T0220899

Annals of Mathematics Studies
Number 189

Multi-parameter Singular Integrals

Brian Street

PRINCETON UNIVERSITY PRESS

PRINCETON AND OXFORD

2014

Published by Princeton University Press, 41 William Street,
Princeton, New Jersey 08540
In the United Kingdom: Princeton University Press, 6 Oxford Street,
Woodstock, Oxfordshire OX20 1TW

press.princeton.edu

Library of Congress Cataloging-in-Publication Data

Street, Brian, 1981- author.
  Multi-parameter singular integrals / Brian Street.
    pages cm
  Includes bibliographical references and index.
  ISBN 978-0-691-16251-5 (hardcover : alk. paper) – ISBN 978-0-691-16252-2 (pbk. :
alk. paper)   1. Singular Integrals. 2. Transformations (Mathematics) I. Title.
  QA329.6.S77 2014
  515'.98–dc23
                                2013045259

British Library Cataloging-in-Publication Data is available

This book has been composed in LATEX.

The publisher would like to acknowledge the author of this volume for providing the
camera-ready copy from which this book was printed.

Printed on acid-free paper ∞

Printed in the United States of America

10 9 8 7 6 5 4 3 2 1

Contents

Preface

This monograph concerns a new theory of an algebra of singular integral operators which we refer to as "multi-parameter singular integral operators." These are operators which act on functions on $\mathbb{R}^n$, and have an underlying geometry which is given by a family of "balls" on $\mathbb{R}^n$ with many "radius" parameters: $B(x, \delta_1, \ldots, \delta_\nu)$. The classical theory of singular integrals corresponds to the case when $\nu = 1$. For higher ν, there are several different theories which have been developed by many authors. Some examples are the product theory of singular integrals, convolution with "flag kernels" on graded groups, and compositions of two singular integrals corresponding to different geometries. The goal of this monograph is to develop a general algebra of singular integrals which generalizes and unifies each of these (and other) examples.

Because our goal is to define a new kind of singular integral, it is best to start with the basic question "what is a singular integral?" For this, we start with the well-understood "single-parameter" case. The most classical example of a singular integral operator is the Hilbert transform. To understand the Hilbert transform, we wish to make sense of $\int f(t)/t \, dt$, for $f \in C_0^\infty(\mathbb{R})$. This integral does not make sense classically, because t^{-1} is not in L^1. However, t^{-1} has "cancellation": t^{-1} is an *odd* function. Since t^{-1} is odd, we make the convention that $\int_{-1}^1 t^{-1} \, dt = 0$. With this convention we have

$$\int f(t)/t \, dt = \int_{-1}^1 (f(t) - f(0))/t \, dt + \int_{|t| \geq 1} f(t)/t \, dt. \tag{1}$$

This allows us to define $\int f(t) \frac{1}{t} \, dt$ for $f \in C_0^\infty(\mathbb{R})$ by (1) and sees t^{-1} as a distribution on $\mathbb{R}$. The Hilbert transform is defined as the operator $Hf = f * t^{-1}$, for $f \in C_0^\infty(\mathbb{R})$; i.e., $Hf(x) = \int f(x - t)/t \, dt$. A classical theorem states that H extends to a bounded operator $H : L^p \to L^p$, $1 < p < \infty$. We call H a "singular integral operator" because it is defined by an integral which does not converge in the classical sense,[1] but can be made sense of using some sort of "cancellation." Some notion of cancellation is central to any theory of singular integrals.

In this single-parameter case, the above ideas have been greatly generalized. The most well-known such generalization is the theory of spaces of homogeneous type as developed by Coifman and Weiss [CW71] (see, also, Stein's book [Ste93]). In this theory, one is given a (quasi-)metric ρ on $\mathbb{R}^n$ and a Borel measure[2] Vol on $\mathbb{R}^n$. Define the balls by $B(x, \delta) := \{y \mid \rho(x, y) < \delta\}$. Provided these balls satisfy certain axioms,[3]

---

[1]But the integral nearly converges in the sense that $|t|^{-\delta} \in L^1([-1, 1])$ for $\delta < 1$.

[2]We are mostly interested in the case when Vol is given by Lebesgue measure.

[3]A key such axiom is the "doubling condition": $\mathrm{Vol}(B(x, 2\delta)) \leq C\mathrm{Vol}(B(x, \delta))$.

a theory of "singular integral operators" can be developed. These operators are given, formally, by $f \mapsto Tf$ where

$$Tf(x) = \int K(x,y) f(y) \ d\mathrm{Vol}(y),$$

where $K(x,y)$ is not necessarily integrable, and instead satisfies estimates like

$$|K(x,y)| \lesssim \frac{1}{\mathrm{Vol}(B(x,\rho(x,y)))}, \tag{2}$$

along with a cancellation condition so that we may make sense of the integral. When the right conditions are imposed, these operators are bounded on L^p ($1 < p < \infty$); again, an important point is to make appropriate sense of the "cancellation condition." Furthermore, in many situations the operators that arise form an algebra: the composition of two singular integral operators is again a singular integral operator.

In Chapters 1 and 2 we discuss two important cases of these single-parameter singular integral operators. Much of the theory in these chapters is well-known, but we present these results as a way to motivate the more general multi-parameter theory discussed later; moreover, some of the results we prove are key tools in studying the multi-parameter situation. In Chapter 1, we discuss the case when ρ is the usual distance on $\mathbb{R}^n$. There, we obtain the most classical theory of singular integrals, which we see is useful for studying elliptic partial differential operators.

In Chapter 1 we are introduced to a running theme of the monograph: singular integral operators can be defined in three equivalent ways. Each way is useful for different purposes. The three ways, roughly speaking, are as follows:

- $Tf(x) = \int K(x,y) f(y) \ dy$, where K satisfies certain estimates like (2); we refer to these estimates as "growth conditions." In addition we need to assume a "cancellation condition." This condition takes the form of bounds for $T\phi$ (and $T^*\phi$) where ϕ ranges over certain test functions. In this "single-parameter" case, the cancellation condition is closely related to the conditions of the $T(1)$ theorem of David and Journé [DJ84]. This type of definition is the most classical type we consider.

- The second equivalent definition introduces a type of "elementary operator." The condition states, roughly, that if E is an elementary operator, then so is TE. This condition is useful for showing that the operators in question form an algebra. In this simplest case, an example of an elementary operator is a Littlewood-Paley cut-off.

- The third equivalent definition sees T as an appropriate sum of elementary operators. This definition is useful for proving the L^p ($1 < p < \infty$) boundedness of singular integral operators. In this Euclidean case, such decompositions of singular integral operators are often called "Littlewood-Paley decompositions."

Already in Chapter 1, we see this trichotomy three times, in increasing generality: Theorems 1.1.23, 1.1.26, and 1.2.10. Results like the ones in Chapter 1 can be found in many sources (e.g., [Ste93]). One way in which the thrust of our presentation in Chapter 1 differs is the emphasis of the above trichotomy. Indeed, we develop it even when the operators are not translation invariant (many authors discuss such ideas only for translation invariant, or nearly translation invariant operators). We also present these ideas in a slightly different way than is usual, which helps to motivate later results and definitions.

In Chapter 2 we remain in the single-parameter case, and turn to the case when the metric is a Carnot-Carathéodory (or sub-Riemannian) metric.[4] We define a class of singular integral operators adapted to this metric. The setting here is an instance of a space of homogeneous type, but we have more structure to work with. Indeed, there is a natural way to discuss "smoothness" with respect to a Carnot-Carathéodory structure. This makes these ideas useful for studying regularity properties of certain partial differential operators.

Chapter 2 has two major themes. The first is a more general reprise of the trichotomy described above (Theorem 2.0.29); this accounts for much of the work in Chapter 2 and paves the way to proving many properties of these operators (e.g., that they form an algebra). The second theme is a generalization of the fact (from Chapter 1) that Euclidean singular integral operators are closely related to elliptic partial differential equations. In fact, there is a far-reaching generalization of ellipticity, known as maximal hypoellipticity, and the singular integrals defined in Chapter 2 are an essential tool in studying this concept. The concept of maximal hypoellipticity was developed by several authors [Hör67, RS76, Koh78, HN85]. The connection between maximal hypoellipticity and singular integrals has also been used by several authors [RS76, NRSW89, CNS92, Koe02, Str09], but this seems to be the first time that the connection was made explicit in full generality.

A major tool we introduce in Chapter 2 is a quantitative version of the classical Frobenius theorem from differential geometry. This "quantitative Frobenius theorem" can be thought of as yielding "scaling maps" which are well adapted to the Carnot-Carathéodory geometry, and is of central use throughout the rest of the monograph. The statement of the result is quite technical so we devote time to carefully explaining it and its uses. We briefly indicate the proof, but refer the reader to the original source [Str11] for the full details.

Chapters 1 and 2 should be thought of as the background and motivation for the main goal of this monograph: to develop a general theory of "multi-parameter" singular integral operators. To understand this concept, consider the fact that one can reconstruct the metric ρ from the metric balls:

$$\rho(x, y) = \inf \left\{ \delta > 0 \mid y \in B(x, \delta) \right\}.$$

Thus, when defining a class of singular integrals, the most basic ingredient is the corresponding family of balls. The remainder of this monograph is concerned with the following questions: what if, instead of balls of the form $B(x, \delta)$, we are given balls with many "radius" parameters $B(x, \delta_1, \ldots, \delta_\nu)$? What should we assume on these balls to develop a notion of a singular integral operator? What is the right definition of

---

[4]We also work on a compact manifold, instead of $\mathbb{R}^n$.

a singular integral operator? A major difficulty is that (2) involves the metric ρ in an essential way, and there is no one natural metric associated to the balls $B\left(x, \delta_1, \ldots, \delta_\nu\right)$. It is therefore not obvious what a natural generalization of bounds like (2) might be. Chapters 3, 4, and 5 are devoted to these questions.

We again restrict attention to Carnot-Carathéodory type balls, and we offer answers to the above questions in that situation. We refer to the corresponding singular integral operators as "multi-parameter singular integral operators."

The most basic and well understood example of a multi-parameter singular integral comes from the so-called product theory of singular integrals. In that case, the ambient space $\mathbb{R}^n$ is decomposed into factors $\mathbb{R}^{n_1} \times \cdots \times \mathbb{R}^{n_\nu}$. On each factor, $\mathbb{R}^{n_\mu}$, one is given a metric ρ_μ. We denote the corresponding metric balls by $B_\mu\left(x_\mu, \delta_\mu\right) \subseteq \mathbb{R}^{n_\mu}$. We obtain multi-parameter balls by

$$B\left(\left(x_1, \ldots, x_\nu\right), \left(\delta_1, \ldots, \delta_\nu\right)\right) = B_1\left(x_1, \delta_1\right) \times \cdots \times B_\nu\left(x_\nu, \delta_\nu\right).$$

Generalizing (2) to this situation is easy; we use estimates like

$$\left|K\left(\left(x_1, \ldots, x_\nu\right), \left(y_1, \ldots y_\nu\right)\right)\right| \lesssim \prod_{\mu=1}^{\nu} \mathrm{Vol}_\mu\left(B_\mu\left(x_\mu, \rho_\mu\left(x_\mu, y_\mu\right)\right)\right)^{-1}.$$

In this situation, it is well known how to develop a theory of singular integrals. See, for example, Section 4.1. These ideas were developed by many authors, beginning with foundational work of Fefferman and Stein [FS82] and Journé [Jou85] (see Section 4.1.2 for more references). The theory we develop in this monograph incorporates this product theory, but the main point is to develop a theory of multi-parameter singular integral operators when the balls are not necessarily of product type.

In Chapter 3 we develop the theory of multi-parameter Carnot-Carathéodory geometry which we need to study these singular integral operators. In the case when the balls are of product type, all of the results from Chapter 3 are simple variants of results in the single-parameter theory. When the balls are not of product type, these ideas become more difficult. What saves the day is the quantitative Frobenius theorem given in Chapter 2. Using this we can estimate certain integrals, and also develop an appropriate maximal function and an appropriate Littlewood-Paley square function, all of which are essential to our study of singular integral operators.

There are a few special cases where such a theory of multi-parameter singular integral operators has already been developed, and we discuss these in Chapter 4. These include the product theory of singular integrals, convolution with flag kernels on graded groups, convolution with both the left and right invariant Calderón-Zygmund singular integral operators on stratified Lie groups, and composition of standard pseudodifferential operators with certain singular integrals corresponding to non-Euclidean geometries. We outline these examples and their applications and relate them to the trichotomy discussed above.

Finally, in Chapter 5, we turn to a general theory which generalizes and unifies all of the above examples. As mentioned above, a main issue is that the first definition from our trichotomy does not generalize to the multi-parameter situation (there is no useful analog of the "growth conditions" in general). To deal with this, we introduce strengthened cancellation conditions. We do this in two different ways, leaving us with four total definitions for singular integral operators (the first two use the strengthened cancellation conditions, while the later two are generalizations of the later two parts of the above trichotomy). Thus we obtain four classes of singular integral operators, which we denote by $\mathcal{A}_1$, $\mathcal{A}_2$, $\mathcal{A}_3$, and $\mathcal{A}_4$. The main theorem of Chapter 5 is $\mathcal{A}_1 = \mathcal{A}_2 = \mathcal{A}_3 = \mathcal{A}_4$; i.e., all four of these definitions are equivalent. This leads to many nice properties of these singular integral operators. For instance, it proves they are an algebra, and proves their boundedness on appropriate non-isotropic Sobolev spaces. We also include several special cases of this algebra, and relate it to the examples from Chapter 4; in particular, we see that this algebra arises naturally in some questions from partial differential equations.

There are three appendices. The first gives the background theory on functional analysis which we need throughout the monograph. This includes the basic aspects of locally convex topological vector spaces, and also categorical limits and tensor products of locally convex topological vector spaces. The second appendix records three results from calculus which are useful to us: the smoothness of exponentials of vector fields, a version of the inverse function theorem which is "uniform on compact sets," and a technical change of variables which is used several times in the monograph. The third appendix is meant to be a quick reference for some notation which is used throughout the monograph which may be somewhat nonstandard.

This work uses the ideas of many authors. We have included several sections titled "Further reading and references" at the end of Chapters 1, 2, and 3, and at the end of each section in Chapter 4. These sections include references to the literature for the results we have used, along with some comments on surrounding ideas and theories of other authors. Especially in Chapters 1, 2, and 4, few of the ideas are new, and we have attempted to give appropriate credit and surrounding history in these final sections.

ACKNOWLEDGMENTS

This monograph would not exist without the suggestions and encouragement of Eli Stein. His comments helped shape many of my ideas, and I am indebted to him. I also thank the anonymous referees who gave detailed suggestions on how to improve the exposition. Finally, I acknowledge support from the NSF (NSF DMS-0802587 and NSF DMS-1066020).

Multi-parameter Singular Integrals

Chapter One

The Calderón-Zygmund Theory I: Ellipticity

Our story begins with a classical situation: convolution with homogeneous, Calderón-Zygmund kernels on $\mathbb{R}^n$. Let $S^{n-1} \hookrightarrow \mathbb{R}^n$ denote the unit sphere in $\mathbb{R}^n$. We let $C_z^\infty \left(S^{n-1} \right) \subset C^\infty \left(S^{n-1} \right)$ denote those $k \in C^\infty \left(S^{n-1} \right)$ with $\int_{S^{n-1}} k \left(\omega \right) \, d\omega = 0$ (where ω denotes the surface area measure on S^{n-1}).

For a function $k \in C_z^\infty \left(S^{n-1} \right)$, and a complex number $c \in \mathbb{C}$, we define a distribution $K = K \left(k, c \right) \in C_0^\infty \left(\mathbb{R}^n \right)'$ by

$$\langle K, f \rangle = cf \left(0 \right) + \lim_{\epsilon \to 0} \int_{|x| > \epsilon} k \left(x / |x| \right) \frac{1}{|x|^n} f \left(x \right) \, dx, \quad f \in C_0^\infty \left(\mathbb{R}^n \right). \quad (1.1)$$

LEMMA 1.0.1. *(1.1) defines a distribution.*[1]

PROOF. Fix $M > 0$, and let $\Gamma := \overline{B^n \left(M \right)}$, where $B^n \left(M \right) \subset \mathbb{R}^n$ is the ball of radius M in $\mathbb{R}^n$, centered at 0. Note that Γ is compact. Let $C^\infty \left(\Gamma \right)$ denote the Fréchet space of those $f \in C_0^\infty \left(\mathbb{R}^n \right)$ with $\mathrm{supp} \left(f \right) \subset \Gamma$. We wish to show that for $f \in C^\infty \left(\Gamma \right)$, the limit in (1.1) exists and that $\langle K, \cdot \rangle : C^\infty \left(\Gamma \right) \to \mathbb{C}$ is continuous. Consider, for $f \in C^\infty \left(\Gamma \right)$,

$$\langle K, f \rangle = cf \left(0 \right) + \lim_{\epsilon \to 0} \int_{\epsilon < |x| \le M} k \left(x / |x| \right) \frac{1}{|x|^n} f \left(x \right) \, dx$$

$$= cf \left(0 \right) + \lim_{\epsilon \to 0} \int_{\epsilon < |x| \le M} k \left(x / |x| \right) \frac{1}{|x|^{n-1}} \frac{\left(f \left(x \right) - f \left(0 \right) \right)}{|x|} \, dx,$$

where we have used that $\int k \left(\omega \right) \, d\sigma \left(\omega \right) = 0$. Since $|f \left(x \right) - f \left(0 \right)| / |x| \le 2 \left\| f \right\|_{C^1}$, and since $1 / |x|^{n-1} \in L^1 \left(\Gamma \right)$, the result follows. $\square$

DEFINITION 1.0.2. *We define* $\mathrm{CZ} \subset C_0^\infty \left(\mathbb{R}^n \right)'$ *to be the vector space of distributions given by*

$$\mathrm{CZ} = \left\{ K \left(k, c \right) \mid k \in C_z^\infty \left(S^{n-1} \right), c \in \mathbb{C} \right\},$$

where $K \left(k, c \right)$ *is given by* (1.1).

Remark 1.0.3 Definition 1.0.2 uses the standard space $C_0^\infty \left(\mathbb{R}^n \right)'$–the space of distributions on $\mathbb{R}^n$. The reader wishing background on this and other standard spaces which appear in this text (e.g., $C^\infty \left(\mathbb{R}^n \right)$ and $C^\infty \left(\mathbb{R}^n \right)'$) is referred to Appendix A for more details and further references.

---

[1] See Appendix A for the basic definitions concerning distributions.

Remark 1.0.4 CZ stands for "Calderón-Zygmund," and the distributions in CZ are called "Calderón-Zygmund kernels."

Corresponding to $K \in$ CZ, we define an operator $\mathrm{Op}\,(K) : C_0^\infty\,(\mathbb{R}^n) \to C^\infty\,(\mathbb{R}^n)$ by

$$\mathrm{Op}\,(K)\,f\,(x) = K * f\,(x)\,.$$

Operators of the form $\mathrm{Op}\,(K)$ are often referred to as "Calderón-Zygmund singular integral operators." The main properties of these operators are outlined in the following theorems.

THEOREM 1.0.5. *For $K \in$ CZ, $\mathrm{Op}\,(K)$ extends to a bounded operator $\mathrm{Op}\,(K)$: $L^p\,(\mathbb{R}^n) \to L^p\,(\mathbb{R}^n)$, $1 < p < \infty$.*

In light of Theorem 1.0.5, if $K_1, K_2 \in$ CZ, it makes sense to consider

$$\mathrm{Op}\,(K_1)\,\mathrm{Op}\,(K_2)$$

as a composition of bounded operators on L^p $(1 < p < \infty)$.

THEOREM 1.0.6. *Let $K_1, K_2 \in$ CZ. There is $K_3 \in$ CZ with*

$$\mathrm{Op}\,(K_3) = \mathrm{Op}\,(K_1)\,\mathrm{Op}\,(K_2)\,.$$

Formally, we may write $K_3 = K_1 * K_2$.

We discuss proofs of the above two theorems in the next section, where we proceed in a more general setting.

Remark 1.0.7 Theorem 1.0.6 can be restated as saying that the space of operators

$$\{\mathrm{Op}\,(K) \mid K \in \mathrm{CZ}\}$$

is a subalgebra of the algebra bounded operators on L^p $(1 < p < \infty)$.

The operators $\mathrm{Op}\,(K)$ are homogeneous in the following sense. For $r > 0$ define a dilation operator by $D_r f\,(x) = f\,(rx)$. Then, $D_{1/r}\mathrm{Op}\,(K)\,D_r = \mathrm{Op}\,(K)$, for $K \in$ CZ. It is this basic dilation invariance that is fundamental to many of the aspects of Calderón-Zygmund singular integral operators. We refer to this theory and generalizations of it as "single-parameter theories" where this refers to the "single-parameter" dilations D_r; in this simple case, this means that D_r depends on only one variable r. One main goal of this monograph is to develop a reasonable, useful, and somewhat general definition of a "multi-parameter singular integral."

In this chapter and the next, we discuss various generalizations and modifications of the above definitions. These ideas are well known, and we refer to all of these situations as being "single-parameter." After, we turn to the "multi-parameter" situation, where new theory is developed.

Remark 1.0.8 Even though all the main ideas of this chapter are well known, we present them with an eye toward the new situation covered in Chapter 5. Results like Theorem 1.1.26 and Theorem 1.2.10, below, provide a main motivation for the new definitions in Chapter 5.

1.1 NON-HOMOGENEOUS KERNELS

It is not essential that we consider only operators which are exactly homogeneous in the sense that $D_{1/r}\mathrm{Op}\,(K)\,D_r = \mathrm{Op}\,(K)$. We merely need the *class* of operators we consider to be scale invariant in an appropriate way, i.e., that $D_{1/r}\mathrm{Op}\,(K)\,D_r$ be "of the same form" as $\mathrm{Op}\,(K)$. In addition, the operators we introduce can be seen in three equivalent ways (Theorem 1.1.23). This trichotomy will be a running theme throughout this monograph. We now turn to making these ideas precise.

Let $K \in C_0^\infty\,(\mathbb{R}^n)'$ be a distribution; we abuse notation and for $\phi \in C_0^\infty\,(\mathbb{R}^n)$ we write $\langle K, \phi \rangle = \int K\,(x)\,\phi\,(x)\ dx$. If, for some set $U \subseteq \mathbb{R}^n$ and for $\phi \in C_0^\infty\,(U)$, we have that $\int K\,(x)\,\phi\,(x)\ dx$ agrees with integration against an L_{loc}^1 function, then we identify $K\big|_U$ with this function.[2]

DEFINITION 1.1.1. *We say $K \in C_0^\infty\,(\mathbb{R}^n)'$ is a Calderón-Zygmund kernel if:*

(i) *(Growth Condition) For every multi-index α, $|\partial_x^\alpha K\,(x)| \le C_\alpha\,|x|^{-n-|\alpha|}$. In particular, we assume that $K\,(x)$ is a C^∞ function for $x \ne 0$.*

(ii) *(Cancellation Condition) For every bounded set $\mathcal{B} \subset C_0^\infty\,(\mathbb{R}^n)$, we assume*

$$\sup_{\substack{\phi \in \mathcal{B} \\ R > 0}} \left| \int K\,(x)\,\phi\,(Rx)\ dx \right| \le C_\mathcal{B}.$$

The reader interested in a simple characterization of bounded subsets of $C_0^\infty\,(\Omega)$ is referred to Corollary A.1.26.

For each multi-index α, we define a semi-norm on the space of Calderón-Zygmund kernels by taking the least C_α in the Growth Condition. For each bounded set $\mathcal{B} \subset C_0^\infty\,(\mathbb{R}^n)$, we define a semi-norm to be the least $C_\mathcal{B}$ in the Cancellation Condition. We give the space of Calderón-Zygmund kernels the coarsest topology with respect to which all of these semi-norms are continuous.[3]

Given a Calderón-Zygmund kernel, we define an operator, $\mathrm{Op}\,(K) : C_0^\infty\,(\mathbb{R}^n) \to C^\infty\,(\mathbb{R}^n)$, by

$$\mathrm{Op}\,(K)\,f = f * K = \int f\,(x - y)\,K\,(y)\ dy.$$

Three of the fundamental properties of Calderón-Zygmund kernels are contained in the following theorem.

THEOREM 1.1.2. *(a) If K is a Calderón-Zygmund kernel, then*

$$\mathrm{Op}\,(K) : L^p\,(\mathbb{R}^n) \to L^p\,(\mathbb{R}^n), \quad 1 < p < \infty.$$

More precisely, $\mathrm{Op}\,(K)$ extends to a bounded operator[4] $L^p \to L^p$, $1 < p < \infty$.

---

[2] See Appendix A for more details on this notation.

[3] For the notion of defining a topology by giving a family of semi-norms, we refer the reader to Appendix A.

[4] In the rest of the monograph, when we are given an operator T, initially defined on some dense subspace of L^p, and we say $T : L^p \to L^p$, we mean that T extends to a bounded operator $L^p \to L^p$.

(b) If K is a Calderón-Zygmund kernel, then $\operatorname{Op}(K)^* = \operatorname{Op}(K')$, where K' is a Calderón-Zygmund kernel and $\operatorname{Op}(K)^*$ denotes the L^2 adjoint of $\operatorname{Op}(K)$.

(c) If K_1 and K_2 are Calderón-Zygmund kernels, then it makes sense to consider $\operatorname{Op}(K_1)\operatorname{Op}(K_2)$–for instance, as a bounded operator on L^2. We have

$$\operatorname{Op}(K_1)\operatorname{Op}(K_2) = \operatorname{Op}(K_3)$$

for a Calderón-Zygmund kernel K_3. Formally, we have $K_3 = K_1 * K_2$.

Part (c) of Theorem 1.1.2 can be restated as saying that

$$\{\operatorname{Op}(K) \mid K \text{ is a Calderón-Zygmund kernel}\} \tag{1.2}$$

is a subalgebra of the algebra of bounded operators on L^2. As motivation for the more complicated situations which arise later, we review some aspects of the proof of Theorem 1.1.2.

The easiest way to see that (1.2) forms an algebra involves the Fourier transform. To use the Fourier transform, we introduce Schwartz space.

DEFINITION 1.1.3. *$S(\mathbb{R}^n)$–the space of Schwartz functions on $\mathbb{R}^n$–is defined to be*

$$S(\mathbb{R}^n) = \left\{ f \in C^\infty(\mathbb{R}^n) \;\middle|\; \forall \alpha, \beta \sup_{x \in \mathbb{R}^n} \left|\partial_x^\alpha x^\beta f(x)\right| < \infty \right\},$$

and we give $S(\mathbb{R}^n)$ the Fréchet topology given by the countable family of semi-norms[5]

$$\|f\|_{\alpha,\beta} = \sup_{x \in \mathbb{R}^n} \left|\partial_x^\alpha x^\beta f(x)\right|.$$

For a function $f \in S(\mathbb{R}^n)$ we define the Fourier transform

$$\hat{f}(\xi) = \int e^{-2\pi i x \cdot \xi} f(x)\, dx,$$

so that $\hat{f} : \mathbb{R}^n \to \mathbb{C}$. A fundamental fact is that the Fourier transform is an automorphism of the Fréchet space $S(\mathbb{R}^n)$.[6] We denote by $\vee : S(\mathbb{R}^n) \to S(\mathbb{R}^n)$ the inverse Fourier transform. Because the Fourier transform is an automorphism of Schwartz space, we may define the Fourier transform on the space of "tempered distributions," $S(\mathbb{R}^n)'$, by duality. Namely, for $K \in S(\mathbb{R}^n)'$ and $f \in S(\mathbb{R}^n)$, we define $\int \hat{K}(x) f(x)\, dx = \int K(x) \hat{f}(x)\, dx$. This extends the Fourier transform to be an automorphism of $S(\mathbb{R}^n)'$.

Remark 1.1.4 The Fourier transform is well-adapted to convolution operators. Indeed, $(K * f)^\wedge = \hat{K}\hat{f}$. This makes the Fourier transform a useful tool when dealing with the convolution operators in this section. Later in the monograph, we work with

---

[5]For the definition of a Fréchet space see Definition A.1.9.
[6]By this we mean that the Fourier transform is a bijective, linear, homeomorphism $S(\mathbb{R}^n) \to S(\mathbb{R}^n)$.

operators which are not translation invariant, and the Fourier transform is not as directly applicable. As motivation for these later applications, we somewhat limit our use of the Fourier transform in this section. Some arguments which follow may have shorter arguments by way of the Fourier transform, but we choose the ones that follow as these are the ones we generalize. We do still use the Fourier transform heavily in this section, but as we move to more generalize situations, it will be used less and less. For a further discussion of this, see the remarks following the proof of Theorem 1.1.23.

In light of the Growth Condition, even though Calderón-Zygmund kernels are a priori elements of $C_0^\infty(\mathbb{R}^n)'$, we see that they may be extended to be elements of $\mathcal{S}(\mathbb{R}^n)'$. This allows us to characterize Calderón-Zygmund kernels in terms of their Fourier transforms. Indeed, we have the following theorem.

THEOREM 1.1.5. *Let $K \in \mathcal{S}(\mathbb{R}^n)'$. The following are equivalent.*

(i) K is a Calderón-Zygmund kernel.

(ii) $\widehat{K}$ is given by a function which satisfies

$$\left| \partial_\xi^\alpha \widehat{K}(\xi) \right| \le C_\alpha |\xi|^{-|\alpha|}, \quad \xi \ne 0. \tag{1.3}$$

Before we prove Theorem 1.1.5, we need to introduce a new notation that will be used throughout the monograph. We write $A \lesssim B$ for $A \le CB$, where C is a constant which is independent of any relevant parameters. Also, we write $A \approx B$ for $A \lesssim B$ and $B \lesssim A$.

PROOF OF THEOREM 1.1.5. We begin with (i)⇒(ii). A priori, $\widehat{K}$ is only a tempered distribution. We will show that $\widehat{K}(\xi)$ agrees with a C^∞ function away from $\xi = 0$, and that for every α

$$\sup_{1 \le |\xi| \le 2} \left| \partial_\xi^\alpha \widehat{K}(\xi) \right| \le C_\alpha, \tag{1.4}$$

where C_α can be chosen to depend only on α and $\|K\|$, and where $\|\cdot\|$ is a continuous semi-norm (depending on α) on the space of Calderón-Zygmund kernels.

First, we see why this yields (ii). Indeed, for $R > 0$ if we replace K by the distribution $K^R(x) := R^n K(Rx)$, then $\{K^R \mid R > 0\}$ is a bounded set[7] of Calderón-Zygmund kernels. If we take the Fourier transform of K^R, we obtain $\widehat{K}(R^{-1}\xi)$. From this homogeneity and (1.4), (1.3) follows. The only remaining issue is to show that $\widehat{K}$ is given by a function; i.e., we need to show that $\widehat{K}(\xi)$ does not have a part supported at $\xi = 0$. The only distributions supported at 0 are finite linear combinations of derivatives of $\delta_0(\xi)$ (where δ_0 denotes the δ function at 0), which are all homogeneous of degree $\ge n$. That is, we know that $\widehat{K}$ is a sum of a function satisfying (1.3) plus a finite linear combination of derivatives of δ_0–and we wish to show that this finite linear combination of derivatives of δ_0 is actually 0. Because $\{K^R \mid R > 0\}$ is a bounded set

---

[7]See Definition A.1.15 for the notion of a bounded set in a locally convex topological vector space.

of Calderón-Zygmund kernels, it is also a bounded set of tempered distributions. Thus, $\left\{ \widehat{K}\left(R^{-1}\xi\right) \mid R > 0 \right\}$ is a bounded set of tempered distributions. Taking $R \to \infty$, we see that if there were any terms supported at $\xi = 0$, then $\left\{ \widehat{K}\left(R^{-1}\xi\right) \mid R > 0 \right\}$ would not be a bounded set of tempered distributions, which shows that $\widehat{K}$ does not have a part which is supported at 0.

Hence, to prove (i)$\Rightarrow$(ii), it suffices to prove (1.4). Now consider ξ with $1 \le |\xi| \le 2$. We have

$$\partial_\xi^\alpha \widehat{K}\left(\xi\right) = \partial_\xi^\alpha \int K\left(x\right) e^{-2\pi i x \cdot \xi}\, dx = \int \left(-2\pi i x\right)^\alpha K\left(x\right) e^{-2\pi i x \cdot \xi}\, dx.$$

Let $\phi \in C_0^\infty\left(\mathbb{R}^n\right)$ be supported in $B^n\left(2\right)$ (the ball of radius 2, centered at 0), and equal to 1 on $B^n\left(3/2\right)$.

We decompose

$$\partial_\xi^\alpha \widehat{K}\left(\xi\right) = \int \left(-2\pi i x\right)^\alpha \phi\left(x\right) e^{-2\pi i x \cdot \xi} K\left(x\right)\, dx$$
$$+ \int \left(1 - \phi\left(x\right)\right)\left(-2\pi i x\right)^\alpha e^{-2\pi i x \cdot \xi} K\left(x\right)\, dx,$$

and we estimate the two terms separately. The estimate

$$\left| \int \left(-2\pi i x\right)^\alpha \phi\left(x\right) e^{-2\pi i x \cdot \xi} K\left(x\right)\, dx \right| \lesssim 1$$

follows immediately from the cancellation condition (recall $1 \le |\xi| \le 2$). For the second term, we integrate by parts to see (for any $L \in \mathbb{N}$)

$$\int \left(-2\pi i x\right)^\alpha \left(1 - \phi\left(x\right)\right) K\left(x\right) e^{i x \cdot \xi}\, dx$$
$$= \left(-|2\pi\xi|^2\right)^L \int \triangle_x^L \left[\left(-2\pi i x\right)^\alpha \left(1 - \phi\left(x\right)\right) K\left(x\right)\right] e^{i x \cdot \xi}\, dx.$$

We take $L = L\left(\alpha, m\right)$ large. If any of the derivatives land on $1 - \phi\left(x\right)$, then the resulting function is supported in $B^n\left(2\right) \setminus B^n\left(3/2\right)$ and the growth condition shows that the integral converges absolutely proving the desired estimate. Otherwise, all but at most $|\alpha|$ derivatives land on K. In this case, the growth condition shows that the resulting distribution falls off like $|x|^{-n-2L+|\alpha|}$. Taking $2L \ge |\alpha| + 1$, the integral converges absolutely, showing

$$\left| \int \left(-2\pi i x\right)^\alpha \left(1 - \phi\left(x\right)\right) K\left(x\right) e^{i x \cdot \xi}\, dx \right| \lesssim 1,$$

as desired. This completes the proof of (1.4) and therefore completes the proof of (i)$\Rightarrow$(ii).

We now turn to (ii)⇒(i), and we assume $\widehat{K}$ is a function satisfying (1.3). We wish to show K is a Calderón-Zygmund kernel. We begin with the growth condition. We show, for every multi-index β,

$$\sup_{1 \le |x| \le 2} \left| \partial_x^\beta K(x) \right| \le D_\beta, \tag{1.5}$$

where D_β is a constant which depends on only a finite number (depending on β) of the constants C_α in (1.3). Because of the relationship between $\widehat{K}(R^{-1}\xi)$ and K^R discussed in the first part of the proof, the growth condition follows immediately from (1.5). (1.5), in turn, follows just as in the estimates for the first part of this proof–indeed, we do not need a "cancellation condition" for $\widehat{K}$, since it is an element of $L^\infty \subset L^1_{\text{loc}}$.

We turn to the cancellation condition. Let $\mathcal{B} \subset C_0^\infty(\mathbb{R}^n)$ be a bounded set. Let $R > 0$ and $\phi \in \mathcal{B}$. We have, letting $\check{\phi}$ denote the inverse Fourier transform of ϕ,

$$\left| \int K(x)\phi(Rx)\,dx \right| = \left| \int \widehat{K}(\xi) R^{-n} \check{\phi}(-R^{-1}\xi)\,d\xi \right|$$

$$= \left| \int \widehat{K}(R\xi) \check{\phi}(\xi)\,d\xi \right|$$

$$\le \int \left| \widehat{K}(R\xi) \check{\phi}(\xi) \right|\,d\xi$$

$$\lesssim 1,$$

where in the last line we have used $\left| \widehat{K}(R\xi) \right| \lesssim 1$ and $\{\check{\phi} \mid \phi \in \mathcal{B}\} \subset \mathcal{S}(\mathbb{R}^n)$ is a bounded set. This completes the proof. $\qquad\square$

Remark 1.1.6 The proof of Theorem 1.1.5 yields something which might seem somewhat surprising. Indeed, it shows that we could have defined the topology on Calderón-Zygmund kernels in another way, by taking the semi-norms to be, for each multi-index α, the least possible C_α from (1.3). This shows that the space of Calderón-Zygmund kernels is, in fact, a Fréchet space, even though we originally defined it with an uncountable collection of semi-norms (indeed, we had one semi-norm for each bounded subset of $C_0^\infty(\mathbb{R}^n)$). We see examples of this sort of phenomenon several times in the sequel. See Remarks 2.8.2 and 5.7.6.

Theorem 1.1.5 immediately implies parts (b) and (c) of Theorem 1.1.2. Notice

$$\text{Op}(K)f = \left(\widehat{K}\widehat{f} \right)^\vee,$$

where $\vee$ denotes the inverse Fourier transform. Therefore $\text{Op}(K)^* f = \left(\overline{\widehat{K}}\widehat{f} \right)^\vee$, where $\overline{\widehat{K}}$ denotes the complex conjugate of $\widehat{K}$. Since $\widehat{K}$ satisfies (1.3) if and only if $\overline{\widehat{K}}$ does, (b) follows.[8] To see (c) notice

$$\text{Op}(K_1)\,\text{Op}(K_2)f = \left(\widehat{K_1}\widehat{K_2}\widehat{f} \right)^\vee,$$

---

[8](b) is also quite easy to see directly from the definitions.

and the result follows from Theorem 1.1.5. Finally, since $\widehat{K}$ is a bounded function, it follows that $\mathrm{Op}\,(K)$ is bounded on L^2. Thus, the remainder of Theorem 1.1.2 follows from the next, conditional, proposition.

PROPOSITION 1.1.7. *If* $K \in C_0^\infty\,(\mathbb{R}^n)'$ *is a distribution which satisfies the Growth Condition (i.e., (i) of Definition 1.1.1) and for which* $\mathrm{Op}\,(K)$ *extends to a bounded operator* $L^2\,(\mathbb{R}^n) \to L^2\,(\mathbb{R}^n)$, *then* $\mathrm{Op}\,(K)$ *extends to a bounded operator* $L^p\,(\mathbb{R}^n) \to L^p\,(\mathbb{R}^n)$, $(1 < p \leq 2)$.

Notice that Proposition 1.1.7 completes the proof that $\mathrm{Op}\,(K)$ is bounded on L^p, $1 < p < \infty$. Indeed, for $p > 2$, we may apply Proposition 1.1.7 to $\mathrm{Op}\,(K)^*$, and the result follows by duality.

The key to proving Proposition 1.1.7 is the "Calderón-Zygmund decomposition," which we state without proof. See [Ste93] for a proof and further details.

LEMMA 1.1.8. *Fix* $f \in L^1\,(\mathbb{R}^n)$ *and* $\alpha > 0$. *There is a countable family of closed cubes* $\{Q_k\}$ *whose interiors are disjoint, and functions* g *and* b *with* $f = g + b$, *such that*

(i) $\sum \mathrm{Vol}\,(Q_k) \leq \frac{1}{\alpha} \int |f|$.

(ii) b *is supported on* $\bigcup Q_k$ *and* $\int_{Q_k} b = 0$, $\frac{1}{\mathrm{Vol}(Q_k)} \int |b| \leq 2^{n+1}\alpha$.

(iii) $|g| \leq 2^n \alpha$.

PROOF SKETCH OF PROPOSITION 1.1.7. The goal is to show that $\mathrm{Op}\,(K)$ is weak-type $(1, 1)$. That is, we want to show,

$$\mathrm{Vol}\,(\{x \mid |\mathrm{Op}\,(K)\,f\,(x)| > \alpha\}) \lesssim \frac{1}{\alpha} \int |f|. \qquad (1.6)$$

The result then follows from the L^2 boundedness of $\mathrm{Op}\,(K)$ and the Marcinkiewicz interpolation theorem.

For a fixed α, we apply Lemma 1.1.8 to obtain $\{Q_k\}$ and a decomposition $f = g + b$ as in the statement of that lemma. (1.6) follows once we show

$$\mathrm{Vol}\,(\{x \mid |\mathrm{Op}\,(K)\,g\,(x)| > \alpha/2\}) + \mathrm{Vol}\,(\{x \mid |\mathrm{Op}\,(K)\,b\,(x)| > \alpha/2\}) \lesssim \frac{1}{\alpha} \int |f|.$$

First notice $g \in L^2$. Indeed, $g = f$ on $(\bigcup Q_k)^c$, and we therefore have

$$\int |g|^2 \leq \int_{(\bigcup Q_k)^c} |f|\,|g| + \int_{\bigcup Q_k} |g|^2.$$

We use that $|g| \leq 2^n \alpha$, to see

$$\int_{\bigcup Q_k} |g|^2 \lesssim \alpha^2 \sum \mathrm{Vol}\,(Q_k) \lesssim \alpha \int |f|,$$

and

$$\int_{(\bigcup Q_k)^c} |f| \, |g| \lesssim \alpha \int |f| \, .$$

It follows that

$$\int |g|^2 \lesssim \alpha \int |f| \, ,$$

and therefore $g \in L^2$. Chebycheff's inequality then completes the proof for g:

$$\mathrm{Vol}\left(\{x \mid |\mathrm{Op}\,(K)\,g| > \alpha/2\}\right) \lesssim \alpha^{-2} \int |\mathrm{Op}\,(K)\,g|^2 \lesssim \alpha^{-2} \int |g|^2 \lesssim \frac{1}{\alpha} \int |f| \, .$$

We now turn to the estimate of $\mathrm{Op}\,(K)\,b$. Let B_k be the smallest ball containing the cube Q_k and let $2B_k$ denote the ball with the same center as B_k but with twice the radius. Notice,

$$\sum \mathrm{Vol}\,(2B_k) \approx \sum \mathrm{Vol}\,(Q_k) \lesssim \frac{1}{\alpha} \int |f| \, .$$

Thus, it suffices to show

$$\mathrm{Vol}\left(\left\{x \in \left(\bigcup 2B_k\right)^c \;\middle|\; |\mathrm{Op}\,(K)\,b\,(x)| > \alpha/2\right\}\right) \lesssim \frac{1}{\alpha} \int |f| \, .$$

We have

$$\int_{(\bigcup 2B_k)^c} |\mathrm{Op}\,(K)\,b| \leq \sum_k \int_{(2B_k)^c} |\mathrm{Op}\,(K)\,b_k| \, ,$$

where $b_k = b$ on Q_k and is 0 on Q_k^c. Let y_k denote the center of the cube Q_k. Using that $\int b_k = 0$, we have

$$\int_{(2B_k)^c} |\mathrm{Op}\,(K)\,b_k| \leq \int_{(2B_k)^c} \int_{B_k} |K\,(x - y) - K\,(x - y_k)| \, |b_k\,(y)| \; dy \, dx. \quad (1.7)$$

For $y \in B_k$, $x \in (2B_k)^c$, it is easy to see that

$$|K\,(x - y) - K\,(x - y_k)| \lesssim |y - y_k| \, |x - y_k|^{-n-1} \, .$$

Using $\int |b_k| \lesssim \alpha \mathrm{Vol}\,(Q_k)$, we have from (1.7),

$$\int_{(2B_k)^c} |\mathrm{Op}\,(K)\,b_k| \lesssim \alpha \mathrm{Vol}\,(Q_k) \, ,$$

and it follows that $\int_{(\bigcup 2B_k)^c} |\mathrm{Op}\,(K)\,b| \lesssim \alpha \sum_k \mathrm{Vol}\,(Q_k) \lesssim \int |f|$. That

$$\mathrm{Vol}\,(\{x \mid |\mathrm{Op}\,(K)\,b\,(x)| > \alpha/2\}) \lesssim \frac{1}{\alpha} \int |f|$$

now follows from another application of Chebycheff's inequality. This completes the proof. $\qquad\square$

Remark 1.1.9 The proof of Proposition 1.1.7 generalizes to a number of settings; for instance see Chapter 2. In fact, these ideas work in the even more general setting of a "space of homogeneous type," as developed by Coifman and Weiss [CW71]. See [Ste93] for more details. These methods will be a useful tool in creating an appropriate Littlewood-Paley theory, which we will use to show the L^p boundedness of certain "multi-parameter" operators. This is discussed in Sections 2.15.4 and 3.4.

Let us now turn to characterizing operators of the form $\mathrm{Op}\,(K)$, where K is a Calderón-Zygmund kernel, in several ways. These ideas will be used as motivation for definitions in later chapters. For this we need to introduce a subspace of Schwartz space which plays a pivotal role.

DEFINITION 1.1.10. $\mathcal{S}_0\,(\mathbb{R}^n)$–*the space of Schwartz functions, all of whose moments vanish–is the closed subspace of $\mathcal{S}\,(\mathbb{R}^n)$ defined by*

$$\mathcal{S}_0\,(\mathbb{R}^n) = \left\{ f \in \mathcal{S}\,(\mathbb{R}^n) \;\Big|\; \forall \alpha, \int x^\alpha f\,(x)\; dx = 0 \right\}.$$

Remark 1.1.11 As a closed subspace of $\mathcal{S}\,(\mathbb{R}^n)$, $\mathcal{S}_0\,(\mathbb{R}^n)$ has the induced subspace topology which turns $\mathcal{S}_0\,(\mathbb{R}^n)$ into a Fréchet space. Notice that $\mathcal{B} \subset \mathcal{S}_0\,(\mathbb{R}^n)$ is a bounded set if and only if $\mathcal{B}$ is a bounded subset of $\mathcal{S}\,(\mathbb{R}^n)$; i.e., $\forall \alpha, \beta, \sup_{f \in \mathcal{B}} \|f\|_{\alpha,\beta} < \infty$.[9]

When working with $\mathcal{S}_0\,(\mathbb{R}^n)$ it is often more convenient to work on the Fourier transform side, as the next few results illustrate.

LEMMA 1.1.12. *Let $f \in \mathcal{S}\,(\mathbb{R}^n)$. Then, $f \in \mathcal{S}_0\,(\mathbb{R}^n)$ if and only if $\forall \alpha$, $\partial_\xi^\alpha \hat{f}\,(0) = 0$.*

PROOF. The result follows immediately from the definitions. $\qquad\square$

COROLLARY 1.1.13. *Fix $s \in \mathbb{R}$. The map $\triangle^s : \mathcal{S}_0\,(\mathbb{R}^n) \to \mathcal{S}_0\,(\mathbb{R}^n)$ given by $\triangle^s : f \mapsto \left(|2\pi\xi|^{2s}\,\hat{f}\,(\xi)\right)^{\vee}$ is an automorphism.*

PROOF. It follows easily from Lemma 1.1.12 that for $f \in \mathcal{S}_0\,(\mathbb{R}^n)$, $\triangle^s f \in \mathcal{S}_0\,(\mathbb{R}^n)$. The closed graph theorem (Theorem A.1.14) shows that $\triangle^s$ is continuous. The continuous inverse of $\triangle^s$ is given by $\triangle^{-s}$. $\qquad\square$

COROLLARY 1.1.14. $\mathcal{S}_0\,(\mathbb{R}^n)$ *is dense in $L^2\,(\mathbb{R}^n)$.*

PROOF. This follows easily from Lemma 1.1.12. $\qquad\square$

Remark 1.1.15 Actually, $\mathcal{S}_0\,(\mathbb{R}^n)$ is dense in $L^p\,(\mathbb{R}^n)$ $(1 < p < \infty)$, but it is dense in neither L^1, nor L^∞.

The next lemma offers a characterization of $\mathcal{S}_0\,(\mathbb{R}^n)$, which we do not use directly, but which motivates some future definitions.

---

[9]For further information regarding bounded sets and their relationship with semi-norms, see Appendix A.

LEMMA 1.1.16. *Consider subsets $\mathcal{G} \subset \mathcal{S}(\mathbb{R}^n)$ satisfying the following condition:* $\forall f \in \mathcal{G}, \exists f_1, \ldots, f_n \in \mathcal{G}, \text{ with } f = \sum_{j=1}^{n} \partial_{x_j} f_j.$ $\mathcal{S}_0(\mathbb{R}^n) \subset \mathcal{S}(\mathbb{R}^n)$ *is the largest subset satisfying this condition.*

PROOF. Suppose $\mathcal{G} \subset \mathcal{S}(\mathbb{R}^n)$ satisfies the above condition. Fix a multi-index α. By repeated applications of the above property, we may write $g = \sum_{|\beta|=|\alpha|+1} \partial_x^\beta g_\beta$, where $g_\beta \in \mathcal{G} \subset \mathcal{S}(\mathbb{R}^n)$. We have,

$$\int x^\alpha g(x) \, dx = \sum_{|\beta|=|\alpha|+1} \int x^\alpha \partial_x^\beta g_\beta(x) \, dx = 0,$$

where the last equality follows by integration by parts. Thus $\mathcal{G} \subset \mathcal{S}_0(\mathbb{R}^n)$.

Conversely, given $g \in \mathcal{S}_0(\mathbb{R}^n)$ we may write $g = \triangle g_0$, where $g_0 = \triangle^{-1} g \in \mathcal{S}_0(\mathbb{R}^n)$ (by Corollary 1.1.13). We have $g = \sum_{j=1}^{n} \partial_{x_j}(-\partial_{x_j} g_0)$. Since $-\partial_{x_j} g_0 \in \mathcal{S}_0(\mathbb{R}^n)$, the result follows. $\square$

Remark 1.1.17 Suppose $\mathcal{B} \subset \mathcal{S}_0(\mathbb{R}^n)$ is a bounded set. In light of Lemma 1.1.16, for $f \in \mathcal{B}$, we may write $f = \sum_{|\alpha|=1} \partial_x^\alpha f_\alpha$, where $f_\alpha \in \mathcal{S}_0(\mathbb{R}^n)$. The proof of Lemma 1.1.16 shows more: we may choose the f_α so that $\{f_\alpha \mid f \in \mathcal{B}, |\alpha| = 1\} \subset \mathcal{S}_0(\mathbb{R}^n)$ is a bounded set.

Remark 1.1.18 Let $K \in \mathcal{S}_0(\mathbb{R}^n)'$. It makes sense to define $\mathrm{Op}(K) : \mathcal{S}_0(\mathbb{R}^n) \to C^\infty(\mathbb{R}^n)$. Since $\mathcal{S}_0(\mathbb{R}^n)$ is dense in L^2, there is at most one continuous extension of $\mathrm{Op}(K) : L^2 \to L^2$ (but there may be no continuous extension). If $\widetilde{K} \in \mathcal{S}(\mathbb{R}^n)'$ is a Calderón-Zygmund kernel such that $\widetilde{K}\big|_{\mathcal{S}_0(\mathbb{R}^n)} = K$, then $\mathrm{Op}(K)$ and $\mathrm{Op}\left(\widetilde{K}\right)$ extend to the same bounded operator on L^2. Thus, the map $\widetilde{K} \mapsto \widetilde{K}\big|_{\mathcal{S}_0(\mathbb{R}^n)}$ from Calderón-Zygmund kernels to elements of $\mathcal{S}_0(\mathbb{R}^n)'$ is injective. It, therefore, makes sense to ask whether an element $K \in \mathcal{S}_0(\mathbb{R}^n)'$ is a Calderón-Zygmund kernel, and if so to identify it with a unique Calderón-Zygmund kernel in $C_0^\infty(\mathbb{R}^n)'$.

For a function $f \in \mathcal{S}(\mathbb{R}^n)$ and $R > 0$, we define $f^{(R)}(x) = R^n f(Rx)$. Note that this is defined in such a way that

$$\int f^{(R)}(x) \, dx = \int f(x) \, dx,$$

and, more generally,

$$\left(f^{(R)}\right)^{\wedge}(\xi) = \hat{f}(\xi/R).$$

LEMMA 1.1.19. *Let $\mathcal{B} \subset \mathcal{S}_0(\mathbb{R}^n)$ be a bounded set. For $R_1, R_2 > 0, \phi_1, \phi_2 \in \mathcal{B}$, define a function $\psi = \psi(R_1, R_2, \phi_1, \phi_2)$ by*

$$\psi^{(R_1)} = \phi_1^{(R_1)} * \phi_2^{(R_2)}.$$

Then, for every N, the set

$$\left\{ \left(\frac{R_1 \vee R_2}{R_1 \wedge R_2} \right)^N \psi \,\middle|\, \phi_1, \phi_2 \in \mathcal{B}, R_1, R_2 > 0 \right\}$$

is a bounded subset of $\mathcal{S}_0 (\mathbb{R}^n)$. Here $\vee$ denotes maximum and $\wedge$ denotes minimum.

PROOF. Fix M large, to be chosen later. By Corollary 1.1.13, each $\phi \in \mathcal{B}$ may be written as $\phi = \triangle^M \widetilde{\phi}$, where $\left\{ \widetilde{\phi} \,\middle|\, \phi \in \mathcal{B} \right\}$ is a bounded set. Fix $\phi_1, \phi_2 \in \mathcal{B}$ and $R_1, R_2 > 0$. Suppose $R_1 \geq R_2$, we have,

$$\phi_1^{(R_1)} * \phi_2^{(R_2)} = \left(\frac{R_2}{R_1} \right)^{2M} \widetilde{\phi}_1^{(R_1)} * \left(\triangle^M \phi_2 \right)^{(R_2)}.$$

If $R_2 > R_1$, we instead have

$$\phi_1^{(R_1)} * \phi_2^{(R_2)} = \left(\frac{R_1}{R_2} \right)^{2M} (\triangle \phi_1)^{(R_1)} * \widetilde{\phi}_2^{(R_2)}.$$

Either way, we have

$$\psi^{(R_1)} = \left(\frac{R_1 \wedge R_2}{R_1 \vee R_2} \right)^{2M} \gamma_1^{(R_1)} * \gamma_2^{(R_2)},$$

where γ_1, γ_2 range over a bounded subset of $\mathcal{S}_0 (\mathbb{R}^n)$. By replacing M with $M + N$, it suffices to prove the result with $N = 0$. For any fixed M, and using the above argument, we may write ψ in the form

$$\psi^{(R_1)} = \left(\frac{R_1 \wedge R_2}{R_1 \vee R_2} \right)^{2M} \gamma_1^{(R_1)} * \gamma_2^{(R_2)}, \tag{1.8}$$

where γ_1 and γ_2 range over a bounded subset of $\mathcal{S}_0 (\mathbb{R}^n)$. Fix a semi-norm $\|\cdot\|_{\alpha,\beta}$. We will take M large in terms of α, β. We wish to show

$$\sup_{\psi} \left\| \widehat{\psi} \right\|_{\alpha,\beta} < \infty,$$

where ψ is defined by (1.8) and the supremum is taken as γ_1, γ_2 range over a bounded subset of $\mathcal{S}_0 (\mathbb{R}^n)$ and R_1, R_2 range over $(0, \infty)$. But,

$$\widehat{\psi}(\xi) = \left(\frac{R_1 \wedge R_2}{R_1 \vee R_2} \right)^{2M} \widehat{\gamma}_1(\xi) \, \widehat{\gamma}_2(R_1 \xi / R_2).$$

Taking M sufficiently large in terms of α, β, the result follows. $\qquad \square$

LEMMA 1.1.20. *Let* $\left\{\varsigma_j \mid j \in \mathbb{Z}\right\} \subset \mathcal{S}_0\left(\mathbb{R}^n\right)$ *be a bounded set. The sum*

$$\sum_{j \in \mathbb{Z}} \varsigma_j^{\left(2^j\right)}$$

converges in the sense of distributions. Furthermore,

$$\sum_{j \in \mathbb{Z}} \mathrm{Op}\left(\varsigma_j^{\left(2^j\right)}\right) = \mathrm{Op}\left(\sum_{j \in \mathbb{Z}} \varsigma_j^{\left(2^j\right)}\right), \tag{1.9}$$

thought of as operators on $\mathcal{S}_0\left(\mathbb{R}^n\right)$. *Here, the sum on the left is taken in the topology of bounded convergence as operators* $\mathcal{S}_0\left(\mathbb{R}^n\right) \to \mathcal{S}_0\left(\mathbb{R}^n\right)$ *(see Definition A.1.28 for the definition of this topology), while the sum on the right is taken in the sense of distributions.*

PROOF. By Corollary 1.1.13, $\varsigma_j = \triangle \widetilde{\varsigma}_j$, where $\left\{\widetilde{\varsigma}_j \mid j \in \mathbb{Z}\right\} \subset \mathcal{S}_0\left(\mathbb{R}^n\right)$ is a bounded set, and so, for $f \in C_0^\infty\left(\mathbb{R}^n\right)$,

$$\left|\int \varsigma_j^{\left(2^j\right)}(x) f(x)\, dx\right| = \left|\int \widetilde{\varsigma}_j^{\left(2^j\right)}(x)\, 2^{-2j} \triangle f(t)\, dt\right|$$

$$\leq 2^{-2j}\left(\int |\widetilde{\varsigma}_j|\right)\left(\int |\triangle f|\right)$$

$$\lesssim 2^{-2j}\left(\int |\triangle f|\right).$$

It follows immediately that the sum $\sum_{j \geq 0} \varsigma_j^{\left(2^j\right)}$ converges in the sense of distributions. For $j < 0$, we have

$$\left|\int \varsigma_j^{\left(2^j\right)}(x) f(x)\, dx\right| \lesssim 2^{jn} \int |f(x)|\, dx,$$

and so $\sum_{j < 0} \varsigma_j^{\left(2^j\right)}$ also converges in the sense of distributions. Combining these, we see that $\sum_{j \in \mathbb{Z}} \varsigma_j^{\left(2^j\right)}$ converges in the sense of distributions. The same proof shows that the sum, in fact, converges in the sense of tempered distributions.

Let $\mathcal{B} \subset \mathcal{S}_0\left(\mathbb{R}^n\right)$ be a bounded set. For $g \in \mathcal{B}$, define $g_j = 2^{|j|} \varsigma_j^{\left(2^j\right)} * g$. Lemma 1.1.19 shows $\left\{g_j \mid j \in \mathbb{Z}, g \in \mathcal{B}\right\} \subset \mathcal{S}_0\left(\mathbb{R}^n\right)$ is a bounded set, and we have

$$\sum_{j \in \mathbb{Z}} \mathrm{Op}\left(\varsigma_j^{\left(2^j\right)}\right) g = \sum_{j \in \mathbb{Z}} 2^{-|j|} g_j,$$

and it follows that $\sum_{j \in \mathbb{Z}} \mathrm{Op}\left(\varsigma_j^{\left(2^j\right)}\right)$ converges in the topology of bounded convergence as operators $\mathcal{S}_0\left(\mathbb{R}^n\right) \to \mathcal{S}_0\left(\mathbb{R}^n\right)$. The equality (1.9) follows easily from the above proofs. $\qquad\square$

DEFINITION 1.1.21. *For $x \in \mathbb{R}^n$, we define a tempered distribution δ_x by*

$$\int \delta_x(y) f(y) \ dy = f(x), \quad f \in \mathcal{S}(\mathbb{R}^n).$$

δ_x *is called the Dirac δ function at x.*

LEMMA 1.1.22. *There is a function $\varsigma \in \mathcal{S}_0(\mathbb{R}^n)$ such that*

$$\delta_0 = \sum_{j \in \mathbb{Z}} \varsigma^{(2^j)},$$

where δ_0 denotes the Dirac δ function at 0.

PROOF. Let $\psi \in C_0^\infty(\mathbb{R}^n)$ equal 1 on a neighborhood of 0. Define

$$\hat{\varsigma}(\xi) = \psi(\xi) - \psi(2\xi).$$

Notice,

$$\sum_{j \in \mathbb{Z}} \hat{\varsigma}(2^{-j}\xi) = 1,$$

in the sense of tempered distributions. Taking the inverse Fourier transform of both sides yields the result. $\qquad \square$

We have the following characterization of Calderón-Zygmund kernels.

THEOREM 1.1.23. *Let $K \in \mathcal{S}_0(\mathbb{R}^n)'$. The following are equivalent:*

(i) *K is a Calderón-Zygmund kernel.*

(ii) *$\mathrm{Op}(K) : \mathcal{S}_0(\mathbb{R}^n) \to \mathcal{S}_0(\mathbb{R}^n)$ and for any bounded set $\mathcal{B} \subset \mathcal{S}_0(\mathbb{R}^n)$, the set*

$$\left\{ g \in \mathcal{S}_0(\mathbb{R}^n) \ \middle| \ \exists R > 0, f \in \mathcal{B}, g^{(R)} = \mathrm{Op}(K) f^{(R)} \right\} \subset \mathcal{S}_0(\mathbb{R}^n)$$

is a bounded set.

(iii) *For each $j \in \mathbb{Z}$ there is a function $\varsigma_j \in \mathcal{S}_0(\mathbb{R}^n)$ with $\{\varsigma_j \mid j \in \mathbb{Z}\} \subset \mathcal{S}_0(\mathbb{R}^n)$ a bounded set and such that*

$$K = \sum_{j \in \mathbb{Z}} \varsigma_j^{(2^j)}.$$

See Lemma 1.1.20 for a discussion of the convergence of this sum.

PROOF. We begin with (i)$\Rightarrow$(ii), and we use Theorem 1.1.5. Let K be a Calderón-Zygmund kernel and let $\mathcal{B} \subset \mathcal{S}_0(\mathbb{R}^n)$ be bounded. Let

$$\mathcal{T} := \left\{ g \mid \exists R > 0, f \in \mathcal{B}, g^{(R)} = \mathrm{Op}(K) f^{(R)} \right\}.$$

We wish to show $\mathcal{T}$ is a bounded subset of $\mathcal{S}(\mathbb{R}^n)$ and moreover $\mathcal{T} \subset \mathcal{S}_0(\mathbb{R}^n)$. Notice that if $g^{(R)} = \mathrm{Op}(K) f^{(R)}$, then

$$\hat{g}(\xi) = \widehat{K}(R\xi) \hat{f}(\xi).$$

Thus, to show $\mathcal{T}$ is a bounded subset of $\mathcal{S}(\mathbb{R}^n)$, and using the fact that the Fourier transform is an automorphism of $\mathcal{S}(\mathbb{R}^n)$, it suffices to show

$$\widehat{\mathcal{T}} := \left\{ \widehat{K}(R\xi) \hat{f}(\xi) \mid R > 0, f \in \mathcal{B} \right\}$$

is a bounded subset of $\mathcal{S}(\mathbb{R}^n)$. But, we have for $f \in \mathcal{S}_0(\mathbb{R}^n)$, using Theorem 1.1.5,

$$\left\| \widehat{K}(R\xi) \hat{f}(\xi) \right\|_{\alpha,\beta} = \left\| \left(|\xi|^{2|\alpha|} \widehat{K}(R\xi) \right) \left(|\xi|^{-2|\alpha|} \hat{f}(\xi) \right) \right\|_{\alpha,\beta}$$

$$\lesssim \sum_{\substack{|\alpha'| \le 3|\alpha| \\ |\beta'| \le |\beta|}} \left\| \triangle^{-2|\alpha|} f \right\|_{\alpha',\beta'}.$$

The right-hand side is a continuous semi-norm on $\mathcal{S}_0(\mathbb{R}^n)$ (by Corollary 1.1.13). Taking the supremum over $f \in \mathcal{B}$ and $R > 0$ shows that $\sup_{\hat{g} \in \widehat{\mathcal{T}}} \|\hat{g}\|_{\alpha,\beta} < \infty$, and it follows that $\mathcal{T}$ is a bounded subset of $\mathcal{S}(\mathbb{R}^n)$. We wish to show that $\mathcal{T} \subset \mathcal{S}_0(\mathbb{R}^n)$. Indeed, let $\hat{g} \in \widehat{\mathcal{T}}$, so that $\hat{g}(\xi) = \widehat{K}(R\xi) \hat{f}(\xi)$. By Lemma 1.1.12, we wish to show $\partial_\xi^\alpha \widehat{K}(R\xi) \hat{f}(\xi) \big|_{\xi=0} = 0$, $\forall \alpha$. But this follows immediately from Lemma 1.1.12 and Theorem 1.1.5.

We turn to (ii)$\Rightarrow$(iii); let K be as in (ii). We apply Lemma 1.1.22 to decompose $\delta_0 = \sum_{j \in \mathbb{Z}} \varsigma^{(2^j)}$. Lemma 1.1.20 shows that $I = \sum_{j \in \mathbb{Z}} \mathrm{Op}\left(\varsigma^{(2^j)} \right)$, where the sum is taken in the topology of bounded convergence as operators $\mathcal{S}_0(\mathbb{R}^n) \to \mathcal{S}_0(\mathbb{R}^n)$, and I denotes the identity operator $\mathcal{S}_0(\mathbb{R}^n) \to \mathcal{S}_0(\mathbb{R}^n)$. Let $\varsigma_j^{(2^j)} = \mathrm{Op}(K) \varsigma^{(2^j)}$, so that our assumption implies $\left\{ \varsigma_j \mid j \in \mathbb{Z} \right\} \subset \mathcal{S}_0(\mathbb{R}^n)$ is bounded. We have

$$\mathrm{Op}(K) = \mathrm{Op}(K) I = \sum_{j \in \mathbb{Z}} \mathrm{Op}(K) \mathrm{Op}\left(\varsigma^{(2^j)} \right) = \sum_{j \in \mathbb{Z}} \mathrm{Op}\left(\varsigma_j^{(2^j)} \right).$$

Lemma 1.1.20 then shows that $K = \sum_{j \in \mathbb{Z}} \varsigma_j^{(2^j)}$.

Finally, we prove (iii)$\Rightarrow$(i); let $K = \sum_{j \in \mathbb{Z}} \varsigma_j^{(2^j)}$ as in (iii), where $\left\{ \varsigma_j \mid j \in \mathbb{Z} \right\} \subset \mathcal{S}(\mathbb{R}^n)$ is bounded. We need to show $\sum_{j \in \mathbb{Z}} \varsigma_j^{(2^j)}$ converges in distribution to a Calderón-Zygmund kernel. We have for $x \ne 0$,

$$\left| \partial_x^\alpha \sum_{j \in \mathbb{Z}} \varsigma_j^{(2^j)}(x) \right| \le \sum_{j \in \mathbb{Z}} 2^{nj + |\alpha|j} \left| (\partial_x^\alpha \varsigma_j)(2^j x) \right|$$

$$\lesssim \sum_{j \in \mathbb{Z}} 2^{nj + |\alpha|j} \left(1 + |2^j x| \right)^{-n - |\alpha| - 1}$$

$$\lesssim |x|^{-n - |\alpha|},$$

which establishes the growth condition.

We now verify the cancellation condition. Let $\mathcal{B} \subset C_0^\infty(\mathbb{R}^n)$ be a bounded set. Write $\varsigma_j = \triangle \widetilde{\varsigma}_j$ where, by Corollary 1.1.13, $\{\widetilde{\varsigma}_j \mid j \in \mathbb{Z}\} \subset \mathcal{S}_0(\mathbb{R}^n)$ is a bounded set. For $\phi \in \mathcal{B}$ and $R > 0$, we have

$$\left| \int \sum_{j \in \mathbb{Z}} \varsigma_j^{(2^j)}(x) \, \phi(Rx) \, dx \right|$$

$$\lesssim \sum_{2^j \geq R} 2^{-2j} R^2 \int \left| \widetilde{\varsigma}_j^{(2^j)}(x) \, (\triangle \phi)(Rx) \right| \, dx + \sum_{2^j < R} \int \left| \phi(2^{-j} Rx) \right| \, dx$$

$$\lesssim \sum_{2^j \geq R} 2^{-2j} R^2 + \sum_{2^j < R} (2^j R^{-1})^n$$

$$\lesssim 1,$$

which completes the proof.   □

The proof of Theorem 1.1.23 used the Fourier transform many times. However, it is the statement of Theorem 1.1.23 (which does not mention the Fourier transform) which is of the most interest to us. Indeed, when we leave the translation invariant Euclidean setting, the Fourier transform will be much less applicable. However, we will be able to prove an analog of Theorem 1.1.23 using other methods.

It is not hard to see that many of the conclusions above concerning Calderón-Zygmund kernels, whose proofs used the Fourier transform, can be reproved using the conclusions of Theorem 1.1.23. For instance, that operators of the form $\mathrm{Op}(K)$ form an algebra follows immediately from (ii). Also, that $\mathrm{Op}(K)$ is bounded on L^2 follows from (iii) and the Cotlar-Stein Lemma (Lemma 1.2.26). Indeed, it follows easily from Lemma 1.1.19 that

$$\left\| \mathrm{Op}\left(\varsigma_j^{(2^j)} \right)^* \mathrm{Op}\left(\varsigma_k^{(2^k)} \right) \right\|_{L^2 \to L^2} \lesssim 2^{-|j-k|},$$

and the L^2 boundedness of $\mathrm{Op}(K)$ follows from the Cotlar-Stein Lemma. Of course, these ideas are somewhat circular: the proof of Theorem 1.1.23 uses the Fourier transform. As we will see, it is possible to prove Theorem 1.1.23, and more general analogs, without direct use of the Fourier transform, whereby the above ideas become much more useful.

The operators discussed in this section are operators of "order 0." When we move to a more general setting, we will work with operators of other orders. We state here the basic definitions and results for operators of orders other than 0. We will prove more general analogs of these results in later chapters.

DEFINITION 1.1.24. *We say $K \in C_0^\infty(\mathbb{R}^n)'$ is a Calderón-Zygmund kernel of order $t \in (-n, \infty)$ if:*

(i) *(Growth Condition) For every multi-index α, $|\partial_x^\alpha K(x)| \leq C_\alpha |x|^{-n-t-|\alpha|}$. In particular, we assume that $K(x)$ is a C^∞ function for $x \neq 0$.*

(ii) (*Cancellation Condition*) *For every bounded set* $\mathcal{B} \subset C_0^\infty (\mathbb{R}^n)$, *we assume*

$$\sup_{\substack{\phi \in \mathcal{B} \\ R > 0}} R^{-t} \left| \int K(x) \, \phi(Rx) \, dx \right| \le C_\mathcal{B}.$$

Remark 1.1.25 When $-n < t < 0$, the Cancellation Condition follows from the Growth Condition and the weaker assumption that K has no part supported at 0. I.e., suppose K is a distribution satisfying the Growth Condition where $t \in (-n, 0)$. Define the function

$$\widetilde{K}(x) = \begin{cases} K(x) & \text{if } x \ne 0, \\ 0 & \text{if } x = 0. \end{cases}$$

If $K - \widetilde{K} = 0$ as a distribution, then K is a Calderón-Zygmund kernel of order t. Conversely, if K is a Calderón-Zygmund kernel of order t, then $K - \widetilde{K} = 0$. In this sense, the Cancellation Condition is superfluous for $-n < t < 0$.

As before, given a Calderón-Zygmund kernel, we define an operator $\mathrm{Op}(K) :$ $C_0^\infty(\mathbb{R}^n) \to C^\infty(\mathbb{R}^n)$ by $\mathrm{Op}(K) f = K * f$. Similar to the case of operators of order 0, a density argument shows that to uniquely determine a Calderón-Zygmund kernel K of any order $t \in (-n, \infty)$, it suffices to consider $K\big|_{\mathcal{S}_0(\mathbb{R}^n)}$. Just as in Theorem 1.1.23, we may use this to characterize Calderón-Zygmund kernels in other ways, as the next theorem shows. We state this theorem without proof.

THEOREM 1.1.26. *Fix* $t \in (-n, \infty)$, *and let* $K \in \mathcal{S}_0(\mathbb{R}^n)'$. *The following are equivalent:*

(i) K *is a Calderón-Zygmund kernel of order* t.

(ii) $\mathrm{Op}(K) : \mathcal{S}_0(\mathbb{R}^n) \to \mathcal{S}_0(\mathbb{R}^n)$ *and for any bounded set* $\mathcal{B} \subset \mathcal{S}_0(\mathbb{R}^n)$, *the set*

$$\left\{ g \in \mathcal{S}_0(\mathbb{R}^n) \mid \exists R > 0, f \in \mathcal{B}, g^{(R)} = R^{-t} \mathrm{Op}(K) f^{(R)} \right\} \subset \mathcal{S}_0(\mathbb{R}^n)$$

is a bounded set.

(iii) *For each* $j \in \mathbb{Z}$ *there is a function* $\varsigma_j \in \mathcal{S}_0(\mathbb{R}^n)$ *with* $\{ \varsigma_j \mid j \in \mathbb{Z} \} \subset \mathcal{S}_0(\mathbb{R}^n)$ *a bounded set and such that*

$$K = \sum_{j \in \mathbb{Z}} 2^{jt} \varsigma_j^{(2^j)}.$$

The above sum converges in distribution, though the equality is taken in the sense of elements of $\mathcal{S}_0(\mathbb{R}^n)'$.

Furthermore, (ii) *and* (iii) *are equivalent for any* $t \in \mathbb{R}$.

For $t \le -n$, we define Calderón-Zygmund kernels in the following way.

DEFINITION 1.1.27. *Let* $K \in \mathcal{S}_0(\mathbb{R}^n)'$ *and* $t \in \mathbb{R}$. *We say* K *is a Calderón-Zygmund kernel of order* t *if either of the two equivalent conditions* (ii) *or* (iii) *of Theorem 1.1.26 holds.*

Remark 1.1.28   Restricting a tempered distribution to $\mathcal{S}_0\left(\mathbb{R}^n\right)$ does not uniquely determine the distribution (polynomials are all 0, when thought of in the dual to $\mathcal{S}_0\left(\mathbb{R}^n\right)$). For Calderón-Zygmund operators of order $t > -n$, we used a density argument to uniquely pick out a distribution, given its values on $\mathcal{S}_0\left(\mathbb{R}^n\right)$. For $t \leq -n$, this procedure does not always work. For now, we satisfy ourselves by working only with operators defined on $\mathcal{S}_0\left(\mathbb{R}^n\right)$. Later in the monograph, (in Chapters 2 and 5) we restrict attention to non-translation invariant operators whose Schwartz kernels have compact support, and use this to avoid this non-uniqueness problem.

Remark 1.1.29   With the above definitions, $\triangle^s$ is an isomorphism of Calderón-Zygmund kernels of order t to Calderón-Zygmund kernels of order $t + 2s$.[10] This gives another (equivalent) way to extend Definition 1.1.24 to kernels of order $t \leq -n$. Indeed, we say $K \in \mathcal{S}_0\left(\mathbb{R}^n\right)'$ is a Calderón-Zygmund kernel of order t if $\triangle^{-t/2}K$ is a Calderón-Zygmund kernel of order 0. As pointed out in the previous remark, though, this only uniquely specifies the kernel as an element of $\mathcal{S}_0\left(\mathbb{R}^n\right)'$, and not as a distribution.

Remark 1.1.30   Notice the scale invariance of conditions in Theorem 1.1.26. For instance, consider (ii). If $\mathrm{Op}\left(K\right) : \mathcal{S}_0\left(\mathbb{R}^n\right) \to \mathcal{S}_0\left(\mathbb{R}^n\right)$,[11] then it follows by continuity that for any bounded set $\mathcal{B} \subset \mathcal{S}_0\left(\mathbb{R}^n\right)$, $\mathrm{Op}\left(K\right)\mathcal{B}$ is also bounded. (ii) takes this automatic fact, and instead assumes a *scale invariant* version of it.

This leads us directly to the first main property of Calderón-Zygmund kernels.

THEOREM 1.1.31.   *Suppose $K_1, K_2 \in \mathcal{S}_0\left(\mathbb{R}^n\right)'$ are Calderón-Zygmund kernels of order $s, t \in \mathbb{R}$, respectively. Then $\mathrm{Op}\left(K_1\right)\mathrm{Op}\left(K_2\right) = \mathrm{Op}\left(K_3\right)$, where $K_3 \in \mathcal{S}_0\left(\mathbb{R}^n\right)'$ is a Calderón-Zygmund kernel of order $s + t$.*

PROOF.   This is an immediate consequence of (ii) of Theorem 1.1.26.   □

To discuss the L^p boundedness of these operators, we need appropriate L^p Sobolev spaces. For $1 < p < \infty$ and $s \in \mathbb{R}$, we define $\mathring{L}^p_s\left(\mathbb{R}^n\right)$ to be the completion of $\mathcal{S}_0\left(\mathbb{R}^n\right)$ under the following norm:

$$\|f\|_{\mathring{L}^p_s(\mathbb{R}^n)} := \left\|\triangle^{s/2}f\right\|_{L^p(\mathbb{R}^n)}.$$

As mentioned before, $\mathring{L}^p_0\left(\mathbb{R}^n\right) = L^p\left(\mathbb{R}^n\right)$ $(1 < p < \infty)$.

Remark 1.1.32   Unlike the usual Sobolev spaces, elements of $\mathring{L}^p_s$ cannot be identified with distributions if $s << 0$; instead they can be thought of as elements of $\mathcal{S}_0\left(\mathbb{R}^n\right)'$.

THEOREM 1.1.33.   *Let K be a Calderón-Zygmund kernel of order t. Then, $\mathrm{Op}\left(K\right)$: $\mathring{L}^p_s\left(\mathbb{R}^n\right) \to \mathring{L}^p_{s-t}\left(\mathbb{R}^n\right)$.*

To prove Theorem 1.1.33, we need a lemma.

---

[10]This follows by combining Lemma 1.1.34 with Theorem 1.1.31.

[11]Here, and in the rest of the monograph, when we are given a linear operator $T : V \to W$, where V and W are topological vector spaces, we always assume that T is continuous.

LEMMA 1.1.34. *For $s \in \mathbb{R}$, the operator $\triangle^s : \mathcal{S}_0(\mathbb{R}^n) \to \mathcal{S}_0(\mathbb{R}^n)$ is of the form* $\mathrm{Op}(K)$ *for a Calderón-Zygmund kernel K of order $2s$.*

PROOF. This follows easily by using (iii) of Theorem 1.1.26. Indeed, let $\widehat{\phi} \in C_0^\infty(\mathbb{R}^n)$ equal 1 on a neighborhood of 0 and set $\widehat{\psi}(\xi) = \widehat{\phi}(\xi) - \widehat{\phi}(2\xi)$ so that $\sum_{j \in \mathbb{Z}} \widehat{\psi}(2^{-j}\xi) = 1$ in the sense of tempered distributions. Let $\widehat{\varsigma} = |2\pi\xi|^{2s}\,\widehat{\psi}$, so that $|2\pi\xi|^{2s} = \sum_{j \in \mathbb{Z}} 2^{2js}\widehat{\varsigma}(2^{-j}\xi)$. Define $\varsigma = (\widehat{\varsigma})^{\vee}$–the inverse Fourier transform of $\widehat{\varsigma}$. Note that $\varsigma \in \mathcal{S}_0(\mathbb{R}^n)$. We have $\triangle^s = \sum_{j \in \mathbb{Z}} 2^{2js}\mathrm{Op}\left(\varsigma^{(2^j)}\right) = \sum_{j \in \mathbb{Z}} 2^{2js}\mathrm{Op}(\varsigma)^{(2^j)}$. Thus (iii) of Theorem 1.1.26 holds and $\triangle^s$ is a Calderón-Zygmund operator of order $2s$. $\qquad\square$

PROOF OF THEOREM 1.1.33. Let K be a Calderón-Zygmund kernel of order t. For $f \in \mathcal{S}_0(\mathbb{R}^n)$, we wish to show

$$\left\| \triangle^{(s-t)/2}\mathrm{Op}(K)\,f \right\|_{L^p} \lesssim \left\| \triangle^{s/2} f \right\|_{L^p}.$$

By Corollary 1.1.13 we may write $f = \triangle^{-s/2}g$, $g \in \mathcal{S}_0(\mathbb{R}^n)$, and we therefore wish to show

$$\left\| \triangle^{(s-t)/2}\mathrm{Op}(K)\,\triangle^{-s/2}\,g \right\|_{L^p} \lesssim \|g\|_{L^p}.$$

This follows from the fact that $\triangle^{(s-t)/2}\mathrm{Op}(K)\,\triangle^{-s/2}$ is bounded on L^p, by Theorem 1.1.5, as it is an operator of order 0 (by Theorem 1.1.31 and Lemma 1.1.34). $\qquad\square$

1.2 NON-TRANSLATION INVARIANT OPERATORS

Above, we discussed translation invariant operators: operators of the form $\mathrm{Op}(K)\,f = f * K$. To understand these operators, we made significant use of the Fourier transform. Even though the Fourier transform is less applicable, many of the ideas have generalizations to the non-translation invariant setting. First, a bit of notation. For an operator $T : C_0^\infty(\mathbb{R}^n) \to C_0^\infty(\mathbb{R}^n)'$, we identify T with its Schwartz kernel, $T(x,y) \in C_0^\infty(\mathbb{R}^n \times \mathbb{R}^n)'$ (see Theorem A.1.30 and the remarks surrounding it for this identification). We use the identification of functions in L^1_{loc} with certain distributions, as in the previous discussion. Thus, given an operator, we often treat it as a function, and when we do, we are assuming that its Schwartz kernel is given by integration against an $L^1_{\mathrm{loc}}(\mathbb{R}^n \times \mathbb{R}^n)$ function. Conversely, given a function in $L^1_{\mathrm{loc}}(\mathbb{R}^n \times \mathbb{R}^n)$ we will treat it as an operator $C_0^\infty(\mathbb{R}^n) \to C_0^\infty(\mathbb{R}^n)'$, by this same identification. See Appendix A.1.1 for more details on this notation.

DEFINITION 1.2.1. *We say $T : C_0^\infty(\mathbb{R}^n) \to C^\infty(\mathbb{R}^n)$ is a Calderón-Zygmund operator of order $t \in (-n, \infty)$ if:*

(i) *(Growth Condition) For every multi-indices α and β,*

$$\left| \partial_x^\alpha \partial_y^\beta T(x,y) \right| \leq C_{\alpha,\beta} |x-y|^{-n-t-|\alpha|-|\beta|}.$$

In particular, we assume $T(x,y)$ is a C^∞ function for $x \neq y$.

(ii) *(Cancellation Condition) For all bounded sets* $\mathcal{B} \subset C_0^\infty(\mathbb{R}^n)$ *and for all* $\phi \in \mathcal{B}$, $R > 0$, *and* $z \in \mathbb{R}^n$, *define* $\phi_{R,z}(x) = \phi(R(x-z))$. *We assume, for every multi-index* α,

$$\sup_{\phi \in \mathcal{B}} \sup_{R > 0} \sup_{x, z \in \mathbb{R}^n} R^{-t-|\alpha|} |\partial_x^\alpha T \phi_{R,z}(x)| \leq C_{\mathcal{B}, \alpha},$$

with the same estimates for T *replaced by* T^*, *the formal* L^2 *adjoint of* T.

Remark 1.2.2   Above we used the formal L^2 adjoint of T. To define this, we first define the transpose, T^t. The Schwartz kernel of T^t is defined by $T^t(x,y) = T(y,x)$; more precisely,

$$\int T^t(x,y) \, \phi(x,y) \, dx \, dy = \int T(x,y) \, \phi(y,x) \, dx \, dy,$$

for $\phi \in C_0^\infty(\mathbb{R}^n \times \mathbb{R}^n)$, where, as usual, we have written the pairing between distributions and test functions as integration. We define the Schwartz kernel of T^* by $T^* = \overline{T^t}$, where $\bar{z}$ denotes the complex conjugate of z. Here, for a distribution λ, we are defining the distribution $\bar{\lambda}$ by $\bar{\lambda}(f) = \overline{\lambda(\bar{f})}$.

A key tool for studying these operators is a characterization similar to Theorem 1.1.26. For this, we need a generalization of operators of the form $\mathrm{Op}(\varsigma)$, where $\varsigma \in \mathcal{S}_0(\mathbb{R}^n)$, to the non-translation invariant setting. We begin with a generalization of $\mathrm{Op}(f)$, where $f \in \mathcal{S}(\mathbb{R}^n)$.

DEFINITION 1.2.3.   *We define* $\mathcal{P} \subset C^\infty(\mathbb{R}^n \times \mathbb{R}^n)$ *to be the Fréchet space of functions* $E(x,y) \in C^\infty(\mathbb{R}^n \times \mathbb{R}^n)$ *satisfying for every* $m \in \mathbb{N}$, *and every multi-indices* α, β,

$$\left| \partial_x^\alpha \partial_y^\beta E(x,y) \right| \leq C_{\alpha,\beta,m} (1 + |x-y|)^{-m}.$$

We give $\mathcal{P}$ *the coarsest topology such that the least possible* $C_{\alpha,\beta,m}$ *defines a continuous semi-norm on* $\mathcal{P}$, *for each choice of* α, β, *and* m.

LEMMA 1.2.4.   *Let* $\mathcal{B} \subset \mathcal{P}$. *For each* $E \in \mathcal{B}$ *define two new functions* $E_1(x,z) = E(x, x-z)$, $E_2(z,y) = E(y-z, y)$. *Fix* α *and define* $\mathcal{B}_\alpha \subset C^\infty(\mathbb{R}^n)$ *by*

$$\mathcal{B}_\alpha := \left\{ \partial_x^\alpha E_1(x, \cdot), \partial_y^\alpha E_2(\cdot, y) \mid x, y \in \mathbb{R}^n, E \in \mathcal{B} \right\}.$$

Then, for every α, $\mathcal{B}_\alpha \subset \mathcal{S}(\mathbb{R}^n)$. *Furthermore,* $\mathcal{B}$ *is a bounded subset of* $\mathcal{P}$ *if and only if* $\mathcal{B}_\alpha$ *is a bounded subset of* $\mathcal{S}(\mathbb{R}^n)$, *for every* α.

PROOF.   This follows immediately from the definitions.   □

Remark 1.2.5   We think of the elements of $\mathcal{P}$ as *operators*, by identifying the function $E \in \mathcal{P}$ with the operator whose Schwartz kernel is given by integration against E,

$$Ef(x) = \int E(x,y) f(y) \, dy.$$

DEFINITION 1.2.6. *We define $\mathcal{P}_0 \subset \mathcal{P}$ to consist of those $E \in \mathcal{P}$ such that $\forall x \in \mathbb{R}^n$, $E(x, \cdot) \in \mathcal{S}_0(\mathbb{R}^n)$ and $\forall y \in \mathbb{R}^n$, $E(\cdot, y) \in \mathcal{S}_0(\mathbb{R}^n)$.*

Remark 1.2.7 $\mathcal{P}_0$ is a closed subspace of $\mathcal{P}$ and we give it the induced Fréchet topology.

LEMMA 1.2.8. *For $s \in \mathbb{R}$, the maps $\triangle_x^s, \triangle_y^s : \mathcal{P}_0 \to \mathcal{P}_0$ given by $(\triangle_x^s E)(\cdot, y) = \triangle^s E(\cdot, y)$ and $(\triangle_y^s E)(x, \cdot) = \triangle^s E(x, \cdot)$ are automorphisms of $\mathcal{P}_0$.*

PROOF. It is easy to verify that $\triangle_x^s$ and $\triangle_y^s$ map $\mathcal{P}_0 \to \mathcal{P}_0$. The closed graph theorem (Theorem A.1.14) then shows that both maps are continuous. Their respective continuous inverses are $\triangle_x^{-s}$ and $\triangle_y^{-s}$. $\qquad \square$

Example 1.2.9 *There is a continuous inclusion $\mathcal{S}(\mathbb{R}^n) \hookrightarrow \mathcal{P}$ given by*

$$f \mapsto f(x - y);$$

or if we think of elements of $\mathcal{P}$ as operators,

$$f \mapsto \mathrm{Op}(f).$$

Under this map, $\mathcal{S}_0(\mathbb{R}^n)$ maps in to $\mathcal{P}_0$. Thus, operators in $\mathcal{P}_0$ should be thought of as a non-translation invariant generalization of $\mathrm{Op}(f)$, where $f \in \mathcal{S}_0(\mathbb{R}^n)$.

We define dilations of $\mathcal{P}$ by, for $R > 0$,

$$E^{(R)}(x, y) = R^n E(Rx, Ry).$$

Using this, for $f \in \mathcal{S}(\mathbb{R}^n)$, we have $\mathrm{Op}(f^{(R)}) = \mathrm{Op}(f)^{(R)}$. We have the following result, which is a generalization of Theorem 1.1.26.

THEOREM 1.2.10. *Fix $t \in (-n, \infty)$, and let $T : \mathcal{S}_0(\mathbb{R}^n) \to C^\infty(\mathbb{R}^n)$. The following are equivalent:*

(i) *T is a Calderón-Zygmund operator of order t.[12]*

(ii) *$T : \mathcal{S}_0(\mathbb{R}^n) \to \mathcal{S}_0(\mathbb{R}^n)$, and for any bounded set $\mathcal{B} \subset \mathcal{P}_0$, the set*

$$\left\{ E \mid \exists R > 0, F \in \mathcal{B}, E^{(R)} = R^{-t} T F^{(R)} \right\} \subset \mathcal{P}_0$$

is a bounded set.

---

[12]As in Theorem 1.1.26, when we say $T : \mathcal{S}_0(\mathbb{R}^n) \to \mathcal{S}_0(\mathbb{R}^n)$ is a Calderón-Zygmund operator of order $t \in (-n, \infty)$ we mean that there is a Calderón-Zygmund operator $\widetilde{T} : C_0^\infty(\mathbb{R}^n) \to C^\infty(\mathbb{R}^n)$ such that $\widetilde{T}\big|_{\mathcal{S}_0(\mathbb{R}^n)} = T$. We will see in the proof that follows that if $\widetilde{T} : C_0^\infty(\mathbb{R}^n) \to C^\infty(\mathbb{R}^n)$ is a Calderón-Zygmund operator of order $t \in (-n, \infty)$, then $\widetilde{T}\big|_{\mathcal{S}_0(\mathbb{R}^n)} : \mathcal{S}_0(\mathbb{R}^n) \to \mathcal{S}_0(\mathbb{R}^n)$.

(iii) *For each $j \in \mathbb{Z}$ there is $E_j \in \mathcal{P}_0$ with $\{E_j \mid j \in \mathbb{Z}\} \subset \mathcal{P}_0$ a bounded set and such that*

$$T = \sum_{j \in \mathbb{Z}} 2^{jt} E_j^{(2^j)},$$

where above sum converges in the topology of bounded convergence as operators $\mathcal{S}_0(\mathbb{R}^n) \to \mathcal{S}_0(\mathbb{R}^n)$.

Furthermore, (ii) and (iii) are equivalent for any $t \in \mathbb{R}$.

Remark 1.2.11   When we move to more general situations, we will call operators analogous to operators like $E^{(R)}$, where $E \in \mathcal{P}_0$, *elementary operators*.

To prove Theorem 1.2.10, we need several lemmas.

LEMMA 1.2.12.   *Let $\mathcal{E} \subset \mathcal{P}$ be a bounded set. For each $E_1, E_2 \in \mathcal{E}$ and $j_1, j_2 \in \mathbb{R}$ define an operator F by*

$$F^{(2^{j_1 \wedge j_2})} = E_1^{(2^{j_1})} E_2^{(2^{j_2})}.$$

Then, $\forall \alpha, \beta, \exists N = N(\alpha, \beta), \forall m, \exists C, \forall E_1, E_2 \in \mathcal{E}, \forall j_1, j_2 \in \mathbb{R}$,

$$2^{-N|j_1 - j_2|} \left| \partial_x^\alpha \partial_z^\beta F(x, z) \right| \le C (1 + |x - y|)^{-m}.$$

PROOF.   The conclusion of the lemma is equivalent to

$$2^{-N|j_1 - j_2|} \left| \left(2^{-j_1 \wedge j_2} \partial_x \right)^\alpha \left(2^{-j_1 \wedge j_2} \partial_z \right)^\beta F^{(2^{j_1 \wedge j_2})}(x, z) \right|$$
$$\le C 2^{n j_1 \wedge j_2} \left(1 + 2^{j_1 \wedge j_2} |x - y| \right)^{-m}.$$

By taking N large in terms of α and β, the previous equation follows from

$$\left| \left(2^{-j_1} \partial_x \right)^\alpha \left(2^{-j_2} \partial_z \right)^\beta F^{(2^{j_1 \wedge j_2})}(x, z) \right| \le C 2^{n j_1 \wedge j_2} \left(1 + 2^{j_1 \wedge j_2} |x - y| \right)^{-m}.$$

$\partial_x^\alpha E_1(x, y)$ and $\partial_z^\beta E_2(y, z)$ are of the same forms as E_1 and E_2 (i.e., as E_1 and E_2 range over $\mathcal{E}$, $\partial_x^\alpha E_1(x, y)$ and $\partial_z^\beta E_2(y, z)$ range over a bounded set in $\mathcal{P}$). Thus, we may replace $\left(2^{-j_1} \partial_x^\alpha \right) E_1^{(2^{j_1})}(x, y)$ and $\left(2^{-j_2} \partial_z^\alpha \right) E_2^{(2^{j_2})}(y, z)$ with $E_1^{(2^{j_1})}$ and $E_2^{(2^{j_2})}$, respectively; i.e., it suffices to prove the case when $\alpha = \beta = 0$ and $N = 0$. From here, the result follows by a straightforward estimate, and is left to the reader. $\square$

LEMMA 1.2.13.   *Let $\mathcal{E} \subset \mathcal{P}$ be a bounded set. For each $E_1, E_2 \in \mathcal{E}$ and $j_1, j_2 \in \mathbb{R}$ define two operators F_1 and F_2 by*

$$F_1^{(2^{j_1})} = E_1^{(2^{j_1})} E_2^{(2^{j_2})}, \quad F_2^{(2^{j_2})} = E_1^{(2^{j_1})} E_2^{(2^{j_2})}.$$

Then, $\forall \alpha, \beta, m, \exists N = N(\alpha, \beta, m), \exists C, \forall E_1, E_2 \in \mathcal{E}, \forall j_1, j_2 \in \mathbb{R}$, for $k = 1, 2$,

$$2^{-N|j_1 - j_2|} \left| \partial_x^\alpha \partial_z^\beta F_k(x, z) \right| \le C (1 + |x - y|)^{-m}.$$

PROOF. We prove the result for $k = 1$; the proof for $k = 2$ is similar. The conclusion of the lemma, in this case, is equivalent to

$$2^{-N|j_1-j_2|} \left| \left(2^{-j_1} \partial_x\right)^\alpha \left(2^{-j_1} \partial_z\right)^\beta F_1^{(2^{j_1})} (x, z) \right| \le C2^{nj_1} \left(1 + 2^{j_1} |x - y|\right)^{-m}.$$

By taking $N = N_1 + N_2$ where $N_1 = N_1(\alpha, \beta, m, n)$ is large and N_2 is to be chosen later, this follows from

$$2^{-N_2|j_1-j_2|} \left| \left(2^{-j_1 \wedge j_2} \partial_x\right)^\alpha \left(2^{-j_1 \wedge j_2} \partial_z\right)^\beta F_1^{(2^{j_1})} (x, z) \right|$$
$$\le C2^{nj_1 \wedge j_2} \left(1 + 2^{j_1 \wedge j_2} |x - y|\right)^{-m}.$$

This is the conclusion of Lemma 1.2.12. □

LEMMA 1.2.14. *Suppose $\mathcal{B} \subset \mathcal{P}_0$ is a bounded set. For $E_1, E_2 \in \mathcal{B}$ and $j_1, j_2 \in \mathbb{R}$, define two operators $F_1 = F_1(E_1, E_2, j_1, j_2)$ and $F_2 = F_2(E_1, E_2, j_1, j_2)$ by*

$$F_1^{(2^{j_1})} = E_1^{(2^{j_1})} E_2^{(2^{j_2})}, \quad F_2^{(2^{j_2})} = E_1^{(2^{j_1})} E_2^{(2^{j_2})}.$$

Then, for every N, the set

$$\left\{ 2^{N|j_1-j_2|} F_1, 2^{N|j_1-j_2|} F_2 \mid E_1, E_2 \in \mathcal{B}, j_1, j_2 \in \mathbb{R} \right\} \subset \mathcal{P}_0$$

is a bounded set.

PROOF. We prove that

$$\left\{ 2^{N|j_1-j_2|} F_1, 2^{N|j_1-j_2|} F_2 \mid E_1, E_2 \in \mathcal{B}, j_1, j_2 \in \mathbb{R} \right\} \subset \mathcal{P} \qquad (1.10)$$

is a bounded set. Once this is done, the result will follow, since it is immediate to verify that if $F_1, F_2 \in \mathcal{P}$, they are in fact in $\mathcal{P}_0$, by using that $E_1, E_2 \in \mathcal{P}_0$. We show

$$\left\{ 2^{N|j_1-j_2|} F_1 \mid E_1, E_2 \in \mathcal{B}, j_1, j_2 \in \mathbb{R} \right\} \subset \mathcal{P}$$

is a bounded set. The proof where F_1 is replaced by F_2 is similar.

Fix N, α, β, and m. Take $M = M(N, \alpha, \beta, m)$ large, to be chosen later. We separate into two cases. If $j_1 \ge j_2$, we define $\tilde{E}_1(x, y) = \triangle_y^M E_1(x, y)$ and $\tilde{E}_2(x, y) = \triangle_x^{-M} E_2(x, y)$. If $j_2 > j_1$ we define $\tilde{E}_1(x, y) = \triangle_y^{-M} E_1$ and $\tilde{E}_2(x, y) = \triangle_x^M E_2(x, y)$. Notice that $\left\{ \tilde{E}_1, \tilde{E}_2 \mid E_1, E_2 \in \mathcal{B}, j_1, j_2 \in \mathbb{R} \right\} \subset \mathcal{P}_0$ is a bounded set and we have

$$F_1^{(2^{j_1})} = 2^{-2M|j_1-j_2|} \tilde{E}_1^{(2^{j_1})} \tilde{E}_2^{(2^{j_2})}.$$

Applying Lemma 1.2.13, and by taking $M = M(N, \alpha, \beta, m)$ sufficiently large, we have

$$\left| \partial_x^\alpha \partial_y^\beta 2^{N|j_1-j_2|} F_1(x, y) \right| \lesssim (1 + |x - y|)^{-m}.$$

This completes the proof. □

LEMMA 1.2.15. *Suppose $\mathcal{B} \subset \mathcal{P}$ and $\mathcal{B}_0 \subset \mathcal{P}_0$ are bounded sets. For $E_1 \in \mathcal{B}_0$, $E_2 \in \mathcal{B}$ and $j_1, j_2 \in \mathbb{R}$ with $j_1 \geq j_2$. Define two operators*

$$F_1^{(2^{j_2})} = E_1^{(2^{j_1})} E_2^{(2^{j_2})}, \quad F_2^{(2^{j_2})} = E_2^{(2^{j_2})} E_1^{(2^{j_1})}.$$

Then, for every N, $\left\{ 2^{N|j_1-j_2|} F_1, 2^{N|j_1-j_2|} F_2 \mid E_1 \in \mathcal{B}_0, E_2 \in \mathcal{B}, j_1 \geq j_2 \right\} \subset \mathcal{P}$ is a bounded set. Note F_1 and F_2 are defined in a slightly different way than the operators of the same name in Lemma 1.2.14.

PROOF. The proof follows in the same way as Lemma 1.2.14 and we leave the details to the interested reader. □

LEMMA 1.2.16. *Let T be a Calderón-Zygmund operator of order $t \in (-n, \infty)$, and let $\mathcal{B} \subset \mathcal{P}_0$ be a bounded set. For $E \in \mathcal{B}$, $j \in \mathbb{R}$, define $F^{(2^j)} = 2^{-jt} T E^{(2^j)}$. Then,*

$$\{ F \mid E \in \mathcal{B}, j \in \mathbb{R} \} \subset \mathcal{P}$$

is a bounded set. The same result holds for $F^{(2^j)} = 2^{-jt} E^{(2^j)} T$.

PROOF. We prove the result for $2^{-jt} T E^{(2^j)}$; the result for $2^{-jt} E^{(2^j)} T$ follows by taking adjoints. Fix multi-indices α, β and fix $m \in \mathbb{N}$. We wish to show

$$\left| \left(2^{-j} \partial_x \right)^\alpha \left(2^{-j} \partial_z \right)^\beta F^{(2^j)} (x, z) \right| \lesssim 2^{nj} \left(1 + 2^j |x - z| \right)^{-m}.$$

As E ranges over $\mathcal{B}$, $\partial_z^\beta E$ ranges over a bounded subset of $\mathcal{P}_0$. Thus we may, without loss of generality, assume that $\beta = 0$.

Fix $\phi \in C_0^\infty (B^n (2))$, with $\phi \equiv 1$ on $B^n (1)$. Take $M = M(\alpha, \beta, m)$ and $m' = m'(m)$ large to be chosen later. Set $\widetilde{E}(x, z) = \triangle_x^{-M} E(x, z)$, so that $\widetilde{E}$ ranges over a bounded subset of $\mathcal{P}_0$ as E ranges over $\mathcal{B}$. We have,

$$\left(2^{-j} \partial_x \right)^\alpha F^{(2^j)} (x, z) = 2^{-2Mj-tj} \left[\left(2^{-j} \partial_x \right)^\alpha T \triangle^M \widetilde{E}^{(2^j)} (\cdot, z) \right] (x)$$

$$= 2^{-2Mj-tj} \left[\left(2^{-j} \partial_x \right)^\alpha T \triangle^M \phi \left(2^j (\cdot - x) \right) \widetilde{E}^{(2^j)} (\cdot, z) \right] (x)$$

$$+ 2^{-2Mj-tj} \left[\left(2^{-j} \partial_x \right)^\alpha T \triangle^M \left(1 - \phi \left(2^j (\cdot - x) \right) \right) \widetilde{E}^{(2^j)} (\cdot, z) \right] (x).$$

We bound these two terms separately.

Using the cancellation condition applied with $\phi_{R,z}$ replaced by

$$2^{-nj-2Mj} \triangle^M \phi \left(2^j (\cdot - x) \right) \widetilde{E}^{(2^j)} (\cdot, z) =: \psi$$

we see

$$\left| 2^{-2Mj-tj} \left[\left(2^{-j} \partial_x \right)^\alpha T \triangle^M \phi \left(2^j (\cdot - x) \right) \widetilde{E}^{(2^j)} (\cdot, z) \right] (x) \right|$$

$$\lesssim 2^{nj-tj} \left| \left(2^{-j} \partial_x \right)^\alpha T \psi \right|$$

$$\lesssim 2^{nj} \left(1 + 2^j |x - z| \right)^{-m};$$

here, we have used the rapid decrease of $\widetilde{E}$ to obtain the factor $(1 + 2^j \, |x - z|)^{-m}$.

Using, now, the growth condition

$$
\left| 2^{-2Mj-tj} \left[\left(2^{-j} \partial_x \right)^\alpha T \triangle^M \left(1 - \phi \left(2^j \left(\cdot - x \right) \right) \right) \widetilde{E}^{(2^j)} \left(\cdot, z \right) \right] (x) \right|
$$

$$
\lesssim 2^{-2Mj-tj} \int_{|y-x|>2^{-j}} \left| \triangle_y^M T \left(x, y \right) \left(1 - \phi \left(2^j \left(y - x \right) \right) \right) \widetilde{E}^{(2^j)} \left(y, z \right) \right| \, dy
$$

$$
\lesssim 2^{-2Mj-tj} \int_{|y-x|>2^{-j}} |x - y|^{-n-t-2M} \, 2^{nj} \left(1 + 2^j \, |y - z| \right)^{-m'} \, dy
$$

$$
\lesssim 2^{nj} \int \left(1 + |2^j x - y| \right)^{-n-t-2M} \left(1 + |y - 2^j z| \right)^{-m'} \, dy
$$

$$
\lesssim 2^{nj} \left(1 + 2^j \, |x - z| \right)^{-m},
$$

provided M and m' are sufficiently large. This completes the proof. $\qquad\square$

LEMMA 1.2.17. *Let T be a Calderón-Zygmund operator of order $t \in (-n, \infty)$, and let $\mathcal{B} \subset \mathcal{P}_0$ be a bounded set. For $E_1, E_2 \in \mathcal{B}$, $j_1, j_2 \in \mathbb{R}$, define $F^{(2^{j_1 \wedge j_2})} = 2^{-j_1 \wedge j_2 t} E_1^{(2^{j_1})} T E_2^{(2^{j_2})}$. Then, for every N, $\left\{ 2^{N|j_1-j_2|} F \mid E_1, E_2 \in \mathcal{B}, j_1, j_2 \in \mathbb{R} \right\} \subset \mathcal{P}_0$ is a bounded set.*

PROOF. We prove that $\left\{ 2^{N|j_1-j_2|} F \mid E_1, E_2 \in \mathcal{B}, j_1, j_2 \in \mathbb{R} \right\} \subset \mathcal{P}$ is a bounded set. The result follows, as it is easy to then show that $F \in \mathcal{P}_0$, given that $E_1, E_2 \in \mathcal{P}_0$.

Suppose that $j_2 \leq j_1$. We apply Lemma 1.2.16 to $G^{(2^{j_2})} := 2^{-tj_2} T E_2^{(2^{j_2})}$ to see that $\left\{ G \mid E_2 \in \mathcal{B}, j_2 \in \mathbb{R} \right\} \subset \mathcal{P}$ is a bounded set. The lemma then follows from Lemma 1.2.15.

If, instead, $j_1 \leq j_2$, we instead apply Lemma 1.2.16 to $2^{-tj_1} E_1^{(2^{j_1})} T$ and then apply Lemma 1.2.15 to complete the proof. $\qquad\square$

LEMMA 1.2.18. *For each $j \in \mathbb{Z}$, let $E_j \in \mathcal{P}_0$. Suppose that $\left\{ E_j \mid j \in \mathbb{Z} \right\} \subset \mathcal{P}_0$ is a bounded set. Fix $t \in \mathbb{R}$. Then,*

$$
\sum_{j \in \mathbb{Z}} 2^{jt} E_j^{(2^j)}
$$

converges in the topology of bounded convergence as operators $\mathcal{S}_0 \left(\mathbb{R}^n \right) \to \mathcal{S}_0 \left(\mathbb{R}^n \right)$.

PROOF. This follows just as in Lemma 1.1.20. $\qquad\square$

LEMMA 1.2.19. *There is an operator $E \in \mathcal{P}_0$ with*

$$
I = \sum_{j \in \mathbb{Z}} E^{(2^j)},
$$

with the convergence of the sum taken in the topology of bounded convergence as operators $\mathcal{S}_0 \left(\mathbb{R}^n \right) \to \mathcal{S}_0 \left(\mathbb{R}^n \right)$.

PROOF. Take ς as in Lemma 1.1.22, and set $E = \mathrm{Op}\,(\varsigma)$. The result follows. $\square$

PROOF OF THEOREM 1.2.10. (i)$\Rightarrow$(iii): Let T be a Calderón-Zygmund operator of order $t \in (-n, \infty)$. Let E be as in Lemma 1.2.19. We have,

$$T = ITI = \sum_{j,k \in \mathbb{Z}} E^{(2^j)} T E^{(2^k)}.$$

Setting $F_{j,k}^{(2^{j \wedge k})} = 2^{-(j \wedge k)t} 2^{|j-k|} E^{(2^j)} T E^{(2^k)}$, Lemma 1.2.17 shows that

$$\left\{ F_{j,k} \mid j, k \in \mathbb{Z} \right\} \subset \mathcal{P}_0$$

is a bounded set. We define

$$F_l = \sum_{j \wedge k = l} 2^{-|j-k|} F_{j,k},$$

so that $\left\{ F_l \mid l \in \mathbb{Z} \right\} \subset \mathcal{P}_0$ is a bounded set and

$$T = \sum_{l \in \mathbb{Z}} 2^{lt} F_l^{(2^l)}.$$

This competes the proof of (i)$\Rightarrow$(iii).

(iii)$\Rightarrow$(i): Let $\left\{ E_j \mid j \in \mathbb{Z} \right\} \subset \mathcal{P}_0$ be a bounded set, and let $t \in (-n, \infty)$. We wish to show that $T = \sum_{j \in \mathbb{Z}} 2^{jt} E_j^{(2^j)}$ is a Calderón-Zygmund operator of order t. First we verify the growth condition. Fix multi-indices α and β and let $m = m\,(\alpha, \beta, t)$ be large.

$$\left| \partial_x^\alpha \partial_z^\beta T\,(x, z) \right| \leq \sum_{j \in \mathbb{Z}} 2^{j(|\alpha|+|\beta|+t)} \left| \left(2^{-j} \partial_x \right)^\alpha \left(2^{-j} \partial_z \right)^\beta 2^{jn} E_j\,\left(2^j x, 2^j z \right) \right|$$

$$\lesssim \sum_{j \in \mathbb{Z}} 2^{j(|\alpha|+|\beta|+t+n)} \left(1 + 2^j\,|x - z| \right)^{-m}.$$

We separate the above sum into two terms: when $|x - z| \geq 2^{-j}$ and when $|x - z| < 2^{-j}$. For the first, we have

$$\sum_{|x-z| \geq 2^{-j}} 2^{j(|\alpha|+|\beta|+t+n)} \left(1 + 2^j\,|x - z| \right)^{-m}$$

$$\lesssim \sum_{|x-z| \geq 2^{-j}} 2^{j(|\alpha|+|\beta|+t+n)} 2^{-mj}\,|x - z|^{-m}$$

$$\lesssim |x - z|^{-|\alpha|-|\beta|-t-n}.$$

For the second, we have

$$\sum_{|x-z|<2^{-j}} 2^{j(|\alpha|+|\beta|+t+n)} \left(1 + 2^j |x - z|\right)^{-m}$$

$$\lesssim \sum_{|x-z|<2^{-j}} 2^{j(|\alpha|+|\beta|+t+n)}$$

$$\lesssim |x - z|^{-|\alpha|-|\beta|-t-n},$$

where the last line uses $t > -n$. We, therefore, have

$$\left|\partial_x^\alpha \partial_z^\beta T(x, z)\right| \lesssim |x - z|^{-n-|\alpha|-|\beta|-t},$$

thereby establishing the growth condition.

We now turn to the cancellation condition. Let $\mathcal{B} \subset C_0^\infty(\mathbb{R}^n)$ be a bounded set. For $\phi \in \mathcal{B}$, define $F \in \mathcal{P}$ by $F(x, y) = \phi(x - y)$. Note that $\{F \mid \phi \in \mathcal{B}\} \subset \mathcal{P}$ is a bounded set, and that $\phi_{R,z}(x) = R^{-n} F^{(R)}(x, z)$. Fix $R > 0$ and let $R = 2^k$ for $k \in \mathbb{R}$. In light of the above remarks, to prove the cancellation condition, we need to show

$$\left|\partial_x^\alpha \left[T F^{(2^k)}\right](x, z)\right| \lesssim 2^{k(n+t+|\alpha|)},$$

with implicit constant depending on α and $\mathcal{B}$, but independent of $x, z \in \mathbb{R}^n$ and $k \in \mathbb{R}$. Fix N large to be chosen later. For $j \geq k$, define $G_j^{(2^k)} = 2^{N|j-k|} E_j^{(2^j)} F^{(2^k)}$. In light of Lemma 1.2.15, we have $\{G_j \mid j \geq k, \phi \in \mathcal{B}\} \subset \mathcal{P}$ is a bounded set. For $j < k$ define $G_j^{(2^j)} = E_j^{(2^j)} F^{(2^k)}$. A simple estimate shows $|\partial_x^\alpha G_j(x, z)| \lesssim 1$. We have, for N sufficiently large,

$$\sum_{j \geq k} 2^{jt} 2^{-N|j-k|} \left|\partial_x^\alpha G_j^{(2^k)}(x, z)\right| \lesssim \sum_{j \geq k} 2^{jt} 2^{-N|j-k|} 2^{k(n+|\alpha|)} \lesssim 2^{k(t+n+|\alpha|)},$$

$$\sum_{j < k} 2^{jt} \left|\partial_x^\alpha G_j^{(2^j)}(x, z)\right| \lesssim \sum_{j < k} 2^{j(t+n+|\alpha|)} \lesssim 2^{k(t+n+|\alpha|)},$$

where in the later sum we have used $t > -n$. Since

$$\partial_x^\alpha T F^{(2^k)} = \sum_{j < k} 2^{jt} \partial_x^\alpha G_j^{(2^j)} + \sum_{j \geq k} 2^{jt} 2^{-N|j-k|} \partial_x^\alpha G_j^{(2^k)},$$

combining the above two estimates completes the proof of the cancellation condition.

(ii)$\Rightarrow$(iii): Let T be as in (ii). Take $E \in \mathcal{P}_0$ as in Lemma 1.2.19 so that $I = \sum_{j \in \mathbb{Z}} E^{(2^j)}$. Define $E_j^{(2^j)} = 2^{-jt} T E^{(2^j)}$. By our assumption, $\{E_j \mid j \in \mathbb{Z}\} \subset \mathcal{P}_0$ is a bounded set. We have

$$T = TI = \sum_{j \in \mathbb{Z}} T E^{(2^j)} = \sum_{j \in \mathbb{Z}} 2^{jt} E_j,$$

completing the proof of (iii).

(iii)$\Rightarrow$(ii): Let $T = \sum_{j \in \mathbb{Z}} 2^{jt} E_j^{(2^j)}$, where $\{E_j \mid j \in \mathbb{Z}\} \subset \mathcal{P}_0$ is a bounded set, as in (iii). Let $\mathcal{B} \subset \mathcal{P}_0$ be a bounded set, and fix N large to be chosen later. For $E \in \mathcal{B}$, $k \in \mathbb{R}$, define $F_j = F_j(E, k)$ by $F_j^{(2^k)} = 2^{N|j-k|} E_j^{(2^j)} E^{(2^k)}$. By Lemma 1.2.14, $\{F_j \mid j \in \mathbb{Z}, E \in \mathcal{B}, k \in \mathbb{R}\} \subset \mathcal{P}_0$ is a bounded set. Define $\widetilde{E} = \widetilde{E}(E, k)$ by $2^{kt} \widetilde{E}^{(2^k)} = \sum_{j \in \mathbb{Z}} 2^{jt - N|j-k|} F_j^{(2^k)}$. We take $N > t$ and have that

$$\left\{ \widetilde{E} \mid E \in \mathcal{B}, k \in \mathbb{R} \right\} \subset \mathcal{P}_0$$

is a bounded set. Consider,

$$TE^{(2^k)} = \sum_{j \in \mathbb{Z}} 2^{jt - N|j-k|} F_j^{(2^k)} = 2^{kt} \widetilde{E}^{(2^k)},$$

completing the proof. $\square$

As in the translation invariant setting, we may extend the definition of Calderón-Zygmund operators to any order in the following way.

DEFINITION 1.2.20. *We say $T : \mathcal{S}_0(\mathbb{R}^n) \to \mathcal{S}_0(\mathbb{R}^n)$ is a Calderón-Zygmund operator of order $t \in \mathbb{R}$ if either of the equivalent conditions (ii) or (iii), of Theorem 1.2.10, holds.*

Remark 1.2.21 Suppose $K \in \mathcal{S}_0(\mathbb{R}^n)'$ is a Calderón-Zygmund kernel of order s. Then $\mathrm{Op}(K)$ is a Calderón-Zygmund operator of order s. Indeed, (iii) of Theorem 1.1.26 shows that $\mathrm{Op}(K)$ may be written as $\mathrm{Op}(K) = \sum_{j \in \mathbb{Z}} \mathrm{Op}(\varsigma_j)^{(2^j)}$, where $\{\varsigma_j \mid j \in \mathbb{Z}\} \subset \mathcal{S}_0(\mathbb{R}^n)$ is a bounded set. We have $\{\mathrm{Op}(\varsigma_j) \mid j \in \mathbb{Z}\} \subset \mathcal{P}_0$ is a bounded set and (iii) of Theorem 1.2.10 shows that $\mathrm{Op}(K)$ is a Calderón-Zygmund operator of order s.

PROPOSITION 1.2.22. *If $T, S : \mathcal{S}_0(\mathbb{R}^n) \to \mathcal{S}_0(\mathbb{R}^n)$ are Calderón-Zygmund operators of orders $t, s \in \mathbb{R}$, respectively, then $TS : \mathcal{S}_0(\mathbb{R}^n) \to \mathcal{S}_0(\mathbb{R}^n)$ is a Calderón-Zygmund operator of order $t + s$.*

PROOF. This follows immediately from (ii) of Theorem 1.2.10. $\square$

LEMMA 1.2.23. *For $s \in \mathbb{R}$, $\triangle^s : \mathcal{S}_0(\mathbb{R}^n) \to \mathcal{S}_0(\mathbb{R}^n)$ is a Calderón-Zygmund operator of order $s/2$.*

PROOF. This follows by combining Lemma 1.1.34 and Remark 1.2.21. $\square$

Remark 1.2.24 Just as in Remark 1.1.29, Lemma 1.2.23 when combined with Proposition 1.2.22 shows that $\triangle^s$ is an isomorphism between Calderón-Zygmund operators of order t and Calderón-Zygmund operators of order $t + 2s$. Thus, T is a Calderón-Zygmund operator of order $t \in \mathbb{R}$ if and only if $\triangle^{-t/2} T$ is a Calderón-Zygmund operator of order 0.

THEOREM 1.2.25. *Let T be a Calderón-Zygmund operator of order t. Then, T :*
$\mathring{L}_s^p \to \mathring{L}_{s-t}^p$.

We prove Theorem 1.2.25 in three steps, and each step involves one of the equivalent characterizations from Theorem 1.2.10. The first is the L^2 boundedness of operators of order 0. This uses (iii) and the Cotlar-Stein Lemma, which we state without proof; see [Ste93] for details.

LEMMA 1.2.26 (Cotlar-Stein Lemma). *Let $\mathcal{H}_1$ and $\mathcal{H}_2$ be Hilbert spaces. For each $j \in \mathbb{N}$, let $T_j : \mathcal{H}_1 \to \mathcal{H}_2$ be such that*

$$\sup_{j \in \mathbb{N}} \sum_{k \in \mathbb{N}} \|T_j^* T_k\|_{\mathcal{H}_1 \to \mathcal{H}_1}^{1/2}, \sup_{j \in \mathbb{N}} \sum_{k \in \mathbb{N}} \|T_j T_k^*\|_{\mathcal{H}_2 \to \mathcal{H}_2}^{1/2} \le C.$$

Then, the sum

$$\sum_{j \in \mathbb{N}} T_j$$

converges in the strong operator topology to a bounded operator $\mathcal{H}_1 \to \mathcal{H}_2$ and

$$\left\| \sum_{j \in \mathbb{N}} T_j \right\|_{\mathcal{H}_1 \to \mathcal{H}_2} \le C.$$

LEMMA 1.2.27. *Let $\mathcal{E} \subset \mathcal{P}$ be a bounded set. Then,*

$$\sup_{E \in \mathcal{E}} \sup_{R>0} \sup_{1 \le p \le \infty} \left\| E^{(R)} \right\|_{L^p \to L^p} < \infty.$$

PROOF. That $\left\| E^{(R)} \right\|_{L^1 \to L^1} \lesssim 1$ and $\left\| E^{(R)} \right\|_{L^\infty \to L^\infty} \lesssim 1$ follows immediately from the definitions. The result follows by interpolation. □

PROPOSITION 1.2.28. *Suppose T is a Calderón-Zygmund operator of order 0. Then $T : L^2 \to L^2$.*

PROOF. We use (iii) and write $T = \sum_{j \in \mathbb{Z}} E_j^{(2^j)}$, where $\{E_j \mid j \in \mathbb{Z}\} \subset \mathcal{P}_0$ is a bounded set. Note that $\{E_j^* \mid j \in \mathbb{Z}\} \subset \mathcal{P}_0$ is also a bounded set, and that $\left(E_j^*\right)^{(2^j)} = \left(E_j^{(2^j)}\right)^*$. We define $F_{j,k}^{(2^j)} = E_j^{(2^j)} \left(E_k^{(2^k)}\right)^*$ and $G_{j,k}^{(2^j)} = \left(E_j^{(2^j)}\right)^* E_k^{(2^k)}$. Lemma 1.2.14 shows that $\left\{2^{|j-k|} F_{j,k}, 2^{|j-k|} G_{j,k} \mid j, k \in \mathbb{Z}\right\} \subset \mathcal{P}_0$ is a bounded set. Lemma 1.2.27 then shows

$$\left\| E_j^{(2^j)} \left(E_k^{(2^k)}\right)^* \right\|_{L^2 \to L^2}, \left\| \left(E_j^{(2^j)}\right)^* E_k^{(2^k)} \right\|_{L^2 \to L^2} \lesssim 2^{-|j-k|}.$$

The result follows from the Cotlar-Stein Lemma. □

Next, we extend the above result to L^p, $1 < p < \infty$. This uses (i).

PROPOSITION 1.2.29. *Let T be a Calderón-Zygmund operator of order 0. Then*
$T : L^p \to L^p$, $1 < p < \infty$.

PROOF SKETCH. Because $T : L^2 \to L^2$ by Proposition 1.2.28, to show $T : L^p \to L^p$, $1 < p \le 2$, it suffices to show T is weak-type $(1,1)$. This follows just as in Proposition 1.1.7. Applying this to T^*, which is also a Calderón-Zygmund operator of order 0, we see that $T^* : L^p \to L^p$, $1 < p \le 2$, and therefore we see $T : L^p \to L^p$, $2 \le p < \infty$, completing the proof. $\qquad\square$

Finally we prove Theorem 1.2.25. Here we use Proposition 1.2.22, and thus implicitly use (ii).

PROOF OF THEOREM 1.2.25. Let T be a Calderón-Zygmund operator of order t. For $f \in \mathcal{S}_0(\mathbb{R}^n)$, we wish to show

$$\left\| \triangle^{(s-t)/2} T f \right\|_{L^p} \lesssim \left\| \triangle^{s/2} f \right\|_{L^p}.$$

By Corollary 1.1.13 we may write $f = \triangle^{-s/2} g$, $g \in \mathcal{S}_0(\mathbb{R}^n)$, and we therefore wish to show

$$\left\| \triangle^{(s-t)/2} T \triangle^{-s/2} g \right\|_{L^p} \lesssim \|g\|_{L^p}.$$

This follows from the fact that $\triangle^{(s-t)/2} \mathrm{Op}(K) \triangle^{-s/2}$ is bounded on L^p, by Proposition 1.2.29, as it is an operator of order 0 (by Proposition 1.2.22 and Lemma 1.2.23). $\qquad\square$

Remark 1.2.30 Let T be a Calderón-Zygmund operator of order t, with $-n < t < 0$. Let $T_0(x, y)$ be the function which equals $T(x, y)$ for $x \ne y$ and which equals 0 for $x = y$. From the Growth Condition it follows that $T_0(x, y) \in L^1_{\mathrm{loc}}(\mathbb{R}^n \times \mathbb{R}^n)$, and we may therefore identify T_0 with an operator $T_0 : C_0^\infty(\Omega) \to C_0^\infty(\Omega)'$. In fact, it is easy to see that $T_0 = T$; i.e., T is itself given by integration against an L^1_{loc} function. To see this, merely use (iii) of Theorem 1.2.10. Indeed, if $\{ E_j \mid j \in \mathbb{Z} \} \subset \mathcal{P}_0$ is a bounded set and $-n < t < 0$, then it is easy to see $\sum_{j \in \mathbb{Z}} 2^{jt} E_j^{(2^j)}$ converges in distribution to an L^1_{loc} function.

1.3 PSEUDODIFFERENTIAL OPERATORS

We now wish to introduce another, closely related, class of singular integral operators: the standard pseudodifferential operators on $\mathbb{R}^n$. To a distribution $a \in \mathcal{S}(\mathbb{R}^n \times \mathbb{R}^n)'$, we may associate an operator, $a(x, D) : \mathcal{S}(\mathbb{R}^n) \to \mathcal{S}(\mathbb{R}^n)'$, by

$$\int g(x)(a(x, D) f)(x)\, dx := \int g(x) a(x, \xi) \hat{f}(\xi) e^{2\pi i x \cdot \xi}\, dx\, d\xi, \text{ for } f, g \in \mathcal{S}(\mathbb{R}^n)$$

where we have, as usual, written the pairing between distributions and test functions as integration. An analog of the Schwartz kernel theorem (see Theorem A.1.30) states that the map $a \mapsto a(x, D)$ is a *bijection* between distributions in $\mathcal{S}(\mathbb{R}^n \times \mathbb{R}^n)'$ and operators $\mathcal{S}(\mathbb{R}^n) \to \mathcal{S}(\mathbb{R}^n)'$. a is called the "symbol" of the operator, and for appropriate choices of a, operators written in this way are referred to as "pseudodifferential operators."

The first example to consider, and the one that justifies the name, is that of linear partial differential operators with smooth coefficients, to which we now turn.

DEFINITION 1.3.1. *We let $C_b^\infty(\mathbb{R}^n)$ denote the Fréchet space consisting of those $f \in C^\infty(\mathbb{R}^n)$ such that, for every multi-index α,*

$$\sup_x |\partial_x^\alpha f(x)| < \infty.$$

We give $C_b^\infty(\mathbb{R}^n)$ the coarsest topology such that the left-hand side of the above equation defines a continuous semi-norm, for each α.

Example 1.3.2 *Let $D = \frac{1}{2\pi i}\frac{\partial}{\partial x}$, so that $(D^\alpha f)^\wedge(\xi) = \xi^\alpha \hat{f}(\xi)$. Consider a linear partial differential operator of order M, $p(x, D)$, defined by*

$$p(x, D) f(x) = \sum_{|\alpha| \leq M} a_\alpha(x) D^\alpha f(x),$$

where $a_\alpha \in C_b^\infty(\mathbb{R}^n)$. $p(x, D)$ is a pseudodifferential operator corresponding to the tempered distribution $p(x, \xi) = \sum_{|\alpha| \leq M} a_\alpha(x) \xi^\alpha$ (where we are identifying the function $p(x, \xi)$ with a distribution in the usual way).

We now define a class of symbols which we study. The class that follows is the most commonly used class of symbols, and we refer to them as the "standard symbols."

DEFINITION 1.3.3. *Fix $m \in \mathbb{R}$. The space of standard symbols of order m, S^m, is the Fréchet space of functions $a \in C^\infty(\mathbb{R}^n \times \mathbb{R}^n)$ which satisfy, for all multi-indices α, β,*

$$\left|\partial_x^\alpha \partial_\xi^\beta a(x, \xi)\right| \leq C_{\alpha,\beta}(1 + |\xi|)^{m-|\beta|}.$$

We give S^m the coarsest topology such that the least possible $C_{\alpha,\beta}$ defines a continuous semi-norm on S^m. Operators of the form $a(x, D)$, with $a \in S^m$, are called "pseudodifferential operators of order m." When we wish to be explicit that we are referring to standard symbols, we call $a(x, D)$ a "standard pseudodifferential operator of order m."

Remark 1.3.4 We defined Calderón-Zygmund operators via estimates on the Schwartz kernel of the operator (the growth condition), along with an extra condition (the cancellation condition). In a similar vein, pseudodifferential operators are defined via estimates on the symbol. This highlights a major convenience of pseudodifferential operators: the estimates on the symbols play the role of both the growth condition *and* the cancellation condition from Calderón-Zygmund operators.

Remark 1.3.5 As we shall see, pseudodifferential operators of order m are closely related to Calderón-Zygmund operators of order m. However, neither class is contained in the other. For instance, if $\phi \in C_b^\infty(\mathbb{R}^n)$, the operator given by $f \mapsto \phi f$ is a pseudodifferential operator of order 0, however it is not a Calderón-Zygmund operator of order 0 except in very special cases (it does not satisfy the cancellation condition for $R \ll 1$). Conversely, we shall see that the Schwartz kernels of pseudodifferential operators satisfy estimates which are somewhat better than those satisfied by Calderón-Zygmund operators.

As with Calderón-Zygmund operators, the two main theorems concerning pseudodifferential operators are that they form an algebra, and that they are bounded on appropriate Sobolev spaces. We now turn to discussing these results in more detail. We do not include proofs of all results in the section. We refer the reader to [Ste93, Chapter VI] for more details and proofs.

THEOREM 1.3.6. *Suppose* $a \in S^{m_1}$ *and* $b \in S^{m_2}$. *We consider the operator* $a(x, D) b(x, D) : S(\mathbb{R}^n) \to S(\mathbb{R}^n)$. *There is* $c \in S^{m_1+m_2}$ *such that* $c(x, D) = a(x, D) b(x, D)$. *Furthermore, for every* N,

$$c - \sum_{|\alpha| < N} \frac{(2\pi i)^{|\alpha|}}{\alpha!} (\partial_\xi^\alpha a) (\partial_x^\alpha b) \in S^{m_1+m_2-N}.$$

Remark 1.3.7   Taking $N = 1$ in Theorem 1.3.6 we see

$$c(x, \xi) \equiv a(x, \xi) b(x, \xi) \mod S^{m_1+m_2-1};$$

and we think of the remainder term in $S^{m_1+m_2-1}$ as a "lower order" term.

Remark 1.3.8   Theorem 1.3.6 has a much better conclusion than Proposition 1.2.22: not only do we have $c \in S^{m_1+m_2}$, but we know exactly what c is, modulo elements of $\bigcap_{m \in \mathbb{R}} S^m$. This makes pseudodifferential operators far easier to work with, but is closely tied to the fact that we may use the Fourier transform extensively.

Remark 1.3.9   Theorem 1.3.6 is often referred to as the "calculus of pseudodifferential operators."

For $s \in \mathbb{R}$, let Λ^s denote the pseudodifferential operator of order s with symbol $\left(1 + |\xi|^2\right)^{s/2}$. It is easy to verify that $\Lambda^s \Lambda^t = \Lambda^{s+t}$. As such, $\Lambda^s : S(\mathbb{R}^n) \to S(\mathbb{R}^n)$ is an automorphism with inverse Λ^{-s}.

DEFINITION 1.3.10. *For* $1 \le p \le \infty$, *we define* $L_s^p(\mathbb{R}^n)$, *the standard* L^p *Sobolev space of order* $s \in \mathbb{R}$, *to be the completion of* $S(\mathbb{R}^n)$ *in the norm*

$$\|f\|_{L_s^p} := \|\Lambda^s f\|_{L^p}.$$

THEOREM 1.3.11. *Let* $a \in S^m$. *Then* $a(x, D) : L_s^p \to L_{s-m}^p$, $1 < p < \infty$.

As in Theorem 1.2.25, the proof of Theorem 1.3.11 separates into three main parts, each of which is proved in a manner closely analogous to the corresponding part of the proof of Theorem 1.2.25. Details of the proofs may be found in [Ste93]. The three parts are:

(I) If $a \in S^0$, then $a(x, D) : L^2 \to L^2$.

(II) If $a \in S^0$, then $a(x, D) : L^p \to L^p$, $1 < p < \infty$.

(III) Using the previous part and Theorem 1.3.6, Theorem 1.3.11 follows.

The proof of (I) relies on a decomposition of $a(x, D)$ analogous to the one used in Proposition 1.2.28. Indeed, we decompose

$$a(x, D) = a(x, D) I = \sum_{j \in \mathbb{Z}} a(x, D) \operatorname{Op}\left(\varsigma^{(2^j)}\right) =: \sum_{j \in \mathbb{Z}} A_j,$$

where $\varsigma \in \mathcal{S}_0(\mathbb{R}^n)$ is as in Lemma 1.1.22. The operators A_j can be shown to satisfy

$$\left\| A_j^* A_k \right\|_{L^2 \to L^2}, \left\| A_j A_k^* \right\|_{L^2 \to L^2} \lesssim 2^{-|j-k|},$$

and the L^2 boundedness of $a(x, D)$ follows from the Cotlar-Stein Lemma (Lemma 1.2.26).

For (II), the key is to show that $a(x, D)$ is weak-type $(1, 1)$. This follows by taking the inverse Fourier transform in the ξ variable. To a distribution $a(x, \xi) \in \mathcal{S}(\mathbb{R}^n \times \mathbb{R}^n)'$ we associate a distribution $\breve{a}(x, z) \in \mathcal{S}(\mathbb{R}^n \times \mathbb{R}^n)'$ by, for $f, g \in \mathcal{S}(\mathbb{R}^n)$,

$$\int \breve{a}(x, z) f(x) g(x - z) \, dx \, dz := \int f(x) (a(x, D) g)(x) \, dx.$$

I.e., $\breve{a}(x, z)$ is the inverse Fourier transform of $a(x, \xi)$ in the ξ variable. It is not hard to see that $a \mapsto \breve{a}$ is an automorphism of $\mathcal{S}(\mathbb{R}^n \times \mathbb{R}^n)'$. To characterize symbols in S^m, we define a class of kernels.

DEFINITION 1.3.12. *We say $K \in \mathcal{S}(\mathbb{R}^n \times \mathbb{R}^n)'$ is a kernel of order m if the following conditions hold.*

- *(Growth Condition) For every multi-indices α, β and every N,*

$$\left| \partial_x^\beta \partial_z^\alpha K(x, z) \right| \leq C_{\alpha, \beta, N} |z|^{-n-m-|\alpha|-N},$$

  in particular, $K(x, z)$ is C^∞ for $z \neq 0$.

- *(Cancellation Condition) For $m \geq 0$, we assume for every bounded set $\mathcal{B} \subset C_0^\infty(\mathbb{R}^n)$, and for every multi-index α,*

$$\sup_{x \in \mathbb{R}^n} \sup_{R \geq 1} \sup_{\phi \in \mathcal{B}} R^{-m} \left| \int \partial_x^\alpha K(x, z) \phi(Rz) \, dz \right| < \infty.$$

- *If $m < 0$, the growth condition implies that for $\phi \in C_0^\infty(\mathbb{R}^n)$,*

$$\lim_{\epsilon \to 0} \int_{|z| > \epsilon} K(x, z) \phi(z) \, dz$$

  converges to a C^∞ function. We assume that $\int K(x, z) \phi(z) \, dz$ agrees with this limit.

PROPOSITION 1.3.13. $a \mapsto \breve{a}$ is a bijection between S^m and kernels of order m.

PROOF. See [Ste93, Chapter VI]. □

Remark 1.3.14   In light of Proposition 1.3.13, we may write pseudodifferential operators in another way. Namely, for $a \in S^m$,

$$a(x, D) f(x) = \int \breve{a}(x, z) f(x - z) \, dz,$$

where $\breve{a}$ is a kernel of order m. In the sequel, we use this idea to generalize some aspects of pseudodifferential operators to settings where we have no immediate analog of the Fourier transform.

PROPOSITION 1.3.15. *If $a \in S^0$, then $a(x, D) : L^p \to L^p$, $1 < p \leq 2$.*

MAIN IDEA OF PROOF. Writing $a(x, D)$ as in Remark 1.3.14, the proof is nearly identical to the proof in Proposition 1.1.7. □

The proof of (II) is then completed by duality using the next result, which we state without proof.

PROPOSITION 1.3.16. *Let $a \in S^m$, then $a(x, D)^*$ is a pseudodifferential operator of order m.*

(III) now follows just as in the proof of Theorem 1.1.33, using (II) and the fact that if $a(x, D)$ is a pseudodifferential operator of order t, then $\Lambda^{s-t} a(x, D) \Lambda^{-s}$ is a pseudodifferential operator of order 0, by Theorem 1.3.6.

1.4 ELLIPTIC EQUATIONS

Theorem 1.3.6 allows us to approximately invert certain pseudodifferential operators while staying in the class of pseudodifferential operators. In what follows, we write $\mathrm{Op}(S^m)$ to denote the class of pseudodifferential operators of order m.

DEFINITION 1.4.1. *We say $a \in S^m$ is elliptic if there exists $R > 0$ such that $|a(x, \xi)| \geq C |\xi|^m$, for some C and all $|\xi| > R$.*

PROPOSITION 1.4.2. *Suppose $a \in S^m$ is elliptic. Then there is $b \in S^{-m}$ such that $a(x, D) b(x, D) \equiv I \mod \mathrm{Op}(S^{-1})$.*

PROOF. Let $a \in S^m$ be elliptic, and take R as in Definition 1.4.1. Let $\eta \in C_0^\infty(\mathbb{R}^n)$ be equal to 1 on a neighborhood of $B^n(R)$. Define

$$b(x, \xi) = (1 - \eta(\xi)) a(x, \xi)^{-1}.$$

It is simple to verify that $b \in S^{-m}$. Furthermore, by Theorem 1.3.6, $b(x, D) a(x, D) \equiv I \mod \mathrm{Op}(S^{-1})$. □

COROLLARY 1.4.3. *Suppose $a \in S^m$ is elliptic. Then, for every N, there is $b_N \in S^{-m}$ such that $b_N(x, D) a(x, D) \equiv I \mod \mathrm{Op}(S^{-N})$.*

PROOF. Let $b \in S^{-m}$ be as in Proposition 1.4.2. Define $R \in \mathrm{Op}(S^{-1})$ by $R = b(x, D) a(x, D) - I$. Then let

$$b_N(x, D) = \sum_{j=0}^{N-1} (-1)^j R^j b(x, D).$$

Theorem 1.3.6 shows that $b_N(x, \xi) \in S^{-m}$ satisfies the conclusions of the corollary. $\qquad \square$

Remark 1.4.4 Actually, one may improve Corollary 1.4.3 by showing that there is $b_\infty \in S^{-m}$ such that $b_\infty(x, D) a(x, D) \equiv I \mod \mathrm{Op}(S^{-\infty})$, where $S^{-\infty} = \bigcap_{m \in \mathbb{R}} S^m$. See [Ste93]. We do not pursue this here.

THEOREM 1.4.5. *Let $a \in S^m$ be elliptic, and suppose $u \in C^\infty(\mathbb{R}^n)'$ satisfies $a(x, D) u \in L_s^p$ for some $s \in \mathbb{R}$ and some $p \in (1, \infty)$. Then, $u \in L_{s+m}^p$.*

Remark 1.4.6 The space $C^\infty(\mathbb{R}^n)'$ is the space of "distributions with compact support." The reader wishing for more details on this space is referred to Appendix A.1.1.

PROOF OF THEOREM 1.4.5. It is classical that u, being a distribution with compact support, is in L_{-M}^p for some M. Take N large, and let b_N be as in Corollary 1.4.3. Define $R_N = b_N(x, D) a(x, D) - I \in \mathrm{Op}(S^{-N})$. We have

$$u = b_N(x, D) a(x, D) u - R_N u.$$

By Theorem 1.3.11, we have $b_N(x, D) a(x, D) u \in L_{s+m}^p$ and $R_N u \in L_{-M+N}^p$. Taking $N \geq M + s + m$, the result follows. $\qquad \square$

The most interesting examples of elliptic psuedodifferential operators come from elliptic differential operators.[13] See Example 1.3.2. In this case, Theorem 1.4.5 gives the optimal L^p regularity of elliptic differential operators.

In the sequel, we use similar ideas in a non-Euclidean setting, where we do not have a convenient analog of the Fourier transform. To motivate our later definitions, we present an equivalent characterization of elliptic pseudodifferential operators, which we state without proof.

THEOREM 1.4.7. *Let $a \in S^m$. The following are equivalent:*

(i) a is elliptic.

---

[13] An elliptic differential operator is a differential operator which is an elliptic pseudodifferential operator.

(ii) For every N there exists C_N such that $\forall f \in \mathcal{S}(\mathbb{R}^n)$,

$$\|f\|_{L^2_m} \leq C_N \left(\|a(x,D)f\|_{L^2} + \|f\|_{L^2_{-N}} \right). \tag{1.11}$$

(iii) For some N with $-N < m$ there exists C_N so that (1.11) holds.

Suppose $a(x,D) \in \mathrm{Op}(S^m)$ is an elliptic pseudodifferential operator of order m. Fix p, $1 < p < \infty$, and suppose $u \in C^\infty(\mathbb{R}^n)'$ satisfies $a(x,D)u \in L^p_s$, for some $s \in \mathbb{R}$. Theorem 1.4.5 shows that $u \in L^p_{s+m}$.

More is true, the above holds "locally." For $\phi_1, \phi_2 \in C^\infty_0(\mathbb{R}^n)$, we write $\phi_1 \prec \phi_2$ to mean that $\phi_2 \equiv 1$ on a neighborhood of the support of ϕ_1. Suppose $\phi_1, \phi_2 \in C^\infty_0(\mathbb{R}^n)$ with $\phi_1 \prec \phi_2$ and suppose $u \in C^\infty(\mathbb{R}^n)'$, with $\phi_2 a(x,D)u \in L^p_s$ (we are still taking $a(x,D) \in \mathrm{Op}(S^m)$ to be elliptic, and fixing $p \in (1,\infty)$). Then, we have $\phi_1 u \in L^p_{s+m}$. To formally discuss this, we state a definition and a theorem.

DEFINITION 1.4.8. *Let $T : C^\infty(\mathbb{R}^n)' \to C^\infty_0(\mathbb{R}^n)'$ be an operator. We say T is L^p-subelliptic of order $\epsilon > 0$ on an open set $V \subseteq \mathbb{R}^n$, if for every $\phi_1 \prec \phi_2 \in C^\infty_0(V)$, we have the following estimate for all $u \in C^\infty_0(V)'$, and all s, N,*

$$\|\phi_1 u\|_{L^p_{s+\epsilon}} \leq C_{\phi_1,\phi_2,s,N} \left(\|\phi_2 T u\|_{L^p_s} + \|\phi_2 u\|_{L^p_s} \right),$$

where if the left-hand side is infinite, the right-hand side is infinite, and if the right-hand side is finite, the left-hand side is finite. We say T is L^p-subelliptic on V, if T is L^p-subelliptic of order $\epsilon > 0$ on V for some ϵ. We say T is subelliptic on V, if T is L^2-subelliptic on V.

We state the following theorem without proof.

THEOREM 1.4.9. *Let $a(x,D) \in \mathrm{Op}(S^m)$, $m > 0$, be elliptic. Then $a(x,D)$ is L^p-subelliptic of order m on $\mathbb{R}^n$, for every $1 < p < \infty$.*

Theorem 1.4.9 shows that ellipticity implies subellipticity. There is a third, even weaker, condition which will also be of interest to us.

DEFINITION 1.4.10. *We say $T : C^\infty(\mathbb{R}^n)' \to C^\infty_0(\mathbb{R}^n)'$ is hypoelliptic on an open set $V \subseteq \mathbb{R}^n$ if the following holds. For every distribution $u \in C^\infty(\mathbb{R}^n)'$ with $Tu|_V \in C^\infty(V)$, we have $u|_V \in C^\infty(V)$.*

THEOREM 1.4.11. *Suppose $T : C^\infty(\mathbb{R}^n)' \to C^\infty_0(\mathbb{R}^n)'$ is L^p-subelliptic on V for some $p \in [1,\infty]$. Then, T is hypoelliptic on V.*

PROOF. Take $u \in C^\infty(\mathbb{R}^n)'$ with $Tu|_V \in C^\infty(V)$. Fix $x_0 \in V$, we wish to show u is C^∞ near x_0. Take $\phi_1 \in C^\infty_0(V)$ with $\phi_1 \equiv 1$ on a neighborhood of x_0 and take $\phi_2 \in C^\infty_0(V)$ with $\phi_1 \prec \phi_2$. Since $u \in C^\infty(\mathbb{R}^n)' \subset \bigcup_N L^p_{-N}$, we have $\phi_2 u \in L^p_{-N}$ for some N. Using that $\phi_2 Tu \in C^\infty_0(V) \subset L^p_s$, $\forall s$, subellipticity shows $\phi_1 u \in L^p_{s+\epsilon}$, $\forall s$. Since $\bigcap_s L^p_s \subset C^\infty$ (by the Sobolev embedding theorem), we have $\phi_1 u \in C^\infty$, and the result follows. $\qquad \square$

One may succinctly restate Theorems 1.4.9 and 1.4.11 as:

$$\text{Ellipticity} \Rightarrow \text{Subellipticity} \Rightarrow \text{Hypoellipticity.}$$

In general, none of the reverse implications hold, even for partial differential operators–see Sections 2.6 and 4.3.1.

1.5   FURTHER READING AND REFERENCES

Our prototypical example of a singular integer operator, the Hilbert transform, first arose in Hilbert's work on what is now known as the Riemann-Hilbert problem where he studied a similar operator on the unit circle (instead of on $\mathbb{R}$). Hilbert's proof was published by Weyl [Wey08]. Schur improved these results and introduced the form of the Hilbert transform mentioned in the introduction [Sch11]. All of these results were restricted to L^2. It was Marcel Riesz who extended these results to L^p ($1 < p < \infty$) [Rie28]. The above citations were focused on "complex analysis methods," and did not generalize to higher dimensions. Besicovitch [Bes26], Titchmarsh [Tit29], and Marcinkiewicz [Mar36] offered a "real-variable" analysis of the Hilbert transform.

This real variable analysis of the Hilbert transform was a main motivating example for Calderón and Zygmund when they introduced the homogeneous Calderón-Zygmund kernels as discussed in the introduction to this chapter [CZ52]. The proof of (a) of Theorem 1.1.2 uses their methods. The interpolation theorem used in that proof (the "Marcinkiewicz Interpolation theorem") was proved by Marcinkiewicz [Mar39].

The concept of a pseudodifferential operator is rooted in the work of Marcinkiewicz [Mar39], later work by Calderón and Zygmund, and the work of Seeley in his thesis [See59]. This was followed by further work of Seeley [See65] and Unterberger and Bokobza [UB64]. This culminated in the work of Kohn and Nirenberg [KN65] and Hörmander [Hör65]. It was these last two references that first exhibited the theory of pseudodifferential operators as covered in Section 1.3. Our presentation more closely follows the one from Chapter VI of [Ste93].

Non-translation invariant, non-homogeneous Calderón-Zygmund operators (as discussed in Section 1.2) was the work of many authors. A systematic approach for operators of order 0, working in more general "spaces of homogeneous type," and working with much less regular kernels was developed by Coifman and Weiss [CW71]. Definition 1.2.1 was taken from much more recent work of Nagel, Rosay, Stein, and Wainger [NRSW89] and Koenig [Koe02] who worked in the more general setting discussed in the next chapter (see Section 2.16 for further comments on their work). The concept of defining the cancellation condition as in Definition 1.2.1 is closely related to the hypothesis of the $T(1)$ theorem of David and Journé [DJ84]; see also the presentation in [Ste93, pages 293-294]. This idea was further championed by E. M. Stein; see, e.g., [Ste93, page 248]. The idea to extend operators to order $\leq -n$ by considering them acting on $\mathcal{S}_0(\mathbb{R}^n)$ appears in the work by Christ, Geller, Głowacki, and Polin [CGGP92] though it has been used in many situations.

The decomposition of a Calderón-Zygmund kernel as a sum of dilates of functions in $\mathcal{S}_0(\mathbb{R}^n)$ (see, e.g, (iii) of Theorem 1.1.23, (iii) of Theorem 1.1.26, and more generally, (iii) of Theorem 1.2.10) is called a Littlewood-Paley decomposition of the operator, named so because of the first place similar decompositions appeared: in the work of Littlewood and Paley on Fourier series [LP31, LP37, LP38]. These ideas were later worked on by Zygmund and Marcinkiewicz, but were moved to higher dimensions and used in greater generality by E. M. Stein (see, e.g., [Ste70b]). A decomposition, which is very similar to the ones we use, appears in a "multi-parameter" situation in the work of Nagel, Ricci, Stein, and Wainger on flag kernels [NRS01, NRSW12] (see Section 4.2 for a discussion of their work). Since our main goal is to generalize such concepts to a multi-parameter setting, these works were of the greatest inspiration to us. It is worth noting that in these papers the authors used bounded subsets of $C_0^\infty (B^n (1))$ with one moment vanishing (in place of bounded subsets of $\mathcal{S}_0 (\mathbb{R}^n)$) to study operators of order 0. For our purposes (here, to study operators of all orders, and later to study a more complicated multi-parameter setting in Chapter 5) we need to have many moments vanish and this is why we moved to the space $\mathcal{S}_0 (\mathbb{R}^n)$. In the non-translation invariant setting, results similar to Theorem 1.2.10 are known to experts but we could not find this exact statement in the literature. It is probably most closely related to the translation invariant settings described above.

Finally, a key idea in this monograph is that one may characterize singular integral operators in terms of their actions on certain special functions or operators. Our simplest example of this is (ii) of Theorem 1.1.23–which leads to a short proof that such operators form an algebra. The author first heard of this characterization in a graduate class given by E. M. Stein at Princeton University in 2007.

Further history on some of the above topics and more references can be found in the expository articles by Stein [Ste99, Ste82a] along with Stein's book [Ste93].

Chapter Two

The Calderón-Zygmund Theory II: Maximal Hypoellipticity

In Chapter 1, we were concerned with $\mathbb{R}^n$ endowed with the usual Euclidean metric. As we saw in Section 1.4, singular integrals associated to the Euclidean metric play an important role in understanding *elliptic* partial differential operators. If a partial differential operator fails to be elliptic, then there is no immediate analog of the theory developed above. However, in certain circumstances, many of the theorems in Chapter 1 do have analogs. In this chapter, we discuss one such situation, where we study a generalization of ellipticity, known as maximal hypoellipticity.

We begin with some notation. Associated to r non-commuting indeterminates, $w = (w_1, \ldots, w_r)$, we use ordered multi-index notation. If $\alpha = (\alpha_1, \ldots, \alpha_L)$ is a list of elements of $\{1, \ldots, r\}$, we define $w^\alpha = w_{\alpha_1} w_{\alpha_2} \cdots w_{\alpha_L}$, and we denote by $|\alpha| = L$ the length of the list. For instance, if $\alpha = (1, 2, 3, 1)$, then $|\alpha| = 4$ and $w^\alpha = w_1 w_2 w_3 w_1$.

The setting is a **compact, connected, smooth manifold** M, without boundary. We are given C^∞ vector fields $W_1, \ldots, W_r$ on M. Suppose $P(w_1, \ldots, w_r)$ is a polynomial in r non-commuting indeterminates, with coefficients in $C^\infty(M)$:

$$P(w_1, \ldots, w_r) = \sum_{|\alpha| \leq L} a_\alpha(x) w^\alpha, \quad a_\alpha \in C^\infty(M).$$

Our goal is to study certain partial differential operators of the form $P(W_1, \ldots, W_r)$. To do so, we often assume the following condition on $W_1, \ldots, W_r$:

DEFINITION 2.0.1. *We say $W_1, \ldots, W_r$ satisfy Hörmander's condition on M if*

$$W_1, \ldots, W_r, \ldots, [W_i, W_j], \ldots, [W_i, [W_j, W_l]], \ldots,$$
$$\ldots, (commutators\ of\ order\ m), \ldots$$

span the tangent space $T_x M$ at every point $x \in M$. I.e., if the Lie algebra generated by $W_1, \ldots, W_r$ spans the tangent space to M at every point.

Remark 2.0.2 Because M is compact, we need only use commutators of $W_1, \ldots, W_r$ up to some fixed finite order m to span the tangent space at each point of M, where m is independent of the point. If we wish to make the choice of m explicit, we say that $W_1, \ldots, W_r$ satisfy Hörmander's condition of order m.

Associated to any finite collection of C^∞ vector fields $W_1, \ldots, W_r$ there is a distance $\rho : M \times M \to [0, \infty]$ defined by

$$\rho(x, y) = \inf \left\{ \delta > 0 \,\middle|\, \exists \gamma : [0, 1] \to M, \gamma(0) = x, \gamma(1) = y, \right.$$

$$\gamma'(t) = \sum_{j=1}^{q} a_j(t) \, \delta W_j(\gamma(t)), a_j \in L^\infty([0, 1]),$$

$$\left. \left\| \sum_{j=1}^{r} |a_j|^2 \right\|_{L^\infty([0,1])} < 1 \right\}.$$

Here, and in all similar places in the rest of this monograph, we have written $\gamma'(t) = Z(t)$ to mean $\gamma(t) = \gamma(0) + \int_0^t Z(s) \, ds$ (in particular, γ need not be differentiable everywhere). We call ρ a *Carnot-Carathéodory*[1] distance.

PROPOSITION 2.0.3. $\rho : M \times M \to [0, \infty]$ *is an extended metric. I.e.,*

(a) $\rho(x, z) \le \rho(x, y) + \rho(y, z)$,

(b) $\rho(x, y) \ge 0$ *and* $\rho(x, y) = 0$ *if and only if* $y = x$,

(c) $\rho(x, y) = \rho(y, x)$.

PROOF. (a): Suppose $\rho(x, y) < \delta_1$ and $\rho(y, z) < \delta_2$. We wish to show that $\rho(x, z) < \delta_1 + \delta_2$ and then (a) follows. Since $\rho(x, y) < \delta_1$ and $\rho(y, z) < \delta_2$, there exists $\gamma_1, \gamma_2 : [0, 1] \to M$, with

- $\gamma_1(0) = x, \gamma_1(1) = y = \gamma_2(0), \gamma_2(1) = z$,

- $\gamma_1'(t) = \sum_{j=1}^{r} a_j(t) \delta_1 W_j(\gamma(t)), \gamma_2'(t) = \sum_{j=1}^{r} b_j(t) \delta_2 W_j(\gamma(t))$,

- $\sum_{j=1}^{r} |a_j|^2, \sum_{j=1}^{r} |b_j|^2 < 1$.

Define

$$\gamma(t) = \begin{cases} \gamma_1\left(\frac{\delta_1 + \delta_2}{\delta_1} t\right), & 0 \le t \le \frac{\delta_1}{\delta_1 + \delta_2}, \\ \gamma_2\left(\frac{\delta_1 + \delta_2}{\delta_2}\left(t - \frac{\delta_1}{\delta_1 + \delta_2}\right)\right), & \frac{\delta_1}{\delta_1 + \delta_2} \le t \le 1. \end{cases}$$

It is immediate to verify that γ satisfies the axioms to show $\rho(x, z) < \delta_1 + \delta_2$, completing the proof of (a).

(b): That $\rho(x, y) \ge 0$ and that $\rho(x, x) = 0$ are immediate from the definition. We now wish to show that $\rho(x, y) = 0 \Rightarrow x = y$. We show a stronger result. Fix $x \in M$ and let V be an open neighborhood of x. We claim that there is $\delta > 0$ such that $\rho(x, y) < \delta \Rightarrow y \in V$. Indeed, let $a_1, \ldots, a_r \in L^\infty([0, 1])$, with $\sum_{j=1}^{r} |a_j|^2 < 1$. By the Picard-Lindelöf theorem, for $\delta > 0$ sufficiently small (δ not depending on the

---

[1] Sometimes these distance functions are referred to as *sub-Riemannian* distances.

particular choices of $a_1, \ldots, a_r$), there is a unique solution $\gamma_0 : [0, \delta] \to V$, $\gamma_0'(t) = \sum_{j=1}^{r} a_j(t) W_j(\gamma(t))$. Set $\gamma_1(t) = \gamma_0(\delta t)$.

Now suppose $\rho(x, y) < \delta$. There exists $\gamma : [0, 1] \to M$, $\gamma(0) = x$, $\gamma(1) = y$, $\gamma'(t) = \sum_{j=1}^{r} a_j(t) \delta W_j(\gamma(t))$, with $\sum_{j=1}^{r} |a_j| < 1$. Let γ_1 be as in the previous paragraph, with this choice of $a_1, \ldots, a_r$. Gronwall's inequality shows that $\gamma = \gamma_1$. We conclude that $y = \gamma(1) = \gamma_1(1) \in V$, which completes the proof of (b).

(c): We show that $\rho(x, y) < \delta \Leftrightarrow \rho(y, x) < \delta$, which completes the proof. By symmetry, we need only show $\rho(x, y) < \delta \Rightarrow \rho(y, x) < \delta$. Suppose $\rho(x, y) < \delta$. Then, there exists $\gamma : [0, 1] \to M$ with

- $\gamma(0) = x$, $\gamma(1) = y$,

- $\gamma'(t) = \sum_{j=1}^{r} a_j(t) \delta W_j(\gamma(t))$,

- $\sum_{j=1}^{r} |a_j|^2 < 1$.

Using $\gamma_0(t) = \gamma(1 - t)$ shows that $\rho(y, x) < \delta$, completing the proof. $\qquad \square$

We state the next theorem of Chow [Cho39] without proof.

THEOREM 2.0.4. *If* $W_1, \ldots, W_r$ *satisfy Hörmander's condition, then* ρ *is a metric. That is,* $\rho(x, y) < \infty$, *for every* $x, y \in M$.

Remark 2.0.5   Suppose M is endowed with a Riemannian metric, and denote by $R(x, y)$ the distance between x and y in this metric. Suppose, also, that $W_1, \ldots, W_r$ satisfy Hörmander's condition of order m on M. Then, it turns out that $R(x, y) \lesssim \rho(x, y) \lesssim R(x, y)^{\frac{1}{m}}$ (see Remark 2.1.2). Thus, the *topology* induced by ρ is the same as the topology on the manifold, however the metrics are not necessarily equivalent.

In Chapter 1, to define our singular integrals, we worked on $\mathbb{R}^n$ and used Lebesgue measure on $\mathbb{R}^n$. Now we turn to our arbitrary compact, connected, smooth manifold M. Unlike in $\mathbb{R}^n$, there is no obvious choice of measure to use. However, there is a class of equivalent measures, all of which give rise to the same definitions. We call these "strictly positive, smooth measures."

DEFINITION 2.0.6. *A smooth measure,* μ, *on* M *is a Borel measure on* M *such that in any local coordinates* x, *we may write* $d\mu = \phi_x dm(x)$, *where* dm *denotes Lebesgue measure, and* ϕ_x *is a* C^∞ *function. We say* μ *is a strictly positive, smooth measure if* $\phi_x > 0$ *in every local coordinate system.*

Remark 2.0.7   The function ϕ_x in Definition 2.0.6 is called a 1-density. If y is any other coordinate system, then ϕ transforms via

$$\phi_x = |\det(\partial y / \partial x)| \phi_y.$$

See [Fol99, p. 361–362] for more details.

Remark 2.0.8   It is clear that a strictly positive smooth measure exists on any manifold (one can, e.g., give the manifold a Riemannian structure and take the Riemannian volume density see [Fol99, p. 362]). Furthermore, if μ_1 and μ_2 are two strictly positive smooth measures, then they are mutually absolutely continuous, and the Radon-Nikodym derivative $d\mu_1/d\mu_2$ is C^∞ with $d\mu_1/d\mu_2 \approx 1$ (this last part uses compactness).

Henceforth, we endow M with a strictly positive, smooth measure. For a measurable set $U \subseteq M$, we write $\mathrm{Vol}\,(U)$ to denote the measure of U with respect to this measure. We also write $\int f\,(x)\,dx$ to denote the integral of f against this measure.

Remark 2.0.9 We make several definitions involving the above measure. However, none of these definitions depend on the choice of a smooth, strictly positive measure. Indeed from the comments in Remark 2.0.8, it is immediate to check that all of the definitions that follow are independent of the choice of measure. For instance, it makes sense to talk about $L^p\,(M)$, with the measure given by Vol. Even though the norm depends on the choice of Vol, different choices yield equivalent norms, and we may therefore unambiguously refer to $\|\cdot\|_{L^p}$, $1 \le p \le \infty$. Furthermore, it makes sense to talk of the standard L^p Sobolev spaces. Indeed, we say $f \in L^p_s$ if $\forall x \in M$, there is a neighborhood U of x which is diffeomorphic to an open subset of $\mathbb{R}^n$, and the pullback of $f\big|_U$ via this diffeomorphism is in L^p_s. This yields a Banach space of distributions. As with the L^p spaces, the norm is not well-defined, but the equivalence class of the norm is, and we may therefore refer to $\|f\|_{L^p_s}$, where this is any choice from this equivalence class. We may now refer to differential operators on M being subelliptic (see Definition 1.4.8), and as before subellipticity implies hypoellipticity.

As in Chapter 1, we abuse notation and write the pairing between $C^\infty\,(M)$ and $C^\infty\,(M)'$ as integration: for $\lambda \in C^\infty\,(M)'$, we write $\lambda\,(f) = \int \lambda\,(x)\,f\,(x)\,dx$. If λ is given by integration against an L^1_{loc} function on some open subset $U \subset M$:

$$\int_U \lambda\,(x)\,f\,(x)\,dx, \quad f \in C^\infty_0\,(U),$$

where dx denotes the above chosen smooth measure, then we identify λ with this function on U. We use similar notation for the pairing between $C^\infty\,(M \times M)$ and $C^\infty\,(M \times M)'$. Furthermore, as in Chapter 1, we identify operators $T : C^\infty\,(M) \to C^\infty\,(M)'$ with their Schwartz kernels in $C^\infty\,(M \times M)'$.

From now on we assume $W_1, \ldots, W_r$ satisfy Hörmander's condition. We let

$$B_W\,(x, \delta) := \{y \in M \mid \rho\,(x, y) < \delta\},$$

the ball of radius δ centered at x in the ρ metric. A key inequality comes from the following result of Nagel, Stein, and Wainger [NSW85].

THEOREM 2.0.10. *There are constants $Q_2 \ge Q_1 > 0$ such that for any $x \in M$, $\delta > 0$,*

$$2^{Q_1}\mathrm{Vol}\,(B_W\,(x, \delta)) \le \mathrm{Vol}\,(B_W\,(x, 2\delta)) \le 2^{Q_2}\mathrm{Vol}\,(B_W\,(x, \delta)).$$

DEFINITION 2.0.11. *The least possible Q_2 satisfying the conclusion of Theorem 2.0.10 is referred to as the homogeneous dimension of (M, ρ, Vol). Depending on the situation, Q_1 or Q_2 often play the role that n played in Chapter 1.*

Remark 2.0.12   Henceforth, we use Q_1 to denote the greatest possible Q_1 and Q_2 to be the least possible Q_2 so that Theorem 2.0.10 holds.

We defer discussion of Theorem 2.0.10 to Section 2.3. For now, we mention that $\mathrm{Vol}\left(B_W\left(x, 2\delta\right)\right) \leq 2^{Q_2}\,\mathrm{Vol}\left(B_W\left(x, \delta\right)\right)$ is the key estimate that allows us to transfer the proof of Proposition 1.1.7 to this setting.[2]

We now introduce the class of singular integral operators which we study in this chapter–these are singular integrals corresponding to the balls $B_W\left(x, \delta\right)$. They were first introduced by Nagel, Rosay, Stein, and Wainger [NRSW89] under the name "NIS (Non-isotropic smoothing) operators." The definitions that follow are closely analogous to the definitions in Section 1.2.

DEFINITION 2.0.13.   *We say $\mathcal{B} \subset C^\infty\left(M\right) \times M \times (0, 1]$ is a bounded set of bump functions if:*

- $\forall\left(\phi, x, \delta\right) \in \mathcal{B}$, $\mathrm{supp}\left(\phi\right) \subset B_W\left(x, \delta\right)$.

- *For every ordered multi-index α, there exists C, such that $\forall\left(\phi, x, \delta\right) \in \mathcal{B}$,*

$$\sup_z \left|\left(\delta W\right)^\alpha \phi\left(z\right)\right| \leq C\mathrm{Vol}\left(B_W\left(x, \delta\right)\right)^{-1}.$$

Example 2.0.14   *In the special case[3] $M = \mathbb{R}^n$ and if $W_1, \ldots, W_r$ are given by $\partial_{x_1}, \ldots, \partial_{x_n}$, bounded sets of bump functions are easy to understand. Indeed $\mathcal{B} \subset C_0^\infty\left(\mathbb{R}^n\right) \times \mathbb{R}^n \times (0, 1]$ is a bounded set of bump functions if and only if*

$$\left\{\delta^n \phi\left(\delta\left(\cdot + x\right)\right) \mid \left(\phi, x, \delta\right) \in \mathcal{B}\right\} \subset C_0^\infty\left(\mathbb{R}^n\right)$$

is a bounded set and $\mathrm{supp}\left(\phi\left(\delta\left(\cdot + x\right)\right)\right) \subset B^n\left(1\right)$, $\forall\left(\phi, x, \delta\right) \in \mathcal{B}$, where $B^n\left(1\right)$ denotes the open ball of radius 1, centered at 0, in $\mathbb{R}^n$.

Remark 2.0.15   Despite the name, bounded sets of bump functions are not bounded subsets of a topological vector space in any obvious way. Instead, the name comes from the analogy given in Example 2.0.14.

DEFINITION 2.0.16.   *We say $T : C^\infty\left(M\right) \to C^\infty\left(M\right)$ is a Calderón-Zygmund operator of order $t \in \left(-Q_1, \infty\right)$ if*

(i) (Growth Condition) For each ordered multi-indices α, β,

$$\left|W_x^\alpha W_z^\beta T\left(x, z\right)\right| \leq C_{\alpha, \beta} \frac{\rho\left(x, z\right)^{-t - |\alpha| - |\beta|}}{\mathrm{Vol}\left(B_W\left(x, \rho\left(x, z\right)\right)\right)},$$

where W_x denotes the list of vector fields $W_1, \ldots, W_r$ thought of as partial differential operators in the x variable and similarly for W_z. In particular, the above implies that the distribution $T\left(x, z\right)$ corresponds with a C^∞ function for $x \neq z$.

---

[2]In fact, the inequality $\mathrm{Vol}\left(B_W\left(x, 2\delta\right)\right) \lesssim \mathrm{Vol}\left(B_W\left(x, \delta\right)\right)$ is the key inequality which turns M, when paired with $\mathrm{Vol}\left(\cdot\right)$, into a *space of homogeneous type*, where the general theory of Coifman and Weiss [CW71] can be applied. See [Ste93] for more details.

[3]Strictly speaking, $M = \mathbb{R}^n$ is not a special case, as $\mathbb{R}^n$ is not compact, but this is not a significant issue.

(ii) *(Cancellation Condition) For each bounded set of bump functions $\mathcal{B} \subset C^\infty (M) \times M \times (0, 1]$ and each ordered multi-index α,*

$$\sup_{(\phi, z, \delta) \in \mathcal{B}} \sup_{x \in M} \delta^{t+|\alpha|} \text{Vol} \left(B_W (z, \delta) \right) |W^\alpha T \phi (x)| \le C_{\mathcal{B}, \alpha},$$

with the same estimates for T^* in place of T. Here, the formal adjoint T^* is taken in the sense of $L^2 (M)$ which is defined in terms of the chosen strictly positive, smooth measure.

Remark 2.0.17   In light of Example 2.0.14, it is immediate to verify that if we extend the definition of bounded sets of bump functions to be subsets of $C^\infty (M) \times M \times (0, \infty)$ in the obvious way, then Definition 2.0.16 is a generalization of Definition 1.2.1. We use $(0, 1]$ instead of $(0, \infty)$ to account for the compactness of M.

The above definition is symmetric in x and z: if T is a Calderón-Zygmund operator of order t, so too is T^*. This follows immediately from the following simple lemma.

LEMMA 2.0.18. $\text{Vol} \left(B_W \left(x, \rho \left(x, z \right) \right) \right) \approx \text{Vol} \left(B_W \left(z, \rho \left(z, x \right) \right) \right)$.

PROOF.   We use Theorem 2.0.10 to see

$$\text{Vol} \left(B_W \left(x, \rho \left(x, z \right) \right) \right) \approx \text{Vol} \left(B_W \left(x, 2\rho \left(x, z \right) \right) \right) \ge \text{Vol} \left(B_W \left(z, \rho \left(z, x \right) \right) \right).$$

The result now follows by symmetry. $\square$

As in Chapter 1, Definition 2.0.16 is not always the most convenient definition. We now turn to stating an analog of Theorem 1.2.10 in this setting. Our first goal is to generalize operators of the form $E^{(2^j)}$, where $E \in \mathcal{P}$ ranges over a bounded set. As in Chapter 1, we often write elements of $(0, 1]$ as 2^{-j}, where $j \in [0, \infty)$.

DEFINITION 2.0.19.   *We say $\mathcal{E} \subset C^\infty (M \times M) \times (0, 1]$ is a bounded set of pre-elementary operators if: $\forall \alpha, \beta, m, \exists C = C (\mathcal{E}, \alpha, \beta, m), \forall \left(E, 2^{-j} \right) \in \mathcal{E}$,*

$$\left| \left(2^{-j} W_x \right)^\alpha \left(2^{-j} W_z \right)^\beta E \left(x, z \right) \right| \le C \frac{\left(1 + 2^j \rho \left(x, z \right) \right)^{-m}}{\text{Vol} \left(B_W \left(x, 2^{-j} \left(1 + 2^j \rho \left(x, z \right) \right) \right) \right)}.$$

Remark 2.0.20   Note that the bound $\frac{\left(1 + 2^j \rho (x, z)\right)^{-m}}{\text{Vol}(B_W (x, 2^{-j}(1 + 2^j \rho (x, z))))}$ from Definition 2.0.19 is essentially symmetric in x and z. Indeed, $1 + \rho (x, z) = 1 + \rho (z, x)$ and it follows by Lemma 2.0.18 that

$$\text{Vol} \left(B_W \left(x, 2^{-j} \left(1 + 2^j \rho \left(x, z \right) \right) \right) \right) \approx \text{Vol} \left(B_W \left(z, 2^{-j} \left(1 + 2^j \rho \left(z, x \right) \right) \right) \right).$$

Hence,

$$\frac{\left(1 + 2^j \rho \left(x, z \right) \right)^{-m}}{\text{Vol} \left(B_W \left(x, 2^{-j} \left(1 + 2^j \rho \left(x, z \right) \right) \right) \right)} \approx \frac{\left(1 + 2^j \rho \left(z, x \right) \right)^{-m}}{\text{Vol} \left(B_W \left(z, 2^{-j} \left(1 + 2^j \rho \left(z, x \right) \right) \right) \right)}.$$

We now turn to an analog of $\left\{ E^{(2^j)} \mid E \in \mathcal{B} \right\}$, where $\mathcal{B} \subset \mathcal{P}_0$ is a bounded set. For motivation, we look to Lemma 1.1.16.

DEFINITION 2.0.21. *We define the **set** of bounded sets of elementary operators, $\mathcal{G}$, to be the largest **set of subsets** of $C^\infty (M \times M) \times (0, 1]$ such that for all $\mathcal{E} \in \mathcal{G}$,*

(i) $\mathcal{E}$ is a bounded set of pre-elementary operators.

(ii) $\forall \left(E, 2^{-j} \right) \in \mathcal{E}$,

$$E = \sum_{|\alpha|, |\beta| \leq 1} 2^{-(2 - |\alpha| - |\beta|)j} \left(2^{-j} W \right)^\alpha E_{\alpha, \beta} \left(2^{-j} W \right)^\beta, \qquad (2.1)$$

where $\left\{ \left(E_{\alpha, \beta}, 2^{-j} \right) \mid \left(E, 2^{-j} \right) \in \mathcal{E} \right\} \in \mathcal{G}$.

We say $\mathcal{E}$ is a bounded set of elementary operators if $\mathcal{E} \in \mathcal{G}$.

Remark 2.0.22 We refer to (ii) from Definition 2.0.21, by saying that we can "pull out" derivatives from elementary operators.

Remark 2.0.23 As in Remark 2.0.15, bounded sets of elementary operators and bounded sets of pre-elementary operators are not bounded sets of a topological vector space in any obvious way. Rather, the names come from the analogies to bounded subsets of $\mathcal{P}_0$ and $\mathcal{P}$, respectively.

We outline some simple properties of elementary operators:

PROPOSITION 2.0.24. *Let $\mathcal{E}$ be a bounded set of elementary operators. Then,*

(a) If $\psi \in C^\infty (M)$, then $\left\{ \left(\psi E, 2^{-j} \right), \left(E \psi, 2^{-j} \right) \mid \left(E, 2^{-j} \right) \in \mathcal{E} \right\}$ is a bounded set of elementary operators. Here, we are identifying ψ with the operator $f \mapsto \psi f$.

(b) $\left\{ \left(E^*, 2^{-j} \right) \mid \left(E, 2^{-j} \right) \in \mathcal{E} \right\}$ is a bounded set of elementary operators.

(c) Fix an ordered multi-index α. Then

$$\left\{ \left(\left(2^{-j} W \right)^\alpha E, 2^{-j} \right), \left(E \left(2^{-j} W \right)^\alpha, 2^{-j} \right) \mid \left(E, 2^{-j} \right) \in \mathcal{E} \right\}$$

is a bounded set of elementary operators.

(d) For every $N \in \mathbb{N}$, each $\left(E, 2^{-j} \right) \in \mathcal{E}$ can be written as

$$E = \sum_{|\alpha| \leq N} 2^{(|\alpha| - N)j} \left(2^{-j} W \right)^\alpha E_\alpha$$

where $\left\{ \left(E_\alpha, 2^{-j} \right) \mid \left(E, 2^{-j} \right) \in \mathcal{E} \right\}$ is a bounded set of elementary operators. Similarly, each $\left(E, 2^{-j} \right) \in \mathcal{E}$ can be written as

$$E = \sum_{|\alpha| \leq N} 2^{(|\alpha| - N)j} \widetilde{E}_\alpha \left(2^{-j} W \right)^\alpha,$$

where $\left\{ \left(\widetilde{E}_\alpha, 2^{-j} \right) \mid \left(E, 2^{-j} \right) \in \mathcal{E} \right\}$ is a bounded set of elementary operators.

PROOF. (a): We prove just the result for ψE, the result for $E\psi$ is similar. It is immediate to verify that $\{(\psi E, 2^{-j}) \mid (E, 2^{-j}) \in \mathcal{E}\}$ is a bounded set of pre-elementary operators. The proof will be complete if we can show

$$\psi E = \sum_{|\alpha|, |\beta| \leq 1} 2^{-(2-|\alpha|-|\beta|)j} \left(2^{-j}W\right)^{\alpha} \widetilde{\psi}_{\alpha} E_{\alpha,\beta} \left(2^{-j}W\right)^{\beta},$$

where $\widetilde{\psi}_{\alpha} \in C^{\infty}(M)$ and $\{(E_{\alpha,\beta}, 2^{-j}) \mid (E, 2^{-j}) \in \mathcal{E}\}$ is a bounded set of elementary operators: i.e., that ψE is an appropriate sum of derivatives of operators of the same form as ψE. We have

$$\psi E = \sum_{|\alpha|, |\beta| \leq 1} 2^{-(2-|\alpha|-|\beta|)j} \psi \left(2^{-j}W\right)^{\alpha} E_{\alpha,\beta} \left(2^{-j}W\right)^{\beta}$$

$$= \sum_{|\alpha|, |\beta| \leq 1} 2^{-(2-|\alpha|-|\beta|)j} \left(2^{-j}W\right)^{\alpha} \psi E_{\alpha,\beta} \left(2^{-j}W\right)^{\beta}$$

$$+ \sum_{|\alpha|=1, |\beta| \leq 1} 2^{-(2-|\alpha|-|\beta|)j} \left[\psi, \left(2^{-j}W\right)^{\alpha}\right] E_{\alpha,\beta} \left(2^{-j}W\right)^{\beta}.$$

Since, for $|\alpha| = 1$, $\left[\psi, \left(2^{-j}W\right)^{\alpha}\right] = 2^{-j}\widetilde{\psi}$, where $\widetilde{\psi} \in C^{\infty}(M)$, the proof is complete.

(b): That $\{(E^{*}, 2^{-j}) \mid (E, 2^{-j}) \in \mathcal{E}\}$ is a bounded set of pre-elementary operators follows immediately from Remark 2.0.20. Thus, we turn to showing that E is an appropriate sum of derivatives of terms of the same form as E as in (2.1).

Taking the adjoint of (2.1), we see

$$E^{*} = \sum_{|\alpha|, |\beta| \leq 1} 2^{-(2-|\alpha|-|\beta|)j} \left(\left(2^{-j}W\right)^{\beta}\right)^{*} E_{\alpha,\beta}^{*} \left(\left(2^{-j}W\right)^{\alpha}\right)^{*}.$$

Since, for $|\alpha| = 1$, $\left(\left(2^{-j}W\right)^{\alpha}\right)^{*} = 2^{-j}\psi - \left(2^{-j}W\right)^{\alpha}$, where $\psi \in C^{\infty}(M)$, this shows that E^{*} is an appropriate sum of derivatives of operators of the same form as E^{*}, except possibly multiplied by functions in $C^{\infty}(M)$. Combining this with the proof of (a) completes the proof.

(c): We prove only the result for $\left(2^{-j}W\right)^{\alpha} E$; the proof for $E \left(2^{-j}W\right)^{\alpha}$ is similar. It is immediate to verify that $\{((2^{-j}W)^{\alpha} E, 2^{-j}) \mid (E, 2^{-j}) \in \mathcal{E}\}$ is a bounded set of pre-elementary operators. We have

$$\left(2^{-j}W\right)^{\alpha} E = \sum_{|\gamma|, |\beta| \leq 1} 2^{(|\gamma|+|\beta|-2)j} \left(2^{-j}W\right)^{\alpha} \left(2^{-j}W\right)^{\gamma} E_{\gamma,\beta} \left(2^{-j}W\right)^{\beta}$$

$$= \sum_{\substack{|\alpha'|=|\alpha| \\ |\gamma|, |\beta| \leq 1}} 2^{(|\gamma|+|\beta|-2)j} \left(2^{-j}W\right)^{\gamma} \left(2^{-j}W\right)^{\alpha'} E_{\gamma,\alpha',\beta} \left(2^{-j}W\right)^{\beta},$$

where $\{(E_{\gamma,\alpha',\beta}, 2^{-j}) \mid (E, 2^{-j}) \in \mathcal{E}\}$ is a bounded set of elementary operators; i.e., $\left(2^{-j}W\right)^{\alpha} E$ is an appropriate sum of derivatives of operators of the same form, which completes the proof.

(d): By repeated applications of (2.1), we see

$$E = \sum_{|\alpha|,|\beta| \leq N} 2^{(|\alpha|+|\beta|-2N)j} \left(2^{-j}W\right)^{\alpha} E_{\alpha,\beta} \left(2^{-j}W\right)^{\beta},$$

where $\left\{\left(E_{\alpha,\beta}, 2^{-j}\right) \mid \left(E, 2^{-j}\right) \in \mathcal{E}\right\}$ is a bounded set of elementary operators. The result now follows by applying (c) to either $E_{\alpha,\beta} \left(2^{-j}W\right)^{\beta}$ or $\left(2^{-j}W\right)^{\alpha} E_{\alpha,\beta}$. $\qquad \square$

(d) of Proposition 2.0.24 has a converse which will be of use to us.

PROPOSITION 2.0.25. *Fix $N \geq 1$. Let $\mathcal{G}_N$ be the largest **set of subsets** of*

$$C^{\infty}\left(M \times M\right) \times (0, 1]$$

such that for all $\mathcal{E} \in \mathcal{G}$,

- *$\mathcal{E}$ is a bounded set of pre-elementary operators.*

- *$\forall \left(E, 2^{-j}\right) \in \mathcal{E}$,*

$$E = \begin{cases} \sum_{|\alpha| \leq N} 2^{-\epsilon_{\alpha} j} \left(2^{-j}W\right)^{\alpha} E_{\alpha,1}, & and \\ \sum_{|\alpha| \leq N} 2^{-\epsilon_{\alpha} j} E_{\alpha,2} \left(2^{-j}W\right)^{\alpha}, \end{cases}$$

where $\epsilon_{\alpha} = 1$ if $|\alpha| = 0$ and 0 otherwise and

$$\left\{\left(E_{\alpha,1}, 2^{-j}\right), \left(E_{\alpha,2}, 2^{-j}\right) \mid \left(E, 2^{-j}\right) \in \mathcal{E}\right\} \in \mathcal{G}_N.$$

Then, $\mathcal{G}_N = \mathcal{G}$ for every N, where $\mathcal{G}$ is the set of bounded sets of elementary operators.

PROOF. $\mathcal{G} \subseteq \mathcal{G}_N$ is a consequence of (d) of Proposition 2.0.24. Proceeding just as in the proof of (c) of Proposition 2.0.24, we see for $\mathcal{E} \in \mathcal{G}_N$ and α fixed,

$$\left\{\left(\left(2^{-j}W\right)^{\alpha} E, 2^{-j}\right), \left(E \left(2^{-j}W\right)^{\alpha}, 2^{-j}\right) \mid \left(E, 2^{-j}\right) \in \mathcal{E}\right\} \in \mathcal{G}_N.$$

Fix $\mathcal{E} \in \mathcal{G}_N$ and $\left(E, 2^{-j}\right) \in \mathcal{E}$. Applying the definition of $\mathcal{G}_N$ twice, we see

$$E = \sum_{|\alpha|,|\beta| \leq N} 2^{-\epsilon_{\alpha} j - \epsilon_{\beta} j} \left(2^{-j}W\right)^{\alpha} E_{\alpha,\beta} \left(2^{-j}W\right)^{\beta},$$

with $\left\{ \left(E_{\alpha,\beta}, 2^{-j} \right) \mid \left(E, 2^{-j} \right) \in \mathcal{E}, |\alpha|, |\beta| \le N \right\} \in \mathcal{G}_N$. Rewriting this sum, we have

$$E = 2^{-2j} E_{0,0} + 2^{-j} \sum_{|\beta_0|=1} \left[\sum_{|\beta_1| \le N-1} E_{0,\beta} \left(2^{-j} W \right)^{\beta_1} \right] \left(2^{-j} W \right)^{\beta_0}$$

$$+ 2^{-j} \sum_{|\alpha_0|=1} \left(2^{-j} W \right)^{\alpha_0} \left[\sum_{|\alpha_1| \le N-1} \left(2^{-j} W \right)^{\alpha_1} E_{\alpha,0} \right]$$

$$+ \sum_{|\alpha_0|=|\beta_0|=1} \left(2^{-j} W \right)^{\alpha_0} \left[\sum_{|\alpha_1|,|\beta_1| \le N-1} \left(2^{-j} W \right)^{\alpha_1} E_{\alpha,\beta} \left(2^{-j} W \right)^{\beta_1} \right] \left(2^{-j} W \right)^{\beta_0}$$

$$=: \sum_{|\alpha_0|,|\beta_0| \le 1} 2^{-(2-|\alpha_0|-|\beta_0|)j} \left(2^{-j} W \right)^{\alpha_0} \widetilde{E}_{\alpha_0,\beta_0} \left(2^{-j} W \right)^{\beta_0}.$$

From the above comments $\left\{ \left(\widetilde{E}_{\alpha_0,\beta_0}, 2^{-j} \right) \mid \left(E, 2^{-j} \right) \in \mathcal{E}, |\alpha_0|, |\beta_0| \le 1 \right\} \in \mathcal{G}_N$. Thus, $\mathcal{G}_N$ satisfies the same axioms as $\mathcal{G}$ and we conclude $\mathcal{G}_N \subseteq \mathcal{G}$, completing the proof. $\qquad\square$

Remark 2.0.26 Notice, when $j = 0$, pre-elementary operators and elementary operators are the same. That is, if $\mathcal{E}$ is a bounded set of pre-elementary operators, then

$$\left\{ \left(E, 2^{-0} \right) \mid \left(E, 2^{-0} \right) \in \mathcal{E} \right\}$$

is a bounded set of elementary operators.

LEMMA 2.0.27. *Let $\mathcal{E}$ be a bounded set of pre-elementary operators. Then,*

$$\sup_{(E,2^{-j}) \in \mathcal{E}} \sup_{y \in M} \int |E(x,y)| \, dx < \infty, \qquad \sup_{(E,2^{-j}) \in \mathcal{E}} \sup_{x \in M} \int |E(x,y)| \, dy < \infty.$$

PROOF. By symmetry (see Remark 2.0.20), it suffices to verify

$$\sup_{(E,2^{-j}) \in \mathcal{E}} \sup_{x \in M} \int |E(x,y)| \, dy < \infty.$$

We have

$$\int |E(x,y)| \, dy \lesssim \int \frac{\left(1 + 2^j \rho(x,y) \right)^{-1}}{\mathrm{Vol}\left(B_W \left(x, 2^{-j} \left(1 + 2^j \rho(x,y) \right) \right) \right)} \, dy$$

$$\lesssim \sum_{k=0}^{\infty} \int_{2^k \le 1 + 2^j \rho(x,y) \le 2^{k+1}} 2^{-k} \mathrm{Vol}\left(B_W \left(x, 2^k \right) \right)^{-1} \, dy$$

$$\lesssim \sum_{k=0}^{\infty} 2^{-k} \int_{B_W(x,2^{k+1})} \mathrm{Vol}\left(B_W \left(x, 2^{k+1} \right) \right)^{-1} \, dy$$

$$\lesssim 1,$$

completing the proof. $\qquad\square$

LEMMA 2.0.28. *Fix* $t \in \mathbb{R}$ *and let* $\{(E_j, 2^{-j}) \mid j \in \mathbb{N}\}$ *be a bounded set of elementary operators. Then the sum*

$$\sum_{j \in \mathbb{N}} 2^{jt} E_j$$

converges in the topology of bounded convergence as operators $C^\infty(M) \to C^\infty(M)$ *(and therefore converges in distribution). See Appendix A.1 for these notions of convergence.*

PROOF. Let $\mathcal{B} \subset C^\infty(M)$ be a bounded set, and fix a multi-index α. We will show

$$\sum_{j \in \mathbb{N}} 2^{jt} W^\alpha E_j f(x)$$

converges uniformly for $x \in M$, $f \in \mathcal{B}$, and the result follows (this uses the fact that $W_1, \ldots, W_r$ satisfy Hörmander's condition). Note that

$$\sum_{j \in \mathbb{N}} 2^{jt} W^\alpha E_j f = \sum_{j \in \mathbb{N}} 2^{j(t+|\alpha|)} \left(2^{-j} W\right)^\alpha E_j f.$$

By Proposition 2.0.24, $\left\{\left(2^{-j} W\right)^\alpha E_j \mid j \in \mathbb{N}\right\}$ is a bounded set of elementary operators, and it therefore suffices to consider only the case $|\alpha| = 0$ (by replacing t with $t + |\alpha|$). By Proposition 2.0.24, we have

$$\sum_{j \in \mathbb{N}} 2^{jt} E_j f = \sum_{j \in \mathbb{N}} \sum_{|\alpha| \le N} 2^{j(t-N)} \widetilde{E}_{j,\alpha} W^\alpha f,$$

where $\left\{\left(\widetilde{E}_{j,\alpha}, 2^{-j}\right) \mid j \in \mathbb{N}, |\alpha| \le N\right\}$ is a bounded set of elementary operators. Using this, by taking $N \ge t + 1$, and by replacing f with $W^\alpha f$, we see that the full result follows from the result in the case when $t = -1$ and $|\alpha| = 0$. We then are considering

$$\sum_{j \in \mathbb{N}} 2^{-j} E_j f(x). \tag{2.2}$$

By Lemma 2.0.27, $\sup_x |E_j f(x)| \le \sup_x \int |E_j(x,y)| |f(y)| \, dy \lesssim \sup_y |f(y)|$. That (2.2) converges uniformly in x now follows. $\square$

We are now prepared to state the main characterization theorem for Calderón-Zygmund operators in this context. We defer the proof to Section 2.7.

THEOREM 2.0.29. *Let* $T : C^\infty(M) \to C^\infty(M)$, *and fix* $t \in (-Q_1, \infty)$. *The following are equivalent.*

(i) *T is a Calderón-Zygmund operator of order* t.

(ii) *For every bounded set of elementary operators* $\mathcal{E}$,

$$\left\{\left(2^{-jt} T E, 2^{-j}\right) \mid (E, 2^{-j}) \in \mathcal{E}\right\}$$

is a bounded set of elementary operators.

(iii) There is a bounded set of elementary operators $\left\{ \left(E_j, 2^{-j} \right) \mid j \in \mathbb{N} \right\}$ such that $T = \sum_{j \in \mathbb{N}} 2^{jt} E_j$. See Lemma 2.0.28 for more on the convergence of this sum.

Furthermore, (ii) and (iii) are equivalent for any $t \in \mathbb{R}$.

In light of Theorem 2.0.29, we extend Definition 2.0.16 in the following way.

DEFINITION 2.0.30. *Fix $t \in \mathbb{R}$. We say $T : C^\infty(M) \to C^\infty(M)$ is a Calderón-Zygmund operator of order t if either of the equivalent conditions (ii) or (iii) of Theorem 2.0.29 holds.*

PROPOSITION 2.0.31. *If T and S are Calderón-Zygmund operators of order t and s, respectively, then TS is a Calderón-Zygmund operator of order $t + s$.*

PROOF. This follows immediately from (ii). $\qquad\qquad\qquad\qquad\qquad\qquad\square$

Remark 2.0.32 It is not hard to see that if T is a Calderón-Zygmund operator of order t, then T is a Calderón-Zygmund operator of order s for all $s \leq t$. This is not true in Chapter 1, when M was replaced with $\mathbb{R}^n$, and is closely related to the compactness of M. In light of this, Proposition 2.0.31 can be rephrased as saying that Calderón-Zgymund operators form a *filtered algebra*.

PROPOSITION 2.0.33. *If T is a Calderón-Zygmund operator of order $s \in \mathbb{R}$, then T^* is also a Calderón-Zygmund operator of order s, where T^* denotes the formal L^2 adjoint of T.*

PROOF. This follows from (iii) of Theorem 2.0.29 using (b) of Proposition 2.0.24. $\qquad\qquad\qquad\qquad\qquad\qquad\qquad\qquad\qquad\qquad\qquad\qquad\qquad\qquad\square$

For a fixed s, the space of Calderón-Zygmund operators of order s is a vector space. It is convenient to give this vector space the structure of a Fréchet space. Each of the characterizations of Calderón-Zygmund operators from Theorem 2.0.29 give rise to a locally convex topology in a usual way. The proof of Theorem 2.0.29 shows that each of these topologies are equivalent.[4] Because of this, we introduce only the characterization of this topology which will be of most use to us: the one that arises from (iii).

DEFINITION 2.0.34. *We say $E \in C^\infty(M \times M)$ is a 2^{-j} elementary operator if $\left\{ \left(E, 2^{-j} \right) \right\}$ is a bounded set of elementary operators.*

We begin by defining semi-norms on 2^{-j} elementary operators. These will be semi-norms and will also depend on two additional parameters $N, M \in \mathbb{N}$ and will take the form $|E|_{2^{-j}, N, M}$ where E is a 2^{-j} elementary operator.

DEFINITION 2.0.35. *We define $|E|_{2^{-j}, N, M}$ recursively on M in the following way.*

---

[4]It is tempting to apply the open mapping theorem (Theorem A.1.13) to deduce that these topologies are equivalent. However, we use the proof of Theorem 2.0.29 as a key part of the proof that the space of Calderón-Zygmund operators of order s is a Fréchet space. Thus, the proof of Theorem 2.0.29 is an essential part of the proof of the equivalence of these topologies, even if we invoke the open mapping theorem.

- $M = 0$: $|E|_{2^{-j},N,0}$ is the least possible C so that

$$\sum_{|\alpha|,|\beta|\leq N} \left| \left(2^{-j}W_x\right)^\alpha \left(2^{-j}W_z\right)^\beta E\left(x,z\right) \right|$$

$$\leq C \frac{\left(1 + 2^j\rho\left(x,z\right)\right)^{-N}}{\mathrm{Vol}\left(B_W\left(x, 2^{-j}\left(1 + 2^j\rho\left(x,z\right)\right)\right)\right)}.$$

- $M \geq 1$: Once we have defined the semi-norm $|\cdot|_{2^{-j},N,M-1}$, then we define $|\cdot|_{2^{-j},N,M}$ in the following way. For E a 2^{-j} elementary operator, we write

$$E = \sum_{|\alpha|,|\beta|\leq 1} 2^{-(2-|\alpha|-|\beta|)j} \left(2^{-j}W\right)^\alpha E_{\alpha,\beta}\left(2^{-j}W\right)^\beta, \qquad (2.3)$$

where each $E_{\alpha,\beta}$ is a 2^{-j} elementary operator. We define a semi-norm

$$|E|_{2^{-j},N,M} := |E|_{2^{-j},N,M-1} + \inf \sum_{|\alpha|,|\beta|\leq 1} |E_{\alpha,\beta}|_{2^{-j},N,M-1}, \qquad (2.4)$$

where the infimum is taken over all representations of E of the form (2.3).

DEFINITION 2.0.36. *For each $N \in \mathbb{N}$, we define a semi-norm on the set of Calderón-Zygmund operators of order $s \in \mathbb{R}$ in the following way. For T a Calderón-Zygmund operator of order s, we write*

$$T = \sum_{j\in\mathbb{N}} 2^{js} E_j, \qquad (2.5)$$

where $\left\{ \left(E_j, 2^{-j}\right) \mid j \in \mathbb{N} \right\}$ is a bounded set of elementary operators. We define a semi-norm

$$|T|_{s,N} = \inf \sup_{j\in\mathbb{N}} |E_j|_{2^{-j},N,N}, \qquad (2.6)$$

where the infimum is taken over all such representations of T of the form (2.5). We give the vector space of Calderón-Zygmund operators of order s the coarsest topology under which all of the above semi-norms are continuous.

PROPOSITION 2.0.37. *With the above topology, the space of Calderón-Zygmund operators of order s is a Fréchet space.*

Definition 2.0.36 gives a countable family of semi-norms which induce the topology on the space of Calderón-Zygmund operators of order s. Thus, the heart of Proposition 2.0.37 is that the space is complete with this topology. We defer the proof to Section 2.8.

Remark 2.0.38 Our proofs will show that all maps we consider are continuous with this topology. For instance, the map $(T, S) \mapsto TS$ taking a pair of Calderón-Zygmund operators of order t and s respectively to a Calderón-Zygmund operator of order $t + s$ is continuous. And the maps taking a Calderón-Zygmund operator of order 0 to a bounded operator on L^p ($1 < p < \infty$), where bounded operators are given the uniform topology, are also continuous. Alternatively, one can apply the closed graph theorem (Theorem A.1.14) to prove the desired continuity.

Remark 2.0.39   Here, and in the rest of the monograph, the topologies we put on the classes of singular integrals we consider are useful for some purposes, but are not central to our theory. The reader uninterested in these topologies can safely ignore them and still understand the proofs of our main results.

2.1   VECTOR FIELDS WITH FORMAL DEGREES

While the above definition of ρ and the corresponding balls $B_W (x, \delta)$ are intuitive, there is an equivalent family of balls with a slightly different definition, which better lends itself to analysis and generalizations. First we notice a particular natural scaling of the balls $B_W (x, \delta)$. Indeed, $B_W (x, \delta) = B_{\delta W} (x, 1)$. Thus, balls of any radius are equal to balls of unit radius, provided we change the vector fields. All of our definitions that follow respect this scaling. Given a finite collection of vector fields $W = W_1, \ldots, W_r$, we write $B_W (x)$ to denote $B_W (x, 1)$, thereby emphasizing the importance of balls of unit radius.

Now suppose $W_1, \ldots, W_r$ satisfy Hörmander's condition of order m. We assign to $W_1, \ldots, W_r$ the *formal degree* 1. To vector fields of the form $[W_i, W_j]$ we assign the formal degree 2. Recursively, if Y has formal degree d_0, we assign to $[W_j, Y]$ the formal degree $d_0 + 1$. Let $(X_1, d_1), \ldots, (X_q, d_q)$ be an enumeration of the above collection of vector fields with formal degrees, which have formal degree $\leq m$. Note that, in light of Hörmander's condition, $X_1, \ldots, X_q$ span $T_x M$ for every x.

The formal degrees encapsulate the above notion of scaling. Indeed, if we replace $W_1, \ldots, W_r$ with $\delta W_1, \ldots, \delta W_r$ in the above, then (X_j, d_j) is replaced by $(\delta^{d_j} X_j, d_j)$. Because this plays a crucial role in what follows, we denote by δX the list of vector fields

$$\delta X = \delta^{d_1} X_1, \ldots, \delta^{d_q} X_q. \tag{2.7}$$

We define

$$B_{(X,d)} (x, \delta) := B_{\delta X} (x). \tag{2.8}$$

It is clear that $B_W (x, \delta) \subseteq B_{(X,d)} (x, \delta)$. The converse is nearly true, and was shown by Nagel, Stein, and Wainger [NSW85]:

PROPOSITION 2.1.1. *There is a constant $c > 0$ such that $B_{(X,d)} (x, c\delta) \subseteq B_W (x, \delta)$, for all $\delta > 0$.*

We defer discussion of Proposition 2.1.1 to Section 2.3. In light of Proposition 2.1.1 and Theorem 2.0.10, we may replace $B_W (x, \delta)$ with $B_{(X,d)} (x, \delta)$ throughout the previous section, and obtain equivalent definitions. We may also replace $\rho (x, z)$ with the equivalent metric

$$\inf \left\{ \delta > 0 \mid z \in B_{(X,d)} (x, \delta) \right\}. \tag{2.9}$$

For the rest of this chapter, we shall proceed with both of these replacements. Henceforth, we write $\rho(x, z)$ to denote (2.9).

Remark 2.1.2   With ρ given by (2.9), Remark 2.0.5 is obvious. In addition, this new choice of ρ is equivalent to our original choice of ρ by Proposition 2.1.1, which proves Remark 2.0.5.

We now turn to discussing the main property of (X, d) we shall use.

PROPOSITION 2.1.3.

$$[X_j, X_k] = \sum_{d_l \leq d_j + d_k} c_{j,k}^l X_l, \tag{2.10}$$

where $c_{j,k}^l \in C^\infty(M)$.

PROOF. We separate into two cases. The first, $d_j + d_k \leq m$. In this case, it follows by the Jacobi identity that $[X_j, X_k]$ is a linear combination, with constant coefficients, of vector fields of the form X_l with $d_l = d_j + d_k$.

If $d_j + d_k > m$, then we use that $[X_j, X_k] = \sum_{l=1}^q c_{j,k}^l X_l$, with $c_{j,k}^l \in C^\infty(M)$. This follows from the fact that $X_1, \dots, X_q$ span $T_x M$ for every x. Since, in this case, $d_l \leq m < d_j + d_k$, $\forall l$, (2.10) follows, completing the proof. $\quad\square$

Actually, Proposition 2.1.3 is not the property which will be most important to us. Rather, a weaker corollary leads us down the correct path.

COROLLARY 2.1.4. *For every $\delta \in (0, 1]$ there are functions $c_{j,k}^{l,\delta} \in C^\infty(M)$ satisfying*

$$\left[\delta^{d_j} X_j, \delta^{d_k} X_k\right] = \sum_{l=1}^q c_{j,k}^{k,\delta} \delta^{d_l} X_l,$$

with $\left\{ c_{j,k}^{l,\delta} \mid \delta \in (0, 1] \right\} \subset C^\infty(M)$ a bounded set.

PROOF. Take $c_{j,k}^l \in C^\infty(M)$ as in Proposition 2.1.3. For $\delta \in (0, 1]$, set

$$c_{j,k}^{l,\delta} = \begin{cases} \delta^{d_j + d_k - d_l} c_{j,k}^l & \text{if } d_l \leq d_j + d_k, \\ 0 & \text{otherwise.} \end{cases}$$

It is immediate to verify that $c_{j,k}^{l,\delta}$ satisfy the conclusions of the corollary. $\quad\square$

2.2   THE FROBENIUS THEOREM

Closely tied to Carnot-Carathéodory balls is a quantitative version of the classical Frobenius theorem about involutive distributions. We begin by stating this classical theorem in the particular form which will be of most use to us. The presentation of the classical Frobenius theorem which follows is a special case of the theory developed by Sussmann [Sus73], and we refer the reader to that paper for more details.

The setting is a connected manifold M (M is not necessarily compact). We remind the reader of a definition which will be useful in what follows.

DEFINITION 2.2.1. *Let S and M be smooth manifolds. An immersion of S in M is a smooth map $S \to M$ whose differential is everywhere injective.*

DEFINITION 2.2.2. *Let M be a manifold. An injectively immersed submanifold S of M is a smooth manifold S along with an injective immersion $i : S \hookrightarrow M$.*

Remark 2.2.3 If S is an injectively immersed submanifold of M, then it is often convenient to identify S with its image under the immersion, and therefore think of S as a *subset* of M. However, S need not be a topological subspace of M. See Example 2.2.15.

Remark 2.2.4 If $i : S \hookrightarrow M$ is an injectively immersed submanifold, for $s \in S$, we may identify $T_s S$ with a subspace of $T_{i(s)} M$: namely we identify $T_s S$ with the image of the injective map $di\,(s) : T_s S \to T_{i(s)} M$. Henceforth, we write $T_s S$ to denote this subspace of $T_{i(s)} M$.

DEFINITION 2.2.5. *A distribution on M is a map $\triangle$ which assigns to each $x \in M$ a subspace $\triangle (x) \subseteq T_x M$.*

DEFINITION 2.2.6. *We let $\Gamma\,(TM)$ denote the space of smooth sections of TM. I.e., $\Gamma\,(TM)$ is the space of all vector fields on M.*

Remark 2.2.7 $\Gamma\,(TM)$ is a C^∞-module: given vector fields $X, Y \in \Gamma\,(TM)$ and $f, g \in C^\infty\,(M)$, $fX + gY \in \Gamma\,(TM)$.

DEFINITION 2.2.8. *Associated to a distribution $\triangle$ on M, there is a C^∞-submodule of $\Gamma\,(TM)$, $\mathcal{D}_\triangle$, defined by $X \in \mathcal{D}_\triangle$ if and only if $X\,(x) \in \triangle\,(x)$, $\forall x$.*

DEFINITION 2.2.9. *We say a distribution $\triangle$ is a C^∞ distribution if $\forall x \in M$, $\triangle\,(x) = \{ X\,(x) \mid X \in \mathcal{D}_\triangle \}$.*

DEFINITION 2.2.10. *We say a C^∞ distribution $\triangle$ is involutive if $X, Y \in \mathcal{D}_\triangle \Rightarrow [X, Y] \in \mathcal{D}_\triangle$. I.e., if $\mathcal{D}_\triangle$ is a Lie subalgebra of $\Gamma\,(TM)$.*

THEOREM 2.2.11 (The Frobenius theorem). *Let $\triangle$ be a C^∞ distribution on the connected manifold M. Suppose:*

- *$\triangle$ is involutive.*

- *$\mathcal{D}_\triangle$ is locally finitely generated as a C^∞ module. That is, $\forall x \in M$, there exists an open neighborhood U of x and a finite collection of vector fields $Z_1, \ldots, Z_q \in \mathcal{D}_\triangle$, such that for every $Y \in \mathcal{D}_\triangle$ there exist $c_1, \ldots, c_q \in C^\infty\,(U)$ with $Y\big|_U = \sum_{j=1}^{q} c_j Z_j \big|_U$.*

Then, for each point $x \in M$ there is a unique, maximal, connected, injectively immersed submanifold $L \hookrightarrow M$, with $x \in L$ and $\forall y \in L$, $T_y L = \triangle\,(y)$. The set of all such L, as $x \in M$ varies, form a partition of M into disjoint injectively immersed submanifolds.

DEFINITION 2.2.12. *In the setting of Theorem 2.2.11, we say L is the leaf passing through x, and refer to the conclusion by saying that $\triangle$ foliates M into leaves.*

Remark 2.2.13 It is possible that the various leaves of the foliation may not each have the same dimension. Some authors refer to this possibility by saying that the foliation may be singular. See Definition 2.2.18 and Example 2.2.19, below.

Example 2.2.14 *Consider $\mathbb{R}^2$ with the usual coordinates (x, y). Let $\triangle(x, y) = \text{span}\left\{\frac{\partial}{\partial y}\right\}$ for every $(x, y) \in \mathbb{R}^2$. Then the leaf passing through (x_0, y_0) is*

$$\left\{(x_0, y) \mid y \in \mathbb{R}\right\}.$$

Thus, in this case, the Frobenius theorem decomposes $\mathbb{R}^2$ into its usual product structure $\mathbb{R} \times \mathbb{R}$.

Example 2.2.15 *Let $M = \mathbb{R}/\mathbb{Z} \times \mathbb{R}/\mathbb{Z}$ be the torus, with inherited coordinates (x, y) from $\mathbb{R} \times \mathbb{R}$. Fix $\theta \in \mathbb{R} \setminus \mathbb{Q}$ an irrational number. Let $\triangle(x, y) = \text{span}\left\{\frac{\partial}{\partial x} + \theta\frac{\partial}{\partial y}\right\}$, for every (x, y). In this case the leaves are one dimensional, dense subsets of M, and carry a topology strictly finer than the subspace topology.*

Example 2.2.16 *For a trivial example, consider the case when $\triangle(x) = T_x M$ for every $x \in M$. In this case, there is only one leaf, and that leaf equals M.*

Remark 2.2.17 The difficulty outlined in Example 2.2.15 is nonlocal: it relied on the global attributes of the space $\mathbb{R}/\mathbb{Z} \times \mathbb{R}/\mathbb{Z}$. In particular, if we fix a point $(x_0, y_0) \in \mathbb{R}/\mathbb{Z} \times \mathbb{R}/\mathbb{Z}$ and replace M with a small, connected, open neighborhood of (x_0, y_0), then the leaves will no longer be dense: they are line segments with slope θ. Thus, *locally* Example 2.2.15 and Example 2.2.14 are quite similar. In this monograph, we are concerned with local questions, and will not need to address issues like those in Example 2.2.15.

Fix $x \in M$, and let L be the leaf passing through x. Notice that $\dim L = \dim T_x L = \dim \triangle(x)$. Nowhere did we assume that $\dim \triangle(x)$ was constant in x, and therefore the dimensions of the leaves may vary from point to point. To discuss this, we introduce a definition:

DEFINITION 2.2.18. *We say $x \in M$ is a singular point of the distribution $\triangle$ if $\dim \triangle(y)$ is not constant on any neighborhood of x.*

Example 2.2.19 *Again we work in the case $M = \mathbb{R}^2$ with the usual coordinates (x, y). We set $\triangle(x, y) = \text{span}\left\{x\frac{\partial}{\partial y}\right\}$. Notice*

$$\dim \triangle(x, y) = \begin{cases} 1 & \text{if } x \neq 0, \\ 0 & \text{if } x = 0. \end{cases}$$

*If $(x_0, y_0) \in \mathbb{R}^2$ has $x_0 \neq 0$, then the leaf passing through (x_0, y_0) is $\{(x_0, y) \mid y \in \mathbb{R}\}$.
On the other hand, the leaf passing through $(0, y_0)$ is a point: $\{(0, y_0)\}$. Thus there is
a significant "discontinuity" between leaves when $x_0 \neq 0$ and leaves when $x_0 = 0$.
Each point $(0, y_0)$ is a singular point in the sense of Definition 2.2.18.*

In proofs of the Frobenius theorem, one usually constructs the leaves. The leaves
are, in particular, manifolds and the goal is to construct an atlas of coordinate charts
which give the leaves the right manifold structure (see, for example, [Lun92]). For
most applications, existence of these charts is enough. However, for our applications,
we will need detailed quantitative control of these charts. It is here that singular points
become an issue. In most proofs of the Frobenius theorem, the coordinate charts "blow
up" as one approaches a singular point. We require coordinate charts that avoid this
blow up, and which give good quantitative control uniformly as one varies over the
leaves. Because of the quantitative nature of what follows, the version of the Frobenius
theorem we now present is useful even when the classical Frobenius theorem is trivial,
as in Example 2.2.16.

We work locally: let $\Omega \subset \mathbb{R}^n$ be an open set. We suppose we are given C^∞ vector
fields $Z_1, \ldots, Z_q$, on Ω. These should be thought of as the generators of $\mathcal{D}_\triangle$ on Ω
from the classical Frobenius theorem (where $\triangle(x) = \text{span}\{Z_1(x), \ldots, Z_q(x)\}$).
We assume,

$$[Z_j, Z_k] = \sum_{l=1}^{q} c_{j,k}^l Z_l, \tag{2.11}$$

for certain functions $c_{j,k}^l$; we will specify the hypotheses on these functions later. Notice, if $\mathcal{D}_\triangle$ is the C^∞ module generated by $Z_1, \ldots, Z_q$, then (2.11) (if we assume
$c_{j,k}^l \in C^\infty$) is equivalent to $\triangle$ being involutive.

Fix $x_0 \in \Omega$, we wish to create a coordinate chart near x_0, on the leaf passing
through x_0, generated by by the distribution $\triangle(x) = \text{span}\{Z_1(x), \ldots, Z_q(x)\}$. Let
n_0 be the dimension of this leaf, so $n_0 = \dim \text{span}\{Z_1(x_0), \ldots, Z_q(x_0)\}$.

We are now in a position to explain in a more detailed way in which sense the coor-
dinate charts do not "blow up." For an $n \times q$ matrix A, let $\det_{n_0 \times n_0} A$ denote the *vector*
whose coordinates are determinants of $n_0 \times n_0$ submatrices of A; it is not important to
us in which order the coordinates are arranged. We write $(Z_1(x) \mid \cdots \mid Z_q(x))$ for the
$n \times q$ matrix whose columns are given by the vectors $Z_1(x), \ldots, Z_q(x)$. That n_0 is
the dimension of the leaf implies

$$\left| \det_{n_0 \times n_0} (Z_1(x_0) \mid \cdots \mid Z_q(x_0)) \right| > 0. \tag{2.12}$$

In the classical proofs of the Frobenius theorem, various quantities depend on a lower
bound for the left-hand side of (2.12); it is essential for our applications that this is not
the case in the version we use. In particular, notice that the left-hand side of (2.12)
tends to 0 as x_0 tends to a singular point.

We need to make the notion of "near" quantitative. Even though $\Omega \subset \mathbb{R}^n$, we
are thinking of Ω as an open subset of a manifold, and the Euclidean metric is not
necessarily a natural choice. Because of this, there is only one natural choice of distance

available to use: the Carnot-Carathéodory distance with respect to $Z_1, \ldots, Z_q$. We make this notion rigorous next, but before we do so, it is worth noting that this Carnot-Carathéodory distance is actually a metric on the leaves of the foliation corresponding to $Z_1, \ldots, Z_q$. Points on different leaves have infinite distance.

Because it will be useful in our applications, we assign to each Z_j a formal degree $d_j \in (0, \infty)$. We let (Z, d) denote the list of vector fields with formal degrees $(Z_1, d_1), \ldots, (Z_q, d_q)$. The most natural choice, from the perspective of the Frobenius theorem, is $1 = d_1 = \cdots = d_q$, however we will need to make other choices in some of our applications. With these formal degrees in hand, it makes sense to talk about the ball $B_{(Z,d)}(x_0, \delta)$, for $\delta > 0$; see (2.8). This ball is an open subset of the leaf passing through x_0.

We need one more piece of notation. For an *arbitrary* set $U \subseteq \mathbb{R}^n$, we write

$$\|f\|_{C^m(U)} := \sum_{|\alpha| \leq m} \sup_{x \in U} |\partial_x^\alpha f(x)|, \tag{2.13}$$

and if we write $\|f\|_{C^m(U)} < \infty$, we assume that all these partial derivatives exist on U and are continuous. If $V = \sum_{j=1}^n a_j \frac{\partial}{\partial x_j}$ is a vector field, we denote by $\|V\|_{C^m(U)} := \sum_{j=1}^n \|a_j\|_{C^m(U)}$.

We now turn to a technical statement of the quantitative Frobenius theorem. It is important for our applications that we make each of our qualitative assumptions quantitative. We assume there there exists $0 < \xi_1 \leq 1$ such that:

(a) For every $a_1, \ldots, a_q \in L^\infty([0,1])$, with $\left\| \sum_{j=1}^q |a_j|^2 \right\|_{L^\infty} < 1$, there exists a solution to the ODE

$$\gamma'(t) = \sum_{j=1}^q a_j(t) \xi_1^{d_j} Z_j(\gamma(t)), \quad \gamma(0) = x_0, \quad \gamma : [0,1] \to \Omega.$$

Notice, by the Picard-Lindelöf theorem for existence of ODEs, this condition holds so long as we take ξ_1 small enough, depending on the C^1 norms of $Z_1, \ldots, Z_q$ and the Euclidean distance from x_0 to $\partial\Omega$.

(b) For each m, there is a constant C_m such that[5]

$$\|Z_j\|_{C^m\left(B_{(Z,d)}(x_0, \xi_1)\right)} \leq C_m. \tag{2.14}$$

(c) $[Z_j, Z_k] = \sum_{l=1}^q c_{j,k}^l Z_l$ on $B_{(Z,d)}(x_0, \xi_1)$, where for every m there is a constant D_m such that

$$\sum_{|\alpha| \leq m} \|Z^\alpha c_{j,k}^l\|_{C^0\left(B_{(Z,d)}(x_0, \xi_1)\right)} \leq D_m. \tag{2.15}$$

---

[5]To be clear, even though $B_{(Z,d)}(x_0, \xi_1)$ is an open subset of the leaf passing through x_0, the C^m norm is not taken in the sense of the manifold structure on the leaf. Rather, we are using (2.13), which gives a larger norm.

Remark 2.2.20   We think of ξ_1 as being some fixed number, and therefore not "small." (a) says the $B_{(Z,d)}(x_0,\xi_1)$ lies "inside" of Ω. Of course, by definition $B_{(Z,d)}(x_0,\xi_1)$ is a subset of Ω, but even if the vector fields Z_j were extended outside of Ω, (a) would insist $B_{(Z,d)}(x_0,\xi_1) \subseteq \Omega$. (b) allows us to discuss the fact that the Z_j are C^∞ in a quantitative way. (c) assumes that the $c_{j,k}^l$ are C^∞, when thought of as functions on the leaf passing through x_0, and assumes this in a quantitative way. This is less than assuming that the $c_{j,k}^l$ are C^∞ on Ω.

For $m \geq 2$, we say C is an m-admissible constant if C can be chosen to depend only on upper bounds for m, n, q, C_m from (2.14), D_m from (2.15), d_j ($1 \leq j \leq q$), and positive lower bounds for ξ_1 and d_j ($1 \leq j \leq q$). For $m < 2$, we say C is an m-admissible constant if C is a 2-admissible constant. We say C is an admissible constant if C is an m-admissible constant, where m can be chosen to depend only on upper bounds for n, q, d_j ($1 \leq j \leq q$), and lower bounds for ξ_1 and d_j ($1 \leq j \leq q$).

Remark 2.2.21   In particular, m-admissible constants and admissible constants do not depend on a lower bound for the left-hand side of (2.12).

We write $A \lesssim_m B$ if $A \leq CB$, where C is an m-admissible constant and $A \approx_m B$ for $A \lesssim_m B$ and $B \lesssim_m A$. We write $A \lesssim B$ if $A \leq CB$ where C is an admissible constant and $A \approx B$ for $A \lesssim B$ and $B \lesssim A$.

We continue to use the notation $B^{n_0}(\eta)$ to denote the usual Euclidean ball of radius η in $\mathbb{R}^{n_0}$, centered at 0. When S is a subset of the leaf passing through x_0, we denote by $\mathrm{Vol}(S)$ the volume of S in the sense of the induced Lebesgue measure on the leaf, thought of as a submanifold of $\mathbb{R}^n$.

We state without proof the quantitative Frobenius theorem. The proof can be found in [Str11], though we do outline some of the main ideas of the proof in Section 2.2.2.

THEOREM 2.2.22 (The quantitative Frobenius theorem).   *There exist 2-admissible constants* $\xi_2, \xi_3, \eta > 0$, $\xi_3 < \xi_2 < \xi_1$ *and a* C^∞ *map*

$$\Phi : B^{n_0}(\eta) \to B_{(Z,d)}(x_0,\xi_2),$$

such that

- $\Phi(0) = x_0.$

- Φ *is injective.*

- $B_{(Z,d)}(x_0,\xi_3) \subseteq \Phi(B^{n_0}(\eta)) \subseteq B_{(Z,d)}(x_0,\xi_2).$

- *For* $u \in B^{n_0}(\eta),$

$$\left| \det_{n_0 \times n_0} d\Phi(u) \right| \approx_2 \left| \det_{n_0 \times n_0} (Z_1(x_0)|\cdots|Z_q(x_0)) \right|$$

$$\approx_2 \mathrm{Vol}\left(B_{(Z,d)}(x_0,\xi_2)\right).$$

(2.16)

Furthermore, if $Y_1, \ldots, Y_q$ are the pullbacks of $Z_1, \ldots, Z_q$ to $B^{n_0}(\eta)$, then

$$\|Y_j\|_{C^m(B^{n_0}(\eta))} \lesssim_m 1, \tag{2.17}$$

and

$$\inf_{u \in B^{n_0}(\eta)} \left| \det_{n_0 \times n_0} (Y_1(u) | \cdots | Y_q(u)) \right| \approx_2 1. \tag{2.18}$$

Φ is the desired coordinate chart. We take a moment to discuss some aspects of Theorem 2.2.22: that $|\det_{n_0 \times n_0} (Z_1(x_0) | \cdots | Z_q(x_0))| \approx_2 \mathrm{Vol}\left(B_{(Z,d)}(x_0, \xi_2)\right)$ is closely related to a change of variables and the Cauchy-Binet formula. Indeed, suppose Ψ is a C^1 diffeomorphism from an open subset U in $\mathbb{R}^{n_0}$ to an n_0 dimensional submanifold of $\mathbb{R}^n$, where this submanifold is given the induced Lebesgue measure dx. Then we have

$$\int_{\Psi(U)} f(x)\, dx = \int_U f(\Psi(t)) \left| \det_{n_0 \times n_0} d\Psi(t) \right| dt. \tag{2.19}$$

In particular, we have

$$\mathrm{Vol}\left(\Phi\left(B^{n_0}(\eta)\right)\right) = \int_{B^{n_0}(\eta)} \left| \det_{n_0 \times n_0} d\Phi(u) \right| du \approx_2 \left| \det_{n_0 \times n_0} (Z_1(x_0) | \cdots | Z_q(x_0)) \right|.$$

Remark 2.2.23 The vector fields $Y_1, \ldots, Y_q$ can often be used instead of the Euclidean vector fields, to define the usual norms. For instance, for $f : B^{n_0}(\eta) \to \mathbb{C}$, we have

$$\|f\|_{C^m(B^{n_0}(\eta))} \approx_{m-1} \sum_{|\beta| \le m} \|Y^\beta f\|_{C^0(B^{n_0}(\eta))}.$$

This follows easily from (2.17) and (2.18).

Remark 2.2.24 As a particular case of the above remark, $\det_{n_0 \times n_0} d\Phi$ is a C^∞ map whose derivatives are admissibly bounded. After all, $d\Phi$ is a linear transformation which takes the spanning set $Y_1, \ldots, Y_q$ to the set $Z_1, \ldots, Z_q$. (2.17) and (2.18) then show

$$\left\| \det_{n_0 \times n_0} d\Phi \right\|_{C^m(B^{n_0}(\eta))} \lesssim_m 1.$$

In fact, more is true. The map $t \mapsto \Phi(t) - x_0$ is C^∞ with C^m norm bounded by an m-admissible constant. The easiest way to see this is the particular form of Φ which we have not yet made precise. It can be found in (2.26); from that formula and standard theorems from ODEs (see Appendix B.1), the smoothness of Φ follows.

2.2.1   Scaling techniques

The power of Theorem 2.2.22 (the quantitative Frobenius theorem) lies in the fact that the coordinate chart Φ allows us to convert questions about the Z vector fields into questions about the Y vector fields; we refer to this as "rescaling." Indeed, $Z_1, \ldots, Z_q$ might be very small, and might be very linearly dependent (in the sense that

$$\left| \det_{n_0 \times n_0} \left(Z_1 \left(x_0 \right) | \cdots | Z_q \left(x_0 \right) \right) \right|$$

might be very small). Neither of these are true about the pulled back vector fields $Y_1, \ldots, Y_q$. A general strategy is as follows: given a question about the Z vector fields, conjugate everything by the pullback via Φ. This changes the question about the Z vector fields to an equivalent question about the Y vector fields. Calculus methods can often be applied to answer the question about the Y vector fields, thereby answering the question about the Z vector fields.

Before we discuss this, we need to introduce one important concept: the exponential of a vector field. Let M be manifold and let X be a smooth vector field on M. For $x \in M$, we define $e^X x$ in the following way. We let $E(t)$ be the unique solution to the differential equation $E'(t) = X(E(t))$, $E(0) = x$. The Picard-Lindelöf theorem shows that the solution exists for small time and is unique. The interval of existence depends on the C^1 norm of X; if the C^1 norm is sufficiently small then the solution will exist up to $t = 1$.[6] Provided the solution exists up to $t = 1$, we define $e^X x = E(1)$. By multiplying X by a real number t, we have a definition for $e^{tX} x$ (provided the solution to the differential equation exists). Let $U \Subset M$ be a relatively open compact set. Standard theorems from ODEs show for $\delta_0 > 0$ small enough that $e^{tX} x$ exists for $|t| < \delta_0$ and $x \in U$, and $e^{tX} x$ is a C^∞ map in (x, t) for $|t| < \delta_0$ and $x \in U$. It is easy to check, by uniqueness, that $e^{tX} x = E(t)$. As a consequence, we have the group property $e^{tX} e^{sX} x = e^{(t+s)X} x$, and in particular $e^{tX} e^{-tX} x = x$–therefore e^X defines a local diffeomorphism of M. Notice, $e^{tX} x$ is defined in such a way that if $f \in C^\infty(M)$, then

$$\frac{d}{dt} f\left(e^{tX} x\right) = (Xf)\left(e^{tX} x\right). \tag{2.20}$$

Finally, if $[X_1, X_2] = 0$, then $e^{X_1} e^{X_2} x = e^{X_1 + X_2} x = e^{X_2} e^{X_1} x$. See Appendix B.1 for more details on this exponential map.

In special cases the scaling procedure referred to above is surely familiar to the reader. Indeed, consider the following very simple example, which helps to elucidate the nature of Φ.

Example 2.2.25   *We work on $\mathbb{R}^3$ with coordinates (x, y, z). Fix $0 < \delta_1, \delta_2, \theta \ll 1$ small numbers. Let Z_1 and Z_2 be the vector fields $Z_1 = \delta_1 \frac{\partial}{\partial x}$, $Z_2 = \delta_2 \left(\frac{\partial}{\partial x} + \theta \frac{\partial}{\partial y} \right)$ (we give both Z_1, Z_2 degree 1). Note that $[Z_1, Z_2] = 0$, and the classical Frobenius theorem applies to foliate $\mathbb{R}^3$ into leaves: the leaf passing through (x_0, y_0, z_0) is given*

---

[6]Actually, the interval of existence can be proven to only depend on the C^0 norm of X, so long as X is Lipschitz; we will not need this, though.

by $\{(x, y, z_0) \,|\, x, y \in \mathbb{R}\}$. $B_{(Z,d)}\left((x_0, y_0, z_0), 1\right)$ is (approximately) an ellipse lying in this leaf of size $\approx \delta_1 + \delta_2$ in the x direction and size $\approx \delta_2 \theta$ in the y direction. If we define $\Phi(u, v) = e^{uZ_1 + vZ_2}(x_0, y_0, z_0)$, it is easy to see that $\Phi : B^2(1) \to \mathbb{R}^3$ maps onto a set which is bounded above and below of balls of the form $B_{(Z,d)}\left((x_0, y_0, z_0), \xi\right)$, where $\xi \approx 1$. Pulling back Z_1 and Z_2 via Φ yields $Y_1 = \frac{\partial}{\partial u}$, $Y_2 = \frac{\partial}{\partial v}$. It is easy to verify that Φ satisfies the conclusions of Theorem 2.2.22. Thus, in this case, Φ "rescales" Z_1 and Z_2 and "straightens" them as well. The difficulty that Theorem 2.2.22 addresses is that this procedure is more difficult when the Z_j do not commute. When they do not commute, more work is required, but Theorem 2.2.22 shows that a similar result is still possible. For another example along these lines, see Section 2.13, in particular Remark 2.13.13.

In this section we apply this idea to prove two simple inequalities which are useful in the sequel. We also hope that this section will provide the reader with a simple situation in which to understand these scaling techniques, which appear in more complicated forms throughout the monograph.

For the first inequality we fix points $x_0, y_0 \in \Omega$ ($\Omega \subseteq \mathbb{R}^n$ an open set as above) and C^∞ vector fields on Ω with formal degrees $(Z_1, d_1), \ldots, (Z_q, d_q)$. We assume that these vector fields satisfy the hypotheses of Theorem 2.2.22, both with the base point x_0 and with the base point y_0 (i.e., with y_0 playing the role of x_0 in the theorem)–we use the same ξ_1 for both x_0 and y_0. Below we use "admissible constants." These may depend on the same parameters as do the admissible constants in Theorem 2.2.22 both with the base point x_0 and with the base point y_0. We take admissible constants ξ_2 and η as in Theorem 2.2.22–note, we may pick the same ξ_2 and η for both x_0 and y_0.

PROPOSITION 2.2.26. *Suppose $K : \Omega \times \Omega \to \mathbb{C}$ is a measurable function such that $\forall \phi : B_{(X,d)}(y_0, \xi_2) \to \mathbb{C}$ with*

$$\sum_{|\beta| \le m} \left\| Z^\beta \phi \right\|_{C^0\left(B_{(Z,d)}(y_0, \xi_2)\right)} < \infty, \quad \forall m$$

we have $\forall \alpha$,

$$\sup_{x \in B_{(Z,d)}(x_0, \xi_2)} \left| \int Z_x^\alpha K(x, y) \, \phi(y) \, dy \right| \le C_\alpha \mathrm{Vol}\left(B_{(Z,d)}(y_0, \xi_2)\right)^{\frac{1}{2}} \|\phi\|_{L^2};$$

with the same estimate with the roles of x and y (and x_0 and y_0) reversed. Here dy and $\mathrm{Vol}(\cdot)$ denote the induced Lebesgue measure on the appropriate leaf, and $\|\phi\|_{L^2}$ is defined with respect to this induced Lebesgue measure. Then, $\forall \beta, \gamma$, there exists an admissible constant $C_{\beta, \gamma}$ (depending on the C_α), such that

$$\left| Z_x^\beta Z_y^\gamma K(x_0, y_0) \right| \le C_{\beta, \gamma}.$$

We will see, after scaling, that Proposition 2.2.26 follows from the next lemma which we prove by calculus methods.

LEMMA 2.2.27. *Fix $n_1, n_2 \in \mathbb{N}$ and suppose $K : B^{n_1}(\eta) \times B^{n_2}(\eta) \to \mathbb{C}$ is a measurable function such that $\forall \alpha$, $\forall \phi \in C_0^\infty(B^{n_2}(\eta))$ we have*

$$\sup_{u \in B^{n_1}(\eta)} \left| \int \partial_u^\alpha K(u, v) \phi(v) \, dv \right| \le C_\alpha \|\phi\|_{L^2} ; \tag{2.21}$$

with the same estimate with the roles of u and v (and n_1 and n_2) reversed. Then, $\forall \beta, \gamma$, there exists a constant $C_{\beta,\gamma}$ (depending on the above C_α and on η) such that

$$\left| \partial_u^\beta \partial_v^\gamma K(0,0) \right| \le C_{\beta,\gamma}.$$

PROOF. (2.21) shows that

$$\sup_{u \in B^{n_1}(\eta)} \left\| \partial_u^\beta K(u, \cdot) \right\|_{L^2(B^{n_2}(\eta))} \le C_\beta, \tag{2.22}$$

$$\sup_{v \in B^{n_2}(\eta)} \left\| \partial_v^\beta K(\cdot, v) \right\|_{L^2(B^{n_1}(\eta))} \le C_\beta. \tag{2.23}$$

Let $\phi_1 \in C_0^\infty(B^{n_1}(\eta))$ and $\phi_2 \in C_0^\infty(B^{n_2}(\eta))$ equal 1 on a neighborhood of 0. Consider the function $\widetilde{K}(u, v) = \phi_1(u) K(u, v) \phi_2(u)$, and let $\widehat{K}(\xi, \eta)$ denote the Fourier transform of $\widetilde{K}$. (2.22) and (2.23) show that

$$\left\| \xi^\beta \widehat{K}(\xi, \eta) \right\|_{L^2(\mathbb{R}^{n_1} \times \mathbb{R}^{n_2})} \le C_\beta',$$

$$\left\| \eta^\beta \widehat{K}(\xi, \eta) \right\|_{L^2(\mathbb{R}^{n_1} \times \mathbb{R}^{n_2})} \le C_\beta'.$$

Thus, for every s,

$$\left\| \widetilde{K} \right\|_{L^2_s(\mathbb{R}^{n_1} \times \mathbb{R}^{n_2})} \le C_s'',$$

for some C_s'' depending on the C_β'. The result now follows immediately from the Sobolev embedding theorem. $\qquad \square$

PROOF OF PROPOSITION 2.2.26. Let

$$\Phi_1 : B^{n_1}(\eta) \to B_{(Z,d)}(x_0, \xi_2), \quad \Phi_2 : B^{n_2}(\eta) \to B_{(Z,d)}(y_0, \xi_2)$$

be the maps given by Theorem 2.2.22 when applied with base points x_0 and y_0 respectively. Define

$$\widehat{K}(u, v) = K(\Phi_1(u), \Phi_2(v)).$$

Let $\widehat{\phi} \in C_0^\infty(B^{n_2}(\eta))$, and define $\phi = \widehat{\phi} \circ \Phi_2^{-1}$; note that $\text{supp}(\phi) \subset B_{(Z,d)}(y_0, \xi_2)$. Let $Y_1, \ldots, Y_q$ be the pullbacks of $Z_1, \ldots, Z_q$ via Φ_1 and let $V_1, \ldots, V_q$ be the pullbacks of $Z_1, \ldots, Z_q$ via Φ_2. By (2.17) (applied to $V_1, \ldots, V_q$) we have, for every m,

$$\sum_{|\beta| \le m} \left\| V^\beta \widehat{\phi} \right\|_{C^0(B^{n_0}(\eta))} < \infty,$$

and it follows that

$$\sup_y \sum_{|\beta| \le m} \left| Z^\beta \phi(y) \right| < \infty.$$

Consider, using a change of variables as in (2.19), and applying (2.16), we have

$$\sup_{u \in B^{n_1}(\eta)} \left| \int Y_u^\beta \widehat{K}(u,v) \, \widehat{\phi}(v) \, dv \right| \le \sup_{x \in B_{(Z,d)}(x_0,\xi_2)} \left| \int Z_x^\beta K(x, \Phi_2(v)) \, \widehat{\phi}(v) \, dv \right|$$

$$= \sup_{x \in B_{(Z,d)}(x_0,\xi_2)} \left| \int Z_x^\beta K(x,y) \, \phi(y) \left| \det_{n_2 \times n_2} d\Phi_2^{-1}(y) \right| \, dy \right|$$

$$\lesssim \mathrm{Vol}\left(B_{(Z,d)}(y_0,\xi_2) \right)^{-\frac{1}{2}} \|\phi\|_{L^2}$$

$$\approx \left\| \widehat{\phi} \right\|_{L^2},$$

where the second-to-last line follows by our hypothesis applied with $\phi(y)$ replaced by $\phi(y) \left| \det_{n_2 \times n_2} d\Phi_2^{-1}(y) \right|$ and the last line follows by another change of variables as in (2.19), using (2.16).

Using (2.17) and (2.18) as in Remark 2.2.23, we have

$$\sup_{u \in B^{n_1}(\eta)} \left| \int \partial_u^\beta \widehat{K}(u,v) \, \widehat{\phi}(v) \, dv \right| \lesssim \left\| \widehat{\phi} \right\|_{L^2},$$

for each β. Similarly, one has for $\widehat{\phi} \in C_0^\infty(B^{n_1}(\eta))$,

$$\sup_{v \in B^{n_2}(\eta)} \left| \int \partial_v^\beta \widehat{K}(u,v) \, \widehat{\phi}(u) \, du \right| \lesssim \left\| \widehat{\phi} \right\|_{L^2},$$

for each β. Lemma 2.2.27 applies to show

$$\left| \partial_u^\beta \partial_v^\gamma \widehat{K}(0,0) \right| \lesssim 1,$$

for each β, γ. Using (2.17) we have

$$\left| Y_u^\beta V_v^\gamma \widehat{K}(0,0) \right| \lesssim 1,$$

for each β, γ. This is equivalent to

$$\left| Z_x^\beta Z_y^\gamma K(x_0,y_0) \right| \lesssim 1,$$

for each β, γ, which completes the proof. $\square$

We now turn to the second inequality. Here, we fix one point $x_0 \in \Omega$ and C^∞ vector fields with formal degrees $(Z_1, d_1), \ldots, (Z_q, d_q)$ satisfying the hypotheses of Theorem 2.2.22. We assume, in addition, that $Z_1(x_0), \ldots, Z_q(x_0)$ span the tangent space $T_{x_0}\Omega$. I.e., we assume $n_0 = n$. We take all constants and notation as in Theorem 2.2.22 (e.g., ξ_2, ξ_3, η, and Φ). In what follows, for $t = (t_1, \ldots, t_q) \in \mathbb{R}^q$ we write $t \cdot Z$ for $t_1 Z_1 + \cdots + t_q Z_q$, and similarly for other similar expressions.

PROPOSITION 2.2.28. *There exists a 2-admissible constant $a > 0$ such that the following holds. Fix M. Suppose that $\varsigma \in C_0^\infty (B^q (a))$ and*

$$\kappa \in C^M \left(B^q (a) \times B_{(Z,d)} (x_0, \xi_2) \right)$$

satisfies

$$\sup_{\substack{x \in B_{(Z,d)}(x_0,\xi_2) \\ t \in B^q(a)}} \sum_{|\alpha|+|\beta| \leq M} \left| \partial_t^\beta Z^\alpha \kappa (t, x) \right| \leq C,$$

for some constant $C > 0$. Then if F is the operator given by

$$Ff (x) = \int f \left(e^{t \cdot Z} x \right) \kappa (t, x) \varsigma (t) \, dt$$

we have

$$\sup_{x \in B_{(Z,d)}\left(x_0,\frac{\xi_3}{2}\right)} \sum_{|\alpha|+|\beta| \leq M} \left| Z_x^\alpha Z_z^\beta F (x, z) \right| \lesssim_M \frac{\chi_{B_{(Z,d)}(x_0,\xi_3)} (z)}{\mathrm{Vol} \left(B_{(Z,d)} (x_0, \xi_3) \right)},$$

where the implicit constant is an M admissible constant which is also allowed to depend on C and $\|\varsigma\|_{C^M(B^q(a))}$.

PROOF. It is easy to see that if $a > 0$ is a sufficiently small 2-admissible constant, then for $x \in B_{(Z,d)} (x_0, \xi_3)$, $F (x, \cdot)$ is supported in $B_{(Z,d)} (x_0, \xi_3)$. The result will follow once we show

$$\sup_{x \in B_{(Z,d)}\left(x_0,\frac{\xi_3}{2}\right)} \sum_{|\alpha|+|\beta| \leq M} \left| Z_x^\alpha Z_z^\beta F (x, z) \right| \lesssim_M \frac{1}{\mathrm{Vol} \left(B_{(Z,d)} (x_0, \xi_3) \right)}. \tag{2.24}$$

Denote by $\Phi^\# f (u) = f \circ \Phi (u)$: the pullback via Φ.

Notice that

$$\sum_{|\alpha|+|\beta| \leq M} \left| Y^\alpha \partial_t^\beta \kappa (t, \Phi (u)) \right| \leq C.$$

Using (2.17) and (2.18) as in Remark 2.2.23, we see

$$\|\kappa (\cdot, \Phi (\cdot))\|_{C^M(B^q(a) \times B^n (\eta))} \lesssim_M 1.$$

Consider,

$$\Phi^\# F \left(\Phi^\# \right)^{-1} g (u) = \int g \left(e^{t \cdot Y} u \right) \kappa (t, \Phi (u)) \varsigma (t) \, dt. \tag{2.25}$$

Using that $|\det_{n \times n} (Y_1 (0) | \cdots | Y_q (0))| \approx_2 1$ and $\|Y_j\|_{C^m(B^q(\eta))} \lesssim_m 1$ we see that if $a > 0$ is a sufficiently small 2-admissible constant, we may apply a change of variables in the t variable to (2.25) as in Appendix B.3 to see

$$\Phi^\# F \left(\Phi^\# \right)^{-1} g (u) = \int g (v) K (u, v) \, dv,$$

where $\sum_{|\alpha|+|\beta|\leq M} \left\|\partial_u^\alpha \partial_v^\beta K(u,v)\right\|_{C^0} \lesssim_M 1$. Hence, by (2.17) and (2.18),

$$\sum_{|\alpha|+|\beta|\leq M} \left|Y_u^\alpha Y_v^\beta K(u,v)\right| \lesssim_M 1.$$

Notice, $K(u,v) = F(\Phi(u), \Phi(v)) \left|\det_{n\times n} d\Phi(v)\right|^{-1}$. Using that $Y_j f\circ \Phi = (Z_j f)\circ \Phi$, $\|\det_{n\times n} d\Phi\|_{C^m} \lesssim_m 1$ (see Remark 2.2.24), and (2.16), (2.24) follows, completing the proof. $\qquad\square$

2.2.2 Ideas in the proof

This section is devoted to a discussion of a few parts of the proof of the quantitative Frobenius theorem (Theorem 2.2.22). The ideas that follow are indicative of the main ideas involved in the proof. Unfortunately, the proof of Theorem 2.2.22 is rather involved and we will not be able to include any significant portion of the proof here; for the whole proof, we refer the reader to [Str11]. Fortunately, we will be able to use Theorem 2.2.22 as a "black box" wherever we need it. Thus understanding the proof is not essential to understanding the rest of this monograph. We do not use the ideas from this section in the rest of the monograph. The uninterested reader may safely skip this section and still understand the rest of the results.

We begin by defining the map Φ. We pick n_0 of the vector fields, $Z_{j_1}, \ldots, Z_{j_{n_0}}$, such that

$$\left|\det_{n_0\times n_0} \left(Z_{j_1}(x_0) | \cdots | Z_{j_{n_0}}(x_0)\right)\right|_\infty = \left|\det_{n_0\times n_0} \left(Z_1(x_0) | \cdots | Z_q(x_0)\right)\right|_\infty.$$

Without loss of generality, we re-order our vector fields so that $j_1 = 1, \ldots, j_{n_0} = n_0$, i.e.,

$$\left|\det_{n_0\times n_0} \left(Z_1(x_0) | \cdots | Z_{n_0}(x_0)\right)\right|_\infty = \left|\det_{n_0\times n_0} \left(Z_1(x_0) | \cdots | Z_q(x_0)\right)\right|_\infty.$$

We define Φ by

$$\Phi(t_1, \ldots, t_n) = e^{t_1 Z_1 + \cdots + t_{n_0} Z_{n_0}} x_0. \tag{2.26}$$

Because

$$\left|\det_{n_0\times n_0} \left(Z_1(x_0) | \cdots | Z_{n_0}(x_0)\right)\right|_\infty = \left|\det_{n_0\times n_0} \left(Z_1(x_0) | \cdots | Z_q(x_0)\right)\right|_\infty > 0,$$

by assumption, and because

$$\partial_{t_j}\big|_{t=0} \Phi(t) = Z_j(x_0),$$

the inverse function theorem shows that Φ is a diffeomorphism on some small neighborhood of 0, $\Phi : B^{n_0}(\epsilon) \to \Phi(B^{n_0}(\epsilon))$ (where $\Phi(B^{n_0}(\epsilon))$ is an open subset of the leaf passing through x_0). Unfortunately, ϵ depends on a lower bound for

$$\left|\det_{n_0\times n_0} \left(Z_1(x_0) | \cdots | Z_q(x_0)\right)\right|_\infty,$$

and is therefore not admissible. This is a main problem that makes the quantitative Frobenius theorem more difficult than the classical Frobenius theorem.

The first step to show is the following proposition.

PROPOSITION 2.2.29. *There is a 2-admissible constant $0 < \xi^1$ such that for $x \in B_{(Z,d)}\left(x_0, \xi^1\right)$,*

$$
\left| \det_{n_0 \times n_0} \left(Z_1\left(x\right) | \cdots | Z_q\left(x\right)\right) \right| \approx_2 \left| \det_{n_0 \times n_0} \left(Z_1\left(x\right) | \cdots | Z_{n_0}\left(x\right)\right) \right|
$$

$$
\approx_2 \left| \det_{n_0 \times n_0} \left(Z_1\left(x_0\right) | \cdots | Z_q\left(x_0\right)\right) \right|.
\tag{2.27}
$$

To prove Proposition 2.2.29 we introduce some new notation. For $n_1 \leq m$ let $\mathcal{I}\left(n_1, m\right)$ denote the set of all lists of integers $\left(i_1, \ldots, i_{n_0}\right)$ with $1 \leq i_1 < i_2 < \cdots < i_{n_0} \leq m$. If A is an $n \times q$ matrix, and $I \in \mathcal{I}\left(n_1, n\right)$, $J \in \mathcal{I}\left(n_1, q\right)$, we denote by $A_{I,J}$ the $n_1 \times n_1$ matrix given by taking the rows from A listed in I and the columns listed in J. Thus,

$$
\left| \det_{n_1 \times n_1} A \right|_\infty = \sup_{\substack{I \in \mathcal{I}(n_1, n) \\ J \in \mathcal{I}(n_1, q)}} \left| \det A_{I,J} \right|.
$$

We let $Z\left(x\right)$ denote the $n \times q$ matrix $\left(Z_1\left(x\right) | \cdots | Z_q\left(x\right)\right)$.

LEMMA 2.2.30. *Fix $1 \leq n_1 \leq n \wedge q$. Then for $1 \leq j \leq q$, $I \in \mathcal{I}\left(n_1, n\right)$, $J \in \mathcal{I}\left(n_1, q\right)$, $x \in B_{(Z,d)}\left(x_0, \xi_1\right)$,*

$$
\left| Z_j \det Z\left(x\right)_{I,J} \right| \lesssim_2 \left| \det_{n_1 \times n_1} Z\left(x\right) \right|.
$$

PROOF. We use the notation $\mathcal{L}_U$ to denote the Lie derivative with respect to the vector field U, and i_V to denote the interior product with the vector field V. If ω is a p-form, then $i_V \omega$ is defined to be the $p-1$ form which satisfies $\left(i_V \omega\right)\left(V_1, \ldots, V_{p-1}\right) := \omega\left(V, V_1, \ldots, V_{p-1}\right)$. $\mathcal{L}_U$ can be defined in many ways, but one way is to define it as $\mathcal{L}_U \omega := i_U d\omega + d i_U \omega$, for any form ω. $\mathcal{L}_U$ and i_V have the following, well-known, properties:

- $\mathcal{L}_U f = U f$, for functions f.

- $[\mathcal{L}_U, i_V] = i_{[U,V]}$.

- $\mathcal{L}_U\left(\omega_1 \wedge \omega_2\right) = \left(\mathcal{L}_U \omega_1\right) \wedge \omega_2 + \omega_1 \wedge \left(\mathcal{L}_U \omega_2\right)$.

- If $U = \sum_k b_k \frac{\partial}{\partial x_k}$, then

$$
\mathcal{L}_U dx_k = d i_U dx_k = db_k = \sum \frac{\partial b_k}{\partial x_j} dx_j.
$$

Fix $I = (i_1, \ldots, i_{n_1}) \in \mathcal{I}(n_1, n)$, $J = (j_1, \ldots, j_{n_1}) \in \mathcal{I}(n_1, q)$ as in the statement of the lemma. Then,

$$\det Z(x)_{I,J} = i_{Z_{j_{n_1}}} i_{Z_{j_{n_1-1}}} \cdots i_{Z_{j_1}} dx_{i_1} \wedge dx_{i_2} \wedge \cdots \wedge dx_{i_{n_1}}.$$

Thus, we see

$$
\begin{aligned}
Z_j \det Z(x)_{I,J} &= \mathcal{L}_{Z_j} i_{Z_{j_{n_1}}} i_{Z_{j_{n_1-1}}} \cdots i_{Z_{j_1}} dx_{i_1} \wedge dx_{i_2} \wedge \cdots \wedge dx_{i_{n_1}} \\
&= i_{[Z_j, Z_{j_{n_1}}]} i_{Z_{j_{n_1-1}}} \cdots i_{Z_{j_1}} dx_{i_1} \wedge dx_{i_2} \wedge \cdots \wedge dx_{i_{n_1}} \\
&\quad + i_{Z_{j_{n_1}}} i_{[Z_j, Z_{j_{n_1-1}}]} \cdots i_{Z_{j_1}} dx_{i_1} \wedge dx_{i_2} \wedge \cdots \wedge dx_{i_{n_1}} \\
&\quad + \cdots \\
&\quad + i_{Z_{j_{n_1}}} i_{Z_{j_{n_1-1}}} \cdots i_{[Z_j, Z_{j_1}]} dx_{i_1} \wedge dx_{i_2} \wedge \cdots \wedge dx_{i_{n_1}} \\
&\quad + i_{Z_{j_{n_1}}} i_{Z_{j_{n_1-1}}} \cdots i_{Z_{j_1}} \mathcal{L}_{Z_j} \left(dx_{i_1} \wedge dx_{i_2} \wedge \cdots \wedge dx_{i_{n_1}} \right).
\end{aligned}
\tag{2.28}
$$

There are two types of terms on the right-hand side of (2.28): the first n_1 terms, and the last term. We deal with these two types separately. First, we bound the first n_1 terms. All of these terms work in the same way, so we bound just the first term as an example.

$$
\begin{aligned}
&\left| i_{[Z_j, Z_{j_{n_1}}]} i_{Z_{j_{n_1-1}}} \cdots i_{Z_{j_1}} dx_{i_1} \wedge dx_{i_2} \wedge \cdots \wedge dx_{i_{n_1}} \right| \\
&= \left| \sum_{k=1}^{q} c_{j,j_{n_1}}^k i_{Z_k} i_{Z_{j_{n_1-1}}} \cdots i_{Z_{j_1}} dx_{i_1} \wedge dx_{i_2} \wedge \cdots \wedge dx_{i_{n_1}} \right| \\
&\lesssim_2 \left| \det_{n_1 \times n_1} Z(x) \right|.
\end{aligned}
$$

In the equality, we have used our main assumption on the Z_j (i.e., (c): $[Z_j, Z_{j_{n_1}}] = \sum_{k=1}^{1} c_{j,j_{n_1}}^k Z_k$). In the $\lesssim_2$, we have the fact that for each k, $i_{Z_k} i_{Z_{j_{n_1-1}}} \cdots i_{Z_{j_1}} dx_{i_1} \wedge dx_{i_2} \wedge \cdots \wedge dx_{i_{n_1}}$ is either 0 or of the form $\pm \det Z(x)_{I,J'}$ for some $J' \in \mathcal{I}(n_1, q)$.

We now turn to the last term on the right-hand side of (2.28). We have

$$
\begin{aligned}
\mathcal{L}_{Z_j} \left(dx_{i_1} \wedge dx_{i_2} \wedge \cdots \wedge dx_{i_{n_1}} \right) &= \left(\mathcal{L}_{Z_j} dx_{i_1} \right) \wedge dx_{i_2} \wedge \cdots \wedge dx_{i_{n_1}} \\
&\quad + dx_{i_1} \wedge \left(\mathcal{L}_{Z_j} dx_{i_2} \right) \wedge \cdots \wedge dx_{i_{n_1}} \\
&\quad + \cdots \\
&\quad + dx_{i_1} \wedge dx_{i_2} \wedge \cdots \wedge \left(\mathcal{L}_{Z_j} dx_{i_{n_1}} \right).
\end{aligned}
$$

Using this, we may separate the last term on the right-hand side of (2.28) into a sum of n_1 terms. Each behaves in a similar way, so we bound just the first as an example. To do this, write $Z_j = \sum_k b_j^k \frac{\partial}{\partial x_k}$. Part of the definition of 2-admissible constants states

$\left\| b_j^k \right\|_{C^2 \left(B_{(Z,d)}(x_0, \xi_1) \right)} \lesssim_2 1$. We have

$$\left| i_{Z_{j_{n_1}}} i_{Z_{j_{n_1-1}}} \cdots i_{Z_{j_1}} \left(\mathcal{L}_{Z_j} dx_{i_1} \right) \wedge dx_{i_2} \wedge \cdots \wedge dx_{i_{n_1}} \right|$$

$$= \left| \sum_{l=1}^{n} \frac{\partial b_j^{i_1}}{\partial x_l} i_{Z_{j_{n_1}}} i_{Z_{j_{n_1-1}}} \cdots i_{Z_{j_1}} dx_l \wedge dx_{i_2} \wedge \cdots \wedge dx_{i_{n_1}} \right|$$

$$\lesssim_2 \left| \det_{n_1 \times n_1} Z(x) \right|,$$

since each of the terms $i_{Z_{j_{n_1}}} i_{Z_{j_{n_1-1}}} \cdots i_{Z_{j_1}} dx_l \wedge dx_{i_2} \wedge \cdots \wedge dx_{i_{n_1}}$ is either 0 or of the form $\pm \det Z(x)_{I',J}$ for some $I' \in \mathcal{I}(n_1, n)$. $\qquad \square$

LEMMA 2.2.31. *For* $y \in B_{(Z,d)}(x_0, \xi_1)$, $1 \le n_1 \le n \wedge q$,

$$\left| \det_{n_1 \times n_1} Z(y) \right| \approx_2 \left| \det_{n_1 \times n_1} Z(x_0) \right|.$$

In particular, $\forall y \in B_{(Z,d)}(x_0, \xi_1)$, $\dim \operatorname{span} \{ Z_1(y), \ldots, Z_q(y) \} = n_0$.

PROOF. Since $y \in B_{(Z,d)}(x_0, \xi_1)$, there exists $\gamma : [0,1] \to B_{(Z,d)}(x_0, \xi_1)$ with

- $\gamma(0) = x_0$, $\gamma(1) = y$,

- $\gamma'(t) = \sum_{j=1}^{q} \xi_1^{d_j} a_j(t) Z_j(\gamma(t))$,

- $a_j \in L^\infty([0,1])$, and $\sum |a_j(t)|^2 < 1$.

Consider,

$$\frac{d}{dt} \left| \det_{n_1 \times n_1} Z(\gamma(t)) \right|^2 = 2 \sum_{\substack{I \in \mathcal{I}(n_1, n) \\ J \in \mathcal{I}(n_1, q)}} \det Z_{I,J}(\gamma(t)) \frac{d}{dt} \det Z_{I,J}(\gamma(t))$$

$$= 2 \sum_{\substack{I \in \mathcal{I}(n_1, n) \\ J \in \mathcal{I}(n_1, q)}} \det Z_{I,J}(\gamma(t)) \sum_{j=1}^{q} \xi_1^{d_j} a_j(t) (Z_j \det Z_{I,J})(\gamma(t))$$

$$\lesssim_2 \left| \det_{n_1 \times n_1} Z(\gamma(t)) \right|^2,$$

where in the last line we have applied Lemma 2.2.30. Gronwall's inequality now shows

$$\left| \det_{n_1 \times n_1} Z(y) \right| = \left| \det_{n_1 \times n_1} Z(\gamma(1)) \right| \lesssim_2 \left| \det_{n_1 \times n_1} Z(\gamma(0)) \right| = \left| \det_{n_1 \times n_1} Z(x_0) \right|.$$

Reversing the path γ and applying the same argument we see

$$\left| \det_{n_1 \times n_1} Z(x_0) \right| \lesssim_2 \left| \det_{n_1 \times n_1} Z(y) \right|,$$

completing the proof. $\qquad \square$

For the remainder of this section, set $J_0 = (1, \ldots, n_0)$ and let $I_0 \in \mathcal{I}(n_0, n)$ be such that $\left| \det_{n_0 \times n_0} Z(x_0) \right|_\infty = \left| \det Z(x_0)_{I_0, J_0} \right|$.

LEMMA 2.2.32. *There exists an admissible constant $\xi^1 > 0$ such that for every $y \in B_{(Z,d)}(x_0, \xi^1)$, we have:*

$$\left| \det Z(y)_{I_0, J_0} \right| \gtrsim_2 \left| \det_{n_0 \times n_0} Z(y) \right|.$$

PROOF. Let $\gamma : [0, 1] \to B_{(Z,d)}(x_0, \xi_1)$ satisfy

- $\gamma'(t) = \sum_{j=1}^q \xi_1^{d_j} a_j(t) Z_j(\gamma(t))$,

- $a_j \in L^\infty([0, 1])$, and $\sum |a_j(t)|^2 < 1$.

As in the proof of Lemma 2.2.31, we have

$$\left| \frac{d}{dt} \det_{n_0 \times n_0} Z(\gamma(t)) \right|^2 \lesssim_2 \left| \det_{n_0 \times n_0} Z(\gamma(t)) \right|^2.$$

Applying Lemma 2.2.31, we therefore have

$$\left| \frac{d}{dt} \det_{n_0 \times n_0} Z(\gamma(t)) \right|^2 \lesssim_2 \left| \det_{n_0 \times n_0} Z(\gamma(t)) \right|^2$$

$$\approx_2 \left| \det_{n_0 \times n_0} Z(x_0) \right|^2$$

$$\approx_2 \left| \det Z(x_0)_{I_0, J_0} \right|^2.$$

I.e., there is a 2-admissible constant C such that

$$\left| \frac{d}{dt} \det_{n_0 \times n_0} Z(\gamma(t)) \right|^2 \leq C \left| \det Z(x_0)_{I_0, J_0} \right|^2.$$

Hence, if $t \leq \frac{1}{2C}$,

$$\left| \det Z(\gamma(t))_{I_0, J_0} \right| \approx_2 \left| \det Z(x_0)_{I_0, J_0} \right|$$

and

$$\left| \det_{n_0 \times n_0} Z(\gamma(t)) \right| \lesssim_2 \left| \det_{n_0 \times n_0} Z(x_0) \right| + \left| \det Z(x_0)_{I_0, J_0} \right|$$

$$\approx_2 \left| \det Z(x_0)_{I_0, J_0} \right|$$

$$\approx_2 \left| \det Z(\gamma(t))_{I_0, J_0} \right|.$$

We complete the proof by noting that there exists a 2-admissible constant $\xi^1 > 0$ such that for every $y \in B_{(Z,d)}(x_0, \xi^1)$ there is a γ of the above form and $t \leq \frac{1}{2C}$ with $y = \gamma(t)$. $\square$

PROOF OF PROPOSITION 2.2.29. This follows by combining Lemmas 2.2.31 and 2.2.32. □

LEMMA 2.2.33. *Let ξ^1 be as in Lemma 2.2.32 and let $I \in \mathcal{I}(n_0, n)$, $J \in \mathcal{I}(n_0, q)$. Then,*

$$\sum_{|\alpha| \le m} \left\| Z^\alpha \frac{\det Z_{I,J}}{\det Z_{I_0,J_0}} \right\|_{C^0(B_{(Z,d)}(x_0,\xi^1))} \lesssim_m 1.$$

PROOF. For $m = 0$, this follows from Lemma 2.2.32. For $m > 0$, we look back to the proof of Lemma 2.2.30. There it was shown that $Z_j \det Z_{I,J}$ could be written as a sum of terms of the form

$$f \det Z_{I',J'},$$

where $I' \in \mathcal{I}(n_0, n)$, $J' \in \mathcal{I}(n_0, q)$, and f was either of the form $c_{i,j}^k$ or f was a derivative of one of the components of Z_l ($1 \le l \le q$). From this, Lemma 2.2.32 and a simple induction, the lemma follows. We leave the details to the reader. □

PROPOSITION 2.2.34. *Let ξ^1 be as in Lemma 2.2.32. For $1 \le i, j, k \le n_0$, there exist functions*

$$\hat{c}_{i,j}^k \in C\left(B_{(X,d)}(x_0, \xi^1)\right)$$

such that, for $1 \le i, j \le n_0$,

$$[Z_i, Z_j] = \sum_{k=1}^{n_0} \hat{c}_{i,j}^k Z_k.$$

These functions satisfy

$$\sum_{|\alpha| \le m} \left\| Z^\alpha \hat{c}_{i,j}^k \right\|_{C^0(B_{(Z,d)}(x_0,\xi^1))} \lesssim_m 1.$$

PROOF. For $1 \le j, k \le q$, let $Z^{(j,k)}$ be the matrix obtained by replacing the jth column of the matrix Z with Z_k. Note that

$$\det Z_{I_0,J_0}^{(j,k)} = \epsilon_{j,k} \det Z_{I_0,J(j,k)}$$

where $\epsilon_{j,k} \in \{0, 1, -1\}$ and $J(j,k) \in \mathcal{I}(n_0, q)$.

We may write, by Cramer's rule,

$$Z_k = \sum_{l=1}^{n_0} \frac{\det Z_{I_0,J_0}^{(l,k)}}{\det Z_{I_0,J_0}} Z_l = \sum_{l=1}^{n_0} \epsilon_{l,k} \frac{\det Z_{I_0,J(l,k)}}{\det Z_{I_0,J_0}} Z_l.$$

Thus, we have for $1 \le i, j \le n_0$,

$$[Z_i, Z_j] = \sum_{k=1}^{q} c_{i,j}^k Z_k = \sum_{l=1}^{n_0} \left(\sum_{k=1}^{q} c_{i,j}^k \epsilon_{l,k} \frac{\det Z_{I_0,J(l,k)}}{\det Z_{I_0,J_0}} \right) Z_l =: \sum_{l=1}^{n_0} \hat{c}_{i,j}^l Z_l.$$

The desired estimates for $\hat{c}_{i,j}^l$ now follow from Lemma 2.2.33. □

Proposition 2.2.34 shows that the vector fields $Z_1, \ldots, Z_{n_0}$ satisfy the same assumptions as the vector fields $Z_1, \ldots, Z_q$ (with ξ_1 replaced by ξ^1). In addition, we know $Z_1, \ldots, Z_{n_0}$ are linearly independent at x_0. From here, it can be shown that Theorem 2.2.22 follows from the special case when $q = n_0$; i.e., when $Z_1, \ldots, Z_q$ are linearly independent at x_0 (see [Str11, Section 4] for further details on this reduction).

Let's now return to our map Φ, defined by

$$\Phi(t_1, \ldots, t_n) = e^{t_1 Z_1 + \cdots + t_{n_0} Z_{n_0}} x_0.$$

$\Phi : B^{n_0}(\eta_0) \to B_{(Z,d)}(x_0, \xi^1)$, where η_0 is some 2-admissible constant. As mentioned before, the inverse function theorem shows $\Phi : B^{n_0}(\epsilon) \to \Phi(B^{n_0}(\epsilon))$ is a diffeomorphism for some $\epsilon > 0$, but the inverse function theorem, alone, does not guarantee ϵ can be taken to be an admissible constant.

Let Y_j be the pullback of Z_j via Φ to $B^{n_0}(\epsilon)$. Write

$$Y_j = \frac{\partial}{\partial t_j} + \sum_k a_j^k(t) \frac{\partial}{\partial t_k}. \tag{2.29}$$

Note $a_j^k(0) = 0$; i.e., $Y_j(0) = \frac{\partial}{\partial t_j}$. Also, $a_j^k(t)$ is C^∞ (though we do not yet know that the C^m norms are admissibly bounded). We define two $n_0 \times n_0$ matrices, $A(t) : B^{n_0}(\epsilon) \to \mathbb{M}_{n_0 \times n_0}$, $C : B^{n_0}(\eta_0) \to \mathbb{M}_{n_0 \times n_0}$, where $\mathbb{M}_{n_0 \times n_0}$ denotes the space of $n_0 \times n_0$ real matrices. Let $A_{i,j}$ denote the i, j component of A, and similarly for C. We let

$$A_{i,j}(t) = a_i^j(t), \quad C_{i,k}(t) = \sum_{j=1}^{n_0} t_j \hat{c}_{i,j}^k \circ \Phi(t).$$

We write t in polar coordinates $t = r\omega$, $\omega \in S^{n_0-1}$, $r > 0$. It can be shown that A satisfies the differential equation

$$\frac{\partial}{\partial r} r A(r\omega) = -A^2 - CA - C, \quad A(0) = 0. \tag{2.30}$$

Using standard methods from ODEs, it can be shown that (2.30) has a unique solution (with $A(r\omega) = O(r)$) for $r < \eta$, where $\eta \le \eta_0$ is an admissible constant; moreover, the C^m norms of the unique solution can be bounded in a useful way. Thus, the matrix valued function A can be extended to $A : B^{n_0}(\eta) \to \mathbb{M}_{n_0 \times n_0}$. This allows us to extend the vector fields $Y_1, \ldots, Y_{n_0}$ from vector fields defined on $B^{n_0}(\epsilon)$ to vector fields defined on $B^{n_0}(\eta)$ by using the formula (2.29), where a_i^j is the (i, j) entry of the matrix A. Using, again, standard methods for ODEs, one can show $d\Phi(Y_j) = Z_j$.

The above constitute some of the main methods of the proof of Theorem 2.2.22. For the remainder of the proof, we refer the reader to [Str11].

2.3 VECTOR FIELDS WITH FORMAL DEGREES REVISITED

With the quantitative Frobenius theorem in hand, we return to the main setting of this chapter, which takes place on a **compact**, connected manifold M. Thus, we have used the vector fields $W_1, \ldots, W_r$ to generate the list of vector fields with formal degrees $(X, d) = (X_1, d_1), \ldots, (X_q, d_q)$; where each X_j is a commutator of order d_j of the W_k, and $X_1, \ldots, X_q$ span the tangent space to M at every point. For $\delta > 0$ define $Z_1 = \delta^{d_1} X_1, \ldots, Z_q = \delta^{d_q} X_q$, and assign to Z_j the formal degree d_j (here we have suppressed the dependance of Z_j on δ). It is immediate to verify that, for $\xi > 0$, $B_{(X,d)}(x, \delta \xi) = B_{(Z,d)}(x, \xi)$. Furthermore, Corollary 2.1.4 tells us that the vector fields (Z, d) satisfy the hypotheses of the quantitative Frobenius theorem, *uniformly* for $\delta \in (0, 1]$ and $x_0 \in M$. That is, we have

$$[Z_j, Z_k] = \sum_{l=1}^{q} c_{j,k}^{l} Z_l,$$

where $c_{j,k}^{l} \in C^\infty(M)$ implicitly depends on δ, but as δ varies, $c_{j,k}^{l}$ varies over a bounded subset of $C^\infty(M)$. In particular, for each $x_0 \in M$ and $\delta \in (0, 1]$, we may apply Theorem 2.2.22 with this choice of x_0 and to $Z_1, \ldots, Z_q$ (which depend implicitly on δ). The conclusions of Theorem 2.2.22 are in terms of admissible and m-admissible constants–these constants may be chosen *independent* of the choice of $x_0 \in M$ and $\delta \in (0, 1]$ because the assumptions of Theorem 2.2.22 hold uniformly in these choices.

We turn now to a more precise description. Since M is a compact manifold, its smooth structure is given by an atlas consisting of a *finite* collection of coordinate charts $\{(\phi_\alpha, U_\alpha)\}$, where α ranges over a finite set, U_α is an open cover for M, and the ϕ_α are diffeomorphisms from U_α to open subsets of $\mathbb{R}^n$. Since we endowed M with a strictly positive, smooth measure (see Definition 2.0.6), when we push forward this measure to $\mathbb{R}^n$ via ϕ_α, we obtain a measure comparable to Lebesgue measure on $\phi_\alpha(U_\alpha) \subset \mathbb{R}^n$. Thus, if we wish to estimate the measure of a subset of U_α (up to constant multiples), it suffices to estimate the Lebesgue volume of its image under ϕ_α. Let $2\xi_1 > 0$ be a Lebesgue number of this cover[7] with respect to the metric ρ (recall, $\rho(x, z)$ is given by (2.9)).

Fix $x_0 \in M$. There is some U_α with $B_{(X,d)}(x_0, \xi_1) \subset U_\alpha$. We abuse notation and write X, Z, and x_0 both for the objects on U_α and on their images on $\phi_\alpha(U_\alpha)$ under the diffeomorphism ϕ_α. Since we are only interested in estimating the volumes of various sets up to constants, we may use Vol $(\cdot)$ to unambiguously mean either Lebesgue measure on $\phi_\alpha(U_\alpha) \subset \mathbb{R}^n$ or the chosen strictly positive, smooth measure on U_α. The quantitative Frobenius theorem applies to $B_{(Z,d)}(x_0, \xi_2)$ to show

$$\text{Vol}\left(B_{(X,d)}(x_0, \delta \xi_2)\right) = \text{Vol}\left(B_{(Z,d)}(x, \xi_2)\right) \approx_2 \left| \det_{n_0 \times n_0} (Z_1(x_0) | \cdots | Z_q(x_0)) \right|$$

$$\approx_2 \left| \det_{n_0 \times n_0} \left(\delta^{d_1} X_1(x_0) | \cdots | \delta^{d_q} X_q(x_0)\right) \right|.$$

Replacing $\delta \xi_2$ with δ, so that $\delta \in (0, \xi_2]$ and using $\xi_2 \approx_2 1$, we have

$$\text{Vol}\left(B_{(X,d)}(x_0, \delta)\right) \approx_2 \left| \det_{n_0 \times n_0} \left(\delta^{d_1} X_1(x_0) | \cdots | \delta^{d_q} X_q(x_0)\right) \right|.$$

---

[7] I.e., every subset of M with ρ diameter less than or equal to $2\xi_1$ is a subset of some U_α.

As a consequence, we see, for $2\delta \leq \xi_2$,

$$\text{Vol}\left(B_{(X,d)}(x_0, 2\delta)\right) \approx_2 \text{Vol}\left(B_{(X,d)}(x_0, \delta)\right). \tag{2.31}$$

Furthermore, for $\delta \geq \frac{\xi_2}{2}$, $\text{Vol}\left(B_{(X,d)}(x_0, \delta)\right) \approx_2 1$ (via the compactness of M), and we see (2.31) in fact holds for all $\delta > 0$. As pointed out before, the implicit constants are independent of $x_0 \in M$ and δ. This establishes Theorem 2.0.10.

The same ideas allow us to prove Proposition 2.1.1; i.e., that $B_{(X,d)}(x_0, c\delta) \subseteq B_W(x_0, \delta)$ for some $c > 0$.

PROOF SKETCH OF PROPOSITION 2.1.1. Proposition 2.1.1 is trivial for δ large (by the compactness of M), and we therefore prove it only for δ small. Fix x_0 and δ, and apply the quantitative Frobenius theorem to $Z_1 = \delta^{d_1} X_1, \ldots, Z_q = \delta^{d_q} X_q$ with this choice of δ and x_0, to obtain the map Φ as in that theorem. Let $Y_1, \ldots, Y_q$ be the pullbacks of $Z_1, \ldots, Z_q$ via Φ, and choose the enumeration of $X_1, \ldots, X_q$ in such a way that $W_1 = X_1, \ldots, W_r = X_r$ and $1 = d_1 = \cdots = d_r$. Thus $Y_1, \ldots, Y_r$ are the pullbacks of $\delta W_1, \ldots, \delta W_r$.

Because of the definition of $Y_1, \ldots, Y_q$, each Y_j $(j > r)$ is an iterated commutator of Y_j, $j \leq r$. Since $Y_1, \ldots, Y_q$ span the tangent space, $Y_1, \ldots, Y_r$ satisfy Hörmander's condition. In light of (2.18), they satisfy Hörmander's condition *uniformly* in δ and x_0. It is easy to see, via (2.17) and (2.18), that there is an admissible constant $\eta_1 > 0$ with $B^n(\eta_1) \subset B_{\xi_3 Y_1, \ldots, \xi_3 Y_r}(0)$. Furthermore, by (2.17), there is an admissible constant $\xi_4 > 0$ such that $B_{(Y,d)}(0, \xi_4) \subset B^n(\eta_1)$. Pushing this forward via Φ yields $B_{(X,d)}(x_0, \xi_4\delta) \subset B_{\xi_3 \delta W_1, \ldots, \xi_3 \delta W_r}(x_0)$. This establishes the result for δ sufficiently small, and completes the proof. $\qquad\square$

2.4 MAXIMAL HYPOELLIPTICITY

In Section 1.4 we saw that singular integrals corresponding to the Euclidean distance on $\mathbb{R}^n$ were useful for studying elliptic partial differential operators. A main motivation for studying the Calderón-Zygmund operators in this chapter is that they are useful for studying certain partial differential operators which generalize elliptic partial differential operators. These operators are known as "maximally hypoelliptic" operators (and are sometimes also called "maximally subelliptic" operators).

Let $P(w_1, \ldots, w_r)$ be a polynomial in r-noncommuting indeterminants $w_1, \ldots, w_r$, with coefficients in $C^\infty(M)$:

$$P(w_1, \ldots, w_r) = \sum_{|\alpha| \leq k} a_\alpha(x) w^\alpha, \quad a_\alpha \in C^\infty(M),$$

where the sum is taken over ordered multi-indices, since the indeterminants do not commute. We wish to study certain differential operators of the form

$$\mathcal{P} = P(W_1, \ldots, W_r),$$

where $W_1, \ldots, W_r$ satisfy Hörmander's condition.

Remark 2.4.1   Because $W_1, \ldots, W_r$ satisfy Hörmander's condition, *every* linear partial differential operator is of the form $P(W_1, \ldots, W_r)$ for some P. However, we will be considering only those that have a strong relationship with the vector fields $W_1, \ldots, W_r$.

Remark 2.4.2 P might be a differential operator of order $< k$. For instance, $[W_i, W_j] = W_i W_j - W_j W_i$ is a differential operator of order ≤ 1, though the corresponding polynomial P has degree 2.

Remark 2.4.3   Many different polynomials, P, may give rise to the same $\mathcal{P}$. For instance in $\mathbb{R}^2$, if $W_1 = \frac{\partial}{\partial_x}$, $W_2 = \frac{\partial}{\partial_y}$, then the polynomial $P(w_1, w_2) = w_1 w_2 - w_2 w_1$ gives rise to $\mathcal{P} = [W_1, W_2] = 0$. In what follows, it is most useful to consider a choice of the polynomial P with minimal degree.

DEFINITION 2.4.4.   *We say* $\mathcal{P} = P(W_1, \ldots, W_r)$ *is maximally hypoelliptic if*

$$\sum_{|\alpha| \leq k} \|W^\alpha f\|_{L^2} \leq C\left(\|\mathcal{P}f\|_{L^2} + \|f\|_{L^2}\right), \quad \forall f \in C^\infty(M).$$

Recall, k is the degree of the polynomial P.

Remark 2.4.5   Note that the definition of maximal hypoellipticity depends on the degree of the chosen polynomial representation of $\mathcal{P}$ (see Remark 2.4.3). It is easy to see, though, that it can only hold for a choice of a polynomial with minimal degree. Thus, the choice of k is unambiguous in the above definition.

A fundamental result concerning maximal hypoellipticity is the following.

THEOREM 2.4.6.   *Suppose $\mathcal{P}$ is maximally hypoelliptic, then $\mathcal{P}$ is subelliptic.*

We prove Theorem 2.4.6 in Section 2.4.1.

Remark 2.4.7   Theorem 2.4.6 helps to justify the name "maximally hypoelliptic." Indeed, if $\mathcal{P}$ is maximally hypoelliptic, then it is subelliptic, and therefore hypoelliptic. Maximally hypoelliptic gives more than subellipticity: u is smoother than $\mathcal{P}u$ by a "maximal" amount. See, for instance, Corollary 2.11.6.

Note that if $\mathcal{P} = P(W_1, \ldots, W_r)$, then $\mathcal{P}^* = Q(W_1, \ldots, W_r)$, where Q is also a polynomial of degree k in r non-commuting indeterminants. Thus, it also makes sense to ask whether $\mathcal{P}^*$ is maximally hypoelliptic. This allows us to state the main connection between maximally hypoelliptic operators and singular integrals in the following theorem.

THEOREM 2.4.8.   *Let $\mathcal{P} = P(W_1, \ldots, W_r)$ be as above. Then, the following are equivalent.*

(i) *$\mathcal{P}$ and $\mathcal{P}^*$ are maximally hypoelliptic.*

(ii) *There is a Calderón-Zygmund operator, $T : C^\infty(M) \to C^\infty(M)$, of order $-k$ such that $\mathcal{P}T, T\mathcal{P} \equiv I \mod C^\infty(M \times M)$.*

To prove Theorem 2.4.8, we must develop a good deal of surrounding theory. The proof concludes in Section 2.11, but requires many of the ideas leading up to that section.

There is another related type of maximal hypoellipticity which is also of use to us.

DEFINITION 2.4.9. *Suppose* $k = \deg P$ *is even; so that* $k = 2l$, $l \in \mathbb{N}$. *We say* $\mathcal{P} = P(W_1, \ldots, W_r)$ *is maximally hypoelliptic of type* 2 *if*

$$\sum_{|\alpha| \leq l} \|W^\alpha f\|_{L^2}^2 \leq C \left(|\langle \mathcal{P} f, f \rangle| + \|f\|_{L^2}^2 \right), \quad \forall f \in C^\infty(M).$$

Here $\langle \cdot, \cdot \rangle$ *denotes the* L^2 *inner product.*

Remark 2.4.10   It is clear that $\mathcal{P}$ is maximally hypoelliptic if and only if $\mathcal{P}^*\mathcal{P}$ is maximally hypoelliptic of type 2. We will see in the proofs that follow that these two concepts are closely related; however, their exact relationship remains unclear.

2.4.1 Subellipticity

This section is devoted to Theorem 2.4.6: maximal hypoellipticity implies subellipticity. In fact, we prove more: we prove a quantitative, local result. This quantitative result is essential to deduce Theorem 2.4.8. We shall also prove similar results for maximal hypoellipticity of type 2.

Fix an open set $\Omega' \subseteq \mathbb{R}^n$ and C^∞ vector fields $Y_1, \ldots, Y_r$ on Ω' and let $\Omega \Subset \Omega'$ be a relatively compact open subset of Ω. Suppose $Y_1, \ldots, Y_r$ satisfy Hörmander's condition of order m on Ω'. Let $\mathcal{Y}_1 = \{Y_1, \ldots, Y_r\}$, and recursively define

$$\mathcal{Y}_j = \{[Y_l, V] \mid 1 \leq l \leq r, V \in \mathcal{Y}_{j-1}\}.$$

Each $\mathcal{Y}_j$ is a finite set and the vector fields in $\bigcup_{j=1}^m \mathcal{Y}_j$ span the tangent space to every point at Ω. In fact, we have

$$\eta := \inf_{x \in \Omega} \sup_{V_1, \ldots, V_n \in \bigcup_{j=1}^m \mathcal{Y}_j} |\det (V_1(x) | \cdots | V_n(x))| > 0. \tag{2.32}$$

Let $P(y_1, \ldots, y_r) = \sum_{|\alpha| \leq k} a_\alpha(x) y^\alpha$ be a polynomial in non-commuting indeterminants $y_1, \ldots, y_r$ with coefficients $a_\alpha \in C^\infty(\Omega')$. Let $k = \deg P$. We consider the differential operator $\mathcal{P} := P(Y)$, and we will assume one of two possible assumptions:

Maximal Hypoellipticity: $\mathcal{P}$ is maximally hypoelliptic. That is, there is a constant $C > 0$ such that $\forall f \in C_0^\infty(\Omega)$,

$$\sum_{|\alpha| \leq k} \|W^\alpha f\|_{L^2} \leq C (\|\mathcal{P} f\|_{L^2} + \|f\|_{L^2}); \tag{2.33}$$

or

Maximal Hypoellipticity of type 2: $\mathcal{P}$ is maximally hypoelliptic of type 2. That is, $k = 2l$ where $l \in \mathbb{N}$ and there is a constant $C > 0$ such that $\forall f \in C_0^\infty(\Omega)$,

$$\sum_{|\alpha| \le l} \|W^\alpha f\|_{L^2}^2 \le C \left(|\langle \mathcal{P}f, f \rangle| + \|f\|_{L^2}^2 \right). \tag{2.34}$$

Here, $\langle \cdot, \cdot \rangle$ denotes the L^2 inner product.

Before we state our main result, we need a few more pieces of notation. If $\phi_1, \phi_2 \in C_0^\infty(\Omega)$, we write $\phi_1 \prec \phi_2$ if $\phi_2 \equiv 1$ on a neighborhood of supp (ϕ_1). As in Section 1.3, for $s \in \mathbb{R}$, we let Λ^s denote the pseudodifferential operator of order s with symbol $\left(1 + |\xi|^2\right)^{s/2}$. It is easy to verify that $\Lambda^s \Lambda^t = \Lambda^{s+t}$ and $(\Lambda^s)^* = \Lambda^s$. Finally, as in Definition 1.3.10, we write $\|f\|_{L_s^p} := \|\Lambda^s f\|_{L^p}$.

We turn to stating the main result of this section, though first we state it without being precise on how the constants depend on our various assumptions.

THEOREM 2.4.11. *Suppose $\mathcal{P}$ is either maximally hypoelliptic or maximally hypoelliptic of type 2. Then, there is $\epsilon_0 > 0$ such that $\forall \phi_1, \phi_2 \in C_0^\infty(\Omega)$ with $\phi_1 \prec \phi_2$, $\forall s, N \in \mathbb{R}$, $\exists D = D(\phi_1, \phi_2, s, N)$, $\forall u \in C_0^\infty(\Omega)'$,*

$$\|\phi_1 u\|_{L^2_{s+\epsilon_0}} \le D \left(\|\phi_2 \mathcal{P}u\|_{L^2_s} + \|\phi_2 u\|_{L^2_{s-N}} \right). \tag{2.35}$$

We may take $\epsilon_0 = k2^{1-m}$ if $\mathcal{P}$ is maximally hypoelliptic or $\epsilon_0 = l2^{1-m}$ if $\mathcal{P}$ is maximally hypoelliptic of type 2.[8] Here, if the right-hand side of (2.35) is finite, then the left-hand side is finite as well, and if the left-hand side is infinite, then the right-hand side is infinite as well.

One main issue we have not addressed is in what way the constant D from (2.35) depends on the various ingredients in our setup. As mentioned in the theorem, D depends on ϕ_1, ϕ_2, s, and N. It also depends on r, m, η from (2.32), and C from either (2.33) or (2.34). In addition, D depends on $\|Y_l\|_{C^L(\Omega)}$, $\|a_\alpha\|_{C^L(\Omega)}$ ($1 \le l \le r$, $|\alpha| \le k$), where L can be chosen to depend only on s, N, and m.

Remark 2.4.12 Note that Theorem 2.4.11 proves Theorem 2.4.6: maximal hypoellipticity implies subellipticity. Indeed, subellipticity is a local property: a partial differential operator $\mathcal{P}$ is subelliptic on a compact manifold M if and only if $\forall x \in M$, $\exists$ a neighborhood U of x such that $\mathcal{P}$ is subelliptic on U. This follows by a simple partition of unity argument. From here, Theorem 2.4.6 follows directly from Theorem 2.4.11. Similarly, we see that maximal hypoellipticity of type 2 implies subellipticity.

Remark 2.4.13 The dependance of D on η is in contrast to many of the other results in this chapter. Indeed, a main point of the quantitative Frobenius theorem (Theorem 2.2.22) is that the conclusions do *not* depend on assumed lower bound for values like

---

[8] These choices of ϵ_0 are not optimal, but are sufficient for our purposes. In fact, one may take $\epsilon = \frac{k}{m}$ in the maximally hypoelliptic case, but this follows from much more detailed considerations. See Remark 2.4.18.

η. In our applications, the role vector fields Y_l from this section will be played by the vector fields of the same name in Theorem 2.2.22. In that setting, one of the *conclusions* of Theorem 2.2.22 is a lower bound for η independent of any relevant parameters. The idea here is that our goal is not to apply Theorem 2.4.11 directly to the vector fields $W_1, \ldots, W_r$, but rather to appropriately scaled versions of these vector fields given by an application of Theorem 2.2.22. One of the first results of this type is given in Theorem 2.4.29, below.

We turn to the proof of Theorem 2.4.11. Throughout the rest of this section we use the notation $A \lesssim B$ to denote $A \leq CB$ where C is a constant which depends only on the appropriate parameters as outlined above. We write $A \approx B$ to denote $A \lesssim B$ and $B \lesssim A$. This is in contrast to the rest of the chapter, where $A \lesssim B$ denotes that the implicit constant is an admissible constant in the sense of the quantitative Frobenius theorem (Theorem 2.2.22).

Notice that all of the assumptions and conclusions of Theorem 2.4.11 depend only on the values of Y_j and $\mathcal{P}$ on Ω. Thus if $\psi \in C_0^\infty(\Omega')$ equals 1 on a neighborhood of $\overline{\Omega}$, then we may replace Y_j with ψY_j and a_α with ψa_α in all of the above, without affecting any of our assumptions or conclusions. We henceforth do this (for the rest of this section). The main use of this is that the vector fields Y_j are now standard pseudodifferential operators of order 1, and we may apply the calculus from Theorem 1.3.6 to help manipulate these operators.

Theorem 2.4.11 was stated $\forall u \in C_0^\infty(\Omega)'$. It is much more convenient to prove (2.35) only for $u \in C_0^\infty(\Omega)$. The next lemma shows that this is sufficient.

LEMMA 2.4.14. *Let* $\phi_1, \phi_2 \in C_0^\infty(\Omega)$, $s_1, s_2, s_3 \in \mathbb{R}$, $\mathcal{P}$ *be any partial differential operator with smooth coefficients, and* $C > 0$ *satisfy*

$$\|\phi_1 u\|_{L_{s_1}^2} \leq C \left(\|\phi_2 \mathcal{P} u\|_{L_{s_2}^2} + \|\phi_2 u\|_{L_{s_3}^2} \right), \quad \forall u \in C_0^\infty(\Omega).$$

Then, the same inequality holds (with the same constant C), $\forall u \in C_0^\infty(\Omega)'$. Furthermore, when $u \in C_0^\infty(\Omega)'$, if the right-hand side is finite, then so is the left-hand side, and if the left-hand side is infinite, then so is the right-hand side.

PROOF. Fix $u \in C_0^\infty(\Omega)'$. Take $\phi_3 \in C_0^\infty(\Omega)$ with $\phi_1, \phi_2 \prec \phi_3$. Let $\psi \in C_0^\infty(\mathbb{R}^n)$ satisfy $\int \psi = 1$. Define, for $N > 0$, $\psi^{(N)} = N^n \psi(Nx)$, and $u_N = (\phi_3 u) * \psi^{(N)}$. Because $\psi_3 \in C_0^\infty(\Omega)$, $u_N \in C_0^\infty(\Omega)$ for N sufficiently large (depending on $\text{supp}(\psi)$). We also have

$$\lim_{N \to \infty} \|\phi_1 u_N\|_{L_{s_1}^2} = \|\phi_1 u\|_{L_{s_1}^2},$$

$$\lim_{N \to \infty} \|\phi_2 \mathcal{P} u_N\|_{L_{s_2}^2} = \|\phi_2 \mathcal{P} u\|_{L_{s_2}^2},$$

$$\lim_{N \to \infty} \|\phi_2 u_N\|_{L_{s_3}^2} = \|\phi_2 u\|_{L_{s_3}^2}.$$

These limits hold even if the limiting value is ∞. The result follows. $\square$

Before we continue with the proof of Theorem 2.4.11, we need a few more pieces of notation. We write (s.c.) to denote "small constant" and (l.c.) to denote "large constant." An equation of the form $A \leq$ (l.c.)$B_1 +$ (s.c.)B_2 denotes: $\forall \epsilon > 0$, $\exists C$ with $A \leq CB_1 + \epsilon B_2$. We use, repeatedly, the inequality $AB \leq$ (s.c.)$A^2 +$ (l.c.)B^2 in what follows. One trick will be quite useful to us. If we have $A \leq$ (l.c.)$B +$ (s.c.)A, then we have $A \lesssim B$.

Also, for $f, g, h \in C_0^\infty(\mathbb{R}^n)$ we write

$$\|f\|_{L^2_{s_1}} \lesssim \|g\|_{L^2_{s_2}} + \|h\|_{-\infty}$$

to mean $\forall N$, $\exists C_N$,

$$\|f\|_{L^2_{s_1}} \leq C_N \left(\|g\|_{L^2_{s_2}} + \|h\|_{L^2_{-N}} \right).$$

Finally, we write $S^{(s)}$ to denote some standard pseudodifferential operator of order s (which may change from line to line). In each case, the operator $S^{(s)}$ can easily be explicitly written down, but what will be important to us is only that it is a pseudodifferential operator of order s (and this is true uniformly in any relevant parameters). Thus, we have inequalities like

$$\left\| S^{(s)} u \right\|_{L^2_{s_1}} \lesssim \|u\|_{L^2_{s_1+s}}.$$

PROPOSITION 2.4.15. *For any $s_1 < s < s_2$, we have*

$$\|u\|_{L^2_s} \leq \text{(s.c.)}\, \|u\|_{L^2_{s_2}} + \text{(l.c.)}\, \|u\|_{L^2_{s_1}}, \quad \forall u \in C_0^\infty(\mathbb{R}^n).$$

To prove Proposition 2.4.15 we prove the following intermediate lemma.

LEMMA 2.4.16. *Let $s < s_2$. Then,*

$$\|u\|_{L^2_s} \leq \text{(s.c.)}\, \|u\|_{L^2_{s_2}} + \text{(l.c.)}\, \|u\|_{L^2_{2s-s_2}}, \quad \forall u \in C_0^\infty(\mathbb{R}^n).$$

PROOF. We have

$$\begin{aligned}
\|u\|_{L^2_s}^2 &= \langle \Lambda^s u, \Lambda^s u \rangle \\
&= \langle \Lambda^{s_2} u, \Lambda^{2s-s_2} u \rangle \\
&\leq \text{(s.c.)}\, \|u\|_{L^2_{s_2}}^2 + \text{(l.c.)}\, \|u\|_{L^2_{2s-s_2}}^2,
\end{aligned}$$

where the last line follows from the Cauchy-Schwartz inequality.   □

PROOF OF PROPOSITION 2.4.15. We claim that for all $j \in \mathbb{N}$,

$$\|u\|_{L^2_s} \leq \text{(s.c.)}\, \|u\|_{L^2_{s_2}} + \text{(l.c.)}\, \|u\|_{L^2_{2^j s-(2^j-1)s_2}}, \quad \forall u \in C_0^\infty(\mathbb{R}^n). \qquad (2.36)$$

The result then follows by taking j so large $2^j s - \left(2^j - 1\right) s_2 \leq s_1$. We prove (2.36) by induction on j. The base case ($j = 0$) is obvious (in that case, the small constant can actually be taken to be 0, and the large constant can be taken to be 1).

We assume we have (2.36) for j and prove it for $j + 1$. Lemma 2.4.16 shows

$$\|u\|_{L^2_{2^j s - (2^j - 1)s_2}} \leq \text{(s.c.)} \|u\|_{L^2_{s_2}} + \text{(l.c.)} \|u\|_{L^2_{2^{j+1} s - (2^{j+1} - 1)s_2}}.$$

Combining this with our inductive hypothesis proves (2.36) with j replaced by $j + 1$. This completes the proof. $\qquad\square$

The next proposition is the key step where Hörmander's condition comes into play.

PROPOSITION 2.4.17. $\forall s \in \mathbb{R}$,

$$\|u\|_{L^2_{s+2^{(1-m)}}} \lesssim \sum_{j=1}^{r} \|Y_j u\|_{L^2_s} + \|u\|_{L^2_s}, \quad \forall u \in C_0^\infty(\Omega). \tag{2.37}$$

Remark 2.4.18 $s + 2^{(1-m)}$ is not the optimal value on the left-hand side of (2.37). The result remains true with $s + \frac{1}{m}$, instead, and this choice is optimal. This is a result of Rothschild and Stein [RS76, Theorem 12] and requires a much more detailed argument. Using this, when $\mathcal{P}$ is maximally hypoelliptic, one may replace ϵ_0 in Theorem 2.4.11 with $\frac{k}{m}$. In fact, in the maximally hypoelliptic case, if one uses Proposition 2.4.17 with $\frac{1}{m}$ in place of $2^{(1-m)}$ in all of the proofs that follow, Theorem 2.4.11 with $\epsilon_0 = \frac{k}{m}$ follows. When $\mathcal{P}$ is maximally hypoelliptic of type 2, one only gets $\epsilon_0 = \frac{l}{m}$ using this argument; one would expect $\epsilon_0 = \frac{k}{m}$, but we do not know of a way to achieve this in general.

To prove Proposition 2.4.17 we use the following lemma.

LEMMA 2.4.19. *Let* $Z_j \in \mathcal{Y}_j$. *Then,*

$$\|Z_j u\|_{L^2_{s+2^{1-j}-1}} \lesssim \sum_{k=1}^{r} \|Y_k u\|_{L^2_s} + \|u\|_{L^2_s}, \quad \forall u \in C_0^\infty(\Omega).$$

PROOF. We prove the result by induction on j. The base case, $j = 1$, is obvious. Fix $j \geq 2$ for which we wish to prove the result, and we assume the result for all lesser values of j. Take $Z_j \in \mathcal{Y}_j$. Write $Z_j = [Y_l, Z_{j-1}]$ for some Y_l and some $Z_{j-1} \in \mathcal{Y}_{j-1}$. We have

$$\|Z_j u\|^2_{L^2_{s+2^{1-j}-1}} = \left\langle Z_j u, \Lambda^{2s+2^{2-j}-2} Z_j u \right\rangle \tag{2.38}$$

$$= \left\langle Y_l Z_{j-1} u, \Lambda^{2s+2^{2-j}-2} Z_j u \right\rangle - \left\langle Z_{j-1} Y_l u, \Lambda^{2s+2^{2-j}-2} Z_j u \right\rangle.$$

We bound the two terms on the right-hand side of the above equation separately. Write $Y_l^* = -Y_l + f$, where $f \in C_0^\infty(\Omega')$. We have, for $u \in C_0^\infty(\Omega)$,

$$\left| \left\langle Y_l Z_{j-1} u, \Lambda^{2s+2^{2-j}-2} Z_j u \right\rangle \right|$$

$$= \left| \left\langle Z_{j-1} u, Y_l \Lambda^{2s+2^{2-j}-2} Z_j u \right\rangle - \left\langle Z_{j-1} u, f \Lambda^{2s+2^{2-j}-2} Z_j u \right\rangle \right| \tag{2.39}$$

$$\leq \left| \left\langle Z_{j-1} u, \Lambda^{2s+2^{2-j}-2} Z_j Y_l u \right\rangle \right| + \left| \left\langle Z_{j-1} u, S^{\left(2^{2s+2^{2-j}-1}\right)} u \right\rangle \right|,$$

where in the last line we have used the calculus of pseudodifferential operators.

Consider

$$\left|\left\langle Z_{j-1}u, S^{\left(2^{2s}+2^{2-j}-1\right)}u\right\rangle\right| = \left|\left\langle S^{\left(s+2^{2-j}-1\right)}Z_{j-1}u, \Lambda^s u\right\rangle\right|$$

$$\lesssim \|Z_{j-1}u\|^2_{L^2_{s+2^{2-j}-1}} + \|u\|^2_{L^2_s} \tag{2.40}$$

$$\lesssim \sum_{k=1}^r \|Y_k u\|^2_{L^2_s} + \|u\|^2_{L^2_s},$$

where the last line follows by our inductive hypothesis. Also,

$$\left|\left\langle Z_{j-1}u, \Lambda^{2s+2^{2-j}-2}Z_j Y_l u\right\rangle\right| \lesssim \|Z_{j-1}u\|^2_{L^2_{s+2^{2-j}-1}} + \|Y_l u\|^2_{L^2_s}$$

$$\lesssim \sum_{k=1}^r \|Y_k u\|^2_{L^2_s} + \|u\|^2_{L^2_s}, \tag{2.41}$$

where in the first line we have used $\Lambda^{s-1}Z_j Y_l u = S^{(s)}Y_l u$, since Z_j is a pseudodifferential operator of order 1, and in the second line we have used our inductive hypothesis. Combining (2.40) and (2.41) with (2.39) shows

$$\left|\left\langle Y_l Z_{j-1}u, \Lambda^{2s+2^{2-j}-2}Z_j u\right\rangle\right| \lesssim \sum_{k=1}^r \|Y_k u\|^2_{L^2_s} + \|u\|^2_{L^2_s}, \quad \forall u \in C_0^\infty(\Omega). \tag{2.42}$$

We now turn to the second term on the right-hand side of (2.38). Recall, $j \geq 2$.

$$\left|\left\langle Z_{j-1}Y_l u, \Lambda^{2s+2^{2-j}-2}Z_j u\right\rangle\right|$$

$$\leq \left|\left\langle Y_l u, Z_{j-1}\Lambda^{2s+2^{2-j}-2}Z_j u\right\rangle\right| + \left|\left\langle Y_l u, S^{\left(2s+2^{2-j}-1\right)}u\right\rangle\right|$$

$$\lesssim \left|\left\langle Y_l u, Z_{j-1}\Lambda^{2s+2^{2-j}-2}Z_j u\right\rangle\right| + \|Y_l u\|^2_{L^2_{s+2^{2-j}-1}} + \|u\|^2_{L^2_s}$$

$$\lesssim \left|\left\langle Y_l u, \Lambda^{2s+2^{2-j}-2}Z_j Z_{j-1}u\right\rangle\right| + \left|\left\langle Y_l u, S^{\left(2s+2^{2-j}-1\right)}u\right\rangle\right|$$

$$\qquad + \|Y_l u\|^2_{L^2_{s+2^{2-j}-1}} + \|u\|^2_{L^2_s} \tag{2.43}$$

$$\lesssim \left|\left\langle Y_l u, \Lambda^{2s+2^{2-j}-2}Z_j Z_{j-1}u\right\rangle\right| + \|Y_l u\|^2_{L^2_{s+2^{2-j}-1}} + \|u\|^2_{L^2_s}$$

$$\lesssim \|Y_l u\|^2_{L^2_s} + \|Z_{j-1}u\|^2_{L^2_{s+2^{2-j}-1}} + \|u\|^2_{L^2_s}$$

$$\lesssim \sum_{k=1}^r \|Y_k u\|^2_{L^2_s} + \|u\|^2_{L^2_s},$$

where the last line follows by our inductive hypothesis. Combining (2.42), (2.43), and (2.38) yields the result. $\square$

PROOF OF PROPOSITION 2.4.17. Notice, for $u \in C_0^\infty(\Omega)$,

$$\|u\|_{L^2_{s+2^{1-m}}} \lesssim \sum_{j=1}^n \|\partial_{x_j} u\|_{L^2_{s+2^{1-m}-1}} + \|u\|_{L^2_s}$$

$$\lesssim \sum_{Z \in \bigcup_{j=1}^m \mathcal{Y}_j} \|Zu\|_{L^2_{s+2^{1-m}-1}} + \|u\|_{L^2_s}$$

$$\lesssim \sum_{j=1}^r \|Y_j u\|_{L^2_s} + \|u\|_{L^2_s},$$

where the last line follows by Lemma 2.4.19. $\qquad\square$

LEMMA 2.4.20. *Let $k_0 \geq 1$ be an integer. Then,*

$$\sum_{|\alpha| \leq k_0 - 1} \|Y^\alpha u\|_{L^2_s} \lesssim \sum_{|\alpha| \leq k_0} \|Y^\alpha u\|_{L^2_{s-2^{1-m}}}, \quad \forall u \in C_0^\infty(\Omega).$$

PROOF. This follows immediately from Proposition 2.4.17. $\qquad\square$

LEMMA 2.4.21. *Let $k_0 \geq 1$ be an integer and $\mathcal{P}$ any partial differential operator with smooth coefficients. Suppose we have $\forall \phi_1, \phi_2 \in C_0^\infty(\Omega)$ with $\phi_1 \prec \phi_2$ and $\forall s \in \mathbb{R}$,*

$$\sum_{|\alpha| \leq k_0} \|Y^\alpha \phi_1 u\|_{L^2_s} \lesssim \|\phi_1 \mathcal{P} u\|_{L^2_s} + \sum_{|\alpha| \leq k_0 - 1} \|Y^\alpha \phi_2 u\|_{L^2_s}, \quad \forall u \in C_0^\infty(\Omega). \quad (2.44)$$

Then, $\forall \phi_1, \phi_2 \in C_0^\infty(\Omega)$ with $\phi_1 \prec \phi_2$ and $\forall s \in \mathbb{R}$,

$$\|\phi_1 u\|_{L^2_{s+k_0 2^{1-m}}} \lesssim \|\phi_2 \mathcal{P} u\|_{L^2_s} + \|\phi_2 u\|_{-\infty}, \quad \forall u \in C_0^\infty(\Omega)'. \quad (2.45)$$

PROOF. Lemma 2.4.14 shows that it suffices to prove (2.45) for $u \in C_0^\infty(\Omega)$. Pick a sequence $\eta_j \in C_0^\infty(\Omega)$ with $\phi_1 = \eta_0 \prec \eta_1 \prec \eta_2 \prec \cdots \prec \phi_2$. We claim, $\forall u \in C_0^\infty(\Omega)$,

$$\sum_{|\alpha| \leq k_0} \|Y^\alpha \eta_0 u\|_{L^2_s} \lesssim \|\eta_{j-1} \mathcal{P} u\|_{L^2_s} + \sum_{|\alpha| \leq k_0 - 1} \|Y^\alpha \eta_j u\|_{L^2_{s-(j-1)2(1-m)}}. \quad (2.46)$$

We prove (2.46) by induction. The base case ($j = 1$) is the assumption (2.44). Assume we have (2.46) for some value of j, and we wish to prove it for $j + 1$. We have $\forall u \in C_0^\infty(\Omega)$

$$\sum_{|\alpha| \leq k_0} \|Y^\alpha \eta_0 u\|_{L^2_s} \lesssim \|\eta_{j-1} \mathcal{P} u\|_{L^2_s} + \sum_{|\alpha| \leq k_0 - 1} \|Y^\alpha \eta_j u\|_{L^2_{s-(j-1)2(1-m)}}$$

$$\lesssim \|\eta_{j-1} \mathcal{P} u\|_{L^2_s} + \sum_{|\alpha| \leq k_0} \|Y^\alpha \eta_j u\|_{L^2_{s-j2(1-m)}},$$

where we have applied Lemma 2.4.20. But by our hypothesis, we have

$$\sum_{|\alpha| \le k_0} \|Y^\alpha \eta_j u\|_{L^2_{s-j2(1-m)}} \lesssim \|\eta_j \mathcal{P}u\|_{L^2_{s-j2(1-m)}} + \sum_{|\alpha| \le k_0 - 1} \|Y^\alpha \eta_{j+1} u\|_{L^2_{s-j2(1-m)}}.$$

Combining these two equations gives

$$\sum_{|\alpha| \le k_0} \|Y^\alpha \eta_0 u\|_{L^2_s} \lesssim \|\eta_j \mathcal{P}u\|_{L^2_s} + \sum_{|\alpha| \le k_0 - 1} \|Y^\alpha \eta_{j+1} u\|_{L^2_{s-j2(1-m)}},$$

completing the proof of (2.46).

Using Proposition 2.4.17 and (2.46), we have

$$\|\eta_0 u\|_{L^2_{s+k_0 2(1-m)}} \lesssim \sum_{|\alpha| \le k_0} \|Y^\alpha \eta_0 u\|_{L^2_s}$$

$$\lesssim \|\eta_{j-1} \mathcal{P}u\|_{L^2_s} + \sum_{|\alpha| \le k_0 - 1} \|Y^\alpha \eta_j u\|_{L^2_{s-(j-1)2(1-m)}}$$

$$\lesssim \|\phi_2 \mathcal{P}u\|_{L^2_s} + \|\phi_2 u\|_{L^2_{s+k_0-1-(j-1)2(1-m)}}.$$

Taking j large, and using $\eta_0 = \phi_1$, completes the proof.   □

LEMMA 2.4.22. *Let $\phi_1, \phi_2 \in C_0^\infty(\Omega)$ with $\phi_1 \prec \phi_2$. Let α be an ordered multi-index and $S^{(s)}$ a pseudodifferential of order s. We can write*

$$\left[S^{(s)}, Y^\alpha \phi_1 \right] = \sum_{|\beta| < |\alpha|} S_\beta^{(s)} Y^\beta \phi_2,$$

where each $S_\beta^{(s)}$ is a pseudodifferential operator of order s.

PROOF. This follows easily from the calculus of pseudodifferential operators.   □

LEMMA 2.4.23. *Let $\phi_1, \phi_2 \in C_0^\infty(\Omega)$ with $\phi_1 \prec \phi_2$, and let $S^{(s)}$ be a pseudodifferential operator of order s. Then,*

$$\left[\mathcal{P}, S^{(s)} \phi_1 \right] = \sum_{|\beta| \le k-1} S_\beta^{(s)} Y^\beta \phi_2,$$

where each $S_\beta^{(s)}$ is a pseudodifferential operator of order s.

PROOF. This follows easily from the calculus of pseudodifferential operators.   □

PROOF OF THEOREM 2.4.11 FOR $\mathcal{P}$ MAXIMALLY HYPOELLIPTIC. Let $\mathcal{P}$ be maximally hypoelliptic, and let $\phi_1, \phi_2 \in C_0^\infty(\Omega)$ with $\phi_1 \prec \phi_2$. Consider, for $u \in$

$C_0^\infty (\Omega)$, applying Lemmas 2.4.22 and 2.4.23 freely,

$$
\begin{aligned}
\sum_{|\alpha|\leq k} \|Y^\alpha \phi_1 u\|_{L_s^2} &\lesssim \sum_{|\alpha|\leq k} \|\phi_2 \Lambda^s Y^\alpha \phi_1 u\|_{L^2} + \|\phi_2 u\|_{L_s^2} \\
&\lesssim \sum_{|\alpha|\leq k} \|Y^\alpha \phi_2 \Lambda^s \phi_1 u\|_{L^2} + \sum_{|\alpha|\leq k-1} \|Y^\alpha \phi_2 u\|_{L_s^2} \\
&\lesssim \|\mathcal{P}\phi_2 \Lambda^s \phi_1 u\|_{L^2} + \sum_{|\alpha|\leq k-1} \|Y^\alpha \phi_2 u\|_{L_s^2} \\
&\lesssim \|\phi_2 \Lambda^s \phi_1 \mathcal{P} u\|_{L^2} + \sum_{|\alpha|\leq k-1} \|Y^\alpha \phi_2 u\|_{L_s^2} \\
&\lesssim \|\phi_1 \mathcal{P} u\|_{L_s^2} + \sum_{|\alpha|\leq k-1} \|Y^\alpha \phi_2 u\|_{L_s^2} .
\end{aligned}
$$

The result now follows from Lemma 2.4.21.   □

PROOF OF THEOREM 2.4.11 FOR $\mathcal{P}$ MAXIMALLY HYPOELLIPTIC OF TYPE 2. Let $\mathcal{P}$ be maximally hypoelliptic of type 2. Consider, for $u \in C_0^\infty (\Omega)$, using Lemma 2.4.22,

$$
\begin{aligned}
\sum_{|\alpha|\leq l} \|Y^\alpha \phi_1 u\|_{L_s^2}^2 &\lesssim \sum_{|\alpha|\leq l} \|\phi_2 \Lambda^s Y^\alpha \phi_1 u\|_{L^2}^2 + \|\phi_2 u\|_{L_s^2}^2 \\
&\lesssim \sum_{|\alpha|\leq l} \|Y^\alpha \phi_2 \Lambda^s \phi_1 u\|_{L^2}^2 + \sum_{|\alpha|\leq l-1} \|Y^\alpha \phi_2 u\|_{L_s^2}^2 \qquad (2.47) \\
&\lesssim |\langle \mathcal{P}\phi_2 \Lambda^s \phi_1 u, \phi_2 \Lambda^s \phi_1 u\rangle| + \sum_{|\alpha|\leq l-1} \|Y^\alpha \phi_2 u\|_{L_s^2}^2 .
\end{aligned}
$$

We have, using Lemma 2.4.23,

$$
\begin{aligned}
|\langle \mathcal{P}&\phi_2 \Lambda^s \phi_1 u, \phi_2 \Lambda^s \phi_1 u\rangle| \\
&\lesssim |\langle \phi_2 \Lambda^s \phi_1 \mathcal{P} u, \phi_2 \Lambda^s \phi_1 u\rangle| + \sum_{|\beta|\leq k-1} \left|\left\langle S_\beta^{(s)} Y^\beta \phi_2 u, \phi_2 \Lambda^s \phi_1 u\right\rangle\right|, \qquad (2.48)
\end{aligned}
$$

where $S_\beta^{(s)}$ is a pseudodifferential operator of order s.

We bound the terms on the right-hand side of (2.48) separately. We have, using Lemma 2.4.22,

$$
\sum_{|\beta|\leq k-1} \left|\left\langle S_\beta^{(s)} Y^\beta \phi_2 u, \phi_2 \Lambda^s \phi_1 u\right\rangle\right| \lesssim \sum_{\substack{|\alpha|\leq l \\ |\beta|\leq l-1}} \left|\left\langle S_\beta^{(s)} Y^\beta \phi_2 u, S_\alpha^{(s)} Y^\alpha \phi_1 u\right\rangle\right|,
$$

where $S_\beta^{(s)}$ and $S_\alpha^{(s)}$ are pseudodifferential operators of order s (possibly different pseudodifferential operators as those from the previous line). We have

$$\sum_{\substack{|\alpha|\leq l \\ |\beta|\leq l-1}} \left|\left\langle S_\beta^{(s)} Y^\beta \phi_2 u, S_\alpha^{(s)} Y^\alpha \phi_1 u\right\rangle\right| \leq (\text{l.c.}) \sum_{|\beta|\leq l-1} \left\|Y^\beta \phi_2 u\right\|_{L_s^2}^2$$

$$+ \sum_{|\alpha|\leq l} (\text{s.c.}) \left\|Y^\alpha \phi_1 u\right\|_{L_s^2}^2.$$

Combining the above two equations, we conclude

$$\sum_{|\beta|\leq k-1} \left|\left\langle S_\beta^{(s)} Y^\beta \phi_2 u, \phi_2 \Lambda^s \phi_1 u\right\rangle\right| \leq (\text{l.c.}) \sum_{|\beta|\leq l-1} \left\|Y^\beta \phi_2 u\right\|_{L_s^2}^2$$

$$+ (\text{s.c.}) \sum_{|\alpha|\leq l} \left\|Y^\alpha \phi_1 u\right\|_{L_s^2}^2. \tag{2.49}$$

Also, we have

$$|\langle \phi_2 \Lambda^s \phi_1 \mathcal{P} u, \phi_2 \Lambda^s \phi_1 u\rangle| \lesssim \|\phi_1 \mathcal{P} u\|_{L_s^2}^2 + \|\phi_1 u\|_{L_s^2}^2. \tag{2.50}$$

Plugging (2.49) and (2.50) into (2.48) we see

$$|\langle \mathcal{P} \phi_2 \Lambda^s \phi_1 u, \phi_2 \Lambda^s \phi_1 u\rangle| \leq (\text{l.c.}) \left(\|\phi_1 \mathcal{P} u\|_{L_s^2}^2 + \sum_{|\beta|\leq l-1} \left\|Y^\beta \phi_2 u\right\|_{L_s^2}^2\right)$$

$$+ (\text{s.c.}) \sum_{|\alpha|\leq l} \left\|Y^\alpha \phi_1 u\right\|_{L_s^2}^2.$$

Combining this with (2.47), we have

$$\sum_{|\alpha|\leq l} \left\|Y^\alpha \phi_1 u\right\|_{L_s^2}^2 \leq (\text{l.c.}) \left(\|\phi_1 \mathcal{P} u\|_{L_s^2}^2 + \sum_{|\beta|\leq l-1} \left\|Y^\beta \phi_2 u\right\|_{L_s^2}^2\right)$$

$$+ (\text{s.c.}) \sum_{|\alpha|\leq l} \left\|Y^\alpha \phi_1 u\right\|_{L_s^2}^2,$$

and therefore

$$\sum_{|\alpha|\leq l} \left\|Y^\alpha \phi_1 u\right\|_{L_s^2}^2 \lesssim \|\phi_1 \mathcal{P} u\|_{L_s^2}^2 + \sum_{|\beta|\leq l-1} \left\|Y^\beta \phi_2 u\right\|_{L_s^2}^2.$$

The result now follows from Lemma 2.4.21. $\qquad\square$

2.4.2 Scale invariance

We now discuss the main aspect of maximal hypoellipticity that leads to Theorem 2.4.8: maximal hypoellipticity is a "scale invariant" condition. That is, we have the following proposition.

PROPOSITION 2.4.24. *Suppose* $\mathcal{P} = P(W)$ *is maximally hypoelliptic. Then so is* $\delta^k P(W)$, *for* $0 < \delta \leq 1$, *uniformly in* δ. *More precisely, there is a constant* C, *independent of* $\delta \in (0, 1]$ *such that*

$$\sum_{|\alpha| \leq k} \|(\delta W)^\alpha f\|_{L^2} \leq C \left(\left\| \delta^k P(W) \right\|_{L^2} + \|f\|_{L^2} \right), \quad \forall f \in C^\infty(M),$$

where $k = \deg P$. *Similarly if* $P(W)$ *is maximally hypoelliptic of type* 2, *then* $\delta^k P(W)$ *is as well for* $0 < \delta \leq 1$, *uniformly in* δ. *More precisely, there is a constant* C, *independent of* $\delta \in (0, 1]$ *such that*

$$\sum_{|\alpha| \leq l} \|(\delta W)^\alpha f\|_{L^2}^2 \leq C \left(\left| \langle \delta^k P(W) f, f \rangle \right| + \|f\|_{L^2}^2 \right), \quad \forall f \in C^\infty(M),$$

where $l = \frac{\deg P}{2}$ *is an integer.*

We devote the remainder of this section to the proof of Proposition 2.4.24 and the important consequences which follow from it. The proof requires several lemmas. We let $Z_1, \ldots, Z_r$ be arbitrary C^∞ vector fields on M, which we do not require to satisfy Hörmander's condition. The implicit constants in the lemmas that follow are allowed to depend on upper bounds for the C^m norms of $Z_1, \ldots, Z_r$, but no lower bounds. In particular, we will take $Z_1 = \delta W_1, \ldots, Z_r = \delta W_r$, and with this choice, the implicit constants are independent of $\delta \in (0, 1]$.

As in the previous section, we write (s.c.) to denote "small constant" and (l.c.) to denote "large constant." An equation of the form $A \leq (\text{l.c.})B_1 + (\text{s.c.})B_2$ denotes: $\forall \epsilon > 0, \exists C$ with $A \leq CB_1 + \epsilon B_2$. Notice, if $A \leq (\text{l.c.})B + (\text{s.c.})A$, then $A \leq CB$ for some constant C. We use, repeatedly, the inequality $AB \leq (\text{s.c.})A^2 + (\text{l.c.})B^2$ in what follows. Also, we use $\langle \cdot, \cdot \rangle$ to denote the $L^2(M)$ inner product: here $L^2(M)$ is defined with respect to our chosen strictly positive, smooth measure. Note that if Z_0 is a C^∞ vector field, then $Z_0^* = -Z_0 + g$, where $g \in C^\infty(M)$, and the adjoint is taken with respect to this inner product.

The key to the proof of Proposition 2.4.24 is the next lemma.

LEMMA 2.4.25. *Fix* $l \in \mathbb{N}$. *For* $f \in C^\infty(M)$,

$$\sum_{|\alpha| \leq l} \|Z^\alpha f\|_{L^2}^2 \lesssim \sum_{|\alpha| = l} \|Z^\alpha f\|_{L^2}^2 + \|f\|_{L^2}^2.$$

We prove Lemma 2.4.25 by proving several simpler lemmas which lead up to it.

LEMMA 2.4.26. *Let* $g \in C^\infty(M)$ *and fix* $j \leq l \in \mathbb{N}$. *Then, for* $f \in C^\infty(M)$,

$$\sum_{\substack{|\alpha|=l \\ |\beta|=j}} \left| \langle Z^\alpha f, g Z^\beta f \rangle \right|$$

$$\leq (\text{l.c.}) \left(\sum_{|\alpha|=l+1} \|Z^\alpha f\|_{L^2}^2 + \|f\|_{L^2}^2 \right) + (\text{s.c.}) \sum_{1 \leq |\beta| \leq l} \left\| Z^\beta f \right\|_{L^2}^2.$$

PROOF. We prove the result by induction on j. The base case, $j = 0$, is trivial. For higher j, we write $Z^\beta = Z_s Z^{\beta_0}$ where $|\beta_0| = j - 1$. We have, letting $g_1, g_2 \in C^\infty(M)$ be functions which are implicitly defined below,

$$
\begin{aligned}
\left| \left(Z^\alpha f, g Z^\beta f \right) \right| &= \left| \left(Z^\alpha f, g Z_s Z^{\beta_0} f \right) \right| \\
&\leq \left| \left(Z^\alpha f, [g, Z_s] Z^{\beta_0} f \right) \right| + \left| \left(Z_s^* Z^\alpha f, g Z^{\beta_0} f \right) \right| \\
&\leq \left| \left(Z^\alpha f, g_1 Z^{\beta_0} f \right) \right| + \left| \left(Z^\alpha f, g_2 Z^{\beta_0} f \right) \right| + \left| \left(Z_s Z^\alpha f, g Z^{\beta_0} f \right) \right| .
\end{aligned}
$$

The first two terms have the desired bound by our inductive hypothesis. The last term satisfies the proper bound by a simple application of the Cauchy-Schwartz inequality. □

LEMMA 2.4.27. *Fix $l \in \mathbb{N}$. For $f \in C^\infty(M)$*

$$
\sum_{|\alpha|=l} \| Z^\alpha f \|_{L^2}^2 \leq (\text{l.c.}) \left(\sum_{|\alpha|=l+1} \| Z^\alpha f \|_{L^2}^2 + \| f \|_{L^2}^2 \right) + (\text{s.c.}) \sum_{1 \leq |\alpha| \leq l-1} \| Z^\alpha f \|_{L^2}^2 .
$$

PROOF. By the previous lemma, we have

$$
\sum_{|\alpha|=l} \| Z^\alpha f \|_{L^2}^2 \leq (\text{l.c.}) \left(\sum_{|\alpha|=l+1} \| Z^\alpha f \|_{L^2}^2 + \| f \|_{L^2}^2 \right) + (\text{s.c.}) \sum_{1 \leq |\alpha| \leq l} \| Z^\alpha f \|_{L^2}^2 .
$$

Subtracting $(\text{s.c.}) \sum_{|\alpha|=l} \| Z^\alpha f \|_{L^2}^2$ from both sides yields the result. □

LEMMA 2.4.28. *Fix $l \in \mathbb{N}$. For $f \in C^\infty(M)$,*

$$
\sum_{|\alpha| \leq l} \| Z^\alpha f \|_{L^2}^2 \lesssim \sum_{|\alpha|=l+1} \| Z^\alpha f \|_{L^2}^2 + \| f \|_{L^2}^2 .
$$

PROOF. We proceed by induction on l. The base case, $l = 0$, is trivial. We assume the result for $l - 1$ and prove it for l.

$$
\begin{aligned}
\sum_{|\alpha| \leq l} \| Z^\alpha f \|_{L^2}^2 &\leq C \left(\sum_{|\alpha|=l} \| Z^\alpha f \|_{L^2}^2 + \| f \|_{L^2}^2 \right) \\
&\leq (\text{l.c.}) \left(\sum_{|\alpha|=l+1} \| Z^\alpha f \|_{L^2}^2 + \| f \|_{L^2}^2 \right) + (\text{s.c.}) \sum_{1 \leq |\alpha| \leq l} \| Z^\alpha f \|_{L^2}^2 ,
\end{aligned}
$$

where the first inequality follows from our inductive hypothesis, and the second follows from the previous lemma. The result follows. □

PROOF OF LEMMA 2.4.25. This follows immediately from Lemma 2.4.28. □

PROOF OF PROPOSITION 2.4.24. Suppose $\mathcal{P}$ is maximally hypoelliptic; i.e.,

$$\sum_{|\alpha|\leq k} \|W^\alpha f\|_{L^2}^2 \lesssim \|\mathcal{P}f\|_{L^2}^2 + \|f\|_{L^2}^2 .$$

By Lemma 2.4.25 this is equivalent to

$$\sum_{|\alpha|=k} \|W^\alpha f\|_{L^2}^2 \lesssim \|\mathcal{P}f\|_{L^2}^2 + \|f\|_{L^2}^2 .$$

Multiplying both sides by δ^{2k} (for $\delta \in (0,1]$), we have

$$\sum_{|\alpha|=k} \|(\delta W)^\alpha f\|_{L^2}^2 \lesssim \left\|\delta^k \mathcal{P}f\right\|_{L^2}^2 + \|f\|_{L^2}^2 .$$

Lemma 2.4.25 shows that this implies

$$\sum_{|\alpha|\leq k} \|(\delta W)^\alpha f\|_{L^2}^2 \lesssim \left\|\delta^k \mathcal{P}f\right\|_{L^2}^2 + \|f\|_{L^2}^2 ,$$

completing the proof in the maximally hypoelliptic case. When $\mathcal{P}$ is maximally hypoelliptic of type 2, a similar proof yields the result. $\qquad\square$

Proposition 2.4.24 shows that maximal hypoellipticity and maximal hypoellipticity of type 2 are "scale invariant" properties. Because of this, we expect Theorem 2.4.6 (maximal hypoellipticity$\Rightarrow$subellipticity) to be true in a "scale invariant" way, as well. This is difficult to make precise at this stage, because the standard Sobolev spaces do not scale in the same way that our vector fields scale. Instead, for the time being, we make do with a consequence of subellipticity which we can make scale invariant.[9]

THEOREM 2.4.29. *Suppose* $P(W)$ *is either maximally hypoelliptic or maximally hypoelliptic of type 2. Then* $\exists \xi_3, \xi_4 > 0$, $\forall \alpha$, $\exists L = L(\alpha)$, $C = C(\alpha)$, $\forall x_0 \in M$, $\forall \delta \in (0,1]$, $\forall f \in C^\infty(M)$,

$$\sup_{x \in B_{(X,d)}(x_0,\xi_4\delta)} |(\delta W)^\alpha f(x)|$$

$$\leq C \sum_{j=0}^{L} \mathrm{Vol}\left(B_{(X,d)}(x_0,\xi_3\delta)\right)^{-\frac{1}{2}} \left\|\delta^{jk} P(W)^j f\right\|_{L^2\left(B_{(X,d)}(x_0,\xi_3\delta)\right)} .$$

We turn to the proof of Theorem 2.4.29. The key is to use the quantitative Frobenius theorem to leverage the scale invariant nature of maximal hypoellipticity. Fix a point $x_0 \in M$ and $\delta \in (0,1]$. The quantitative Frobenius theorem applies to the vector fields $\delta^{d_1}X_1,\ldots,\delta^{d_q}X_q$ to give a map $\Phi : B^n(\eta) \to B_{(X,d)}(x_0,\xi_2\delta)$, satisfying the conclusions of Theorem 2.2.22, uniformly for $x_0 \in M$ and $\delta \in (0,1]$. Let $Y_1,\ldots,Y_q$ be the pullbacks of $\delta^{d_1}X_1,\ldots,\delta^{d_q}X_q$ via Φ, so that $Y_1,\ldots,Y_q$ are C^∞ vector fields

---

[9]See Corollary 2.11.6 for a scale invariant version of subellipticity.

on $B^n(\eta)$. The vector fields with formal degrees $(W_1, 1), \ldots, (W_r, 1)$ appear in the list $(X_1, d_1), \ldots, (X_q, d_q)$. Without loss of generality, we assume that we have enumerated (X, d) so that these are the first r elements of the list. Thus, $Y_1, \ldots, Y_r$ are the pullbacks, via Φ, of $\delta W_1, \ldots, \delta W_r$. For clarity, define also $V_1 = Y_1, \ldots, V_r = Y_r$. Each $\delta^{d_j} X_j$ is an iterated commutator of the δW_j (where the number of commutators is $\leq m$, where $W_1, \ldots, W_r$ satisfy Hörmander's condition of order m). Pulling this back via Φ, each Y_j $1 \leq j \leq q$ is an iterated commutator (of order $\leq m$) of the V_j $1 \leq j \leq r$. (2.18) shows that $Y_1, \ldots, Y_q$ span the tangent space, and we conclude $V_1, \ldots, V_r$ satisfy Hörmander's condition of order m on $B^n(\eta)$, *uniformly for $x_0 \in M$, $\delta \in (0, 1]$,* in the sense that (2.18) holds with implicit constant independent of $x_0 \in M$, $\delta \in (0, 1]$.

From Theorem 2.4.11 we immediately get Theorem 2.4.29 for $\delta = 1$, which we apply to the vector fields $V_1, \ldots, V_r$:

COROLLARY 2.4.30. *Let $P(V)$ be maximally hypoelliptic or maximally hypoelliptic of type 2 on $B^n(\eta)$. Fix $0 < \eta_3 < \eta_2 < \eta$, $\forall \alpha$, $\exists L = L(\alpha)$, $C = C(\alpha)$,*

$$\sup_{u \in B^n(\eta_3)} |V^\alpha f(u)| \leq C \sum_{j=0}^{L} \left\| P(V)^j f \right\|_{L^2(B^n(\eta_2))}.$$

C and L depend only on η_2, η_3 and the same parameters that D does in Theorem 2.4.11 (where L depends on m and α).

PROOF. Fix $\eta_3 < \eta_2 < \eta$ and take $\phi_1 \in C_0^\infty(B^n(\eta_2))$ with $\phi_1 \equiv 1$ on $B^n(\eta_3)$. Let $\phi_2^0 = \phi_1$ and recursively choose $\phi_2^j \in C_0^\infty(B^n(\eta_2))$ with $\phi_2^{j-1} \prec \phi_2^j$. By the Sobolev embedding theorem, we have for some $s = s(\alpha)$,

$$\sup_{u \in B^n(\eta_3)} |V^\alpha f(u)| \lesssim \|\phi_1 f\|_{L_s^2}.$$

We have, by repeated applications of Theorem 2.4.11 (using $V_1, \ldots, V_r$ to play the role of $Y_1, \ldots, Y_r$ in that theorem),

$$\|\phi_2 f\|_{L_s^2} \lesssim \left\| \phi_2^1 P(V) f \right\|_{L_{s-\epsilon_0}^2} + \|f\|_{L^2(B^n(\eta_2))}$$

$$\lesssim \left\| \phi_2^2 P(V)^2 f \right\|_{L_{s-2\epsilon_0}^2} + \sum_{j=0}^{1} \left\| P(V)^j f \right\|_{L^2(B^n(\eta_2))}$$

$$\lesssim \cdots$$

$$\lesssim \sum_{j=0}^{L} \left\| P(V)^j f \right\|_{L^2(B^n(\eta_2))},$$

which completes the proof. $\qquad \square$

PROOF OF THEOREM 2.4.29. We exhibit the proof when $P(W)$ is maximally hypoelliptic; the proof when $P(W)$ is maximally hypoelliptic of type 2 is nearly identical. Fix $x_0 \in M$ and $\delta \in (0, 1]$. Let $V_1, \ldots, V_r$ be the pullbacks, via Φ, of $\delta W_1, \ldots, \delta W_r$

as discussed above, so that $V_1, \ldots, V_r$ satisfy Hörmander's condition uniformly in δ. Define an operator

$$P_\Phi (V) := \sum_{|\alpha| \leq k} \delta^{k - |\alpha|} a_\alpha \circ \Phi (u) \, V^\alpha.$$

We claim that since $P(W)$ is maximally hypoelliptic, $P_\Phi (V)$ is maximally hypoelliptic (uniformly for $x_0 \in M$, $\delta \in (0, 1]$). Indeed, using a change of variables as in $(2.19)^{10}$ combined with the estimate (2.16), we have

$$\|f\|_{L^2 (\Phi (B^n (\eta)))} \approx \mathrm{Vol} \left(B_{(X, d)} (x_0, \xi_2 \delta) \right)^{\frac{1}{2}} \|f \circ \Phi\|_{L^2 (B^n (\eta))}.$$

Applying Proposition 2.4.24 with $f = g \circ \Phi^{-1}$, for $g \in C_0^\infty (B^n (\eta))$, and using this change of variables, we have

$$\sum_{|\alpha| \leq k} \|V^\alpha g\|_{L^2} \lesssim \|P_\Phi (V) g\|_{L^2} + \|g\|_{L^2},$$

where the implicit constant is independent of $x_0 \in M$ and $\delta \in (0, 1]$. This shows that $P_\Phi (V)$ is maximally hypoelliptic, uniformly for $x_0 \in M$, $\delta \in (0, 1].^{11}$

From the discussion above, and using Theorem 2.2.22, it follows that the hypotheses of Corollary 2.4.30 are satisfied, uniformly for $\delta \in (0, 1]$, $x_0 \in M$. We conclude, for whichever $\eta_2 > \eta_3 > 0$ we pick, with $\eta_2 < \eta$,

$$\sup_{u \in B^n (\eta_3)} |V^\alpha g (u)| \leq C_{\eta_2, \eta_3} \sum_{j=0}^{L} \left\| P_\Phi (V)^j g \right\|_{L^2 (B^n (\eta_2))}.$$

Let ξ_3 be as in Theorem 2.2.22 and take $\eta_2 \gtrsim 1$ so small $B^n (\eta_2) \subset B_{(Y, d)} (0, \xi_3)$, which we may do since $Y_1, \ldots, Y_q$ span the tangent space, uniformly in the relevant parameters. Take $\eta_3 = \eta_2 / 2$. Using $\|Y_j\|_{C^1 (B^n (\eta))} \lesssim 1$ (which is contained in Theorem 2.2.22), we may pick $\xi_4 \gtrsim 1$ such that $B_{(Y, d)} (0, \xi_4) \subset B^n (\eta_3)$. We therefore have

$$\sup_{u \in B_{(Y, d)} (0, \xi_4)} |V^\alpha g (u)| \leq C \sum_{j=0}^{L} \left\| P_\Phi (V)^j g \right\|_{L^2 \left(B_{(Y, d)} (0, \xi_3) \right)}.$$

Taking $g = f \circ \Phi$, and applying a change of variables as in (2.19), the conclusion follows. $\qquad \square$

---

[10](2.19) uses that the range space of Φ is a subset of $\mathbb{R}^n$, given Lebesgue measure. Here, though, the measure is given by the chosen strictly positive, smooth measure. This yields a measure which is locally equivalent to Lebesgue measure, and the change of variables in (2.19) then works, provided we are only estimating the various integrals up to constant multiples, which is sufficient for our purposes.

[11]Similarly, when $P(W)$ is maximally hypoelliptic of type 2, the same proof shows $P_\Phi (V)$ is maximally hypoelliptic of type 2, uniformly for $x_0 \in M$, $\delta \in (0, 1]$.

2.5   SMOOTH METRICS AND BUMP FUNCTIONS

To prove Theorem 2.4.8, we need to better understand the nature of the metric ρ, as it relates to the smooth structure on M. In this section, we develop partitions of unity at each scale, with respect to the metric ρ, which behave well with respect to derivatives in the direction of the vector fields W. We use this to then develop a "smooth" version of the metric ρ. We begin by a discussion of bump functions. In this section, we use the notation $A \lesssim B$ to denote $A \leq CB$ where C is a constant that does not depend on certain parameters. More specifically, the constant will be an admissible constant in the sense discussed in Section 2.3. For instance, there is often a fixed point $x_0 \in M$ and a fixed scale $\delta \in (0, 1]$ and the constant C is to be independent of these two parameters.

PROPOSITION 2.5.1. *Let $x_0 \in M$, $\delta \in (0, 1]$. There is $1 \lesssim \xi_4 < 1$, and a function $\phi \in C^\infty(M)$ with $\phi \equiv 1$ on $B_{(X,d)}(x_0, \xi_4 \delta)$ and $\phi \equiv 0$ outside $B_{(X,d)}(x_0, \delta)$, $0 \leq \phi \leq 1$, and ϕ and satisfies, $\forall \alpha$,*

$$|(\delta W)^\alpha \phi| \lesssim 1,$$

where the implicit constant depends on α, but on neither x_0 nor δ.

PROOF. As in Section 2.3, we may apply the quantitative Frobenius theorem to the vector fields $\delta^{d_1} X_1, \ldots, \delta^{d_q} X_q$ to obtain a map $\Phi = \Phi_{x_0, \delta} : B^n(\eta) \to B_{(X,d)}(x_0, \xi_2 \delta)$ as in that theorem. Fix $\psi \in C_0^\infty(B^n(\eta))$, with $\psi \equiv 1$ on $B^n(\eta/2)$. Let $Y_1, \ldots, Y_q$ denote the pullbacks, via Φ, of $\delta^{d_1} X_1, \ldots, \delta^{d_q} X_q$. By Theorem 2.2.22, $\|Y_j\|_{C^l(B^n(\eta))} \lesssim 1$, $\forall l$. The Picard-Lindelöf theorem shows that there is $\xi_4 \gtrsim 1$ with $B_{(Y,d)}(0, \xi_4) \subset B^n(\eta/2)$. Define

$$\phi(x) = \begin{cases} \psi \circ \Phi^{-1}(x) & \text{if } x \in \Phi(B^n(\eta)), \\ 0 & \text{otherwise.} \end{cases}$$

Note that $\phi \in C^\infty(M)$. Moreover, for any ordered multi-index, we have (for $x \in \Phi(B^n(\eta))$)

$$|(\delta X)^\alpha \phi(x)| = |(Y^\alpha \psi) \circ \Phi^{-1}(x)| \lesssim 1.$$

The result follows, since each δW_j is of the form $\delta^{d_l} X_l$ for some (X_l, d_l) by our choice of (X, d). $\square$

LEMMA 2.5.2. *Fix $K > 1$. There is a constant $N = N(K) \lesssim 1$ such that if $\delta \in (0, 1]$ and $\mathcal{C} \subset M$ is such that for $x, y \in \mathcal{C}$, $B_{(X,d)}(x, \delta) \cap B_{(X,d)}(y, \delta) = \emptyset$, then no point lies in more than N of the balls $B_{(X,d)}(x_0, K\delta)$, $x_0 \in \mathcal{C}$.*

PROOF. Let $\mathcal{C}$ be as above and suppose $y \in M$ is such that $y \in B_{(X,d)}(x_j, K\delta)$ for $x_1, \ldots, x_N \in \mathcal{C}$. We wish to show $N \lesssim 1$. We have

$$B_{(X,d)}(x_j, \delta) \subset B_{(X,d)}(x_j, K\delta) \subset B_{(X,d)}(y, 2K\delta).$$

By disjointness of the collection of balls, we have

$$\text{Vol}\left(B_{(X,d)}\left(y, 2K\delta\right)\right) \geq \text{Vol}\left(\bigcup_{l=1}^{N} B_{(X,d)}\left(x_l, \delta\right)\right) = \sum_{l=1}^{N} \text{Vol}\left(B_{(X,d)}\left(x_l, \delta\right)\right).$$

However, by (2.31),

$$\text{Vol}\left(B_{(X,d)}\left(x_l, \delta\right)\right) \gtrsim \text{Vol}\left(B_{(X,d)}\left(x_l, 4K\delta\right)\right) \geq \text{Vol}\left(B_{(X,d)}\left(y, 2K\delta\right)\right).$$

Combining these two estimates we have

$$\text{Vol}\left(B_{(X,d)}\left(y, 2K\delta\right)\right) \gtrsim N\text{Vol}\left(B_{(X,d)}\left(y, 2K\delta\right)\right),$$

and we conclude $N \lesssim 1$. $\square$

THEOREM 2.5.3 (Partition of Unity). *Fix $\delta \in (0, 1]$ and let $\xi_4 > 0$ be as in Proposition 2.5.1. There is a finite set $\mathcal{C} \subset M$ and functions $\psi_x(y) \in C^\infty(M)$, $x \in \mathcal{C}$, satisfying*

- $\sum_{x \in \mathcal{C}} \psi_x(y) \equiv 1.$
- *$\forall m$,*

$$\sum_{|\alpha| \leq m} \|(\delta W)^\alpha \psi_x\|_{L^\infty} \lesssim 1.$$

- *For $x \in \mathcal{C}$, $\text{supp}(\psi_x) \subset B_{(X,d)}(x, \delta)$.*
- *For $x \in \mathcal{C}$, $\psi_x(y) \gtrsim 1$ for $y \in B_{(X,d)}(x, \xi_4\delta)$.*
- *$0 \leq \psi_x \leq 1$.*
- *$\bigcup_{x \in \mathcal{C}} B_{(X,d)}(x, \xi_4\delta/2) = M$.*
- *For $x \in M$, $1 \leq \#\{x_0 \in \mathcal{C} \mid \psi_{x_0}(x) \neq 0\} \lesssim 1$.*

PROOF. Cover M by a finite collection of balls of the form $B_{(X,d)}(x, \xi_4\delta/8)$. Pick from this collection a maximal disjoint collection $B_{(X,d)}(x, \xi_4\delta/8)$, $x \in \mathcal{C}$, $\mathcal{C}$ a finite set. By maximality, $B_{(X,d)}(x, \xi_4\delta/2)$, $x \in \mathcal{C}$, is a cover for M, as if y were in none of the balls, then any element of our original cover $B_{(X,d)}(x, \xi_4\delta/8)$ which contains y would be disjoint from our collection of balls, contradicting maximality.

For each $x_0 \in \mathcal{C}$, let ϕ_{x_0} be as in Proposition 2.5.1, with $\phi_{x_0} \equiv 1$ on $B_{(X,d)}(x_0, \xi_4\delta)$ and $\phi_{x_0} \equiv 0$ outside $B_{(X,d)}(x_0, \delta)$, $0 \leq \phi_{x_0} \leq 1$, and $|(\delta W)^\alpha \phi_{x_0}| \lesssim 1$, $\forall\alpha$. We claim $1 \leq \sum_{x \in \mathcal{C}} \phi_x(y) \leq N$, where $N = N\left(\frac{8}{\xi_4}\right)$ is from Lemma 2.5.2. That $1 \leq \sum_{x \in \mathcal{C}} \phi_x(y)$ follows by the fact that $B_{(X,d)}(x, \xi_4\delta)$ is a cover for M, and that $\sum_{x \in \mathcal{C}} \phi_x(y) \leq N$ follows from the fact that no point lies in more than N of the balls $B_{(X,d)}(x, \delta)$, $x \in \mathcal{C}$, by Lemma 2.5.2.

Define

$$\psi_x(y) = \frac{\phi_x(y)}{\sum_{z \in \mathcal{C}} \phi_z(y)}.$$

The desired properties for ψ_x follow easily from the properties of ϕ_x, completing the proof. $\square$

LEMMA 2.5.4. *Fix $\delta \in (0, 1]$, and let $\xi_4 \gtrsim 1$ be as in Proposition 2.5.1. There is a function $\omega_\delta \in C^\infty (M \times M)$ such that for $x, y \in M$,*

- $0 \leq \omega_\delta (x, y) \lesssim 1.$

- $\omega_\delta (x, y) \gtrsim 1$ *if $\rho (x, y) \leq \xi_4 \delta/2$.*

- $\omega_\delta (x, y) = 0$ *if $\rho (x, y) \geq 2\delta$.*

- *For any ordered multi-indices α and β,*

$$\left| (\delta W_x)^\alpha (\delta W_y)^\beta \omega_\delta (x, y) \right| \lesssim 1.$$

PROOF. Fix $\delta \in (0, 1]$ and let $\mathcal{C}$ be as in Theorem 2.5.3, and let $\psi_x \in C^\infty (M)$, $x \in \mathcal{C}$ be the corresponding partition of unity. Define $\omega_\delta (x, y) = \sum_{x_0 \in \mathcal{C}} \psi_{x_0} (x) \psi_{x_0} (y) \in C^\infty (M \times M)$.

Suppose $\rho (x, y) < \xi_4 \delta/2$. Since $\bigcup_{x_0 \in \mathcal{C}} B_{(X,d)} (x_0, \xi_4 \delta/2) = M$, there is some $x_0 \in \mathcal{C}$ with $x \in B_{(X,d)} (x_0, \xi_4 \delta/2)$. We have $x, y \in B_{(X,d)} (x_0, \xi_4 \delta)$, and it follows that $\psi_{x_0} (x) \psi_{x_0} (y) \gtrsim 1$, and therefore $\omega_\delta (x, y) \gtrsim 1$. Also, since $\mathrm{supp} (\psi_{x_0}) \subset B_{(X,d)} (x_0, \delta)$, we have that if $\rho (x, y) \geq 2\delta$, $\psi_{x_0} (x) \psi_{x_0} (y) = 0$.

The remaining properties of ω_δ follow easily from the properties of the ψ_{x_0}. $\square$

THEOREM 2.5.5. *There is a function $\tilde{\rho} : M \times M \to [0, \infty)$, with $\tilde{\rho} (x, y) \approx \rho (x, y)$, $\tilde{\rho} \in C^\infty (M \times M \setminus \{x = y\})$, and for all ordered multi-indices α, β, for $x \neq y$,*

$$\left| W_x^\alpha W_y^\beta \tilde{\rho} (x, y) \right| \lesssim \tilde{\rho} (x, y)^{1 - |\alpha| - |\beta|}.$$

PROOF. Set $\tilde{\rho} (x, x) = 0$ and for $x \neq y$ let

$$\tilde{\rho} (x, y) = \left[1 + \sum_{j=0}^\infty 2^j \omega_{2^{-j}} (x, y) \right]^{-1},$$

where $\omega_{2^{-j}}$ is the function from Lemma 2.5.4. The result now follows easily from the properties of the $\omega_{2^{-j}}$. $\square$

Remark 2.5.6 Let $\tilde{\rho}$ be as in Theorem 2.5.5. By replacing $\tilde{\rho} (x, y)$ with $\tilde{\rho} (x, y) + \tilde{\rho} (y, x)$, we may assume $\tilde{\rho} (x, y) = \tilde{\rho} (y, x)$. Furthermore, by multiplying $\tilde{\rho}$ by a fixed constant we may assume for all $|\alpha| = 1$,

$$\sup_{x \neq y} \left| W_y^\alpha \tilde{\rho} (x, y) \right| \leq \frac{1}{r}.$$

Recall, r is the number of vector fields $W_1, \ldots, W_r$. From now on, we assume both of these properties for $\tilde{\rho}$.

2.6   THE SUB-LAPLACIAN

We now turn to a special maximally hypoelliptic operator, Hörmander's sub-Laplacian:

$$\mathcal{L} := W_1^* W_1 + \cdots + W_r^* W_r.$$

It is evident that $\mathcal{L}$ is maximally hypoelliptic of type 2 (see Definition 2.4.9), and therefore by Theorem 2.4.11, it is subelliptic.[12] We will see that $\mathcal{L}$ is, in fact, maximally hypoelliptic (Remark 2.6.9), but this fact is not obvious.

Remark 2.6.1   Note that $\mathcal{L}$ depends on the particular choice of the strictly positive, smooth measure we picked on M (because the definition of W_j^* depends on this choice). Any two such choices differ by terms of the form $f_j W_j$ and g, where $f_j, g \in C^\infty(M)$. We will think of these terms as "lower order," and they do not affect our analysis.

LEMMA 2.6.2.   *Suppose $u \in C^\infty(M)'$ with $\mathcal{L}u = 0$. Then u is constant.*

PROOF.   Because $\mathcal{L}$ is subelliptic, and therefore hypoelliptic, if $\mathcal{L}u = 0$, then $u \in C^\infty(M)$. Using that M is compact, without boundary, integration by parts shows that (for $u \in C^\infty(M)$), if $\mathcal{L}u = 0$ then $\langle \mathcal{L}u, u \rangle = 0$ and therefore $\|W_j u\|_{L^2(M)} = 0$, $j = 1, \ldots, r$. We conclude $W_j u = 0$, $j = 1, \ldots, r$. Since $W_1, \ldots, W_r$ satisfy Hörmander's condition, this shows $Yu = 0$ for every vector field Y. Thus u is locally constant. The result follows, since M is connected.   □

One may think of $\mathcal{L}$ as an operator defined on $C^\infty(M)$, and think of $C^\infty(M)$ as a dense subspace of $L^2(M)$. $\mathcal{L}$ is clearly symmetric: for $f, g \in C^\infty(M)$, we have $\langle \mathcal{L}f, g \rangle = \langle f, \mathcal{L}g \rangle$. However, $\mathcal{L}$, thought of as an operator on $C^\infty(M)$ is not *self-adjoint*. A priori such a densely defined symmetric operator may have many, or no, self-adjoint extensions. Fortunately, $\mathcal{L}$ has precisely one self-adjoint extension. In fact, the closure of $\mathcal{L}$ is self-adjoint; i.e., $\mathcal{L}$ is *essentially self-adjoint*. For more details on these concepts, see [RS80, Chapter VIII].

LEMMA 2.6.3.   *$\mathcal{L}$, with domain $C^\infty(M)$, is an essentially self-adjoint operator on $L^2(M)$.*

PROOF.   It suffices to show that the range of $\mathcal{L} \pm i$ is dense (see [RS80, page 257]). We prove the result for $\mathcal{L}+i$, the proof of the result for $\mathcal{L}-i$ is nearly identical. Suppose $u \in L^2(M)$ is orthogonal to the range of $\mathcal{L} + i$. We have $\langle u, (\mathcal{L}+i) f \rangle = 0$, for $f \in C^\infty(M)$. We conclude that $(\mathcal{L} - i) u = 0$ as a distribution. We know $u \in L^2(M)$ and we conclude that $\mathcal{L}u = iu \in L^2(M)$. Subellipticity implies $u \in L^2_\epsilon$, which in turn implies $\mathcal{L}u = iu \in L^2_\epsilon$. Subellipiticity now implies $u \in L^2_{2\epsilon}$. Continuing this process, we conclude $u \in L^2_N$ for every N. The Sobolev embedding theorem implies $u \in C^\infty(M)$.

Integrating by parts, and using $u \in C^\infty(M)$, we have $0 = \langle (\mathcal{L} - i) u, u \rangle = \sum_{j=1}^r \|W_j u\|_{L^2}^2 + i \|u\|_{L^2}^2$. We conclude $\|u\|_{L^2} = 0$ and therefore $u = 0$, completing the proof.   □

---

[12]The subellipticity of $\mathcal{L}$ was first shown by Hörmander [Hör67].

We henceforth identify the operator $\mathcal{L}$ with its closure which is a self-adjoint operator on $L^2(M)$. It is easy to see that, for $u \in L^2(M)$ in the domain of the self-adjoint operator $\mathcal{L}$, $\mathcal{L}u$ agrees with the usual distributional derivative. We are therefore free to think of $\mathcal{L}$ either as a self-adjoint operator, or in the distributional sense.

Let E be the spectral decomposition of $\mathcal{L}$, so that for a Borel measurable function $m : [0, \infty) \to \mathbb{C}$,

$$m(\mathcal{L}) = \int m(\lambda)\, dE(\lambda).$$

In light of Lemma 2.6.2, $E(0)$ is projection onto the constant functions. At best, we can only hope to invert $\mathcal{L}$ relative to the constant functions. Formally, we may define a candidate inverse to $\mathcal{L}$, relative to the constant functions:

$$\mathcal{L}^{-1} = \int_{(0,\infty)} \lambda^{-1}\, dE(\lambda).$$

In fact, this defines $\mathcal{L}^{-1}$ as a bounded operator on L^2. To see this, we prove the following lemma.

LEMMA 2.6.4. *The spectrum of $\mathcal{L}$ is discrete.*

COROLLARY 2.6.5. *$\mathcal{L}^{-1}$ is a bounded operator on $L^2(M)$, and is an inverse to $\mathcal{L}$ on the orthocomplement of the constant functions (and takes constant functions to 0).*

PROOF OF COROLLARY 2.6.5. This follows from Lemma 2.6.4. $\qquad\qquad\square$

PROOF OF LEMMA 2.6.4. We define the bounded operator

$$(I + \mathcal{L})^{-1} = \int (1 + \lambda)^{-1}\, dE(\lambda).$$

We will show $(I + \mathcal{L})^{-1} : L^2(M) \to L^2(M)$ is compact. Notice, once this is shown, it follows that (as $(I + \mathcal{L})^{-1}$ is a compact self-adjoint operator) the spectrum of $(I + \mathcal{L})^{-1}$ is countable and can accumulate only at 0. Since λ is in the spectrum of $\mathcal{L}$ if and only if $(1 + \lambda)^{-1}$ is in the spectrum of $(I + \mathcal{L})^{-1}$, the result will then follow.

The compactness of $(I + \mathcal{L})^{-1}$ follows from the fact that $(I + \mathcal{L})^{-1} : L^2(M) \to L^2_\epsilon(M)$ for some $\epsilon > 0$, and the well-known fact that the inclusion $L^2_\epsilon(M) \hookrightarrow L^2(M)$ is compact. Let $f \in L^2(M)$. To see that $(I + \mathcal{L})^{-1} f \in L^2_\epsilon(M)$, it suffices to note $\mathcal{L}(I + \mathcal{L})^{-1} f \in L^2(M)$ and the result follows by the subellipticity of $\mathcal{L}$. $\qquad\square$

Our main theorem concerning $\mathcal{L}$ is the following.

THEOREM 2.6.6. *Let $\lambda_0 > 0$ be the least nonzero eigenvalue of $\mathcal{L}$ (this is possible by Lemma 2.6.4). Fix $t \in \mathbb{R}$ and let $m : [0, \infty) \to \mathbb{C}$ be a function with $m\big|_{(\lambda_0/2,\infty)} \in C^\infty((\lambda_0/2, \infty))$ and which satisfies*

$$\sup_{\lambda \in (\lambda_0/2,\infty)} \lambda^{-t} |(\lambda\partial_\lambda)^a\, m(\lambda)| < \infty, \quad \forall a \in \mathbb{N}.$$

Then $m(\mathcal{L})$ is a Calderón-Zygmund operator of order $2t$, in the sense that there is a bounded set of elementary operators $\{(E_j, 2^{-j}) \mid j \in \mathbb{N}\}$ with $m(\mathcal{L}) = \sum_{j \in \mathbb{N}} 2^{2tj} E_j$.

COROLLARY 2.6.7. *For $s \in \mathbb{R}$ define*

$$\mathcal{L}^s = \int_{(0,\infty)} \lambda^s \, dE(\lambda).$$

$\mathcal{L}^s$ *is a Calderón-Zygmund operator of order $2s$. In particular, $\mathcal{L}^{-1}$ is a Calderón-Zygmund operator of order -2.*

PROOF. It is evident that the function

$$m(\lambda) = \begin{cases} \lambda^s & \text{if } \lambda > 0, \\ 0 & \text{if } \lambda = 0 \end{cases}$$

satisfies the hypotheses of Theorem 2.6.6 with $t = s$. □

Remark 2.6.8   Note that our notation in Corollary 2.6.7 may be a little counterintuitive. Indeed, $\mathcal{L}^0 = I - E(0)$ is projection onto the orthocomplement of the constant functions, and thus $\mathcal{L}^0$ is not the identity operator. We do have $\mathcal{L}^s \mathcal{L}^t = \mathcal{L}^{s+t}$, which will be useful to us.

Remark 2.6.9   In light of Corollary 2.6.7 and the (ii)⇒(i) part of Theorem 2.4.8, we see that $\mathcal{L}$ is maximally hypoelliptic.

COROLLARY 2.6.10. *For $s \in \mathbb{R}$, $(I + \mathcal{L})^s$ is a Calderón-Zygmund operator of order $2s$.*

PROOF. The function $m(\lambda) = (1 + \lambda)^s$ satisfies the hypotheses of Theorem 2.6.6 with $t = s$. □

Let $\widehat{\mathcal{S}_0}(\mathbb{R}) = \left\{ f \in \mathcal{S}(\mathbb{R}) \mid \left(\partial_\xi^j f \right)(0) = 0, \forall j \right\}$: the space of Schwartz functions which vanish to infinite order at 0. $\widehat{\mathcal{S}_0}(\mathbb{R})$ is a closed subspace of $\mathcal{S}(\mathbb{R})$ and we give it the subspace topology, endowing it with the structure of a Fréchet space. Note that the Fourier transform is an isomorphism $\mathcal{S}_0(\mathbb{R}) \to \widehat{\mathcal{S}_0}(\mathbb{R})$, justifying the notation. The key to Theorem 2.6.6 is the following proposition.

PROPOSITION 2.6.11. *Let $\mathcal{B} \subset \widehat{\mathcal{S}_0}(\mathbb{R})$ be a bounded set. Then,*

$$\left\{ \left(m\left(s^2 \mathcal{L}\right), s\right) \mid s \in (0,1], m \in \mathcal{B} \right\}$$

is a bounded set of elementary operators.

Before we prove Proposition 2.6.11, we discuss how it yields Theorem 2.6.6.

PROOF OF THEOREM 2.6.6. Let $\psi \in C_0^\infty(\mathbb{R})$ be a nonnegative function which equals 1 on $(-3\lambda_0/4, 3\lambda_0/4)$ and which equals 0 outside $(-7\lambda_0/8, 7\lambda_0/8)$. Define $\phi(\lambda) = \psi(\lambda) - \psi(4\lambda)$. Note $1 = \psi(\lambda) + \sum_{j=1}^\infty \phi(2^{-2j}\lambda)$ and therefore

$$m(\mathcal{L}) = m(0)\psi(\mathcal{L}) + \sum_{j=1}^\infty \phi(2^{-2j}\mathcal{L}) m(\mathcal{L}).$$

Because $\psi(\mathcal{L})$ is projection onto the constant functions, it immediately follows that $\{(m(0)\psi(\mathcal{L}), 1)\}$ is a bounded set of elementary operators (see Remark 2.0.26). The result then follows from Proposition 2.6.11 once we show

$$\{2^{-j2t} m(2^{2j}\lambda) \phi(\lambda) \mid j \in \mathbb{N}\} \subset \widehat{\mathcal{S}_0}(\mathbb{R})$$

is a bounded set. This follows immediately from the properties of m and ϕ, completing the proof.   □

LEMMA 2.6.12. *Let $\mathcal{B} \subset \mathcal{S}(\mathbb{R})$ be a bounded set. Then,*

$$\{(m(s^2\mathcal{L}), s) \mid s \in (0, 1], m \in \mathcal{B}\}$$

is a bounded set of pre-elementary operators.

PROOF OF PROPOSITION 2.6.11 GIVEN LEMMA 2.6.12. Let $m \in \mathcal{B}$ and $s \in (0, 1]$. We have $m(s^2\mathcal{L}) = (s^2\mathcal{L}) \widetilde{m}(s^2\mathcal{L}) (s^2\mathcal{L})$, where $\widetilde{m}(\lambda) = \lambda^{-2}m(\lambda)$. Note that $\{\widetilde{m} \mid m \in \mathcal{B}\} \subset \widehat{\mathcal{S}_0}(\mathbb{R})$ is a bounded set; i.e., $\widetilde{m}(s^2\mathcal{L})$ is of the same form as $m(s^2\mathcal{L})$. The result follows from Lemma 2.6.12, as sets of the form

$$\{(\widetilde{m}(s^2\mathcal{L}), s) \mid s \in (0, 1], m \in \mathcal{B}\}$$

are bounded sets of pre-elementary operators. Here we have used Proposition 2.0.25 for our characterization of elementary operators.   □

We close this section with the proof of Lemma 2.6.12, which is somewhat involved and requires that we prove several results first. We begin with a weaker result. Recall, we identify operators with their Schwartz kernels, throughout.

LEMMA 2.6.13. *Let $s > 0$. For every ordered multi-index α, there exists $m = m(\alpha)$ such that*

$$\left\| W_x^\alpha (I + s\mathcal{L})^{-m}(x, \cdot) \right\|_{L^2(M)} \lesssim \frac{s^{-\frac{|\alpha|}{2}}}{\mathrm{Vol}\left(B_{(X,d)}(x, \sqrt{s})\right)^{\frac{1}{2}}},$$

where the implicit constant depends on neither $s > 0$ nor $x \in M$.

PROOF. Fix α and take $m = m(\alpha)$ large, to be chosen later. Let $\phi \in C^\infty(M)$. The result will follow once we show

$$\left| W^\alpha (I + s\mathcal{L})^{-m} \phi(x) \right| \lesssim \mathrm{Vol}\left(B_{(X,d)}\left(x, \sqrt{s}\right) \right)^{-\frac{1}{2}} s^{-\frac{|\alpha|}{2}} \|\phi\|_{L^2(M)}. \qquad (2.51)$$

We separate the proof of (2.51) into two cases. The first is when $s \in \left(0, \xi_3^2\right]$ (where $\xi_3 > 0$ is as in Theorem 2.4.29). We apply Theorem 2.4.29 with $\delta = \sqrt{s}/\xi_3$ to see

$$\left| W^\alpha (I + s\mathcal{L})^{-m} \phi(x) \right|$$

$$\lesssim \mathrm{Vol}\left(B_{(X,d)}\left(x, \sqrt{s}\right) \right)^{-\frac{1}{2}} \sum_{j=0}^{L(\alpha)} s^{-\frac{|\alpha|}{2}} \left\| (s\mathcal{L})^j (1 + s\mathcal{L})^{-m} \phi \right\|_{L^2(M)}$$

$$\lesssim \mathrm{Vol}\left(B_{(X,d)}\left(x, \sqrt{s}\right) \right)^{-\frac{1}{2}} s^{-\frac{|\alpha|}{2}} \|\phi\|_{L^2(M)},$$

where in the last line we have take $m = m(\alpha)$ large and used

$$\left\| (s\mathcal{L})^j (I + s\mathcal{L})^{-m} \right\|_{L^2 \to L^2} \lesssim 1 \text{ if } m \geq j.$$

If $s \geq \xi_3^2$, we use the fact (by the compactness of M) that $\mathrm{Vol}\left(B_{(X,d)}\left(x, \sqrt{s}\right) \right) \approx 1 \approx \mathrm{Vol}\left(B_{(X,d)}(x, 1) \right)$. We then have, applying Theorem 2.4.29 with $\delta = 1$,

$$\left| W^\alpha (I + s\mathcal{L})^{-m} \phi(x) \right| \lesssim \mathrm{Vol}\left(B_{(X,d)}(x, \xi_3) \right)^{-\frac{1}{2}} \sum_{j=0}^{L(\alpha)} \left\| \mathcal{L}^j (1 + s\mathcal{L})^{-m} \phi \right\|_{L^2(M)}.$$

If $|\alpha| = 0$, (2.51) follows by taking m large and using $\left\| \mathcal{L}^j (1 + s\mathcal{L})^{-m} \right\|_{L^2 \to L^2} \lesssim 1$, if $m \geq j$. If $|\alpha| > 0$, we use that $W^\alpha E(0) = 0$, as $E(0)$ is projection onto the constant functions. We have, using the identity $\mathcal{L}^N \mathcal{L}^{-N} = \mathcal{L}^0 = I - E(0)$, for any N and $m \geq N$, and using $\mathrm{Vol}\left(B_{(X,d)}(x, \xi_3) \right) \approx 1 \approx \mathrm{Vol}\left(B_{(X,d)}\left(x, \sqrt{s}\right) \right)$,

$$\left| W^\alpha (I + s\mathcal{L})^{-m} \phi(x) \right| = \left| W^\alpha (I + s\mathcal{L})^{-m} (I - E(0)) \phi(x) \right|$$

$$\lesssim \mathrm{Vol}\left(B_{(X,d)}\left(x, \sqrt{s}\right) \right)^{-\frac{1}{2}} \sum_{j=0}^{L(\alpha)} \left\| \mathcal{L}^j (1 + s\mathcal{L})^{-m} (I - E(0)) \phi \right\|_{L^2(M)}$$

$$= \mathrm{Vol}\left(B_{(X,d)}\left(x, \sqrt{s}\right) \right)^{-\frac{1}{2}} \sum_{j=0}^{L(\alpha)} \left\| \mathcal{L}^{-N+j} \mathcal{L}^N (1 + s\mathcal{L})^{-m} \phi \right\|_{L^2(M)}$$

$$\lesssim \mathrm{Vol}\left(B_{(X,d)}\left(x, \sqrt{s}\right) \right)^{-\frac{1}{2}} \left\| \mathcal{L}^N (1 + s\mathcal{L})^{-m} \phi \right\|_{L^2(M)}$$

$$\approx \mathrm{Vol}\left(B_{(X,d)}\left(x, \sqrt{s}\right) \right)^{-\frac{1}{2}} s^{-N} \left\| (s\mathcal{L})^N (1 + s\mathcal{L})^{-m} \phi \right\|_{L^2(M)}$$

$$\lesssim \mathrm{Vol}\left(B_{(X,d)}\left(x, \sqrt{s}\right) \right)^{-\frac{1}{2}} s^{-N} \|\phi\|_{L^2(M)}.$$

Taking $N \geq |\alpha|/2$, $m \geq N$, and using that $s \geq \xi_3^2 \gtrsim 1$, the result follows. $\qquad \square$

To use Lemma 2.6.13 to prove the stronger results we need, we turn to the wave equation:

$$\frac{\partial^2}{\partial t^2} u\left(t, x\right) = -\mathcal{L}u\left(t, x\right),$$

where $u : [0, \infty) \times M \to \mathbb{R}$ and $\mathcal{L}$ acts in the x variable. The spectral theorem allows us to easily write down $u\left(t, x\right)$ given $u\left(0, x\right) = u_0\left(x\right)$, for $u_0\left(x\right) \in C^\infty\left(M\right)$, and assuming $\partial_t u\left(0, x\right) \equiv 0$. Namely, $u\left(t, x\right) = \left(\cos\left(t\sqrt{\mathcal{L}}\right) u_0\right)\left(x\right)$. The key fact we use about the wave equation is that it has finite propagation speed with respect to the metric ρ, which is a result of Melrose [Mel86].

THEOREM 2.6.14 (Finite speed of propagation). *There is $\kappa > 0$ such that $\forall t > 0$,*

$$\mathrm{supp}\left(\cos\left(t\sqrt{\mathcal{L}}\right)\right) \subseteq \left\{(x, y) \in M \times M \mid \rho\left(x, y\right) \le \kappa t\right\}.$$

To prove Theorem 2.6.14, we use the following lemma.

LEMMA 2.6.15. *Suppose $u\left(t, x\right) \in C^2\left([0, T] \times M\right)$ satisfies $\partial_t^2 u + \mathcal{L}u = 0$ and $u = \partial_t u = 0$ on the ball*

$$\left\{(0, y) \in [0, T] \times M \mid \tilde{\rho}\left(x_0, y\right) \le t_0\right\},$$

where $\tilde{\rho}$ is as in Theorem 2.5.5 and Remark 2.5.6, $x_0 \in M$, and $0 < t_0 \le T$. Then u vanishes in the region

$$\Omega = \left\{(t, y) \mid 0 \le t \le t_0, \tilde{\rho}\left(x_0, y\right) \le (t_0 - t)\right\}.$$

PROOF. Given $\delta > 0$, let $\chi_\delta \in C^\infty\left(\mathbb{R}\right)$ be such that $\chi_\delta\left(s\right) = 1$ if $s \le 1$, $\chi_\delta\left(s\right) = 0$ for $s \ge 1 + \delta$, and $\chi_\delta' \le 0$. Let χ_0 denote the characteristic function of $(-\infty, 1]$. Note that $\lim_{\delta \to 0} \chi_\delta = \chi_0$, pointwise. For $\delta \ge 0$ define

$$E_\delta\left(t\right) = \frac{1}{2}\int\left(|\partial_t u\left(t, y\right)|^2 + \sum_{j=1}^{r}|W_j u\left(t, y\right)|^2\right)\chi_\delta\left(\frac{\tilde{\rho}\left(x_0, y\right)}{t_0 - t}\right) dy,$$

where each W_j is acting in the y variable.

Consider, writing $\langle a, b \rangle$ for $a\bar{b}$, we have for $\delta > 0$,

$$\frac{dE_\delta}{dt}\left(t\right) = \mathrm{Re}\int\left(\langle u_{tt}, u_t \rangle + \sum_{j=1}^{r}\langle W_j u, W_j u_t \rangle\right)\chi_\delta\left(\frac{\tilde{\rho}\left(x_0, y\right)}{t_0 - t}\right) dy$$

$$+ \frac{1}{2}\int\left(|u_t|^2 + \sum_{j=1}^{r}|W_j u|^2\right)\chi_\delta'\left(\frac{\tilde{\rho}\left(x_0, y\right)}{t_0 - t}\right)\frac{\tilde{\rho}\left(x_0, y\right)}{(t_0 - t)^2} dy$$

$$=: I + II.$$

Since $\chi_\delta' \le 0$, II is clearly non-positive. We will show (for $t < t_0$), $|I| \le |II|$, and it will follow that $\frac{dE_\delta}{dt}\left(t\right) \le 0$.

We use the fact that $W_j^* = -W_j + g_j$, where $g_j \in C^\infty(M)$ to see

$$|I| \leq \left| \int \left(\langle u_{tt}, u_t \rangle + \sum_{j=1}^{r} \langle W_j^* W_j u, u_t \rangle \right) \chi_\delta \left(\frac{\tilde{\rho}(x_0, y)}{t_0 - t} \right) dy \right|$$

$$+ \sum_{j=1}^{r} \left| \int \langle W_j u, u_t \rangle \left(W_j \left(\chi_\delta \left(\frac{\tilde{\rho}(x_0, y)}{t_0 - t} \right) \right) \right) dy \right|$$

$$=: III + IV.$$

The integrand of III contains the term

$$\langle u_{tt}, u_t \rangle + \sum_{j=1}^{r} \langle W_j^* W_j u, u_t \rangle = \langle u_{tt} + \mathcal{L}u, u_t \rangle = 0$$

and thus $III = 0$. In the next equation and in what follows, W_j acts in the y variable. To bound IV note, for $j = 1, \ldots, r$ (see Remark 2.5.6)

$$\left| W_j \left(\chi_\delta \left(\frac{\tilde{\rho}(x_0, y)}{t_0 - t} \right) \right) \right| \leq -\chi_\delta' \left(\frac{\tilde{\rho}(x_0, y)}{t_0 - t} \right) \left| \frac{W_j \tilde{\rho}(x_0, y)}{t_0 - t} \right|$$

$$\leq -\chi_\delta' \left(\frac{\tilde{\rho}(x_0, y)}{t_0 - t} \right) r^{-1} |t - t_0|^{-1}$$

$$\leq -\chi_\delta' \left(\frac{\tilde{\rho}(x_0, y)}{t_0 - t} \right) \frac{1}{r} \frac{\tilde{\rho}(x_0, y)}{(t_0 - t)^2},$$

where in the last line we used that $\frac{\tilde{\rho}(x_0, y)}{t_0 - t} \geq 1$ on the support of $\chi_\delta' \left(\frac{\tilde{\rho}(x_0, y)}{t_0 - t} \right)$.

Thus, we have

$$IV \leq \int \sum_{j=1}^{r} r^{-1} |\langle W_j u, u_t \rangle| \left(-\chi_\delta' \left(\frac{\tilde{\rho}(x_0, y)}{t_0 - t} \right) \right) \frac{\tilde{\rho}(x_0, y)}{(t_0 - t)^2} dy$$

$$\leq \frac{1}{2} \int \left(|u_t|^2 + \sum_{j=1}^{r} |W_j u|^2 \right) \left(-\chi_\delta' \left(\frac{\tilde{\rho}(x_0, y)}{t_0 - t} \right) \right) \frac{\tilde{\rho}(x_0, y)}{(t_0 - t)^2} dy$$

$$= |II|.$$

Hence, $|I| \leq |II|$, and $\frac{dE_\delta}{dt}(t) \leq 0$ for $0 \leq t < t_0$. We conclude

$$E_\delta(t) \leq E_\delta(0), \tag{2.52}$$

for $0 \leq t \leq t_0$. Taking the limit of both sides of (2.52) as $\delta \to 0$ and applying the dominated convergence theorem, we see

$$E_0(t) \leq E_0(0),$$

for $0 \leq t \leq t_0$.

Our assumptions on u imply $E_0(0) = 0$ and it follows that $E_0(t) = 0$ for $0 \leq t \leq t_0$, and in particular $\partial_t u = 0$ on Ω. Thus $u(t, y) = 0$ on Ω. $\qquad \square$

Theorem 2.6.14 follows immediately from the next corollary.

COROLLARY 2.6.16. *For $t > 0$,*

$$\text{supp}\left(\cos\left(t\sqrt{\mathcal{L}}\right)\right) \subseteq \{(x,y) \in M \times M \mid \tilde{\rho}(x,y) \le t\}.$$

PROOF. Let $B_{\tilde{\rho}}(x,\delta) = \{y \in M \mid \tilde{\rho}(x,y) < \delta\}$. Fix $x_0, y_0 \in M$ with $x_0 \ne y_0$ and take $t_1 > 0$ such that $\tilde{\rho}(x_0,y_0) > t_1$. Fix $\epsilon > 0$ so small that for all $x \in B_{\tilde{\rho}}(x_0,\epsilon)$ and all $y \in B_{\tilde{\rho}}(y_0,\epsilon)$ we have $\tilde{\rho}(x,y) > t_1 + \epsilon$. We will show for every $\phi \in C_0^\infty(B_{\tilde{\rho}}(x_0,\epsilon))$, $\psi \in C_0^\infty(B_{\tilde{\rho}}(y_0,\epsilon))$ we have

$$\int \psi(z)\left(\cos\left(t_1\sqrt{\mathcal{L}}\right)\phi\right)(z)\,dz = 0, \tag{2.53}$$

and it will follow that $(x_0,y_0) \notin \text{supp}\left(\cos\left(t_1\sqrt{\mathcal{L}}\right)\right)$ and the claim follows.

Define $u(t,x) = \left(\cos\left(t\sqrt{\mathcal{L}}\right)\phi\right)(x)$. We claim $u \in C^\infty(\mathbb{R} \times M)$. We show $\forall R > 0$, $u \in C^\infty([-R,R] \times M)$. Indeed, for every N_1, N_2, we have

$$\mathcal{L}^{N_1}\partial_t^{2N_2}u(t,x) = \cos\left(t\sqrt{\mathcal{L}}\right)(-1)^{N_2}\mathcal{L}^{N_1+N_2}\phi \in L^2([-R,R] \times M).$$

Subellipticity of $\mathcal{L}$ shows $\partial_t^{2N_2}u(t,x) \in L^2([-R,R];C^\infty(M))$, $\forall N_2$–this space is the space of functions which are $L^2([-R,R])$ functions in the t variable, taking values in the Fréchet space $C^\infty(M)$. It follows, by the Sobolev embedding theorem, that $u \in C^\infty([-R,R] \times M)$. As R was arbitrary, $u \in C^\infty(\mathbb{R} \times M)$.

We are in a position to apply Lemma 2.6.15 to u. Note, for all $y \in B_{\tilde{\rho}}(y_0, t_1 + \epsilon)$, $u(0,y) = 0 = \partial_t u(0,y)$. Taking $t_0 = t_1 + \epsilon$ in Lemma 2.6.15 and taking those points in Ω with t component equal to t_1, we see $u(t_1,y) = 0$ for $y \in B_{\tilde{\rho}}(y_0,\epsilon)$. Hence, $\int \psi(y)u(t_1,y)\,dy = 0$, completing the proof. $\qquad\square$

The finite propagation speed of the wave equation will be used by way of the next corollary.

COROLLARY 2.6.17. *Suppose $\widehat{F}$ is the Fourier transform[13] of an even, bounded Borel function with $\text{supp}\left(\widehat{F}\right) \subseteq [-s,s]$. Then,*

$$\text{supp}\left(F\left(\sqrt{\mathcal{L}}\right)\right) \subseteq \{(x,y) \mid \rho(x,y) \le \kappa s\}.$$

PROOF. Because F is even, we have

$$F\left(\sqrt{\mathcal{L}}\right) = \frac{1}{2\pi}\int_{\mathbb{R}}\widehat{F}(t)\cos\left(t\sqrt{\mathcal{L}}\right)\,dt = \frac{1}{2\pi}\int_{-s}^{s}\widehat{F}(t)\cos\left(|t|\sqrt{\mathcal{L}}\right)\,dt.$$

The result now follows from Theorem 2.6.14. $\qquad\square$

---

[13]In this section, we define the Fourier transform with a slightly different normalization than in the rest of the monograph. Namely, $\hat{f}(\xi) = \frac{1}{2\pi}\int f(t)e^{-it\xi}\,dt$, for $f \in \mathcal{S}(\mathbb{R})$. This difference in normalization is not essential, but it does make the statement of some results cleaner (e.g., Corollary 2.6.17).

LEMMA 2.6.18. *Suppose $S_1, S_2 : L^2(M) \to L^2(M)$. Fix an ordered multi-index α. Suppose for some open set $U \subseteq M$, $W_x^\alpha(S_1 S_2)(x, y) \in L_{loc}^1(U \times M)$ and that $\sup_{x \in U} \|W_x^\alpha S_1(x, \cdot)\|_{L^2(M)} < \infty$. Then, for $x \in U$,*

$$\|W_x^\alpha(S_1 S_2)(x, \cdot)\|_{L^2(M)} \le \|S_2\|_{L^2 \to L^2} \|W_x^\alpha S_1(x, \cdot)\|_{L^2(M)}. \qquad (2.54)$$

If instead we have two neighborhoods $U, V \subset M$ and two ordered multi-indices α and β and if

$$W_x^\alpha S_1(x, y), W_y^\beta S_2(x, y) \in L_{loc}^1(M \times M)$$

with

$$\sup_{x \in U} \|W_x^\alpha S_1(x, \cdot)\|_{L^2(M)} + \sup_{y \in V} \|W_y^\beta S_2(\cdot, y)\|_{L^2(M)} < \infty$$

then for $x \in U$, $y \in V$,

$$|W_x^\alpha W_y^\beta(S_1 S_2)(x, y)| \le \|W_x^\alpha S_1(x, \cdot)\|_{L^2(M)} \|W_y^\beta S_2(\cdot, y)\|_{L^2(M)}. \qquad (2.55)$$

PROOF. To prove (2.54) it suffices to note that, if $T = S_1 S_2$ or $T = S_1$, we have

$$\|W_x^\alpha T(x, \cdot)\|_{L^2(M)} = \sup_{\|\phi\|_{L^2} = 1} |(W^\alpha T \phi)(x)|.$$

(2.54) follows easily.

To see (2.55) note

$$W_x^\alpha W_y^\beta(S_1 S_2)(x, y) = \int W_x^\alpha S_1(x, z) W_y^\beta S_2(z, y) \, dz.$$

The result now follows from the Cauchy-Schwartz inequality. $\square$

LEMMA 2.6.19. *Let $\mathcal{B}_1 \subset \mathcal{S}(\mathbb{R})$ and $\mathcal{B}_2 \subset C_b^\infty(\mathbb{R})$ be bounded sets. Suppose, further, for $\phi \in \mathcal{B}_2$, $\phi(\xi) = 0$ for $\xi \in [-\frac{1}{4}, \frac{1}{4}]$. For $s, t > 0$, $\psi \in \mathcal{B}_1$, $\phi \in \mathcal{B}_2$, define a function $F_{s,t}$ by*

$$\widehat{F_{s,t}}(\xi) = \phi\left(\frac{\xi}{t}\right) \frac{1}{s} \psi\left(\frac{\xi}{s}\right).$$

For all $a, b > 0$ there is $C = C(a, b, \mathcal{B}_1, \mathcal{B}_2) > 0$ with

$$\sup_{\substack{s > 0, t > 0, \lambda > 0 \\ \psi \in \mathcal{B}_1, \phi \in \mathcal{B}_2}} (1 + \lambda t)^b \left(1 + \frac{t}{s}\right)^a |F_{s,t}(\lambda)| \le C.$$

PROOF. Notice,

$$2\pi F_{s,t}(\lambda) = \frac{t}{s} \int \phi(\xi) \psi\left(\frac{t}{s}\xi\right) e^{i\lambda t \xi} \, d\xi.$$

We claim that it suffices to prove the result when $b = 0$. Indeed, if $\lambda t \leq 1$ this is immediate. Suppose $\lambda t \geq 1$. Fix $N \in \mathbb{N}$ with $N \geq b$. We have, integrating by parts,

$$2\pi F_{s,t}(\lambda) = (\lambda t)^{-N}(-1)^N \frac{t}{s} \int \partial_\xi^N \left(\phi(\xi) \psi\left(\frac{t}{s}\xi\right) \right) e^{i\lambda t \xi}\, d\xi$$

$$= \sum_{N_1+N_2=N} (-\lambda t)^{-N} \left(\frac{t}{s}\right)^{1+N_2} \binom{N}{N_1} \int \left(\partial_\xi^{N_1}\phi\right)(\xi) \left(\partial_\xi^{N_2}\psi\right)\left(\frac{t}{s}\xi\right) e^{i\lambda t \xi}\, d\xi.$$

Thus, the result for a and b is implied by the result when $b = 0$ and a is replaced by $a + N$, where $N \geq b$, $N \in \mathbb{N}$ (this uses that $\partial_\xi^{N_1}\phi$ and $\partial_\xi^{N_2}\psi$ are of the same form as ϕ and ψ, respectively).

We now assume $b = 0$. We have, for any $m \geq a + 2$,

$$\left| \frac{t}{s} \int \phi(\xi)\psi\left(\frac{t}{s}\xi\right) e^{i\lambda t \xi}\, d\xi \right| \lesssim \frac{t}{s} \int_{|\xi|\geq \frac{1}{4}} \left(1 + \left|\frac{t}{s}\xi\right| \right)^{-m}\, d\xi$$

$$\lesssim \frac{t}{s}\left(1+\frac{t}{s}\right)^{-a} \int \left(1+\left|\frac{t}{s}\xi\right|\right)^{-m+a}\, d\xi$$

$$\lesssim \left(1+\frac{t}{s}\right)^{-a} \int (1+|\xi|)^{-m+a}\, d\xi$$

$$\lesssim \left(1+\frac{t}{s}\right)^{-a}.$$

This completes the proof. $\square$

PROOF OF LEMMA 2.6.12. Let $m \in \mathcal{B}$ and $s > 0$. Fix $x_0, y_0 \in M$ and ordered multi-indices α and β. We wish to bound

$$\left. (sW_x)^\alpha (sW_y)^\beta m(s^2\mathcal{L})(x,y) \right|_{\substack{x=x_0 \\ y=y_0}}.$$

We begin with the case $\rho(x_0, y_0) \geq s$. Define $\psi(\lambda) = m(\lambda^2)$. It is easy to see $\{\psi \mid m \in \mathcal{B}\} \subset \mathcal{S}(\mathbb{R})$ is a bounded set. Let $\psi_s(\lambda) = \psi(s\lambda)$ so that $\psi_s\left(\sqrt{\mathcal{L}}\right) = m(s^2\mathcal{L})$, and $\widehat{\psi_s}(\xi) = \frac{1}{s}\widehat{\psi}\left(\frac{\xi}{s}\right)$. Let $\phi \in C_b^\infty(\mathbb{R})$ be such that

$$\phi(\xi) = \begin{cases} 0 & \text{if } |\xi| \leq \frac{1}{4}, \\ 1 & \text{if } |\xi| \geq \frac{1}{2}. \end{cases}$$

For $t > 0$ define a function $F_{s,t}$ by

$$\widehat{F_{s,t}}(\xi) = \phi\left(\frac{\xi}{t}\right)\frac{1}{s}\widehat{\psi}\left(\frac{\xi}{s}\right).$$

Lemma 2.6.19 shows, for all $a, b > 0$,

$$(1 + \lambda t)^b \left(1 + \frac{t}{s}\right)^a |F_{s,t}(\lambda)| \lesssim 1, \tag{2.56}$$

where the implicit constant is independent of any relevant parameters.

Set $t = \frac{\rho(x_0, y_0)}{\kappa}$, where κ is as in Theorem 2.6.14, so that $\frac{t}{s} \gtrsim 1$. By Corollary 2.6.17 and the definition of $F_{s,t}$, we have

$$W_x^\alpha W_y^\beta m \left(s^2 \mathcal{L}\right) (x, y) \Big|_{\substack{x=x_0 \\ y=y_0}} = W_x^\alpha W_y^\beta F_{s,t} \left(\sqrt{\mathcal{L}}\right) (x, y) \Big|_{\substack{x=x_0 \\ y=y_0}}.$$

Let $J_{s,t}(\lambda)$ be a measurable function with $J_{s,t}(\lambda)^2 = F_{s,t}(\lambda)$. (2.56) can be rephrased as, for all $a, b > 0$,

$$\left(1 + \lambda^2 t^2\right)^b \left(1 + \frac{t}{s}\right)^a |J_{s,t}(\lambda)| \lesssim 1. \tag{2.57}$$

From the spectral theorem, this implies, for every $a, b > 0$,

$$\left\| \left(1 + \frac{t}{s}\right)^a \left(1 + t^2 \mathcal{L}\right)^b J_{s,t} \left(\sqrt{\mathcal{L}}\right) \right\|_{L^2 \to L^2} \lesssim 1.$$

Using

$$J_{s,t} \left(\sqrt{\mathcal{L}}\right) = \left(1 + \frac{t}{s}\right)^{-a} \left(1 + t^2 \mathcal{L}\right)^{-b} \left(1 + \frac{t}{s}\right)^a \left(1 + t^2 \mathcal{L}\right)^b J_{s,t} \left(\sqrt{\mathcal{L}}\right),$$

and by using Lemma 2.6.18 we have

$$\left\| (sW_x)^\alpha J_{s,t} \left(\sqrt{\mathcal{L}}\right) (x, \cdot) \Big|_{x=x_0} \right\|_{L^2(M)}$$

$$\lesssim \left(1 + \frac{t}{s}\right)^{-a} s^{|\alpha|} \left\| W_x^\alpha \left(1 + t^2 \mathcal{L}\right)^{-b} (x, \cdot) \Big|_{x=x_0} \right\|_{L^2(M)}$$

$$\lesssim \left(1 + \frac{\rho(x_0, y_0)}{s}\right)^{-a} \frac{s^{|\alpha|} \rho(x_0, y_0)^{-|\alpha|}}{\text{Vol} \left(B_{(X,d)}(x_0, \rho(x_0, y_0))\right)^{\frac{1}{2}}}$$

$$\leq \left(1 + \frac{\rho(x_0, y_0)}{s}\right)^{-a} \text{Vol} \left(B_{(X,d)}(x_0, \rho(x_0, y_0))\right)^{-\frac{1}{2}},$$

where in the second-to-last line we have used Lemma 2.6.13 (by taking b sufficiently large), and that $t = \frac{\rho(x_0, y_0)}{\kappa}$, and in the last line we have used $s \leq \rho(x_0, y_0)$. Similarly, we have

$$\left\| (sW_y)^\beta J_{s,t} \left(\sqrt{\mathcal{L}}\right) (\cdot, y) \Big|_{y=y_0} \right\|_{L^2(M)}$$

$$\lesssim \left(1 + \frac{\rho(x_0, y_0)}{s}\right)^{-a} \text{Vol} \left(B_{(X,d)}(y_0, \rho(x_0, y_0))\right)^{-\frac{1}{2}}.$$

Using $F_{s,t}\left(\sqrt{\mathcal{L}}\right) = J_{s,t}\left(\sqrt{\mathcal{L}}\right) J_{s,t}\left(\sqrt{\mathcal{L}}\right)$ and applying Lemma 2.6.18, we have for all $a > 0$,

$$
\left| (sW_x)^\alpha (sW_y)^\beta F_{s,t}\left(\sqrt{\mathcal{L}}\right) (x,y) \right|_{\substack{x=x_0 \\ y=y_0}}
$$
$$
\lesssim \left\| (sW_x)^\alpha J_{s,t}\left(\sqrt{\mathcal{L}}\right)(x,\cdot) \right|_{x=x_0} \right\|_{L^2(M)} \left\| (sW_y)^\beta J_{s,t}\left(\sqrt{\mathcal{L}}\right)(\cdot,y) \right|_{y=y_0} \right\|_{L^2(M)}
$$
$$
\lesssim \mathrm{Vol}\left(B_{(X,d)}\left(x_0, \rho\left(x_0, y_0\right)\right)\right)^{-\frac{1}{2}} \mathrm{Vol}\left(B_{(X,d)}\left(y_0, \rho\left(x_0, y_0\right)\right)\right)^{-\frac{1}{2}}
$$
$$
\times \left(1 + s^{-1}\rho\left(x_0, y_0\right)\right)^{-a}.
$$

We have $\mathrm{Vol}\left(B_{(X,d)}\left(x_0, \rho\left(x_0, y_0\right)\right)\right) \approx \mathrm{Vol}\left(B_{(X,d)}\left(y_0, \rho\left(x_0, y_0\right)\right)\right)$ (this follows just as in Lemma 2.0.18), and since $s \leq \rho\left(x_0, y_0\right)$, we also have

$$
\mathrm{Vol}\left(B_{(X,d)}\left(x_0, s + \rho\left(x_0, y_0\right)\right)\right) \approx \mathrm{Vol}\left(B_{(X,d)}\left(x_0, \rho\left(x_0, y_0\right)\right)\right).
$$

Putting these together, we conclude

$$
\left| (sW_x)^\alpha (sW_y)^\beta m\left(s^2\mathcal{L}\right)(x,y) \right|_{\substack{x=x_0 \\ y=y_0}}
$$
$$
= \left| (sW_x)^\alpha (sW_y)^\beta F_{s,t}\left(\sqrt{\mathcal{L}}\right)(x,y) \right|_{\substack{x=x_0 \\ y=y_0}}
$$
$$
\lesssim \frac{\left(1 + s^{-1}\rho\left(x_0, y_0\right)\right)^{-a}}{\mathrm{Vol}\left(B_{(X,d)}\left(x_0, s + \rho\left(x_0, y_0\right)\right)\right)}.
$$

Since $a > 0$ was arbitrary, this is the desired bound for a bounded set of pre-elementary operators.

We now turn to the case $\rho\left(x_0, y_0\right) \leq s$. Let $j_s\left(\lambda\right)$ be a measurable function such that $j_s\left(\lambda\right)^2 = m\left(s^2\lambda\right)$. Note, by rapid decrease of m, we have for every N, $\sup_{\lambda > 0}\left(1 + s^2\lambda\right)^N |j_s\left(\lambda\right)| \lesssim 1$, and as a consequence $\left\|\left(1 + s^2\mathcal{L}\right)^N j_s\left(\mathcal{L}\right)\right\|_{L^2} \lesssim 1$. Using this, the fact that $m\left(s^2\mathcal{L}\right) = j_s\left(\mathcal{L}\right) j_s\left(\mathcal{L}\right)$, and Lemmas 2.6.18 and 2.6.13 we have

$$\left| (sW_x)^\alpha (sW_y)^\beta \, m \left(s^2 \mathcal{L} \right) (x, y) \right|_{\substack{x=x_0 \\ y=y_0}}$$

$$\leq \left\| (sW_x)^\alpha j_s \left(\mathcal{L} \right) (x, \cdot) \right|_{x=x_0} \right\|_{L^2(M)} \left\| (sW_y)^\beta j_s \left(\mathcal{L} \right) (\cdot, y) \right|_{y=y_0} \right\|_{L^2(M)}$$

$$\lesssim s^{|\alpha|+|\beta|} \left\| W_x^\alpha \left(1 + s^2 \mathcal{L} \right)^{-N} (x, \cdot) \right|_{x=x_0} \right\|_{L^2(M)} \left\| W_y^\beta \left(1 + s^2 \mathcal{L} \right)^{-N} (\cdot, y) \right|_{y=y_0} \right\|_{L^2(M)}$$

$$\lesssim \mathrm{Vol} \left(B_{(X,d)} (x_0, s) \right)^{-\frac{1}{2}} \mathrm{Vol} \left(B_{(X,d)} (y_0, s) \right)^{-\frac{1}{2}}$$

$$\approx \mathrm{Vol} \left(B_{(X,d)} (x_0, s) \right)^{-1}$$

$$\approx \frac{\left(1 + s^{-1} \rho (x_0, y_0) \right)^{-m}}{\mathrm{Vol} \left(B_{(X,d)} (x_0, s + \rho (x_0, y_0)) \right)},$$

for any m, where in the last two estimates we have used $s \geq \rho (x_0, y_0)$. This completes the proof. $\qquad\square$

2.7   THE ALGEBRA OF SINGULAR INTEGRALS

In this section we prove Theorem 2.0.29, thereby characterizing the algebra of Calderón-Zygmund singular integrals. A key lemma is the following.

LEMMA 2.7.1. *There is a bounded set of elementary operators* $\left\{ (E_j, 2^{-j}) \mid j \in \mathbb{N} \right\}$ *such that* $I = \sum_{j \in \mathbb{N}} E_j$. *See Lemma 2.0.28 for details on this convergence of this infinite sum.*

PROOF. The constant function $m (\lambda) = 1$ satisfies the assumptions of Theorem 2.6.6 with $t = 0$. Since $m (\mathcal{L}) = I$, the conclusion of Theorem 2.6.6 in this case is exactly the statement of the lemma. $\qquad\square$

PROOF OF (ii)$\Rightarrow$(iii) OF THEOREM 2.0.29. Let $T : C^\infty (M) \to C^\infty (M)$ be as in (ii) (where t is any element of $\mathbb{R}$), and let $I = \sum_{j \in \mathbb{N}} E_j$ where $\left\{ (E_j, 2^{-j}) \mid j \in \mathbb{N} \right\}$ is a bounded set of elementary operators as in Lemma 2.7.1. Let $\widetilde{E}_j = 2^{-jt} T E_j$, so that $\left\{ \left(\widetilde{E}_j, 2^{-j} \right) \mid j \in \mathbb{N} \right\}$ is a bounded set of elementary operators by our assumption. We have $T = TI = \sum_{j \in \mathbb{N}} T E_j = \sum_{j \in \mathbb{N}} 2^{jt} \widetilde{E}_j$, completing the proof of (iii). $\qquad\square$

We now turn to (iii)$\Rightarrow$(ii). For this, we prove the following result.

PROPOSITION 2.7.2. *Let* $\mathcal{E}$ *be a bounded set of elementary operators. Then, for every* N, *the set*

$$\left\{ \left(2^{N|j_1 - j_2|} E_1 E_2, 2^{-j_1} \right), \left(2^{N|j_1 - j_2|} E_1 E_2, 2^{-j_2} \right) \mid (E_1, 2^{-j_1}), (E_2, 2^{-j_2}) \in \mathcal{E} \right\}$$

is a bounded set of elementary operators.

To prove Proposition 2.7.2, we need several preliminary lemmas.

LEMMA 2.7.3. *Fix a constant $D > 0$. $\forall m > Q_1$, $\forall j \in \mathbb{N}$,*

$$\sum_{-\infty < k \leq j} 2^{-m(j-k)} \frac{\chi_{\{\rho(x,z) < D2^{-k}\}}}{\text{Vol}\left(B_{(X,d)}\left(x, 2^{-k}\right)\right)}$$

$$\approx \frac{\left(1 + 2^j \rho\left(x, z\right)\right)^{-m}}{\text{Vol}\left(B_{(X,d)}\left(x, 2^{-j} + \rho\left(x, z\right)\right)\right)}, \tag{2.58}$$

where the implicit constants depend on D and m, but not on $j \in \mathbb{N}$.

PROOF. Fix $x, z \in M$. The sum on the left-hand side of (2.58) is bounded termwise by a geometric sum, and therefore is comparable to its largest nonzero term. If $\rho\left(x, z\right) < D2^{-j}$, then the largest nonzero term is when $k = j$, and both sides are comparable to

$$\text{Vol}\left(B_{(X,d)}\left(x, 2^{-j}\right)\right)^{-1}.$$

Otherwise, the largest nonzero term is when $\rho\left(x, z\right) < D2^{-k}$ and $\rho\left(x, z\right) \geq D2^{-k-1}$ for some $k < j$. In this case, $\rho\left(x, z\right) \approx 2^{-k}$, and both sides of (2.58) are comparable to

$$2^{-m(j-k)} \text{Vol}\left(B_{(X,d)}\left(x, 2^{-k}\right)\right)^{-1}.$$

$\square$

LEMMA 2.7.4. *Let $j, k \in \mathbb{R}$. Then,*

$$\int \frac{\chi_{\{\rho(x,y) < 2^{-j}\}}}{\text{Vol}\left(B_{(X,d)}\left(x, 2^{-j}\right)\right)} \frac{\chi_{\{\rho(y,z) < 2^{-k}\}}}{\text{Vol}\left(B_{(X,d)}\left(y, 2^{-k}\right)\right)} dy$$

$$\lesssim \frac{\chi_{\{\rho(x,z) < 2^{-j} + 2^{-k}\}}}{\text{Vol}\left(B_{(X,d)}\left(x, 2^{-k} + 2^{-j}\right)\right)}$$

$$\lesssim \frac{\chi_{\{\rho(x,z) < 2 \cdot 2^{-j \wedge k}\}}}{\text{Vol}\left(B_{(X,d)}\left(x, 2^{-j \wedge k}\right)\right)}.$$

PROOF. The second $\lesssim$ is clear, so we concentrate only on the first. It is clear that the integral is only nonzero for $\rho\left(x, z\right) < 2^{-j} + 2^{-k}$. Thus, we fix $x, z \in M$ with $\rho\left(x, z\right) < 2^{-j} + 2^{-k}$. There are two possibilities: either $j \leq k$ or $k \leq j$. We treat the case $j \leq k$, the other case being similar. When $j \leq k$, we have

$$\int \frac{\chi_{\{\rho(x,y) < 2^{-j}\}}}{\text{Vol}\left(B_{(X,d)}\left(x, 2^{-j}\right)\right)} \frac{\chi_{\{\rho(y,z) < 2^{-k}\}}}{\text{Vol}\left(B_{(X,d)}\left(y, 2^{-k}\right)\right)} dy$$

$$\lesssim \text{Vol}\left(B_{(X,d)}\left(x, 2^{-j}\right)\right)^{-1} \int \frac{\chi_{\{\rho(y,z) < 2^{-k}\}}}{\text{Vol}\left(B_{(X,d)}\left(z, 2^{-k}\right)\right)} dy$$

$$\lesssim \text{Vol}\left(B_{(X,d)}\left(x, 2^{-j}\right)\right)^{-1}$$

$$\approx \text{Vol}\left(B_{(X,d)}\left(x, 2^{-j} + 2^{-k}\right)\right)^{-1},$$

where in the second line we used that for $\rho\left(y, z\right) < 2^{-k}$, $\text{Vol}\left(B_{(X,d)}\left(z, 2^{-k}\right)\right) \approx \text{Vol}\left(B_{(X,d)}\left(y, 2^{-k}\right)\right)$, and in the last line we have used $j \leq k$. This completes the proof. $\square$

LEMMA 2.7.5. *For every $m > Q_1$, and $\forall j_1, j_2 \in [0, \infty)$,*

$$\int \frac{\left(1 + 2^{j_1} \rho(x, y)\right)^{-m}}{\operatorname{Vol}\left(B_{(X,d)}\left(x, 2^{-j_1} + \rho(x, y)\right)\right)} \frac{\left(1 + 2^{j_2} \rho(y, z)\right)^{-m}}{\operatorname{Vol}\left(B_{(X,d)}\left(y, 2^{-j_2} + \rho(y, z)\right)\right)} \, dy$$

$$\lesssim \frac{\left(1 + 2^{j_1 \wedge j_2} \rho(x, z)\right)^{-m}}{\operatorname{Vol}\left(B_{(X,d)}\left(x, 2^{-j_1 \wedge j_2} + \rho(x, z)\right)\right)},$$

where the implicit constant depends on m, but not on $j_1, j_1 \in [0, \infty)$.

PROOF. We apply Lemmas 2.7.3 and 2.7.4 to see

$$\int \frac{\left(1 + 2^{j_1} \rho(x, y)\right)^{-m}}{\operatorname{Vol}\left(B_{(X,d)}\left(x, 2^{-j_1} + \rho(x, y)\right)\right)} \frac{\left(1 + 2^{j_2} \rho(y, z)\right)^{-m}}{\operatorname{Vol}\left(B_{(X,d)}\left(y, 2^{-j_2} + \rho(y, z)\right)\right)} \, dy$$

$$\lesssim \sum_{\substack{-\infty < l_1 \leq j_1 \\ -\infty < l_2 \leq j_2}} 2^{-m(j_1 - l_2) - m(j_2 - l_2)} \int \frac{\chi_{\{\rho(x, y) < 2^{-l_1}\}}}{\operatorname{Vol}\left(B_{(X,d)}\left(x, 2^{-l_1}\right)\right)} \frac{\chi_{\{\rho(y, z) < 2^{-l_2}\}}}{\operatorname{Vol}\left(B_{(X,d)}\left(y, 2^{-l_2}\right)\right)} \, dy$$

$$\lesssim \sum_{\substack{-\infty < l_1 \leq j_1 \\ -\infty < l_2 \leq j_2}} 2^{-m(j_1 - l_1) - m(j_2 - l_2)} \frac{\chi_{\{\rho(x, z) < 2^{2 - l_1 \wedge l_2}\}}}{\operatorname{Vol}\left(B_{(X,d)}\left(x, 2^{-l_1 \wedge l_2}\right)\right)}$$

$$\lesssim \sum_{-\infty < l \leq j_1 \wedge j_2} 2^{-m(j_1 \wedge j_2 - l)} \frac{\chi_{\{\rho(x, z) < 2^{2 - l}\}}}{\operatorname{Vol}\left(B_{(X,d)}\left(x, 2^{-l}\right)\right)},$$

where in the last line we have set $l = l_1 \wedge l_2$ and summed over $l_1 \vee l_2$. The result now follows from one final application of Lemma 2.7.3. □

LEMMA 2.7.6. *Let $\mathcal{E}$ be a bounded set of pre-elementary operators. Then, $\forall m$, $\exists C$, $\forall \left(F_1, 2^{-j_1}\right), \left(F_2, 2^{-j_2}\right) \in \mathcal{E}$,*

$$|F_1 F_2(x, z)| \leq C \frac{\left(1 + 2^{j_1 \wedge j_2} \rho(x, z)\right)^{-m}}{\operatorname{Vol}\left(B_{(X,d)}\left(x, 2^{-j_1 \wedge j_2} + \rho(x, z)\right)\right)}.$$

PROOF. This is an immediate consequence of Lemma 2.7.5. □

LEMMA 2.7.7. *$\forall m$, $\exists N$, $\forall j_1, j_2 \in [0, \infty)$, $\forall x, y \in M$,*

$$2^{-N|j_1 - j_2|} \frac{\left(1 + 2^{j_1} \rho(x, y)\right)^{-m}}{\operatorname{Vol}\left(B_{(X,d)}\left(x, 2^{-j_1} + \rho(x, y)\right)\right)} \leq \frac{\left(1 + 2^{j_2} \rho(x, y)\right)^{-m}}{\operatorname{Vol}\left(B_{(X,d)}\left(x, 2^{-j_2} + \rho(x, y)\right)\right)}.$$

PROOF. It is clear that

$$2^{-m|j_1 - j_2|} \left(1 + 2^{j_1} \rho(x, y)\right)^{-m} \leq \left(1 + 2^{j_2} \rho(x, y)\right)^{-m}. \tag{2.59}$$

By (2.31), we have,

$$\operatorname{Vol}\left(B_{(X,d)}\left(x, 2^{-j_2} + \rho(x, y)\right)\right) \leq \operatorname{Vol}\left(B_{(X,d)}\left(x, 2^{|j_1 - j_2|}\left(2^{-j_1} + \rho(x, y)\right)\right)\right)$$

$$\leq 2^{Q_2|j_1 - j_2|} \operatorname{Vol}\left(B_{(X,d)}\left(x, 2^{-j_1} + \rho(x, y)\right)\right).$$

Combining this with (2.59) completes the proof. □

LEMMA 2.7.8. *Let $\mathcal{E}$ be a bounded set of pre-elementary operators. Then, $\forall m$, α, and β, $\exists N, C$, such that $\forall \left(F_1, 2^{-j_1}\right), \left(F_2, 2^{-j_2}\right) \in \mathcal{E}$, and letting $k = j_1$ or $k = j_2$, we have*

$$2^{-N|j_1 - j_2|} \left| \left(2^{-k} W_x\right)^{\alpha} \left(2^{-k} W_z\right)^{\beta} [F_1 F_2](x, z) \right|$$

$$\leq C \frac{\left(1 + 2^k \rho(x, z)\right)^{-m}}{\mathrm{Vol}\left(B_{(X,d)}\left(x, 2^{-k} + \rho(x, z)\right)\right)}.$$

PROOF. The result with $k = j_2$ follows from the result with $k = j_1$ and taking adjoints. We, therefore, prove only the result with $k = j_1$. By letting $N_1 = N - |\beta|$, we wish to show, for some N_1,

$$2^{-N_1|j_1 - j_2|} \left| \left(2^{-j_1} W_x\right)^{\alpha} \left(2^{-j_2} W_z\right)^{\beta} [F_1 F_2](x, z) \right|$$

$$\lesssim \frac{\left(1 + 2^{j_1} \rho(x, z)\right)^{-m}}{\mathrm{Vol}\left(B_{(X,d)}\left(x, 2^{-j_1} + \rho(x, z)\right)\right)}. \tag{2.60}$$

By the definition of bounded sets of pre-elementary operators, we have

$$\left\{ \left(\left(2^{-j_1} W_x\right)^{\alpha} F_1(x, z), 2^{-j_1} \right), \left(\left(2^{-j_2} W_z\right)^{\beta} F_2(x, z), 2^{-j_2} \right) \right.$$

$$\left. \left| \left(F_1, 2^{-j_1}\right), \left(F_2, 2^{-j_2}\right) \in \mathcal{E} \right\} \right.$$

is a bounded set of pre-elementary operators. Thus, it suffices to prove (2.60) in the case when $\alpha = 0 = \beta$. Lemma 2.7.6 shows

$$|[F_1 F_2](x, z)| \lesssim \frac{\left(1 + 2^{j_1 \wedge j_2} \rho(x, z)\right)^{-m}}{\mathrm{Vol}\left(B_{(X,d)}\left(x, 2^{-j_1 \wedge j_2} + \rho(x, z)\right)\right)}.$$

(2.60) (with $\alpha = 0 = \beta$) now follows from Lemma 2.7.7, completing the proof. $\square$

LEMMA 2.7.9. *Let $\mathcal{E}$ be a bounded set of elementary operators. Then, for every N, the set*

$$\left\{ \left(2^{N|j_1 - j_2|} E_1 E_2, 2^{-j_1}\right), \left(2^{N|j_1 - j_2|} E_1 E_2, 2^{-j_2}\right) \mid \left(E_1, 2^{-j_1}\right), \left(E_2, 2^{-j_2}\right) \in \mathcal{E} \right\}$$

is a bounded set of pre-elementary operators.

PROOF. Fix N. We prove that

$$\left\{ \left(2^{N|j_1 - j_2|} E_1 E_2, 2^{-j_1}\right) \mid \left(E_1, 2^{-j_1}\right), \left(E_2, 2^{-j_2}\right) \in \mathcal{E} \right\}$$

is a bounded set of pre-elementary operators. The result with j_1 replaced with j_2 then follows by taking adjoints (see (b) of Proposition 2.0.24).

Fix m, N_0, α, and β. We wish to show

$$2^{N_0|j_1-j_2|}\left|\left(2^{-j_1}W_x\right)^\alpha\left(2^{-j_1}W_z\right)^\beta[E_1E_2](x,z)\right|$$

$$\lesssim \frac{\left(1+2^{j_1}\rho(x,y)\right)^{-m}}{\mathrm{Vol}\left(B_{(X,d)}\left(x,2^{-j_1}+\rho(x,z)\right)\right)}. \tag{2.61}$$

Let $N = N(m,\alpha,\beta)$ be as in Lemma 2.7.7. We separate into two cases. The first case is $j_1 \leq j_2$. We use (d) of Proposition 2.0.24 to write

$$E_2 = \sum_{|\alpha|\leq N+N_0} 2^{j_2(|\alpha|-N-N_0)}\left(2^{-j_2}W\right)^\alpha E_{2,\alpha},$$

where $\left\{\left(E_{2,\alpha},2^{-j_2}\right)\mid\left(E_2,2^{-j_2}\right)\in\mathcal{E},|\alpha|\leq N+N_0\right\}$ is a bounded set of elementary operators. Thus,

$$E_1E_2 = \sum_{|\alpha|\leq N+N_0} 2^{-j_2(N+N_0-|\alpha|)}2^{-|\alpha||j_1-j_2|}\left(E_1\left(2^{-j_1}W\right)^\alpha\right)E_{2,\alpha}.$$

Let $E_{1,\alpha} = 2^{-j_1(N+N_0-|\alpha|)}E_1\left(2^{-j_1}W\right)^\alpha$, so that (by Proposition 2.0.24)

$$\left\{\left(E_{1,\alpha},2^{-j_1}\right)\mid\left(E_1,2^{-j_1}\right)\in\mathcal{E},|\alpha|\leq N+N_0\right\}$$

is a bounded set of elementary operators, and we have

$$E_1E_2 = \sum_{|\alpha|\leq N+N_0} 2^{-(N+N_0)|j_1-j_2|}E_{1,\alpha}E_{2,\alpha}.$$

Plugging this into (2.61), we see that it suffices to prove (2.61) with N_0 replaced by $-N$. Now the result follows from Lemma 2.7.8.

When $j_1 \geq j_2$, the proof is similar except we now apply (d) of Proposition 2.0.24 to E_1 to write

$$E_1 = \sum_{|\alpha|\leq N+N_0} 2^{j_1(|\alpha|-N-N_0)}E_{1,\alpha}\left(2^{-j_1}W\right)^\alpha.$$

From here, (2.61) follows in a similar manner, and we leave the details to the reader. $\square$

PROOF OF PROPOSITION 2.7.2. We prove that

$$\mathcal{E}_N := \left\{\left(2^{N|j_1-j_2|}E_1E_2,2^{-j_1}\right)\mid\left(E_1,2^{-j_1}\right),\left(E_2,2^{-j_2}\right)\in\mathcal{E}\right\}$$

is a bounded set of elementary operators. The corresponding result with j_1 replaced by j_2 follows by taking adjoints. We already know that $\mathcal{E}_N$ is a bounded set of pre-elementary operators (by Lemma 2.7.9). The result will follow once we show for

$\left(E_1, 2^{-j_1}\right), \left(E_2, 2^{-j_2}\right) \in \mathcal{E}$ we have $2^{N|j_1-j_2|} E_1 E_2$ is a sum of derivatives of operators of the same form, as in the definition of elementary operators. But we have, using Proposition 2.0.24,

$$
\begin{aligned}
& 2^{N|j_1-j_2|} E_1 E_2 \\
&= \sum_{|\alpha|,|\beta|\leq 1} 2^{N|j_1-j_2|-(1-|\alpha|)j_1-(1-|\beta|)j_2} \left(2^{-j_1}W\right)^{\alpha} E_{1,\alpha} E_{2,\beta} \left(2^{-j_2}W\right)^{\beta},
\end{aligned}
$$

where $\left\{\left(E_{1,\alpha}, 2^{-j_1}\right), \left(E_{2,\beta}, 2^{-j_2}\right) \mid \left(E_1, 2^{-j_1}\right), \left(E_2, 2^{-j_2}\right) \in \mathcal{E}\right\}$ is a bounded set of elementary operators. And therefore,

$$
\begin{aligned}
& 2^{N|j_1-j_2|} E_1 E_2 \\
&= \sum_{|\alpha|,|\beta|\leq 1} 2^{-(2-|\alpha|-|\beta|)j_1} \left(2^{-j_1}W\right)^{\alpha} \left[2^{N|j_1-j_2|+(j_1-j_2)} E_{1,\alpha} E_{2,\alpha}\right] \left(2^{-j_1}W\right)^{\beta}.
\end{aligned}
$$

This completes the proof, since $2^{N|j_1-j_2|+(j_1-j_2)} E_{1,\alpha} E_{2,\alpha}$ is of the same form as $2^{N|j_1-j_2|} E_1 E_2$ (with a different choice of N). $\qquad\square$

PROOF OF (iii)$\Rightarrow$(ii) OF THEOREM 2.0.29. Fix $t \in \mathbb{R}$ and let $T = \sum_{j\in\mathbb{N}} 2^{jt} E_j$, where $\left\{\left(E_j, 2^{-j}\right) \mid j \in \mathbb{N}\right\}$ is a bounded set of elementary operators, as in (iii). Let $\mathcal{E}$ be a bounded set of elementary operators. For $\left(E, 2^{-k}\right) \in \mathcal{E}$ and $j \in \mathbb{N}$ define $\widetilde{E}_j = 2^{t(j-k)+|j-k|} E_j E$, so that $\left\{\left(\widetilde{E}_j, 2^{-k}\right) \mid j \in \mathbb{N}, \left(E, 2^{-k}\right) \in \mathcal{E}\right\}$ is a bounded set of elementary operators, by Proposition 2.7.2. Set $\widetilde{E} = \sum_{j\in\mathbb{N}} 2^{-|j-k|} \widetilde{E}_j$ so that $\left\{\left(\widetilde{E}, 2^{-k}\right) \mid \left(E, 2^{-k}\right) \in \mathcal{E}\right\}$ is a bounded set of elementary operators. We have

$$
2^{-kt} T E = \sum_{j\in\mathbb{N}} 2^{(j-k)t} T E_j = \sum_{j\in\mathbb{N}} 2^{-|j-k|} \widetilde{E}_j = \widetilde{E},
$$

which completes the proof. $\qquad\square$

Next, we turn to (i)$\Rightarrow$(iii). In what follows, when we refer to a "Calderón-Zygmund operator of order $t \in (-Q_1, \infty)$" we are referring to Definition 2.0.16.

LEMMA 2.7.10. Suppose $\mathcal{E}_1$ is a bounded set of elementary operators and $\mathcal{E}_2$ is a bounded set of pre-elementary operators. Then, for every N, the set

$$
\left\{ \left(2^{N|j_1-j_2|} E_1 E_2, 2^{-j_2}\right), \left(2^{N|j_1-j_2|} E_2 E_1, 2^{-j_2}\right) \right.
$$
$$
\left. \quad \middle| \left(E_1, 2^{-j_1}\right) \in \mathcal{E}_1, \left(E_2, 2^{-j_2}\right) \in \mathcal{E}_2, j_1 \geq j_2 \right\}
$$

is a bounded set of pre-elementary operators.

PROOF. This follows just as in the proof of Lemma 2.7.9 using that $j_1 \geq j_2$, and we leave the details to the reader. $\qquad\square$

LEMMA 2.7.11. *Fix $t \in \mathbb{R}$. Suppose $T : C^\infty(M) \to C^\infty(M)$ is such that for every bounded set of elementary operators $\mathcal{E}$,*

$$\left\{ \left(2^{-jt}ET, 2^{-j} \right), \left(2^{-jt}TE, 2^{-j} \right) \mid (E, 2^{-j}) \in \mathcal{E} \right\},$$

is a bounded set of pre-elementary operators. Then there is a bounded set of elementary operators $\left\{ (E_j, 2^{-j}) \mid j \in \mathbb{N} \right\}$ with $T = \sum_{j \in \mathbb{N}} 2^{jt} E_j$.

PROOF. Let $\mathcal{E}$ be a bounded set of elementary operators. We will show that for every N, the set

$$\mathcal{E}_N := \left\{ \left(2^{N|j_1 - j_2|} 2^{-j_1 \wedge j_2 t} E_1 T E_2, 2^{-j_1 \wedge j_2} \right) \mid (E_1, 2^{-j_1}), (E_2, 2^{-j_2}) \in \mathcal{E} \right\}$$

is a bounded set of elementary operators. First we see how this completes the proof. Indeed, we use Lemma 2.7.1 to write $I = \sum_{j \in \mathbb{N}} E_j$, where $\left\{ (E_j, 2^{-j}) \mid j \in \mathbb{N} \right\}$ is a bounded set of elementary operators. Fix $l \in \mathbb{N}$. Let

$$\widetilde{E}_l := \sum_{\substack{j_1, j_2 \in \mathbb{N} \\ j_1 \wedge j_2 = l}} 2^{-lt} E_{j_1} T E_{j_2} = \sum_{\substack{j_1, j_2 \in \mathbb{N} \\ j_1 \wedge j_2 = l}} 2^{-|j_1 - l| - |j_2 - l|} \left(2^{-j_1 \wedge j_2 t} 2^{|j_1 - j_2|} E_{j_1} T E_{j_2} \right).$$

From the fact (which we have yet to prove) that $\mathcal{E}_1$ is a bounded set of elementary operators, we have that $\left\{ \left(\widetilde{E}_l, 2^{-l} \right) \mid l \in \mathbb{N} \right\}$ is a bounded set of elementary operators. We also have

$$T = ITI = \sum_{j_1, j_2 \in \mathbb{N}} E_{j_1} T E_{j_2} = \sum_{l \in \mathbb{N}} 2^{lt} \widetilde{E}_l,$$

which is the claim of the lemma.

We now turn to showing that $\mathcal{E}_N$ is a bounded set of elementary operators. First, we show that it is a bounded set of pre-elementary operators. We separate $\mathcal{E}_N$ into two sets The first, when $j_1 \geq j_2$. We have, by hypothesis, $\left\{ \left(2^{-j_2 t} T E_2, 2^{-j_2} \right) \mid (E_2, 2^{-j_2}) \in \mathcal{E} \right\}$ is a bounded set of pre-elementary operators, and it follows from Lemma 2.7.10 that, for every N,

$$\left\{ \left(2^{N|j_1 - j_2|} 2^{-t j_1 \wedge j_2} E_1 T E_2, 2^{-j_1 \wedge j_2} \right) \mid (E_1, 2^{-j_1}), (E_2, 2^{-j_2}) \in \mathcal{E}, j_1 \geq j_2 \right\}$$

is a bounded set of pre-elementary operators. We next need to show that

$$2^{N|j_1 - j_2|} 2^{-t j_1 \wedge j_2} E_1 T E_2$$

is an appropriate sum of derivatives of terms of the same form, as in the definition of elementary operators. Proposition 2.0.24 allows us to write

$$2^{N|j_1 - j_2|} 2^{-t j_1 \wedge j_2} E_1 T E_2$$

$$= \sum_{|\alpha|, |\beta| \leq 1} 2^{-(1 - |\beta|) j_2 - (1 - |\alpha|) j_1} \left(2^{-j_1} W \right)^\alpha \left[2^{N|j_1 - j_2| - t j_1 \wedge j_2} \widetilde{E}_{1,\alpha} T \widetilde{E}_{2,\beta} \right] \left(2^{-j_2} W \right)^\beta,$$

where

$$\left\{\left(\tilde{E}_{1,\alpha}, 2^{-j_1}\right), \left(\tilde{E}_{2,\beta}, 2^{-j_2}\right) \mid (E_1, 2^{-j_1}), (E_2, 2^{-j_2}) \in \mathcal{E}, j_1 \geq j_2\right\}$$

is a bounded set of elementary operators. Thus, we have

$$2^{N|j_1 - j_2|} 2^{-t j_1 \wedge j_2} E_1 T E_2$$

$$= \sum_{|\alpha|, |\beta| \leq 1} 2^{-(1-|\beta|)j_2 - (1-|\alpha|)j_1} \left(2^{-j_1} W\right)^\alpha \left[2^{N|j_1-j_2| - t j_1 \wedge j_2} \tilde{E}_{1,\alpha} T \tilde{E}_{2,\beta}\right] \left(2^{-j_2} W\right)^\beta$$

$$=: \sum_{|\alpha|, |\beta| \leq 1} 2^{-(2-|\alpha|-|\beta|)j_2} \left(2^{-j_2} W\right)^\alpha \left[2^{(N+1)|j_1-j_2| - t j_1 \wedge j_2} E_{1,\alpha} T E_{2,\beta}\right] \left(2^{-j_2} W\right)^\beta,$$

where

$$\left\{\left(E_{1,\alpha}, 2^{-j_1}\right), \left(E_{2,\beta}, 2^{-j_2}\right) \mid (E_1, 2^{-j_1}), (E_2, 2^{-j_2}) \in \mathcal{E}, j_1 \geq j_2\right\}$$

is a bounded set of elementary operators. Since $2^{(N+1)|j_1-j_2| - t j_1 \wedge j_2} E_{1,\alpha} T E_{2,\beta}$ is of the same form as $2^{N|j_1-j_2|} 2^{-t j_1 \wedge j_2} E_1 T E_2$ (with a different choice of N), this completes the proof that

$$\left\{\left(2^{N|j_1-j_2|} 2^{-t j_1 \wedge j_2} E_1 T E_2, 2^{-j_1 \wedge j_2}\right) \mid (E_1, 2^{-j_1}), (E_2, 2^{-j_2}) \in \mathcal{E}, j_1 \geq j_2\right\}$$

is a bounded set of elementary operators.

The proof when $j_1 \leq j_2$ is similar, reversing the roles of j_1 and j_2 in the argument. $\qquad\square$

LEMMA 2.7.12. *Suppose T is a Calderón-Zygmund operator of order $t \in (-Q_1, \infty)$, in the sense of Definition 2.0.16, and suppose $\mathcal{E}$ is a bounded set of elementary operators. Then,*

$$\left\{\left(2^{-jt} T E, 2^{-j}\right) \mid (E, 2^{-j}) \in \mathcal{E}\right\}$$

is a bounded set of pre-elementary operators.

PROOF. Fix ordered multi-indices α, β and fix $m \in \mathbb{N}$. We wish to show, for $(E, 2^{-j}) \in \mathcal{E}$,

$$2^{-jt} \left|\left(2^{-j} W_x\right)^\alpha \left(2^{-j} W_z\right)^\beta [TE](x, z)\right|$$

$$\lesssim \frac{\left(1 + 2^j \rho(x, z)\right)^{-m}}{\mathrm{Vol}\left(B_{(X,d)}(x, 2^{-j} + \rho(x, z))\right)}. \tag{2.62}$$

Using the fact that $\left\{\left(\left(2^{-j} W_z\right)^\beta E(x, z), 2^{-j}\right) \mid (E, 2^{-j}) \in \mathcal{E}\right\}$ is a bounded set of elementary operators (Proposition 2.0.24) and that $W^\alpha T$ is a Calderón-Zygmund operator of order $t + |\alpha|$, we see that it suffices to prove (2.62) in the case when $\alpha = 0 = \beta$, by replacing t with $t + |\alpha|$.

Fix $N = N(m)$ large, to be chosen later. Applying Proposition 2.0.24 to E, we see

$$E = \sum_{|\alpha| \leq N} 2^{-(N-|\alpha|)j} \left[\left(2^{-j} W \right)^{\alpha} \right]^{*} E_{\alpha},$$

where $\left\{ \left(E_{\alpha}, 2^{-j} \right) \mid \left(E, 2^{-j} \right) \in \mathcal{E} \right\}$ is a bounded set of elementary operators. Here, we have used that $\left[\left(2^{-j} W \right)^{\alpha} \right]^{*} = \sum_{|\beta| \leq \alpha} \left(2^{-j} W \right)^{\beta} g_{\beta}$, where $g_{\beta} \in C^{\infty}(M)$ and $g_{\alpha} \equiv (-1)^{|\alpha|}$. Plugging this into (2.62), we see that it suffices to show, for each $|\alpha| \leq N$,

$$2^{-jt} 2^{-(N-|\alpha|)j} \left| \left[T \left[\left(2^{-j} W \right)^{\alpha} \right]^{*} E_{\alpha} \right] (x, z) \right|$$
$$\lesssim \frac{\left(1 + 2^{j} \rho(x, z) \right)^{-m}}{\mathrm{Vol}\left(B_{(X,d)} \left(x, 2^{-j} + \rho(x, z) \right) \right)}. \tag{2.63}$$

Fix $x, z \in M$. Let $\phi \in C^{\infty}(M)$ satisfy the conclusions of Proposition 2.5.1, with $\delta = 2^{-j}$ and $x_0 = x$. In particular, $\mathrm{supp}(\phi) \subseteq B_{(X,d)} \left(x, 2^{-j} \right)$ and $\phi \equiv 1$ on $B_{(X,d)} \left(x, \xi_4 2^{-j} \right)$ where $\xi_4 > 0$ is some constant independent of any relevant parameters. We also have $\left| \left(2^{-j} W \right)^{\gamma} \phi \right| \lesssim 1$, for all γ. Set $\psi(y) = \phi(y) E_{\alpha}(y, z)$. Note that, $\forall \gamma$,

$$\left| \left(2^{-j} W \right)^{\gamma} \psi(y) \right| \lesssim \frac{\left(1 + 2^{j} \rho(x, z) \right)^{-m}}{\mathrm{Vol}\left(B_{(X,d)} \left(x, 2^{j} + \rho(x, z) \right) \right)}, \tag{2.64}$$

where we have used the rapid decrease of $E_{\alpha}(y, z)$ and that $\mathrm{supp}(\Phi) \subseteq B_{(X,d)} \left(x, 2^{-j} \right)$.

Consider,

$$2^{-jt} 2^{-(N-|\alpha|)j} \left| \left[T \left[\left(2^{-j} W \right)^{\alpha} \right]^{*} E_{\alpha} \right] (x, z) \right|$$
$$\leq 2^{-jt} 2^{-(N-|\alpha|)j} \left| \left[T \left[\left(2^{-j} W \right)^{\alpha} \right]^{*} \psi \right] (x) \right|$$
$$+ 2^{-jt} 2^{-Nj} \int \left| \left[T \left(W^{\alpha} \right)^{*} \right] (x, y) \left(1 - \phi(y) \right) E(y, z) \right| \, dy.$$

We bound the two terms separately. For the first term, we use the cancellation condition and (2.64) to see

$$2^{-jt} 2^{-(N-|\alpha|)j} \left| \left[T \left[\left(2^{-j} W \right)^{\alpha} \right]^{*} \psi \right] (x) \right|$$
$$\lesssim 2^{-(N-|\alpha|)j} \frac{\left(1 + 2^{j} \rho(x, z) \right)^{-m}}{\mathrm{Vol}\left(B_{(X,d)} \left(x, 2^{-j} + \rho(x, z) \right) \right)},$$

which implies the desired bound.

For the second term, we use the growth condition, and the fact that $T(W^{\alpha})^{*}$ is a Calderón-Zygmund operator of order $t + |\alpha|$. We have, for $\rho(x, y) \geq 2^{-j} \xi_4$ (and using

$\rho(x, y) \lesssim 1, \forall x, y \in M)$,

$$2^{-jt}2^{-Nj} \left| [T(W^\alpha)^*](x, y) \right| \lesssim 2^{-jt-Nj} \frac{\rho(x, y)^{-t-|\alpha|}}{\mathrm{Vol}\left(B_{(X,d)}(x, \rho(x, y))\right)}$$

$$\lesssim 2^{-jt-Nj} \frac{\rho(x, y)^{-t-N}}{\mathrm{Vol}\left(B_{(X,d)}(x, \rho(x, y))\right)}$$

$$\approx \frac{\left(1 + 2^j \rho(x, y)\right)^{-t-N}}{\mathrm{Vol}\left(B_{(X,d)}(x, 2^{-j} + \rho(x, y))\right)}.$$

Because $1 - \phi(y)$ is supported on $\rho(x, y) \geq 2^{-j}\xi_4$, we may use the above bound to conclude

$$2^{-jt}2^{-(N-|\alpha|)j} \int \left| T(x, y) \left[(2^{-j}W)^\alpha \right]^* (1 - \phi(y)) E(y, z) \right| dy$$

$$\lesssim \int \frac{\left(1 + 2^j \rho(x, y)\right)^{-N-t}}{\mathrm{Vol}\left(B_{(X,d)}(x, 2^{-j} + \rho(x, y))\right)} \frac{\left(1 + 2^j \rho(y, z)\right)^{-N-t}}{\mathrm{Vol}\left(B_{(X,d)}(y, 2^{-j} + \rho(y, z))\right)} dy.$$

Choosing N so that $N + t > m \vee Q_1$, (2.63) follows from Lemma 2.7.5, completing the proof. $\square$

PROOF OF (i)$\Rightarrow$(iii) OF THEOREM 2.0.29. Let T be a Calderón-Zygmund operator of order $t \in (-Q_1, \infty)$, and let $\mathcal{E}$ be a bounded set of elementary operators. Lemma 2.7.12 shows that $\left\{ \left(2^{-jt}TE, 2^{-j} \right) \mid (E, 2^{-j}) \in \mathcal{E} \right\}$ is a bounded set of pre-elementary operators. Since T^* is also a Calderón-Zygmund kernel of order t, Lemma 2.7.12 combined with Proposition 2.0.24 shows

$$\left\{ \left(2^{-jt}T^*E^*, 2^{-j} \right) \mid (E, 2^{-j}) \in \mathcal{E} \right\}$$

is a bounded set of pre-elementary operators. It follows that

$$\left\{ \left(2^{-jt}ET, 2^{-j} \right) \mid (E, 2^{-j}) \in \mathcal{E} \right\}$$

is a bounded set of pre-elementary operators. Lemma 2.7.11 now completes the proof. $\square$

Finally, we prove (iii)$\Rightarrow$(i).

LEMMA 2.7.13. *Let $\mathcal{B}$ be a bounded set of bump functions and $\mathcal{E}$ be a bounded set of pre-elementary operators. $\forall m, \exists C$, such that $\forall (E, 2^{-j}) \in \mathcal{E}, (\phi, z, 2^{-k}) \in \mathcal{B}$,*

$$|E\phi(x)| \lesssim \frac{\left(1 + 2^{j\wedge k}\rho(x, z)\right)^{-m}}{\mathrm{Vol}\left(B_{(X,d)}(x, 2^{-j\wedge k} + \rho(x, z))\right)}.$$

PROOF. Since ϕ satisfies

$$
\begin{aligned}
|\phi(y)| &\lesssim \frac{\chi_{\{\rho(y,z)<2^{-k}\}}}{\mathrm{Vol}\left(B_{(X,d)}\left(z, 2^{-k}\right)\right)} \\
&\approx \frac{\chi_{\{\rho(y,z)<2^{-k}\}}}{\mathrm{Vol}\left(B_{(X,d)}\left(y, 2^{-k}\right)\right)} \\
&\lesssim \frac{\left(1 + 2^{k}\rho(y,z)\right)^{-m}}{\mathrm{Vol}\left(B_{(X,d)}\left(y, 2^{-k} + \rho(y,z)\right)\right)},
\end{aligned}
$$

this is an immediate consequence of Lemma 2.7.5. $\qquad\square$

LEMMA 2.7.14. *Let $\mathcal{B}$ be a bounded set of bump functions and $\mathcal{E}$ be a bounded set of elementary operators. $\forall N, m, \exists C$ such that $\forall\left(E, 2^{-j}\right) \in \mathcal{E}, \left(\phi, z, 2^{-k}\right) \in \mathcal{B}$,*

$$
|E\phi(x)| \lesssim 2^{-N(j-j\wedge k)} \frac{\left(1 + 2^{j\wedge k}\rho(x,z)\right)^{-m}}{\mathrm{Vol}\left(B_{(X,d)}\left(x, 2^{-j\wedge k} + \rho(x,z)\right)\right)}.
$$

PROOF. When $j \leq k$, this bound follows from Lemma 2.7.13. For $j > k$ apply Proposition 2.0.24 to write $E = \sum_{|\alpha|\leq N} 2^{-j(N-|\alpha|)} E_{\alpha}\left(2^{-j}W\right)^{\alpha}$, where

$$
\left\{\left(E_{\alpha}, 2^{-j}\right) \mid j \in \mathbb{N}\right\}
$$

is a bounded set of elementary operators. We have

$$
E\phi(x) = \sum_{|\alpha|\leq N} 2^{-j(N-|\alpha|)} 2^{-(j-k)|\alpha|} E_{\alpha}\left(2^{-k}W\right)^{\alpha}\phi(x).
$$

Using that $2^{-j(N-|\alpha|)} 2^{-(j-k)|\alpha|} \leq 2^{-N(j-j\wedge k)}$ and that

$$
\left\{\left(\left(2^{-k}W\right)^{\alpha}\phi, z, 2^{-k}\right) \mid \left(\phi, z, 2^{-k}\right) \in \mathcal{B}, |\alpha| \leq N\right\}
$$

is a bounded set of bump functions, the result now follows from Lemma 2.7.13. $\qquad\square$

PROOF OF (iii)$\Rightarrow$(i) OF THEOREM 2.0.29. Let $\left\{\left(E_j, 2^{-j}\right) \mid j \in \mathbb{N}\right\}$ be a bounded set of elementary operators, and fix $t \in (-Q_1, \infty)$. Let $T = \sum_{j\in\mathbb{N}} 2^{jt} E_j$. We wish to show that T is a Calderón-Zygmund operator of order t. We begin by verifying the growth condition. Fix ordered multi-indices α and β. Consider, for $x \neq y \in M$, and for any m,

$$
\begin{aligned}
\left|W_x^{\alpha} W_y^{\beta} T(x,y)\right| &\leq \sum_{j\in\mathbb{N}} 2^{j(t+|\alpha|+|\beta|)}\left|\left(2^{-j}W_x\right)^{\alpha}\left(2^{-j}W_y\right)^{\beta} E_j(x,y)\right| \\
&\lesssim \sum_{j\in\mathbb{N}} 2^{j(t+|\alpha|+|\beta|)} \frac{\left(1 + 2^{j}\rho(x,y)\right)^{-m}}{\mathrm{Vol}\left(B_{(X,d)}\left(x, 2^{-j} + \rho(x,y)\right)\right)} \\
&\lesssim \sum_{2^{-j}\leq\rho(x,y)} 2^{j(t+|\alpha|+|\beta|)-jm}\rho(x,y)^{-m}\,\mathrm{Vol}\left(B_{(X,d)}\left(x, \rho(x,y)\right)\right)^{-1} \\
&\quad + \sum_{\rho(x,y)<2^{-j}} 2^{j(t+|\alpha|+|\beta|)}\,\mathrm{Vol}\left(B_{(X,d)}\left(x, 2^{-j}\right)\right)^{-1}.
\end{aligned}
$$

Using that, for $\delta > 0$, $2^{jQ_1} \text{Vol}\left(B_{(X,d)}\left(x, 2^{-j}\delta\right)\right) \geq \text{Vol}\left(B_{(X,d)}\left(x, \delta\right)\right)$ and that $t > -Q_1$, we see, by taking $m > |\alpha| + |\beta| + t$, that both terms are bounded termwise by a geometric series and are therefore bounded by a constant times their largest term. We conclude

$$\left|W_x^\alpha W_y^\beta T\left(x, y\right)\right| \lesssim \frac{\rho\left(x, y\right)^{-t-|\alpha|-|\beta|}}{\text{Vol}\left(B_{(X,d)}\left(x, \rho\left(x, y\right)\right)\right)},$$

completing the proof of the growth condition.

We turn to the cancellation condition. Let $\mathcal{B}$ be a bounded set of bump functions and fix an ordered multi-index α. For $\left(\phi, z, 2^{-k}\right) \in \mathcal{B}$ consider

$$\left|2^{-kt}\left(2^{-k}W\right)^\alpha T\phi\left(x\right)\right| \leq \sum_{j \in \mathbb{N}} 2^{(j-k)t} 2^{(j-k)|\alpha|} \left|\left(2^{-j}W\right)^\alpha E_j\phi\left(x\right)\right|.$$

Proposition 2.0.24 shows

$$\left\{\left(\left(2^{-j}W\right)^\alpha E_j, 2^{-j}\right) \mid j \in \mathbb{N}\right\}$$

is a bounded set of elementary operators, and we may therefore apply Lemma 2.7.14 to $\left(2^{-j}W\right)^\alpha E_j\phi$ to see, for every N, m,

$$\left|2^{-kt}\left(2^{-k}W\right)^\alpha T\phi\left(x\right)\right|$$

$$\lesssim \sum_{j \in \mathbb{N}} 2^{(j-k)t} 2^{(j-k)|\alpha|} 2^{-N(j-j\wedge k)} \frac{\left(1 + 2^{j\wedge k}\rho\left(x, z\right)\right)^{-m}}{\text{Vol}\left(B_{(X,d)}\left(x, 2^{-j\wedge k} + \rho\left(x, z\right)\right)\right)}$$

$$\lesssim \sum_{j \in \mathbb{N}} 2^{(j-k)t} 2^{(j-k)|\alpha|} 2^{-N(j-j\wedge k)} \text{Vol}\left(B_{(X,d)}\left(x, 2^{-j\wedge k}\right)\right)^{-1}.$$

We separate the above sum into two parts. The first,

$$\sum_{0 \leq j \leq k} 2^{(j-k)t} 2^{(j-k)|\alpha|} \text{Vol}\left(B_{(X,d)}\left(x, 2^{-j}\right)\right)^{-1}.$$

Using that, for $\delta > 0$, $2^{jQ_1} \text{Vol}\left(B_{(X,d)}\left(x, 2^{-j}\delta\right)\right) \geq \text{Vol}\left(B_{(X,d)}\left(x, \delta\right)\right)$ and that $t > -Q_1$, we see that this sum is termwise bounded by a geometric sum, and therefore bounded by a constant times its largest term. We obtain

$$\sum_{0 \leq j \leq k} 2^{(j-k)t} 2^{(j-k)|\alpha|} \text{Vol}\left(B_{(X,d)}\left(x, 2^{-j}\right)\right)^{-1} \lesssim \text{Vol}\left(B_{(X,d)}\left(x, 2^{-k}\right)\right)^{-1}.$$

The second sum is

$$\sum_{k \leq j} 2^{(j-k)t} 2^{(j-k)|\alpha|} 2^{-N(j-k)} \text{Vol}\left(B_{(X,d)}\left(x, 2^{-k}\right)\right)^{-1}.$$

By taking N large, this is a geometric sum, and therefore bounded by a constant times its largest term. We obtain

$$\sum_{k \leq j} 2^{(j-k)t} 2^{(j-k)|\alpha|} 2^{-N(j-k)} \text{Vol}\left(B_{(X,d)}\left(x, 2^{-k}\right)\right)^{-1} \lesssim \text{Vol}\left(B_{(X,d)}\left(x, 2^{-k}\right)\right)^{-1}.$$

Combining all of the above, yields

$$\left| 2^{-kt} \left(2^{-k} W \right)^{\alpha} T \phi \left(x \right) \right| \lesssim \mathrm{Vol} \left(B_{(X,d)} \left(x, 2^{-k} \right) \right)^{-1},$$

which is the desired bound for the cancellation condition for T.

$T^* = \sum_{j \in \mathbb{N}} 2^{jt} E_j^*$, and Proposition 2.0.24 shows that $\left\{ \left(E_j^*, 2^{-j} \right) \mid j \in \mathbb{N} \right\}$ is a bounded set of elementary operators. The above proof, therefore, applies to prove the cancellation condition for T^*. $\qquad \square$

2.7.1 More on the cancellation condition

Proposition 2.5.1 shows that there are appropriate "cut-off" functions adapted to the Carnot-Carathéodory geometry. In Definition 2.0.16, we defined Calderón-Zygmund operators of order $t \in (-Q_1, \infty)$ by using a "Growth Condition" and a "Cancellation Condition." Using the cut-off functions we have developed, we will see that these two conditions overlap quite a bit. In fact, we will be able to equivalently define Calderón-Zygmund operators in terms of some strengthened "Cancellation Conditions." More precisely, have the following theorem.

THEOREM 2.7.15. *Fix $t \in (-Q_1, \infty)$, and let $T : C^{\infty} (M) \to C^{\infty} (M)$. The following are equivalent.*

(i) *T is a Calderón-Zygmund operator of order t.*

(ii) *For every bounded set of bump functions $\mathcal{B}$:*

$\forall \alpha, \beta, \exists C = C \left(\mathcal{B}, \alpha, \beta \right), \forall \left(\phi, z, 2^{-j} \right) \in \mathcal{B},$

$$\left| W^{\alpha} T W^{\beta} \phi \left(x \right) \right| \le C \frac{\left(2^{-j} + \rho \left(x, z \right) \right)^{-t - |\alpha| - |\beta|}}{\mathrm{Vol} \left(B_{(X,d)} \left(x, 2^{-j} + \rho \left(x, z \right) \right) \right)}, \qquad \forall x \in M.$$

We also assume the same estimate for T replaced by T^*.

(iii) *For every bounded set of bump functions $\mathcal{B}$:*

$\forall \alpha, \beta, \exists C = C \left(\mathcal{B}, \alpha, \beta \right), \forall \left(\phi_1, x, 2^{-j_1} \right), \left(\phi_2, z, 2^{-j_2} \right) \in \mathcal{B},$

$$\left| \left\langle W^{\alpha} \phi_1, T W^{\beta} \phi_2 \right\rangle \right| \le C \frac{\left(2^{-j_1 \wedge j_2} + \rho \left(x, z \right) \right)^{-t - |\alpha| - |\beta|}}{\mathrm{Vol} \left(B_{(X,d)} \left(x, 2^{-j_1 \wedge j_2} + \rho \left(x, z \right) \right) \right)}.$$

The key to the proof of Theorem 2.7.15 is the following lemma.

LEMMA 2.7.16. *There is a bounded set of bump functions $\mathcal{B}$ such that $\forall x \in M$, there is a sequence $\left(\psi_j, x, 2^{-j} \right) \in \mathcal{B}$ $(j \in \mathbb{N})$ with $\psi_j \to \delta_x$ in the sense of distributions, where δ_x denotes the Dirac delta functions at the point x.*

PROOF. For $j \in \mathbb{N}$, $x \in M$, let ϕ_j be the function given by Proposition 2.5.1 with $\delta = 2^{-j}$ and $x_0 = x$. Define

$$\psi_j = \left(\int \phi_j \right)^{-1} \phi_j.$$

By the properties of ϕ_j, it immediately follows that $\psi_j \to \delta_x$, and that

$$\left\{ \left(\phi_j, x, 2^{-j} \right) \mid x \in M, j \in \mathbb{N} \right\}$$

is a bounded set of bump functions. This completes the proof. $\qquad \square$

We also need one additional result to prove Theorem 2.7.15.

LEMMA 2.7.17. *For* $t \in [-Q_1, \infty)$,

$$\frac{\left(2^{-j} + \rho\left(x, y \right) \right)^{-t}}{\operatorname{Vol}\left(B_{(X,d)}\left(x, 2^{-j} + \rho\left(x, y \right) \right) \right)} \lesssim \frac{2^{jt}}{\operatorname{Vol}\left(B_{(X,d)}\left(x, 2^{-j} \right) \right)}. \tag{2.65}$$

We prove Lemma 2.7.17 by introducing two simpler lemmas.

LEMMA 2.7.18. *For* $R \geq 1$ *and* $\delta > 0$,

$$R^{Q_1} \operatorname{Vol}\left(B_{(X,d)}\left(x, \delta \right) \right) \lesssim \operatorname{Vol}\left(B_{(X,d)}\left(x, R\delta \right) \right).$$

PROOF. When $R = 2^j$ for some $j \in \mathbb{N}$, this follows by repeated applications of Theorem 2.0.10 (in fact with $\leq$ in place of $\lesssim$). Otherwise, take j so that $2^j \leq R < 2^{j+1}$. Then,

$$\begin{aligned} R^{Q_1} \operatorname{Vol}\left(B_{(X,d)}\left(x, \delta \right) \right) &\approx 2^{jQ_1} \operatorname{Vol}\left(B_{(X,d)}\left(x, \delta \right) \right) \\ &\leq \operatorname{Vol}\left(B_{(X,d)}\left(x, 2^j \delta \right) \right) \\ &\approx \operatorname{Vol}\left(B_{(X,d)}\left(x, R\delta \right) \right), \end{aligned}$$

completing the proof. $\qquad \square$

LEMMA 2.7.19. *For* $R \geq 1$ *and* $t \in [-Q_1, \infty)$,

$$R^{-t} \operatorname{Vol}\left(B_{(X,d)}\left(x, \delta \right) \right) \lesssim \operatorname{Vol}\left(B_{(X,d)}\left(x, R\delta \right) \right).$$

PROOF. Because $R \geq 1$ and $t \in [-Q_1, \infty)$,

$$\begin{aligned} R^{-t} \operatorname{Vol}\left(B_{(X,d)}\left(x, \delta \right) \right) &\leq R^{Q_1} \operatorname{Vol}\left(B_{(X,d)}\left(x, \delta \right) \right) \\ &\lesssim \operatorname{Vol}\left(B_{(X,d)}\left(x, R\delta \right) \right), \end{aligned}$$

where in the last line, we have applied Lemma 2.7.18. $\qquad \square$

PROOF OF LEMMA 2.7.17. Notice, (2.65) is equivalent to

$$\frac{\left(1 + 2^j \rho(x, y)\right)^{-t}}{\mathrm{Vol}\left(B_{(X,d)}\left(x, 2^{-j} + \rho(x, y)\right)\right)} \lesssim \frac{1}{\mathrm{Vol}\left(B_{(X,d)}\left(x, 2^{-j}\right)\right)}. \tag{2.66}$$

When $2^j \rho(x, y) \le 1$, (2.66) is obvious. Thus we assume $R := 2^j \rho(x, y) \ge 1$. In this case, (2.66) is equivalent to showing

$$R^{-t} \mathrm{Vol}\left(B_{(X,d)}\left(x, 2^{-j}\right)\right) \lesssim \mathrm{Vol}\left(B_{(X,d)}\left(x, \rho(x, y)\right)\right). \tag{2.67}$$

Since $R2^{-j} = \rho(x, y)$, and we are assuming $R \ge 1$, (2.67) follows from Lemma 2.7.19. $\qquad\square$

PROOF OF THEOREM 2.7.15. (iii)⇒(ii): Assume that (iii) holds. Let $\mathcal{B}$ be a bounded set of bump functions. Because $2^{-j_1 \wedge j_2} + \rho(x, z) \lesssim 1$ $(\forall j_1, j_2 \in [0, \infty)$, $x, z \in M)$, and because $(W^\alpha)^* = \sum_{|\alpha'| \le |\alpha|} W^{\alpha'} g_{\alpha, \alpha'}$, where $g_{\alpha, \alpha'} \in C^\infty(M)$, we have $\forall \left(\phi_1, x, 2^{-j_1}\right), \left(\phi_2, z, 2^{-j_2}\right) \in \mathcal{B}$,

$$\left|\left\langle \phi_1, W^\alpha T W^\beta \phi_2 \right\rangle\right| \le C_{\mathcal{B}, \alpha, \beta} \frac{\left(2^{-j_1 \wedge j_2} + \rho(x, z)\right)^{-t - |\alpha| - |\beta|}}{\mathrm{Vol}\left(B_{(X,d)}\left(x, 2^{-j_1 \wedge j_2} + \rho(x, z)\right)\right)}.$$

Replacing $\left(\phi_1, x, 2^{-j_1}\right)$ with $\left(\psi_j, x, 2^{-j}\right)$ where $\left(\psi_j, x, 2^{-j}\right)$ is as in Lemma 2.7.16, we have

$$\left|\left\langle \psi_j, W^\alpha T W^\beta \phi_2 \right\rangle\right| \le C'_{\mathcal{B}, \alpha, \beta} \frac{\left(2^{-j \wedge j_2} + \rho(x, z)\right)^{-t - |\alpha| - |\beta|}}{\mathrm{Vol}\left(B_{(X,d)}\left(x, 2^{-j \wedge j_2} + \rho(x, z)\right)\right)}.$$

Taking $j \to \infty$ yields

$$\left|W^\alpha T W^\beta \phi_2(x)\right| \le C'_{\mathcal{B}, \alpha, \beta} \frac{\left(2^{-j_2} + \rho(x, z)\right)^{-t - |\alpha| - |\beta|}}{\mathrm{Vol}\left(B_{(X,d)}\left(x, 2^{-j_2} + \rho(x, z)\right)\right)},$$

as desired. Because (iii) is symmetric in T and T^*, the same estimate holds for T^*, which completes the proof of (ii).

(ii)⇒(i): In light of Lemma 2.7.17, the cancellation condition follows immediately from (ii), and so we need only prove the growth condition. Fix $x, z \in M$, $x \ne z$. Take $\left(\psi_j, z, 2^{-j}\right)$ as in lemma 2.7.16 so that $\psi_j \to \delta_z$. We have

$$\left|W^\alpha T W^\beta \psi_j(x)\right| \lesssim \frac{\left(2^{-j} + \rho(x, z)\right)^{-t - |\alpha| - |\beta|}}{\mathrm{Vol}\left(B_{(X,d)}\left(x, 2^{-j} + \rho(x, z)\right)\right)}.$$

Taking $j \to \infty$, we see

$$\left|\left[W^\alpha T W^\beta\right](x, z)\right| \lesssim \frac{\rho(x, z)^{-t - |\alpha| - |\beta|}}{\mathrm{Vol}\left(B_{(X,d)}\left(x, \rho(x, z)\right)\right)},$$

and the growth condition follows.

(i)$\Rightarrow$(iii): If T is a Calderón-Zygmund operator of order t, then it is easy to see $(W^\alpha)^* T W^\beta$ is a Calderón-Zygmund operator of order $t + |\alpha| + |\beta|$. Thus we prove (iii) only in the case $|\alpha| = |\beta| = 0$, and the full result then follows. I.e., we assume T is a Calderón-Zygmund operator of order $t > -Q_1$, $\mathcal{B}$ is a bounded set of bump functions, and we wish to show

$$|\langle \phi_1, T\phi_2 \rangle| \lesssim \frac{\left(2^{-j_1 \wedge j_2} + \rho(x,z)\right)^{-t}}{\mathrm{Vol}\left(B_{(X,d)}\left(x, 2^{-j_1 \wedge j_2} + \rho(x,z)\right)\right)}, \qquad (2.68)$$

$\forall \left(\phi_1, x, 2^{-j_1}\right), \left(\phi_2, z, 2^{-j_2}\right) \in \mathcal{B}$.

If $\rho(x,z)/10 < 2^{-j_1 \wedge j_2}$, then we use the cancellation condition to prove (2.68). Because our assumptions are symmetric in T and T^* we may assume without loss of generality that $j_2 \le j_1$. We apply the cancellation condition to see

$$|T\phi_2(y)| \lesssim \frac{\left(2^{-j_2}\right)^{-t}}{\mathrm{Vol}\left(B_{(X,d)}\left(x, 2^{-j_2}\right)\right)} \approx \frac{\left(2^{-j_1 \wedge j_2} + \rho(x,z)\right)^{-t}}{\mathrm{Vol}\left(B_{(X,d)}\left(x, 2^{-j_1 \wedge j_2} + \rho(x,z)\right)\right)}.$$

Using that $\int |\phi_1| \lesssim 1$, (2.68) follows.

Now suppose $\rho(x,z) \ge 10\left(2^{-j_1 \wedge j_2}\right)$. Then the supports of ϕ_1 and ϕ_2 are disjoint, and (2.68) can be estimated using only the growth condition. For $y_1 \in \mathrm{supp}(\phi_1) \subset B_{(X,d)}\left(x, 2^{-j_1}\right)$ and $y_2 \in \mathrm{supp}(\phi_2) \subset B_{(X,d)}\left(x, 2^{-j_2}\right)$, we have

$$|T(y_1, y_2)| \lesssim \frac{\rho(y_1, y_2)^{-t}}{\mathrm{Vol}\left(B_{(X,d)}\left(y_1, \rho(y_1, y_2)\right)\right)} \approx \frac{\rho(x,z)^{-t}}{\mathrm{Vol}\left(B_{(X,d)}\left(x, \rho(x,z)\right)\right)}.$$

Using that $\int |\phi_1|, \int |\phi_2| \lesssim 1$, we have

$$\begin{aligned}
|\langle \phi_1, T\phi_2 \rangle| &\lesssim \int |\phi_1(y_1) T(y_1, y_2) \phi_2(y_2)| \, dy_1 dy_2 \\
&\lesssim \frac{\rho(x,z)^{-t}}{\mathrm{Vol}\left(B_{(X,d)}\left(x, \rho(x,z)\right)\right)} \int |\phi_1(y_1)| |\phi_2(y_2)| \, dy_1 \, dy_2 \\
&\lesssim \frac{\rho(x,z)^{-t}}{\mathrm{Vol}\left(B_{(X,d)}\left(x, \rho(x,z)\right)\right)},
\end{aligned}$$

proving (2.68), and completing the proof. $\qquad \square$

The same proof as Theorem 2.7.15 gives the following slight modification of that theorem.

THEOREM 2.7.20. *Fix* $t \in (-Q_1, \infty)$, *and let* $T : C^\infty(M) \to C^\infty(M)$. *The following are equivalent.*

(i) *T is a Calderón-Zygmund operator of order t.*

(ii) *For every bounded set of bump functions $\mathcal{B}$:*

$\forall \alpha, \beta, \exists C = C(\mathcal{B}, \alpha, \beta), \forall (\phi, z, 2^{-j}) \in \mathcal{B}$,

$$\left| W^\alpha T (W^\beta)^* \phi(x) \right| \le C \frac{\left(2^{-j} + \rho(x, z) \right)^{-t - |\alpha| - |\beta|}}{\mathrm{Vol}\left(B_{(X,d)} (x, 2^{-j} + \rho(x, z)) \right)}, \quad \forall x \in M.$$

We also assume the same estimate for T replaced by T^*.

(iii) *For every bounded set of bump functions $\mathcal{B}$:*

$\forall \alpha, \beta, \exists C = C(\mathcal{B}, \alpha, \beta), \forall (\phi_1, x, 2^{-j_1}), (\phi_2, z, 2^{-j_2}) \in \mathcal{B}$,

$$\left| \left\langle (W^\alpha)^* \phi_1, T (W^\beta)^* \phi_2 \right\rangle \right| \le C \frac{\left(2^{-j_1 \wedge j_2} + \rho(x, z) \right)^{-t - |\alpha| - |\beta|}}{\mathrm{Vol}\left(B_{(X,d)} (x, 2^{-j_1 \wedge j_2} + \rho(x, z)) \right)}.$$

We have the following corollary.

COROLLARY 2.7.21. *Fix $t \in \mathbb{R}$, and let $T : C^\infty(M) \to C^\infty(M)$. The following are equivalent.*

(i) *T is a Calderón-Zygmund operator of order t.*

(ii) *For every bounded set of bump functions $\mathcal{B}$, $\forall \alpha, \beta$ with $|\alpha| + |\beta| > -Q_1 - t$, $\exists C = C(\mathcal{B}, \alpha, \beta), \forall (\phi, z, 2^{-j}) \in \mathcal{B}$,*

$$\left| W^\alpha T (W^\beta)^* \phi(x) \right| \le C \frac{\left(2^{-j} + \rho(x, z) \right)^{-t - |\alpha| - |\beta|}}{\mathrm{Vol}\left(B_{(X,d)} (x, 2^{-j} + \rho(x, z)) \right)}, \quad \forall x \in M.$$

We also assume the same estimate for T replaced by T^*.

(iii) *For every bounded set of bump functions $\mathcal{B}$, $\forall \alpha, \beta$ with $|\alpha| + |\beta| > -Q_1 - t$, $\exists C = C(\mathcal{B}, \alpha, \beta), \forall (\phi_1, x, 2^{-j_1}), (\phi_2, z, 2^{-j_2}) \in \mathcal{B}$,*

$$\left| \left\langle (W^\alpha)^* \phi_1, T (W^\beta)^* \phi_2 \right\rangle \right| \le C \frac{\left(2^{-j_1 \wedge j_2} + \rho(x, z) \right)^{-t - |\alpha| - |\beta|}}{\mathrm{Vol}\left(B_{(X,d)} (x, 2^{-j_1 \wedge j_2} + \rho(x, z)) \right)}.$$

To deduce Corollary 2.7.21 we need a lemma.

LEMMA 2.7.22. *Fix $t \in \mathbb{R}$ and let $T : C^\infty(M) \to C^\infty(M)$. For $N_1, N_2 \in \mathbb{N}$, the following are equivalent.*

(i) *T is a Calderón-Zygmund operator of order t.*

(ii) *$W^\alpha T (W^\beta)^*$ is a Calderón-Zygmund operator of order $t + N_1 + N_2$, $\forall |\alpha| = N_1, |\beta| = N_2$.*

PROOF. (i)$\Rightarrow$(ii): If $|\alpha| = N_1, |\beta| = N_2$, then W^α is a Calderón-Zygmund operator of order N_1 and $\left(W^\beta\right)^*$ is a Calderón-Zygmund operator of order N_2. Thus, $W^\alpha T \left(W^\beta\right)^*$ is a Calderón-Zygmund operator of order $t + N_1 + N_2$.

(ii)$\Rightarrow$(i): The result for all N_1, N_2 follows from the result with $N_1 = 1$ and $N_2 = 0$ and the result when $N_1 = 0$ and $N_2 = 1$ (by a straightforward induction argument). The result when $N_1 = 0$ and $N_2 = 1$ follows from the result when $N_1 = 1$ and $N_2 = 0$ by taking adjoints. Thus, we prove (ii)$\Rightarrow$(i) only in the case when $N_1 = 1$ and $N_2 = 0$. Suppose $W_j T$ is a Calderón-Zygmund operator of order $t + 1$ for every j. Then, $\sum_{j=1}^r W_j^* W_j T = \mathcal{L} T$ is a Calderón-Zygmund operator of order $t + 2$. It follows that $(I - E(0)) T = \mathcal{L}^{-1} \mathcal{L} T$ is a Calderón-Zygmund operator of order t where $E(0)$ is the orthogonal projection onto the constant functions and $\mathcal{L}^{-1}$ is a Calderón-Zygmund operator of order -2 (by Corollary 2.6.7).

Because $T^* \mathcal{L}$ is a Calderón-Zygmund kernel of order $t + 2$, we have

$$T^* \mathcal{L} : C^\infty (M) \to C^\infty (M)$$

and therefore $\mathcal{L} T$ extends to a continuous mapping $C^\infty (M)' \to C^\infty (M)'$. Because $\mathcal{L}$ is subelliptic, we have T extends to an operator $C^\infty (M)' \to C^\infty (M)'$. Thus, $E(0) T = T - \mathcal{L}^{-1} \mathcal{L} T : C^\infty (M)' \to C^\infty (M)$; i.e., $E(0) T \in C^\infty (M \times M)$. This shows that (in particular) $E(0) T$ is also a Calderón-Zygmund operator of order t, completing the proof. $\square$

PROOF OF COROLLARY 2.7.21. (i)$\Rightarrow$(ii): If T is a Calderón-Zygmund operator of order t, then $\forall \alpha, \beta$ with $|\alpha| + |\beta| > -Q_1 - t$, $W^\alpha T \left(W^\beta\right)^*$ is a Calderón-Zygmund operator of order $t + |\alpha| + |\beta| \in (-Q_1, \infty)$. (ii) follows from the corresponding implication in Theorem 2.7.20.

(ii)$\Rightarrow$(i): Suppose (ii) holds. Take N so large $t + N > -Q_1$. Let $|\alpha_0| = N$ and set $S_{\alpha_0} = W^{\alpha_0} T$. (ii) shows that for every bounded set of bump functions $\mathcal{B}$, $\forall \alpha, \beta$, $\exists C = C(\mathcal{B}, \alpha, \beta), \forall \left(\phi, z, 2^{-j}\right) \in \mathcal{B}$,

$$\left| W^\alpha S_{\alpha_0} \left(W^\beta\right)^* \phi(x) \right| \le C \frac{\left(2^{-j} + \rho(x, z)\right)^{-t-N-|\alpha|-|\beta|}}{\mathrm{Vol}\left(B_{(X,d)}\left(x, 2^{-j} + \rho(x, z)\right)\right)}, \quad \forall x \in M.$$

The same estimate holds for S_{α_0} replaced by $S_{\alpha_0}^*$. Theorem 2.7.20 now shows that S_{α_0} is a Calderón-Zygmund operator of order $t + N$. From here, Lemma 2.7.22 shows that T is a Calderón-Zygmund operator of order t.

The proof of (i)$\Leftrightarrow$(iii) follows from the same ideas, and we leave the details to the reader. $\square$

Remark 2.7.23 In light of Theorem 2.7.20, we think of (ii) and (iii) of that theorem as encapsulating both the growth conditions and the cancellation conditions. When we move to the setting $t \le -Q_1$, where we do not have a direct analog of the growth conditions, we like to think of (ii) and (iii) of Corollary 2.7.21 as containing an analog of the growth conditions. When we move to the multi-parameter setting in Chapter 5, we no longer define singular integrals in terms of growth conditions. However, we will use an analog of Corollary 2.7.21 to help define these singular integrals. Thus, in the multi-parameter setting, we will have at least a partial analog of the growth and cancellation conditions. See Remark 5.1.18.

2.8   THE TOPOLOGY

In this section, we discuss the topology on the space of Calderón-Zygmund operators of order s, and prove that it is a Fréchet space (i.e., prove Proposition 2.0.37). In Definitions 2.0.35 and 2.0.36 we define the semi-norms which induce the locally convex topology on the space of Calderón-Zygmund operators of order s. These are clearly a countable family of semi-norms so the only thing to verify is that the space is complete.

Proposition 2.0.37 would be straightforward were it not for the infima in (2.4) and (2.6). For instance, if we are able to pick a particular way to decompose T as in (2.5) and then a particular way to decompose each E_j as in (2.3) and so forth, so that the infima could be avoided, then it would easily follow that the space of Calderón-Zygmund operators of order s is a Fréchet space. In fact, we will be able to choose such a particular decomposition. The key is the next lemma.

LEMMA 2.8.1. *Let E_0 be a 2^{-j_0} elementary operator and let $\mathcal{E}$ be a bounded set of elementary operators. Then $\forall \left(E_1, 2^{-j_1}\right), \left(E_2, 2^{-j_2}\right) \in \mathcal{E}$ if we define $j = j_1 \wedge j_2$ and set*

$$E = E_1 E_0 E_2$$

then $\forall N, \exists M$, with

$$|E|_{2^{-j}, N, 0} \le C_{\mathcal{E}} \, |E_0|_{2^{-j_0}, M, M} \, 2^{-N \mathrm{diam}\{j_0, j_1, j_2\}},$$

where $\mathrm{diam}\{j_0, j_1, j_2\} = \max_{0 \le l_1, l_2 \le 2} |j_{l_1} - j_{l_2}|$, and $C_{\mathcal{E}}$ depends on $\mathcal{E}$ but not on E_0.

COMMENTS ON THE PROOF.   That

$$|E|_{2^{-j}, N, 0} \le C 2^{-N \mathrm{diam}\{j_0, j_1, j_2\}}$$

for some constant C follows immediately from Lemma 2.7.9. To see that C is of the form

$$C_{\mathcal{E}} \, |E_0|_{2^{-j_0}, M, M},$$

for some M, one can trace through the proof of Lemma 2.7.9 keeping track of constants. We leave the details to the interested reader.   □

Let T be a Calderón-Zygmund operator of order s, and decompose the identity operator $I = \sum_{j \in \mathbb{N}} D_j$, where $\left\{ (D_j, 2^{-j}) \mid j \in \mathbb{N} \right\}$ is a bounded set of elementary operators (see Lemma 2.7.1). Note that

$$T = ITI = \sum_{j_1, j_2 \in \mathbb{N}} D_{j_1} T D_{j_2}.$$

For $j \in \mathbb{N}$ define

$$\widetilde{E}_j := 2^{-js} \sum_{j_1 \wedge j_2 = j} D_{j_1} T D_{j_2}.$$

Proposition 2.7.2 can be used to easily show that $\left\{ \left(\widetilde{E}_j, 2^{-j} \right) \mid j \in \mathbb{N} \right\}$ is a bounded set of elementary operators, and we have

$$T = \sum_{j \in \mathbb{N}} 2^{js} \widetilde{E}_j.$$

This will be our particular choice of decomposition in (2.5).

Suppose we take any decomposition of T as in (2.5): $T = \sum_{j \in \mathbb{N}} 2^{js} E_j$. We have

$$\widetilde{E}_j = 2^{(j_0 - j)s} \sum_{j_1 \wedge j_2 = j} \sum_{j_0 \in \mathbb{N}} D_{j_1} E_{j_0} D_{j_2}.$$

Lemma 2.8.1 shows that $\forall N, \exists M,$

$$\left| \widetilde{E}_j \right|_{2^{-j}, N, 0} \lesssim \sup_{j_0 \in \mathbb{N}} |E_{j_0}|_{2^{-j_0}, M, M} .$$

Taking the infimum over all such representations of T, we get

$$\left| \widetilde{E}_j \right|_{2^{-j}, N, 0} \lesssim |T|_{s, M} .$$

Next, we discuss "pulling out derivatives." We use Proposition 2.0.25 (in the case $N = 1$) to write

$$D_j = \begin{cases} \sum_{|\alpha| \leq N} 2^{-(1-|\alpha|)j} \left(2^{-j} W \right)^{\alpha} D_{j,\alpha,1}, & \text{and} \\ \sum_{|\alpha| \leq N} 2^{-(1-|\alpha|)j} D_{j,\alpha,2} \left(2^{-j} W \right)^{\alpha}, & \end{cases}$$

where $\left\{ \left(D_{j,\alpha,1}, 2^{-j} \right), \left(D_{j,\alpha,2}, 2^{-j} \right) \mid j \in \mathbb{N}, |\alpha| \leq 1 \right\}$ is a bounded set of elementary operators. We have for $T = \sum_{j \in \mathbb{N}} 2^{js} E_j,$

$$\widetilde{E}_j = \sum_{|\alpha|, |\beta| \leq 1} \sum_{j_1 \wedge j_2 = j} \sum_{j_0 \in \mathbb{N}} 2^{-(1-|\alpha|)j_1 - (1-|\beta|)j_2}$$

$$\times \left(2^{-j_1} W \right)^{\alpha} 2^{(j_0 - j)s} D_{j_1,\alpha,2} E_{j_0} D_{j_2,\beta,1} \left(2^{-j_2} W \right)^{\beta}$$

$$=: \sum_{|\alpha|, |\beta| \leq 1} 2^{-(2-|\alpha|-|\beta|)j} \left(2^{-j} W \right)^{\alpha} \widetilde{E}_{j,\alpha,\beta} \left(2^{-j} W \right)^{\beta},$$

where, by two applications of Proposition 2.7.2,

$$\left\{ \left(\widetilde{E}_{j,\alpha,\beta}, 2^{-j} \right) \mid j \in \mathbb{N}, |\alpha|, |\beta| \leq 1 \right\}$$

is a bounded set of elementary operators (here we are using that, in the sum, $j_1, j_2 \geq j$). Moreover, Lemma 2.8.1 shows $\forall N, \exists M,$

$$\left| \tilde{E}_j \right|_{2^{-j}, N, 0} + \sum_{|\alpha|, |\beta| \leq 1} \left| \tilde{E}_{j, \alpha, \beta} \right|_{2^{-j}, N, 0} \lesssim \sup_{j \in \mathbb{N}} |E_j|_{2^{-j}, M, M}.$$

Taking the infimum over all such representations of T we have

$$\left| \tilde{E}_j \right|_{2^{-j}, N, 0} + \sum_{|\alpha|, |\beta| \leq 1} \left| \tilde{E}_{j, \alpha, \beta} \right|_{2^{-j}, N, 0} \lesssim |T|_{s, M}.$$

Inductively continuing this process, pulling out more and more derivatives, we see that we can define the topology on Calderón-Zygmund operators of order s equivalently in terms of the $\tilde{E}_j$, $\tilde{E}_{j, \alpha, \beta}$, etc., thereby removing the infima from (2.4) and (2.6).

PROOF OF PROPOSITION 2.0.37. Clearly, the space of Calderón-Zygmund operators of order $s \in \mathbb{R}$ is a locally convex topological vector space whose topology is given by a countable family of semi-norms. The only thing to verify is that this space is complete. Let $\{T_n\}$ be a Cauchy sequence of Calderón-Zygmund operators of order s. For each n, we decompose

$$T_n = \sum_{j \in \mathbb{N}^\nu} 2^{js} \tilde{E}_j^n,$$

where $\tilde{E}_j^n$ is as above. By the above remarks, we have $\forall N, \exists M,$

$$\left| \tilde{E}_j^n - \tilde{E}_j^m \right|_{2^{-j}, N, 0} \lesssim |T_n - T_m|_{s, M}.$$

Thus, $\left\{ \tilde{E}_j^n \right\}$ is Cauchy in the $|\cdot|_{2^{-j}, N, 0}$ norm (uniformly in j), and it follows that $\tilde{E}_j^n$ converges (uniformly in j) to some $\tilde{E}_j^\infty$ in the $|\cdot|_{2^{-j}, N, 0}$ norm.

Furthermore, we showed that we can canonically pull out derivatives so that, for instance,

$$\tilde{E}_j^n = \sum_{|\alpha|, |\beta| \leq 1} 2^{-(2 - |\alpha| - |\beta|)j} \left(2^{-j} W \right)^\alpha \tilde{E}_{j, \alpha, \beta}^n \left(2^{-j} W \right)^\beta,$$

and $\forall N, \exists M$ with

$$\sum_{|\alpha|, |\beta| \leq 1} \left| \tilde{E}_{j, \alpha, \beta}^n - \tilde{E}_{j, \alpha, \beta}^m \right|_{2^{-j}, N, 0} \lesssim |T_n - T_m|_{s, M}.$$

Thus, $\tilde{E}_{j, \alpha, \beta}^n$ converge, uniformly in j, to some $\tilde{E}_{j, \alpha, \beta}^\infty$ in the $|\cdot|_{2^{-j}, N, 0}$ norm, for every N. We have

$$\tilde{E}_j^\infty = \sum_{|\alpha|, |\beta| \leq 1} 2^{-(2 - |\alpha| - |\beta|)j} \left(2^{-j} W \right)^\alpha \tilde{E}_{j, \alpha, \beta}^\infty \left(2^{-j} W \right)^\beta.$$

Inductively continuing this process, we have that $\left\{ \left(\widetilde{E}_j^\infty, 2^{-j} \right) \mid j \in \mathbb{N} \right\}$ is a bounded set of elementary operators and $T_\infty = \sum_{j \in \mathbb{N}} 2^{js} \widetilde{E}_j^\infty$ is a Calderón-Zygmund operator of order s. Finally, we have $T_n \to T_\infty$ in the topology of Calderón-Zygmund operators of order s. This proves that the space of Calderón-Zygmund operators of order s is complete. $\qquad\qquad\qquad\qquad\qquad\qquad\qquad\qquad\qquad\qquad\qquad\qquad\qquad\quad$ $\square$

Remark 2.8.2   For $s \in (-Q_1, \infty)$, we could have used Definition 2.0.16 to define a locally convex topology on the space of Calderón-Zygmund operators of order s. I.e., we would take the least $C_{\alpha,\beta}$ in the growth condition as a semi-norm (for each α, β), and let least $C_{\mathcal{B},\alpha}$ for T and T^* in the cancellation condition as a semi-norm (for each $\mathcal{B}$ and α). This gives an uncountable family of semi-norms (for uncountably many choices of $\mathcal{B}$). The proof of Theorem 2.0.29 shows that this topology is the same as the one defined above. Thus, even though this process gives rise to uncountably many semi-norms, the induced space is still a Fréchet space, but this does not seem to be obvious from the definition. This is analogous to Remark 1.1.6.

2.9 THE MAXIMAL FUNCTION

Closely related to Calderón-Zygmund singular integrals is the maximal function corresponding to the balls $B_{(X,d)}(x, \delta)$. Indeed, for $f \in C^\infty(M)$, we define

$$\mathcal{M}f(x) = \sup_{\delta > 0} \frac{1}{\mathrm{Vol}\left(B_{(X,d)}(x, \delta) \right)} \int_{B_{(X,d)}(x,\delta)} |f(y)| \, dy.$$

The main theorem concerning $\mathcal{M}$ is the following.

THEOREM 2.9.1. $\|\mathcal{M}f\|_{L^p} \leq C_p \|f\|_{L^p}$, for $1 < p \leq \infty$.

The proof of Theorem 2.9.1 is quite standard, and we do not prove it here. See [Ste93, Chapter I] for a proof in the more general setting of spaces of homogeneous type.

One of our main uses of $\mathcal{M}$ is to prove bounds for certain vector valued operators. For $p, q \in [1, \infty]$, we define the Banach space $L^p(\ell^q(\mathbb{N}))$ to be the Banach space of sequences of measurable functions $f_j : M \to \mathbb{C}$, $j \in \mathbb{N}$, such that the following norm is finite.

$$\|f\|_{L^p(\ell^q(\mathbb{N}))} := \begin{cases} \left\| \left(\sum_{j \in \mathbb{N}} |f_j|^q \right)^{\frac{1}{q}} \right\|_{L^p} & \text{if } q < \infty, \\ \left\| \sup_{j \in \mathbb{N}} |f_j| \right\|_{L^p} & \text{if } q = \infty. \end{cases}$$

LEMMA 2.9.2. *Let $\mathcal{E}$ be a bounded set of pre-elementary operators. Let*

$$\left\{ (E_j, 2^{-j}) \mid j \in \mathbb{N} \right\} \subset \mathcal{E}.$$

Define the vector valued operator $\mathcal{T} \{f_j\}_{j \in \mathbb{N}} := \{E_j f_j\}_{j \in \mathbb{N}}$. Then, for $1 < p < \infty$, there is a constant $C_{p,\mathcal{E}}$ such that

$$\|\mathcal{T}\|_{L^p(\ell^2(\mathbb{N})) \to L^p(\ell^2(\mathbb{N}))} \leq C_{p,\mathcal{E}}.$$

We separate the proof of Lemma 2.9.2 into several smaller results.

LEMMA 2.9.3. *Let $\mathcal{E}$ be a bounded set of pre-elementary operators. Then there is a constant $C_{\mathcal{E}}$ such that for all $\left(E, 2^{-j}\right) \in \mathcal{E}$, $|Ef(x)| \leq C_{\mathcal{E}} \mathcal{M} f(x)$.*

PROOF. For $\left(E, 2^{-j}\right) \in \mathcal{E}$, we have $|E(x, y)| \lesssim \frac{\left(1+2^j \rho(x,y)\right)^{-1}}{\operatorname{Vol}\left(B_{(X,d)}(x, 2^{-j}+\rho(x,y))\right)}$. Thus,

$$
\begin{aligned}
|Ef(x)| &\lesssim \int \frac{\left(1+2^j \rho(x, y)\right)^{-1}}{\operatorname{Vol}\left(B_{(X,d)}\left(x, 2^{-j}+\rho(x, y)\right)\right)} |f(y)| \, dy \\
&\lesssim \int_{B_{(X,d)}(x,2^{-j})} \frac{1}{\operatorname{Vol}\left(B_{(X,d)}\left(x, 2^{-j}\right)\right)} |f(y)| \, dy \\
&\quad + \sum_{k=1}^{\infty} \int_{B_{(X,d)}(x,2^{-j+k}) \setminus B_{(X,d)}(x,2^{-j+k-1})} \frac{\left(2^j \rho(x, y)\right)^{-1}}{\operatorname{Vol}\left(B_{(X,d)}\left(x, \rho(x, y)\right)\right)} |f(y)| \, dy \\
&\lesssim \sum_{k=0}^{\infty} 2^{-k} \operatorname{Vol}\left(B_{(X,d)}\left(x, 2^{-j+k}\right)\right)^{-1} \int_{B_{(X,d)}(x,2^{-j+k})} |f(y)| \, dy \\
&\lesssim \mathcal{M} f(x).
\end{aligned}
$$

$\square$

LEMMA 2.9.4. *Suppose $\mathcal{E}$ is a bounded set of pre-elementary operators. Then, for $1 \leq p \leq \infty$,*

$$
\sup_{(E,2^{-j}) \in \mathcal{E}} \|E\|_{L^p \to L^p} < \infty.
$$

PROOF. The results for $p = 1$ and $p = \infty$ follow from Lemma 2.0.27. For $1 < p < \infty$, the result follows by interpolation. $\square$

LEMMA 2.9.5. *Let $\mathcal{E}$ be a bounded set of pre-elementary operators, and let*

$$
\left\{\left(E_j, 2^{-j}\right) \mid j \in \mathbb{N}\right\} \subset \mathcal{E}.
$$

Define the vector valued operator $\mathcal{T} \{f_j\}_{j \in \mathbb{N}} := \{E_j f_j\}_{j \in \mathbb{N}}$. Then, there is a constant $C_{\mathcal{E}}$ such that for $1 \leq p < \infty$,

$$
\|\mathcal{T}\|_{L^p(\ell^p(\mathbb{N})) \to L^p(\ell^p(\mathbb{N}))} \leq C_{\mathcal{E}}.
$$

PROOF. By Lemma 2.9.4, we have $\|E_j\|_{L^p \to L^p} \lesssim 1$. Notice,

$$
\left\| \left(\sum_{j \in \mathbb{N}} |E_j f_j|^p\right)^{\frac{1}{p}} \right\|_{L^p}^p = \sum_{j \in \mathbb{N}} \|E_j f\|_{L^p}^p \lesssim \sum_{j \in \mathbb{N}} \|f\|_{L^p}^p = \left\| \left(\sum_{j \in \mathbb{N}} |f_j|^p\right)^{\frac{1}{p}} \right\|_{L^p}^p,
$$

completing the proof. $\square$

LEMMA 2.9.6. *Let $\mathcal{E}$ be a bounded set of pre-elementary operators, and let*

$$\{(E_j, 2^{-j}) \mid j \in \mathbb{N}\} \subset \mathcal{E}.$$

Define the vector valued operator $\mathcal{T}\{f_j\}_{j\in\mathbb{N}} := \{E_j f_j\}_{j\in\mathbb{N}}$. Then, for $1 < p < \infty$, there is a constant $C_{p,\mathcal{E}}$ such that

$$\|\mathcal{T}\|_{L^p(\ell^\infty(\mathbb{N})) \to L^p(\ell^\infty(\mathbb{N}))} \le C_{p,\mathcal{E}}.$$

PROOF. We use Lemma 2.9.3 and Theorem 2.9.1 to see

$$\left\| \sup_{j\in\mathbb{N}} |E_j f_j| \right\|_{L^p} \lesssim \left\| \sup_j \mathcal{M} f_j \right\|_{L^p} \lesssim \left\| \mathcal{M} \sup_j |f_j| \right\|_{L^p} \lesssim \left\| \sup_j |f_j| \right\|_{L^p},$$

completing the proof. ◻

PROOF OF LEMMA 2.9.2. If q is the dual exponent to p ($1 < p < \infty$), then the dual of $L^p(\ell^2(\mathbb{N}))$ is $L^q(\ell^2(\mathbb{N}))$. Moreover, the adjoint of $\mathcal{T}$ is given by $\mathcal{T}^*\{f_j\} = \{E_j^* f_j\}$. Thus, the result for $p > 2$ follows from the result for $p < 2$. For $p \le 2$ the result follows by interpolating Lemmas 2.9.5 and 2.9.6. ◻

The main consequence of Lemma 2.9.2 that we will use is the next proposition.

PROPOSITION 2.9.7. *Suppose $\mathcal{E}$ is a bounded set of elementary operators and let $\{(E_j, 2^{-j}), (F_j, 2^{-j}) \mid j \in \mathbb{N}\} \subset \mathcal{E}$. For $j \in \mathbb{Z} \setminus \mathbb{N}$ set $F_j = 0$. For $l \in \mathbb{Z}$ define the vector valued operator $\mathcal{T}_l\{f_j\}_{j\in\mathbb{N}} = \{E_j F_{j+l} f_j\}_{j\in\mathbb{N}}$. Then for $1 < p < \infty$, and $N \in \mathbb{N}$, there is a constant $C_{p,N,\mathcal{E}}$ such that $\|\mathcal{T}_l\|_{L^p(\ell^2(\mathbb{N})) \to L^p(\ell^2(\mathbb{N}))} \le C_{p,N,\mathcal{E}} 2^{-N|l|}$.*

PROOF. Proposition 2.7.2 shows that, for every N, the set

$$\left\{ \left(2^{N|j_1 - j_2|} E_1 E_2, 2^{-j_1} \right) \mid (E_1, 2^{-j_1}), (E_2, 2^{-j_2}) \in \mathcal{E} \right\}$$

is a bounded set of elementary operators. The result now follows from Lemma 2.9.2. ◻

2.10   NON-ISOTROPIC SOBOLEV SPACES

For $p \in (1, \infty)$, $s \in \mathbb{R}$, let NL_s^p be the completion of $C^\infty(M)$ under the norm

$$\|f\|_{\mathrm{NL}_s^p} := \left\| (I + \mathcal{L})^{s/2} f \right\|_{L^p}.$$

In particular, note $L^p = \mathrm{NL}_0^p$.

THEOREM 2.10.1. *Let T be a Calderón-Zygmund operator of order $t \in \mathbb{R}$. Then, for $p \in (1, \infty)$, $s \in \mathbb{R}$, $T : \mathrm{NL}_s^p \to \mathrm{NL}_{s-t}^p$.*

Just as in Section 1.2, the proof separates into three main parts:

- If T is a Calderón-Zygmund operator of order 0, then $T : L^2 \to L^2$.

- If T is a Calderón-Zygmund operator of order 0, then $T : L^p \to L^p, 1 < p < \infty$.

- Theorem 2.10.1 follows from the above, and the algebra properties of Calderón-Zygmund operators.

We now turn to presenting these three parts in more detail.

LEMMA 2.10.2. *Let $\mathcal{E}$ be a bounded set of elementary operators.*
$\forall N, \exists C, \forall \left(E_1, 2^{-j_1} \right), \left(E_2, 2^{-j_2} \right) \in \mathcal{E}$,

$$\left\| E_1^* E_2 \right\|_{L^2 \to L^2}, \left\| E_2 E_1^* \right\|_{L^2 \to L^2} \leq C 2^{-N|j_1 - j_2|}.$$

PROOF. Proposition 2.0.24 shows $\left\{ \left(E^*, 2^{-j} \right) \mid \left(E, 2^{-j} \right) \in \mathcal{E} \right\}$ is a bounded set of elementary operators. Using this, it follows from Proposition 2.7.2 that

$$\left\{ \left(2^{N|j_1 - j_2|} E_1^* E_2, 2^{-j_1 \wedge j_2} \right), \left(2^{N|j_1 - j_2|} E_2 E_1^*, 2^{-j_1 \wedge j_2} \right) \right.$$

$$\left. \mid \left(E_1, 2^{-j_1} \right), \left(E_2, 2^{-j_2} \right) \in \mathcal{E} \right\}$$

is a bounded set of elementary operators. The result now follows from Lemma 2.9.4.
□

PROPOSITION 2.10.3. *Suppose T is a Calderón-Zygmund operator of order 0. Then $T : L^2 \to L^2$. Moreover, if we decompose $T = \sum_{j \in \mathbb{N}} E_j$, where $\left\{ \left(E_j, 2^{-j} \right) \mid j \in \mathbb{N} \right\}$ is a bounded set of elementary operators, then the sum $\sum_{j \in \mathbb{N}} E_j$ converges to T in the strong operator topology as bounded operators $L^2 \to L^2$.*

PROOF. Lemma 2.10.2 shows $\left\| E_j E_k^* \right\|_{L^2 \to L^2}, \left\| E_j^* E_k \right\|_{L^2 \to L^2} \lesssim 2^{-|j-k|}$ and the result follows from the Cotlar-Stein Lemma (Lemma 1.2.26). □

PROPOSITION 2.10.4. *Suppose T is a Calderón-Zygmund operator of order 0. Then $T : L^p \to L^p, 1 < p < \infty$.*

PROOF SKETCH. The main goal is to show that T is weak-type $(1, 1)$. The proof of this is similar to the proof of Proposition 1.1.7 (using Proposition 2.10.3). See [Ste93] for more details. Once this is done, interpolation shows $T : L^p \to L^p, 1 < p \leq 2$. Since T is a Calderón-Zygmund operator of order 0 if and only if T^* is, the result for $2 < p < \infty$ follows by duality.

In fact, this proof works in the more general setting of "spaces of homogeneous type." See [Ste93, Chatper I] for details on this. □

Remark 2.10.5 One does not need to proceed via Proposition 2.10.3 to conclude Proposition 2.10.4. Indeed, T satisfies the hypotheses of the so-called $T(1)$ theorem. We refer the reader to [Ste93] for more details on this approach.

PROOF OF THEOREM 2.10.1. Let T be a Calderón-Zygmund operator of order $t \in \mathbb{R}$ and fix $s \in \mathbb{R}$. We wish to show

$$\|Tf\|_{\mathrm{NL}^p_{s-t}} \lesssim \|f\|_{\mathrm{NL}^p_s}, \quad f \in C^\infty(M). \tag{2.69}$$

Let $g = (I + \mathcal{L})^{s/2} f \in C^\infty(M)$ so that $f = (I + \mathcal{L})^{-s/2} g$. (2.69) is equivalent to showing

$$\left\| (I + \mathcal{L})^{(s-t)/2} T (I + \mathcal{L})^{-s/2} g \right\|_{L^p} \lesssim \|g\|_{L^p}. \tag{2.70}$$

Using that $(I + \mathcal{L})^{s_0}$ is a Calderón-Zygmund operator of order $2s_0$ (Corollary 2.6.10), we have that $(I + \mathcal{L})^{(s-t)/2} T (I + \mathcal{L})^{-s/2}$ is a Calderón-Zygmund operator of order 0 (by Proposition 2.0.31). (2.70) now follows from Proposition 2.10.4. $\qquad\square$

We now present another characterization of the space NL^p_s. This is useful for motivating our later definitions. Note that $I : C^\infty(M) \to C^\infty(M)$ (the identity operator) is clearly a Calderón-Zygmund operator of order 0 (one may see this by applying Lemma 2.7.1 or by using either (i) or (ii) of Theorem 2.0.29). As such, there is a bounded set of elementary operators $\left\{ \left(D_j, 2^{-j} \right) \mid j \in \mathbb{N} \right\}$ with $I = \sum_{j \in \mathbb{N}} D_j$. This allows us to characterize NL^p_s via a "Littlewood-Paley square function":

THEOREM 2.10.6. *Fix* $1 < p < \infty$, $s \in \mathbb{R}$.

$$\|f\|_{\mathrm{NL}^p_s} \approx \left\| \left(\sum_{j \in \mathbb{N}} \left| 2^{sj} D_j f \right|^2 \right)^{\frac{1}{2}} \right\|_{L^p}, \quad f \in C^\infty(M),$$

where the implicit constants depend on p, s, and the particular decomposition $I = \sum_{j \in \mathbb{N}} D_j$.

Remark 2.10.7   In particular, note that while $\left\| \left(\sum_{j \in \mathbb{N}} \left| 2^{sj} D_j f \right|^2 \right)^{\frac{1}{2}} \right\|_{L^2}$ clearly depends on the choice of D_j in the decomposition $I = \sum_{j \in \mathbb{N}} D_j$, the *equivalence class* of the norm does not because $\|\cdot\|_{\mathrm{NL}^p_s}$ does not depend on any such choice.

To prove Theorem 2.10.6, we begin with the case $s = 0$.

PROPOSITION 2.10.8. *Fix* $1 < p < \infty$, *then*

$$\|f\|_{L^p} \approx \left\| \left(\sum_{j \in \mathbb{N}} |D_j f|^2 \right)^{\frac{1}{2}} \right\|_{L^p}, \quad f \in C^\infty(M),$$

where the implicit constants depend on p and the particular decomposition of $I = \sum_{j \in \mathbb{N}} D_j$.

We begin with the $\gtrsim$ side of the inequality.

LEMMA 2.10.9. *For $1 < p < \infty$,*

$$\left\|\left(\sum_{j\in\mathbb{N}}|D_jf|^2\right)^{\frac{1}{2}}\right\|_{L^p} \lesssim \|f\|_{L^p}, \quad f \in C^\infty(M).$$

The same result holds with D_j^* in place of D_j.

We include the proof of Lemma 2.10.9, as these techniques are fundamental to the ideas in later chapters. However, the proof is very standard; see, for instance, [Ste93, page 267] and [Ste70a, Chapter 4, Section 5]. We begin by introducing the Khintchine inequality, which we state without proof.

THEOREM 2.10.10 (The Khintchine inequality–see [Haa81]). *Let $\{\epsilon_j\}_{j=0}^\infty$ be a sequence of i.i.d. random variables with mean zero taking values ± 1. Let $x_0, x_1, \ldots$ be a sequence of complex numbers, and fix $0 < p < \infty$. There are constant A_p, B_p, depending only on p such that*

$$A_p\left(\sum_{j=0}^\infty |x_j|^2\right)^{\frac{1}{2}} \le \left(\mathbb{E}\left|\sum_{j=0}^\infty \epsilon_j x_j\right|^p\right)^{\frac{1}{p}} \le B_p\left(\sum_{j=0}^\infty |x_j|^2\right)^{\frac{1}{2}}.$$

PROOF OF LEMMA 2.10.9. Let $\{(E_j, 2^{-j}) \mid j \in \mathbb{N}\}$ be a bounded set of elementary operators. We will show

$$\left\|\left(\sum_{j\in\mathbb{N}}|E_jf|^2\right)^{\frac{1}{2}}\right\|_{L^p} \lesssim \|f\|_{L^p}. \tag{2.71}$$

Let $\epsilon_j \in \{\pm 1\}$ be a sequence. It suffices to show $\left\|\sum_{j\in\mathbb{N}}\epsilon_j E_j f\right\|_{L^p} \lesssim \|f\|_{L^p}$, with implicit constant independent of the sequence; (2.71) then follows from the Khintchine inequality. However, $\sum_{j\in\mathbb{N}}\epsilon_j E_j$ is a Calderón-Zygmund operator of order 0 (uniformly in the choice of the sequence) and is therefore bounded on L^p ($1 < p < \infty$) by Proposition 2.10.4. (2.71) follows. The result now follows for D_j. The result with D_j replaced with D_j^* follows from this and the fact that $\{(D_j^*, 2^{-j}) \mid j \in \mathbb{N}\}$ is a bounded set of elementary operators (by (b) of Proposition 2.0.24). $\square$

For $l \in \mathbb{Z} \setminus \mathbb{N}$, define $D_l = 0$. For $M > 0$ define

$$U_M = \sum_{\substack{j\in\mathbb{N}\\|l|\le M}} D_j D_{j+l}, \quad R_M = \sum_{\substack{j\in\mathbb{N}\\|l|>M}} D_j D_{j+l}.$$

We have the following Calderón reproducing-type formula.

PROPOSITION 2.10.11. *Fix p, $1 < p < \infty$. There exists $M = M(p)$ such that $U_M : L^p \to L^p$ is an isomorphism. I.e., there is a bounded operator $V_M : L^p \to L^p$ with $U_M V_M = V_M U_M = I$.*

To prove Proposition 2.10.11 we use the following lemma.

LEMMA 2.10.12. *Fix p, $1 < p < \infty$. $\lim_{M \to \infty} \|R_M\|_{L^p \to L^p} = 0$.*

PROOF OF PROPOSITION 2.10.11 GIVEN LEMMA 2.10.12. Note that $U_M = I - R_M$. We apply Lemma 2.10.12 to take $M = M(p)$ so large $\|R_M\|_{L^p \to L^p} < 1$. Set $V_M = \sum_{m=0}^{\infty} R_M^m$, with convergence in the uniform operator topology as operators $L^p \to L^p$. It is direct to verify that V_M satisfies the conclusions of the proposition. $\square$

Lemma 2.10.12 follows immediately by interpolating the next two lemmas.

LEMMA 2.10.13. *For $1 < p < \infty$, $M \geq 1$,*

$$\|R_M\|_{L^p \to L^p} \leq C_p M.$$

LEMMA 2.10.14. *For every $N > 0$, $\|R_M\|_{L^2 \to L^2} \leq C_N 2^{-NM}$.*

PROOF OF LEMMA 2.10.13. Since $R_M = I - U_M$, it suffices to prove the result with U_M in place of R_M. Let L^q be dual to L^p. Fix $f \in L^p$ and $g \in L^q$ with $\|g\|_{L^q} = 1$. Consider,

$$
\begin{aligned}
|\langle g, U_M f \rangle| &\leq \sum_{|l| \leq M} \left| \left\langle g, \sum_{j \in \mathbb{N}} D_j D_{j+l} f \right\rangle \right| \\
&= \sum_{|l| \leq M} \left| \sum_{j \in \mathbb{N}} \langle D_j^* g, D_{j+l} f \rangle \right| \\
&\leq \sum_{|l| \leq M} \int \left(\sum_{j \in \mathbb{N}} |D_j^* g|^2 \right)^{\frac{1}{2}} \left(\sum_{j \in \mathbb{N}} |D_j f|^2 \right)^{\frac{1}{2}} \\
&\lesssim M \left\| \left(\sum_{j \in \mathbb{N}} |D_j^* g|^2 \right)^{\frac{1}{2}} \right\|_{L^q} \left\| \left(\sum_{j \in \mathbb{N}} |D_j f|^2 \right)^{\frac{1}{2}} \right\|_{L^p} \\
&\lesssim M \|g\|_{L^q} \|f\|_{L^p} \\
&= M \|f\|_{L^p},
\end{aligned}
$$

where in the second-to-last line we applied Lemma 2.10.9. Taking the supremum over all such g yields the result. $\square$

PROOF OF LEMMA 2.10.14. We wish to apply the Cotlar-Stein Lemma to the sum $R_M = \sum_{\substack{j \in \mathbb{N} \\ |l| > M}} D_j D_{j+l}$. We know $\{(D_j, 2^{-j}) \mid j \in \mathbb{N}\}$ and $\{(D_j^*, 2^{-j}) \mid j \in \mathbb{N}\}$ are bounded sets of elementary operators (see (b) of Proposition 2.0.24). For four numbers j_1, j_2, j_3, j_4, we let diam $\{j_1, j_2, j_3, j_4\} = \max_{1 \leq l_1, l_2 \leq 4} |j_{l_1} - j_{l_2}|$. Repeated applications of Proposition 2.7.2 show, for every N, the sets

$$\left\{ \left(2^{N \text{diam}\{j_1, j_1+l_1, j_2, j_2+l_2\}} D_{j_1} D_{j_1+l_1} D_{j_2+l_2}^* D_{j_2}^*, 2^{-j_1} \right) \mid j_1, j_2 \in \mathbb{N}, l_1, l_2 \in \mathbb{Z} \right\},$$

$$\left\{ \left(2^{N\operatorname{diam}\{j_1,j_1+l_1,j_2,j_2+l_2\}} D^*_{j_1+l_1} D^*_{j_1} D_{j_2} D_{j_2+l_2}, 2^{-j_1} \right) \mid j_1, j_2 \in \mathbb{N}, l_1, l_2 \in \mathbb{Z} \right\},$$

are bounded sets of elementary operators. Lemma 2.9.4 shows, for every N,

$$\left\| D_{j_1} D_{j_1+l_1} D^*_{j_2+l_2} D^*_{j_2} \right\|_{L^2 \to L^2} \lesssim 2^{-N\operatorname{diam}\{j_1,j_1+l_1,j_2,j_2+l_2\}},$$

$$\left\| D^*_{j_1+l_1} D^*_{j_1} D_{j_2} D_{j_2+l_2} \right\|_{L^2 \to L^2} \lesssim 2^{-N\operatorname{diam}\{j_1,j_1+l_1,j_2,j_2+l_2\}}.$$

The Cotlar-Stein Lemma (Lemma 1.2.26) now shows

$$\|R_M\|_{L^2 \to L^2} \lesssim \sup_{\substack{j_1 \in \mathbb{N} \\ |l_1| > M}} \sum_{\substack{j_2 \in \mathbb{N} \\ |l_2| > M}} 2^{-N\operatorname{diam}\{j_1,j_1+l_1,j_2,j_2+l_2\}/2} \lesssim 2^{-NM/2},$$

completing the proof with N replaced by $N/2$. $\qquad\square$

PROOF OF PROPOSITION 2.10.8. The $\gtrsim$ part of the result follows from Lemma 2.10.9, so it suffices to prove only $\lesssim$. Fix p, $1 < p < \infty$, and take $M = M(p)$ and V_M as in Proposition 2.10.11. Let L^q be dual to L^p so that $V^*_M : L^q \to L^q$. Fix $g \in L^q$ with $\|g\|_{L^q} = 1$. We have, for $f \in L^p$,

$$|\langle g, f \rangle| = |\langle V^*_M g, U_M f \rangle|$$

$$\leq \sum_{|l| \leq M} \left| \sum_{j \in \mathbb{N}} \langle D^*_j V_M g, D_{j+l} f \rangle \right|$$

$$\leq \sum_{|l| \leq M} \int \left(\sum_{j \in \mathbb{N}} |D^*_j V_M g|^2 \right)^{\frac{1}{2}} \left(\sum_{j \in \mathbb{N}} |D_j f|^2 \right)^{\frac{1}{2}}$$

$$\leq \sum_{|l| \leq M} \left\| \left(\sum_{j \in \mathbb{N}} |D^*_j V_M g|^2 \right)^{\frac{1}{2}} \right\|_{L^q} \left\| \left(\sum_{j \in \mathbb{N}} |D_j f|^2 \right)^{\frac{1}{2}} \right\|_{L^p}$$

$$\lesssim M \|V^*_M g\|_{L^q} \left\| \left(\sum_{j \in \mathbb{N}} |D_j f|^2 \right)^{\frac{1}{2}} \right\|_{L^p}$$

$$\lesssim \left\| \left(\sum_{j \in \mathbb{N}} |D_j f|^2 \right)^{\frac{1}{2}} \right\|_{L^p},$$

where in the second-to-last line, we have applied Lemma 2.10.9, and in the last line used that M is fixed since p is, $V^*_M : L^q \to L^q$, and $\|g\|_{L^q} = 1$. Taking the supremum over all such g completes the proof. $\qquad\square$

Temporarily, for $1 < p < \infty$, $s \in \mathbb{R}$, define a new Banach space, $\widetilde{\mathrm{NL}}_s^p$ to be the completion of $C^\infty(M)$ under the norm

$$\|f\|_{\widetilde{\mathrm{NL}}_s^p} := \left\| \left(\sum_{j \in \mathbb{N}} \left| 2^{js} D_j f \right|^2 \right)^{\frac{1}{2}} \right\|_{L^p}.$$

The content of Theorem 2.10.6 (which we have yet to complete the proof of) is that $\mathrm{NL}_s^p = \widetilde{\mathrm{NL}}_s^p$. We have already established the case $s = 0$ (Proposition 2.10.8). We need the following analog of Theorem 2.10.1.

PROPOSITION 2.10.15. *Let T be a Calderón-Zygmund operator of order $t \in \mathbb{R}$. Then for $p \in (1, \infty)$, $s \in \mathbb{R}$, we have $T : \widetilde{\mathrm{NL}}_s^p \to \widetilde{\mathrm{NL}}_{s-t}^p$.*

To prove Proposition 2.10.15 we need a technical result. Let $\mathcal{E}$ be a bounded set of elementary operators, and fix $p \in (1, \infty)$ and $s \in \mathbb{R}$. Define the norm, for $f \in C^\infty(M)$,

$$\|f\|_{p,s,\mathcal{E}} := \sup_{\{(E_j, 2^{-j}) \mid j \in \mathbb{N}\} \subseteq \mathcal{E}} \left\| \left(\sum_{j \in \mathbb{N}^\nu} \left| 2^{js} E_j f \right|^2 \right)^{\frac{1}{2}} \right\|_{L^p}.$$

If we take $\mathcal{E} := \{(D_j, 2^{-j}) \mid j \in \mathbb{N}\}$, we see $\|\cdot\|_{\widetilde{\mathrm{NL}}_s^p} = \|\cdot\|_{p,s,\mathcal{E}}$. The next result shows that $\|\cdot\|_{\widetilde{\mathrm{NL}}_s^p}$ dominates all other choices of $\mathcal{E}$.

PROPOSITION 2.10.16. *Let $\mathcal{E}$ be a bounded set of elementary operators, and let $p \in (1, \infty)$, $s \in \mathbb{R}$. Then, there exists $C = C(p, s, \mathcal{E})$ such that for all $f \in C^\infty(M)$,*

$$\|f\|_{p,s,\mathcal{E}} \leq C \|f\|_{\widetilde{\mathrm{NL}}_s^p}.$$

PROOF. Fix $s \in \mathbb{R}$. For $j, k, l \in \mathbb{Z}$ with $|l| \geq |k|$ define a new operator by

$$F_{j,k,l} := 2^{-ks+|k|+|l|} D_{j+k} D_{j+k+l},$$

where we recall that for $m \in \mathbb{Z} \setminus \mathbb{N}$, we define $D_m = 0$. Define

$$\mathcal{E}' := \mathcal{E} \bigcup \left\{ (F_{j,k,l}, 2^{-j-k}) \mid j, k, l \in \mathbb{Z}, j + k \in \mathbb{N}, |l| \geq |k| \right\}.$$

Proposition 2.7.2 shows $\mathcal{E}'$ is a bounded set of elementary operators. Notice that $\|f\|_{p,s,\mathcal{E}} \leq \|f\|_{p,s,\mathcal{E}'}$, and so it suffices to prove the result with $\mathcal{E}$ replaced by $\mathcal{E}'$. Let $\{(E_j, 2^{-j}) \mid j \in \mathbb{N}\} \subseteq \mathcal{E}'$. Consider, where $I : C^\infty(M) \to C^\infty(M)$ denotes the identity operator,

$$
\left\| \left(\sum_{j \in \mathbb{N}} |2^{js} E_j f|^2 \right)^{\frac{1}{2}} \right\|_{L^p} = \left\| \left(\sum_{j \in \mathbb{N}} |2^{js} E_j II f|^2 \right)^{\frac{1}{2}} \right\|_{L^p}
$$

$$
= \left\| \left(\sum_{j \in \mathbb{N}} \left| \sum_{k,l \in \mathbb{Z}} 2^{js} E_j D_{j+k} D_{j+k+l} f \right|^2 \right)^{\frac{1}{2}} \right\|_{L^p} \tag{2.72}
$$

$$
\leq \sum_{k,l \in \mathbb{Z}} \left\| \left(\sum_{j \in \mathbb{N}} |2^{js} E_j D_{j+k} D_{j+k+l} f|^2 \right)^{\frac{1}{2}} \right\|_{L^p},
$$

where the last line uses the triangle inequality.

Fix M large, to be chosen later. We separate the right-hand side of (2.72) into three terms,

$$
(I) := \sum_{\substack{k,l \in \mathbb{Z} \\ |l| \leq M}} \left\| \left(\sum_{j \in \mathbb{N}} |2^{js} E_j D_{j+k} D_{j+k+l} f|^2 \right)^{\frac{1}{2}} \right\|_{L^p},
$$

$$
(II) := \sum_{\substack{k,l \in \mathbb{Z} \\ |l| > M \\ |k| \geq |l|}} \left\| \left(\sum_{j \in \mathbb{N}} |2^{js} E_j D_{j+k} D_{j+k+l} f|^2 \right)^{\frac{1}{2}} \right\|_{L^p},
$$

$$
(III) := \sum_{\substack{k,l \in \mathbb{Z} \\ |l| > M \\ |l| > |k|}} \left\| \left(\sum_{j \in \mathbb{N}} |2^{js} E_j D_{j+k} D_{j+k+l} f|^2 \right)^{\frac{1}{2}} \right\|_{L^p}.
$$

Notice that the right-hand side of (2.72) is exactly $(I)+(II)+(III)$. We bound each of these terms separately. In what follows, implicit constants may depend on $p \in (1, \infty)$, $s \in \mathbb{R}$, and $\mathcal{E}'$, but not on k, l, M, f, or the particular choice of $\{ (E_j, 2^{-j}) \mid j \in \mathbb{N} \} \subseteq \mathcal{E}'$.

We begin with (I). Define a vector valued operator, for $k \in \mathbb{Z}$,

$$
T_k^1 \{f_j\}_{j \in \mathbb{N}} := \left\{ 2^{|k|-ks} E_j D_{j+k} f_j \right\}_{j \in \mathbb{N}}.
$$

Proposition 2.9.7 shows $\|\mathcal{T}_k\|_{L^p(\ell^2(\mathbb{N}))\to L^p(\ell^2(\mathbb{N}))} \lesssim 1$. We have

$$(I) = \sum_{\substack{k,l\in\mathbb{Z}\\|l|\leq M}} 2^{-ls-|k|} \left\|\left(\sum_{j\in\mathbb{N}} \left|\left(2^{|k|-ks}E_jD_{j+k}\right)\left(2^{(j+k+l)s}D_{j+k+l}\right)f\right|^2\right)^{\frac{1}{2}}\right\|_{L^p}$$

$$= \sum_{\substack{k,l\in\mathbb{Z}\\|l|\leq M}} 2^{-ls-|k|} \left\|\mathcal{T}_k^1\left\{2^{(j+k+l)s}D_{j+k+l}f\right\}_{j\in\mathbb{N}}\right\|_{L^p(\ell^2(\mathbb{N}))}$$

$$\lesssim \sum_{\substack{k,l\in\mathbb{Z}\\|l|\leq M}} 2^{-ls-|k|} \|f\|_{\widetilde{\mathrm{NL}}_s^p}$$

$$\lesssim 2^{M|s|} \|f\|_{\widetilde{\mathrm{NL}}_s^p}.$$

We now bound (II). For $k,l\in\mathbb{Z}$ with $|k|\geq|l|$ define the vector valued operator

$$\mathcal{T}_{k,l}^2\{f_j\}_{j\in\mathbb{N}} := \left\{2^{|k|+|l|-(k+l)s}E_jD_{j+k}f_j\right\}_{j\in\mathbb{N}}.$$

Proposition 2.9.7 shows $\left\|\mathcal{T}_{k,l}^2\right\|_{L^p(\ell^2(\mathbb{N}))\to L^p(\ell^2(\mathbb{N}))} \lesssim 1$ (this uses that $|k|\geq|l|$). We have

$$(II) = \sum_{\substack{k,l\in\mathbb{Z}\\|l|>M\\|k|\geq|l|}} 2^{-|k|-|l|} \left\|\left(\sum_{j\in\mathbb{N}} \left|\left(2^{|k|+|l|-(k+l)s}E_jD_{j+k}\right)\left(2^{(j+k+l)s}D_{j+k+l}\right)f\right|^2\right)^{\frac{1}{2}}\right\|_{L^p}$$

$$= \sum_{\substack{k,l\in\mathbb{Z}\\|l|>M\\|k|\geq|l|}} 2^{-|k|-|l|} \left\|\mathcal{T}_{k,l}^2\left\{2^{(j+k+l)s}D_{j+k+l}f\right\}_{j\in\mathbb{N}}\right\|_{L^p(\ell^2(\mathbb{N}))}$$

$$\lesssim \sum_{\substack{k,l\in\mathbb{Z}\\|l|>M\\|k|\geq|l|}} 2^{-|k|-|l|} \|f\|_{\widetilde{\mathrm{NL}}_s^p}$$

$$\lesssim 2^{-M} \|f\|_{\widetilde{\mathrm{NL}}_s^p}.$$

We now bound (III). Define a vector valued operator

$$\mathcal{T}^3\{f_j\}_{j\in\mathbb{N}} := \{E_jf_j\}_{j\in\mathbb{N}}.$$

Lemma 2.9.2 shows $\left\|\mathcal{T}^3\right\|_{L^p(\ell^2(\mathbb{N}))\to L^p(\ell^2(\mathbb{N}))} \lesssim 1$. Consider,

$$
(III) = \sum_{\substack{k,l\in\mathbb{Z} \\ |l|>M \\ |l|>|k|}} 2^{-|k|-|l|} \left\| \left(\sum_{j\in\mathbb{N}} \left| E_j \left(2^{(j+k)s} 2^{-ks+|k|+|l|} D_{j+k} D_{j+k+l} \right) f \right|^2 \right)^{\frac{1}{2}} \right\|_{L^p}
$$

$$
= \sum_{\substack{k,l\in\mathbb{Z} \\ |l|>M \\ |l|>|k|}} 2^{-|k|-|l|} \left\| \mathcal{T}^3 \left\{ 2^{(j+k)s} F_{j,k,l} f \right\}_{j\in\mathbb{N}^\nu} \right\|_{L^p(\ell^2(\mathbb{N}))}
$$

$$
\lesssim \sum_{\substack{k,l\in\mathbb{Z} \\ |l|>M \\ |l|>|k|}} 2^{-|k|-|l|} \left\| \left\{ 2^{(j+k)s} F_{j,k,l} f \right\}_{j\in\mathbb{N}^\nu} \right\|_{L^p(\ell^2(\mathbb{N}))}
$$

$$
\leq \sum_{\substack{k,l\in\mathbb{Z} \\ |l|>M \\ |l|>|k|}} 2^{-|k|-|l|} \|f\|_{p,s,\mathcal{E}'}
$$

$$
\lesssim 2^{-M} \|f\|_{p,s,\mathcal{E}'}.
$$

Plugging these estimates into (2.72), we have

$$
\left\| \left(\sum_{j\in\mathbb{N}} |2^{js} E_j f|^2 \right)^{\frac{1}{2}} \right\|_{L^p} \leq (I) + (II) + (III) \lesssim 2^{M|s|} \|f\|_{\widetilde{\mathrm{NL}}_s^p} + 2^{-M} \|f\|_{p,s,\mathcal{E}'}.
$$

Taking the supremum over all $\left\{ (E_j, 2^{-j}) \mid j\in\mathbb{N} \right\} \subseteq \mathcal{E}'$ we have that there exists $C = C(p,s,\mathcal{E}')$ (but independent of M and f) such that

$$
\|f\|_{p,s,\mathcal{E}'} \leq C 2^{M|s|} \|f\|_{\widetilde{\mathrm{NL}}_s^p} + C 2^{-M} \|f\|_{p,s,\mathcal{E}'}.
$$

Taking M so large that $C 2^{-M} \leq \frac{1}{2}$ we have

$$
\|f\|_{p,s,\mathcal{E}'} \leq 2C 2^{M|s|} \|f\|_{\widetilde{\mathrm{NL}}_s^p},
$$

completing the proof. $\square$

LEMMA 2.10.17. *Let T be a Calderón-Zygmund operator of order $t\in\mathbb{R}$, and let $\mathcal{E}$ be a bounded set of elementary operators. Then,*

$$
\left\{ \left(2^{-jt} ET, 2^{-j} \right) \mid (E, 2^{-j}) \in \mathcal{E} \right\}
$$

is a bounded set of elementary operators.

PROOF. Using (b) of Proposition 2.0.24, it suffices to show

$$\left\{ \left(2^{-jt} T^* E^*, 2^{-j} \right) \mid (E, 2^{-j}) \in \mathcal{E} \right\}$$

is a bounded set of elementary operators. In light of Proposition 2.0.33, T^* is a Calderón-Zygmund operator of order t and in light of (b) of Proposition 2.0.24

$$\left\{ \left(E^*, 2^{-j} \right) \mid (E, 2^{-j}) \in \mathcal{E} \right\}$$

is a bounded set of elementary operators. The result now follows from (ii) of Theorem 2.0.29. $\qquad\square$

PROOF OF PROPOSITION 2.10.15. For $j \in \mathbb{N}$, let $E_j = 2^{-jt} D_j T$, so that (by Lemma 2.10.17) $\left\{ (E_j, 2^{-j}) \mid j \in \mathbb{N} \right\}$ is a bounded set of elementary operators. For $f \in C^\infty (M)$, we have

$$
\begin{aligned}
\|Tf\|_{\widetilde{\mathrm{NL}}^p_{s-t}} &= \left\| \left(\sum_{j \in \mathbb{N}} \left| 2^{js-jt} D_j T f \right|^2 \right)^{\frac{1}{2}} \right\|_{L^p} \\
&= \left\| \left(\sum_{j \in \mathbb{N}} \left| 2^{js} E_j f \right|^2 \right)^{\frac{1}{2}} \right\|_{L^p} \\
&\lesssim \|f\|_{\widetilde{\mathrm{NL}}^p_s},
\end{aligned}
$$

where the last line follows from Proposition 2.10.16, which completes the proof. $\qquad\square$

PROOF OF THEOREM 2.10.6. Fix $1 < p < \infty$, $s \in \mathbb{R}$. Proposition 2.10.15 shows that $(I + \mathcal{L})^{-s/2} : \widetilde{\mathrm{NL}}^p_0 \to \widetilde{\mathrm{NL}}^p_s$ is continuous, with continuous inverse $(I + \mathcal{L})^{s/2} : \widetilde{\mathrm{NL}}^p_s \to \widetilde{\mathrm{NL}}^p_0$. Thus, $(I + \mathcal{L})^{-s/2} : \widetilde{\mathrm{NL}}^p_0 \to \widetilde{\mathrm{NL}}^p_s$ is an isomorphism. But, $\widetilde{\mathrm{NL}}^p_0 = L^p$ (Proposition 2.10.8) and $(I + \mathcal{L})^{-s/2} : L^p \to \mathrm{NL}^p_s$ is an isomorphism by the definition of NL^p_s. We conclude that $\mathrm{NL}^p_s = \widetilde{\mathrm{NL}}^p_s$, with equivalence of norms. $\qquad\square$

We now turn to another characterization of NL^p_s, which only makes sense for $s \in \mathbb{N}$.

THEOREM 2.10.18. *Fix* $1 < p < \infty$ *and* $N \in \mathbb{N}$. *Then,*

$$\|f\|_{\mathrm{NL}^p_N} \approx \sum_{|\alpha| \leq N} \|W^\alpha f\|_{L^p}, \quad f \in C^\infty (M),$$

where the implicit constants depend on p *and* N.

Theorem 2.10.18 follows from repeated applications of the next lemma.

LEMMA 2.10.19. *Fix* $1 < p < \infty$ *and* $s \in \mathbb{R}$. *Then,*

$$\|f\|_{\mathrm{NL}^p_s} \approx \sum_{|\alpha| \leq 1} \|W^\alpha f\|_{\mathrm{NL}^p_{s-1}}.$$

PROOF. For $|\alpha| \leq 1$, W^α is a Calderón-Zygmund operator of order 1, and Theorem 2.10.1 shows $\sum_{|\alpha|\leq 1} \|W^\alpha f\|_{\mathrm{NL}^p_s} \lesssim \|f\|_{\mathrm{NL}^p_s}$. For the reverse inequality, note that $(I + \mathcal{L})^{1/2} : \mathrm{NL}^p_s \to \mathrm{NL}^p_{s-1}$ is an isometry. We have

$$
\begin{aligned}
\|f\|_{\mathrm{NL}^p_s} &= \left\| (I + \mathcal{L})^{1/2} f \right\|_{\mathrm{NL}^p_{s-1}} = \left\| (I + \mathcal{L})^{-1/2} (I + \mathcal{L}) f \right\|_{\mathrm{NL}^p_{s-1}} \\
&\leq \left\| (I + \mathcal{L})^{-1/2} f \right\|_{\mathrm{NL}^p_{s-1}} + \sum_{j=1}^{r} \left\| (I + \mathcal{L})^{-1/2} W_j^* W_j f \right\|_{\mathrm{NL}^p_{s-1}} \\
&\lesssim \sum_{|\alpha|\leq 1} \|W^\alpha f\|_{\mathrm{NL}^p_{s-1}},
\end{aligned}
$$

where we have used in the last line for $|\alpha| \leq 1$, $(I + \mathcal{L})^{-1/2} W^\alpha$ is a Calderón-Zygmund operator of order 0 and therefore bounded on NL^p_{s-1} (Theorem 2.10.1). $\qquad\square$

2.11 MAXIMAL HYPOELLIPTICITY REVISITED

In this section, we prove Theorem 2.4.8: that for $\mathcal{P} = P(W)$, where $\deg P = k$, the following are equivalent:

(i) $\mathcal{P}$ and $\mathcal{P}^*$ are maximally hypoelliptic.

(ii) There is a Calderón-Zygmund operator, $T : C^\infty(M) \to C^\infty(M)$, of order $-k$ such that $\mathcal{P}T, T\mathcal{P} \equiv I \mod C^\infty(M \times M)$.

We also discuss some related results.

PROOF OF (ii)$\Rightarrow$(i) OF THEOREM 2.4.8. Suppose that (ii) holds, so that there is a Calderón-Zygmund operator T of order $-k$ with $\mathcal{P}T, T\mathcal{P} \equiv I \mod C^\infty(M \times M)$. Let $T\mathcal{P} = I + R$, so that $R \in C^\infty(M \times M)$. For $|\alpha| \leq k$, it is immediate to verify that W^α is a Calderón-Zygmund operator of order $|\alpha|$. Thus, for every $|\alpha| \leq k$, $W^\alpha T$ is a Calderón-Zygmund operator of order ≤ 0, and therefore is a Calderón-Zygmund operator of order 0. Using that operators of order 0 are bounded on L^2, we have

$$
\sum_{|\alpha|\leq k} \|W^\alpha T f\|_{L^2} \lesssim \|f\|_{L^2}, \quad f \in C^\infty(M).
$$

Replacing f with $\mathcal{P}g$, for $g \in C^\infty(M)$, and using $T\mathcal{P} = I + R$, we have,

$$
\sum_{|\alpha|\leq k} \|W^\alpha g\|_{L^2} \lesssim \|\mathcal{P}g\|_{L^2} + \sum_{|\alpha|\leq k} \|W^\alpha R g\|_{L^2}, \quad g \in C^\infty(M).
$$

Since $R \in C^\infty(M \times M)$, $W^\alpha R : L^2 \to L^2$, $\forall \alpha$, and we conclude,

$$
\sum_{|\alpha|\leq k} \|W^\alpha g\|_{L^2} \lesssim \|\mathcal{P}g\|_{L^2} + \|g\|_{L^2}, \quad g \in C^\infty(M),
$$

and therefore $\mathcal{P}$ is maximally hypoelliptic.

Taking adjoints of the fact that $\mathcal{P}T \equiv I \mod C^\infty (M \times M)$, we see $T^*\mathcal{P}^* \equiv I \mod C^\infty (M \times M)$. Since T^* is a Calderón-Zygmund operator of order $-k$, the same proof applies to show that $\mathcal{P}^*$ is maximally hypoelliptic, thereby establishing (i). $\quad\square$

We now turn to (i)$\Rightarrow$(ii). For this, we use the following result.

THEOREM 2.11.1. *Suppose $\mathcal{P}$ is a differential operator and $\mathcal{P}$ and $\mathcal{P}^*$ are both hypoelliptic. Then there is a two sided parametrix $T : C^\infty (M) \to C^\infty (M)$ satisfying $\mathcal{P}T, T\mathcal{P} \equiv I \mod C^\infty (M \times M)$; i.e., the Schwartz kernels of $T\mathcal{P}$ and $\mathcal{P}T$ differ from the Schwartz kernel of I by an element of $C^\infty (M \times M)$. Moreover, the distribution $T (x, y)$ is C^∞ for $x \neq y$.*

PROOF. See the proof of Theorem 52.3 of [Trè67]. $\quad\square$

Henceforth, we work in the setting of (i) of Theorem 2.4.8, thus we assume $\mathcal{P}$ and $\mathcal{P}^*$ are both maximally hypoelliptic, and therefore hypoelliptic. In light of Theorem 2.11.1 there is an operator $T : C^\infty (M) \to C^\infty (M)$ satisfying $\mathcal{P}T, T\mathcal{P} \equiv I \mod C^\infty (M \times M)$. Note that T^* is a priori defined as an operator $T^* : C^\infty (M)' \to C^\infty (M)'$, and satisfies $\mathcal{P}^*T^*, T^*\mathcal{P}^* \equiv I \mod C^\infty (M \times M)$. Using that $\mathcal{P}^*$ is hypoelliptic, we see that T^* restricts to an operator $T^* : C^\infty (M) \to C^\infty (M)$. By duality, this extends $T : C^\infty (M)' \to C^\infty (M)'$. Hence, it makes sense to consider both T and T^* when applied to any distribution.

LEMMA 2.11.2. $T, T^* : L^2 \to L^2$.

PROOF. Let $u \in L^2$. We have $\mathcal{P}Tu \equiv u \mod C^\infty (M)$, and therefore $\mathcal{P}Tu \in L^2$. Since $\mathcal{P}$ is maximally hypoelliptic, it is subelliptic, and therefore, for some $\epsilon > 0$, $Tu \in L^2_\epsilon \hookrightarrow L^2$. We conclude $Tu \in L^2$. The closed graph theorem (Theorem A.1.14) implies $T : L^2 \to L^2$. The same proof with $\mathcal{P}$ replaced by $\mathcal{P}^*$ proves $T^* : L^2 \to L^2$. $\quad\square$

LEMMA 2.11.3. *For $|\alpha| \leq k$, we have $W^\alpha T, W^\alpha T^* : L^2 \to L^2$.*

PROOF. Because $\mathcal{P}$ is maximally hypoelliptic, we have for $f \in C^\infty (M)$,

$$\sum_{|\alpha| \leq k} \|W^\alpha Tf\|_{L^2} \lesssim \|f\|_{L^2} + \|Tf\|_{L^2} .$$

Lemma 2.11.2 shows $\|Tf\|_{L^2} \lesssim \|f\|_{L^2}$, and the result follows for T. The same proof works for T^*, using that $\mathcal{P}^*$ is maximally hypoelliptic. $\quad\square$

LEMMA 2.11.4. *For $|\alpha| \leq k$, we have $TW^\alpha, T^*W^\alpha : L^2 \to L^2$.*

PROOF. This follows by taking adjoints of the result in Lemma 2.11.3. $\quad\square$

LEMMA 2.11.5. *Fix $k \in \mathbb{N}$, $s \in \mathbb{R}$. Suppose $S : C^\infty (M) \to C^\infty (M)$ is such that for all $|\alpha|, |\beta| \leq k$, $W^\alpha SW^\beta$ is a Calderón-Zygmund operator of order s. Then, S is a Calderón-Zygmund operator of order $s - 2k$.*

PROOF. Suppose, first, $k = 2l$ is even. Then, we have

$$S = (I + \mathcal{L})^{-l} \left[(I + \mathcal{L})^l S (I + \mathcal{L})^l \right] (I + \mathcal{L})^{-l} .$$

Our assumption shows that $(I + \mathcal{L})^l S (I + \mathcal{L})^l$ is a Calderón-Zygmund operator of order s. Proposition 2.0.31 combined with Corollary 2.6.10 then completes the proof. If k is odd, we use Proposition 2.0.31 and the fact that each W_j is an operator of order 1 to see for $|\alpha|, |\beta| \leq k + 1$, $W^\alpha S W^\beta$ is an operator of order $s + 2$. The result now follows from the even case. $\qquad\qquad\square$

PROOF OF (i)⇒(ii) OF THEOREM 2.4.8. For $|\alpha_0|, |\beta_0| \leq k$ we will verify that $W^{\alpha_0} T W^{\beta_0}$ is a Calderón-Zygmund operator of order k in the sense of Definition 2.0.16, and the result follows from Lemma 2.11.5. Hence, fix α_0, β_0 with $|\alpha_0|, |\beta_0| \leq k$ and set $S = W^{\alpha_0} T W^{\beta_0}$.

We begin with the growth condition. Fix $x_0 \neq z_0 \in M$ and let $s = \rho(x_0, z_0)$; notice by the compactness of M, $s \lesssim 1$. Take $\delta = \frac{s}{2}$, so that $B_{(X,d)}(x_0, \xi_2 \delta) \cap B_{(X,d)}(x_0, \xi_2 \delta) = \emptyset$ (here $0 < \xi_2 \leq 1$ is as in Theorem 2.2.22). Let $\mathcal{P} = P(W)$ so that $\mathcal{P} T = I + R$, where $R \in C^\infty(M \times M)$. Let $\phi \in C_0^\infty \left(B_{(X,d)}(z_0, \xi_2 \delta) \right)$ and consider, using Theorem 2.4.29,

$$\sup_{x \in B_{(X,d)}(x_0, \xi_4 \delta)} \left| (\delta W)^\alpha W^{\alpha_0} T W^{\beta_0} \phi(x) \right|$$

$$\lesssim \sum_{j=0}^{L(\alpha)} \delta^{-|\alpha_0|} \mathrm{Vol}\left(B_{(X,d)}(x_0, \xi_2 \delta) \right)^{-\frac{1}{2}} \left\| \delta^{jk} \mathcal{P}^j T W^{\beta_0} \phi \right\|_{L^2 \left(B_{(X,d)}(x_0, \xi_2 \delta) \right)}$$

$$\lesssim \delta^{-k} \mathrm{Vol}\left(B_{(X,d)}(x_0, \xi_2 \delta) \right)^{-\frac{1}{2}} \left(\left\| T W^{\beta_0} \phi \right\|_{L^2} + \sum_{j=0}^{L(\alpha)-1} \left\| \mathcal{P}^j R W^{\beta_0} \phi \right\|_{L^2} \right),$$

where we have used $\left\| \mathcal{P}^j W^{\beta_0} \phi \right\|_{L^2 \left(B_{(X,d)}(x_0, \xi_2 \delta) \right)} = 0, \forall j$, due to the support of ϕ. Lemma 2.11.4 and the fact that $R \in C^\infty(M \times M)$ now show

$$\sup_{x \in B_{(X,d)}(x_0, \xi_4 \delta)} \left| (\delta W)^\alpha W^{\alpha_0} T W^{\beta_0} \phi(x) \right| \lesssim \delta^{-k} \mathrm{Vol}\left(B_{(X,d)}(x_0, \xi_2 \delta) \right)^{-\frac{1}{2}} \left\| \phi \right\|_{L^2} .$$

For every j, $\delta^{d_j} X_j$ can be written as a linear combination of commutators of vector fields for the form δW_l. Thus, we have for every α,

$$\sup_{x \in B_{(X,d)}(x_0, \xi_4 \delta)} \left| (\delta X)^\alpha W^{\alpha_0} T W^{\beta_0} \phi(x) \right| \lesssim \delta^{-k} \mathrm{Vol}\left(B_{(X,d)}(x_0, \xi_2 \delta) \right)^{-\frac{1}{2}} \left\| \phi \right\|_{L^2} .$$

The same result holds with T replaced by T^* and the roles of x_0 and z_0 reversed. Note that, since $\delta \approx \rho(x_0, z_0)$, $\mathrm{Vol}\left(B_{(X,d)}(x_0, \xi_2 \delta) \right) \approx \mathrm{Vol}\left(B_{(X,d)}(z_0, \xi_2 \delta) \right)$. Proposition 2.2.26 shows (using ξ_4 in place of ξ_1 in Theorem 2.2.22), for every α and β,

$$\left| (\delta X_x)^\alpha (\delta X_z)^\beta S(x_0, z_0) \right| \lesssim \delta^{-k} \mathrm{Vol}\left(B_{(X,d)}(x_0, \xi_2 \delta) \right)^{-1} .$$

Using that $\delta \approx \rho(x_0, z_0)$, the growth condition follows.

We now turn to the cancellation condition. Let $\mathcal{B}$ be a bounded set of bump functions, and let $(\phi, x_0, \delta) \in \mathcal{B}$. We wish to show that $\forall \alpha$,

$$|(\delta W)^\alpha S_{a_0} \phi(x)| \lesssim \delta^{-k} \mathrm{Vol}\left(B_{(X,d)}(x_0, \delta)\right)^{-1}.$$

For $\rho(x, x_0) >> \delta$, this follows easily from the growth condition. Thus, we need only verify the cancellation condition when $\rho(x, x_0) \lesssim \delta$; i.e., for $x \in B_{(X,d)}(x_0, C\xi_4 \delta)$, where C is some admissible constant independent of x_0 and δ. We have, using Theorem 2.4.29,

$$\sup_{x \in B_{(X,d)}(x_0, \xi_4 C\delta)} \left|(C\delta W)^\alpha W^{\alpha_0} T W^{\beta_0} \phi(x)\right|$$

$$\lesssim \sum_{j=0}^{L(\alpha)} \delta^{-k} \mathrm{Vol}\left(B_{(X,d)}(x_0, \xi_2 C\delta)\right)^{-\frac{1}{2}} \left\|\delta^{jk} \mathcal{P}^j T W^{\beta_0} \phi\right\|_{L^2}$$

$$\lesssim \mathrm{Vol}\left(B_{(X,d)}(x_0, \xi_2 C\delta)\right)^{-\frac{1}{2}} \left[\delta^{-k} \left\|T W^{\beta_0} \phi\right\|_{L^2} \right.$$

$$\left. + \sum_{j=0}^{L(\alpha)-1} \left(\left\|\delta^{jk} \mathcal{P}^j W^{\beta_0} \phi\right\|_{L^2} + \left\|\delta^{jk} \mathcal{P}^j R W^{\beta_0} \phi\right\|_{L^2}\right)\right]$$

$$\lesssim \delta^{-k} \mathrm{Vol}\left(B_{(X,d)}(x_0, \xi_2 C\delta)\right)^{-\frac{1}{2}} \left(\|\phi\|_{L^2} + \sum_{j=0}^{L(\alpha)-1} \left\|\delta^{jk} \mathcal{P}^j (\delta W)^{\beta_0} \phi\right\|_{L^2}\right),$$

where in the last line we have used Lemma 2.11.4, that $0 < \delta \le 1$, and that $\mathcal{P}^j R W^{\beta_0}$ is bounded on L^2 (since $R \in C^\infty(M \times M)$). The definition of bounded sets of bump functions easily shows

$$\|\phi\|_{L^2}, \left\|\delta^{jk} \mathcal{P}^j (\delta W)^{\beta_0} \phi\right\|_{L^2} \lesssim \mathrm{Vol}\left(B_{(X,d)}(x_0, \delta)\right)^{-\frac{1}{2}}.$$

Since $\mathrm{Vol}\left(B_{(X,d)}(x_0, \xi_2 C\delta)\right) \approx \mathrm{Vol}\left(B_{(X,d)}(x_0, \delta)\right)$, we have

$$\sup_{x \in B_{(X,d)}(x_0, \xi_4 C\delta)} \left|(\delta W)^\alpha W^{\alpha_0} T W^{\beta_0} \phi(x)\right| \lesssim \delta^{-k} \mathrm{Vol}\left(B_{(X,d)}(x_0, \delta)\right)^{-1},$$

which proves the cancellation condition for S. The cancellation condition for S^* follows by the same proof, with $\mathcal{P}$ replaced with $\mathcal{P}^*$.

All together, we conclude that S is a Calderón-Zygmund operator of order k, which completes the proof. $\square$

COROLLARY 2.11.6. *Suppose $\mathcal{P}$ and $\mathcal{P}^*$ are maximally hypoelliptic. Fix $p \in (1, \infty)$ and $s \in \mathbb{R}$. Let $u \in C^\infty(M)'$ be a distribution with $\mathcal{P}u \in \mathrm{NL}_s^p$. Then $u \in \mathrm{NL}_{s+k}^p$.*

PROOF. Let T be a Calderón-Zygmund operator of order $-k$ satisfying $T\mathcal{P} \equiv I$ mod $C^{\infty}(M \times M)$, as guaranteed by Theorem 2.11.1. It follows that $u \equiv T\mathcal{P}u$ mod $C^{\infty}(M)$; applying Theorem 2.10.1 yields the result.   □

Remark 2.11.7 When $p = 2$ and $s = 0$, the conclusion of Corollary 2.11.6 is essentially the hypothesis that $\mathcal{P}$ is maximally hypoelliptic. Thus, Theorem 2.11.1 allows us to turn a weak assumption ($p = 2$, $s = 0$) to a strong conclusion ($p \in (1, \infty)$, $s \in \mathbb{R}$).

2.11.1   The Kohn Laplacian

While the singular integrals from Definition 2.0.16 had been implicitly studied in several different situations, they were first explicitly defined by Nagel, Rosay, Stein, and Wainger [NRSW89]. In that paper, these operators were used to study questions from several complex variables, and we take a moment to briefly discuss some of these and related results.

We work on an open, relatively compact subset $\Omega \Subset \mathbb{C}^2$ with smooth boundary, $\partial\Omega$. Let U be a neighborhood of $\partial\Omega$ and let $r : U \to \mathbb{R}$ be a defining function so that $\Omega \cap U = \{z \in \mathbb{C}^2 \mid r(z) > 0\}$, and $\bigtriangledown r(z) \neq 0$ for $r(z) = 0$. On $\mathbb{C}^n$ we have the antiholomorphic exterior derivative $\bar{\partial}$.

We choose a smooth orthonormal basis for $(0, 1)$ forms on U given by $\bar{\omega}_1$ and $\bar{\omega}_2$ where $\bar{\omega}_2 = \sqrt{2}\,\bar{\partial}r$. We let $\bar{L}_1$ and $\bar{L}_2$ be the dual basis of antiholomorphic vector fields on U. Then, L_1 and $\bar{L}_1$ are tangential on $\partial\Omega$. We define a real vector field by $T = \frac{1}{2i}(L_2 - \bar{L}_2)$, so that T is also tangential on $\partial\Omega$. We assume that Ω is "weakly pseudoconvex." That is, we assume that if we expand $[L_1, \bar{L}_1]$ into the basis of tangential vector fields $L_1, \bar{L}_1, iT$, then the coefficient of iT is nonnegative everywhere.

Set $W_1 = \text{Re}(L_1)$ and $W_2 = \text{Im}(L_1)$. We assume that Ω is "finite type." I.e., we assume that W_1 and W_2 satisfy Hörmander's condition on $\partial\Omega$. We use W_1 and W_2 to create a list of vector fields with formal degrees $(X, d) = (X_1, d_2), \ldots, (X_q, d_q)$ as in Section 2.1. The relevant geometry on $\partial\Omega$ corresponds to the balls $B_{(X,d)}(x, \delta)$. We therefore have the algebra of Calderón-Zygmund operators on $\partial\Omega$, as in Definitions 2.0.16 and 2.0.30.

On $\partial\Omega$ we have the boundary exterior derivative $\bar{\partial}_b$ which takes (p, q) forms to $(p, q + 1)$ forms. The Kohn Laplacian is defined as

$$\Box_b = \bar{\partial}_b^* \bar{\partial}_b + \bar{\partial}_b \bar{\partial}_b^*,$$

taking (p, q) forms to (p, q) forms. It turns out that the value of p does not play a role in what follows, and so we consider only $\Box_b$ acting on $(0, q)$ forms. We denote by $\Box_b^1$ the Kohn Laplacian acting on $(0, 1)$ forms and $\Box_b^0$ the Kohn Laplacian acting on $(0, 0)$ forms. Since $\partial\Omega$ is 3-dimensional, we may identify $(0, 1)$ forms with functions. Thus, we may think of both $\Box_b^1$ and $\Box_b^0$ as partial differential operators acting on scalar valued functions. The theory for both operators is similar, and we only discuss the theory for $\Box_b^0$ here.

$\Box_b^0$ is *not* hypoelliptic. In fact, $\Box_b^0$ has an infinite dimensional null space. We let S denote the orthogonal projection onto the L^2 null space of $\Box_b^0$; S is known as the Szegö projection. While $\Box_b^0$ is not hypoelliptic (and therefore not maximally hypoelliptic[14]), it is maximally hypoelliptic "relative to its null space." I.e., on the orthogonal comple- ment of the L^2 null space of $\Box_b^0$, the maximal hypoelliptic estimates hold. This can be used to show (using methods similar to those in Section 2.11), that S is a Calderón- Zygmund operator of order 0 and there is a Calderón-Zygmund operator K of order -2 such that $\Box_b^0 K = K \Box_b^0 = I - S$.

All of the ideas described above were generalized to higher dimensions by Koenig [Koe02]. For further details and proofs, we refer the reader to [NRSW89, Koe02].

A slightly different approach can be used to achieve the same results: one may use the same methods we used to study the sub-Laplacian in Section 2.6 to study $\Box_b$ in this setting. See [Str09] for the additional ideas needed to deal with the infinite dimensional null space of $\Box_b$.

2.12 EXPONENTIAL MAPS

In Section 2.2.1, we introduced the concept of the exponential of a vector field e^Y, where Y is a C^∞ vector field. When two vector fields, Y and Z, commute (i.e., when $[Y, Z] = 0$) we have $e^Y e^Z = e^{Y+Z} = e^Z e^Y$, and this can simplify many computations involving these vector fields. However, when $[Y, Z] \neq 0$, considerations sometimes become much more difficult.

For instance, when $[Y, Z] = 0$, it is easy to compute $Y \left(f \left(e^{sZ} x \right) \right)$. Indeed,

$$
\begin{aligned}
Y \left(f \left(e^{sZ} x \right) \right) &= \left. \frac{d}{dt} \right|_{t=0} f \left(e^{sZ} e^{tY} x \right) = \left. \frac{d}{dt} \right|_{t=0} f \left(e^{tY + sZ} x \right) \\
&= \left. \frac{d}{dt} \right|_{t=0} f \left(e^{tY} e^{sZ} x \right) = (Yf) \left(e^{sZ} x \right).
\end{aligned}
$$

If $[Y, Z] \neq 0$, the situation becomes more complicated. To compute this, we introduce the Baker-Campbell-Hausdorff formula. We work slightly more generally. Let $Y = (Y_1, \ldots, Y_{r_1})$ be a list of vector fields and similarly $Z = (Z_1, \ldots, Z_{r_2})$ another list. For $t \in \mathbb{R}^{r_1}$ define $t \cdot Y = t_1 Y_1 + \cdots + t_{r_1} Y_{r_1}$; similarly define $s \cdot Z$ for $s \in \mathbb{R}^{r_2}$. The Baker-Campbell-Hausdorff formula is a formal series which states:

$$
\begin{aligned}
e^{t \cdot Y} e^{s \cdot Z} x \sim \exp \Bigg(& t \cdot Y + s \cdot Z + \frac{1}{2} [t \cdot Y, s \cdot Z] + \frac{1}{12} [t \cdot Y, [t \cdot Y, s \cdot Z]] \\
& - \frac{1}{12} [s \cdot Z, [t \cdot Y, s \cdot Z]] + \cdots \\
& + \{\text{certain commutators of order } m\} + \cdots \Bigg) x.
\end{aligned}
$$

(2.73)

Here $\sim$ denotes as formal power series in (s, t). E.g.,

$$
e^{t \cdot Y} e^{s \cdot Z} x \sim \exp \left(t \cdot Y + s \cdot Z + \frac{1}{2} [t \cdot Y, s \cdot Z] \right) x + O \left(|(s,t)|^3 \right),
$$

---

[14] $\Box_b$ is a second order polynomial in L_1 and $\overline{L}_1$, and it therefore makes sense to ask if it is maximally hypoelliptic.

and similarly for higher order sums. Furthermore, if sufficiently high order commutators of $t \cdot Y$ and $s \cdot Z$ are 0 (i.e., if there are only finitely many nonzero terms on the right-hand side of (2.73)), then $\sim$ may be replaced with $=$.

For $x_0 \in M$, let $\theta_{x_0}(t) := e^{t \cdot Y} x_0$, so that $\theta_{x_0} : B^{r_1}(\epsilon) \to M$, for ϵ sufficiently small. It is important for some of our applications to understand $d\theta_{x_0}\left(\frac{\partial}{\partial t_j}\right)$. The Baker-Campbell-Hausdorff formula gives full understanding of the Taylor series, as shown in the next lemma.

LEMMA 2.12.1. *There are universal constants $b_1, b_2, \ldots$ such that for any integer m,*

$$d\theta_{x_0}\left(\frac{\partial}{\partial t_j}\right) - \left(Y_j + \sum_{k=1}^{m} b_k \operatorname{ad}(t \cdot Y)^k Y_j\right) = O\left(|t|^{m+1}\right), \qquad (2.74)$$

where the vector fields above are evaluated at the point $\theta_{x_0}(t)$. Furthermore, if the commutators of order $> m+1$ of the vector fields $Y_1, \ldots, Y_{r_1}$ are zero, then the right-hand side is, in fact, zero.

PROOF. We write $\partial_{t_j} e^{t \cdot Y} x$ as $\partial_s\big|_{s=0} e^{t \cdot Y + s Y_j} x$. If it were true that $e^{t \cdot Y + s Y_j} x = e^{s W_j(t)} e^{t \cdot Y} x$, then we would have $d\theta_x(\partial_{t_j}) = W_j(t)$. While we do not have this identity, the Baker-Campbell-Hausdorff formula allows us to compute such a vector field W_j, up to an appropriate error term. Using the Baker-Campbell-Hausdorff formula we have for any m

$$e^{t \cdot Y + s Y_j} e^{-t \cdot Y} x - e^{s\left(Y_j + \sum_{k=1}^{m} b_k \operatorname{ad}(t \cdot Y)^k Y_j\right)} x = O\left(|t|^{n+1}|s| + |s|^2\right). \qquad (2.75)$$

Letting $x = e^{t \cdot Y} x_0$ completes the proof of (2.74). Noting that the right-hand side of (2.75) equals zero if commutators of order $> m+1$ of the Y_js are 0, we get that in this case the right-hand side of (2.74) is zero as well. □

We use Lemma 2.12.1 in the following way. Suppose $Y_1, \ldots, Y_{r_1}$ satisfy an involutivity condition of the form

$$[Y_j, Y_k] = \sum_l c_{j,k}^l Y_l,$$

with $c_{j,k}^l \in C^\infty$. If we then apply (2.74) we have

$$d\theta_{x_0}(\partial_{t_j})(\theta_{x_0}(t)) = Y_j(\theta_{x_0}(t)) + \sum_{\substack{1 \le l \le r_1 \\ 0 < |\alpha| \le m}} t^\alpha c_{\alpha,l}(\theta_{x_0}(t)) Y_l(\theta_{x_0}(t)) + O\left(|t|^{m+1}\right),$$

where $c_{\alpha,l}$ are functions which we can explicitly compute in terms of $c_{j,k}^l$ (in particular if the $c_{j,k}^l$ are constants, so are the $c_{\alpha,l}$). Note, then,

$$d\theta_{x_0}\left(\partial_{t_j} - \sum_{\substack{l=1 \le r_1 \\ |\alpha|=1}} t^\alpha c_{\alpha_1} \partial_{t_l}\right)(\theta_{x_0}(t)) = Y_j(\theta_{x_0}(t)) + O\left(|t|^2\right).$$

And more generally, for each m we can solve for a vector field X_m in the t space which satisfies

$$d\theta_{x_0}\left(X_m\right)\left(\theta_{x_0}\left(t\right)\right) = Y_j\left(\theta_{x_0}\left(t\right)\right) + O\left(|t|^{m+1}\right).$$

Finally, if the iterated commutators of the Y_js are eventually 0, we may solve exactly for an X which satisfies $d\theta_{x_0}\left(X\right) = Y_j$.

2.13   NILPOTENT LIE GROUPS

An important model case for a manifold endowed with vector fields satisfying Hörmander's condition comes from that of a (stratified) nilpotent Lie group. Let $\mathfrak{g}$ be a Lie algebra. Define $\mathfrak{g}^{(0)} = \mathfrak{g}$ and recursively, $\mathfrak{g}^{(k+1)} = \left[\mathfrak{g}, \mathfrak{g}^{(k)}\right]$.

DEFINITION 2.13.1. *We say $\mathfrak{g}$ is nilpotent of step k if $\mathfrak{g}^{(k+1)} = \{0\}$. We say $\mathfrak{g}$ is nilpotent if it is nilpotent of step k of some k.*

DEFINITION 2.13.2. *If G is Lie a group whose Lie algebra is nilpotent, we say G is a nilpotent Lie group.*

It is well known that if $\mathfrak{g}$ is a nilpotent Lie algebra and if G is the corresponding connected, simply connected, Lie group, then the exponential map $\exp : \mathfrak{g} \to G$ is a diffeomorphism [Puk67]. In particular, as a manifold $G \cong \mathbb{R}^{\dim G}$.

DEFINITION 2.13.3. *We say a nilpotent Lie algebra $\mathfrak{g}$ is graded if $\mathfrak{g} = \oplus_{\mu=1}^{\nu} V_\mu$ with $[V_{\mu_1}, V_{\mu_2}] \subseteq V_{\mu_1+\mu_2}$, where we take $V_\mu = \{0\}$ for $\mu > \nu$. We say a connected, simply connected, nilpotent Lie group is a graded Lie group if its Lie algebra is graded.*

DEFINITION 2.13.4. *We say a graded Lie algebra $\mathfrak{g} = \oplus_{\mu=1}^{\nu} V_\mu$ is stratified if $[V_1, V_\mu] = V_{\mu+1}$. We say a connected, simply connected, nilpotent Lie group is a stratified Lie group if its Lie algebra is stratified.*

Suppose $\mathfrak{g}$ is stratified, and suppose $W_1, \ldots, W_r$ are a basis for V_1. We may think of $W_1, \ldots, W_r$ as left invariant vector fields on G. By the definition of a stratified Lie group, $W_1, \ldots, W_r$ satisfy Hörmander's condition. This is an important model case for general vector fields which satisfy Hörmander's condition.

DEFINITION 2.13.5. *Let $\mathfrak{g}$ be a nilpotent Lie algebra. A family of dilations $\delta_t : \mathfrak{g} \to \mathfrak{g}$, $t > 0$, is a family of automorphisms defined by $\delta_t X_j = t^{d_j} X_j$, where $X_1, \ldots, X_{\dim \mathfrak{g}}$ is a basis for $\mathfrak{g}$, and $0 \neq d_j \in (0, \infty)$.*

Example 2.13.6   *Let $\mathfrak{g}$ be a graded Lie algebra; $\mathfrak{g} = \oplus_{\mu=1}^{\nu} V_\mu$. Define a dilation by, for $X \in V_\mu$, $\delta_t X = t^\mu X$; and extend by linearity. This defines a family of dilations on $\mathfrak{g}$. It is this case which we will mostly be concerned with. Henceforth, when we are given a graded group, we use this dilation structure.*

DEFINITION 2.13.7. *A connected, simply connected, nilpotent Lie group whose Lie algebra is endowed with a family of dilations is called a homogeneous Lie group.*

Remark 2.13.8   Not every nilpotent Lie group can be given the structure of a homogeneous Lie group [Dye70]. However, we will be focusing on graded Lie groups, and as we saw in Example 2.13.6, these groups have a natural homogeneous structure.

Let G be a homogeneous group. Since $\exp : \mathfrak{g} \to G$ is a diffeomorphism, we may identify $\mathfrak{g}$ with G; henceforth we do this. In particular, 0 is the identity of G, and our dilations δ_t can be thought of as functions $\delta_t : G \to G$. From here, it is easy to see $\delta_t(xy) = \delta_t x \delta_t y$; i.e., δ_t is a group automorphism. One way to think about this is as follows. Every $x, y \in G$ can be written as $e^X = x$ and $e^Y = y$ for unique choices of $X, Y \in \mathfrak{g}$. Then $xy = e^X e^Y = e^Z$, for some $Z \in \mathfrak{g}$. We can explicitly compute Z using the Baker-Campbell-Hausdorff formula, which is an identity in this case, since the commutators of X and Y of order greater than k are 0. We may therefore give $\mathfrak{g}$ a group structure via the Baker-Campbell-Hausdorff formula, which is isomorphic to the group G. In what follows, we think of the group G under this identification with $\mathfrak{g} \cong \mathbb{R}^{\dim G}$. With this identification $0 \in \mathfrak{g}$ is the identity element of G.

DEFINITION 2.13.9. *Let G be a homogeneous group. A homogeneous norm*

$$| \cdot | : G \to [0, \infty)$$

is a continuous function, smooth away from the identity, with $|x| = 0 \Leftrightarrow x = 0$, and $|\delta_t x| = t|x|$ for $t > 0$.

Henceforth, we work only on graded groups, with the natural homogeneous structure given by Example 2.13.6. The next example shows that these groups also have a homogeneous norm.

Example 2.13.10   *Let $\mathfrak{g} = \oplus_{\mu=1}^{\nu} V_\mu$ be a graded group. For $\mu = 1, \dots, \nu$, fix a basis for V_μ: $X_1^\mu, \dots, X_{l_\mu}^\mu$. Each $X \in \mathfrak{g}$ can be written as $X = \sum_{\mu=1}^{\nu} \sum_{k=1}^{l_\mu} t_{\mu,k} X_k^\mu$. We may define a homogeneous norm by*

$$|X| = \left(\sum_{\mu=1}^{\nu} \sum_{k=1}^{l_\mu} |t_{\mu,k}|^{2(\nu!)/j} \right)^{1/(2(\nu!))} .$$

Recall, we are identifying the group with the Lie algebra via the exponential map.

Remark 2.13.11   For our purposes, all choices of a homogeneous norm are equivalent (see Remark 2.13.12). Since we will be working only with graded groups, we may always work with the norm given in Example 2.13.10.

With a fixed choice of homogeneous norm on a homogeneous group G, there is a natural left invariant metric on G, namely the distance between $x, y \in G$ is given by $\rho(x, y) := |x^{-1}y|$. This metric is also homogeneous: $\rho(\delta_t x, \delta_t y) = t\rho(x, y)$. For $r > 0$, $x \in G$, let $B(x, r) = \{y \mid \rho(x, y) < r\}$.

Remark 2.13.12   Suppose $\rho_1(x, y)$ is a continuous left invariant metric on G, which is homogeneous with respect to the dilations δ_t. Then ρ_1 is equivalent to ρ. Indeed, we have $\rho_1(x,y) = \rho_1(0, x^{-1}y)$ and similarly for ρ, by left invariance–here, $0 \in G$ is the identity element. Thus it suffices to see $\rho_1(0,y) \approx \rho(0,y)$. By dilation invariance, both $\rho_1(0,y)$ and $\rho(0,y)$ are determined by their values for y on the unit sphere centered at 0 in a usual Euclidean metric. Compactness then shows $\rho_1(0,y) \approx \rho(0,y)$ for y in the sphere, completing the proof.

In light of the previous sections, one might think that a Carnot-Carathéodory metric would be the most natural kind of metric to use–indeed, it turns out ρ essentially is a Carnot-Carathéodory metric. Let $X_1^\mu, \ldots, X_{l_\mu}^\mu$ be a basis for V_μ as in Example 2.13.10. Think of each X_k^μ as a left invariant vector field, and assign X_k^μ the formal degree d_μ. Let $(X, d) = (X_1, d_1), \ldots, (X_q, d_q)$ be an enumeration of this list of vector fields with formal degrees, and let $B_{(X,d)}(x, r)$ be the corresponding balls. If we define the metric $\rho_1(x,y) = \inf\{r > 0 \mid y \in B_{(X,d)}(x,r)\}$, then ρ_1 is a left invariant homogeneous metric, and by Remark 2.13.12 is equivalent to ρ. Hence, ρ is equivalent to a Carnot-Carathéodory metric.

Remark 2.13.13 Let (X, d) be the list of left invariant vector fields with formal degrees from the previous paragraph. For $r > 0$, it is immediate to verify that

$$\left[r^{d_j} X_j, r^{d_k} X_k\right] = \sum_{d_j+d_k=d_l} c_{j,k}^l r^{d_l} X_l,$$

where $c_{j,k}^l$ are constants, independent of r. Thus the quantitative Frobenius theorem (Theorem 2.2.22) applies, uniformly in r. Here, though, the scaling maps are quite easy to understand. Indeed, for a fixed $x \in G$ and $r > 0$, consider the map $\Phi_{x,r} : B(0,1) \to B(x,r)$ given by

$$\Phi_{x,r}(y) = (\delta_r y)\, x.$$

$\Phi_{x,r}$ is in fact a bijection with constant Jacobian: $\det d\Phi_{x,r} = r^Q$, where $Q = \sum_{\mu=1}^\nu \mu \dim V_\mu$. Furthermore, the pullback of $r^\mu X_k^\mu$ via $\Phi_{x,r}$ equals X_k^μ. This $\Phi_{x,r}$ satisfies all of the conclusions of Theorem 2.2.22.

Fix a graded group G, $\mathfrak{g} = \oplus_{\mu=1}^\nu V_\mu$, and let $q = \dim G$; decompose $\mathbb{R}^q = \mathbb{R}^{\dim V_1} \times \cdots \times \mathbb{R}^{\dim V_\nu}$. For $r > 0$ we define dilations on $\mathbb{R}^q$ by $r(t_1, \ldots, t_\nu) = (rt_1, r^2 t_2, \ldots, r^\nu t_\nu)$. Notice, if we identify $G \cong \mathfrak{g}$ with $\mathbb{R}^q$ (as a manifold) by identifying V_μ with $\mathbb{R}^{\dim V_\mu}$, these are the dilations given by δ_r, though now we have suppressed the δ. With these dilations $d(rt)/dt = r^Q$, where $d(rt)/dt$ denotes the Radon-Nikodym derivative, and $Q = \sum_{\mu=1}^\nu \mu \dim V_\mu$ is the so-called "homogeneous dimension." Furthermore, we use this identification with G to define $|t|$ for $t \in \mathbb{R}^q$, where $|\cdot|$ denotes a homogeneous norm. With the above notations, we have $|rx| = r|x|$, for $r > 0$. Finally, in the above identification, Lebesgue measure on $\mathbb{R}^q$ corresponds with the two-sided Haar measure on G. Henceforth, integration on G will always be with respect to this measure.

Using these dilations, we can generalize the Calderón-Zygmund kernels of Definition 1.1.24. For a multi-index $\alpha = (\alpha_1, \ldots, \alpha_\nu) \in \mathbb{N}^q = \mathbb{N}^{\dim V_1} \times \cdots \times \mathbb{N}^{\dim V_\nu}$, we define $\deg(\alpha) = \sum_{\mu=1}^\nu \mu |\alpha_\mu|$, where $|\alpha_\mu|$ denotes the usual length of the multi-index, i.e., the ℓ^1 norm.

DEFINITION 2.13.14. *We say* $K \in C_0^\infty (\mathbb{R}^q)'$ *is a Calderón-Zygmund kernel of order* $s \in (-Q, \infty)$ *if:*

(i) *(Growth Condition) For every multi-index* α, $|\partial_t^\alpha K(t)| \leq C_\alpha |t|^{-Q-s-\deg(\alpha)}$.

(ii) *(Cancellation Condition) For every bounded set* $\mathcal{B} \subset C_0^\infty (\mathbb{R}^q)$, *we assume*

$$\sup_{\substack{\phi \in \mathcal{B} \\ R>0}} R^{-s} \left| \int K(t) \, \phi(Rt) \, dt \right| < \infty.$$

Given $K \in C_0^\infty (\mathbb{R}^q)' = C_0^\infty (G)'$ we may define a left invariant operator Op (K) : $C_0^\infty (G) \to C^\infty (G)$ by Op $(K) f(x) = f * K(x) = \int f(xy^{-1}) K(y) \, dy$. For K a Calderón-Zygmund kernel, the theory here in many ways follows the special case in Section 1.1, where $G = \mathbb{R}^n$ was used (and the usual addition gave the group structure). The part which will be of most interest to us is the following generalization of Theorem 1.1.26. For a function $f \in C^\infty (\mathbb{R}^q)$ and $R > 0$, we define $f^{(R)}(t) = R^Q f(Rt)$, where Rt is defined by the above dilations, and therefore $f^{(R)}$ is defined to preserve the L^1 norm: $\int f^{(R)}(t) \, dt = \int f(t) \, dt$.

THEOREM 2.13.15. *Fix* $s \in (-Q, \infty)$, *and let* $K \in \mathcal{S}_0 (\mathbb{R}^q)'$. *The following are equivalent:*

(i) K *is a Calderón-Zygmund kernel of order* s.

(ii) Op $(K) : \mathcal{S}_0 (\mathbb{R}^q) \to \mathcal{S}_0 (\mathbb{R}^q)$ *and for any bounded set* $\mathcal{B} \subset \mathcal{S}_0 (\mathbb{R}^q)$, *the set*

$$\left\{ g \in \mathcal{S}_0 (\mathbb{R}^q) \mid \exists R > 0, f \in \mathcal{B}, g^{(R)} = R^{-s} \text{Op} (K) f^{(R)} \right\} \subset \mathcal{S}_0 (\mathbb{R}^q)$$

is a bounded set.

(iii) *For each* $j \in \mathbb{Z}$, *there is a function* $\varsigma_j \in \mathcal{S}_0 (\mathbb{R}^q)$ *with* $\{\varsigma_j \mid j \in \mathbb{Z}\} \subset \mathcal{S}_0 (\mathbb{R}^q)$ *a bounded set and such that*

$$K = \sum_{j \in \mathbb{Z}} 2^{js} \varsigma_j^{(2^j)}.$$

The above sum converges in distribution, and the equality is taken in the sense of elements of $\mathcal{S}_0 (\mathbb{R}^q)'$.

Furthermore, (ii) and (iii) are equivalent for any $s \in \mathbb{R}$.

DEFINITION 2.13.16. *For* $s \leq -Q$, *we say* $K \in \mathcal{S}_0 (\mathbb{R}^q)'$ *is a Calderón-Zygmund kernel of order* s *if either of the equivalent conditions (ii) or (iii) of Theorem 2.13.15 hold.*

Remark 2.13.17   Definition 2.13.14 and parts (i) and (iii) of Theorem 2.13.15 do not mention the group structure at all. In fact, if we are given dilations on $\mathbb{R}^q$:

$$R(t_1, \ldots, t_q) = \left(R^{d_1} t_1, \ldots, R^{d_q} t_q \right) \text{ where } 0 < d_1, \ldots, d_q \in \mathbb{N},$$

then we define Calderón-Zygmund kernels on $\mathbb{R}^n$ as in Definition 2.13.14. In this context, both of the conclusions (i) and (iii) of Theorem 2.13.15 hold. This can be seen by taking $G = \mathbb{R}^n$ with the usual group structure, but with the above non-standard dilations.

Theorem 2.13.15 follows from many of the same ideas as Theorem 1.1.26. Furthermore, we present a more general version of Theorem 2.13.15 in Section 5.2, and therefore do not discuss the proof here. Using the same ideas as in Chapter 1 it is not hard to prove the following theorem.

THEOREM 2.13.18. *If K is a Calderón-Zygmund kernel of order 0, then $\mathrm{Op}\,(K)$ extends to a bounded operator $L^p\,(G) \to L^p\,(G)$ ($1 < p < \infty$). If K_1 and K_2 are Calderón-Zygmund kernel of orders s_1 and s_2, respectively, then $\mathrm{Op}\,(K_1)\,\mathrm{Op}\,(K_2) = \mathrm{Op}\,(K_3)$, where K_3 is a Calderón-Zygmund kernel of order $s_1 + s_2$.*

DEFINITION 2.13.19. *For $K \in C_0^\infty\,(\mathbb{R}^q)'$ and $r > 0$, we define the distribution $K\,(rt)$ by*

$$\int K\,(rt)\,f\,(t)\,dt = r^{-Q} \int K\,(t)\,f\,(r^{-1}t)\,dt,$$

where $r^{-1}t$ is defined by the dilations. Note, when $K\,(t)$ is a function, then $K\,(rt)$ agrees with precomposition with dilation by r.

DEFINITION 2.13.20. *We say a distribution is homogeneous of degree $s \in \mathbb{R}$ if $K\,(rx) = r^s K\,(x)$, for $r > 0$.*

PROPOSITION 2.13.21. *Fix $s \leq Q$ and suppose $K \in C_0^\infty\,(\mathbb{R}^q)'$ is a distribution which is homogeneous of degree $s - Q$ such that $K\,(x)$ is C^∞ for $x \neq 0$. Then, K is a Calderón-Zygmund kernel of order $-s$.*

PROOF. We begin by verifying the growth condition. Away from $x = 0$, $\partial_t^\alpha K\,(t)$ is a C^∞ function, homogeneous of degree $s - Q - \deg\,(\alpha)$. Thus, $|\partial_t^\alpha K\,(t)| \leq \left[\sup_{|u|=1} |\partial_t^\alpha K\,(u)|\right] |t|^{-Q+s-\deg(\alpha)}$, yielding the growth condition.

For the cancellation condition, let $\mathcal{B} \subset C_0^\infty\,(\mathbb{R}^q)$ be a bounded set, and let $\phi \in \mathcal{B}$. We have $R^s \int K\,(t)\,\phi\,(Rt)\,dt = \int K\,(t)\,\phi\,(t)\,dt$. $\sup_{\phi \in \mathcal{B}} \left|\int K\,(t)\,\phi\,(t)\,dt\right| < \infty$ is true for any distribution, completing the proof. $\square$

DEFINITION 2.13.22. *We say a differential operator $\mathcal{P} : C^\infty\,(\mathbb{R}^q) \to C^\infty\,(\mathbb{R}^q)$ is a homogeneous differential operator of degree $s \in \mathbb{R}$ if $\mathcal{P}\,(f\,(rx)) = r^s\,(\mathcal{P}f)\,(rx)$, for $r > 0$.*

We present, without proof, a theorem of Folland [Fol75].

THEOREM 2.13.23. *Suppose $\mathcal{P} : C^\infty\,(\mathbb{R}^q) \to C^\infty\,(\mathbb{R}^q)$ is a homogeneous differential operator of degree $s < Q$ such that $\mathcal{P}$ and $\mathcal{P}^*$ are hypoelliptic. Then there is a distribution $K \in C_0^\infty\,(\mathbb{R}^q)'$, homogeneous of degree $s - Q$ and C^∞ away from 0, such that $\mathcal{P}K = \delta_0$, where δ_0 denotes the δ function at 0.*

If $\mathcal{P}$ is a left invariant operator on G, then $\mathrm{Op}\,(K)$ yields a fundamental solution for $\mathcal{P}$: $\mathcal{P}\mathrm{Op}\,(K) = I = \mathrm{Op}\,(K)\,\mathcal{P}$. As a consequence, the proof of Theorem 2.4.8 shows that if $\mathcal{P}$ is a left invariant differential operator on G, satisfying the hypotheses of Theorem 2.13.23, then $\mathcal{P}$ and $\mathcal{P}^*$ are maximally hypoelliptic.

We now specialize to the case of a stratified Lie group G. We let $W_1,\ldots,W_l$ be a basis for V_1 (the vector fields homogeneous of degree 1). We think of $W_1,\ldots,W_l$ as left invariant vector fields on G. By the definition of a stratified Lie group, $W_1,\ldots,W_l$ satisfy Hörmander's condition on G. We have identified G with $\mathbb{R}^q \cong \mathfrak{g}$ via the exponential map, and we think of writing $W_1,\ldots,W_l$ in these coordinates. In this coordinate system, $W_1,\ldots,W_l$ can be explicitly computed using the Baker-Campbell Hausdorff formula, since the identification of $\mathbb{R}^q$ and G is given by $t \mapsto e^{t_1 X_1 + \cdots + t_q X_q} 0$, where 0 denotes the identity. See the computation at the end of Section 2.12 for more details.

We define $\mathcal{L} = -W_1^2 - \cdots - W_l^2 = W_1^*W_1 + \cdots + W_l^*W_l$: the sub-Laplacian.[15] Each W_j is a homogeneous differential operator of degree 1 and as a consequence $\mathcal{L}$ is a homogeneous differential operator of degree 2. $\mathcal{L}$ is clearly maximally hypoelliptic of type 2 and therefore hypoelliptic. Since $\mathcal{L}^* = \mathcal{L}$, and under the mild assumption that $\dim G > 2$, Theorem 2.13.23 and the above comments imply that $\mathcal{L}$ is maximally hypoelliptic. This was already discussed in Remark 2.6.9, but that proof used the spectral theorem in an essential way. The proof outlined above does not, but is restricted to the group setting. In the next section, we remove this restriction by a generalization of pseudodifferential operators.

In the next section, we will use a special stratified Lie algebra: the free nilpotent Lie algebra of step $k \in \mathbb{N}$ on r generators. Informally, we take a Lie algebra, $\mathfrak{n}_{k,r}$ of step k, which is generated as a Lie algebra by r elements $\widehat{W}_1,\ldots,\widehat{W}_r$, but otherwise has as few relations as possible. Formally $\mathfrak{n}_{k,r}$ is the unique[16] nilpotent Lie algebra of step k satisfying the following universal property: we let S be a set of r elements and let $\phi : S \to \mathfrak{n}_{k,r}$ be a map which takes these r elements bijectively to $\widehat{W}_1,\ldots,\widehat{W}_r$. If $\mathfrak{g}$ is another Lie algebra of step at most k and $\psi : S \to \mathfrak{g}$ is any map, then there is a unique Lie algebra homomorphism which makes the following diagram commute:

$$\begin{array}{ccc} & & \mathfrak{n}_{k,r} \\ & \overset{\phi}{\nearrow} & \Big\downarrow {\exists !} \\ S & \underset{\psi}{\longrightarrow} & \mathfrak{g} \end{array} \qquad (2.76)$$

It is easy to show that such a Lie algebra exists. Furthermore, this Lie algebra can be given the structure of a stratified Lie algebra. Indeed, we take $\widehat{W}_1,\ldots,\widehat{W}_r$ to be a basis for $\widehat{V}_1$ and recursively define $\widehat{V}_j = \left[\widehat{V}_1, \widehat{V}_{j-1}\right]$, thereby giving $\mathfrak{n}_{k,r}$ the structure of a stratified Lie algebra. For more information on $\mathfrak{n}_{k,r}$ see [CG90, page 2] and [RS76].

To understand how we use $\mathfrak{n}_{k,r}$, we need to unravel our definitions. Let $N_{k,r}$ be the connected, simply connected, nilpotent Lie group with Lie algebra $\mathfrak{n}_{k,r}$, with

---

[15]Here we have used that, for a homogeneous vector field Y, $Y^* = -Y$. For any vector field, we have $Y^* = -Y + f$, where $f \in C^\infty\,(\mathbb{R}^q)$, but homogeneity forces f to be 0.

[16]The universal property determines $\mathfrak{n}_{n,k}$ uniquely up to isomorphisms of Lie algebras.

generators $\widehat{W}_1, \ldots, \widehat{W}_r$. We give $\mathfrak{n}_{k,r}$ the above defined structure of a stratified Lie algebra, and pick a basis $\widehat{X}_1, \ldots, \widehat{X}_q$ of $\mathfrak{n}_{k,r}$ with each $\widehat{X}_j \in \widehat{V}_{d_j}$; i.e., $\widehat{X}_j$ homogeneous of degree d_j. Furthermore, we choose each $\widehat{X}_j$ to be of the form $\widehat{X}_j = \mathrm{ad}\left(\widehat{W}_{l_1^j}\right) \cdots \mathrm{ad}\left(\widehat{W}_{l_{d_j-1}^j}\right) \widehat{W}_{l_{d_j}^j}$, for some choice of $l_1^j, l_2^j, \ldots, l_{d_j}^j$.

We think of all elements of $\mathfrak{n}_{k,r}$ as left invariant vector fields. As before, we identify $N_{k,r}$ with $\mathfrak{n}_{k,r} \cong \mathbb{R}^q$, and we may write an element $\widehat{Y} \in \mathfrak{n}_{k,r}$ in these coordinates. To make this clear, when we write an element $\widehat{Y}$ in these coordinates, we denote it by $\widetilde{Y}$ (technically speaking these are the same operators, but we do this to make the choice of coordinate system clear). $\widetilde{Y}$ acts in the t variable. Recall, we can explicitly solve for $\widetilde{Y}$ using the Baker-Campbell-Hausdorff formula as in Section 2.12.

Let $\mathfrak{g}$ be a stratified Lie group, $\mathfrak{g} = \oplus_{j=1}^k V_j$, with $\dim V_1 \leq r$. Let $W_1, \ldots, W_{\dim V_1}$ be a basis for V_1, and let $W_{\dim V_1+1} = \ldots = W_r = 0$ (if $\dim V_1 < r$). (2.76) guarantees a Lie algebra homomorphism $\lambda : \mathfrak{n}_{k,r} \to \mathfrak{g}$ taking $\widehat{W}_j \mapsto W_j$. Let X_j be the image under λ of $\widehat{X}_j$, so that if $\widehat{X}_j = \mathrm{ad}\left(\widehat{W}_{l_1}\right) \cdots \mathrm{ad}\left(\widehat{W}_{l_{d_j-1}}\right) \widehat{W}_{l_{d_j}}$, then $X_j = \mathrm{ad}(W_{l_1}) \cdots \mathrm{ad}\left(W_{l_{d_j-1}}\right) W_{l_{d_j}}$. For a left invariant vector field $\widehat{Y}$ on N, if $Y = \lambda\left(\widehat{Y}\right)$, we have

$$\widetilde{Y}\left(f\left(e^{t_1 X_1 + \cdots + t_q X_q} x\right)\right) = (Yf)\left(e^{t_1 X_1 + \cdots + t_q X_q} x\right), \qquad (2.77)$$

by a computation using the Baker-Campbell-Hausdorff formula. Here, $\widetilde{Y}$ acts in the t-variable and Y acts in the x-variable. Indeed, the Baker-Campbell-Hausdorff formula dictates how to write $\widehat{Y}$ in the t-coordinates (which we denote by $\widetilde{Y}$), and this same computation applies to show (2.77).

Consider now the sub-Laplacian on G, $\mathcal{L} = -W_1^2 - \cdots - W_r^2$. If $\widetilde{\mathcal{L}} = -\widetilde{W}_1^2 - \cdots - \widetilde{W}_r^2$, then we have

$$\widetilde{\mathcal{L}}\left(f\left(e^{t_1 X_1 + \cdots + t_q X_q} x\right)\right) = (\mathcal{L}f)\left(e^{t_1 X_1 + \cdots + t_q X_q} x\right).$$

Here, again, $\widetilde{\mathcal{L}}$ acts in the t-variables, while $\mathcal{L}$ acts in the x-variable. Let $\widetilde{K}(t) \in C_0^\infty(\mathbb{R}^q)$ be the homogeneous solution to $\widetilde{\mathcal{L}}\widetilde{K} = \delta_0$ and define a map $T : C_0^\infty(G) \to C^\infty(G)$ by

$$Tf(x) = \int f\left(e^{t_1 X_1 + \cdots + t_q X_q} x\right) \widetilde{K}(t) \, dt.$$

Notice, then,

$$T\mathcal{L}f(x) = \int (\mathcal{L}f)\left(e^{t_1 X_1 + \cdots + t_q X_q} x\right) \widetilde{K}(t) \, dt$$

$$= \int \widetilde{\mathcal{L}}f\left(e^{t_1 X_1 + \cdots + t_q X_q} x\right) \widetilde{K}(t) \, dt = \int f\left(e^{t_1 X_1 + \cdots + t_q X_q} x\right) \widetilde{\mathcal{L}}\widetilde{K}(t) \, dt$$

$$= f(x).$$

Thus, T is a fundamental solution to $\mathcal{L}$. It is not hard to see that T is, in fact, a Calderón-Zygmund operator of order -2 (a point we return to more generally in the next section–see Theorem 2.14.1). Such a fundamental solution was already guaranteed by Theorem 2.13.23, however we will use the same ideas in the next section where Theorem 2.13.23 does not apply. The same argument works for more general homogeneous hypoelliptic operators on N, whose adjoints are also hypoelliptic, yielding corresponding hypoelliptic operators on G.

2.14   PSEUDODIFFERENTIAL OPERATORS

We now return to the case of a connected, compact manifold M, endowed with vector fields $W_1, \ldots, W_r$ satisfying Hörmander's condition. We define the sub-Laplacian as before $\mathcal{L} = W_1^* W_1 + \cdots + W_r^* W_r$. In light of Corollary 2.6.7, $\mathcal{L}^{-1}$ is a Calderón-Zygmund operator of order -2 and satisfies $\mathcal{L}\mathcal{L}^{-1} = \mathcal{L}^{-1}\mathcal{L} = I - E(0) \equiv I$ mod $C^\infty(M \times M)$,[17] where $E(0)$ is the orthogonal projection onto the constant functions (see Remark 2.6.8). It follows that $\mathcal{L}$ is maximally hypoelliptic (Remark 2.6.9). The key to this proof (and many of the methods in Section 2.6) was that $\mathcal{L}$ is self-adjoint, and we had access to the spectral theorem. In Section 2.13, however, in the case of a stratified Lie group, we were able to construct a fundamental solution[18] without resorting to the spectral theorem. In this section, we discuss a method, first developed by Rothschild and Stein [RS76] and later studied by Goodman [Goo76], to use the knowledge we have of the group situation to prove results about $\mathcal{L}$ on M.

The basic idea is to generalize some aspects of pseudodifferential operators to this setting. Informally, we wish to view the tangent space of M at each point as (a quotient of) a stratified Lie group, instead of as $\mathbb{R}^n$, and to then generalize pseudodifferential operators using this more general notion of a tangent space. Unfortunately, following this idea directly leads to several difficulties–one particular difficulty is that if the tangent bundle is defined in the most natural way, it does not have an appropriate smooth structure. However, for many purposes, a slightly weaker version of this idea works– we instead view all the tangent spaces as quotients of one fixed free nilpotent Lie group on r generators.

For the purposes of this section, we need to be slightly more careful about how we generate the list (X, d) from the list $W_1, \ldots, W_r$. Suppose $W_1, \ldots, W_r$ satisfy Hörmander's condition of order k, and let $\mathfrak{n}_{k,r}$ be the free nilpotent Lie algebra of step k on r generators, and we denote by $\widehat{W}_1, \ldots, \widehat{W}_r$ r generators for $\mathfrak{n}_{k,r}$. As in the

---

[17]When we write $T \equiv S \mod C^\infty(M \times M)$, where T and S are operators, we are once again identifying T and S with their Schwartz kernels, and this equation means the Schwartz kernel of $T - S$ is an element of $C^\infty(M \times M)$.

[18]On a stratified Lie group, we were actually able to obtain a fundamental solution to $\mathcal{L}$: $\mathcal{L}\mathcal{L}^{-1} = I$. On a compact manifold we only obtained $\mathcal{L}\mathcal{L}^{-1} = I - E(0)$. This is related to the fact that the sub-Laplacian is not injective on $C^\infty(M)$ (it annihilates constant functions), while it is injective on $C_0^\infty(G)$, where G is a stratified nilpotent Lie group (constant functions do not have compact support).

previous section, let $\widehat{X}_1, \ldots, \widehat{X}_q$ be a basis for $\mathfrak{n}_{k,r}$ with

$$\widehat{X}_j = \mathrm{ad}\left(\widehat{W}_{l_1^j}\right) \mathrm{ad}\left(\widehat{W}_{l_2^j}\right) \cdots \mathrm{ad}\left(\widehat{W}_{l_{d_j-1}^j}\right) \widehat{W}_{l_{d_j}^j}$$

for some choice of $l_1^j, \ldots, l_{d_j}^j$. Note that $\widehat{X}_j$ is homogeneous of degree d_j. On M we define the corresponding vector fields

$$X_j = \mathrm{ad}\left(W_{l_1^j}\right) \mathrm{ad}\left(W_{l_2^j}\right) \cdots \mathrm{ad}\left(W_{l_{d_j-1}^j}\right) W_{l_{d_j}^j},$$

and we assign to X_j the formal degree d_j. Because $\mathfrak{n}_{k,r}$ is the *free* nilpotent Lie group of step k on r generators, it follows that every commutator of $W_1, \ldots, W_r$ of order $\leq k$ can be written as a linear combination, with constant coefficients, of $X_1, \ldots, X_q$. Thus, $X_1, \ldots, X_q$ span the tangent space at every point of M, since $W_1, \ldots, W_r$ satisfy Hörmander's condition of order k. We let $(X, d) = (X_1, d_1), \ldots, (X_q, d_q)$. This choice of (X, d) is slightly different than the one used in previous sections: the one in previous sections had more vector fields with formal degrees. However, each vector field with a formal degree in the choice from the previous sections can be written as a linear combination with constant coefficients of vector fields X_j, with the same degree. These superfluous vector fields may therefore be ignored, and the balls $B_{(X,d)}(x, \delta)$ are equivalent to the balls we used in the previous sections.

On $\mathbb{R}^q$ we define dilations as in the previous section, for $\delta > 0$, $\delta(t_1, \ldots, t_q) = (\delta^{d_1} t_1, \ldots, \delta^{d_q} t_q)$. Let $K \in C_0^\infty(\mathbb{R}^q)'$ be a Calderón-Zygmund kernel of order $s > -Q$ as in Definition 2.13.14 (see Remark 2.13.17). Consider the operator $T : C^\infty(M) \to C^\infty(M)$ defined by

$$Tf(x) = \int f\left(e^{t_1 X_1 + \cdots + t_q X_q} x\right) K(t) \, dt. \tag{2.78}$$

Our first main result of this section is

THEOREM 2.14.1. *If K is supported on a sufficiently small neighborhood of 0, then T is a Calderón-Zygmund operator of order s, as in Definition 2.0.16.*

The restriction that $s > -Q$ in Theorem 2.14.1 is not essential. However, Definition 2.13.16 does not offer the right generalization of Calderón-Zygmund kernels.

DEFINITION 2.14.2. *Fix $a > 0$, $s \in \mathbb{R}$. The class $\mathrm{CZ}(s, a) \subset C_0^\infty(\mathbb{R}^q)'$ consists of those distributions $K \in C_0^\infty(\mathbb{R}^q)'$, such that there exists $\eta \in C_0^\infty(B^q(a))$ and a bounded set $\{\varsigma_j \mid j \in \mathbb{N}\} \subset \mathcal{S}(\mathbb{R}^q)$ with $\varsigma_j \in \mathcal{S}_0(\mathbb{R}^q)$ for $j \neq 0$, with*

$$K(t) = \eta(t) \sum_{j \in \mathbb{N}} 2^{js} \varsigma_j^{(2^j)}(t). \tag{2.79}$$

Remark 2.14.3 Notice that the definition of $\mathrm{CZ}(s, a)$ depends on the choice of dilations on $\mathbb{R}^q$. Also, all distributions in $\mathrm{CZ}(s, a)$ are supported in $B^q(a)$. Finally, every sum of the form (2.79) converges in the sense of distributions–this can be shown in a manner similar to the proof of Lemma 1.1.20.

Definition 2.14.2 is justified by the next proposition, which we will prove later in this section.

PROPOSITION 2.14.4. *Let $s > -Q$ and let $K \in C_0^\infty (\mathbb{R}^q)'$ be a Calderón-Zygmund kernel of order s which is supported in $B^q (a)$. Then, $K \in CZ (s, a)$.*

It is perhaps not obvious that $CZ (s, a)$ is a vector space. This follows from the following proposition.

PROPOSITION 2.14.5. *Fix $s \in \mathbb{R}$ and $a > 0$. For $K \in S (\mathbb{R}^q)'$ the following are equivalent:*

(i) $K \in CZ (s, a)$.

(ii) $\operatorname{supp} (K) \subset B^q (a)$ *and there is a bounded set $\{\widetilde{\varsigma}_j \mid j \in \mathbb{N}\} \subset S (\mathbb{R}^q)$ with $\widetilde{\varsigma}_j \in S_0 (\mathbb{R}^q)$ for $j > 0$ such that $K = \sum_{j \in \mathbb{N}} 2^{js} \widetilde{\varsigma}_j^{(2^j)}$.*

Remark 2.14.6 Propositions 2.14.4 and 2.14.5 and the definition of a Calderón-Zygmund kernel all used the dilations $\delta (t_1, \ldots, t_q) = \left(\delta^{d_1} t_1, \ldots, \delta^{d_q} t_q\right)$, and are not associated to the vector fields $(X_1, d_1), \ldots, (X_q, d_q)$ in any other way.

To prove Proposition 2.14.5, we need two lemmas.

LEMMA 2.14.7. *Let $B_1 \subset S (\mathbb{R}^q)$, $B_2 \subset S_0 (\mathbb{R}^q)$, and $B_3 \subset C_0^\infty (B^q (a))$ be bounded sets. For $\varsigma_1 \in B_1$, $\varsigma_2 \in B_2$, $\eta \in B_3$, and $0 \leq j_1 \leq j_2$, define $\varsigma_3^{(2^{j_1})} = \varsigma_1^{(2^{j_1})} * \left(\eta \varsigma_2^{(2^{j_2})}\right)$, $\varsigma_4^{(2^{j_1})} = \left(\eta \varsigma_1^{(2^{j_1})}\right) * \varsigma_2^{(2^{j_2})}$, where the convolution is taken with respect to the usual Euclidean group structure on $\mathbb{R}^q$. Then, for every N, the set*

$$\left\{ 2^{N|j_1 - j_2|} \varsigma_3, 2^{N|j_1 - j_2|} \varsigma_4 \mid \varsigma_1 \in B_1, \varsigma_2 \in B_2, \eta \in B_3, 0 \leq j_1 \leq j_2 \right\} \subset S (\mathbb{R}^q)$$

is bounded.

PROOF. We prove only

$$\left\{ 2^{N|j_1 - j_2|} \varsigma_4 \mid \varsigma_1 \in B_1, \varsigma_2 \in B_2, \eta \in B_3, 0 \leq j_1 \leq j_2 \right\} \subset S (\mathbb{R}^q)$$

is bounded, for every N. The proof with ς_4 replaced by ς_3 is similar.

Fix N. Using repeated applications of Lemma 1.1.16 (and Remark 1.1.17), we may write $\varsigma_2 = \sum_{|\alpha| = N} \partial_t^\alpha \varsigma_\alpha$, where $\{\varsigma_\alpha \mid \varsigma_2 \in B_2\} \subset S_0 (\mathbb{R}^q)$ is a bounded set. We have,

$$\varsigma_4 (t) = \int \varsigma_1 (t) \eta \left(2^{-j_1} (t)\right) \varsigma_2^{(2^{j_2 - j_1})} (t - s) \, dt$$

$$= \sum_{|\alpha| = N} 2^{(j_1 - j_2) \deg(\alpha)} (-1)^{|\alpha|} \int \partial_t^\alpha \left[\varsigma_1 (t) \eta \left(2^{-j_1} (t)\right)\right] \varsigma_\alpha^{(2^{j_2 - j_2})} (t - s) \, dt.$$

For $|\alpha| = N$, $\deg(\alpha) \geq N$. Also,

$$\left\{ \partial_t^\alpha \left[\varsigma_1(t)\, \eta\left(2^{-j_1}t\right)\right] \mid |\alpha| = N, \varsigma_1 \in \mathcal{B}_1, \eta \in \mathcal{B}_3, j_1 \geq 0\right\} \subset \mathcal{S}\left(\mathbb{R}^q\right)$$

is a bounded set. The result now follows from the elementary fact that if $\mathcal{B} \subset \mathcal{S}\left(\mathbb{R}^q\right)$ is a bounded set, then $\left\{ f * g^{\left(2^j\right)} \mid f, g \in \mathcal{B}, j \geq 0\right\} \subset \mathcal{S}\left(\mathbb{R}^q\right)$ is also a bounded set–this can be easily seen by using the Fourier transform (recall, we are using the Euclidean group structure on $\mathbb{R}^q$). $\qquad\square$

LEMMA 2.14.8. *Suppose* $\mathcal{B}_1 \subset \mathcal{S}_0\left(\mathbb{R}^q\right)$ *and* $\mathcal{B}_2 \subset C_0^\infty\left(B^q(a)\right)$ *are bounded sets. For* $\varsigma \in \mathcal{B}_1$, $\eta \in \mathcal{B}_2$, *and* $j \in \mathbb{N}$, *we may write*

$$\eta\varsigma^{\left(2^j\right)} = \sum_{\substack{k \leq j \\ k \in \mathbb{N}}} \varsigma_{j,k}^{\left(2^k\right)},$$

where, for every N, $\mathcal{B}_N := \left\{ 2^{N|j-k|}\varsigma_{j,k} \mid \varsigma \in \mathcal{B}_1, \eta \in \mathcal{B}_2, j \in \mathbb{N}, k \leq j\right\} \subset \mathcal{S}\left(\mathbb{R}^q\right)$ *is bounded, and* $\varsigma_{j,k} \in \mathcal{S}_0\left(\mathbb{R}^q\right)$ *if* $k > 0$.

PROOF. We decompose δ_0–the Dirac δ function at 0. Indeed, let $\widehat{\phi} \in C_0^\infty\left(\mathbb{R}^q\right)$ equal 1 on a neighborhood of 0. Define $\widehat{\psi} = \widehat{\phi}(\xi) - \widehat{\phi}(2\xi)$, where 2ξ is defined in terms of the dilations on $\mathbb{R}^q$–i.e., $2\left(\xi_1, \ldots, \xi_q\right) = \left(2^{d_1}\xi_1, \ldots, 2^{d_q}\xi_q\right)$. Note, $1 \equiv \widehat{\phi}(\xi) + \sum_{\substack{k \geq 1 \\ k \in \mathbb{N}}} \widehat{\psi}\left(2^{-k}\xi\right)$; here, again, $2^{-k}\xi$ is defined in terms of the chosen dilations on $\mathbb{R}^q$. Taking the inverse Fourier transform, we have $\delta_0 = \phi + \sum_{k \geq 1} \psi^{\left(2^k\right)}$. For $\varsigma \in \mathcal{B}_1$, $\eta \in \mathcal{B}_2$ and $j \geq 0$, define[19]

$$\widetilde{\varsigma}_{j,k}^{\left(2^{j \wedge k}\right)} := \begin{cases} \psi^{\left(2^k\right)} * \left(\eta\varsigma^{\left(2^j\right)}\right) & \text{if } k \geq 1, \\ \phi * \left(\eta\varsigma^{\left(2^j\right)}\right) & \text{if } k = 0, \end{cases}$$

so that $\eta\varsigma^{\left(2^j\right)} = \delta_0 * \left(\eta\varsigma^{\left(2^j\right)}\right) = \sum_{k \geq 0} \widetilde{\varsigma}_{j,k}^{\left(2^{j \wedge k}\right)}$ and by Lemma 2.14.7 for every N,

$$\left\{ 2^{N|j-k|}\widetilde{\varsigma}_{j,k} \mid \varsigma \in \mathcal{B}_1, \eta \in \mathcal{B}_2, j, k \in \mathbb{N}\right\} \subset \mathcal{S}\left(\mathbb{R}^q\right)$$

is a bounded set. If $k > 0$, $\psi \in \mathcal{S}_0\left(\mathbb{R}^q\right)$ implies $\psi^{\left(2^k\right)} * \left(\eta\varsigma^{\left(2^j\right)}\right) \in \mathcal{S}_0\left(\mathbb{R}^q\right)$. For $k \leq j$, define

$$\varsigma_{j,k}^{\left(2^k\right)} = \begin{cases} \widetilde{\varsigma}_{j,k}^{\left(2^k\right)} & \text{if } k < j, \\ \sum_{l \geq j} \widetilde{\varsigma}_{j,l}^{\left(2^k\right)} & \text{if } k = j. \end{cases}$$

By the above remarks, $\varsigma_{j,k}$ satisfy the conclusions of the lemma. $\qquad\square$

---

[19]Here, we are using the usual Euclidean group structure on $\mathbb{R}^q$ to define the convolution.

PROOF OF PROPOSITION 2.14.5. (ii)⇒(i): Suppose $K \in \mathcal{S}(\mathbb{R}^q)'$ has supp $(K) \subset \overline{B^q}(a)$ and there is a bounded set $\{\widetilde{\varsigma}_j \mid j \in \mathbb{N}^\nu\} \subset \mathcal{S}(\mathbb{R}^q)$ with $\varsigma_j \in \mathcal{S}_0(\mathbb{R}^q)$ for $j > 0$ such that $K = \sum_{j \in \mathbb{N}} 2^{js} \varsigma_j^{(2^j)}$. Since supp (K) is a closed subset of $B^q(a)$, we may take $\eta \in C_0^\infty(B^q(a))$ with $\eta \equiv 1$ on a neighborhood of supp (K). Then, $K = \eta K = \eta \sum_{j \in \mathbb{N}} 2^{j \cdot s} \varsigma_j^{(2^j)}$, thereby proving $K \in CZ(s, a)$.

(i)⇒(ii): Let $K \in CZ(s, a)$, so that there is $\eta \in C_0^\infty(B^q(a))$ and a bounded set $\{\varsigma_j \mid j \in \mathbb{N}\} \subset \mathcal{S}(\mathbb{R}^q)$ with $\varsigma_j \in \mathcal{S}_0(\mathbb{R}^q)$ for $j > 0$ such that $K = \eta \sum_{j \in \mathbb{N}} \varsigma_j^{(2^j)}$. For each j apply Lemma 2.14.8 to decompose $\eta \varsigma_j^{(2^j)} := \sum_{k \leq j} \varsigma_{j,k}^{(2^k)}$, where, for every N, $\{2^{N|j-k|} \varsigma_{j,k} \mid j \in \mathbb{N}, k \leq j\} \subset \mathcal{S}(\mathbb{R}^q)$ is bounded, and $\varsigma_{j,k} \in \mathcal{S}_0(\mathbb{R}^q)$ for $k > 0$. For $k \in \mathbb{N}$ define $2^{ks} \widetilde{\varsigma}_k = \sum_{j \geq k} 2^{js} \varsigma_{j,k}$, so that $\{\widetilde{\varsigma}_k \mid k \in \mathbb{N}\} \subset \mathcal{S}(\mathbb{R}^q)$ is a bounded set, $\widetilde{\varsigma}_k \in \mathcal{S}_0(\mathbb{R}^q)$ for $k > 0$, and $K = \sum_{k \in \mathbb{N}} 2^{ks} \widetilde{\varsigma}_k^{(2^k)}$, thus proving (ii). ☐

It follows from Proposition 2.14.5 that $CZ(s, a)$ is a vector space. It will be useful to define a locally convex topology on $CZ(s, a)$. This topology will only be tangentially useful, and the reader uninterested may skip the definition and corresponding results, without missing any main ideas. We define this topology in the following way.

DEFINITION 2.14.9. *Fix* $s \in \mathbb{R}$. *For* $0 < b < a$, *define the vector space* $CZ_0(s, b)$ *to be the space of those* $K \in CZ(s, a)$ *with* supp $(K) \subseteq \overline{B^q}(b)$. *For each* $K \in CZ_0(s, b)$, *Proposition 2.14.5 shows that we may write* $K = \sum_{j \in \mathbb{N}} 2^{js} \varsigma_j^{(2^j)}$, *where* $\{\varsigma_j \mid j \in \mathbb{N}^\nu\} \subset \mathcal{S}(\mathbb{R}^q)$ *is a bounded set and* $\varsigma_j \in \mathcal{S}_0(\mathbb{R}^q)$ *for* $j > 0$. *For each continuous semi-norm* $|\cdot|$ *on* $\mathcal{S}(\mathbb{R}^q)$ *we define a semi-norm on* $CZ_0(s, b)$ *by* $|K| :=$ inf $\sup_{j \in \mathbb{N}} |\varsigma_j|$, *where the infimum is taken over all such representations of* K. *We give* $CZ_0(s, b)$ *the coarsest topology such that all of these semi-norms are continuous.*

LEMMA 2.14.10. $CZ_0(s, b)$ *is a Fréchet space.*

To prove Lemma 2.14.10 we need another lemma.

LEMMA 2.14.11. *Let* $\mathcal{B} \subset \mathcal{S}(\mathbb{R}^q)$ *be a bounded set. For* $\varsigma_1, \varsigma_2 \in \mathcal{B}$, $j_1, j_2 \in [0, \infty)$ *with* $0 \leq j_1 \leq j_2$ *and* $\varsigma_2 \in \mathcal{S}_0(\mathbb{R}^q)$ *if* $j_2 > 0$, *we define*

$$\varsigma_3^{(2^{j_1})} = \varsigma_1^{(2^{j_1})} * \varsigma_2^{(2^{j_2})}, \quad \varsigma_4^{(2^{j_2})} = \varsigma_1^{(2^{j_1})} * \varsigma_2^{(2^{j_2})},$$

then for every N *the set*

$$\left\{ 2^{N|j_1 - j_2|} \varsigma_3, 2^{N|j_1 - j_2|} \varsigma_4 \mid \varsigma_1, \varsigma_2 \in \mathcal{B}, j_1, j_2 \in [0, \infty), \right.$$

$$\left. 0 \leq j_1 \leq j_2, \varsigma_2 \in \mathcal{S}_0(\mathbb{R}^q) \text{ if } j_2 > 0 \right\} \subset \mathcal{S}(\mathbb{R}^q)$$

is a bounded set and $\varsigma_3, \varsigma_4 \in \mathcal{S}_0(\mathbb{R}^q)$ *if* $j_2 > 0$.

PROOF. This follows just as in Lemma 1.1.19. ☐

PROOF SKETCH OF LEMMA 2.14.10. Because the topology of $S(\mathbb{R}^q)$ is defined by a countable family of semi-norms, Definition 2.14.9 gives a countable family of semi-norms generating the topology of $CZ_0(s, b)$, so the content of the lemma is that $CZ_0(s, b)$ is complete. As in Section 2.8, the main difficulty is the infimum defining the semi-norms in Definition 2.14.9. To avoid this, we pick a particular way to decompose K, and we do this in such a way that the topology can be equivalently defined by the particular decomposition, thereby avoiding the infimum.

For this, we look back to the proof of Lemma 2.14.8. There we decomposed $\delta_0 = \phi + \sum_{k \geq 1} \psi^{(2^k)}$, where $\phi \in S(\mathbb{R}^q)$ and $\psi \in S_0(\mathbb{R}^q)$. Define

$$
\widetilde{\varsigma}_j^{(2^j)} := \begin{cases} K * \phi & \text{if } j = 0, \\ 2^{-js} K * \psi^{(2^j)} & \text{if } j > 0. \end{cases}
$$

I.e., if $K = \sum_{j \in \mathbb{N}} 2^{js} \varsigma_j^{(2^j)}$, with $\{\varsigma_j \mid j \in \mathbb{N}^\nu\} \subset S(\mathbb{R}^q)$ a bounded set and $\varsigma_j \in S_0(\mathbb{R}^q)$ for $j \neq 0$, then

$$
\widetilde{\varsigma}_j^{(2^j)} := \begin{cases} \sum_{k \in \mathbb{N}} 2^{ks} \varsigma_k^{(2^k)} * \phi & \text{if } j = 0, \\ \sum_{k \in \mathbb{N}} 2^{-(j-k)s} \varsigma_k^{(2^k)} * \psi^{(2^j)} & \text{if } j > 0. \end{cases}
$$

We have $K = \sum_{j \in \mathbb{N}} 2^{js} \widetilde{\varsigma}_j^{(2^j)}$. It follows from Lemma 2.14.11 that $\{\widetilde{\varsigma}_j \mid j \in \mathbb{N}\} \subset S(\mathbb{R}^q)$ is a bounded set and $\widetilde{\varsigma}_j \in S_0(\mathbb{R}^q)$ for $j > 0$. Furthermore, it follows by the same proof that for every continuous semi-norm $|\cdot|_1$ on $S(\mathbb{R}^q)$ there is a continuous semi-norm $|\cdot|_2$ on $S(\mathbb{R}^q)$ such that if $K = \sum_{j \in \mathbb{N}^\nu} 2^{js} \varsigma_j^{(2^j)}$ is another such representation with $\{\varsigma_j \mid j \in \mathbb{N}\} \subset S(\mathbb{R}^q)$ a bounded set and $\varsigma_j \in S_0(\mathbb{R}^q)$ for $j > 0$, then

$$
\sup_{j \in \mathbb{N}} |\widetilde{\varsigma}_j|_1 \lesssim \sup_{j \in \mathbb{N}} |\varsigma_j|_2 .
$$

Taking the infimum over all such representations of K, we have that for every continuous semi-norm $|\cdot|_1$ on $S(\mathbb{R}^q)$, there is a continuous semi-norm $|\cdot|_2$ on $CZ_0(s, b)$ such that

$$
\sup_{j \in \mathbb{N}} |\widetilde{\varsigma}_j|_1 \lesssim |K|_2 .
$$

Thus, for each continuous semi-norm $|\cdot|$ on $S(\mathbb{R}^q)$ we obtain a continuous semi-norm on $CZ_0(s, b)$ given by $|K| := \sup_{j \in \mathbb{N}} |\widetilde{\varsigma}_j|$, where $\widetilde{\varsigma}_j$ is chosen as above. It follows that this family of semi-norms generates the topology of $CZ_0(s, b)$. Since $S(\mathbb{R}^q)$ is a Fréchet space, we can pick a countable collection of such semi-norms which define the topology of $CZ_0(s, b)$.

Now suppose $\{K_n\}_{n \in \mathbb{N}} \subset CZ_0(s, b)$ is a Cauchy sequence. Decompose each K_n as $K_n = \sum_{j \in \mathbb{N}} 2^{js} \widetilde{\varsigma}_{n,j}^{(2^j)}$ as above. Because $\{K_n\}_{n \in \mathbb{N}}$ is Cauchy, it follows that for each j, $\{\widetilde{\varsigma}_{n,j}\} \subset S(\mathbb{R}^q)$ is Cauchy uniformly in j. Thus, $\widetilde{\varsigma}_{n,j} \to \widetilde{\varsigma}_{\infty,j}$ in $S(\mathbb{R}^q)$ uniformly for $j \in \mathbb{N}$, where $\{\widetilde{\varsigma}_{\infty,j} \mid j \in \mathbb{N}\} \subset S(\mathbb{R}^q)$ is a bounded set and $\widetilde{\varsigma}_{\infty,j} \in S_0(\mathbb{R}^q)$ for $j > 0$. Set $K_\infty = \sum_{j \in \mathbb{N}} 2^{js} \widetilde{\varsigma}_{\infty,j}^{(2^j)} \in CZ_0(s, b)$. We have $K_n \to K_\infty$ in $CZ_0(s, b)$ proving that $CZ_0(s, b)$ is complete, and therefore it is a Fréchet space. □

Note that if $0 < b_1 < b_2 < a$, Proposition 2.14.5 shows that $CZ_0(s, b_1) \hookrightarrow CZ_0(s, b_2)$. Furthermore, it follows directly from the definitions that the topology on $CZ_0(s, b_1)$ is exactly the topology $CZ_0(s, b_1)$ inherits as a subspace of $CZ_0(s, b_2)$.

DEFINITION 2.14.12. *We define a locally convex topology on* $CZ(s, a)$ *by seeing it as* $CZ(s, a) = \varinjlim CZ(s, b)$ *where the inductive limit is taken over* $0 < b < a$. *See Definition A.1.21 and Theorem A.1.22 for the notion of an inductive limit.*

Remark 2.14.13 Pick a countable sequence $b_1 < b_2 < \cdots < a$ with $\lim_{j \to \infty} b_j = a$. We have $CZ(s, a) = \varinjlim CZ_0(s, b_j)$. This sees $CZ(s, a)$ as an LF space (Definition A.1.24) and we therefore have $CZ(s, a)$ is complete (Remark A.2.4).

Now that we have a complete locally convex topology on $CZ(s, a)$ it makes sense to consider $C^\infty(M; CZ(s, a))$: $C^\infty(M)$ functions taking values in $CZ(s, a)$–see Appendix A.2 for details.

DEFINITION 2.14.14. *Fix* $a > 0$ *a small number to be chosen later. For* $K(x, t) \in C^\infty(M; CZ(s, a))$, *we call operators,* $T : C^\infty(M) \to C^\infty(M)$, *of the form*

$$Tf(x) = \int f\left(e^{t_1 X_1 + \ldots + t_q X_q} x\right) K(x, t) \, dt$$

pseudodifferential operators of order s.

Remark 2.14.15 We think of the operator T from Definition 2.14.14 as a generalization of a pseudodifferential operator from the Euclidean setting, as covered in Section 1.3. Indeed, consider the formula for a pseudodifferential operator given in Remark 1.3.14. Noting that $x - z = e^{-z_1 \partial_{x_1} - \cdots - z_n \partial_{x_n}} x$, we see that the operator $a(x, D)$ in Remark 1.3.14 is of a closely related form to T, where in T we have replaced $\partial_{x_1}, \ldots, \partial_{x_n}$ with $X_1, \ldots, X_q$. The analogy with pseudodifferential operators does not end here, though. See Remark 2.14.29.

THEOREM 2.14.16. *Let* T *be a pseudodifferential operator of order* s, *then* T *is a Calderón-Zygmund operator of order* s.

We turn to proving Proposition 2.14.4 and Theorem 2.14.16 (which together imply Theorem 2.14.1). We begin with Proposition 2.14.4 and we start with a lemma.

LEMMA 2.14.17. *Let* $\{\varsigma_j \mid j \in \mathbb{Z}, j \leq 0\} \subset \mathcal{S}(\mathbb{R}^q)$ *be a bounded set,* $\eta' \in C_0^\infty(\mathbb{R}^q)$, *and let* $s > -Q$. *Then,* $\sum_{j \leq 0} \eta' 2^{js} \varsigma_j^{(2^j)} \in C_0^\infty(\mathbb{R}^q)$.

PROOF. Since $\eta'(t) 2^{js} \varsigma_j^{(2^j)}(t) = \eta'(t) 2^{j(Q+s)} \varsigma_j(2^j t)$, the result follows easily (this requires $s > -Q$). □

PROOF OF PROPOSITION 2.14.4. Suppose K is a Calderón-Zygmund kernel of order $s > -Q$ with $\mathrm{supp}(K) \subset B^q(a)$. Fix $\eta, \eta' \in C_0^\infty(B^q(a))$ with $\eta \equiv 1$ on a neighborhood of the support of K and $\eta' \equiv 1$ on a neighborhood of the support of η.

Apply Theorem 2.13.15 to decompose K: $K = \sum_{j\in\mathbb{Z}} 2^{js}\widehat{\varsigma}_j^{(2^j)}$, with $\{\varsigma_j \mid j\in\mathbb{Z}\} \subset \mathcal{S}_0(\mathbb{R}^q)$ a bounded set. For $j > 0$ let $\varsigma_j = \widetilde{\varsigma}_j$ and let $\varsigma_0 = \eta'(t)\sum_{j\le 0} 2^{js}\widetilde{\varsigma}_j$. By Lemma 2.14.17, $\{\varsigma_j \mid j\in\mathbb{N}\} \subset \mathcal{S}(\mathbb{R}^q)$ is a bounded set. We have $K = \eta K = \eta\sum_{j\in\mathbb{N}} \varsigma_j^{(2^j)}$, completing the proof that $K \in \mathrm{CZ}(s,a)$. $\qquad\square$

We now turn to the proof of Theorem 2.14.16. Instead of proving Theorem 2.14.16 directly, we prove the following lemma.

LEMMA 2.14.18. *Let* $\kappa \in C^\infty(\mathbb{R}^q \times M)$ *and* $K \in \mathrm{CZ}(s,a)$. *Define the operator*

$$Tf(x) = \int f\left(e^{t\cdot X}x\right) \kappa(t,x) K(t)\, dt.$$

Then T *is a Calderón-Zygmund operator of order* s. *Moreover, the map* $(\kappa, K) \mapsto T$ *is continuous.*

PROOF OF THEOREM 2.14.16 GIVEN LEMMA 2.14.18. We use the theory outlined in Appendix A.2. We restrict attention to $\kappa(x) \in C^\infty(M)$; i.e., κ independent of $t \in \mathbb{R}^q$. Lemma 2.14.18 shows that the map $(\kappa, K) \mapsto T$ is a continuous, bilinear, map taking $C^\infty(M) \times \mathrm{CZ}(s,a)$ to the space of Calderón-Zygmund operators of order s. By the universal property (Proposition A.2.7), this extends to a continuous map from $C^\infty(M) \widehat{\otimes}_\pi \mathrm{CZ}(s,a)$ to the space of Calderón-Zygmund operators of order s, in the canonical way. However, $C^\infty(M) \widehat{\otimes}_\pi \mathrm{CZ}(s,a) \cong C^\infty(M; \mathrm{CZ}(s,a))$ (see Example A.2.15), which completes the proof. $\qquad\square$

We now turn to proving Lemma 2.14.18. Let T be as in the statement of the lemma:

$$Tf(x) = \int f\left(e^{t\cdot X}x\right) \kappa(t,x) K(t)\, dt,$$

where $\kappa \in C^\infty(\mathbb{R}^q \times M)$ and $K \in \mathrm{CZ}(s,a)$ (for $a > 0$ small). By the definition of $\mathrm{CZ}(s,a)$ there exists a bounded set $\{\varsigma_j \mid j\in\mathbb{N}\} \subset \mathcal{S}(\mathbb{R}^q)$ with $\varsigma_j \in \mathcal{S}_0(\mathbb{R}^q)$ for $j > 0$ and $\eta \in C_0^\infty(B^q(a))$ with $K = \eta\sum_{j\in\mathbb{N}} 2^{js}\varsigma_j^{(2^j)}$. For $j \in \mathbb{N}$ define the operator $E_j : C^\infty(M) \to C^\infty(M)$ by

$$E_j f(x) = \int f\left(e^{t_1 X_1 + \cdots + t_q X_q}x\right) \kappa(t,x)\, \eta(t)\, \varsigma_j^{(2^j)}(t)\, dt,$$

so that $T = \sum_{j\in\mathbb{N}^\nu} 2^{js}E_j$. Lemma 2.14.18 will follow from Theorem 2.0.29 once we show $\{(E_j, 2^{-j}) \mid j\in\mathbb{N}\}$ is a bounded set of elementary operators (the desired continuity is an immediate consequence of the proof that follows). We state this result in a slightly more general way.

Fix $a > 0$ to be a small number to be chosen later. For $\varsigma \in \mathcal{S}(\mathbb{R}^q)$, $j \in [0,\infty)$, $\eta \in C_0^\infty(B^q(a))$, $\kappa \in C^\infty(\mathbb{R}^q \times M)$, define $E = E(\varsigma, 2^{-j}, \eta, \kappa) : C^\infty(M) \to C^\infty(M)$ by

$$Ef(x) = \int f\left(e^{t_1 X_1 + \cdots + t_q X_q}x\right) \kappa(t,x)\, \eta(t)\, \varsigma^{(2^j)}(t)\, dt.$$

PROPOSITION 2.14.19. *Let*

$$\mathcal{B}_1 \subset \mathcal{S}\left(\mathbb{R}^q\right),\ \mathcal{B}_2 \subset C_0^\infty\left(B^q\left(a\right)\right),\ \text{and } \mathcal{B}_3 \subset C^\infty\left(\mathbb{R}^q \times M\right)\ \text{be bounded sets.}$$

The following sets are bounded sets of elementary operators:

$$\mathcal{E}_1 := \left\{\left(E\left(\varsigma, 2^{-0}, \eta, \kappa\right), 2^{-0}\right) \mid \varsigma \in \mathcal{B}_1, \eta \in \mathcal{B}_2, \kappa \in \mathcal{B}_3\right\},$$

$$\mathcal{E}_2 := \left\{\left(E\left(\varsigma, 2^{-j}, \eta, \kappa\right), 2^{-j}\right) \mid \varsigma \in \mathcal{B}_1 \cap \mathcal{S}_0\left(\mathbb{R}^q\right), j \in [0, \infty), \eta \in \mathcal{B}_2, \kappa \in \mathcal{B}_3\right\}.$$

To prove this proposition, we need to further decompose $E\left(\varsigma, 2^{-j}, \eta, \kappa\right)$. To do this, we decompose ς.

LEMMA 2.14.20. *Let* $\mathcal{B}_1 \subset \mathcal{S}\left(\mathbb{R}^q\right)$ *and* $\mathcal{B}_2 \subset C_0^\infty\left(B^q\left(a\right)\right)$ *be bounded sets. Then, for* $\eta \in \mathcal{B}_2, \varsigma \in \mathcal{B}_1,$ *and* $j \in [0, \infty),$ *we have*

$$\eta\left(t\right)\varsigma^{\left(2^j\right)}\left(t\right) = \sum_{\substack{k \leq j \\ k \in \mathbb{N}}} \widetilde{\varsigma}_k^{\left(2^k\right)}\left(t\right),$$

where $\widetilde{\varsigma}_k \in C_0^\infty\left(B^q\left(a\right)\right)$ *and, moreover, for every* N, *the set*

$$\left\{2^{N|j-k|}\widetilde{\varsigma}_k \mid \varsigma \in \mathcal{B}_1, \eta \in \mathcal{B}_2, j \in [0, \infty), k \in [0, j] \cap \mathbb{N}\right\} \subset C_0^\infty\left(B^q\left(a\right)\right) \quad (2.80)$$

is a bounded set.

PROOF. Fix $\eta' \in C_0^\infty\left(B^q\left(a\right)\right)$ such that $\eta' \equiv 1$ on a neighborhood of the closure of $\cup_{\eta \in \mathcal{B}_2}\text{supp}\left(\eta\right)$. Fix $j \in [0, \infty)$. For $k \in \mathbb{N}, k \leq j$, define

$$\delta_k\left(t\right) = \begin{cases} \eta'\left(2^k t\right) - \eta'\left(2^{k+1} t\right), & \text{if } k+1 \leq j, \\ \eta'\left(2^k t\right), & \text{if } j < k+1 \leq j+1. \end{cases}$$

Note, $\eta'\left(t\right) = \sum_{0 \leq k \leq j}\delta_k\left(t\right)$ and $\delta_k\left(t\right) = 0$ if $k+1 \leq j$ and $\left|2^k t\right| \leq \epsilon_0$, where $\epsilon_0 > 0$ depends on η', but does not depend on any of the parameters which vary. Define $\widetilde{\varsigma}_k^{\left(2^k\right)}\left(t\right) := \delta_k\left(t\right)\eta\left(t\right)\varsigma^{\left(2^j\right)}\left(t\right)$. Note, $\widetilde{\varsigma}_k\left(t\right) = 2^{(j-k)Q}\delta_k\left(2^{-k}t\right)\eta\left(2^{-k}t\right)\varsigma\left(2^{j-k}t\right)$. If $k+1 \leq j$, then $\delta_k\left(2^{-k}t\right)$ is supported on $|t| \approx 1$. The rapid decrease of ς and all its derivatives shows, for every N, the set in (2.80) is bounded. The claim follows. $\qquad\square$

LEMMA 2.14.21. *Let* $\mathcal{B}_1 \subset \mathcal{S}\left(\mathbb{R}^q\right), \mathcal{B}_2 \subset C_0^\infty\left(B^q\left(a\right)\right),$ *and* $\mathcal{B}_3 \subset C^\infty\left(\mathbb{R}^q \times M\right)$ *be bounded sets. The set*

$$\left\{\left(E\left(\varsigma, 2^{-j}, \eta, \kappa\right), 2^{-j}\right) \mid \varsigma \in \mathcal{B}_1, j \in [0, \infty), \eta \in \mathcal{B}_2, \kappa \in \mathcal{B}_3\right\}$$

is a bounded set of pre-elementary operators.

PROOF. Let $\varsigma \in \mathcal{B}_1$, $\eta \in \mathcal{B}_2$, $j \in [0, \infty)$, and $\kappa \in \mathcal{B}_3$. Decompose $\eta \varsigma^{(2^j)} = \sum_{\substack{k \leq j \\ k \in \mathbb{N}}} \tilde{\varsigma}_k^{(2^k)}$ as in Lemma 2.14.20. Define the operator

$$
\begin{aligned}
\tilde{E}_k f(x) &= \int f\left(e^{t_1 X_1 + \cdots + t_q X_q} x\right) \kappa(t, x) \tilde{\varsigma}_k^{(2^k)}(t) \, dt \\
&= \int f\left(e^{t_1 2^{-kd_1} X_1 + \cdots + t_q 2^{-kd_q} X_q} x\right) \kappa\left(2^{-k} t, x\right) \tilde{\varsigma}_k(t) \, dt \\
&= \int f\left(e^{t \cdot (2^{-k} X)} x\right) \kappa\left(2^{-k} t, x\right) \tilde{\varsigma}_k(t) \, dt,
\end{aligned}
$$

so that $E = E\left(\varsigma, 2^{-j}, \eta, \kappa\right) = \sum_{\substack{k \leq j \\ k \in \mathbb{N}}} \tilde{E}_k$. Fix ordered multi-indices α and β and fix $m \in \mathbb{N}$. We wish to show

$$
\left|\left(2^{-j} X_x\right)^\alpha \left(2^{-j} X_z\right)^\beta E(x, z)\right| \lesssim \frac{\left(1 + 2^j \rho(x, z)\right)^{-m}}{\operatorname{Vol}\left(B_{(X,d)}\left(x, 2^{-j} + \rho(x, z)\right)\right)}. \tag{2.81}
$$

Let $N = N(m, \alpha, \beta)$ be a large integer to be chosen later. We have

$$
\left\{2^{N|j-k|} \tilde{\varsigma}_k \mid \varsigma \in \mathcal{B}_1, \eta \in \mathcal{B}_2, j \in [0, \infty), k \in [0, j] \cap \mathbb{N}\right\} \subset C_0^\infty\left(B^q(a)\right)
$$

is a bounded set, and therefore by Proposition 2.2.28 (taking $a > 0$ sufficiently small) applied with (Z, d) replaced by $\left(2^{-k} X, d\right)$ we have,

$$
\begin{aligned}
\left|\left(2^{-k} X_x\right)^\alpha \left(2^{-k} X_z\right)^\beta \tilde{E}_k(x, z)\right| &\lesssim 2^{-N|j-k|} \frac{\chi_{B_{(X,d)}(x, \xi_3 2^{-k-1})}(z)}{\operatorname{Vol}\left(B_{(X,d)}\left(x, \xi_3 2^{-k-1}\right)\right)} \\
&\lesssim 2^{-N|j-k|} \frac{\left(1 + 2^k \rho(x, z)\right)^{-m}}{\operatorname{Vol}\left(B_{(X,d)}\left(x, 2^{-k} + \rho(x, z)\right)\right)},
\end{aligned}
$$

where $\xi_3 > 0$ is as in the quantitative Frobenius theorem (Theorem 2.2.22). By taking $N = N(m, \alpha, \beta)$ sufficiently large, we have by Lemma 2.7.7,

$$
\left|\left(2^{-j} X_x\right)^\alpha \left(2^{-j} X_z\right)^\beta \tilde{E}_k(x, z)\right| \lesssim 2^{-|j-k|} \frac{\left(1 + 2^j \rho(x, z)\right)^{-m}}{\operatorname{Vol}\left(B_{(X,d)}\left(x, 2^{-j} + \rho(x, z)\right)\right)}.
$$

Since $E = \sum_{\substack{k \leq j \\ k \in \mathbb{N}}} \tilde{E}_k$, (2.81) follows, which completes the proof. $\qquad\square$

LEMMA 2.14.22. *Let $\mathcal{B}_1 \subset \mathcal{S}(\mathbb{R}^q)$, $\mathcal{B}_2 \subset C_0^\infty(B^q(a))$, and $\mathcal{B}_3 \subset C^\infty(\mathbb{R}^q \times M)$ be bounded sets. For $\varsigma \in \mathcal{B}_1$, $\eta \in \mathcal{B}_2$, $j \in [0, \infty)$, and $\kappa \in \mathcal{B}_3$, we have*

$$
E(\varsigma, \eta, j, \kappa)^* = E\left(\bar{\varsigma}(-t), \bar{\eta}(-t), j, \kappa^*\right),
$$

where $\kappa^* = \kappa^*(\kappa)$ and $\{\kappa^* \mid \kappa \in \mathcal{B}_3\} \subset C^\infty(\mathbb{R}^q \times M)$ is a bounded set and $\bar{\varsigma}(-t)$ and $\bar{\eta}(-t)$ denote the respective complex conjugates of ς and η composed with the map $t \mapsto -t$. In short, E^* is "of the same form" as E. Here we use that we may take $a > 0$ small, depending on $X_1, \ldots, X_q$.

PROOF. Note that $\frac{\partial}{\partial x} e^{t \cdot X} x \big|_{t=0} = I$, because $e^{0 \cdot X} x = x$. Thus,

$$\det \frac{\partial}{\partial x} e^{t \cdot X} x \big|_{t=0} = 1.$$

Hence, if we take $a > 0$ small enough, we have $\det \frac{\partial}{\partial x} e^{t \cdot X} x \approx 1$ for $|t| < a$. The result now follows by a simple change of variables–see Appendix B.3. $\qquad\square$

LEMMA 2.14.23. *Let* $\mathcal{B}_1 \subset \mathcal{S}_0 (\mathbb{R}^q)$, $\mathcal{B}_2 \subset C_0^\infty (B^q (a))$, *and* $\mathcal{B}_3 \subset C^\infty (\mathbb{R}^q \times M)$ *be bounded sets. There is a constant* $R = R(q, d_1, \ldots, d_q)$ *such that for* $\varsigma \in \mathcal{B}_1$, $\eta \in \mathcal{B}_2$, $j \in [0, \infty)$, *and* $\kappa \in \mathcal{B}_3$,

$$
E(\varsigma, \eta, j, \kappa)
$$
$$
= \begin{cases} \sum_{|\alpha| \leq 1} 2^{-(1-|\alpha|)} \sum_{k=1}^R E(\varsigma_{\alpha,1,k}, \eta_{\alpha,1,k}, j, \kappa_{\alpha,1,k}) (2^{-j} X)^\alpha, & and \quad (2.82) \\ \sum_{|\alpha| \leq 1} 2^{-(1-|\alpha|)} \sum_{k=1}^R (2^{-j} X)^\alpha E(\varsigma_{\alpha,2,k}, \eta_{\alpha,2,k}, j, \kappa_{\alpha,2,k}) \end{cases}
$$

where the following sets are bounded:

$$\left\{ \varsigma_{\alpha,1,k}, \varsigma_{\alpha,2,k} \mid \varsigma \in \mathcal{B}_1, \eta \in \mathcal{B}_2, j \in [0, \infty), \kappa \in \mathcal{B}_3, |\alpha| \leq 1, 1 \leq k \leq R \right\} \subset \mathcal{S}_0 (\mathbb{R}^q),$$

$$\left\{ \eta_{\alpha,1,k}, \eta_{\alpha,2,k} \mid \varsigma \in \mathcal{B}_1, \eta \in \mathcal{B}_2, j \in [0, \infty), \kappa \in \mathcal{B}_3, |\alpha| \leq 1, 1 \leq k \leq R \right\} \subset C_0^\infty (B^q (a)),$$

$$\left\{ \kappa_{\alpha,1,k}, \kappa_{\alpha,2,k} \mid \varsigma \in \mathcal{B}_1, \eta \in \mathcal{B}_2, j \in [0, \infty), \kappa \in \mathcal{B}_3, |\alpha| \leq 1, 1 \leq k \leq R \right\} \subset C^\infty (\mathbb{R}^q \times M).$$

PROOF. The second line of (2.82) follows by taking adjoints of the first (Lemma 2.14.22), and so we prove only the first. For $\varsigma \in \mathcal{B}_1$, write $\varsigma = \sum_{l=1}^q \partial_{t_l} \varsigma_l$ (e.g., by writing $\varsigma = \triangle \triangle^{-1} \varsigma$ and using Corollary 1.1.13), such that $\left\{ \varsigma_l \mid \varsigma \in \mathcal{B}_1, 1 \leq l \leq q \right\} \subset \mathcal{S}_0 (\mathbb{R}^q)$ is a bounded set. Thus, $E(\varsigma, \eta, j, \kappa)$ can be written as a sum of terms of the form

$$E_l f (x) := \int f(e^{t \cdot X} x) \kappa(t, x) \eta(t) (\partial_{t_l} \varsigma_l)^{(2^j)} (t) \, dt$$
$$= 2^{-jd_l} \int \partial_{t_l} \left[f(e^{t \cdot X} x) \kappa(t, x) \eta(t) \right] \varsigma_l^{(2^j)} (t) \, dt.$$

We apply the product rule to the differentiation on the right-hand side of the above equation, and there are three terms. When ∂_{t_l} lands on κ we obtain $2^{-jd_l} E(\varsigma_l, \eta, j, \partial_{t_l} \kappa)$ and when ∂_{t_l} lands on η we obtain $2^{-jd_l} E(\varsigma_l, \partial_{t_l} \eta, j, \kappa)$, which are of the desired form. We are therefore left only with the case when ∂_{t_l} lands on $f(e^{t \cdot X} x)$. To compute this, we use Lemma 2.12.1. Fix M large, we have

$$\partial_{t_l} f(e^{t \cdot X} x) = \sum_{k=1}^q \sum_{\substack{\deg(\alpha) + d_l \geq d_k \\ |\alpha| \leq M}} \tilde{c}_{l,\alpha} (t, x) t^\alpha (X_k f) (e^{t \cdot X} x) + O\left(|t|^{M+1}\right),$$

where $\tilde{c}_{l,\alpha} \in C^\infty (\mathbb{R}^q \times M)$ come from an application of Proposition 2.1.3 to the formula in Lemma 2.12.1. Using that $X_1, \ldots, X_q$ span the tangent space at each point, we may write the remainder term in terms of $X_1, \ldots, X_q$. This yields

$$\partial_{t_l} f \left(e^{t \cdot X} x \right) = \sum_{k=1}^{q} \sum_{\substack{\deg(\alpha) + d_l \geq d_k \\ |\alpha| \leq M}} c_{l,\alpha} (t,x) \, t^\alpha \, (X_k f) \left(e^{t \cdot X} x \right),$$

with $c_{l,\alpha} \in C^\infty (\mathbb{R}^q \times M)$. We therefore have

$$\int \left[\partial_{t_l} f \left(e^{t \cdot X} x \right) \right] \kappa (t,x) \, \eta (t) \, \varsigma_l^{(2^j)} (t) \; dt$$

$$= \sum_{k=1}^{q} \sum_{\substack{\deg(\alpha) + d_l \geq d_k \\ |\alpha| \leq M}} 2^{-j(\deg(\alpha) - d_k)}$$

$$\times \int \left(2^{-j d_k} X_k f \right) \left(e^{t \cdot X} x \right) \kappa (t,x) \, c_{l,\alpha} (t,x) \, \eta (t) \, (t^\alpha \varsigma_l)^{(2^j)} (t) \; dt,$$

which is a sum of terms of the desired form, completing the proof. $\square$

PROOF OF PROPOSITION 2.14.19. That $\mathcal{E}_1$ is a bounded set of elementary operators follows from Lemma 2.14.21 (see Remark 2.0.26). That $\mathcal{E}_2$ is a bounded set of elementary operators follows by combining Lemmas 2.14.23 and 2.14.21, and using the characterization of elementary operators given in Proposition 2.0.24 (where we have used that, for each l, $2^{-j d_l} X_l$ is a sum of terms of the form $\left(2^{-j} W \right)^\alpha$ for various α, since $2^{-j d_l} X_l$ is an iterated commutator of the vector fields $2^{-j} W$). $\square$

Remark 2.14.24   In Theorems 1.1.23 and 1.1.26, convolution with an element of $\mathcal{S} (\mathbb{R}^n)$ played the role of a "pre-elementary operator" while convolution with an element of $\mathcal{S}_0 (\mathbb{R}^n)$ played the role of an "elementary operator." Lemma 2.14.21 and Proposition 2.14.19 show that pre-elementary operators and elementary operators are more than just formal generalizations of the Euclidean case: they are, in fact, closely related.

We now use Theorem 2.14.16 to prove the sub-Laplacian $\mathcal{L}$ is maximally hypoelliptic. As in Section 2.13, we identify $N_{k,r}$ with $\mathbb{R}^q \cong \mathfrak{n}_{k,r}$ via the exponential map. Here, $\mathfrak{n}_{k,r}$ is the free nilpotent Lie algebra of step k with r generators, and $N_{k,r}$ is the corresponding connected, simply connected, Lie group; the congruence $\mathbb{R}^q \cong \mathfrak{n}_{k,r}$ is taken in the sense of manifolds and vector spaces and this yields a global coordinate system on $\mathfrak{n}_{k,r} \cong N_{k,r}$. When working in this coordinate system, we write $\widetilde{X}_j$ for $\widehat{X}_j$. In what follows, we use the variable t to denote an element of $\mathbb{R}^q \cong \mathfrak{n}_{k,r}$, so that $\widetilde{X}_j$ is a differential operator in the t-variable.

LEMMA 2.14.25. For $f \in C^\infty (M)$,

$$\widetilde{X}_j f \left(e^{t \cdot X} x \right) = (X_j f) \, f \left(e^{t \cdot X} x \right)$$

$$+ \sum_{2k \geq \deg(\alpha) \geq k} \sum_{l=1}^{q} t^\alpha c_{l,\alpha} (t,x) \, (X_q f) \left(e^{t \cdot X} x \right), \tag{2.83}$$

with $c_{l,\alpha} \in C^\infty$.

PROOF SKETCH. Using the Baker-Campbell-Hausdorff formula, we may compute the Taylor series of $\widetilde{X}_j f \left(e^{t \cdot X} x\right)$. Because of the definition of $\widetilde{X}_j$, if the commutators of order $> k$ of the vector fields $W_1, \ldots, W_r$ were 0, then (2.83) would instead be $\widetilde{X}_j f \left(e^{t \cdot X} x\right) = (X_j f) f \left(e^{t \cdot X} x\right)$. In general, we have this equality only for the Taylor series involving terms t^α with $\deg(\alpha) \le k - 1$. The second term on the right-hand side of (2.83) is the remainder term for this Taylor approximation where we have used that $X_1, \ldots, X_q$ span the tangent space to write the remainder term in terms of $X_1, \ldots, X_q$. See [Goo76] for more details. $\qquad\square$

LEMMA 2.14.26. *Let $K \in CZ(s, a)$ and α be a multi-index. Then, $t^\alpha K(t) \in CZ(s - \deg(\alpha), a)$.*

PROOF. For $K \in CZ(s, a)$, we may decompose $K = \eta \sum_{j \in \mathbb{N}} 2^{js} \varsigma_j^{(2^j)}$, where η and ς_j are as in Definition 2.14.2. Setting $\widetilde{\varsigma}_j(t) = t^\alpha \varsigma_j$, we have $\{\widetilde{\varsigma}_j \mid j \in \mathbb{N}\} \subset \mathcal{S}(\mathbb{R}^q)$ is a bounded set and $\widetilde{\varsigma}_j \in \mathcal{S}_0(\mathbb{R}^q)$ for $j \ne 0$. We also have

$$t^\alpha K = \eta \sum_{j \in \mathbb{N}} 2^{j(s - \deg(\alpha))} \widetilde{\varsigma}_j^{(2^j)},$$

completing the proof. $\qquad\square$

LEMMA 2.14.27. *Let T be a pseudodifferential operator of order s of the form:*

$$Tf(x) = \int f \left(e^{t \cdot X} x\right) \kappa(t, x) \eta(t) K(t) \, dt.$$

Then,

$$TW_j f(x) = -\int f \left(e^{t \cdot X} x\right) \kappa(t, x) \eta(t) \widetilde{W}_j K(t) \, dt + R$$

where R is Calderón-Zygmund operator of order s.

PROOF. By Lemma 2.14.25,

$$\begin{aligned} TW_j f(x) = & \int \left[\widetilde{W}_j f \left(e^{t \cdot X} x\right)\right] \kappa(t, x) \eta(t) K(t) \, dt \\ & + \sum_{\substack{2k \ge \deg(\alpha) \ge k \\ 1 \le l \le q}} \int (X_j f) \left(e^{t \cdot X} x\right) \kappa_{l,\alpha}(t, x) t^\alpha K(t) \, dt, \end{aligned} \tag{2.84}$$

where $\kappa_{l,\alpha} = \kappa c_{l,\alpha}$. The operator

$$f \mapsto \int (X_j f) \left(e^{t \cdot X} x\right) \kappa_{l,\alpha} t^\alpha K(t) \, dt$$

is a composition of two operators. The first operator, $f \mapsto \int f \left(e^{t \cdot X} x\right) \kappa_{l,\alpha} t^\alpha K(t) \, dt$, is by Lemma 2.14.26 a pseudodifferential operator of order $s - \deg(\alpha)$, and is therefore a Calderón-Zygmund operator of order $s - \deg(\alpha)$, by Theorem 2.14.16. The second operator is X_j which is a Calderón-Zygmund operator of order $d_j \leq k$. Since the sum in (2.84) is over $\deg(\alpha) \geq k$, Proposition 2.0.31 shows that each term in the sum is a Calderón-Zygmund operator of order s. $\square$

Now consider the sub-Laplacian on $N_{k,r}$, which we identify with $\mathfrak{n}_{k,r} \cong \mathbb{R}^q$ via the exponential map. $\widetilde{\mathcal{L}} = \widetilde{W}_1^* \widetilde{W}_1 + \cdots + \widetilde{W}_r^* \widetilde{W}_r = -\widetilde{W}_1^2 - \cdots - \widetilde{W}_r^2$. By Theorem 2.13.23 and Proposition 2.13.21, there is a Calderón-Zygmund kernel K of order -2 with $\widetilde{\mathcal{L}} K = \delta_0$. Fix $\eta \in C_0^\infty(B^q(a))$ with $\eta \equiv 1$ on a neighborhood of 0. Define a pseudodifferential operator of order -2 by

$$Tf(x) = \int f\left(e^{t \cdot X} x\right) \eta(t) K(t) \, dt.$$

We have the following theorem.

THEOREM 2.14.28. $T\mathcal{L} = I + R$, where R is a Calderón-Zygmund operator of order -1.

PROOF. First we use that $\mathcal{L} \equiv -W_1^2 - \cdots - W_r^2$, modulo a Calderón-Zygmund operator of order 1, and therefore, $T\mathcal{L} \equiv T\left(-W_1^2 - \cdots - W_r^2\right)$ modulo a Calderón-Zygmund operator of order -1. Consider, by the previous lemma,

$$TW_l^2 f \equiv \int f\left(e^{t \cdot X} x\right) \widetilde{W}_l^2 \left[\eta(t) K(t)\right] \, dt,$$

modulo an operator of order -1 applied to f, and therefore,

$$TW_l^2 f \equiv \int f\left(e^{t \cdot X} x\right) \eta(t) \widetilde{W}_l^2 K(t) \, dt,$$

modulo an operator of order -1 applied to f. It follows that

$$T\mathcal{L} f \equiv \int f\left(e^{t \cdot X} x\right) \eta(t) \widetilde{\mathcal{L}} K(t) \, dt = \int f\left(e^{t \cdot X} x\right) \eta(t) \delta_0(t) \, dt = f(x),$$

modulo an operator of order -1 applied to f. This completes the proof. $\square$

Remark 2.14.29   The proof in Theorem 2.14.28 is closely related to the proof of Proposition 1.4.2. One can think of it as a generalization of the ideas from Proposition 1.4.2, to a situation where we do not have access to the Fourier transform.

COROLLARY 2.14.30. *For every N, there is a Calderón-Zygmund operator T_N of order -2 such that $T_N \mathcal{L} \equiv I$ modulo a Calderón-Zygmund operator of order $-N$.*

PROOF. Let T and R be as in Theorem 2.14.28. Set $T_N = \sum_{n=0}^{N-1} (-1)^n R^n T$. The result follows from the algebra properties of Calderón-Zygmund operators. $\square$

COROLLARY 2.14.31. *$\mathcal{L}$ is maximally hypoelliptic.*

PROOF. Let T_N be as in Corollary 2.14.30. We know, for $u \in C^\infty$,

$$\sum_{|\alpha| \leq 2} \|W^\alpha T_2 u\|_{L^2} \lesssim \|u\|_{L^2}.$$

Replacing u with $\mathcal{L}u$, we have,

$$\sum_{|\alpha| \leq 2} \|W^\alpha u + R_0 u\|_{L^2} \lesssim \|\mathcal{L}u\|_{L^2},$$

where R_0 is a Calderón-Zygmund operator of order 0. We therefore have

$$\sum_{|\alpha| \leq 2} \|W^\alpha u\|_{L^2} \lesssim \|\mathcal{L}u\|_{L^2} + \|u\|_{L^2},$$

which completes the proof.   □

2.15   BEYOND HÖRMANDER'S CONDITION

Up until this point, in this chapter we have worked with vector fields $W_1, \ldots, W_r$ which satisfy Hörmander's condition. From these, we generated the list of vector fields with formal degrees $(X, d) = (X_1, d_1), \ldots, (X_q, d_q)$, which satisfied the conclusion of Proposition 2.1.3,[20] and such that $X_1, \ldots, X_q$ span the tangent space at each point. With this list of vector fields in hand, we developed a theory of singular integrals. It turns out much of this theory can be re-created when we are just given appropriate vector fields with formal degrees $(X, d) = (X_1, d_1), \ldots, (X_q, d_q)$, where these vector fields need not be generated by vector fields satisfying Hörmander's condition.[21] First we discuss the case when (X, d) satisfies a generalization of the conclusion of Proposition 2.1.3, along with the assumption that $X_1, \ldots, X_q$ span the tangent space at each point. Then we turn to the more general case where we still assume a generalization of the conclusion of Proposition 2.1.3, but we do not assume that $X_1, \ldots, X_q$ span the tangent space.

Instead of a compact manifold, we work on an open set $U \subseteq \mathbb{R}^n$. We assume we are given vector fields with formal degrees $(X, d) = (X_1, d_1), \ldots, (X_q, d_q)$ with $0 \neq d_j \in \mathbb{N}$. For $\delta > 0$, we denote by δX the list of vector fields $\delta^{d_1} X_1, \ldots, \delta^{d_q} X_q$. For $x \in U$, we obtain the balls $B_{(X,d)}(x, \delta) := B_{\delta X}(x)$–if $X_1, \ldots, X_q$ do not span the tangent space, these balls may not be open sets. Let $U_1 \Subset U$ be a relatively compact open subset of U and let $\Omega \Subset U_1$ be a relatively compact open subset of U_1. Informally, we assume that the hypotheses of the quantitative Frobenius theorem (Theorem 2.2.22) hold uniformly for $x_0 \in \Omega$ with the choice of vector fields $Z_1 = \delta^{d_1} X_1, \ldots, Z_q = \delta^{d_q} X_q$. More precisely, we assume that there is $0 < \xi_1 \leq 1$ (independent of δ) such that

---

[20]Recall, the conclusion of Proposition 2.1.3 was that $[X_j, X_k] = \sum_{d_l \leq d_j + d_k} c_{j,k}^l X_l$, with $c_{j,k}^l \in C^\infty$.

[21]With the notable exception that there is no analog of the sub-Laplacian to this situation.

(a) For every $x_0 \in \Omega$ and $a_1, \ldots, a_q \in L^\infty([0,1])$, with $\left\|\sum_{j=1}^q |a_j|^2\right\|_{L^\infty} < 1$, there exists a solution to the ODE

$$\gamma'(t) = \sum_{j=1}^q a_j(t) \xi_1^{d_j} \delta^{d_j} X_j(\gamma(t)), \quad \gamma(0) = x_0, \quad \gamma : [0,1] \to U_1.$$

(b) $\forall \delta \in (0,1], \forall x_0 \in \Omega, \left[\delta^{d_j} X_j, \delta^{d_k} X_k\right] = \sum_{l=1}^q c_{j,k}^{l,x_0,\delta} \delta^{d_l} X_l$ on $B_{(X,d)}(x, \xi_1 \delta)$, where for every m, there is a constant D_m such that

$$\sup_{x_0 \in \Omega, \delta \in (0,1]} \sum_{|\alpha| \le m} \left\|(\delta X)^\alpha c_{j,k}^{l,x_0,\delta}\right\|_{C^0\left(B_{(X,d)}(x_0,\xi_1\delta)\right)} \le D_m.$$

Remark 2.15.1 In light of the fact that $[X_j, X_k]$ is spanned by $X_1, \ldots, X_q$, the classical Frobenius theorem applies to foliate the ambient space, Ω, into leaves. The ball $B_{(X,d)}(x, \delta)$ is an open subset of the leaf passing through x_0.

Notice, if we take m-admissible (respectively, admissible) constants as in Theorem 2.2.22 when applied to the list δX at the point x_0, then the above assumptions imply that these constants may be taken to be independent of $x_0 \in \Omega$ and $\delta \in (0,1]$. For each $x_0 \in \Omega$, let $n_0(x_0) = \dim \operatorname{span}\{X_1(x_0), \ldots, X_q(x_0)\}$. We have the following restatement of Theorem 2.2.22 in this situation, where m-admissible constants are independent of $x_0 \in \Omega$, $\delta \in (0,1]$ (and $\lesssim_m$, $\lesssim$, $\approx_m$, and $\approx$ are defined as in that theorem; and therefore all the implicit constants can be taken to be independent of $x_0 \in \Omega$ and $\delta \in (0,1]$).

THEOREM 2.15.2. *There exist 2-admissible constants* $\xi_2, \xi_3, \eta > 0$, $\xi_3 < \xi_2 < \xi_1$ *such that for every* $x_0 \in \Omega$ *and* $\delta \in (0,1]$ *there exists a* C^∞ *map*

$$\Phi_{x_0,\delta} : B^{n_0(x_0)}(\eta) \to B_{(X,d)}(x_0, \xi_2 \delta),$$

such that

- $\Phi_{x_0,\delta}(0) = x_0$.

- $\Phi_{x_0,\delta}$ *is injective.*

- $B_{(X,d)}(x_0, \xi_3) \subseteq \Phi_{x_0,\delta}\left(B^{n_0(x_0)}(\eta)\right) \subseteq B_{(X,d)}(x_0, \xi_2)$.

- *For* $u \in B^{n_0(x_0)}(\eta)$,

$$\left|\det_{n_0(x_0) \times n_0(x_0)} d\Phi_{x_0,\delta}(u)\right|$$
$$\approx_2 \left|\det_{n_0(x_0) \times n_0(x_0)} \left(\delta^{d_1} X_1(x_0) | \cdots | \delta^{d_q} X_q(x_0)\right)\right| \tag{2.85}$$
$$\approx_2 \operatorname{Vol}\left(B_{(X,d)}(x_0, \xi_2 \delta)\right),$$

where $\operatorname{Vol}(\cdot)$ *denotes the induced Lebesgue measure on the leaf passing through* x_0.

Furthermore, if $Y_1^{x_0,\delta}, \ldots, Y_q^{x_0,\delta}$ are the pullbacks of $\delta^{d_1} X_1, \ldots, \delta^{d_q} X_q$ via $\Phi_{x_0,\delta}$ to $B^{n_0(x_0)}(\eta)$, *then*

$$\left\| Y_j^{x_0,\delta} \right\|_{C^m\left(B^{n_0(x_0)}(\eta)\right)} \lesssim_m 1, \tag{2.86}$$

and

$$\inf_{\substack{u \in B^{n_0(x_0)}(\eta)}} \left| \det_{n_0(x_0) \times n_0(x_0)} \left(Y_1^{x_0,\delta}(u) | \cdots | Y_q^{x_0,\delta}(u) \right) \right| \approx_2 1. \tag{2.87}$$

2.15.1 More on the assumptions

The assumptions on the vector fields with formal degrees, (X, d), in this section are clearly the minimal assumptions needed to apply the quantitative Frobenius theorem and obtain Theorem 2.15.2. They are also somewhat weaker than the conclusion of Corollary 2.1.4. In this section, we discuss these assumptions, to better understand what they imply.

Suppose we are given vector fields with formal degrees

$$(X, d) = (X_1, d_1), \ldots, (X_q, d_q)$$

satisfying the assumptions from this section.

DEFINITION 2.15.3. *Let X_{q+1} be a C^∞ vector field on U, with a corresponding formal degree $0 < d_{q+1} \in \mathbb{N}$. We say (X, d) controls (X_{q+1}, d_{q+1}) if $\forall x_0 \in \Omega$, $\delta \in (0, 1]$, $\delta^{d_{q+1}} X_{q+1} = \sum_{j=1}^{q} c_j^{x_0,\delta} \delta^{d_j} X_j$ on $B_{(X,d)}(x_0, \xi_1 \delta)$, where $\forall m$,*

$$\sup_{\substack{x_0 \in \Omega \\ \delta \in (0,1]}} \sum_{|\alpha| \le m} \left\| (\delta X)^\alpha \, c_j^{x_0,\delta} \right\|_{C^0\left(B_{(X,d)}(x_0, \xi_1 \delta)\right)} < \infty. \tag{2.88}$$

Remark 2.15.4   Notice, our main assumption can be rephrased as:
(X, d) controls $([X_j, X_k], d_j + d_k), \forall 1 \le j, k \le q.$

PROPOSITION 2.15.5. *Suppose (X, d) controls (X_{q+1}, d_{q+1}). Then, X_{q+1} is tangent to the leaves of the foliation generated by $X_1, \ldots, X_q$. For each $x_0 \in \Omega$, $\delta \in (0, 1]$, let $Y_{q+1}^{x_0,\delta}$ be the pullback of $\delta^{d_{q+1}} X_{q+1}$ via $\Phi_{x_0,\delta}$ to $B^{n_0(x_0)}(\eta)$. Then, $Y_{q+1}^{x_0,\delta}$ is C^∞ uniformly in x_0 and δ. More precisely, for all $m \in \mathbb{N}$,*

$$\sup_{\substack{x_0 \in \Omega \\ \delta \in (0,1]}} \left\| Y_{q+1}^{x_0,\delta} \right\|_{C^m\left(B^{n_0(x_0)}(\eta)\right)} < \infty. \tag{2.89}$$

PROOF. Since (X, d) controls (X_{q+1}, d_{q+1}), X_{q+1} can be written as a linear combination of $X_1, \ldots, X_q$ and is therefore tangent to the leaves of the foliation generated by $X_1, \ldots, X_q$. Fix $x_0 \in \Omega$ and $\delta \in (0, 1]$. We have $\delta^{d_{q+1}} X_{q+1} = \sum_{j=1}^{q} c_j^{x_0,\delta} \delta^{d_j} X_j,$

where $c_j^{x_0,\delta}$ satisfies (2.88). Let $\tilde{c}_j^{x_0,\delta} = c_j^{x_0,\delta} \circ \Phi_{x_0,\delta}$. Notice, $\left(Y^{x_0,\delta}\right)^\alpha \tilde{c}_j^{x_0,\delta} = \left((\delta X)^\alpha c_j^{x_0,\delta}\right) \circ \Phi_{x_0,\delta}$. It follows from (2.88) that, $\forall \alpha$,

$$\sup_{\substack{x_0 \in \Omega \\ \delta \in (0,1] \\ u \in B^{n_0(x_0)}(\eta)}} \left|\left(Y^{x_0,\delta}\right)^\alpha \tilde{c}_j^{x_0,\delta}(u)\right| < \infty. \tag{2.90}$$

Using (2.86) and (2.87), we see that we may write $\partial_{u_j} = \sum_{l=1}^q f_{j,l}^{x_0,\delta} Y_l^{x_0,\delta}$ ($1 \leq j \leq n_0(x_0)$), where $f_{j,l}^{x_0,\delta}$ is C^∞, uniformly in x_0, δ. From (2.90) it follows that, $\forall \alpha$,

$$\sup_{\substack{x_0 \in \Omega \\ \delta \in (0,1] \\ u \in B^{n_0(x_0)}(\eta)}} \left|(\partial_u)^\alpha \tilde{c}_j^{x_0,\delta}(u)\right| < \infty.$$

The result now follows from the fact that $Y_{q+1}^{x_0,\delta} = \sum_{j=1}^q \tilde{c}_j^{x_0,\delta} Y_j^{x_0,\delta}$ and (2.86). $\qquad \square$

Proposition 2.15.5 has a partial converse.

PROPOSITION 2.15.6. *Suppose X_{q+1} is a C^∞ vector field with associated formal degree $0 < d_{q+1} \in \mathbb{N}$, and suppose that X_{q+1} is tangent to the leaves of the foliation generated by $X_1, \ldots, X_q$. For each $x_0 \in \Omega$, $\delta \in (0,1]$, let $Y_{q+1}^{x_0,\delta}$ be the pullback of $\delta^{d_{q+1}} X_{q+1}$ via $\Phi_{x_0,\delta}$ to $B^{n_0(x_0)}(\eta)$ and suppose $Y_{q+1}^{x_0,\delta}$ is C^∞ uniformly in x_0 and δ in the sense that (2.89) holds. Then, $\forall x_0 \in \Omega$, $\delta \in (0,1]$, $\delta^{d_{q+1}} X_{q+1} = \sum_{j=1}^q c_j^{x_0,\delta} \delta^{d_j} X_j$ on $B_{(X,d)}(x_0, \xi_3 \delta)$, where $\forall m$,*

$$\sup_{\substack{x_0 \in \Omega \\ \delta \in (0,1]}} \sum_{|\alpha| \leq m} \left\|(\delta X)^\alpha c_j^{x_0,\delta}\right\|_{C^0\left(B_{(X,d)}(x_0,\xi_3\delta)\right)} < \infty.$$

PROOF. By (2.86) and (2.87), we may write $Y_{q+1}^{x_0,\delta} = \sum_{j=1}^q \tilde{c}_j^{x_0,\delta} Y_j^{x_0,\delta}$, with $\tilde{c}_j^{x_0,\delta}$ C^∞, uniformly in x_0 and δ. By (2.86), for all α,

$$\sup_{\substack{x_0 \in \Omega \\ \delta \in (0,1] \\ u \in B^{n_0(x_0)}(\eta)}} \left|\left(Y^{x_0,\delta}\right)^\alpha \tilde{c}_j^{x_0,\delta}(u)\right| < \infty.$$

For $x \in B_{(X,d)}(x_0, \xi_3\delta)$ define $c_j^{x_0,\delta}(x) = \tilde{c}_j^{x_0,\delta} \circ \Phi_{x_0,\delta}^{-1}(x)$. We have $\delta^{d_{q+1}} X_{q+1} = \sum_{j=1}^q c_j^{x_0,\delta} \delta^{d_j} X_j$. Since $(\delta X)^\alpha c_j^{x_0,\delta} = \left(\left(Y^{x_0,\delta}\right)^\alpha \tilde{c}_j^{x_0,\delta}\right) \circ \Phi_{x_0,\delta}^{-1}$, the result follows.
$\qquad \square$

PROPOSITION 2.15.7. *Suppose (X,d) controls (X_{q+1}, d_{q+1}). Define a new list of vector fields with formal degrees $(X', d') = (X_1, d_1), \ldots, (X_{q+1}, d_{q+1})$. There is a constant $c > 0$ such that, for $\delta \in (0,1]$, $x_0 \in \Omega$,*

$$B_{(X',d')}(x_0, c\delta) \subseteq B_{(X,d)}(x_0, \delta) \subseteq B_{(X',d')}(x_0, \delta).$$

PROOF. The containment $B_{(X,d)}(x_0, \delta) \subseteq B_{(X',d')}(x_0, \delta)$ is trivial. The containment $B_{(X',d')}(x_0, c\delta) \subseteq B_{(X,d)}(x_0, \delta)$ follows immediately from the definitions, by taking c small. ☐

Remark 2.15.8 The above propositions can be rephrased informally as saying that if (X, d) controls (X_{q+1}, d_{q+1}), then one "does not get anything new" by adding (X_{q+1}, d_{q+1}) to the list (X, d). In fact, in what follows, our definitions will remain unchanged if we add (X_{q+1}, d_{q+1}) to the list (X, d).

LEMMA 2.15.9. *Suppose (X, d) controls (X_{q+1}, d_{q+1}). Then, (X, d) controls $([X_j, X_{q+1}], d_j + d_{q+1})$ for all $1 \le j \le q$.*

PROOF. This follows immediately from the definitions. ☐

Additional Assumption–spanning: We now include the additional assumption that $X_1, \ldots, X_q$ span the tangent space at every point of U. I.e., we assume that $n_0(x_0) = n, \forall x_0 \in U$.

Let $M = \max_{1 \le j \le q} d_j$. We recursively define vector fields with formal degrees. We are given the vector fields with formal degrees $(X_1, d_1), \ldots, (X_q, d_q)$. If Z is a vector field with formal degree e, then we assign to $[X_j, Z]$ the formal degree $d_j + e$. We enumerate all vector fields with formal degrees $\le M$ to obtain the list of vector fields with formal degrees

$$(X', d') := (X_1, d_1), \ldots, (X_{q'}, d_{q'}),$$

so that (X, d) is a sublist of (X', d'). Moreover, by repeated applications of Lemma 2.15.9, for every $q + 1 \le j \le q'$, we have (X_j, d_j) is controlled by (X, d). Thus, by repeated applications of Proposition 2.15.7, we have for some $c > 0$

$$B_{(X',d')}(x_0, c\delta) \subseteq B_{(X,d)}(x_0, \delta) \subseteq B_{(X',d')}(x_0, \delta).$$

Furthermore, for $1 \le j, k \le q'$, we have

$$[X_j, X_k] = \sum_{\substack{d_l \le d_j + d_k \\ 1 \le l \le q'}} c_{j,k}^l X_l, \quad c_{j,k}^l \in C^\infty(U). \tag{2.91}$$

Indeed, for $d_j + d_k \le M$, this follows immediately from the Jacobi identity. For $d_j + d_k > M$, we merely use the fact that $X_1, \ldots, X_q$ span the tangent space at every point of U, as in the proof of Proposition 2.1.3.

In conclusion, when $X_1, \ldots, X_q$ span the tangent space at every point of U, we may replace (X, d) with an *equivalent* list (X', d'), which satisfies (2.91). All of our main definitions that follow remain unchanged when replacing (X, d) with (X', d') and so we may use (2.91) in many of our proofs.

2.15.2   When the vector fields span

We take $(X, d) = (X_1, d_1), \ldots, (X_q, d_q)$ satisfying the above assumptions, and we assume in addition that $X_1, \ldots, X_q$ span the tangent space at every point: i.e., $n_0(x) = n$, $\forall x$.[22] In particular, let

$$\gamma := \inf_{x \in \Omega} \left| \det_{n \times n} (X_1(x) | \cdots | X_q(x)) \right| ;$$

we assume $\gamma > 0$. In this case, the balls $B_{(X,d)}(x, \delta)$ are open sets. In what follows, we write $A \lesssim B$, if $A \leq CB$, where C is an admissible constant in the sense of Theorem 2.15.2, and C is also allowed to depend on a positive lower bound for γ and on an upper bound for diam $\{\Omega\}$ (the diameter in the usual Euclidean distance of the set Ω). We write $A \approx B$ for $A \lesssim B$ and $B \lesssim A$. We have the crucial doubling estimate:

PROPOSITION 2.15.10.   *Let* Vol $(\cdot)$ *denote the usual Lebesgue measure on* $\mathbb{R}^n$. *There are constants* $Q_2 \geq Q_1 > 0$ *($Q_2 \approx Q_1 \approx 1$), such that for any* $x \in \Omega$ *and* $\delta \in (0, 1]$,

$$2^{Q_1} \text{Vol}\left(B_{(X,d)}(x, \delta)\right) \leq \text{Vol}\left(B_{(X,d)}(x, 2\delta)\right) \leq 2^{Q_2} \text{Vol}\left(B_{(X,d)}(x, \delta)\right).$$

PROOF.   For $\delta \leq \frac{\xi_2}{2}$, this follows immediately from (2.85). For $1 \geq \delta > \frac{\xi_2}{2}$, we use the fact that Vol $\left(B_{(X,d)}(x, \delta)\right) \approx 1$–here, the implicit constants depend on a lower bound for γ. This completes the proof. $\square$

Remark 2.15.11   We take Q_1 to be the largest number and Q_2 to be the smallest number such that Proposition 2.15.10 holds.

We turn a series of definitions, which culminate in a generalization of Definition 2.0.16.

DEFINITION 2.15.12.   *For* $x, y \in \Omega$, *we define the metric*

$$\rho(x, y) = \inf \left\{ \delta > 0 \mid y \in B_{(X,d)}(x, \delta) \right\}.$$

DEFINITION 2.15.13.   *For an ordered multi-index* α, *thought of as an ordered list of numbers with repetitions from* $\{1, \ldots, q\}$, *we write* $\deg(\alpha) = \sum_{j=1}^q k_j d_j \in \mathbb{N}$, *where* k_j *denotes the number of times* j *appears in the list* α; *i.e.,* k_j *is the number of times* X_j *appears in the operator* X^α.

DEFINITION 2.15.14.   *We say* $\mathcal{B} \subset C_0^\infty(\Omega) \times \Omega \times (0, 1]$ *is a bounded set of bump functions if:*

- $\forall (\phi, x, \delta) \in \mathcal{B}$, supp $(\phi) \subset B_{(X,d)}(x, \delta)$.

---

[22] As a special case, Corollary 2.1.4 shows that if (X, d) is generated by vector fields satisfying Hörmander's condition, then (X, d) satisfies the assumptions of this section.

- *For every ordered multi-index α, there exists $C = C(\alpha)$, such that $\forall (\phi, x, \delta) \in \mathcal{B}$,*

$$\sup_{z \in \Omega} |(\delta X)^\alpha \phi(z)| \leq C \mathrm{Vol}\left(B_{(X,d)}(x, \delta)\right)^{-1}.$$

DEFINITION 2.15.15. *We say $T : C^\infty(\Omega) \to C_0^\infty(\Omega)$ is a Calderón-Zygmund operator of order $t \in (-Q_1, \infty)$ if:*

- *(Growth Condition) For each ordered multi-indices α, β,*

$$\left| X_x^\alpha X_z^\beta T(x, z) \right| \leq C_{\alpha,\beta} \frac{\rho(x, y)^{-t - \deg(\alpha) - \deg(\beta)}}{\mathrm{Vol}\left(B_{(X,d)}(x, \rho(x, y))\right)}.$$

- *(Cancellation Condition) For each bounded set of bump functions $\mathcal{B} \subset C_0^\infty(\Omega) \times \Omega \times (0, 1]$ and each ordered multi-index α,*

$$\sup_{(\phi, z, \delta) \in \mathcal{B}} \sup_{x \in M} \delta^t \mathrm{Vol}\left(B_{(X,d)}(z, \delta)\right) |(\delta X)^\alpha T\phi(x)| \leq C_{\mathcal{B},\alpha},$$

with the same estimate for T replaced by T^*.

Remark 2.15.16 Notice, since we are taking $T : C^\infty(\Omega) \to C_0^\infty(\Omega)$ in Definition 2.15.15, we have $T \in C^\infty(\Omega \times \Omega)'$ and therefore $\mathrm{supp}(T) \Subset \Omega \times \Omega$. See Appendix A.1.1 for more details.

Many of the results from this chapter extend to this more general situation. There are two obvious difficulties in carrying the proofs over to this situation. The first is that our assumptions involve much less regularity on the $c_{j,k}^{l,x_0,\delta}$ than is guaranteed by Corollary 2.1.4–however, as discussed at the end of Section 2.15.1, this is not an essential point: we may replace (X, d) with an equivalent, larger list (X', d'), Definition 2.15.15 remains unchanged, while (X', d') satisfies the conclusions of Corollary 2.1.4. The second difficulty is that we do not have a sub-Laplacian with which to work. Previously, using the spectral theorem, we were able to use the sub-Lapacian to write the identity operator as $I = \sum_{j \in \mathbb{N}} E_j$, where $\left\{ (E_j, 2^{-j}) \mid j \in \mathbb{N} \right\}$ is a bounded set of elementary operators (Lemma 2.7.1). We are in need of a different way to create this decomposition.

We state the main definitions and theorems, and discuss the changes needed to adapt the previous proofs to this situation. In what follows, (X, d) denotes our original list of vector fields, while (X', d') denotes the equivalent list (from the end of Section 2.15.1) satisfying the conclusion of Corollary 2.1.4.

Remark 2.15.17 In what follows, only sketches of proofs are provided, referring to previous proofs if those proofs already contain all the difficulties. Moreover, all of the results here are special cases of results in the more general theory developed in Chapter 5, where the reader may turn for detailed proofs.

DEFINITION 2.15.18. *We say $\mathcal{E} \subset C_0^\infty(\Omega \times \Omega) \times (0, 1]$ is a bounded set of pre-elementary operators if:*

- $\bigcup_{(E,2^{-j})\in\mathcal{E}} \mathrm{supp}\,(E) \Subset \Omega \times \Omega$. *Recall, $A \Subset B$ denotes that A is a relatively compact subset of B.*

- $\forall \alpha, \beta, \exists C, \forall (E, 2^{-j}) \in \mathcal{E}$,

$$\left| \left(2^{-j} X_x\right)^\alpha \left(2^{-j} X_z\right)^\beta E\,(x, z) \right| \leq C \frac{\left(1 + 2^j \rho\,(x, z)\right)^{-m}}{\mathrm{Vol}\left(B_{(X,d)}\left(x, 2^{-j}\left(1 + 2^j \rho\,(x, z)\right)\right)\right)}.$$

DEFINITION 2.15.19. *We define the set of bounded sets of elementary operators, $\mathcal{G}$, to be the largest set of subsets of $C_0^\infty\,(\Omega \times \Omega) \times (0, 1]$ such that for all $\mathcal{E} \in \mathcal{G}$,*

- *$\mathcal{E}$ is a bounded set of pre-elementary operators.*

- $\forall (E, 2^{-j}) \in \mathcal{E}$,

$$E = \sum_{|\alpha|, |\beta| \leq 1} 2^{-(2 - |\alpha| - |\beta|)} \left(2^{-j} X\right)^\alpha E_{\alpha,\beta} \left(2^{-j} X\right)^\beta,$$

where $\left\{ \left(E_{\alpha,\beta}, 2^{-j}\right) \mid (E, 2^{-j}) \in \mathcal{E} \right\} \in \mathcal{G}$.

We say $\mathcal{E}$ is a bounded set of elementary operators if $\mathcal{E} \in \mathcal{G}$.

THEOREM 2.15.20. *Let $T : C^\infty\,(\Omega) \to C_0^\infty\,(\Omega)$, and fix $t \in (-Q_1, \infty)$. The following are equivalent.*

(i) *T is a Calderón-Zygmund operator of order t.*

(ii) *For every bounded set of elementary operators $\mathcal{E}$,*

$$\left\{ \left(2^{-jt} T E, 2^{-j}\right) \mid (E, 2^{-j}) \in \mathcal{E} \right\}$$

is a bounded set of elementary operators.

(iii) *There is a bounded set of elementary operators $\left\{ (E_j, 2^{-j}) \mid j \in \mathbb{N} \right\}$ such that $T = \sum_{j\in\mathbb{N}} 2^{jt} E_j$. (Any such sum converges in the topology of bounded convergence as operators $C^\infty\,(\Omega) \to C_0^\infty\,(\Omega)$, as in Lemma 2.0.28.)*

Furthermore, (ii) and (iii) are equivalent for every $t \in \mathbb{R}$.

DEFINITION 2.15.21. *Fix $t \in \mathbb{R}$. We say $T : C^\infty\,(\Omega) \to C_0^\infty\,(\Omega)$ is a Calderón-Zygmund operator of order t if either of the equivalent conditions (ii) or (iii) of Theorem 2.15.20 holds.*

PROPOSITION 2.15.22. *If T and S are Calderón-Zygmund operators of order t and s, respectively, then TS is a Calderón-Zygmund operator of order $t + s$.*

PROOF. This follows immediately from (ii) of Theorem 2.15.20. □

THEOREM 2.15.23. *Suppose $T : C^\infty\,(\Omega) \to C_0^\infty\,(\Omega)$ is a Calderón-Zygmund operator of order 0. Then, T extends to a bounded operator $T : L^p\,(\mathbb{R}^n) \to L^p\,(\mathbb{R}^n)$ $(1 < p < \infty)$.*

Define dilations on $\mathbb{R}^q$ by, for $\delta > 0$, $\delta\left(t_1, \ldots, t_q\right) = \left(\delta^{d_1} t_1, \ldots, \delta^{d_q} t_q\right)$. Recall the class of kernels $\mathrm{CZ}\left(s, a\right) \subset C_0^\infty\left(\mathbb{R}^q\right)'$ from Definition 2.14.2, defined for $s \in \mathbb{R}$, $a > 0$–these are defined in terms of the above dilations on $\mathbb{R}^q$.

DEFINITION 2.15.24. *Let $a > 0$ be a small number. For $s \in \mathbb{R}$, $K\left(x, t\right) \in C^\infty\left(\Omega; \mathrm{CZ}\left(s, a\right)\right)$, $\psi_1, \psi_2 \in C_0^\infty\left(\Omega\right)$, we call operators $T : C^\infty\left(\Omega\right) \to C_0^\infty\left(\Omega\right)$ of the form*

$$Tf\left(x\right) = \psi_1\left(x\right) \int f\left(e^{t_1 X_1 + \cdots + t_q X_q} x\right) \psi_2\left(e^{t_1 X_1 + \cdots + t_q X_q} x\right) K\left(x, t\right)\, dt$$

pseudodifferential operators of order s.

THEOREM 2.15.25. *For $a > 0$ sufficiently small, if T is a pseudodifferential operator of order s, then T is a Calderón-Zygmund operator of order s, in the sense of (iii) of Theorem 2.15.20.*

COMMENTS ON THE PROOF. Notice, in the definition of pseudodifferential operators, we used the list X, not the larger list X'; either choice will work, but it is the list X that we use later. The reader can easily check that the proof of Theorem 2.14.16 goes through in this situation to prove this theorem. See also Theorem 5.1.37. $\qquad\square$

PROPOSITION 2.15.26. *Let $\psi \in C_0^\infty\left(\Omega\right)$. There is a bounded set of elementary operators $\left\{\left(D_j, 2^{-j}\right) \mid j \in \mathbb{N}\right\}$ such that $\sum_{j \in \mathbb{N}} D_j = \psi^2$ (where we have identified the function ψ^2 with the operator given by multiplication by ψ^2).*

PROOF. Notice, for any $a > 0$, we have $\delta_0\left(t\right) \in \mathrm{CZ}\left(0, a\right) \hookrightarrow C^\infty\left(\Omega; \mathrm{CZ}\left(0, a\right)\right)$. Taking $\psi_1 = \psi_2 = \psi$ and $K\left(x, t\right) = \delta_0\left(t\right)$, the result follows from Theorem 2.15.25. $\qquad\square$

COMMENTS ON THE PROOF OF THEOREM 2.15.20. The proofs of (iii)⇒(ii) and (iii)⇒(i) follow just as in the corresponding implications of Theorem 2.0.29.

(ii)⇒(iii): Since $T : C^\infty\left(\Omega\right) \to C_0^\infty\left(\Omega\right)$, we have $\mathrm{supp}\left(T\right) \Subset \Omega \times \Omega$. Take $\psi \in C_0^\infty\left(\Omega\right)$ so that $\psi \otimes \psi \equiv 1$ on a neighborhood of $\mathrm{supp}\left(T\right)$. If we identify ψ with the operator given by multiplication by ψ, we have $T\psi^2 = T$. We apply Proposition 2.15.26 to decompose $\psi^2 = \sum_{j \in \mathbb{N}} D_j$, where $\left\{\left(D_j, 2^{-j}\right) \mid j \in \mathbb{N}\right\}$ is a bounded set of elementary operators. We have $T = T\psi^2 = \sum_{j \in \mathbb{N}} TD_j$. The result now follows, just as in the corresponding implication of Theorem 2.0.29.

(i)⇒(iii): Since $T : C^\infty\left(\Omega\right) \to C_0^\infty\left(\Omega\right)$, we have $\mathrm{supp}\left(T\right) \Subset \Omega \times \Omega$. Take $\psi \in C_0^\infty\left(\Omega\right)$ so that $\psi \otimes \psi \equiv 1$ on a neighborhood of $\mathrm{supp}\left(T\right)$. If we identify ψ with the operator given by multiplication by ψ, we have $\psi^2 T \psi^2 = T$. We apply Proposition 2.15.26 to decompose $\psi^2 = \sum_{j \in \mathbb{N}} D_j$, where $\left\{\left(D_j, 2^{-j}\right) \mid j \in \mathbb{N}\right\}$ is a bounded set of elementary operators. We have $T = \psi^2 T \psi^2 = \sum_{j_1, j_2 \in \mathbb{N}} D_{j_1} T D_{j_2}$. The result now follows just as in the corresponding implication of Theorem 2.0.29. $\qquad\square$

COMMENTS ON THE PROOF OF THEOREM 2.15.23. From here, the result follows just as in the proof of Proposition 2.10.4. $\qquad\square$

We can also create non-isotropic Sobolev spaces. Again, we have no sub-Laplacian with which to work, and so we turn to Theorem 2.10.6 for motivation. Fix an open, relatively compact subset of Ω: $\Omega_0 \Subset \Omega$. We wish to only consider functions supported on Ω_0. Let $\psi \in C_0^\infty(\Omega)$ equal 1 on a neighborhood of the closure of Ω_0. Apply Proposition 2.15.26 to decompose $\psi^2 = \sum_{j \in \mathbb{N}} D_j$, where $\left\{ (D_j, 2^{-j}) \mid j \in \mathbb{N} \right\}$ is a bounded set of elementary operators.

DEFINITION 2.15.27. *For $1 < p < \infty$, $s \in \mathbb{R}$, define*

$$\|f\|_{\mathrm{NL}_s^p} := \left\| \left(\sum_{j \in \mathbb{N}} \left| 2^{js} D_j f \right|^2 \right)^{\frac{1}{2}} \right\|_{L^p}.$$

All of the relevant aspects of NL_s^p are contained in the next theorem; the proof of which (in a more general setting) can be found in Section 5.8.

THEOREM 2.15.28.   • *Let $\psi_1 \in C_0^\infty(\Omega)$ be another function which equals 1 on a neighborhood of the closure of Ω_0, and let $\psi_1^2 = \sum_{j \in \mathbb{N}} \widetilde{D}_j$, where $\left\{ \left(\widetilde{D}_j, 2^{-j} \right) \mid j \in \mathbb{N} \right\}$ is a bounded set of elementary operators. Then, for $1 < p < \infty$, $s \in \mathbb{R}$,*

$$\|f\|_{\mathrm{NL}_s^p} \approx \left\| \left(\sum_{j \in \mathbb{N}} \left| 2^{js} \widetilde{D}_j f \right|^2 \right)^{\frac{1}{2}} \right\|_{L^p}, \quad \forall f \in C_0^\infty(\Omega_0),$$

where the implicit constants depend on p, s, ψ, ψ_1, and the particular choice of decompositions of ψ_1^2 and ψ^2. In short, the equivalence class of the norm $\|\cdot\|_{\mathrm{NL}_s^p}$ does not depend on any of the choices we made.

• *Let T be a Calderón-Zygmund operator of order $t \in \mathbb{R}$. Then,*

$$\|Tf\|_{\mathrm{NL}_s^p} \lesssim \|f\|_{\mathrm{NL}_{s+t}^p},$$

for $1 < p < \infty$, $s \in \mathbb{R}$.

• *For $1 < p < \infty$, and f supported on Ω_0,*

$$\|f\|_{L^p} \approx \|f\|_{\mathrm{NL}_0^p}.$$

We consider the maximal function.

DEFINITION 2.15.29. *Fix $\psi \in C_0^\infty(\Omega)$ with $\psi \geq 0$. We consider the maximal function*

$$\mathcal{M}f(x) = \psi(x) \sup_{0 < \delta \leq 1} \frac{1}{\mathrm{Vol}\left(B_{(X,d)}(x, \delta) \right)} \int_{B_{(X,d)}(x,\delta)} |f(y)| \, \psi(y) \, dy.$$

THEOREM 2.15.30. *We have*

$$\|\mathcal{M}f\|_{L^p} \lesssim \|f\|_{L^p},$$

$1 < p \le \infty$, *where the implicit constant depends on p.*

PROOF. This follows from the classical theory of spaces of homogeneous type; see [Ste93, Chapter I]. □

2.15.3   When the vector fields do not span

We return to the assumptions outlined at the beginning of this section, and no longer assume $X_1, \ldots, X_q$ span the tangent space. Indeed, we even allow the function $n_0(x_0)$ to possibly be non-constant. The Frobenius theorem applies to foliate the ambient space, Ω, into leaves, where $X_1, \ldots, X_q$ span the tangent space to each leaf. It may seem likely that we can just apply the theory developed in the previous section to each leaf; putting this together properly would then yield a theory which does not require $X_1, \ldots, X_q$ to span the tangent space. If the distribution has no singular points,[23] this is not difficult to carry out. Unfortunately, near a singular point, this procedure becomes much more complicated. Because of these complications, we discuss only some special operators in this context, and do not discuss the more general operators covered in Definitions 2.15.15 and 2.15.21. Even for these special operators, singular points create significant difficulties. It is the quantitative Frobenius theorem (Theorem 2.15.2) which saves the day, and allows us to study these operators.

The main point of this section is the study of two operators, which we now introduce.

DEFINITION 2.15.31. *Fix $\psi_1, \psi_2 \in C_0^\infty(\Omega)$. Let $a > 0$ be a small 2-admissible constant, to be chosen later. For $s \in \mathbb{R}$ and $K \in C^\infty(\Omega; CZ(s, a))$, we call operators $T : C^\infty(\Omega) \to C_0^\infty(\Omega)$ of the form*

$$Tf(x) = \psi_1(x) \int f\left(e^{t_1 X_1 + \cdots + t_q X_q} x\right) \psi_2\left(e^{t_1 X_1 + \cdots + t_q X_q} x\right) K(x, t) \, dt$$

pseudodifferential operators of order s.

THEOREM 2.15.32. *Let T be a pseudodifferential operator of order 0. Then, $T : L^p \to L^p$ $(1 < p < \infty)$.*

DEFINITION 2.15.33. *Let $\psi \in C_0^\infty(\Omega)$ be nonnegative and let $a > 0$ be a small 2-admissible constant to be chosen later. We define the maximal function*

$$\mathcal{M}f(x) = \psi(x) \sup_{0 < \delta \le a} \frac{1}{\mathrm{Vol}\left(B_{(X,d)}(x, \delta)\right)} \int_{B_{(X,d)}(x,\delta)} |f(y)| \, \psi(y) \, dy,$$

where $\mathrm{Vol}(\cdot)$ and dy denote the induced Lebesgue measure on the leaf passing through x, generated by $X_1, \ldots, X_q$.

---

[23]Recall, a point $x_0 \in \Omega$ is said to be singular if $n_0(x_0)$ is not constant on any neighborhood of x_0.

THEOREM 2.15.34. $\|\mathcal{M}f\|_{L^p} \lesssim \|f\|_{L^p}$, for $1 < p < \infty$, where the implicit constant depends on p.

Remark 2.15.35 In Theorem 2.15.32 and 2.15.34 we restrict to $a > 0$ small. For instance, in Theorem 2.15.32 this forces us to only consider operators T such that $T(x,y)$ is supported for x close to y, where "close" is in the sense of the Carnot-Carathéodory metric induced by $(X_1, d_1), \ldots, (X_q, d_q)$. Of course, if $(x, y) \in \text{supp}(T)$, then y is in the same leaf as x, but this requires more. Indeed, consider Example 2.2.15 on the torus. Even though the torus is compact, the leaves in Example 2.2.15 have points which are arbitrarily far away from each other in the sense of the Carnot-Carathéodory metric. Such problems did not arise when we were working on a compact manifold with vector fields which span the tangent space. The quantitative Frobenius theorem (Theorem 2.15.2) only applies to small distances, which is why we must restrict to $a > 0$ small.

In what follows, we use the notation from Theorem 2.15.2. Fix a compact subset $\mathcal{K} \Subset \Omega$; all implicit constants which follow are allowed to depend on the choice of $\mathcal{K}$. The next proposition is the main result needed to prove the above two theorems. In it, we use the function $\Phi_{x,1}$ from Theorem 2.15.2.

PROPOSITION 2.15.36. *There exists a 2-admissible constant $0 < \eta_1 \le \eta$ such that for all $0 < \eta' \le \eta_1$, and all $f \ge 0$ with $\text{supp}(f) \subset \mathcal{K}$,*

$$\int_{\mathcal{K}} f(x)\, dx \approx_2 \int_\Omega \int_{B^{n_0(x)}(\eta')} f(\Phi_{x,1}(u))\, du\, dx,$$

where the implicit constants depend on a lower bound for η'.

To prove Proposition 2.15.36, we need some preliminary lemmas.

LEMMA 2.15.37. *There exists a 2-admissible constant $\eta_2 > 0$ such that for every $0 < \eta' \le \eta_2$, and all $f \ge 0$ with $\text{supp}(f) \subset \mathcal{K}$,*

$$\int_{\mathcal{K}} f(x)\, dx \approx_2 \frac{1}{(2\eta')^q} \int_\Omega \int_{B^q(\eta')} f(e^{t \cdot X} x)\, dt\, dx,$$

where $t \cdot X = t_1 X_1 + \cdots + t_q X_q$.

PROOF. For $\eta_2 = \eta_2(\mathcal{K}) > 0$ a sufficiently small 2-admissible constant, we have for $|t| < \eta_2$, $e^{t \cdot X} \Omega \supset \mathcal{K}$. Furthermore, since $\left\| e^{t \cdot X} x \right\|_{C^2} \lesssim_2 1$ (see Appendix B.1) and since $\frac{\partial}{\partial x} e^{t \cdot X} x \big|_{t=0} = I$, we may apply a change of variables as in Theorem B.3.1 to see that if we take η_2 to be a sufficiently small 2-admissible constant we have for $|t| \le \eta_2$,

$$\int_{\mathcal{K}} f(x)\, dx \approx_2 \int_\Omega f(e^{t \cdot X} x)\, dx.$$

Averaging both sides over the set $|t| \le \eta'$ yields the result. $\qquad\square$

LEMMA 2.15.38. *Let η_2 be as in Lemma 2.15.37. There exists a 2-admissible constant $\eta_1 > 0$ such that for every $0 < \eta' \leq \eta_1$ there exist 2-admissible constants $0 < \eta_4 = \eta_4(\eta')$ and η_3 with $\eta_4 < \eta_3 < \eta_2$, such that for every $f \geq 0$ with $\operatorname{supp}(f) \subset \mathcal{K}$, we have (for $x \in \Omega$)*

$$\int_{B^q(\eta_4)} f\left(e^{t\cdot X}x\right) dt \lesssim_2 \int_{B^{n_0(x)}(\eta')} f\circ\Phi_{x,1}(u)\, du \lesssim_2 \int_{B^q(\eta_3)} f\left(e^{t\cdot X}x\right) dt. \quad (2.92)$$

PROOF. Let $Y_j^{x,1}$ be the pullback, via $\Phi_{x,1}$, of X_j to $B^{n_0(x)}(\eta)$ as in Theorem 2.15.2. We use that $\left\|Y_j^{x,1}\right\|_{C^2(B^q(\eta))} \lesssim_2 1$ to see that $\left\|e^{t\cdot Y^{x,1}}0\right\|_{C^2(B^q(\eta))} \lesssim_2 1$ (see Appendix B.1). We take $0 < \eta_3 \leq \eta_2$ to be a 2-admissible constant so small for $t \in B^q(\eta_3)$, $e^{t\cdot Y^{x_0,1}}0 \in B^q(\eta)$. We also take $\eta_3 > 0$ small enough that

$$\left|\det_{n_0(x)\times n_0(x)} \frac{\partial}{\partial t} e^{t\cdot Y^{x,1}}0\right| \geq \frac{1}{2}, \quad \forall t \in B^q(\eta_3).$$

Using (2.87), and the fact that $\partial_{t_j}|_{t=0} e^{t\cdot Y^{x,1}}0 = Y_j^{x,1}(0)$, the inverse function theorem (Theorem B.2.1) implies that we may take $\eta_1 > 0$ to be a 2-admissible constant so small $B^{n_0(x)}(\eta_1) \subseteq \left\{e^{t\cdot Y^{x,1}}0 \mid t \in B^q(\eta_3)\right\}$. Given $0 < \eta' \leq \eta_1$, we take $0 < \eta_4 = \eta_4(\eta') \leq \eta_3$ to be a 2-admissible constant so small for $t \in B^q(\eta_4)$, $e^{t\cdot Y^{x,1}}u \in B^{n_0(x)}(\eta')$.

Since $x = \Phi_{x,1}(0)$, (2.92) is equivalent to

$$\int_{B^q(\eta_4)} f\circ\Phi_{x,1}\left(e^{t\cdot Y^{x,1}}0\right) dt \lesssim_2 \int_{B^{n_0(x)}(\eta')} f\circ\Phi_{x,1}(u)\, du$$

$$\lesssim_2 \int_{B^q(\eta_3)} f\circ\Phi_{x,1}\left(e^{t\cdot Y^{x,1}}0\right) dt.$$

But this formula follows from a change of variables $u = e^{t\cdot Y^{x,1}}0$, and the relevant properties described above. See Appendix B.3 for more details on this change of variables. □

PROOF OF PROPOSITION 2.15.36. Let η_1 be as in Lemma 2.15.38, and let $0 < \eta' \leq \eta_1$. The implicit constants which follow may depend on a lower bound for η'. Letting η_3 be as in Lemma 2.15.38, we have by Lemmas 2.15.38 and 2.15.37, for f as in the statement of the proposition,

$$\int_\Omega \int_{B^{n_0(x)}(\eta')} f\circ\Phi_{x,1}(u)\, du\, dx \lesssim_2 \int_\Omega \int_{B^q(\eta_3)} f\left(e^{t\cdot X}x\right) dt\, dx \lesssim_2 \int_\mathcal{K} f(x)\, dx,$$

proving the $\gtrsim_2$ part of the proposition.

For the $\lesssim_2$ part, letting $\eta_4 = \eta_4(\eta')$ be as in Lemma 2.15.38, we have by Lemmas 2.15.38 and 2.15.37, for f as in the statement of the proposition,

$$\int_\mathcal{K} f(x)\, dx \lesssim_2 \int_\Omega \int_{B^q(\eta_4)} f\left(e^{t\cdot X}x\right) dt\, dx \lesssim_2 \int_\Omega \int_{B^{n_0(x)}(\eta')} f\circ\Phi_{x,1}(u)\, du\, dx,$$

completing the proof. □

We now turn to Theorem 2.15.32.

LEMMA 2.15.39. *Let T be as in Definition 2.15.31, and let η_1 be as in Proposition 2.15.36. If the choice of $a > 0$ in Definition 2.15.31 is a sufficiently small 2-admissible constant, we have, for $1 < p < \infty$, $x \in \Omega$,*

$$\int_{B^{n_0(x)}(\eta_1/2)} |(Tf) \circ \Phi_{x,1}(u)|^p \, du \lesssim \int_{B^{n_0(x)}(\eta_1)} |f \circ \Phi_{x,1}(u)|^p \, du,$$

where the implicit constant depends on p.

PROOF. Let $Y_j^{x,1}$ be the pullback, via $\Phi_{x,1}$, of X_j to $B^{n_0(x)}(\eta)$. We take $a > 0$ to be a 2-admissible constant so small that for $t \in B^q(a)$ and $u \in B^{n_0(x)}(\eta_1/2)$, $e^{t \cdot Y^{x,1}} u \in B^{n_0(x)}(\eta_1)$. Consider,

$$(Tf) \circ \Phi_{x,1}(u)$$

$$= \psi_1 \circ \Phi_{x,1}(u) \int f \circ \Phi_{x,1}\left(e^{t \cdot Y^{x,1}} u\right) \psi_2 \circ \Phi_{x,1}\left(e^{t \cdot Y_x,1} u\right) K\left(\Phi_{x,1}(u), t\right) \, dt.$$

Let $\widetilde{\psi} \in C_0^\infty\left(B^{n_0(x)}(\eta)\right)$ be equal to 1 on a neighborhood of the closure of $B^{n_0(x)}(\eta_1)$. Define the operator

$$S_x g(u) := \widetilde{\psi}(u) \psi_1 \circ \Phi_{x,1}(u) \int g\left(e^{t \cdot Y^{x,1}} u\right) \psi \circ \Phi_{x,1}(u) \widetilde{\psi}(u) K\left(\Phi_{x,1}(u), t\right) \, dt.$$

S_x is a pseudodifferential operator of order 0 in the sense of Definition 2.15.24, with respect to the vector fields $Y_1^{x,1}, \ldots, Y_q^{x,1}$. Because of (2.86) and (2.87), the hypotheses of Section 2.15.2 hold uniformly for $x \in \Omega$. Theorem 2.15.25 combined with Theorem 2.15.23 shows that S_x is bounded on L^p ($1 < p < \infty$). Moreover, this is true uniformly in x, in the sense that the L^p operator norm of S_x is bounded independent of x. Since we are considering $a > 0$ small enough that $e^{t \cdot Y^{x,1}} u \in B^{n_0(x)}(\eta_1)$ for $u \in B^{n_0(x)}(\eta_1/2)$ and $|t| < a$, we have that the value of $S_x g(u)$ with $u \in B^{n_0(x)}(\eta_1/2)$ depends only on the values of $g(u)$ for $u \in B^{n_0(x)}(\eta_1)$. Thus,

$$\int_{B^{n_0(x)}(\eta_1/2)} |S_x g(u)|^p \, du \lesssim \int_{B^{n_0(x)}(\eta_1)} |g(u)|^p \, du.$$

Replacing g with $f \circ \Phi_{x_0,1}$, and using that $\widetilde{\psi} \equiv 1$ on $B^{n_0(x)}(\eta_1)$, the result follows. □

PROOF OF THEOREM 2.15.32. Let $\mathcal{K} \Subset \Omega$ be a compact set such that supp (ψ_1) and supp (ψ_2) are compact subsets of the interior of $\mathcal{K}$. Let $\psi_3 \in C_0^\infty(\Omega)$ be supported in $\mathcal{K}$ with $\psi_3 \equiv 1$ on a neighborhood of the support of ψ_2. Notice that $T\psi_3 f = Tf$.

We have, using Lemma 2.15.39 and Proposition 2.15.36, for $1 < p < \infty$,

$$
\begin{aligned}
\|Tf\|_{L^p}^p &= \|T\psi_3 f\|_{L^p}^p \\
&= \int_K |T\psi_3 f(x)|^p \, dx \\
&\approx_2 \int_\Omega \int_{B^{n_0(x)}(\eta_1/2)} |(T\psi_3 f)(\Phi_{x,1}(u))|^p \, du \, dx \\
&\lesssim \int_\Omega \int_{B^{n_0(x)}(\eta_1)} |(\psi_3 f) \circ \Phi_{x,1}(u)|^p \, du \, dx \\
&\approx_2 \int_K |\psi_3(x) f(x)|^p \, dx \\
&\lesssim \int_K |f(x)|^p \, dx,
\end{aligned}
$$

for all $f \in C_0^\infty(\Omega)$. This completes the proof. $\qquad\square$

The proof of Theorem 2.15.34 is very similar to the proof of Theorem 2.15.32. We need the following analog of Lemma 2.15.39.

LEMMA 2.15.40. *Let $\mathcal{M}$ be as in Definition 2.15.33, and let η_1 be as in Proposition 2.15.36. If $a > 0$ is a sufficiently small 2-admissible constant, we have, for $1 < p < \infty$, $x \in \Omega$,*

$$
\int_{B^{n_0(x)}(\eta_1/2)} |(\mathcal{M}f) \circ \Phi_{x,1}(u)|^p \, du \lesssim \int_{B^{n_0(x)}(\eta_1)} |f \circ \Phi_{x,1}(u)|^p \, du,
$$

where the implicit constant depends on p.

PROOF. As in Lemma 2.15.39, we let $Y_j^{x,1}$ be the pullback, via $\Phi_{x,1}$ of X_j to $B^{n_0(x)}(\eta)$. We associate to $Y_j^{x,1}$ the formal degree d_j creating a list of vector fields with formal degrees $(Y^{x,1}, d) = \left(Y_1^{x,1}, d_1\right), \dots, \left(Y_q^{x,1}, d_q\right)$. We take $a > 0$ to be a sufficiently small 2-admissible constant so that for $u \in B^{n_0(x)}(\eta_1/2)$, $B_{(Y^{x,1},d)}(u,a) \subset B^{n_0(x)}(\eta_1)$. Notice, directly from the definitions,

$$
\Phi_{x,1}\left(B_{(Y^{x,1},d)}(u,\delta)\right) = B_{(X,d)}(\Phi_{x,1}(u),\delta).
$$

(2.85) then shows (for $\delta < a$),

$$
\left| \det_{n_0(x) \times n_0(x)} X(x) \right| \operatorname{Vol}\left(B_{(Y^{x,1},d)}(u,\delta)\right) \approx_2 \operatorname{Vol}\left(B_{(X,d)}(\Phi_{x,1}(u),\delta)\right).
$$

We have, for $u \in B^{n_0(x)}(\eta_1/2)$ and $\delta < a$ and $g \geq 0$,

$$\frac{1}{\mathrm{Vol}\left(B_{(X,d)}\left(\Phi_{x,1}\left(u\right),\delta\right)\right)} \int_{B_{(X,d)}\left(\Phi_{x,1}(u),\delta\right)} g \circ \Phi_{x,1}^{-1}\left(y\right) \, dy$$

$$= \frac{1}{\mathrm{Vol}\left(B_{(X,d)}\left(\Phi_{x,1}\left(u\right),\delta\right)\right)} \int_{B_{(Y^x,1,d)}\left(u,\delta\right)} g\left(u\right)$$

$$\times \left| \det_{n_0(x) \times n_0(x)} d\Phi_{x,1}\left(u\right) \right| \, du \qquad (2.93)$$

$$\approx_2 \frac{\left| \det_{n_0(x) \times n_0(x)} X\left(x\right) \right|}{\mathrm{Vol}\left(B_{(X,d)}\left(\Phi_{x,1}\left(u\right),\delta\right)\right)} \int_{B_{(Y^x,1,d)}\left(u,\delta\right)} g\left(u\right) \, du$$

$$\approx_2 \frac{1}{\mathrm{Vol}\left(B_{(Y^x,1,d)}\left(u,\delta\right)\right)} \int_{B_{(Y^x,1,d)}\left(u,\delta\right)} g\left(u\right) \, du.$$

For x fixed, let $\psi \in C_0^\infty\left(B^{n_0(x)}\left(\eta\right)\right)$ equal 1 on a neighborhood of the closure of $B^{n_0(x)}\left(\eta_1\right)$. Define a maximal function,

$$\widetilde{\mathcal{M}}_x f\left(u\right) = \widetilde{\psi}\left(u\right) \psi \circ \Phi_{x,1}\left(u\right) \sup_{0 < \delta \leq a} \frac{1}{\mathrm{Vol}\left(B_{(Y^x,1,d)}\left(u,\delta\right)\right)}$$

$$\times \int_{B_{(Y^x,1,d)}\left(u,\delta\right)} \left|f\left(v\right)\right| \psi \circ \Phi_{x,1}\left(v\right) \widetilde{\psi}\left(v\right) \, dv.$$

The maximal function $\widetilde{\mathcal{M}}_x$ is of the form covered in Definition 2.15.29 (with respect to the vector fields $Y_1^{x,1}, \ldots, Y_q^{x,1}$) and in light of (2.86) and (2.87) the hypotheses of Section 2.15.2 hold uniformly in x. Theorem 2.15.30 applies to show $\left\|\widetilde{\mathcal{M}}_x g\right\|_{L^p} \lesssim \|g\|_{L^p}$, $1 < p \leq \infty$, with implicit constant not depending on x.

For $u \in B^{n_0(x)}\left(\eta_1/2\right)$, using (2.93) and the fact that $\widetilde{\psi}\left(v\right) = 1$ for all v such that $v \in B_{(Y^x,1,d)}\left(u,a\right) \subset B^{n_0(x)}\left(\eta_1\right)$ for $u \in B^{n_0(x)}\left(\eta_1/2\right)$ (by our choice of a), we have

$$(\mathcal{M}f) \circ \Phi_{x,1}(u)$$

$$= \psi \circ \Phi_{x,1}(u) \sup_{0 < \delta \leq a} \frac{1}{\mathrm{Vol}\left(B_{(X,d)}\left(\Phi_{x,1}(u), \delta\right)\right)}$$

$$\times \int_{B_{(X,d)}(\Phi_{x,1}(u),\delta)} |f(y)| \psi(y) \, dy$$

$$\lesssim_2 \psi \circ \Phi_{x,1}(u) \sup_{0 < \delta \leq a} \frac{1}{\mathrm{Vol}\left(B_{(Y^{x,1},d)}(u, \delta)\right)}$$

$$\times \int_{B_{(Y^{x,1},d)}(u,\delta)} |f \circ \Phi_{x,1}(v)| \, \psi \circ \Phi_{x,1}(v) \, dv$$

$$= \psi \circ \Phi_{x,1}(u) \, \widetilde{\psi}(u) \sup_{0 < \delta \leq a} \frac{1}{\mathrm{Vol}\left(B_{(Y^{x,1},d)}(u, \delta)\right)}$$

$$\times \int_{B_{(Y^{x,1},d)}(u,\delta)} |f \circ \Phi_{x,1}(v)| \, \psi \circ \Phi_{x,1}(v) \, \widetilde{\psi}(v) \, dv$$

$$= \widetilde{\mathcal{M}}_x \left(f \circ \Phi_{x,1}\right)(u).$$

The L^p boundedness of $\widetilde{\mathcal{M}}_x$ then shows

$$\int_{B^{n_0(x)}(\eta_1/2)} |(\mathcal{M}f) \circ \Phi_{x,1}(u)|^p \, du \lesssim \int_{B^{n_0(x)}(\eta_1)} |f \circ \Phi_{x,1}(u)|^p \, du,$$

where we have used that the value of $(\mathcal{M}f) \circ \Phi_{x,1}(u)$ (for $u \in B^{n_0(x)}(\eta_1/2)$) depends only on the values of $f \circ \Phi_{x,1}(u)$ on $B_{(Y^{x,1},d)}(u, a) \subset B^{n_0(x)}(\eta_1)$. This completes the proof. $\square$

PROOF OF THEOREM 2.15.34. The proof of Theorem 2.15.32 goes through to prove Theorem 2.15.34, once Lemma 2.15.39 is replaced by Lemma 2.15.40. $\square$

We now turn to a technical result, which we use in the next section to create a Littlewood-Paley theory. Let $a > 0$ be a small 2-admissible constant (to be chosen in the proofs of the next results). For $\varsigma \in \mathcal{S}(\mathbb{R}^q)$, $j \in [0, \infty)$, $\eta \in C_0^\infty(B^q(a))$, $\kappa \in C^\infty(\mathbb{R}^q \times \Omega)$, and $\psi_1, \psi_2 \in C_0^\infty(\Omega)$, define $E = E\left(\varsigma, 2^{-j}, \eta, \kappa, \psi_1, \psi_2\right) : C^\infty(\Omega) \to C_0^\infty(\Omega)$ by

$$Ef(x) = \psi_1(x) \int f\left(e^{t \cdot X} x\right) \psi_2\left(e^{t \cdot X} x\right) \kappa(t, x) \eta(t) \varsigma^{(2^j)}(t) \, dt,$$

where $\varsigma^{(2^j)}$ is defined in terms of the dilations on $\mathbb{R}^q$.

PROPOSITION 2.15.41. Let $\mathcal{B}_1 \subset \mathcal{S}(\mathbb{R}^q)$, $\mathcal{B}_2 \subset C_0^\infty(B^q(a))$, $\mathcal{B}_3 \subset C^\infty(\mathbb{R}^q \times \Omega)$, and $\mathcal{B}_4 \subset C_0^\infty(\Omega)$ be bounded sets. For $l = 1, \ldots, 4$, let $\varsigma_l \in \mathcal{B}_1$, $j_l \in [0, \infty)$, $\varsigma_l \in \mathcal{S}_0(\mathbb{R}^q)$ if $j_l > 0$, $\eta_l \in \mathcal{B}_2$, $\kappa_l \in \mathcal{B}_3$, and $\psi_1^l, \psi_2^l \in \mathcal{B}_4$. Define

$$E_l = E\left(\varsigma_l, 2^{-j_l}, \eta_l, \kappa_l, \psi_1^l, \psi_2^l\right).$$

Then, for every N, there is an admissible constant $C = C(N, \mathcal{B}_1, \mathcal{B}_2, \mathcal{B}_3, \mathcal{B}_4)$, such that

$$\|E_1 E_2 E_3^* E_4^*\|_{L^2 \to L^2}, \|E_1^* E_2^* E_3 E_4\|_{L^2 \to L^2} \leq C 2^{-N \text{diam}\{j_1, j_2, j_3, j_4\}},$$

where $\text{diam}\{j_1, j_2, j_3, j_4\} = \max_{1 \leq l, k \leq 4} |j_l - j_k|$.

Remark 2.15.42 Let E_1, E_2 be as in Proposition 2.15.41. By the same methods that prove Proposition 2.15.41, we can prove $\|E_1 E_2^*\|_{L^2}, \|E_1^* E_2\|_{L^2} \lesssim 2^{-N|j_1 - j_2|}$, for every N. This can be combined with the Cotlar-Stein Lemma (Lemma 1.2.26) to prove the $p = 2$ part of Theorem 2.15.32. The real difficulty in Theorem 2.15.32 was, therefore, the case $p \neq 2$. To deal with this, we used the Frobenius theorem to reduce the question to a situation where the Calderón-Zygmund theory applied. In particular, we pulled back the operators via the coordinate charts given by the Frobenius theorem; the pulled back operators were weak-type $(1, 1)$, which is the key to the proof of Theorem 2.15.32.

To prove Proposition 2.15.41, we need several lemmas.

LEMMA 2.15.43. *For $a > 0$ a sufficiently small 2-admissible constant, we have the following. Let $\mathcal{B}_1 \subset \mathcal{S}(\mathbb{R}^q)$, $\mathcal{B}_2 \subset C_0^\infty(B^q(a))$, $\mathcal{B}_3 \subset C^\infty(\mathbb{R}^q \times \Omega)$, and $\mathcal{B}_4 \subset C_0^\infty(\Omega)$ be bounded sets. Let $\varsigma \in \mathcal{B}_1$, $j \in [0, \infty)$, $\eta \in \mathcal{B}_2$, $\kappa \in \mathcal{B}_3$, and $\psi_1, \psi_2 \in \mathcal{B}_4$. Define $E = E(\varsigma, 2^{-j}, \eta, \kappa, \psi_1, \psi_2)$. Then, there is a 2-admissible constant $C = C(\mathcal{B}_1, \mathcal{B}_2, \mathcal{B}_3, \mathcal{B}_4)$ such that*

$$\|E\|_{L^p \to L^p} \leq C,$$

$1 \leq p \leq \infty$.

PROOF. For $p = \infty$, this is trivial. For $p = 1$, note that $e^{0 \cdot X} x = x$ and therefore $\frac{\partial}{\partial x} e^{t \cdot X} x \big|_{t=0} = I$. By taking $a > 0$ small enough, the result for $p = 1$ follows by a change of variables in the x-variable–see Appendix B.3. Interpolation proves the result for all p. $\square$

LEMMA 2.15.44. *Let $\mathcal{B}_1 \subset \mathcal{S}(\mathbb{R}^q)$, $\mathcal{B}_2 \subset C_0^\infty(B^q(a))$, $\mathcal{B}_3 \subset C^\infty(\mathbb{R}^q \times \Omega)$, and $\mathcal{B}_4 \subset C_0^\infty(\Omega)$ be bounded sets. For $l = 1, 2$, let $\widetilde{\varsigma}_l \in \mathcal{B}_1$, $j_l \in [0, \infty)$, $\widetilde{\varsigma}_l \in \mathcal{S}_0(\mathbb{R}^q)$ if $j_l > 0$, $\eta_l \in \mathcal{B}_2$, $\kappa_l \in \mathcal{B}_3$, and $\psi_1^l, \psi_2^l \in \mathcal{B}_4$. Define $E_l = E(\widetilde{\varsigma}_l, 2^{-j_l}, \eta_l, \kappa_l, \psi_1^l, \psi_2^l)$. Then, for every N, there is an admissible constant $C = C(N, \mathcal{B}_1, \mathcal{B}_2, \mathcal{B}_3, \mathcal{B}_4)$, such that*

$$\|E_1 E_2\|_{L^\infty \to L^\infty} \leq C 2^{-N|j_1 - j_2|}.$$

PROOF OF PROPOSITION 2.15.41 GIVEN LEMMA 2.15.44. It is easy to see that E_l^* is of the same form as E_l (see Lemma 2.14.22), so it suffices to prove

$$\|E_1 E_2 E_3 E_4\|_{L^2 \to L^2} \lesssim 2^{-N \text{diam}\{j_1, j_2, j_3, j_4\}}, \quad \forall N.$$

There is a $1 \leq l \leq 3$ such that $|j_l - j_{l+1}| \geq \frac{1}{3} \text{diam}\{j_1, j_2, j_3, j_4\}$. Lemma 2.15.43 shows $\|E_k\|_{L^2 \to L^2} \lesssim 1$, and so the result will follow once we show

$$\|E_l E_{l+1}\|_{L^2 \to L^2} \lesssim 2^{-N|j_l - j_{l+1}|} \quad \forall N.$$

Lemma 2.15.43 shows $\|E_l E_{l+1}\|_{L^1 \to L^1} \lesssim 1$, and therefore, by interpolation it suffices to show $\forall N$, $\|E_l E_{l+1}\|_{L^\infty \to L^\infty} \lesssim 2^{-N|j_l - j_{l+1}|}$. This is exactly the conclusion of Lemma 2.15.44. $\qquad\square$

We turn to the proof of Lemma 2.15.44. To do this, we need the following decomposition result for $S(\mathbb{R}^q)$.

LEMMA 2.15.45. *Fix $a > 0$. Let $\mathcal{B}_1 \subset S(\mathbb{R}^q)$ and $\mathcal{B}_2 \subset C_0^\infty(B^q(a))$ be bounded sets. For every $M \in \mathbb{N}$, $\tilde{\varsigma} \in \mathcal{B}_1$, $\eta \in \mathcal{B}_2$, $j \in [0, \infty)$, with $\tilde{\varsigma} \in S_0(\mathbb{R}^q)$ if $j > 0$, there exists $\{\varsigma_k\}_{\substack{k \in \mathbb{N} \\ k \leq j}} \subset C_0^\infty(B^q(a))$ such that*

$$\varsigma_k = \begin{cases} \sum_{|\alpha| = 2N} \partial_t^\alpha \gamma_{k,\alpha} & \text{if } k > 0, \\ \gamma_0 & \text{if } k = 0, \end{cases}$$

where $\gamma_{k,\alpha}, \gamma_0 \in C_0^\infty(B^q(a))$, and

$$\eta(t) \tilde{\varsigma}^{(2^j)}(t) = \sum_{\substack{k \in \mathbb{N} \\ k \leq j}} \eta(t) \varsigma_k^{(2^k)}.$$

Moreover, for every M, N, the sets

$$\left\{ 2^{M|j|} \gamma_0 \mid \tilde{\varsigma} \in \mathcal{B}_1, \eta \in \mathcal{B}_2, j \in [0, \infty) \right\} \subset C_0^\infty(B^q(a))$$

$$\left\{ 2^{M|k-j|} \gamma_{k,\alpha} \mid \tilde{\varsigma} \in \mathcal{B}_1, \eta \in \mathcal{B}_2, j \in [0, \infty), k \in \mathbb{N}, k \leq j, |\alpha| = 2N \right\} \subset C_0^\infty(B^q(a))$$

are bounded.

PROOF. If $j = 0$, the result is trivial. Indeed, take $\eta' \in C_0^\infty(B^q(a))$ with $\eta' \equiv 1$ on a neighborhood of the closure of $\cup_{\eta \in \mathcal{B}_2} \text{supp}(\eta)$. Setting $\gamma_0 = \eta'\tilde{\varsigma}$ yields the result for $j = 0$.

We turn to the case when $j > 0$. Let $\tilde{\varsigma} \in \mathcal{B}_1 \cap S_0(\mathbb{R}^q)$ and $\eta \in \mathcal{B}_2$. We use Corollary 1.1.13 to write $\varsigma = \triangle^{-N}\tilde{\varsigma}$, so that $\tilde{\varsigma} = \triangle^N \varsigma$, and $\{\varsigma \mid \tilde{\varsigma} \in \mathcal{B}_1 \cap S_0(\mathbb{R}^q)\} \subset S_0(\mathbb{R}^q)$ is a bounded set. Let $\eta' \in C_0^\infty(B^q(a))$ with $\eta' \equiv 1$ on a neighborhood of the closure of $\cup_{\eta \in \mathcal{B}_2} \text{supp}(\eta)$. We apply Lemma 2.14.20 to $\eta'\varsigma^{(2^j)}$ to write

$$\eta'\varsigma^{(2^j)} = \sum_{k \leq j} \eta' \tilde{\gamma}_k^{(2^k)},$$

where for every M,

$$\left\{ 2^{M|j-k|} \tilde{\gamma}_k \mid j \in (0, \infty), k \leq j, k \in \mathbb{N}, \tilde{\varsigma} \in \mathcal{B}_1 \cap S_0(\mathbb{R}^q) \right\} \subset C_0^\infty(B^q(a))$$

is a bounded set.

Define $2^{-2j}\triangle = \sum_{l=1}^{q} -2^{-2jd_l}\partial_{t_l}^2$, so that $\tilde{\varsigma}^{(2^j)} = \left(2^{-2j}\triangle\right)^N \varsigma^{(2^j)}$. Consider,

$$\eta\tilde{\varsigma}^{(2^j)} = \eta\left(2^{-2j}\triangle\right)^N \varsigma^{(2^j)}$$

$$= \sum_{\substack{k\leq j \\ k\in\mathbb{N}}} \eta\left(2^{-2j}\triangle\right)^N \tilde{\gamma}_k^{(2^k)}.$$

We expand $\triangle^N = \sum_{|\alpha|=2N} c_\alpha \partial_t^\alpha$ where the c_α are explicit constants. Recall, we write $2^{-j}\partial_t$ to denote $\left(2^{-jd_1}\partial_{t_1}, \ldots, 2^{-jd_q}\partial_{t_q}\right)$. Consider,

$$\left(2^{-j}\partial_t\right)^\alpha \tilde{\gamma}_k^{(2^k)} = \left[\partial_t^\alpha 2^{(k-j)\alpha\cdot d}\tilde{\gamma}_k\right]^{(2^k)}.$$

Setting $\gamma_{k,\alpha} = c_\alpha 2^{(k-j)\alpha\cdot d}\tilde{\gamma}_k$, the result follows by the properties of the $\tilde{\gamma}_k$. □

LEMMA 2.15.46. *Let $\mathcal{B}_1 \subset C_0^\infty\left(B^q\left(a\right)\right)$, $\mathcal{B}_2 \subset C^\infty\left(\mathbb{R}^q \times \Omega\right)$, and $\mathcal{B}_3 \subset C_0^\infty\left(\Omega\right)$ be bounded sets. Fix $N \in \mathbb{N}$. For $l = 1,2$, let $\eta_l, \gamma_l \in \mathcal{B}_1$, $j_l \in [0,\infty)$, $\kappa_l \in \mathcal{B}_2$, $\psi_1^l, \psi_2^l \in \mathcal{B}_3$, and let α_1 and α_2 be ordered multi-indices with $|\alpha_1| = |\alpha_2| = 2N$. Let*

$$\varsigma_l = \begin{cases} \partial_t^{\alpha_l}\gamma_l & \text{if } j_l > 0, \\ \gamma_l & \text{if } j_l = 0. \end{cases}$$

Define operators

$$F_l f\left(x\right) = \psi_1^l\left(x\right) \int f\left(e^{t\cdot X}x\right) \psi_2^l\left(e^{t\cdot X}x\right) \kappa_l\left(t,x\right) \eta_l\left(t\right) \varsigma_l^{\left(2^{j_l}\right)}\left(t\right) dt.$$

Then, there is an admissible constant $C = C\left(N, \mathcal{B}_1, \mathcal{B}_2, \mathcal{B}_3\right)$ such that

$$\|F_1 F_2\|_{L^\infty \to L^\infty} \leq C2^{-2N|j_1-j_2|}.$$

PROOF. Fix $x_0 \in \Omega$. We wish to show $|F_1 F_2 f\left(x_0\right)| \lesssim 2^{-2N|j_1-j_2|}\|f\|_{L^\infty}$. If $j_1 = j_2$, the result is trivial. There are two remaining cases, either $j_1 > j_2$ or $j_2 < j_1$. The proofs for the two cases are nearly identical, and we prove only the case $j_1 > j_2$, leaving the other case to the reader. Notice, since $j_1 > j_2 \geq 0$, we have $j_1 > 0$. Consider, letting $\tilde{\kappa}\left(t_1, t_2, x\right), \hat{\kappa}\left(t_1, t_2, x\right) \in C^\infty\left(\mathbb{R}^q \times \mathbb{R}^q \times \Omega\right)$ be functions which range over a bounded set as the various parameters in the lemma change,

$$F_1 F_2 f\left(x_0\right) = \psi_1^1\left(x_0\right) \int f\left(e^{t_2\cdot X}e^{t_1\cdot X}x_0\right) \tilde{\kappa}\left(t_1, t_2, x_0\right) \varsigma_1^{\left(2^{j_1}\right)}\left(t_1\right) \varsigma_2^{\left(2^{j_2}\right)}\left(t_2\right) dt_1 dt_2$$

$$= \psi_1^1\left(x_0\right) \int f\left(e^{t_2\cdot\left(2^{-j_2}X\right)}e^{t_1\cdot\left(2^{-j_2}X\right)}x_0\right) \hat{\kappa}\left(t_1, t_2, x_0\right) \varsigma_1^{\left(2^{j_1-j_2}\right)}\left(t_1\right) \varsigma_2\left(t_2\right) dt_1 dt_2,$$

where $\hat{\kappa}\left(t_1, t_2, x\right) = \tilde{\kappa}\left(2^{-j_2}t_1, 2^{-j_2}t_2, x\right)$. We take $a > 0$ to be a sufficiently small 2-admissible constant so that

$$e^{t_2\cdot\left(2^{-j_2}X\right)}e^{t_1\cdot\left(2^{-j_2}X\right)}x_0 \in B_{\left(X,d\right)}\left(x_0, \xi_3 2^{-j_2}\right), \quad \forall t_1, t_2 \in B^q\left(a\right).$$

Notice, the value of $F_1 F_2 f(x_0)$ depends only on the values of f on $B_{(X,d)}\left(x_0, \xi_3 2^{-j_2}\right)$ $\subseteq \Phi_{x_0, 2^{-j_2}}\left(B^{n_0(x_0)}(\eta)\right)$. Using $x_0 = \Phi_{x_0, 2^{-j_2}}(0)$ and that $e^{t \cdot (2^{-j} X)} \Phi_{x_0, 2^{-j}}(u) = \Phi_{x_0, 2^{-j}}\left(e^{t \cdot Y^{x_0, 2^{-j}}} u\right)$, we have

$$F_1 F_2 f(x_0) = \psi_1^1(x_0) \int f \circ \Phi_{x_0, 2^{-j_2}}\left(e^{t_2 \cdot Y^{x_0, 2^{-j}}} e^{t_1 \cdot Y^{x_0, 2^{-j}}} 0\right)$$
$$\widehat{\kappa}(t_1, t_2, x_0) \varsigma_1^{\left(2^{j_1 - j_2}\right)}(t_1) \varsigma_2(t_2) \, dt_1 dt_2.$$

By (2.87), we have

$$\left| \det_{n_0(x_0) \times n_0(x_0)} \frac{\partial}{\partial t_2} e^{t_2 \cdot Y^{x_0, 2^{-j}}} e^{t_1 \cdot Y^{x_0, 2^{-j}}} 0 \right|_{t_1 = t_2 = 0} \gtrsim 1.$$

We apply a change of variables in the t_2 variable $u = e^{t_2 \cdot Y^{x_0, 2^{-j}}} e^{t_1 \cdot Y^{x_0, 2^{-j}}} 0$, as in Appendix B.3–by taking $a > 0$ to be a sufficiently small 2-admissible constant. With this change of variables, we have

$$F_1 F_2 f(x_0) = \psi_1^1(x_0) \int_{u \in B^{n_0(x)}(\eta)} f \circ \Phi_{x_0, 2^{-j_2}}(u) \, g(u, t_1) \varsigma_1^{\left(2^{j_1 - j_2}\right)}(t_1) \, du \, dt_1,$$

where $\|g\|_{C^m\left(B^{n_0(x_0)}(\eta) \times B^q(a)\right)} \lesssim 1$ for every m.

Since $j_1 > 0$, $\varsigma_1 = \partial_t^{\alpha_1} \gamma_1$ with $|\alpha_1| = 2N$. As before, define

$$\delta \partial_{t_1} = \left(\delta^{d_1} \partial_{t_1^1}, \ldots, \delta^{d_q} \partial_{t_1^q}\right).$$

Integration by parts shows

$$F_1 F_2 f(x_0) = \psi_1^1(x_0) \int_{u \in B^{n_0(x)}(\eta)} f \circ \Phi_{x_0, 2^{-j_2}}(u) \left[\left(2^{-(j_1 - j_2)} \partial_{t_1}\right)^{\alpha_1} g(u, t_1)\right]$$
$$\gamma_1^{\left(2^{j_1 - j_2}\right)}(t_1) \, du \, dt_1.$$

Since $\left|\left(2^{-(j_1 - j_2)} \partial_{t_1}\right)^{\alpha_1} g(u, t_1)\right| \lesssim 2^{-2N|j_1 - j_2|}$ and since $\int \left|\gamma_1^{\left(2^{j_1 - j_2}\right)}(t_1)\right| dt_1 \lesssim 1$, it follows that

$$|F_1 F_2 f(x_0)| \lesssim 2^{-2N|j_1 - j_2|} \sup_{u \in B^{n_0(x_0)}(\eta)} \left|f \circ \Phi_{x_0, 2^{-j_2}}(u)\right| \leq 2^{-2N|j_1 - j_2|} \|f\|_{L^\infty}.$$

As $x_0 \in \Omega$ was arbitrary, this completes the proof. $\square$

PROOF OF LEMMA 2.15.44. For $l = 1, 2$, let $\widetilde{\varsigma}_l$, j_l, η_l, κ_l, and ψ_1^l, ψ_2^l be as in the statement of the lemma. Fix N and take M large, to be chosen later. Apply Lemma 2.15.45 to decompose

$$\eta_l(t) \widetilde{\varsigma}_l^{\left(2^{j_l}\right)}(t) = \sum_{\substack{k \in \mathbb{N} \\ k \leq j_l}} \eta_l(t) \varsigma_{l,k}^{\left(2^k\right)},$$

where

$$\varsigma_{l,k} = \begin{cases} \sum_{|\alpha|=2N} \partial_t^\alpha \gamma_{l,k,\alpha} & \text{if } k > 0, \\ \gamma_0 & \text{if } k = 0, \end{cases}$$

and for every M the set

$$\left\{ 2^{M|j_l-k|} \gamma_{l,k,\alpha} \mid l = 1, 2, \widetilde{\varsigma}_l \in \mathcal{B}_1, \eta_l \in \mathcal{B}_2, j_l \in [0, \infty), k \in \mathbb{N}, k \le j_l, |\alpha| = 2N \right\}$$
$$\subset C_0^\infty \left(B^q(a) \right)$$

(2.94)

is bounded. For $k \le j_l$, $k \in \mathbb{N}$, define operators

$$F_{l,k} f(x) = \psi_1^l(x) \int f\left(e^{t \cdot X} x\right) \psi_2^l\left(e^{t \cdot X} x\right) \kappa_l(t, x) \eta_l(t) \varsigma_{l,k}^{(2^k)}(t) \, dt,$$

so that $E_l = \sum_{\substack{k \le j_l \\ k \in \mathbb{N}}} F_{k,l}$. Combining (2.94) and Lemma 2.15.46, we have

$$\|F_{1,k_1} F_{2,k_2}\|_{L^\infty \to L^\infty} \lesssim 2^{-M|j_1-k_1|-M|j_2-k_2|-2N|k_1-k_2|}$$
$$\le 2^{-|j_1-k_1|-|j_2-k_2|-2N|j_1-j_2|},$$

provided we choose $M = M(N)$ sufficiently large (where the implicit constants depend on N and M, but not on j_1, j_2, k_1, k_2). We have,

$$\|E_1 E_2\|_{L^\infty \to L^\infty} \le \sum_{\substack{k_1 \le j_1 \\ k_2 \le j_2 \\ k_1, k_2 \in \mathbb{N}}} \|F_{1,k_1} F_{2,k_2}\|_{L^\infty \to L^\infty}$$

$$\lesssim \sum_{\substack{k_1 \le j_1 \\ k_2 \le j_2 \\ k_1, k_2 \in \mathbb{N}}} 2^{-|j_1-k_1|-|j_2-k_2|-2N|j_1-j_2|}$$

$$\lesssim 2^{-2N|j_1-j_2|},$$

completing the proof.   □

2.15.4   A Littlewood-Paley theory

We take the same setup as in Section 2.15.3, and therefore we have a list of vector fields with formal degrees $(X, d) = (X_1, d_1), \ldots, (X_q, d_q)$ satisfying the assumptions of this section, and we do *not* assume $X_1, \ldots, X_q$ span the tangent space. The goal of this section is to create a Littlewood-Paley theory adapted to the geometry $B_{(X,d)}(x, \delta)$. We use the same dilations on $\mathbb{R}^q$: for $\delta > 0$

$$\delta(t_1, \ldots, t_q) = \left(\delta^{d_1} t_1, \ldots, \delta^{d_q} t_q\right).$$

(2.95)

Remark 2.15.47   The Littlewood-Paley theory developed in this section is used in Chapter 5 to prove the L^p boundedness of multi-parameter singular integral operators.

Let $a > 0$ be a small 2-admissible constant so that all of the results in Section 2.15.3 apply. Fix $\Omega_0 \Subset \Omega$ an open, relatively compact subset of Ω, and let $\psi_0 \in C_0^\infty(\Omega)$ equal 1 on a neighborhood of the closure of Ω_0. Consider the Dirac δ function at 0: $\delta_0 \in C_0^\infty(\mathbb{R}^q)'$. δ_0 is clearly a Calderón-Zygmund kernel of order 0 in the sense of Definition 2.13.14. Since $\operatorname{supp}(\delta_0) = \{0\} \subset B^q(a)$, Proposition 2.14.4 shows $\delta_0 \in CZ(0, a)$ (see Definition 2.14.2). In all of these definitions, we are using the dilations (2.95).

Since $\delta_0 \in CZ(0, a)$ there is $\eta \in C_0^\infty(B^q(a))$ and $\{\varsigma_j \mid j \in \mathbb{N}\} \subset \mathcal{S}(\mathbb{R}^q)$ a bounded set with $\varsigma_j \in \mathcal{S}_0(\mathbb{R}^q)$ if $j > 0$, such that $\delta_0(t) = \sum_{j \in \mathbb{N}} \eta(t) \varsigma_j^{(2^j)}(t)$. For $j \in \mathbb{N}$ define an operator

$$D_j f(x) = \psi_0(x) \int f(e^{t \cdot X} x) \psi_0(e^{t \cdot X} x) \eta(t) \varsigma_j^{(2^j)}(t) \, dt.$$

Notice $\psi_0^2 = \sum_{j \in \mathbb{N}} D_j$, where we have identified ψ_0^2 with the operator given by multiplication by the function ψ_0^2. This section is devoted to the next theorem.

THEOREM 2.15.48. *For $1 < p < \infty$ and $f \in C^\infty(\Omega)$, we have*

$$\left\| \left(\sum_{j \in \mathbb{N}} |D_j f|^2 \right)^{\frac{1}{2}} \right\|_{L^p} \leq C_p \|f\|_{L^p}, \tag{2.96}$$

for some constant C_p. Conversely, suppose $\operatorname{supp}(f) \subset \Omega_0$. Then, for $1 < p < \infty$,

$$\|f\|_{L^p} \leq C_p' \left\| \left(\sum_{j \in \mathbb{N}} |D_j f|^2 \right)^{\frac{1}{2}} \right\|_{L^p}, \tag{2.97}$$

for some constant C_p'.

The proof of Theorem 2.15.48 is very similar to the proof of Proposition 2.10.8, however there are a few added difficulties. We present the proof of Theorem 2.15.48 in detail: this result is essential to our later proofs of the L^p boundedness of multiparameter singular integrals. The proof requires a Calderón reproducing-type formula. For $j \in \mathbb{Z} \setminus \mathbb{N}$, define $D_j = 0$. For $M > 0$ define

$$U_M = \sum_{\substack{j \in \mathbb{N} \\ |l| \leq M}} D_j D_{j+l}, \quad R_M = \sum_{\substack{j \in \mathbb{N} \\ |l| > M}} D_j D_{j+l}.$$

Note that $U_M + R_M = \psi_0^4$.

PROPOSITION 2.15.49 (Calderón reproducing-type fomula). *Fix p, $1 < p < \infty$, and $\psi_1 \in C_0^\infty(\Omega)$ with $\psi_0 \equiv 1$ on a neighborhood of the support of ψ_1. There exists $M = M(p)$ and $V_M : L^p \to L^p$ such that $\psi_1 U_M V_M = V_M U_M \psi_1 = \psi_1$.*

To prove Proposition 2.15.49, we need the next lemma.

LEMMA 2.15.50. *Fix* p, $1 < p < \infty$. $\lim_{M \to \infty} \|R_M\|_{L^p \to L^p} = 0$.

PROOF OF PROPOSITION 2.15.49 GIVEN LEMMA 2.15.50. Recall $U_M = \psi_0^4 - R_M$. Take $\psi \in C_0^\infty(\Omega)$ with $\psi \equiv 1$ on a neighborhood of the support of ψ_1, but with $\psi_0 \equiv 1$ on a neighborhood of the support of ψ. Take $M = M(p)$ so large $\|R_M \psi\|_{L^p \to L^p} < 1$. Define

$$V_M = \sum_{l=0}^{\infty} \psi (R_M \psi)^l ,$$

with convergence in the uniform operator topology as operators $L^p \to L^p$. It is direct to verify that V_M satisfies the conclusions of the proposition. $\square$

Lemma 2.15.50 follows immediately by interpolating the next two lemmas.

LEMMA 2.15.51. *For* $1 < p < \infty$, $M \geq 1$,

$$\|R_M\|_{L^p \to L^p} \leq C_p M.$$

LEMMA 2.15.52. *For every* $N > 0$,

$$\|R_M\|_{L^2 \to L^2} \leq C_N 2^{-NM}.$$

To prove Lemma 2.15.51, we need another lemma, which implies (2.96).

LEMMA 2.15.53. *For* $f \in C^\infty(\Omega)$ *and* $1 < p < \infty$,

$$\left\| \left(\sum_{j \in \mathbb{N}} |D_j f|^2 \right)^{\frac{1}{2}} \right\|_{L^p} \lesssim \|f\|_{L^p} .$$

The same result holds with D_j *replaced by* D_j^*.

PROOF. Let $\{\epsilon_j\}_{j \in \mathbb{N}} \subseteq \{\pm 1\}$ be a sequence. Notice, $\sum_{j \in \mathbb{N}} \epsilon_j D_j$ is an operator of the form covered in Theorem 2.15.32, and this is true uniformly in the choice of the sequence. Thus, for $1 < p < \infty$, $\left\| \sum_{j \in \mathbb{N}} \epsilon_j D_j \right\|_{L^p \to L^p} \lesssim 1$, where the implicit constant does not depend on the choice of the sequence.

Now let $\{\epsilon_j\}_{j \in \mathbb{N}}$ be a sequence of i.i.d. random variables of mean 0 taking values ± 1. We use the Khintchine inequality (Theorem 2.10.10) to see

$$\left\| \left(\sum_{j \in \mathbb{N}} |D_j f|^2 \right)^{\frac{1}{2}} \right\|_{L^p} \approx \left\| \left(\mathbb{E} \left| \sum_{j \in \mathbb{N}} \epsilon_j D_j f \right|^p \right)^{\frac{1}{p}} \right\|_{L^p}$$

$$= \left(\mathbb{E} \left\| \sum_{j \in \mathbb{N}} \epsilon_j D_j f \right\|_{L^p}^p \right)^{\frac{1}{p}}$$

$$\lesssim (\mathbb{E} \|f\|_{L^p}^p)^{\frac{1}{p}}$$

$$= \|f\|_{L^p} .$$

The same proof works with D_j replaced by D_j^*.   □

PROOF OF LEMMA 2.15.51. With Lemma 2.15.53 in hand, the proof is exactly the same as the proof of Lemma 2.10.13.   □

PROOF OF LEMMA 2.15.52. Proposition 2.15.41 shows, for every N,

$$\left\| D_{j_1} D_{j_1+l_1} D_{j_2+l_2}^* D_{j_2}^* \right\|_{L^2 \to L^2} \le C 2^{-N \operatorname{diam}\{j_1, j_1+l_1, j_2, j_2+l_2\}},$$

$$\left\| D_{j_1+l_1}^* D_{j_1}^* D_{j_2} D_{j_2+l_2} \right\|_{L^2 \to L^2} \le C 2^{-N \operatorname{diam}\{j_1, j_1+l_1, j_2, j_2+l_2\}},$$

for $j_1, j_2 \in \mathbb{N}$, $l_1, l_2 \in \mathbb{Z}$. The Cotlar-Stein Lemma (Lemma 1.2.26) applies to show

$$\| R_M \|_{L^2 \to L^2} \lesssim \sup_{\substack{j_1 \in \mathbb{N} \\ |l_1| > M}} \sum_{\substack{j_2 \in \mathbb{N} \\ |l_2| > M}} 2^{-N \operatorname{diam}\{j_1, j_1+l_1, j_2, j_2+l_2\}/2} \lesssim 2^{-NM/2},$$

completing the proof with N replaced by $N/2$.   □

PROOF OF THEOREM 2.15.48. (2.96) is contained in Lemma 2.15.53, and so we only prove (2.97). Fix $\psi_1 \in C_0^\infty(\Omega)$, with $\psi_0 \equiv 1$ on a neighborhood of the support of ψ_1 and $\psi_1 \equiv 1$ on a neighborhood of the closure of Ω_0. Thus, if $\operatorname{supp}(f) \subset \Omega_0$, we have $\psi_1 f = f$. Fix p $(1 < p < \infty)$, and take $M = M(p)$ and V_M as in Proposition 2.15.49, so that $\psi_1 U_M V_M = \psi_1 = V_M U_M \psi_1$. Let q be dual to p so that $V_M^* : L^q \to$

L^q. Fix $g \in L^q$ with $\|g\|_{L^q} = 1$. We have, for f with $\psi_1 f = f$,

$$
\begin{aligned}
|\langle g, f \rangle| &= |\langle g, \psi_1 f \rangle| \\
&= |\langle V_M^* g, U_M \psi_1 f \rangle| \\
&= |\langle V_M^* g, U_M f \rangle| \\
&\leq \sum_{|l| \leq M} \left| \sum_{j \in \mathbb{N}} \langle D_j^* V_M^* g, D_{j+l} f \rangle \right| \\
&\leq \sum_{|l| \leq M} \int \sum_{j \in \mathbb{N}} |(D_j^* V_M^* g)(D_{j+l} f)| \\
&\leq \sum_{|l| \leq M} \int \left(\sum_{j \in \mathbb{N}} |D_j^* V_M^* g|^2 \right)^{\frac{1}{2}} \left(\sum_{j \in \mathbb{N}} |D_j f|^2 \right)^{\frac{1}{2}} \\
&\leq \sum_{|l| \leq M} \left\| \left(\sum_{j \in \mathbb{N}} |D_j^* V_M^* g|^2 \right)^{\frac{1}{2}} \right\|_{L^q} \left\| \left(\sum_{j \in \mathbb{N}} |D_j f|^2 \right)^{\frac{1}{2}} \right\|_{L^p} \\
&\lesssim M \|V_M^* g\|_{L^q} \left\| \left(\sum_{j \in \mathbb{N}} |D_j f|^2 \right)^{\frac{1}{2}} \right\|_{L^p} \\
&\approx \left\| \left(\sum_{j \in \mathbb{N}} |D_j f|^2 \right)^{\frac{1}{2}} \right\|_{L^p},
\end{aligned}
$$

where in the second-to-last line we applied Lemma 2.15.53. In the last line, we have used that M is fixed since p is, that $V_M^* : L^q \to L^q$, and $\|g\|_{L^q} = 1$. Taking the supremum over all such g yields the result. $\qquad\square$

COROLLARY 2.15.54. *Let* $\{\epsilon_j\}_{j \in \mathbb{N}}$ *be a sequence of i.i.d. random variables of mean* 0 *taking values* ± 1, *and fix* p, $1 < p < \infty$. *For* $f \in C_0^\infty(\Omega_0)$,

$$
\|f\|_{L^p} \approx \left\| \left(\sum_{j \in \mathbb{N}} |D_j f|^2 \right)^{\frac{1}{2}} \right\|_{L^p} \approx \left(\mathbb{E} \left\| \sum_{j \in \mathbb{N}} \epsilon_j D_j f \right\|_{L^p}^p \right)^{\frac{1}{p}}.
$$

PROOF. The first $\approx$ is contained in Theorem 2.15.48. The second $\approx$ follows immediately from the Khintchine inequality (Theorem 2.10.10). $\qquad\square$

COROLLARY 2.15.55. *Let $\{\epsilon_j\}_{j\in\mathbb{N}}$ be a sequence of i.i.d. random variables of mean 0 taking values ± 1, and fix p, $1 < p < \infty$. For $f \in C_0^\infty(\Omega)$,*

$$\left(\mathbb{E}\left\|\sum_{j\in\mathbb{N}}\epsilon_j D_j f\right\|_{L^p}^p\right)^{\frac{1}{p}} \approx \left\|\left(\sum_{j\in\mathbb{N}}|D_j f|^2\right)^{\frac{1}{2}}\right\|_{L^p} \lesssim \|f\|_{L^p}.$$

PROOF. The $\approx$ follows immediately from the Khintchine inequality (Theorem 2.10.10). The $\lesssim$ follows from Theorem 2.15.48. $\qquad\square$

2.15.5 The role of real analyticity

At the start of this section, we introduced assumptions on the list of vector fields $(X, d) = (X_1, d_1), \ldots, (X_q, d_q)$. As discussed earlier, Corollary 2.1.4 shows that if this list were generated by Hörmander vector fields, then these assumptions are satisfied. The reader might wonder, though, what other ways vector fields satisfying these assumptions might arise where the vector fields do not necessarily span the tangent space. A particularly compelling case involves real analytic vector fields.

Let $U \subset \mathbb{R}^n$ be an open set, and let $U_2 \Subset U$ be a relatively compact open subset of U. Suppose we are given real analytic vector fields $W_1, \ldots, W_r$ on U, and corresponding to each W_j we are given a formal degree $0 \neq d_j \in \mathbb{N}$. We recursively assign formal degrees to vector fields in the following way: if Y_1 has formal degree e_1 and Y_2 has formal degree e_2, we assign to $[Y_1, Y_2]$ the formal degree $e_1 + e_2$. Let M be a large integer, and let $(X, d) = (X_1, d_1), \ldots, (X_q, d_q)$ be an enumeration of all of the above vector fields with formal degree $\leq M$.

THEOREM 2.15.56. *If M is sufficiently large, then on U_2, we have*

$$[X_i, X_j] = \sum_{d_i + d_j \leq d_k} c_{i,j}^k X_k, \quad c_{i,j}^k \in C^\infty(U_2).$$

COROLLARY 2.15.57. *The vector fields (X, d) satisfy the hypotheses of this section.*

PROOF. This follows just as in Corollary 2.1.4. $\qquad\square$

We turn to the proof of Theorem 2.15.56. Let $\mathcal{S}$ denote the (infinite) set of all of the above vector fields with formal degrees; i.e., all iterated commuters of the W_j paired with the appropriate formal degrees. We prove

PROPOSITION 2.15.58. *For each point $x \in \overline{U_2}$ there is a finite set $\mathcal{F}_x \subset \mathcal{S}$ such that for each $(Y, e) \in \mathcal{S}$ there is a neighborhood $V_{Y,e,x}$ of x with $Y = \sum_{\substack{(Z,f)\in\mathcal{F}_x \\ f\leq e}} c_{Y,e,Z,f} Z$ on $V_{Y,e,x}$ with $c_{Y,e,Z,f} \in C^\infty(V_{Y,e,x})$.*

PROOF OF THEOREM 2.15.56 GIVEN PROPOSITION 2.15.58. Let $\mathcal{F}_x$ be as in Proposition 2.15.58, and let $M_x = \max\{f \mid (Z, f) \in \mathcal{F}_x\} \vee d_1 \vee d_2 \vee \cdots \vee d_r$.

Let $\mathcal{F}'_x = \{(Z, f) \in \mathcal{S} \mid f \leq M_x\}$. Notice $\mathcal{F}_x \subseteq \mathcal{F}'_x$ and $\mathcal{F}'_x$ is a finite set. Let $V_{Y,e,x}$ be as in Proposition 2.15.58 (for $(Y, e) \in \mathcal{S}$). Define

$$V_x = \bigcap_{(Z_1,e_1),(Z_2,e_2)\in\mathcal{F}'_x} V_{([Z_1,Z_2],e_1+e_2)}.$$

On V_x, for $(Z_1, e_1), (Z_2, e_2) \in \mathcal{F}'_x$,

$$[Z_1, Z_2] = \sum_{\substack{(Z_3,e_3)\in\mathcal{F}'_x \\ e_3 \leq e_1+e_2}} c_{Z_1,e_1,Z_2,e_2,Z_3,e_3} Z_3, \quad c_{Z_1,e_1,Z_2,e_2,Z_3,e_3} \in C^\infty(V_x).$$

Inductively, the same is true for higher order commutators; for instance $[Z_1, [Z_2, Z_3]]$, $[Z_1, [Z_2, [Z_3, Z_4]]]$, etc. But every element of $\mathcal{S}$ is an iterated commutator of elements of $\mathcal{F}'_x$, due to the fact that $(W_1, d_1), \ldots, (W_r, d_r) \in \mathcal{F}'_x$. Thus, $\forall (Y, e) \in \mathcal{S}$, $Y = \sum_{\substack{(Z,f)\in\mathcal{F}'_x \\ f \leq e}} c_{Y,e,Z,f} Z$ on V_x with $c_{Y,e,Z,f} \in C^\infty(V_x)$.

$\{V_x \mid x \in \overline{U_2}\}$ is an open cover of the compact set $\overline{U_2}$. We extract a finite subcover, $V_{x_1}, \ldots, V_{x_n}$. Let $M = \max\{M_{x_j} \mid j = 1, \ldots N\}$, and set $\mathcal{F} = \{(Z, e) \mid e \leq M\}$. Notice that $\mathcal{F}'_{x_j} \subseteq \mathcal{F}$ for every x_j and $\mathcal{F}$ is a finite set. Let

$$(X, d) = (X_1, d_1), \ldots, (X_q, d_q)$$

be an enumeration of $\mathcal{F}$.

A simple partition of unity argument shows $\forall (Y, e) \in \mathcal{S}$, $Y = \sum_{\substack{(Z,f)\in\mathcal{F} \\ f \leq e}} c_{Y,e,Z,f} Z$ on U_2 with $c_{Y,e,Z,f} \in C^\infty(U_2)$. Since each $([X_j, X_k], d_j + d_k) \in \mathcal{S}$, the result now follows. $\qquad\square$

Fix a point in $\mathbb{R}^n$, which we denote by 0 (but it is not important that this is the usual 0 element of $\mathbb{R}^n$). We write $f : \mathbb{R}_0^n \to \mathbb{R}^m$ to denote that f is a germ of a function defined near $0 \in \mathbb{R}^n$. Let

$$\mathcal{A}_n = \{f : \mathbb{R}_0^n \to \mathbb{R} \mid f \text{ is real analytic}\},$$

$$\mathcal{A}_n^m = \{f : \mathbb{R}_0^n \to \mathbb{R}^m \mid f \text{ is real analytic}\}.$$

Notice that $\mathcal{A}_n^m$ can be identified with the m-fold Cartesian product of $\mathcal{A}_n$, justifying our notation.

LEMMA 2.15.59. *The ring $\mathcal{A}_n$ is Noetherian.*

COMMENTS ON THE PROOF. This is a simple consequence of the Weierstrass preparation theorem. See page 148 of [ZS60]. The proof in [ZS60] is for the formal power series ring, however, as mentioned on page 130 of [ZS60], the proof also works for the ring of convergent power series–i.e., the ring of power series with some positive radius of convergence. This is exactly the ring $\mathcal{A}_n$. $\qquad\square$

LEMMA 2.15.60. *The module $\mathcal{A}_n^m$ is a Noetherian $\mathcal{A}_n$ module.*

COMMENTS ON THE PROOF. It is a standard fact that for any Noetherian ring R, the R-module R^m is Noetherian. $\qquad\square$

LEMMA 2.15.61. *Let $\mathcal{S} \subset \mathcal{A}_n^n \times \mathbb{N}$. Then, there exists a finite subset $\mathcal{F} \subseteq \mathcal{S}$ such that for every $(g, e) \in \mathcal{S}$,*

$$g(x) = \sum_{\substack{(f,d)\in\mathcal{F} \\ d\le e}} c_{(f,d)}(x) f(x),$$

with $c_{f,d} \in \mathcal{A}_n$.

PROOF. Define a map $\iota : \mathcal{A}_n^n \times \mathbb{N} \to \mathcal{A}_{n+1}^n$ by $\iota(f, d) = t^d f(x)$, where $t \in \mathbb{R}$. Let M be the submodule of $\mathcal{A}_{n+1}^n$ generated by $\iota\mathcal{S}$. M is finitely generated by Lemma 2.15.60. Let $\mathcal{F} \subseteq \mathcal{S}$ be a finite subset so that $\iota\mathcal{F}$ generates M. We show that $\mathcal{F}$ satisfies the conclusions of the lemma.

Let $(g, e) \in \mathcal{S}$. Since $t^e g \in M$, we have

$$t^e g(x) = \sum_{(f,d)\in\mathcal{F}} \hat{c}_{(f,d)}(t, x) t^d f(x),$$

on a neighborhood of $(0, 0) \in \mathbb{R} \times \mathbb{R}^n$. Applying $\frac{1}{e!}\partial_t^e\big|_{t=0}$ to both sides and using that $\frac{1}{e!}\partial_t^e\big|_{t=0} t^d c_{(f,d)}(t, x) = 0$ if $d > e$, we have

$$g(x) = \sum_{\substack{(f,d)\in\mathcal{F} \\ d\le e}} \left[\frac{1}{e!}\partial_t^e\big|_{t=0}\hat{c}_{(f,d)}(t, x) t^d \right] f(x).$$

The result follows. $\qquad\square$

PROOF OF PROPOSITION 2.15.58. For $x \in \overline{U_2}$ fixed, we relabel x to be 0 and apply Lemma 2.15.61. Thinking of vector fields as functions from $\mathbb{R}^n \to \mathbb{R}^n$, the result follows immediately from Lemma 2.15.61. $\qquad\square$

2.16 FURTHER READING AND REFERENCES

The study of Carnot-Carathéodory (or sub-Riemannian) geometry has a very long history. It began with the work of Carathéodory [Car09] and the work of Chow [Cho39]. The *quantitative* study of Carnot-Carathéodory geometry–precursors to the quantitative Frobenius theorem (Theorem 2.2.22)–began with the work of Nagel, Stein, and Wainger [NSW85]. It was they who introduced the more general formulation involving vector fields with formal degrees as in Section 2.1. Also, Nagel, Stein, and Wainger introduced the map Φ as defined in (2.26). The work of Nagel, Stein, and Wainger amounts to a special case of Theorem 2.2.22 used to study single-parameter balls in the case when the vector fields span the tangent space (so in a setting where the classical Frobenius theorem is trivial). Nagel, Stein, and Wainger's methods are not sufficient for a study of multi-parameter balls. Fortunately, Tao and Wright [TW03] improved on these methods. They studied two parameter balls $B(x, \delta_1, \delta_2)$ under a weakly comparable condition

$$\delta_1^N \lesssim \delta_2 \lesssim \delta_1^{1/N},$$

for some large N–still under the assumption that the vector fields span the tangent space. Perhaps most importantly, Tao and Wright made use of the ODE (2.30) in this situation (the ODE itself is classical; see [Che46, page 155]). Tao and Wright also showed how useful Gronwall's inequality could be in this situation. Theorem 2.2.22, itself, was proved by the author building on both of these works [Str11].

The study of partial differential operators defined by vector fields began with the work of Hörmander [Hör67], where he proved that the sub-Laplacian is subelliptic. A simplified version of Hörmander's proof was provided by Kohn [Koh78], and it is this proof method which is used in Section 2.4.1. At this point, the connection to Carnot-Carathéodory geometry was still unclear. Hörmander's and Kohn's results took place in L^2 and did not prove the maximal order of subellipticity. To extend these results to L^p and to obtain the sharp gain, Rothschild and Stein [RS76] (building on work of Folland and Stein [FS74] and Folland [Fol75]) introduced the pseudodifferential operators from Section 2.14. The connection between these pseudodifferential operators and Carnot-Carathéodory geometry was then solidified by the work of Nagel, Stein, and Wainger [NSW85]. This set the stage for the singular integrals which were discussed in this chapter to be introduced–where they were, in fact, introduced to study questions in several complex variables. They appeared first under the name NIS operators in the work of Nagel, Rosay, Stein, and Wainger [NRSW89], and were later used by Chang, Nagel, and Stein [CNS92] and Koenig [Koe02].[24] The pseudodifferential operators of [RS76] where further clarified by Goodman [Goo76] and were also studied by Christ, Geller, Glowacki, and Polin [CGGP92] and Rothschild [Rot79].

Remark 2.16.1 In [NRSW89, CNS92, Koe02], an extra hypothesis was assumed in Definition 2.0.16. In [NRSW89, CNS92] it was assumed that there is a sequence $T_j \in C^\infty(M \times M)$ satisfying the hypotheses of Definition 2.0.16 uniformly in j such that $T_j \to T$ in distribution. In [Koe02], this was replaced with an a priori estimate. It turns out that these additional assumptions are superfluous: they follow from the assumptions of Definition 2.0.16 as we have stated it. Indeed, the implication (iii)⇒(i) gives such a sequence of T_j (namely the partial sums of $\sum_{j \in \mathbb{N}} 2^{jt} E_j$).

There have been many other works connecting Carnot-Carathéodory geometry with partial differential operators. Two important papers are those of Fefferman and Sánchez-Calle [FSC86] and Fefferman and Phong [FP83].

Maximal hypoellipticity first appeared in the work of Folland and Stein [FS74, Theorems 9.4 and 16.6], where it is proved for the Kohn-Laplacian on strictly pseudoconvex domains. The concept later appeared in the work of Rothschild and Stein [RS76], who proved that the sub-Laplacian is maximally hypoelliptic. Rockland [Roc78] introduced a conjecture about when a left invariant partial differential operator on a nilpotent Lie group would be maximally hypoelliptic in terms of group representation theory (and proved his conjecture in the case of the Heisenberg group). This conjecture was

---

[24]These papers all worked with Definition 2.0.16; the extension to Definition 2.0.30 seems to be new.

proved in full by Helffer and Nourrigat [HN79] who, in turn, conjectured an extension of this result to operators on manifolds like the ones discussed in this chapter [HN85]. This more general conjecture remains open, though some progress was obtained by Rothschild [Rot79]. Helffer and Nourrigat's result concerning Rockland's conjecture was extended to non-differential operators on nilpotent Lie groups by Christ, Geller, Glowacki, and Polin [CGGP92].

The proof of Theorem 2.4.8 uses the ideas of [NRSW89, CNS92, Koe02] and is known (in some form or another) to experts, but we do not know of a reference where Theorem 2.4.8 is explicitly stated.

The smooth metrics and bump functions from Section 2.5 were first constructed by Nagel and Stein [NS01]–the work in Section 2.5 closely follows their methods.

Spectral multipliers $m(\mathcal{L})$ like the ones in Section 2.6 have been studied by many authors. In the group situation, see [HJ83, Mül89, Ale94, Chr91]. A very general result of Stein [Ste70b] deals with the case when m is of Laplace transform type, but this is too restrictive for our purposes.

The finite propagation speed of the $\mathcal{L}$ wave equation follows from a more general result of Melrose [Mel86]. The proof we present, though, is a slight generalization of a proof of Müller [Mül04], which is based on an idea of Folland [Fol95]. This proof appeared in [Str09].

The connection between the finite propagation speed of the wave equation and spectral multipliers has been exploited by many authors. Our presentation follows closely the ideas of Sikora [Sik04], which were inspired by the work of Coulhon and Duong [CD99]. Results very much like the ones in Section 2.6, using the same methods, were obtained by the author in [Str09], where a more complicated situation was considered: $\mathcal{L}$ was replaced with an operator which could possibly have an infinite dimensional null space, so long as the projection onto that null space were well behaved.

The non-isotropic Sobolev spaces from Section 2.10 were introduced by Folland and Stein [FS74] and then more generally by Rothschild and Stein [RS76]. They have been used by many authors, including [NRSW89, CNS92, Koe02]. All of these references proved results like Theorem 2.10.1.

The theory we discuss concerning nilpotent Lie groups was laid out by Folland [Fol75]. The reader should also consult [Goo76] for further information. For general information on nilpotent Lie groups, the textbook by Corwin and Greenleaf [CG90] is an excellent reference.

The theory in Section 2.15.3, when the vector fields do not span the tangent space, is based on the work of the author in [Str11]. Because the distribution considered in that section may have singular points, the quantitative Frobenius theorem is necessary to develop any theory like the one discussed there. For the same reason, the Littlewood-Paley theory discussed in Section 2.15.4 is also quite recent. Some of the ideas there first appeared in the joint work of the author with Stein on singular Radon transforms [SS13]. The results in Section 2.15.5 are based on ideas of Lobry [Lob70]–see [SS12] for further information and generalizations.

Chapter Three

Multi-parameter Carnot-Carathéodory Geometry

In the previous chapters, we focused on "single-parameter" singular integrals. By this, we mean that the singular integrals are defined in terms of an underlying family of balls $B(x, \delta)$ where $\delta > 0$. The main focus of this monograph is a more general setting, where the underlying balls have many "parameters" $B(x, \delta_1, \ldots, \delta_\nu)$. We focus exclusively on the case of Carnot-Carathéodory balls. In this chapter, we develop the theory necessary to deal with these multi-parameter balls. Much as in the single-parameter case, the quantitative Frobenius theorem (Theorem 2.2.22) is a key tool.

We begin by introducing the multi-parameter Carnot-Carathéodory balls. Fix $1 \leq \nu \in \mathbb{N}$–the number of "parameters." Let M be a connected (not necessarily compact) manifold. For each μ let $X_1^\mu, \ldots, X_{q_\mu}^\mu$ be C^∞ vector fields on M. Pair each X_j^μ with a formal degree $0 \neq \hat{d}_j^\mu \in \mathbb{N}$. As in (2.7), we denote by $\left(X^\mu, \hat{d}^\mu\right)$ this list of vector fields with formal degrees. For $\delta_\mu > 0$ we define the list of vector fields $\delta_\mu X^\mu := \delta_\mu^{\hat{d}_1^\mu} X_1^\mu, \ldots, \delta_\mu^{\hat{d}_{q_\mu}^\mu} X_{q_\mu}^\mu$. Then, following (2.8), we obtain *single*-parameter balls $B_{\left(X^\mu, \hat{d}^\mu\right)}(x, \delta_\mu) := B_{\delta_\mu X^\mu}(x)$–if $X_1^\mu, \ldots, X_{q_\mu}^\mu$ do not span the tangent space, then these balls might not be open subsets of M. As we saw in Chapter 2, under appropriate assumptions on $\left(X^\mu, \hat{d}^\mu\right)$, we may develop a theory of singular integrals based on the balls $B_{\left(X^\mu, \hat{d}^\mu\right)}(x, \delta_\mu)$.

Corresponding to $\hat{d}_j^\mu$ define a multi-index $0 \neq d_j^\mu \in \mathbb{N}^\nu$, by letting d_j^μ equal $\hat{d}_j^\mu$ in the μth component, and 0 in all other components. Let $(X, d) = (X_1, d_1), \ldots, (X_q, d_q)$ be an enumeration of all of these vector fields with ν-parameter formal degrees: i.e., (X, d) is an enumeration of

$$\left(X_1^1, d_1^1\right), \ldots, \left(X_{q_1}^1, d_{q_1}^1\right), \ldots, \ldots, \left(X_1^\nu, d_1^\nu\right), \ldots, \left(X_{q_\nu}^\nu, d_{q_\nu}^\nu\right),$$

and $q = q_1 + \cdots + q_\nu$. For $\delta \in [0, \infty)^\nu$, and $\gamma \in \mathbb{N}^\nu$, define δ^γ via standard multi-index notation: $\delta^\gamma = \prod_\mu \delta_\mu^{\gamma_\mu}$. This defines δ^{d_j} for $1 \leq j \leq q$. We define δX to be the list of vector fields $\delta X := \delta^{d_1} X_1, \ldots, \delta^{d_q} X_q$. From here, we may define multi-parameter balls, for $x \in M$,

$$B_{(X,d)}(x, \delta) := B_{\delta X}(x), \quad \delta \in [0, \infty)^\nu.$$

We will see that under a natural generalization of the assumptions on the vector fields in Chapter 2, we are able to create a theory of singular integrals with these underlying balls.

3.1 ASSUMPTIONS ON THE VECTOR FIELDS

We now turn to a more precise description of our assumptions. For this we begin again, and ignore the setup in the introduction to this chapter. Fix an open set $U \subseteq \mathbb{R}^n$ and let $0 \neq \nu \in \mathbb{N}$ be the number of "parameters." We assume we are given a list of C^∞ vector fields on U, each paired with a "formal degree": $(X, d) := (X_1, d_1), \ldots, (X_q, d_q)$, where $0 \neq d_j \in \mathbb{N}^\nu$ (we think of each d_j as a multi-index[1]). As above, for $\delta \in [0, 1]^\nu$, we write δX to denote the list of vector fields $\delta^{d_1} X_1, \ldots, \delta^{d_q} X_q$.

Remark 3.1.1   When we introduce the class of singular integrals we study, the list (X, d) will be the same list from the introduction to this chapter. A key point in later sections is that each d_j from the list of vector fields in the introduction is nonzero in only one component. We do not need this added assumption in this section, though.

We fix a distinguished point in U, which we call 0.[2] We will only be considering operators whose Schwartz kernels are supported on a small neighborhood of $(0, 0) \in U \times U$. Let $U_1 \Subset U_2 \Subset U$ be neighborhoods of 0.

Assumptions on the vector fields: Let $\xi > 0$ be small enough that for every $x \in U_1$ and every $a = (a_1, \ldots, a_q) : [0, 1] \to \mathbb{R}^q$ with $\|a\|_{L^\infty} < 1$, there exists a unique solution $\gamma : [0, 1] \to U_2$ to the ODE

$$\gamma'(t) = \sum_{j=1}^{q} a_j(t) \, \xi^{|d_j|} X_j(\gamma(t)), \quad \gamma(0) = x. \tag{3.1}$$

It is easy to see that such an ξ exists, by the contraction mapping principle.[3] We assume that for each $m \in \mathbb{N}$, there is a constant C_m such that for every $\delta \in [0, 1]^\nu$ with $|\delta|_\infty < \xi$ and for every $x \in U_1$ we have

$$\left[\delta^{d_j} X_j, \delta^{d_k} X_k \right] = \sum_{l=1}^{q} c_{j,k}^{l,\delta,x} \delta^{d_l} X_l \text{ on } B_{(X,d)}(x, \delta),$$

with

$$\sup_{\substack{\delta \in [0,1]^\nu \\ |\delta|_\infty < \xi \\ x \in U_1}} \sum_{|\alpha| \leq m} \left\| (\delta X)^\alpha \, c_{j,k}^{l,\delta,x} \right\|_{C^0 \left(B_{(X,d)}(x,\delta) \right)} \leq C_m,$$

for every m. This concludes our assumptions on the vector fields.

The point of the above assumptions can be seen by looking at the quantitative Frobenius theorem (Theorem 2.2.22). Indeed, for each $\delta \in [0, 1]^\nu$, if we define the vector fields $Z_1 = \delta^{d_1} X_1, \ldots, Z_q = \delta^{d_q} X_1$, and the single-parameter formal degrees $\hat{d}_j = |d_j|$, then for each $x_0 \in U_1$, the assumptions of Theorem 2.2.22 hold uniformly

---

[1] In particular, when we write $|d_j|$ we mean the ℓ^1 norm of the coefficients, as in standard multi-index notation.

[2] It is not important that this point actually be the usual element $0 \in \mathbb{R}^n$.

[3] By standard uniqueness theorems, if a solution to (3.1) exists, it must be unique.

in $x_0 \in U_1$, $\delta \in [0,1]^\nu$, with this choice of $Z_1, \ldots, Z_q$. In particular, we may take admissible (and m-admissible) constants in the sense of that theorem to be independent of $x_0 \in U_1$ and $\delta \in [0,1]^\nu$.

More precisely, for $m \geq 2$, we say C is an m-admissible constant if C can be chosen to depend only on upper bounds for m, n, q, C_m, $\|X_j\|_{C^m(U_2)}$, d_j ($1 \leq j \leq q$), and a positive lower bound for ξ. For $m < 2$, we say C is an m-admissible constant if it is a 2-admissible constant. We say C is an admissible constant if C is an m-admissible constant where m can be chosen to depend only on upper bounds for n, q, and d_j ($1 \leq j \leq q$). We write $A \lesssim_m B$ (respectively, $A \lesssim B$) if $A \leq CB$ where C is an m-admissible (respectively, admissible) constant. We write $A \approx_m B$ (respectively, $A \approx B$) if $A \lesssim_m B$ and $B \lesssim_m A$ (respectively, $A \lesssim B$ and $B \lesssim A$).

The key point is that m-admissible constants (and admissible constants) do not depend on $x_0 \in U_1$ or $\delta \in [0,1]^\nu$. By Theorem 2.2.22, we have the following result.

THEOREM 3.1.2. *There exist 2-admissible constants* $\xi_2, \xi_3, \eta > 0$ *with* $\xi_3 < \xi_2 < \xi$ *such that the following holds. For each* $x_0 \in U_1$ *and* $\delta \in [0,1]^\nu$ *let* $n_0(x_0, \delta) = \dim \mathrm{span}\left\{ \delta^{d_1} X_1(x_0), \ldots, \delta^{d_q} X_q(x_0) \right\}$.[4] *Then, there is a* C^∞ *map*

$$\Phi_{x_0,\delta} : B^{n_0(x_0,\delta)}(\eta) \to B_{(X,d)}(x_0, \xi_2 \delta)$$

such that

- $\Phi_{x_0,\delta}(0) = x_0$.

- $\Phi_{x_0,\delta}$ *is injective.*

- $B_{(X,d)}(x_0, \xi_3 \delta) \subseteq \Phi_{x_0,\delta}\left(B^{n_0(x_0,\delta)}(\eta) \right) \subseteq B_{(X,d)}(x_0, \xi_2 \delta)$.

- *For* $u \in B^{n_0(x_0,\delta)}(\eta)$,

$$\left| \det_{n_0(x_0,\delta) \times n_0(x_0,\delta)} d\Phi_{x_0,\delta}(u) \right|$$
$$\approx_2 \left| \det_{n_0(x_0,\delta) \times n_0(x_0,\delta)} \left(\delta^{d_1} X_1(x_0) | \cdots | \delta^{d_q} X_q(x_0) \right) \right|$$
$$\approx_2 \mathrm{Vol}\left(B_{(X,d)}(x_0, \delta \xi_2) \right).$$

Where $\mathrm{Vol}(\cdot)$ *denotes the induced Lebesgue measure on the leaf passing through* x_0, *generated by* $\delta^{d_1} X_1, \ldots, \delta^{d_q} X_q$.

Furthermore if $Y_1^{x_0,\delta}, \ldots, Y_q^{x_0,\delta}$ *are the pullbacks of* $\delta^{d_1} X_1, \ldots, \delta^{d_q} X_q$ *via* $\Phi_{x_0,\delta}$ *to* $B^{n_0(x_0,\delta)}(\eta)$ *then we have*

$$\sup_{\substack{x_0 \in U_1 \\ \delta \in [0,1]^\nu}} \left\| Y_j^{x_0,\delta} \right\|_{C^m(B^{n_0(x_0,\delta)}(\eta))} \lesssim_m 1, \tag{3.2}$$

---

[4]For $\delta \in (0,1]^\nu$, $n_0(x_0, \delta)$ does not depend on δ. In fact the dependance of $n_0(x_0, \delta)$ on δ depends only on which coordinates of δ are equal to 0.

and

$$\inf_{\substack{x_0 \in U_1 \\ \delta \in [0,1]^\nu}} \left| \det_{n_0(x_0,\delta) \times n_0(x_0,\delta)} \left(Y_1^{x_0,\delta}(u) | \cdots | Y_q^{x_0,\delta}(u) \right) \right| \gtrsim_2 1. \tag{3.3}$$

PROOF. The conditions of this section can be rephrased exactly as saying that the conditions of Theorem 2.2.22 hold for the vector fields $Z_1 = \delta^{d_1} X_1, \ldots, Z_q = \delta^{d_q} X_q$, uniformly for $x_0 \in U_1$, $\delta \in [0,1]^\nu$, where we have taken $\xi_1 = \xi$ and given Z_j the formal degree $|d_j|$ (recall, this denotes the ℓ^1 norm of the multi-index d_j). The conclusions of Theorem 2.2.22 are exactly the conclusions of this theorem in this context. This uses the fact (which follows immediately from the definitions) that for $r > 0$ and $\delta \in [0,1]^\nu$

$$B_{(X,d)}(x_0, r\delta) = B_{(Z,|d|)}(x_0, r),$$

where $(Z, |d|)$ denotes the list of vector fields with single-parameter formal degrees $(Z_1, |d_1|), \ldots, (Z_q, |d_q|)$. □

PROPOSITION 3.1.3. *Suppose $\delta \in [0,1]^\nu$ with $|\delta|_\infty < \xi_2/2$. Then, for $x \in U_1$,* $\mathrm{Vol}\left(B_{(X,d)}(x, 2\delta)\right) \lesssim_2 \mathrm{Vol}\left(B_{(X,d)}(x, \delta)\right).$

PROOF. Theorem 3.1.2 shows that we have

$$\mathrm{Vol}\left(B_{(X,d)}(x, \delta)\right)$$

$$\approx_2 \left| \det_{n_0(x_0,\delta) \times n_0(x_0,\delta)} \left(\left(\xi_2^{-1}\delta\right)^{d_1} X_1(x_0) | \cdots | \left(\xi_2^{-1}\delta\right)^{d_q} X_q(x_0) \right) \right|,$$

with a similar result for δ replaced by 2δ. The result immediately follows. □

Remark 3.1.4 Proposition 3.1.3 is a multi-parameter analog of the crucial doubling condition given in the second inequality of Theorem 2.0.10. It is the first indication that a general theory of singular integrals may be in reach in this context.

In any particular application, the choice of (X, d) will be fixed. We will use $A \lesssim B$ to denote $A \leq CB$, where C is "independent of any relevant parameters"; in particular C will be allowed to depend on anything admissible constants may depend on–it will not be allowed to depend on $x_0 \in U_1$ or $\delta \in [0,1]^\nu$. Essential to our methods is that we may use Theorem 3.1.2 freely, and our constants will remain independent of $x_0 \in U_1$ and $\delta \in [0,1]^\nu$.

3.2 SOME PRELIMINARY ESTIMATES

In this section, we further our study of Carnot-Carathéodory geometry. The results in this section are the technical estimates we need to study the singular integrals which are discussed in the next chapters.

We work with **all the same assumptions** as Section 3.1, and so we are given a list of vector fields on an open set $U \subseteq \mathbb{R}^n$, with formal degrees

$$(X, d) = (X_1, d_1), \ldots, (X_q, d_q), \quad 0 \neq d_j \in \mathbb{N}^\nu$$

and we are given open sets $0 \in U_1 \Subset U_2 \Subset U$. We assume that this list satisfies all the assumptions from Section 3.1. Furthermore, we assume that $X_1, \ldots, X_q$ **span the tangent space** at each point of U (so that $n_0(x_0, \delta) = n$, for every x_0 and $\delta \in (0, 1]^{\nu}$)– and so the balls $B_{(X,d)}(x, \delta)$, for $\delta \in (0, 1]^{\nu}$, are *open* sets.[5]

Let $0 < \xi_0 \leq \frac{\xi_3}{2q}$ be a constant to be chosen later (it will be chosen in Lemma 3.2.17), where $\xi_3 > 0$ is the 2-admissible constant from Theorem 3.1.2. Define $\Omega :=$ $B_{(X,d)}\left(0, \frac{\xi_0}{2}(1, \ldots, 1)\right)$. In what follows, all points x, z will be elements of Ω. For $j = (j_1, \ldots, j_{\nu}) \in [0, \infty)^{\nu}$, we write $2^{-j} = \left(2^{-j_1}, \ldots, 2^{-j_{\nu}}\right) \in (0, 1]^{\nu}$–thus instead of studying balls of radius $\delta \in (0, 1]^{\nu}$, we may equivalently study balls of radius $2^{-j} \in (0, 1]^{\nu}$.

Associated to the balls $B_{(X,d)}\left(x, 2^{-j}\right)$, it is useful to define a family of metrics: for $j \in [0, \infty)^{\nu}$ and $x, y \in \Omega$, define

$$\rho_{2^{-j}}(x, y) := \inf \left\{ \delta > 0 \;\middle|\; \exists \gamma : [0, 1] \to \Omega, \gamma(0) = x, \gamma(1) = y, \right.$$

$$\gamma'(t) = \sum_{l=1}^{q} a_l(t) \left(\delta 2^{-j}\right)^{d_l} X_l(\gamma(t)), \tag{3.4}$$

$$\left. \sum_{l=1}^{q} |a_l(t)|^2 < 1, \sum_{l=1}^{q} \left|\left(\delta 2^{-j}\right)^{d_l} a_l(t)\right|^2 < \xi_0^2 \right\}.$$

Since $\Omega = B_{(X,d)}\left(0, \frac{\xi_0}{2}(1, \ldots, 1)\right)$, $\rho_{2^{-j}}(x, z) < \infty$ for every $x, z \in \Omega$. For $\delta > 0$ define the open set

$$B_{2^{-j}}(x, \delta) := \left\{ y \in \Omega \;\middle|\; \rho_{2^{-j}}(x, y) < \delta \right\}. \tag{3.5}$$

Note that if $\delta 2^{-j}$ is small in every coordinate, then $B_{2^{-j}}(x, \delta) = B_{(X,d)}\left(x, \delta 2^{-j}\right)$. However, if $\delta 2^{-j}$ is large in some coordinate, then these balls might differ. In fact, the ball $B_{2^{-j}}(x, \delta)$ is comparable to the ball $B_{(X,d)}\left(x, \left(\delta 2^{-j_1} \wedge \xi_0, \ldots, \delta 2^{-j_{\nu}} \wedge \xi_0\right)\right)$; this may differ from the ball $B_{(X,d)}\left(x, \delta 2^{-j}\right)$ if some of the coordinates of $\delta 2^{-j}$ are large. Henceforth, in this section, x, z will always be elements of Ω and j, k will always be elements of $[0, \infty)^{\nu}$.

Before we discuss the results, we introduce a few pieces of notation. If $j = (j_1, \ldots, j_{\nu}), k = (k_1, \ldots, k_{\nu}) \in [0, \infty)^{\nu}$, we write $j \wedge k = (j_1, \wedge k_1, \ldots, j_{\nu} \wedge k_{\nu})$–the coordinatewise minimum of j and k. Also, we write $j \leq k$ if $j_{\mu} \leq k_{\mu}$ for every μ.

PROPOSITION 3.2.1. *For $\delta > 0$, $\mathrm{Vol}\left(B_{2^{-j}}(x, 2\delta)\right) \lesssim_2 \mathrm{Vol}\left(B_{2^{-j}}(x, \delta)\right)$.*

PROOF. This follows immediately from Proposition 3.1.3 and the fact that the ball $B_{2^{-j}}(x, \delta)$ is comparable to the ball $B_{(X,d)}\left(x, \left(\delta 2^{-j_1} \wedge \xi_0, \ldots, \delta 2^{-j_{\nu}} \wedge \xi_0\right)\right)$. $\square$

---

[5]In the introduction to this chapter, (X, d) was associated to ν different single-parameter lists $\left(X^{\mu}, \hat{d}^{\mu}\right), 1 \leq \mu \leq \nu$–though we do not assume that now. When we return to that assumption, we will continue assuming that $X_1, \ldots, X_q$ span the tangent space at each point, but we will *not* assume $X_1^{\mu}, \ldots, X_{q_{\mu}}^{\mu}$ spans the tangent space for any μ.

PROPOSITION 3.2.2. *For $x, z \in \Omega$, $j \in [0, \infty)^{\nu}$, $m \in \mathbb{N}$,*

$$\frac{(1 + \rho_{2-j}(x, z))^{-m}}{\mathrm{Vol}\,(B_{2-j}(x, 1 + \rho_{2-j}(x, z)))} \approx_2 \frac{(1 + \rho_{2-j}(z, x))^{-m}}{\mathrm{Vol}\,(B_{2-j}(z, 1 + \rho_{2-j}(z, x)))}.$$

PROOF. First, note that $\rho_{2-j}(x, z) = \rho_{2-j}(z, x)$. In addition, we have

$$\mathrm{Vol}\,(B_{2-j}(x, 1 + \rho_{2-j}(x, z))) \leq \mathrm{Vol}\,(B_{2-j}(z, 2(1 + \rho_{2-j}(x, z))))$$
$$\lesssim_2 \mathrm{Vol}\,(B_{2-j}(z, 1 + \rho_{2-j}(z, x))).$$

By symmetry, we have

$$\mathrm{Vol}\,(B_{2-j}(x, 1 + \rho_{2-j}(x, z))) \approx_2 \mathrm{Vol}\,(B_{2-j}(z, 1 + \rho_{2-j}(z, x))).$$

The result follows. $\qquad\square$

LEMMA 3.2.3. $2^{-|k-j|}\rho_{2-j}(x, z) \leq \rho_{2-k}(x, z)$.

PROOF. Directly from the definition, for $\delta > 0$, $r \in [0, 1]^{\nu}$, we have $\delta\rho_r(x, z) = \rho_{\delta^{-1}r}(x, z)$. Also, if $r_1, r_2 \in [0, 1]^{\nu}$ with $r_1 \geq r_2$ (coordinatewise), then we have $\rho_{r_1}(x, z) \leq \rho_{r_2}(x, z)$. It follows that

$$2^{-|k-j|}\rho_{2-j}(x, z) = \rho_{2^{|k-j|}2-j}(x, z) \leq \rho_{2-k}(x, z).$$

$\qquad\square$

LEMMA 3.2.4. *There exists a 2-admissible constant C such that*

$$2^{-C|k-j|}\mathrm{Vol}\,(B_{2-j}(x, 1 + \rho_{2-j}(x, z))) \leq \mathrm{Vol}\,(B_{2-k}(x, 1 + \rho_{2-k}(x, z))).$$

PROOF. For $\delta > 0$, denote by $\tilde{B}_{(X,d)}(x, 2^{-j}) := B_{2-j}(x, 1)$. Proposition 3.2.1 can be restated as saying for $r \in [0, \infty)^{\nu}$,

$$\mathrm{Vol}\,\left(\tilde{B}_{(X,d)}(x, 2r)\right) \leq 2^C \mathrm{Vol}\,\left(\tilde{B}_{(X,d)}(x, r)\right),$$

for some 2-admissible constant C. We have, using Lemma 3.2.3,

$$\mathrm{Vol}\,(B_{2-j}(x, 1 + \rho_{2-j}(x, z))) = \mathrm{Vol}\,\left(\tilde{B}_{(X,d)}\left(x, (1 + \rho_{2-j}(x, z))\,2^{-j}\right)\right)$$
$$\leq \mathrm{Vol}\,\left(\tilde{B}_{(X,d)}\left(x, (1 + \rho_{2-j}(x, z))\,2^{|k-j|}2^{-k}\right)\right)$$
$$\leq \mathrm{Vol}\,\left(\tilde{B}_{(X,d)}\left(x, \left(1 + 2^{|k-j|}\rho_{2-k}(x, z)\right)2^{|k-j|}2^{-k}\right)\right)$$
$$\leq \mathrm{Vol}\,\left(\tilde{B}_{(X,d)}\left(x, (1 + \rho_{2-k}(x, z))\,2^{2|k-j|}2^{-k}\right)\right)$$
$$\leq 2^{2C|k-j|}\mathrm{Vol}\,\left(\tilde{B}_{(X,d)}\left(x, (1 + \rho_{2-k}(x, z))\,2^{-k}\right)\right),$$

which completes the proof with C replaced by $2C$. $\qquad\square$

LEMMA 3.2.5. $\forall m$, there is a 2-admissible constant $M = M(m)$ such that

$$2^{-M|k-j|} \frac{(1 + \rho_{2^{-k}}(x,z))^{-m}}{\mathrm{Vol}\,(B_{2^{-k}}(x, 1 + \rho_{2^{-k}}(x,z)))} \leq \frac{(1 + \rho_{2^{-j}}(x,z))^{-m}}{\mathrm{Vol}\,(B_{2^{-j}}(x, 1 + \rho_{2^{-j}}(x,z)))}.$$

PROOF. We use Lemmas 3.2.3 and 3.2.4 to see for M sufficiently large, depending on m,

$$2^{-M|k-j|} \frac{(1 + \rho_{2^{-k}}(x,z))^{-m}}{\mathrm{Vol}\,(B_{2^{-k}}(x, 1 + \rho_{2^{-k}}(x,z)))}$$

$$\leq \frac{\left(1 + 2^{|k-j|}\rho_{2^{-k}}(x,z)\right)^{-m}}{2^{(M-m)|k-j|}\mathrm{Vol}\,(B_{2^{-k}}(x, 1 + \rho_{2^{-k}}(x,z)))}$$

$$\leq \frac{(1 + \rho_{2^{-j}}(x,z))^{-m}}{\mathrm{Vol}\,(B_{2^{-j}}(x, 1 + \rho_{2^{-j}}(x,z)))}.$$

$\square$

PROPOSITION 3.2.6. $\forall m$, there exists a 2-admissible constant $M = M(m)$ such that for every $j \in [0, \infty)^\nu$,

$$\sum_{k \in \mathbb{N}^\nu} 2^{-M|k-j|} \frac{(1 + \rho_{2^{-k}}(x,z))^{-m}}{\mathrm{Vol}\,(B_{2^{-k}}(x, 1 + \rho_{2^{-k}}(x,z)))}$$

$$\leq \nu 2^\nu \frac{(1 + \rho_{2^{-j}}(x,z))^{-m}}{\mathrm{Vol}\,(B_{2^{-j}}(x, 1 + \rho_{2^{-j}}(x,z)))}.$$

PROOF. Take M as in Lemma 3.2.5. Then,

$$2^{-(M+1)|k-j|} \frac{(1 + \rho_{2^{-k}}(x,z))^{-m}}{\mathrm{Vol}\,(B_{2^{-k}}(x, 1 + \rho_{2^{-k}}(x,z)))}$$

$$\leq 2^{-|k-j|} \frac{(1 + \rho_{2^{-j}}(x,z))^{-m}}{\mathrm{Vol}\,(B_{2^{-j}}(x, 1 + \rho_{2^{-j}}(x,z)))}.$$

The result now follows with M replaced by $M + 1$, by summing in k. $\square$

We will have a use for a partition of unity like the one used in Section 2.5. To discuss this, fix a compact set $\mathcal{K}$ with $\Omega \Subset \mathcal{K} \Subset U_1$.

THEOREM 3.2.7 (Partition of Unity). There is a 2-admissible constant $\xi_4 > 0$ such that for every $\delta \in (0, 1]^\nu$ there is a finite set $\mathcal{C} = \mathcal{C}(\delta) \subset \mathcal{K}$ and functions ψ_x, $x \in \mathcal{C}$, satisfying

- $\sum_{x \in \mathcal{C}} \psi_x(y) \equiv 1, \forall y \in \mathcal{K}.$

- $\forall m$,

$$\sum_{|\alpha| \leq m} \|(\delta X)^\alpha \psi_x\|_{L^\infty} \lesssim 1.$$

- *For $x \in \mathcal{C}$, supp $(\psi_x) \subset B_{(X,d)}(x, \delta)$.*

- *For $x \in \mathcal{C}$, $\psi_x(y) \gtrsim 1$, for $y \in B_{(X,d)}(x, \xi_4\delta)$.*

- $0 \leq \psi_x \leq 1$.

- $\bigcup_{x \in \mathcal{C}} B_{(X,d)}(x, \xi_4\delta/2) \supset \mathcal{K}$.

- *For $x \in \mathcal{K}$, $1 \leq \# \{x_0 \in \mathcal{C} \mid \psi_{x_0}(x) \neq 0\} \lesssim 1$, where the implicit constant is independent of $\delta \in (0,1]^\nu$.*

PROOF. This follows from a line by line reprise of the proof of Theorem 2.5.3. □

Recall our assumptions on the vector fields (X, d). We assumed that, for $x \in U_1$,

$$[\delta^{d_j} X_j, \delta^{d_k} X_k] = \sum_{l=1}^{q} c_{j,k}^{l,\delta,x} \delta^{d_l} X_l$$

on $B_{(X,d)}(x, \delta)$; in particular, this assumption only holds near each point. The next lemma shows that this local result is equivalent to a, seemingly stronger, global version.

LEMMA 3.2.8. *For $\delta \in (0,1]^\nu$, $[\delta^{d_j} X_j, \delta^{d_k} X_k] = \sum_{l=1}^{q} c_{j,k}^{l,\delta} \delta^{d_l} X_l$ on Ω, where $\sum_{|\alpha| \leq m} \left| (\delta X)^\alpha c_{j,k}^{l,\delta} \right| \lesssim_m 1$, for every m.*

PROOF. First note that it suffices to prove the result for $|\delta|_\infty < \xi$, as the result for any $\delta \in (0,1]^\nu$ follows from the corresponding result for δ replaced with $(\delta_1 \wedge (\xi/2), \ldots, \delta_\nu \wedge (\xi/2))$. For $|\delta|_\infty < \xi$, we have

$$[\delta^{d_j} X_j, \delta^{d_k} X_k] = \sum_{l=1}^{q} c_{j,k}^{l,\delta,x_0} \delta^{d_l} X_l, \text{ on } B_{(X,d)}(x_0, \delta),$$

where $\sum_{|\alpha| \leq m} \left| (\delta X)^\alpha c_{j,k}^{l,\delta,x_0} \right| \lesssim_m 1$. Define $c_{j,k}^{l,\delta} = \sum_{x_0 \in \mathcal{C}} \psi_{x_0} c_{j,k}^{l,\delta,x_0}$, where $\mathcal{C}$ and ψ_{x_0} are as in Theorem 3.2.7. The result follows. □

DEFINITION 3.2.9. *We define ν-parameter dilations on $\mathbb{R}^q$ as follows. For $\delta \in [0, \infty)^\nu$ and $t = (t_1, \ldots, t_q) \in \mathbb{R}^q$, we define $\delta t = (\delta^{d_1} t_1, \ldots, \delta^{d_q} t_q)$. Similarly, we define $\delta\partial_t = (\delta^{d_1} \partial_{t_1}, \ldots, \delta^{d_q} \partial_{t_q})$. If $t = (t_1, \ldots, t_r) \in [\mathbb{R}^q]^r$, for $\delta \in [0, \infty)^\nu$ we define $\delta t = (\delta t_1, \ldots, \delta t_r)$ where each δt_j is defined in terms of the above dilations on $\mathbb{R}^q$. We define $\delta\partial_t$, for $t \in [\mathbb{R}^q]^r$, in a similar way.*

PROPOSITION 3.2.10. *Fix $r \in \mathbb{N}$. There exists a 2-admissible constant $\eta_3 = \eta_3(r) > 0$ ($\eta_3 < \eta$) such that for $(t, x) = (t_1, \ldots, t_r, x) \in [B^q(\eta_3)]^r \times \Omega$, for a multi-index σ with $|\sigma| = 1$, and for $j \in [0, \infty)^\nu$,*

$$(2^{-j}\partial_t)^\sigma f(e^{t_1 \cdot X} e^{t_2 \cdot X} \cdots e^{t_r \cdot X} x) = \sum_{l=1}^{q} c_{l,\sigma}^j(t, x)(2^{-j \cdot d_l} X_l f)(e^{t_1 \cdot X} \cdots e^{t_r \cdot X} x),$$

where for every M,[6]

$$\sum_{|\alpha|,|\beta|\leq M} \left| \left(2^{-j}\partial_t\right)^{\beta} \left(2^{-j}X\right)^{\alpha} c_{l,\sigma}^{j}\left(t,x\right) \left(1+\left|2^{j}t\right|^2\right)^{-\frac{1}{2}} \right| \lesssim 1.$$

We prove Proposition 3.2.10 in the case $r = 1$; the proofs in other cases are similar, only with more complicated notation. For the proof, we need several lemmas.

LEMMA 3.2.11. *There exists a 2-admissible constant $\eta_2 > 0$ ($\eta_2 < \eta$, where η is as in Theorem 3.1.2) such that for every $x_0 \in \Omega$ and $|t| < \eta_2$, we have for $f \in C^{\infty}$, $j \in [0,\infty)^{\nu}$,*

$$\partial_{t_l} f\left(e^{t\cdot 2^{-j}X}x\right) = \sum_{r=1}^{q} c_{r,l}^{j,x_0}\left(t,x\right) \left(2^{-j\cdot d_r}X_r f\right) \left(e^{t\cdot 2^{-j}X}x\right),$$

on $B^q\left(\eta_2\right) \times B_{(X,d)}\left(x_0, \frac{\xi_3}{2}2^{-j}\right)$, where for every m,

$$\sum_{|\alpha|,|\beta|\leq m} \left| \left(2^{-j}X\right)^{\alpha} \partial_t^{\beta} c_{r,l}^{j,x_0} \right| \lesssim 1.$$

PROOF. We use the map $\Phi = \Phi_{x_0,2^{-j}}$ from Theorem 3.1.2. We denote by Y_l the pullback of $2^{-j\cdot d_l}X_l$ to $B^n\left(\eta\right)$, via Φ. The vector fields $Y_1,\ldots,Y_q$ span the tangent space (uniformly in x_0, j) and we therefore have

$$\partial_{t_l} g\left(e^{t\cdot Y}u\right) = \sum_{r=1}^{q} \tilde{c}_{r,l}^{j,x_0}\left(t,y\right) \left(Y_r g\right) \left(e^{t\cdot Y}u\right),$$

where $\tilde{c}_{r,l}^{j,x_0} \in C^{\infty}\left(B^q\left(\eta_2\right) \times B_{(Y,\tilde{d})}\left(0,\frac{\xi_3}{2}\right)\right)$, provided we take η_2 sufficiently small depending on the C^1 norms of the Y_j. The above is true uniformly in any relevant parameters. Thus, we have

$$\sum_{|\alpha|,|\beta|\leq m} \left| \left(2^{-j}Y\right)^{\alpha} \partial_t^{\beta} \tilde{c}_{r,l}^{j,x_0} \right| \lesssim 1, \quad \forall m,$$

where $2^{-j}Y$ denotes the list of vector fields $2^{-j\cdot d_1}Y_1,\ldots,2^{-j\cdot d_q}Y_q$. Setting

$$c_{r,l}^{j,x_0}\left(t,x\right) := \tilde{c}_{r,l}^{j,x_0}\left(t,\Phi^{-1}\left(x\right)\right)$$

completes the proof. $\square$

---

[6]The factor of $\left(1+\left|2^{j}t\right|\right)^{-\frac{1}{2}}$ is possibly an artifact of the proof. However, for our purposes, its presence is irrelevant.

LEMMA 3.2.12. *Fix* $j, k \in [0, \infty)^\nu$ *with* $k \leq j$ *and* $x_0 \in \Omega$. *Let* η_2 *be as in Lemma 3.2.11 and set* $\eta_3 = 2^{-\max_l |d_l|} \eta_2 / \nu$. *Let* t *be such that* $|2^{k-j}t| < \eta_2$ *and for any* μ *such that* $k_\mu < j_\mu$ *we assume* $\eta_3 \leq \left| (2^{k-j}t)_\mu \right|$. *We have*

$$\partial_{t_p} f \left(e^{t \cdot 2^{-j} X} x \right) = \sum_{l=1}^q c_{l,p}^{j,k,x_0} (t, x) \left(2^{-j \cdot d_l} X_l f \right) \left(e^{t \cdot 2^{-j} X} x \right),$$

for $x \in B_{(X,d)} \left(x_0, 2^{-k} \frac{\xi_3}{2} \right)$, *where for all* m,

$$\sum_{|\alpha|, |\beta| \leq m} \left| \partial_t^\beta \left(2^{-j} X \right)^\alpha \left(1 + |t|^2 \right)^{-\frac{1}{2}} c_{l,p}^{j,k,x_0} \right| \lesssim 1.$$

PROOF. Applying Lemma 3.2.11 with j replaced by k, we have

$$\partial_{t_p} f \left(e^{t \cdot 2^{-j} X} x \right) = \partial_{t_p} f \left(e^{2^{k-j} t \cdot 2^{-k} X} x \right)$$

$$= 2^{(k-j) \cdot d_p} \sum_{l=1}^q c_{l,p}^{k,x_0} \left(2^{k-j} t, x \right) \left(2^{-k \cdot d_l} X_l f \right) \left(e^{t \cdot 2^{-j} X} x \right)$$

$$= 2^{(k-j) \cdot d_p} \sum_{l=1}^q \left(2^{(j-k) \cdot d_l} c_{l,p}^{k,x_0} \left(2^{k-j} t, x \right) \right) \left(2^{j \cdot d_l} X_l f \right) \left(e^{t \cdot 2^{-j} X} x \right),$$

where $c_{l,p}^{k,x_0}$ *are the functions given in Lemma 3.2.11. By our choice of* t, *we have* $2^{(j-k) \cdot d_l} \left(1 + |t|^2 \right)^{-\frac{1}{2}} \lesssim 1, \forall l$. *The result now follows by taking* $c_{l,p}^{j,k,x_0} (t, x) = 2^{(k-j) \cdot d_p} \left(2^{(j-k) \cdot d_l} c_{l,p}^{k,x_0} \left(2^{k-j} t, x \right) \right)$; *the desired estimates for* $c_{l,p}^{j,k,x_0}$ *follow from those for* $c_{l,p}^{k,x_0}$ *given in Lemma 3.2.11. This uses that* $k \leq j$, *and we have used* $\left(1 + |t|^2 \right)^{-\frac{1}{2}}$ *to cancel out the factor of* $2^{(j-k) \cdot d_l}$. $\square$

LEMMA 3.2.13. *Let* $j \in \mathbb{N}^\nu$. *On* $\left(2^j B^q (\eta_3) \right) \times B_{(X,d)} \left(x_0, 2^{-j} \frac{\xi_3}{2} \right)$ *(where* $2^j B^q (\eta_3) = \left\{ 2^j t \mid t \in B^q (\eta_3) \right\}$ *and* $2^j t$ *is as in Definition 3.2.9) we have*

$$\partial_{t_p} f \left(e^{t \cdot 2^{-j} X} x \right) = \sum_{l=1}^q c_{l,p}^{j,x_0} (t, x) \left(2^{-j \cdot d_l} X_l f \right) \left(e^{t \cdot 2^{-j} X} x \right),$$

where for every m,

$$\sum_{|\alpha|, |\beta| \leq m} \left| \partial_t^\beta \left(2^{-j} X \right)^\alpha c_{l,p}^{j,x_0} (t, x) \left(1 + |t|^2 \right)^{-\frac{1}{2}} \right| \lesssim 1.$$

PROOF. Let $\kappa_0(t) \in C_0^\infty (B^q(\eta_2))$ be equal to 1 on $B^q(\nu\eta_3)$. For $k \in \mathbb{N}^\nu$ with $k \le j$, define

$$\kappa_k(t) = \sum_{\substack{p \in \{0,1\}^\nu \\ k+p \le j}} (-1)^{\sum_\mu p_\mu} \kappa_0 \left(2^{k+p} t \right),$$

so that

$$\sum_{\substack{k \le j \\ k \in \mathbb{N}^\nu}} \kappa_k \left(2^{k-j} t \right) = \kappa_0 \left(2^{-j} t \right).$$

Note that if $\kappa_k \left(2^{k-j} t \right)$ is nonzero, then $\left| 2^{k-j} t \right| < \eta_2$, and if in addition $k_\mu < j_\mu$ then $\eta_3 \le \left| \left(2^{k-j} t \right)_\mu \right|$. We define

$$c_{l,p}^{j,x_0}(t,x) := \sum_{\substack{k \le j \\ k \in \mathbb{N}^\nu}} \kappa_k \left(2^{k-j} t \right) c_{l,p}^{j,k,x_0}(t,x),$$

where $c_{l,p}^{j,k,x_0}$ is as in Lemma 3.2.12. The desired estimates follow from the corresponding estimates in Lemma 3.2.12. $\square$

PROOF OF PROPOSITION 3.2.10. In the case $r = 1$, let ψ_{x_0}, $x_0 \in \mathcal{C}$ be the partition of unity from Theorem 3.2.7 with $\delta = 2^{-j}$, and let $c_{l,p}^{j,x_0}$ be as in Lemma 3.2.13. Defining $c_{l,p}^{j}(t,x) = \sum_{x_0 \in \mathcal{C}} \psi_{x_0}(x) c_{l,p}^{j,x_0} \left(2^j t, x \right)$, the result follows. For higher r, the proof is the same, by proving appropriate analogs of the above lemmas. There are no additional difficulties, and only the notation is more complicated. We leave the details to the reader. $\square$

LEMMA 3.2.14. *For $j \in [0,\infty)^\nu$,*

$$\sup_{x \in \Omega} \int_{z \in \Omega} \frac{(1 + \rho_{2^{-j}}(x,z))^{-1}}{\mathrm{Vol}\left(B_{2^{-j}}(x, 1 + \rho_{2^{-j}}(x,z)) \right)} \, dz \lesssim_2 1,$$

and

$$\sup_{z \in \Omega} \int_{x \in \Omega} \frac{(1 + \rho_{2^{-j}}(x,z))^{-1}}{\mathrm{Vol}\left(B_{2^{-j}}(x, 1 + \rho_{2^{-j}}(x,z)) \right)} \, dx \lesssim_2 1.$$

PROOF. In light of Proposition 3.2.2, the above two bounds are equivalent, and we prove only the first. We use Proposition 3.2.1 freely in the following. For $x \in \Omega$,

$$\int_{z \in \Omega} \frac{(1 + \rho_{2-j}(x, z))^{-1}}{\mathrm{Vol}(B_{2-j}(x, 1 + \rho_{2-j}(x, z)))} \, dz$$

$$\lesssim_2 \int_{\rho_{2-j}(x,z) \leq 1, z \in \Omega} \mathrm{Vol}(B_{2-j}(x, 1))^{-1} \, dz$$

$$+ \sum_{l \geq 0} \int_{2^l \leq \rho_{2-j}(x,z) \leq 2^{l+1}, z \in \Omega} 2^{-l} \mathrm{Vol}\left(B_{2-j}\left(x, 2^{-l}\right)\right)^{-1} \, dz$$

$$\lesssim_2 1 + \sum_{l \geq 0} 2^{-l}$$

$$\lesssim_2 1,$$

completing the proof. □

We now add an additional assumption, which is needed for the rest of the results in this section.

Additional Assumption: In addition to the above assumptions on (X, d), we assume each $0 \neq d_j \in \mathbb{N}^\nu$ is nonzero in only one component.

LEMMA 3.2.15. *Fix $c \in (0, 1)$ and $\zeta \in (c\xi_0, \xi_0]$. Let $j \in \mathbb{N}^\nu$, $x, z \in \Omega$, and suppose $\rho_{2-j}(x, z) = 2^m \zeta$ for some $m \in [0, \infty)$. Set*

$$k_\mu = \begin{cases} j_\mu - m & \text{if } j_\mu \geq m, \\ 0 & \text{otherwise.} \end{cases}$$

Then, $\rho_{2-k}(x, z) \approx_2 1$, where the implicit constants depend on c and ξ_0. In short, if $\rho_{2-j}(x, z) \approx 2^m$, then $\rho_{2-k}(x, z) \approx 1$.

PROOF. Suppose, for contradiction, that $\rho_{2-k}(x, z) = \delta < \zeta$. Then, for every $\epsilon > 0$, there exists $\gamma : [0, 1] \to \Omega$, $\gamma(0) = x$, $\gamma(1) = z$, $\gamma'(t) = \sum_{l=1}^{q} a_l(t) \left(\delta 2^{-k}\right)^{d_l} X_l(\gamma(t))$, with $\sum_{l=1}^{q} |a_l(t)| < 1 + \epsilon$ and $\sum_{l=1}^{q} \left|\left(\delta 2^{-k}\right)^{d_l} a_l(t)\right|^2 < \xi_0^2$. Each d_l is nonzero in only one component; let e_l be the vector which equals 1 in the component in which d_l is nonzero, and which equals 0 in all other components. Then,

$$\gamma'(t) = \sum_{l=1}^{q} a_l(t) \left(\delta 2^{(j-k) \cdot e_l} 2^{-j}\right)^{d_l} X_l(\gamma(t)).$$

Since $|j - k|_\infty \leq m$, this shows that $\rho_{2-j}(x, z) \leq 2^m \delta < 2^m \zeta$, a contradiction. We conclude $\rho_{2-k}(x, z) \gtrsim_2 1$.

To complete the proof, we show $\rho_{2-k}(x, z) \lesssim_2 1$. Since $\rho_{2-j}(x, z) = 2^m \zeta$, for every $\epsilon > 0$, there exists $\gamma : [0, 1] \to \Omega$ with

- $\gamma(0) = x$, $\gamma(1) = z$,

- $\gamma'(t) = \sum_{l=1}^{q} a_l(t) \left(2^m \zeta 2^{-j}\right)^{d_l} X_l(\gamma(t))$,

- $\sum_{l=1}^{q} |a_l(t)|^2 < 1 + \epsilon$,

- $\sum_{l=1}^{q} \left|\left(2^m \zeta 2^{-j}\right)^{d_l} a_l(t)\right|^2 < \xi_0^2$.

Let $E \subseteq \{1, \ldots, q\}$ be the set of those $l \in \{1, \ldots, q\}$ such that d_l is nonzero in the μth component and $k_\mu = 0$. For $l \in E$, define $b_l(t) = \left(2^m \zeta 2^{-j}\right)^{d_l} a_l(t)$–our assumptions imply $|b_l| < \xi_0$. For $l \notin E$, $j_\mu - k_\mu = m$, and so we have

$$\gamma'(t) = \sum_{l \in E} b_l(t) X_l(\gamma(t)) + \sum_{l \notin E} a_l(t) \left(\zeta 2^{-k}\right)^{d_l} X_l(\gamma(t))$$

$$= \sum_{l \in E} b_l(t) \left(2^{-k}\right)^{d_l} X_l(\gamma(t)) + \sum_{l \notin E} a_l(t) \left(\zeta 2^{-k}\right)^{d_l} X_l(\gamma(t)).$$

We conclude $\rho_{2^{-k}}(x, z) \lesssim_2 1$. $\square$

LEMMA 3.2.16. *Fix ζ with $\xi_0 \le \zeta \le \xi_3$ and $j \in \mathbb{N}^\nu$. $\forall m$, there exists a 2-admissible constant $M = M(m)$ such that for $x, z \in \Omega$,*

$$\sum_{\substack{k \le j \\ k \in \mathbb{N}^\nu}} 2^{-M|k-j|} \frac{\chi_{\{\rho_{2^{-k}}(x,z) < \zeta\}}}{\mathrm{Vol}\left(B_{2^{-k}}(x,1)\right)} \lesssim_2 \frac{\left(1 + \rho_{2^{-j}}(x,z)\right)^{-m}}{\mathrm{Vol}\left(B_{2^{-j}}(x, 1 + \rho_{2^{-j}}(x,z))\right)}. \tag{3.6}$$

Conversely, $\forall M$, there exists a 2-admissible constant $m = m(M)$ such that for all $x, z \in \Omega$,

$$\sum_{\substack{k \le j \\ k \in \mathbb{N}^\nu}} 2^{-M|k-j|} \frac{\chi_{\{\rho_{2^{-k}}(x,z) < \zeta\}}}{\mathrm{Vol}\left(B_{2^{-k}}(x,1)\right)} \gtrsim_2 \frac{\left(1 + \rho_{2^{-j}}(x,z)\right)^{-m}}{\mathrm{Vol}\left(B_{2^{-j}}(x, 1 + \rho_{2^{-j}}(x,z))\right)}. \tag{3.7}$$

PROOF. To show (3.6) consider,

$$\sum_{\substack{k \le j \\ k \in \mathbb{N}^\nu}} 2^{-M|k-j|} \frac{\chi_{\{\rho_{2^{-k}}(x,z) < \zeta\}}}{\mathrm{Vol}\left(B_{2^{-k}}(x,1)\right)} \lesssim_2 \sum_{\substack{k \le j \\ k \in \mathbb{N}^\nu}} 2^{-M|k-j|} \frac{\left(1 + \rho_{2^{-k}}(x,z)\right)^{-m}}{\mathrm{Vol}\left(B_{2^{-k}}(x, 1 + \rho_{2^{-k}}(x,z))\right)}$$

$$\lesssim_2 \frac{\left(1 + \rho_{2^{-j}}(x,z)\right)^{-m}}{\mathrm{Vol}\left(B_{2^{-j}}(x, 1 + \rho_{2^{-j}}(x,z))\right)},$$

where in the last line we have taken $M = M(m)$ large and applied Proposition 3.2.6.

To prove (3.7), fix $x, z \in \Omega$. If $\rho_{2^{-j}}(x, z) < \zeta$, the result is obvious. Otherwise, $\rho_{2^{-j}}(x, z) = 2^l \xi_0$ for some $l \in [0, \infty)$. Define k as in Lemma 3.2.15 so that $\rho_{2^{-k}}(x, z) \approx_2 1$. Notice that $|k - j| \approx_2 l$. Indeed, $|k - j| \lesssim l$ by the definition of k. If $|k - j| \ll l$, then we would have $\rho_{2^{-j}}(x, z) \ll 2^l$, contradicting our assumption. By decreasing the coordinates of k by a fixed 2-admissible constant, we may ensure

$\rho_{2^{-k}}(x, z) < \xi_0$, while still maintaining $|j - k| \approx_2 l$. We then have, using Proposition 3.2.1 and Lemma 3.2.4, for all m,

$$\frac{(1 + \rho_{2^{-j}}(x, z))^{-m}}{\text{Vol}(B_{2^{-j}}(x, 1 + \rho_{2^{-j}}(x, z)))} \approx_2 \frac{(2^l \rho_{2^{-k}}(x, z))^{-m}}{\text{Vol}(B_{2^{-j}}(x, 1 + \rho_{2^{-j}}(x, z)))}$$

$$\lesssim_2 2^{-(cm-C)|j-k|} \frac{\chi\{\rho_{2^{-k}}(x,z)<\varsigma\}}{\text{Vol}(B_{2^{-k}}(x, 1 + \rho_{2^{-k}}(x, z)))}$$

$$\approx_2 2^{-(cm-C)|j-k|} \frac{\chi\{\rho_{2^{-k}}(x,z)<\varsigma\}}{\text{Vol}(B_{2^{-k}}(x, 1))},$$

for some c, C, where we have used the particular chosen x, z and that k was chosen in terms of x, z. The result follows by taking m large.   □

LEMMA 3.2.17. *If we take $\xi_0 > 0$ to be a sufficiently small 2-admissible constant, then there exists a 2-admissible constant $\eta_1 > 0$, such that the following holds. For $j \in [0, \infty)^\nu$ and $x \in \Omega$, define*

$$U_{j,x} = \left\{t \in B^q(\eta_1) \mid e^{t \cdot 2^{-j} X} x \in B_{2^{-j}}(x, \xi_0)\right\}$$

and define the operator

$$B_j f(x) = \int_{U_{j,x}} f\left(e^{t \cdot 2^{-j} X} x\right) dt,$$

then we have, for $x, z \in U_1$,

$$\frac{\chi\{\rho_{2^{-j}}(x,z)<\xi_0\}}{\text{Vol}(B_{2^{-j}}(x, \xi_0))} \lesssim_2 B_j(x, z), \tag{3.8}$$

and for $j, k \in [0, \infty)^\nu$ and $x, z \in U_1$,

$$[B_j B_k](x, z) \lesssim_2 \frac{\chi\{\rho_{2^{-j \wedge k}}(x,z)<2\xi_0\}}{\text{Vol}(B_{2^{-j \wedge k}}(x, 2\xi_0))}. \tag{3.9}$$

PROOF. Let $Y_l = Y_l^{x_0, 2^{-j}}$ be the pullback of $2^{-j \cdot d_l} X_l$ via the map $\Phi_{x_0, 2^{-j}}$ to $B^n(\eta)$ as in Theorem 3.1.2. The C^2 norms of the Y_l are bounded by a 2-admissible constant, and therefore the same is true for the C^2 norm of the map $t \mapsto e^{t \cdot Y} 0$ (for $t \in \mathbb{R}^q$ with $|t|$ sufficiently small–see Appendix B.1). If η_1 is a sufficiently small 2-admissible constant, for $|t| < \eta_1$ we have $e^{t \cdot Y} 0 \in B^n(\eta/2)$. Furthermore, since $\partial_{t_l}\big|_{t=0} e^{t \cdot Y} 0 = Y_l(0)$, we have $\left|\det_{n \times n} d_t e^{t \cdot Y} 0\big|_{t=0}\right| \approx_2 1$, and by making η_1 a possibly smaller 2-admissible constant, we have using that the C^2 norm of $e^{t \cdot Y} 0$ is bounded by a 2-admissible constant, for $t \in B^q(\eta_1)$, $\left|\det_{n \times n} d_t e^{t \cdot Y} 0\right| \approx_2 1$.

Now let $j, k \in [0, \infty)^\nu$, and we use the map $\Phi = \Phi_{x_0, 2^{-j \wedge k}}$ from Theorem 3.1.2. Letting $V_1, \ldots, V_q$ be the pullbacks of $2^{-j \wedge k \cdot d_1} X_1, \ldots, 2^{-j \wedge k \cdot d_q} X_q$ to $B^n(\eta)$, we have that $|\det_{n \times n}(V_1(0)|\cdots|V_q(0))| \approx_2 1$. Consider the matrix

$$\left(2^{(j \wedge k - j) \cdot d_1} V_1(0)|\cdots|2^{(j \wedge k - j) \cdot d_q} V_q(0)|2^{(j \wedge k - k) \cdot d_1} V_1(0)|\cdots|2^{(j \wedge k - k) \cdot d_q} V_q(0)\right),$$

call this matrix $\mathcal{V}$. We have

$$\left|\det_{n \times n} \mathcal{V}\right| \approx_2 1,$$

since each d_l is nonzero in only one component and therefore either $(j \wedge k - j) \cdot d_l = 0$ or $(j \wedge k - k) \cdot d_l = 0$. Consider the map $\Psi(t_1, t_2)$, for $t_1 = \left(t_1^1, \dots, t_1^q\right)$, $t_2 = \left(t_2^1, \dots, t_2^q\right) \in B^q(\eta_1)$ given by

$$
\begin{aligned}
&\Psi(t_1, t_2) \\
&= e^{t_1^1 2^{(j \wedge k - k) \cdot d_1} V_1 + \dots + t_1^q 2^{(j \wedge k - k) \cdot d_q} V_q} e^{t_2^1 2^{(j \wedge k - j) \cdot d_1} V_1 + \dots + t_2^q 2^{(j \wedge k - j) \cdot d_q} V_q} 0.
\end{aligned}
\tag{3.10}
$$

Since $\partial_{t_1^l} \Psi(0,0) = 2^{(j \wedge k - k) \cdot d_l} V_l(0)$ and $\partial_{t_2^l} \Psi(0,0) = 2^{(j \wedge k - j) \cdot d_l} V_l(0)$, we see

$$\left|\det_{n \times n} d\Psi(0,0)\right| \approx_2 1.$$

Since the C^2 norm of Ψ is bounded by a 2-admissible constant, we see that if we make η_1 sufficiently small (while still keeping it 2-admissible),[7] we have

$$\left|\det_{n \times n} d\Psi(t_1, t_2)\right| \approx_2 1, \quad \forall t_1, t_2 \in B^q(\eta_1).$$

We take ξ_0 to be a 2-admissible constant so small that $\Phi^{-1}_{x_0, 2^{-j}}(B_{2-j}(x, \xi_0))$ is contained in the image of $e^{t \cdot Y} 0$ as t ranges over $B^q(\eta_1)$. Pushing this forward via $\Phi_{x_0, 2^{-j}}$, we have that the map $t \mapsto e^{t \cdot 2^{-j} X} x$, $U_{j,x} \to B_{2-j}(x, \xi_0)$ is surjective. We also take ξ_0 to be small enough (while keeping it 2-admissible) that $\Phi^{-1}_{x_0, 2^{-j \wedge k}}(B_{2-j \wedge k}(x, \xi_0))$ is contained in the image of $\Psi(t_1, t_2)$ as t_1 and t_2 range over $B^q(\eta_1)$.

Fix $x_0 \in U_1$ and $j \in [0, \infty)^\nu$ and let $Y_l = Y_l^{x_0, 2^{-j}}$ be the pullbacks of $2^{-j \cdot d_l} X_l$ via $\Phi_{x_0, 2^{-j}}$. Consider the integral

$$I_{x_0, j} f = \int_{U_{j,x_0}} f\left(e^{t \cdot Y} 0\right) dt.$$

Since $\left|\det_{n \times n} d_t e^{t \cdot Y} 0\right| \approx_2 1$, we have by changing variables in the t variable,

$$I_{x_0, j} f = \int_{\Phi^{-1}(B_{2-j}(x_0, \xi_0))} f(u) b_{x_0, j}(u) \, du,$$

where $b_{x_0, j}(u) \approx_2 1$; see Appendix B.3 for more details on this change of variables. Consider,

$$B_j\left(g \circ \Phi^{-1}\right)(x_0) = \int_{U_{j,x_0}} g\left(e^{t \cdot Y} 0\right) dt = \int_{\Phi^{-1}(B_{2-j}(x_0, \xi_0))} g(u) b_{x_0, j}(u) \, du.$$

---

[7]Throughout the argument, we may shrink $\eta_1 > 0$, so long as it remains a 2-admissible constant.

Applying a change of variables, $v = \Phi_{x_0, 2^{-j}}(u)$, we have

$$B_j f(x_0) = \int_{B_{2^{-j}}(x_0, \xi_0)} f(v) \, b_{x_0, j}\left(\Phi_{x_0, 2^{-j}}^{-1}(v)\right) \left|\det d\Phi_{x_0, 2^{-j}}^{-1}(v)\right| dv.$$

Since $b_{j, x_0} \approx_2 1$, and since $\left|\det_{n \times n} d\Phi_{x_0, 2^{-j}}^{-1}(v)\right| \approx_2 \mathrm{Vol}\left(B_{(X, d)}\left(x_0, \xi_2 2^{-j}\right)\right)^{-1} \approx_2$
$\mathrm{Vol}\left(B_{2^{-j}}(x, \xi_0)\right)^{-1}$, (3.8) follows.

We now turn to (3.9). Since $B_j(x, y)$ is supported on $\rho_{2^{-j}}(x, y) < \xi_0$ and $B_k(y, z)$ is supported on $\rho_{2^{-k}}(y, z) < \xi_0$ it follows by the triangle inequality that $[B_j B_k](x, z)$ is supported on $\rho_{2^{-j \wedge k}}(x, z) < 2\xi_0$. To complete the proof, we need only show

$$[B_j B_k](x, z) \lesssim_2 \frac{1}{\mathrm{Vol}\left(B_{2^{-j \wedge k}}(x, 2\xi_0)\right)}. \tag{3.11}$$

Fix $x_0 \in U_1$, and $j, k \in [0, \infty)^\nu$. Let $\Phi = \Phi_{x_0, 2^{-j \wedge k}}$ be the map from Theorem 3.1.2. Define the integral

$$J_{x_0, 2^{-j \wedge k}} g = \int_{B^q(\eta_1) \times B^q(\eta_1)} g(\Psi(t_1, t_2)) \, dt_1 dt_2,$$

where Ψ is as in (3.10). By the above discussion of Ψ, we have by a change of variables $u = \Psi(t_1, t_2)$ as in Appendix B.3,

$$J_{x_0, 2^{-j \wedge k}} g = \int g(u) \, b_{j, k, x_0}(u) \, du,$$

where $b_{j, k, x_0}(u) \lesssim_2 1$, and $b_{j, k, x_0}(u)$ is supported on $B^n(\eta)$.

Consider, for $g \geq 0$,

$$B_j B_k \left(g \circ \Phi^{-1}\right)(x_0) \leq \int_{B^q(\eta_1) \times B^q(\eta_1)} \int g \circ \Phi^{-1}\left(e^{t_1 \cdot 2^{-k} X} e^{t_2 \cdot 2^{-j} X} x_0\right) dt_1 \, dt_2$$

$$= \int_{B^q(\eta_1) \times B^q(\eta_1)} g(\Psi(t_1, t_2)) \, dt_1 \, dt_2$$

$$= \int_{B^n(\eta)} g(u) \, b_{j, k, x_0}(u) \, du.$$

Applying a change of variables $v = \Phi(u)$, we have

$$B_j B_k f(x_0) \leq \int f(v) \, b_{j, k, x_0}\left(\Phi^{-1}(v)\right) \left|\det d\Phi^{-1}(v)\right| dv.$$

Using that $\left|\det d\Phi^{-1}(v)\right| \approx_2 \mathrm{Vol}\left(B_{2^{-j \wedge k}}(x_0, 2\xi_0)\right)^{-1}$ completes the proof of (3.11) as desired. $\qquad \square$

LEMMA 3.2.18. *For $j \in [0, \infty)^\nu$ and $\zeta > 0$ define the operator A_j^ζ, to be the operator with Schwartz kernel*

$$A_j^\zeta(x, z) := \frac{\chi_{\{\rho_{2^{-j}}(x, z) < \zeta\}}}{\mathrm{Vol}\left(B_{2^{-j}}(x, \zeta)\right)}.$$

If $\xi_0 > 0$ is a sufficiently small 2-admissible constant,

$$\left[A_j^{\xi_0} A_k^{\xi_0} \right] (x, z) \lesssim_2 A_{j \wedge k}^{2\xi_0} (x, z).$$

PROOF. This follows immediately from Lemma 3.2.17.   □

PROPOSITION 3.2.19. $\forall m$, there exist 2-admissible constants $C = C(m)$, $m' = m'(m)$ such that for all $j, k \in [0, \infty)^\nu$ and all $x, z \in \Omega$,

$$\int_\Omega \frac{(1 + \rho_{2^{-j}} (x, y))^{-m'}}{\mathrm{Vol}\,(B_{2^{-j}} (x, 1 + \rho_{2^{-j}} (x, y)))} \frac{(1 + \rho_{2^{-k}} (y, z))^{-m'}}{\mathrm{Vol}\,(B_{2^{-k}} (y, 1 + \rho_{2^{-k}} (y, z)))} \, dy$$

$$\leq C \frac{(1 + \rho_{2^{-j \wedge k}} (x, z))^{-m}}{\mathrm{Vol}\,(B_{2^{-j \wedge k}} (x, 1 + \rho_{2^{-j \wedge k}} (x, z)))}.$$

PROOF. First note that it suffices to prove the result for $j, k \in \mathbb{N}^\nu$, as the general result follows by replacing j, k with the closest elements of $\mathbb{N}^\nu$. Let A_j^ς be as in Lemma 3.2.18. Fix m and take $M = M(m)$ to be a 2-admissible constant as in the first part of Lemma 3.2.16, and take $m' = m'(M)$ to be a 2-admissible constant as in the second part of Lemma 3.2.16. We have,

$$\frac{(1 + \rho_{2^{-j}} (x, z))^{-m}}{\mathrm{Vol}\,(B_{2^{-j}} (x, 1 + \rho_{2^{-j}} (x, z)))} \lesssim_2 \sum_{\substack{l \leq j \\ l \in \mathbb{N}^\nu}} 2^{-M|l - j|} A_l^{\xi_0} (x, z),$$

with a similar result with j replaced by k. Applying Lemma 3.2.18, we have

$$\int_\Omega \frac{(1 + \rho_{2^{-j}} (x, y))^{-m'}}{\mathrm{Vol}\,(B_{2^{-j}} (x, 1 + \rho_{2^{-j}} (x, y)))} \frac{(1 + \rho_{2^{-k}} (y, z))^{-m'}}{\mathrm{Vol}\,(B_{2^{-k}} (y, 1 + \rho_{2^{-k}} (y, z)))} \, dy$$

$$\lesssim_2 \sum_{\substack{l_1 \leq j, l_2 \leq k \\ l_1, l_2 \in \mathbb{N}^\nu}} 2^{-M|l_1 - j| - M|l_2 - k|} \left[A_{l_1}^{\xi_0} A_{l_2}^{\xi_0} \right] (x, z)$$

$$\lesssim_2 \sum_{\substack{l_1 \leq j, l_2 \leq k \\ l_1, l_2 \in \mathbb{N}^\nu}} 2^{-M|l_1 - j| - M|l_2 - k|} A_{l_1 \wedge l_2}^{2\xi_0} (x, z)$$

$$\lesssim_2 \sum_{\substack{l \leq j \wedge k \\ l \in \mathbb{N}^\nu}} 2^{-M|j \wedge k - l|} A_l^{2\xi_0} (x, z)$$

$$\lesssim_2 \frac{(1 + \rho_{2^{-j \wedge k}} (x, z))^{-m}}{\mathrm{Vol}\,(B_{2^{-j \wedge k}} (x, 1 + \rho_{2^{-j \wedge k}} (x, z)))},$$

where in the first and last inequality, we used Lemma 3.2.16.   □

LEMMA 3.2.20. Let $j, k \in [0, \infty)^\nu$, and fix $1 \leq l \leq q$. Then,

$$2^{-k \cdot d_l} X_l \left(2^{-j} X \right)^\alpha = \sum_{\substack{|\gamma| \leq 1 \\ |\beta| \leq |\alpha|}} \left(2^{-j} X \right)^\beta c_{\beta, \gamma}^l \left(2^{-k} X \right)^\gamma,$$

on Ω, *where for every* m,

$$\sum_{|\sigma| \le m} \left| \left(2^{-j \wedge k} X \right)^{\sigma} c_{\beta, \gamma}^{l} \right| \lesssim 1.$$

Remark 3.2.21 In Lemma 3.2.20, we have used admissible constants, and have not been explicit about for what m' these are m'-admissible constants. This can be kept track of, but it slightly complicates the proof, and we will not use this additional information. We remark that this m' does depend on m and α.

PROOF. We proceed by induction on $|\alpha|$. The base case, $|\alpha| = 0$, is trivial. For $|\alpha| > 0$, we may write $\left(2^{-j} X \right)^{\alpha} = 2^{-j \cdot d_p} X_p \left(2^{-j} X \right)^{\alpha_0}$, where $|\alpha_0| + 1 = |\alpha|$. We have

$$2^{-k \cdot d_l} X_l 2^{-j \cdot d_p} X_p \left(2^{-j} X \right)^{\alpha_0}$$
$$= \left[2^{-k \cdot d_l} X_l, 2^{-j \cdot d_p} X_p \right] \left(2^{-j} X \right)^{\alpha_0} + 2^{-j \cdot d_p} X_p 2^{-k \cdot d_l} X_l \left(2^{-j} X \right)^{\alpha_0}.$$

The second term on the right-hand side is of the desired form by our inductive hypothesis. We therefore consider only the first term.
Define

$$\gamma = 2^{(j \wedge k - j) \cdot d_p + (j \wedge k - k) \cdot d_l},$$

so that $0 < \gamma \le 1$. We have,

$$\left[2^{-k \cdot d_l} X_l, 2^{-j \cdot d_p} X_p \right] \left(2^{-j} X \right)^{\alpha_0} = \gamma \left[2^{-j \wedge k \cdot d_l} X_l, 2^{-(j \wedge k) \cdot d_p} X_p \right] \left(2^{-j} X \right)^{\alpha_0}$$
$$= \gamma \left[\sum_{r=1}^{q} c_{l,p}^{r, 2^{-j \wedge k}} 2^{-j \wedge k \cdot d_r} X_r \right] \left(2^{-j} X \right)^{\alpha_0},$$

$$(3.12)$$

where $c_{l,p}^{r, 2^{-j \wedge k}}$ is as in Lemma 3.2.8. Using that each d_r is nonzero in only one component, we have $j \wedge k \cdot d_r = j \cdot d_r$ or $j \wedge k \cdot d_r = k \cdot d_r$; and there are therefore two corresponding types of terms on the right-hand side of (3.12). We treat the two possibilities separately.
If $j \wedge k \cdot d_r = j \cdot d_r$, we have

$$c_{l,p}^{r, 2^{-j \wedge k}} 2^{-j \cdot d_r} X_r \left(2^{-j} X \right)^{\alpha_0} = c_{l,p}^{r, 2^{-j \wedge k}} \left(2^{-j} X \right)^{\alpha'},$$

where $|\alpha'| = |\alpha|$. Using the differential inequalities satisfied by $c_{l,p}^{r, 2^{-j \wedge k}}$ as shown in Lemma 3.2.8, it follows that this is of the desired form.
When $j \wedge k \cdot d_r = k \cdot d_r$, we apply the inductive hypothesis to see

$$c_{l,p}^{r, 2^{-j \wedge k}} 2^{-k \cdot d_r} X_r \left(2^{-j} X \right)^{\alpha_0} = c_{l,p}^{r, 2^{-j \wedge k}} \sum_{\substack{|\gamma| \le 1 \\ |\beta| \le |\alpha| - 1}} \left(2^{-j} X \right)^{\beta} c_{\beta, \gamma}^{r} \left(2^{-k} X \right)^{\gamma},$$

where $c_{\beta, \gamma}^{r}$ satisfies the inequalities of the lemma. Using the inequalities satisfied by $c_{l,p}^{r, 2^{-j \wedge k}}$ from Lemma 3.2.8, the result easily follows. $\square$

PROPOSITION 3.2.22. *Let α be an ordered multi-index, and $j, k \in [0, \infty)^\nu$. Then,*

$$\left(2^{-j \wedge k} X\right)^\alpha = \sum_{|\beta| + |\gamma| \leq |\alpha|} \left(2^{-j} X\right)^\beta c_{\beta,\gamma}^{\alpha,j,k} \left(2^{-k} X\right)^\gamma,$$

where $c_{\beta,\gamma}^{\alpha,j,k} \in C^\infty(\Omega)$ and satisfies for every m,

$$\sum_{|\sigma| \leq m} \left\| \left(2^{-j \wedge k} X\right)^\sigma c_{\beta,\gamma}^{\alpha,j,k} \right\|_{C^0(\Omega)} \lesssim 1.$$

PROOF. We proceed by induction on $|\alpha|$. The base case, $|\alpha| = 1$, follows immediately from the fact that each d_l is nonzero in only one component (and so $j \wedge k \cdot d_l = j \cdot d_l$ or $j \wedge k \cdot d_l = k \cdot d_l$). We assume the result for $|\alpha| < m_0$ and prove the result for $|\alpha| = m_0$. Notice,

$$\left(2^{-j \wedge k} X\right)^\alpha = \begin{cases} \left(2^{-j} X\right)^{\alpha_1} \left(2^{-j \wedge k} X\right)^{\alpha_2} & \text{or} \\ \left(2^{-k} X\right)^{\alpha_1} \left(2^{-j \wedge k} X\right)^{\alpha_2}, \end{cases}$$

where $|\alpha_1| = 1$ and $|\alpha_2| = |\alpha| - 1$. In the first case, our inductive hypothesis immediately shows that $\left(2^{-j \wedge k} X\right)^\alpha$ is of the desired form. In the second case, we apply the inductive hypothesis to see

$$\left(2^{-j \wedge k} X\right)^\alpha = \left(2^{-k} X\right)^{\alpha_1} \sum_{|\beta| + |\gamma| \leq m_0 - 1} \left(2^{-j} X\right)^\beta c_{\beta,\gamma}^{\alpha_2,j,k} \left(2^{-k} X\right)^\gamma,$$

where each $c_{\beta,\gamma}^{\alpha_2,j,k}$ satisfies the estimates in the statement of the lemma. Applying Lemma 3.2.20 to each term of the form $\left(2^{-k} X\right)^{\alpha_1} \left(2^{-j} X\right)^\beta$, the result follows easily. $\square$

3.3   THE MAXIMAL FUNCTION

We return to the setting of Section 3.1, and we do *not* assume each d_j is nonzero in only one component. We also do not assume that $X_1, \ldots, X_q$ span the tangent space (but we do assume all the assumptions from Section 3.1). Let $\psi \in C_0^\infty(U_1)$ be nonnegative.

THEOREM 3.3.1. *There is a constant $\xi_6 > 0$ such that if we define the maximal operator*

$$\mathcal{M}f(x) = \psi(x) \sup_{\substack{\delta \in (0,1]^\nu \\ |\delta| < \xi_6}} \frac{1}{\mathrm{Vol}\left(B_{(X,d)}(x,\delta)\right)} \int_{B_{(X,d)}(x,\delta)} |f(y)|\, \psi(y)\, dy,$$

then $\|\mathcal{M}f\|_{L^p} \leq C_p \|f\|_{L^p}$ for $1 < p \leq \infty$.

For each j, we obtain single-parameter balls $B_{(X_j,1)}(x,\delta)$; these balls are either 1-dimensional (if $X_j(x) \neq 0$) or 0-dimensional (if $X_j(x) = 0$).[8] The key to the proof of Theorem 3.3.1 is the following result, which is closely related to Lemma 3.2.17.

---

[8]For a 0-dimensional set, we define $\mathrm{Vol}\left(\{x_0\}\right) = 1$; i.e., $\mathrm{Vol}(\cdot)$ is counting measure.

PROPOSITION 3.3.2. *Let $a > 0$ be given. There exist constants $\xi_5, \xi_4, \eta_1 > 0$ ($\xi_4 \leq a$), depending on a, such that $\forall x_0 \in U_1$, $\delta \in (0,1]^\nu$, $f \geq 0$,*

$$\frac{1}{\text{Vol}\left(B_{(X,d)}(x_0,\xi_5\delta)\right)} \int_{B_{(X,d)}(x_0,\xi_5\delta)} f(y)\, dy$$

$$\lesssim \int_{|t|_\infty < \eta_1} f\left(e^{t_q\delta^{d_q}X_q} e^{t_{q-1}\delta^{d_{q-1}}X_{q-1}} \cdots e^{t_1\delta^{d_1}X_1} x_0\right) dt$$

$$\sim \frac{1}{\text{Vol}\left(B_{(X_1,1)}(x_0,\xi_4\delta^{d_1})\right)} \int_{B_{(X_1,1)}(x_0,\xi_4\delta^{d_1})} \frac{1}{\text{Vol}\left(B_{(X_2,1)}(x_1,\xi_4\delta^{d_2})\right)}$$

$$\int_{B_{(X_2,1)}(x_1,\xi_4\delta^{d_2})} \cdots \frac{1}{\text{Vol}\left(B_{(X_q,1)}(x_{q-1},\xi_4\delta^{d_q})\right)} \int_{B_{(X_q,1)}(x_{q-1},\xi_4\delta^{d_q})}$$

$$f(x_q)\, dx_q dx_{q-1} \cdots dx_1.$$

PROOF. First we verify the second inequality. This follows from the inequality, for $f \geq 0$, $x \in U_2$,

$$\int_{|t_j|_\infty < \eta_1} f\left(e^{t_j\delta^{d_j}X_j}x\right) dt_j \lesssim \frac{1}{\text{Vol}\left(B_{(X_j,1)}(x,\xi_4\delta^{d_j})\right)} \int_{B_{(X_j,1)}(x,\xi_4\delta^{d_j})} f(y)\, dy. \tag{3.13}$$

If $X_j(x) = 0$, this is trivial, so we may work in the case when $X_j(x) \neq 0$ and $B_{(X_j,1)}(x,\xi_4\delta^{d_j})$ is 1-dimensional. We apply the quantitative Frobenius theorem (Theorem 2.2.22) to the *single vector field* $\delta^{d_j}X_j$ with formal degree 1.[9] We obtain a map $\Phi_j : B^1(\eta') \to B_{(X_j,1)}(x,\xi_2'\delta^{d_j})$ with $B_{(X_j,1)}(x,\xi_3'\delta^{d_j}) \subset \Phi_j\left(B^1(\eta')\right)$, for some $\eta', \xi_2', \xi_3' > 0$, where η' and ξ_3' can be chosen independent of any relevant parameters. Let $\xi_4 \leq \xi_3' \wedge a$. Let Y be the pullback of $\delta^{d_j}X_j$ via Φ_j, so that $\|Y\|_{C^m} \approx 1$ (notice Y is a 1-dimensional vector field, and with a choice of a coordinate can be identified with a function). Applying a change of variables $y = \Phi_j(u)$ and using (2.16), we have

$$\frac{1}{\text{Vol}\left(B_{(X_j,1)}(x,\xi_4\delta^{d_j})\right)} \int_{B_{(X_j,1)}(x,\xi_4\delta^{d_j})} f(y)\, dy$$

$$\approx \int_{B_{(Y,1)}(x,\xi_4)} f \circ \Phi_j(u)\, du. \tag{3.14}$$

Because $|Y| \gtrsim 1$ and $\|Y\|_{C^2} \lesssim 1$, we may pick η_1 (using the discussion in Appendix B) so that

$$\int_{|t_j| < \eta_1} f \circ \Phi_j\left(e^{t_j Y}0\right) dt_j \lesssim \int_{B_{(Y,1)}(x,\xi_4)} f \circ \Phi_j(u)\, du. \tag{3.15}$$

But $\Phi_j\left(e^{t_j Y}0\right) = e^{t_j\delta^{d_j}X_j}x$. (3.13) follows by combining (3.15) and (3.14).

---

[9]The main "involutivity" condition of Theorem 2.2.22 (i.e., assumption (c) of that theorem) is trivial when there is only one vector field.

We turn to the first inequality. We apply the quantitative Frobenius theorem in this context (Theorem 3.1.2) to obtain a map $\Phi_{x_0,\delta}$ as in that theorem. Applying the change of variables $y = \Phi_{x_0,\delta}(u)$, for ξ_5 fixed (where the implicit constants may depend on the choice of ξ_5) we have

$$
\frac{1}{\mathrm{Vol}\left(B_{(X,d)}\left(x_0, \xi_5\delta\right)\right)} \int_{B_{(X,d)}(x_0,\xi_5\delta)} f(y) \, dy
$$
$$
\approx \int_{B_{\left(Y^{x_0,\delta},d\right)}(0,\xi_5)} f \circ \Phi_{x_0,\delta}(u) \, du. \tag{3.16}
$$

Using that

$$
\Phi_{x_0,\delta}\left(e^{t_q Y_q^{x_0,\delta}} e^{t_{q-1} Y_{q-1}^{x_0,\delta}} \cdots e^{t_1 Y_1^{x_0,\delta}} 0\right) = e^{t_q \delta^{d_q} X_q} e^{t_{q-1} \delta^{d_{q-1}} X_{q-1}} \cdots e^{t_1 \delta^{d_1} X_1} x_0,
$$

we have

$$
\int_{|t|_\infty < \eta_1} f\left(e^{t_q \delta^{d_q} X_q} e^{t_{q-1} \delta^{d_{q-1}} X_{q-1}} \cdots e^{t_1 \delta^{d_1} X_1} x_0\right) dt
$$
$$
= \int_{|t|_\infty < \eta_1} f \circ \Phi_{x_0,\delta}\left(e^{t_q Y_q^{x_0,\delta}} e^{t_{q-1} Y_{q-1}^{x_0,\delta}} \cdots e^{t_1 Y_1^{x_0,\delta}} 0\right) dt. \tag{3.17}
$$

Notice $\frac{\partial}{\partial t_j} e^{t_q Y_q^{x_0,\delta}} e^{t_{q-1} Y_{q-1}^{x_0,\delta}} \cdots e^{t_1 Y_1^{x_0,\delta}} 0 \Big|_{t=0} = Y_j^{x_0,\delta}(0)$.

Because $\left\| Y_j^{x_0,\delta} \right\|_{C^2} \lesssim 1$, we have $\left| e^{t_q Y_q^{x_0,\delta}} e^{t_{q-1} Y_{q-1}^{x_0,\delta}} \cdots e^{t_1 Y_1^{x_0,\delta}} 0 \right| \lesssim 1, \forall |t| <$ η_1 (see Appendix B.1), and a change of variables as in Appendix B.3 applies to show that there is $\eta_2 > 0$ with

$$
\int_{|u| < \eta_2} f \circ \Phi_{x_0,\delta}(u) \, du
$$
$$
\lesssim \int_{|t|_\infty < \eta_1} f \circ \Phi_{x_0,\delta}\left(e^{t_q Y_q^{x_0,\delta}} e^{t_{q-1} Y_{q-1}^{x_0,\delta}} \cdots e^{t_1 Y_1^{x_0,\delta}} 0\right) dt. \tag{3.18}
$$

Using that $\left\| Y_j^{x_0,\delta} \right\|_{C^2} \lesssim 1$, we take $\xi_5 > 0$ small enough that $B_{\left(Y^{x_0,\delta},d\right)}(0, \xi_5) \subset$ $B^{n_0(x)}(\eta_2)$ and it follows that

$$
\int_{B_{\left(Y^{x_0,\delta},d\right)}(0,\xi_5)} f \circ \Phi_{x_0,\delta}(u) \, du \le \int_{|u| < \eta_2} f \circ \Phi_{x_0,\delta}(u) \, du. \tag{3.19}
$$

Combining (3.16), (3.17), (3.18), and (3.19) completes the proof. $\square$

Let $a > 0$ be a small number to be chosen in a moment, and let ξ_4, ξ_5, and η_1 be as in Proposition 3.3.2. Let $\psi_0 \in C_0^\infty(U)$ equal 1 on a neighborhood of the closure of U_2. For each $j = 1, \ldots, q$, we define two maximal functions:

$$
\widetilde{M}_j f(x) = \psi_0(x) \sup_{\delta \in (0,1]} \int_{|t_j| < \eta_1} \left| f\left(e^{t_j \delta X_j} x\right) \right| \psi_0\left(e^{t_j \delta X_j} x\right),
$$

$$\mathcal{M}_j f(x) = \psi_0(x) \sup_{0 < \delta \le \xi_4} \frac{1}{\operatorname{Vol}\left(B_{(X_j,1)}(x,\delta)\right)} \int_{B_{(X_j,1)}(x,\delta)} |f(y)| \psi_0(y) \, dy.$$

PROOF OF THEOREM 3.3.1. The result is trivial for $p = \infty$, and so we focus on $1 < p < \infty$. We take $a > 0$ so small that Theorem 2.15.34 applies to each $\mathcal{M}_j$, and shows $\|\mathcal{M}_j f\|_{L^p} \lesssim \|f\|_{L^p}$, $1 < p < \infty$. Proposition 3.3.2 shows we have the pointwise inequalities $\mathcal{M}f(x) \lesssim \widetilde{\mathcal{M}}_1 \cdots \widetilde{\mathcal{M}}_q f(x) \lesssim \mathcal{M}_1 \cdots \mathcal{M}_q f(x)$, which completes the proof. □

Recall the set $\Omega = B_{(X,d)}\left(0, \frac{\xi_0}{2}(1,\dots,1)\right)$, where $\xi_0 > 0$ is a small constant.

PROPOSITION 3.3.3. *Suppose* $\xi_0 \le \xi_6/q$ *and let* $j \in [0,\infty)^\nu$. *If* $F(x,y)$ *has* $\operatorname{supp}(F) \subset \Omega \times \Omega$ *and for every* m,

$$|F(x,y)| \le \frac{(1 + \rho_{2^{-j}}(x,y))^{-m}}{\operatorname{Vol}(B_{2^{-j}}(x, 1 + \rho_{2^{-j}}(x,y)))},$$

then we have the pointwise inequality $|Ff(x)| \lesssim \mathcal{M}f(x)$.

PROOF. We choose ψ in the definition of $\mathcal{M}$ to be equal to 1 on a neighborhood of the closure of $B_{(X,d)}(0, 2\xi_0(1,\dots,1)) \Subset U_1$. We have, for $x \in \Omega$, by applying Lemma 3.2.16 and taking m large,

$$|Ff(x)| \le \int_{y \in \Omega} \frac{(1 + \rho_{2^{-j}}(x,y))^{-m}}{\operatorname{Vol}(B_{2^{-j}}(x, 1 + \rho_{2^{-j}}(x,y)))} |f(y)| \, dy$$

$$\lesssim \sum_{\substack{k \le j \\ k \in \mathbb{N}^\nu}} \frac{2^{-|j-k|}}{\operatorname{Vol}(B_{2^{-k}}(x,1))} \int_{\rho_{2^{-k}}(x,y) < \xi_0} |f(y)| \, dy$$

$$\lesssim \sum_{\substack{k \le j \\ k \in \mathbb{N}^\nu}} \frac{2^{-|j-k|}}{\operatorname{Vol}\left(B_{(X,d)}(x, 2^{-k}\xi_0)\right)} \int_{B_{(X,d)}(x, 2^{-k}\xi_0)} |f(y)| \, dy$$

$$\lesssim \mathcal{M}f(x),$$

completing the proof. □

3.4 A LITTLEWOOD-PALEY THEORY

We take again the setting in Section 3.1, and we assume in addition that each d_j is nonzero in only one component. Our goal is to create a Littlewood-Paley square function adapted to the multi-parameter geometry $B_{(X,d)}(x,\delta)$, $\delta \in (0,1]^\nu$. Because each d_j is nonzero in only one component, we may decompose (X,d) into ν single-parameter lists. That is, for each μ, let $\left(X^\mu, \hat{d}^\mu\right) = \left(X_1^\mu, \hat{d}_1^\mu\right), \dots, \left(X_{q_\mu}^\mu, \hat{d}_{q_\mu}^\mu\right)$ denote the list of those vector fields X_j such that d_j is nonzero in only the μth component, and we assign to that vector field the single-parameter degree $|d_j|$.

Fix an open set $U_0 \Subset U_1$. Let $\psi_\nu \in C_0^\infty(U_1)$ be equal to 1 on a neighborhood of the closure of U_0. Recursively, for $1 \le \mu \le \nu - 1$, let $\psi_\mu \in C_0^\infty(U_1)$ equal 1 on a neighborhood of the support of $\psi_{\mu+1}$. On $\mathbb{R}^{q_\mu}$ we define single-parameter dilations

$$\delta_\mu(t_1, \ldots, t_{q_\mu}) = \left(\delta_\mu^{\hat{d}_1^\mu} t_1, \ldots, \delta_\mu^{\hat{d}_{q_\mu}^\mu} t_{q_\mu}\right). \text{ For } t_\mu \in \mathbb{R}^{q_\mu} \text{ and a function } \varsigma : \mathbb{R}^{q_\mu} \to \mathbb{C},$$

we define $\varsigma^{(2^{j_\mu})}(t_\mu)$ in the usual way: $\varsigma^{(2^{j_\mu})}(t_\mu) = 2^{j_\mu(\sum_l \hat{d}_l^\mu)}\varsigma(2^{j_\mu}t_\mu)$.

As in Section 2.15.4, we decompose $\delta_0(t_\mu) = \sum_{j \in \mathbb{N}} \varsigma_{j,\mu}^{(2^j)}(t_\mu)$, where $\{\varsigma_{j,\mu} \mid j \in \mathbb{N}\} \subset \mathcal{S}(\mathbb{R}^{q_\mu})$ is a bounded set with $\varsigma_{j,\mu} \in \mathcal{S}_0(\mathbb{R}^{q_\mu})$ if $j > 0$, and $\delta_0(t_\mu)$ denotes the Dirac delta function at 0 in the t_μ variable. Define, for $j \in \mathbb{N}$,

$$D_j^\mu f(x) = \psi_\mu(x) \int f\left(e^{t_\mu \cdot X^\mu} x\right) \psi_\mu\left(e^{t_\mu \cdot X^\mu} x\right) \varsigma_{j,\mu}^{(2^j)}(t_\mu)\, dt_\mu.$$

Corollary 2.15.54 shows that for f with $\psi_\mu \equiv 1$ on $\mathrm{supp}(f)$, we have for $\{\epsilon_j\}_{j \in \mathbb{N}}$ a sequence of i.i.d. random variables of mean 0 taking values ± 1, and $1 < p < \infty$,

$$\|f\|_{L^p} \approx \left(\mathbb{E}\left\|\sum_{j \in \mathbb{N}} \epsilon_j D_j^\mu f\right\|_{L^p}^p\right)^{\frac{1}{p}}. \tag{3.20}$$

Moreover, Corollary 2.15.55 shows for all f, we have

$$\left(\mathbb{E}\left\|\sum_{j \in \mathbb{N}} \epsilon_j D_j^\mu f\right\|_{L^p}^p\right)^{\frac{1}{p}} \lesssim \|f\|_{L^p}. \tag{3.21}$$

For $j = (j_1, \ldots, j_\nu) \in \mathbb{N}^\nu$ define

$$D_j = D_{j_1}^1 D_{j_2}^2 \cdots D_{j_\nu}^\nu.$$

THEOREM 3.4.1. *We have, for $1 < p < \infty$,*

$$\left\|\left(\sum_{j \in \mathbb{N}^\nu} |D_j f|^2\right)^{\frac{1}{2}}\right\|_{L^p} \lesssim \|f\|_{L^p}, \tag{3.22}$$

and if $\mathrm{supp}(f) \subset U_0$, *we have*

$$\left\|\left(\sum_{j \in \mathbb{N}^\nu} |D_j f|^2\right)^{\frac{1}{2}}\right\|_{L^p} \approx \|f\|_{L^p}. \tag{3.23}$$

Here, the implicit constants depend on $p \in (1, \infty)$.

PROOF. Let $\left\{\epsilon_j^\mu\right\}_{j\in\mathbb{N},1\le\mu\le\nu}$ be i.i.d. random variables of mean 0 taking values ± 1. For $j=(j_1,\ldots,j_\nu)\in\mathbb{N}^\nu$ define $\epsilon_j=\epsilon_{j_1}^1\epsilon_{j_2}^2\cdots\epsilon_{j_\nu}^\nu$. Notice $\{\epsilon_j\}_{j\in\mathbb{N}^\nu}$ are i.i.d. random variables of mean 0 taking values ± 1. The Khintchine inequality (Theorem 2.10.10) applies to show

$$\left\|\left(\sum_{j\in\mathbb{N}^\nu}|D_j f|^2\right)^{\frac{1}{2}}\right\|_{L^p}\approx\left(\mathbb{E}\left\|\sum_{j\in\mathbb{N}^\nu}\epsilon_j D_j f\right\|_{L^p}^p\right)^{\frac{1}{p}}.$$

But,

$$\sum_{j\in\mathbb{N}^\nu}\epsilon_j D_j=\left(\sum_{j_1\in\mathbb{N}}\epsilon_{j_1}^1 D_{j_1}^1\right)\cdots\left(\sum_{j_\nu\in\mathbb{N}}\epsilon_{j_\nu}^\nu D_{j_\nu}^\nu\right)$$

and therefore ν applications of (3.21) apply to prove (3.22). Using the same argument and using the choice of the ψ_μ, ν applications of (3.20) proves (3.23). □

3.5 FURTHER READING AND REFERENCES

As mentioned before, the quantitative study of Carnot-Carathéodory balls began with the work of Nagel, Stein, and Wainger [NSW85] in the single-parameter case. When one moves to the multi-parameter setting, their methods only apply under rather strong assumptions on the vector fields.[10] For instance, suppose we start with the ν lists of vector fields with *single*-parameter formal degrees, from the introduction to this chapter $\left(X^\mu,\hat{d}^\mu\right)=\left(X_1^\mu,\hat{d}_1^\mu\right),\ldots,\left(X_{q_\mu}^\mu,\hat{d}_{q_\mu}^\mu\right)$, $1\le\mu\le\nu$, and from these we generate a list of vector fields with ν-parameter formal degrees, $(X,d)=(X_1,d_1),\ldots,(X_q,d_q)$, as in the introduction to this chapter. If one assumes the rather strong hypothesis that, for each μ, $X_1^\mu,\ldots,X_{q_\mu}^\mu$ span the tangent space at each point, then the methods of [NSW85] apply to yield and analog of Theorem 3.1.2 in this case. However, we are interested in the case where these lists do not necessarily each individually span, and the methods of [NSW85] are insufficient in this case. See [Str11, Sections 1.2.1, 5.2.2, 5.2.3, and 5.2.4] for more details on this issue.

There is another important case where the methods of Nagel, Stein, and Wainger apply: the so-called "weakly-comparable" case.[11] Suppose $W_1,\ldots,W_\nu$ are vector fields on a compact manifold which satisfy Hörmander's condition. We consider the balls $B(x,\delta_1,\ldots,\delta_\nu):=B_{\delta_1 W_1,\ldots,\delta_\nu W_\nu}(x)$. If one wishes to study $B(x,\delta)$ for all $\delta\in[0,1]^\nu$ in the manner described in this monograph, then one needs to assume extra assumptions on $W_1,\ldots,W_\nu$. However, if one is willing to fix a large number N, and restrict attention to only those $\delta=(\delta_1,\ldots,\delta_\nu)\in[0,1]^\nu$ with $\delta_\mu^N\lesssim\delta_{\mu'}$, $\forall\mu,\mu'$ then one needs no additional assumptions on the vector fields. One refers to these δ as "weakly-comparable." Tao and Wright [TW03] were the first to explicitly note that Nagel, Stein, and Wainger's methods apply to the weakly-comparable situation. See

---

[10]In fact, the methods in [NSW85] require strictly stronger assumptions than those in [Str11].

[11]This case does not play a direct role in this monograph.

[Str11, Section 5.2.1] for more details on this. Despite the fact, as Tao and Wright pointed out, that Nagel, Stein, and Wainger's methods apply to the weakly-comparable case, Tao and Wright introduced a new proof method: the key new idea was to exploit the classical ODE (2.30). Also, they made heavy use of Gronwall's inequality, which Nagel, Stein, and Wainger did not. It was by combining the methods of Nagel, Stein, and Wainger [NSW85] with those of Tao and Wright [TW03] and with classical methods, that the author proved the quantitative Frobenius theorem (Theorem 2.2.22) in [Str11], which allowed for the study of multi-parameter balls in this section.

The idea of bounding a multi-parameter maximal function by compositions of single-parameter maximal functions (and thereby proving the L^p boundedness) dates back to the work of Jessen, Marcinkiewicz, and Zygmund [JMZ35], and has been used by many authors since then. A special case of Theorem 3.3.1 was proved by the author using similar techniques in [Str11, Section 6]–the proof of Theorem 3.3.1 simplifies and strengthens those methods.

Multi-parameter Littlewood-Paley square functions (like the ones developed in Section 3.4) have been used by many authors to prove the L^p boundedness of various kinds of multi-parameter singular integrals. Of particular inspiration to us were the papers [NS04, NRS01, NRSW12]. See also [HLL13, HL08, HLY07]. One common theme of the above references is that the quantitative Frobenius theorem (or an appropriate analog) was trivial. Indeed, the corresponding distributions involved no singular points, and the difficulties dealt with in Theorem 2.2.22 were not present. Because of this, the methods used to construct the Littlewood-Paley square function developed in Section 3.4 are slightly different than the above referenced papers. The new ideas involved first appeared in a joint work of the author with Stein [SS13].

Chapter Four

Multi-parameter Singular Integrals I: Examples

In Chapters 1 and 2, we focused on "single-parameter" singular integrals. I.e., the singular integrals are defined in terms of an underlying family of balls $B(x, \delta)$ where $\delta > 0$. The main focus of this monograph is a more general setting, where the underlying balls have many "parameters," $B(x, \delta_1, \ldots, \delta_\nu)$, where $B(x, \delta_1, \ldots, \delta_\nu)$ is the sort of Carnot-Carathéodory ball studied in Chapter 3. In Chapter 5 we introduce a general theory of these multi-parameter singular integrals. In this chapter, we present several examples (many of which already appear in the literature) which fall under this more general theory.

In each case, the general setup is the same–and is the same as the setting of the introduction to Chapter 3. We study singular integrals corresponding to ν parameter balls, where $0 \neq \nu \in \mathbb{N}$. For each μ, $1 \leq \mu \leq \nu$ we are given a list of C^∞ vector fields $X_1^\mu, \ldots, X_{q_\mu}^\mu$; each X_j^μ paired with a formal degree $0 \neq \hat{d}_j^\mu \in \mathbb{N}$. We denote by $\left(X^\mu, \hat{d}^\mu\right)$ the list of vector fields with formal degrees $\left(X_1^\mu, \hat{d}_1^\mu\right), \ldots, \left(X_{q_\mu}^\mu, \hat{d}_{q_\mu}^\mu\right)$. We associate to $\hat{d}_l^\mu$ the multi-index $0 \neq d_l^\mu \in \mathbb{N}^\nu$, which is equal to $\hat{d}_l^\mu$ in the μth component, and 0 in all other components.[1] We let $(X, d) = (X_1, d_1), \ldots, (X_1, d_q)$ be an enumeration of

$$\left(X_1^1, d_1^1\right), \ldots, \left(X_{q_1}^1, d_{q_1}^1\right), \ldots, \ldots, \left(X_1^\nu, d_1^\nu\right), \ldots, \left(X_{q_\nu}^\nu, d_{q_\nu}^\nu\right),$$

so that $q = q_1 + \cdots + q_\nu$. We assume that $X_1, \ldots, X_q$ span the tangent space at each point of the ambient space, so that the balls $B_{(X,d)}(x, \delta)$, $\delta \in (0, 1]^\nu$ are open. The most important property we assume is the following: $\forall \delta \in [0, 1]^\nu$,

$$\left[\delta^{d_j} X_j, \delta^{d_k} X_l\right] = \sum_{l=1}^{q} c_{j,k}^{l,\delta} \delta^{d_l} X_l, \tag{4.1}$$

where the function $c_{j,k}^{l,\delta}$ is "bounded uniformly in δ"; see Section 3.1 for a more detailed description of this assumption.

4.1 THE PRODUCT THEORY OF SINGULAR INTEGRALS

When one hears of a "multi-parameter singular integral," usually it is the product theory of singular integrals to which is being referred. We will see that this product theory is the simplest special case of the general theory developed in Chapter 5.

---

[1] Note that each $d_j \in \mathbb{N}^\nu$ is nonzero in only one component, as was required for the results at the end of Section 3.2.

The setting is as follows. We are given ν connected, **compact** manifolds, without boundary:

$$M_1, \ldots, M_\nu.$$

The singular integrals we define are operators on $M = M_1 \times \cdots \times M_\nu$. For each μ we suppose we are given C^∞ vector fields $W_1^\mu, \ldots, W_{r_\mu}^\mu$ on M_μ satisfying Hörmander's condition on M_μ. As in Section 2.1, we use these vector fields to create a list of vector fields with single-parameter formal degrees, denote them by

$$\left(X^\mu, \hat{d}^\mu\right) = \left(X_1^\mu, \hat{d}_1^\mu\right), \ldots, \left(X_{q_\mu}^\mu, \hat{d}_{q_\mu}^\mu\right);$$

so that, in particular, X_j^μ is an iterated commutator of order $\hat{d}_j^\mu$ of $W_1^\mu, \ldots, W_{r_\mu}^\mu$ and $X_1^\mu, \ldots, X_{q_\mu}^\mu$ span the tangent space to M_μ at every point. Furthermore, these vector fields satisfy the crucial involutivity condition from Proposition 2.1.3:

$$[X_i^\mu, X_j^\mu] = \sum_{\hat{d}_k^\mu \leq \hat{d}_i^\mu + \hat{d}_j^\mu} c_{i,j}^{k,\mu} X_k^\mu, \quad c_{i,j}^{k,\mu} \in C^\infty(M_\mu). \tag{4.2}$$

Corresponding to these vector fields with formal degrees we obtain single-parameter Carnot-Carathéodory balls on M_μ: $B_{\left(X^\mu, \hat{d}^\mu\right)}(x_\mu, \delta_\mu) \subseteq M_\mu$, for $x_\mu \in M_\mu$ and $\delta_\mu > 0$. From here, we define ν parameter balls on M: $B\left((x_1, \ldots, x_\nu), (\delta_1, \ldots, \delta_\nu)\right) := B_{\left(X^1, \hat{d}^1\right)}(x_1, \delta_1) \times \cdots \times B_{\left(X^\nu, \hat{d}^\nu\right)}(x_\nu, \delta_\nu)$ (we will see that these balls are comparable to the balls studied in Chapter 3; see Proposition 4.1.1). As in Chapter 2 we give each M_μ a strictly positive, smooth measure (see Definition 2.0.6), so that it makes sense to talk about $\mathrm{Vol}\left(B_{\left(X^\mu, \hat{d}^\mu\right)}(x_\mu, \delta_\mu)\right)$. We give M the strictly positive, smooth measure corresponding to the product measure on $M_1 \times \cdots \times M_\nu$ so that $\mathrm{Vol}\left(B\left((x_1, \ldots, x_\nu), (\delta_1, \ldots, \delta_\nu)\right)\right) = \prod_\mu \mathrm{Vol}\left(B_{\left(X^\mu, \hat{d}^\mu\right)}(x_\mu, \delta_\mu)\right)$ (as in Chapter 2, none of the definitions that follow depend on the choice of strictly positive, smooth measures).

On each factor, the balls $B_{\left(X^\mu, \hat{d}^\mu\right)}(x_\mu, \delta_\mu)$ induce a Carnot-Carathéodory metric: $\rho_\mu(x_\mu, z_\mu) := \inf\left\{\delta_\mu > 0 \mid z_\mu \in B_{\left(X^\mu, \hat{d}^\mu\right)}(x_\mu, \delta_\mu)\right\}$. We turn these metrics into a "vector valued metric" on M:

$$\rho\left((x_1, \ldots, x_\nu), (z_1, \ldots, z_\nu)\right) := \left(\rho_1(x_1, z_1), \ldots, \rho_\nu(x_\nu, z_\nu)\right).$$

Fix points $x_\mu \in M_\mu$, $1 \leq \mu \leq \nu$. There is a canonical decomposition

$$T_{(x_1, \ldots, x_\nu)} M \cong T_{x_1} M_1 \times \cdots \times T_{x_\nu} M_\nu.$$

We define the usual inclusion $T_{x_\mu} M_\mu \hookrightarrow T_{(x_1, \ldots, x_\nu)} M$ by $V \mapsto (0, \ldots, 0, V, 0, \ldots, 0)$: the tangent vector which is 0 in every coordinate except the μ coordinate where it equals V. Thus we may consider each X_j^μ as a vector field on M, and henceforth we use X_j^μ to denote both this vector field on M and the original vector field on M_μ (which one we mean will be clear from context). Using the procedure in the introduction to this chapter, we combine the ν lists of vector fields with single-parameter formal degrees $\left(X^\mu, \hat{d}^\mu\right)$ to create a list of vector fields with ν-parameter formal degrees $(X_1, d_1), \ldots, (X_q, d_q)$, each $0 \neq d_j \in \mathbb{N}^\nu$. We therefore have multi-parameter balls $B_{(X,d)}(x, \delta)$, $x \in M$, $\delta \in [0, \infty)^\nu$.

PROPOSITION 4.1.1. *There is a constant $C > 0$ such that*

$$B(x, \delta/C) \subseteq B_{(X,d)}(x, \delta) \subseteq B(x, C\delta), \quad \forall x \in M, \delta \in [0, \infty)^\nu.$$

PROOF. This is an immediate consequence of the definitions. □

Remark 4.1.2 (2.31) shows that

$$\mathrm{Vol}\left(B_{(X^\mu, \hat{d}^\mu)}(x_\mu, 2\delta_\mu)\right) \approx \mathrm{Vol}\left(B_{(X^\mu, \hat{d}^\mu)}(x_\mu, \delta_\mu)\right)$$

and as a consequence, $\mathrm{Vol}(B(x, \delta/C)) \approx \mathrm{Vol}(B(x, C\delta))$. Using Proposition 4.1.1, we have

$$\mathrm{Vol}(B(x, \delta)) \approx \mathrm{Vol}\left(B_{(X,d)}(x, \delta)\right).$$

In light of this and Proposition 4.1.1, all of the definitions that follow are equivalent if we use the balls $B_{(X,d)}(x, \delta)$ or the balls $B(x, \delta)$. We proceed by using the balls $B(x, \delta)$ as that is what is most common in the literature.

For an ordered multi-index α, X^α is an $|\alpha|$ order partial differential operator, in the classical sense of the order of a differential operator. We define $\deg(\alpha) = \sum_{j=1}^q k_j d_j \in \mathbb{N}^\nu$, where k_j is the number of times j appears in the list α. We wish to think of X^α as a differential operator of "order" $\deg(\alpha) \in \mathbb{N}^\nu$. Notice, this treats X_j as a differential operator of "order" $d_j \in \mathbb{N}^\nu$.

We think of $\deg(\alpha)$ as a multi-index. Thus, standard multi-index notation gives

$$\rho\left((x_1, \ldots, x_\nu), (z_1, \ldots, z_\nu)\right)^{-\deg(\alpha)} = \prod_{\mu=1}^\nu \rho_\mu(x_\mu, z_\mu)^{-\deg(\alpha)_\mu}.$$

With this notation, we may formally re-write Definition 2.0.16 to obtain a new class of singular integrals, which we call the product theory of singular integrals. We begin bounded sets of bump functions.

DEFINITION 4.1.3. *We say $\mathcal{B} \subset C^\infty(M) \times M \times (0, 1]^\nu$ is a "bounded set of bump functions" if:*

- *$\forall (\phi, x, \delta) \in \mathcal{B}$, $\mathrm{supp}(\phi) \subset B(x, \delta)$.*

- *For every ordered multi-index α, there exists C, such that $\forall (\phi, x, \delta) \in \mathcal{B}$,*

$$\sup_{z \in M} |(\delta X)^\alpha \phi(z)| \leq C \mathrm{Vol}(B(x, \delta))^{-1}.$$

(2.31) shows for each μ there are Q_1^μ and Q_2^μ satisfying

$$2^{Q_1^\mu} \mathrm{Vol}\left(B_{\left(X^\mu,\hat{d}^\mu\right)}\left(x_\mu,\delta_\mu\right)\right) \leq \mathrm{Vol}\left(B_{\left(X^\mu,\hat{d}^\mu\right)}\left(x_\mu,2\delta_\mu\right)\right)$$
$$\leq 2^{Q_2^\mu} \mathrm{Vol}\left(B_{\left(X^\mu,\hat{d}^\mu\right)}\left(x_\mu,\delta_\mu\right)\right).$$

We take Q_1^μ to be the maximal number satisfying the above, and Q_2^μ to be the minimal one.

DEFINITION 4.1.4. *We say* $T : C^\infty(M) \to C^\infty(M)$ *is a product singular integral operator of order* $t = (t_1, \ldots, t_\nu) \in (-Q_1^1, \infty) \times \cdots \times (-Q_1^\nu, \infty) \subset \mathbb{R}^\nu$ *if*

(i) *(Growth Condition) For all ordered multi-indices* α, β,

$$\left|X_x^\alpha X_z^\beta T(x,z)\right| \leq C_{\alpha,\beta} \frac{\rho(x,z)^{-t-\deg(\alpha)-\deg(\beta)}}{\mathrm{Vol}(B(x,\rho(x,z)))}, \tag{4.3}$$

where X_x *denotes the list of vector fields* $X_1, \ldots, X_q$ *acting in the* x *variable, and similarly for* X_z. *In particular, the above implies that* $T(x,z)$ *agrees with a* C^∞ *function on the set* $x_1 \neq z_1, \ldots, x_\nu \neq z_\nu$.

(ii) *(Cancellation Condition) For each bounded set of bump functions* $\mathcal{B} \subset C^\infty(M) \times M \times (0,1]^\nu$ *and each ordered multi-index* α,

$$\sup_{(\phi,z,\delta)\in\mathcal{B}} \sup_{x\in M} \delta^{t+\deg(\alpha)} \mathrm{Vol}(B(z,\delta)) \left|X^\alpha T\phi(x)\right| \leq C_{\mathcal{B},\alpha}, \tag{4.4}$$

with the same estimate for T *replaced by* T^*.

Remark 4.1.5 Just as in Definition 2.0.16, T is a product singular integral operator if and only if T^* is.

Example 4.1.6 *Suppose for each* μ *we are given a Calderón-Zygmund operator,* $T_\mu : C^\infty(M_\mu) \to C^\infty(M_\mu)$, *of order* $t_\mu \in (-Q_1^\mu, \infty)$ *(in the sense of Definition 2.0.16, with respect to the vector fields* $W_1^\mu, \ldots, W_{r_\mu}^\mu$). *Then, the operator* $T = T_1 \otimes \cdots \otimes T_\nu : \left[C^\infty(M) \cong C^\infty(M_1) \widehat{\otimes} \cdots \widehat{\otimes} C^\infty(M_\nu)\right] \to C^\infty(M)$ *is a product singular integral operator of order* $t = (t_1, \ldots, t_\nu)$ *(this is obvious if one uses the equivalent Definition 4.1.8, below). One may think of the space of product singular integral operators of order* t *as a "tensor product" of the spaces of Calderón-Zygmund singular integral operators in each factor (of order* t_μ *in the* μth *factor). It turns out that the space of product singular integral operators is the injective tensor product of the spaces of Calderón-Zygmund singular integral operators. The space of Calderón-Zygmund singular integral operators is not nuclear, and this tensor product does not agree with the projective tensor product, and therefore does not enjoy the universal property—we cannot use the universal property to trivially lift results from the single-parameter case to the product case. It is possible to do something along these lines,* *though; see Remark 4.1.20. See Appendix A.2 for the tensor product terminology used here.*

There are several more equivalent definitions for product singular integral operators. The next is the one most commonly seen in the literature (see, e.g., [NS04]). For this, we need to introduce bounded sets of bump functions on each factor.

DEFINITION 4.1.7. *We say $\mathcal{B}_\mu \subset C^\infty(M_\mu) \times M_\mu \times (0,1]$ is a bounded set of bump functions on M_μ if:*

- $\forall (\phi, x_\mu, \delta_\mu) \in \mathcal{B}_\mu$, $\mathrm{supp}(\phi) \subset B_{(X^\mu, \hat{d}^\mu)}(x_\mu, \delta_\mu)$.

- *For every ordered multi-index α, there exists C, such that $\forall (\phi, x_\mu, \delta_\mu) \in \mathcal{B}$,*

$$\sup_{z_\mu \in M_\mu} |(\delta_\mu X^\mu)^\alpha \phi(z_\mu)| \leq C \mathrm{Vol}\left(B_{(X^\mu, \hat{d}^\mu)}(x_\mu, \delta_\mu)\right)^{-1}.$$

To define product singular integrals in the next way, we instead look to the topology on the space of product singular integral operators. There is an obvious locally convex topology on the space of product singular integral operators of order t induced from Definition 4.1.4 (i.e., semi-norms are defined by the least possible $C_{\alpha,\beta}$ in (4.3), the least possible $C_{\mathcal{B},\alpha}$ in (4.4), and the least possible $C_{\mathcal{B},\alpha}$ in (4.4) when T is replaced by T^*). For the next equivalent definition, we need to use this topology as part of the definition. This definition is recursive.

DEFINITION 4.1.8. *When $\nu = 0$, we define the space of product singular integral operators to be $\mathbb{C}$ with the usual topology. For $\nu \geq 1$ the space of product singular integrals of order $t = (t_1, \ldots, t_\nu) \in (-Q_1^1, \infty) \times \cdots \times (-Q_1^\nu, \infty)$ is the space of those distributions $T \in C^\infty(M \times M)'$ such that the following holds.*

(i) *(Growth Condition) For all ordered multi-indices α, β,*

$$\left|(X_x)^\alpha (X_z)^\beta T(x,z)\right| \leq C_{\alpha,\beta} \frac{\rho(x,z)^{-t-\deg(\alpha)-\deg(\beta)}}{\mathrm{Vol}(B(x, \rho(x,z)))},$$

in particular we assume $T((x_1, \ldots, x_\nu), (z_1, \ldots, z_\nu))$ agrees with a C^∞ function on the set $x_1 \neq z_1, \ldots, x_\nu \neq z_\nu$. We define a semi-norm to be the least possible $C_{\alpha,\beta}$ such that the above holds.

(ii) *(Cancellation Condition) For each μ, $1 \leq \mu \leq \nu$, we assume that the following holds. For every bounded set of bump functions $\mathcal{B}_\mu$ on M_μ, we have the following. For every $x_\mu \in M_\mu$, $(\phi_\mu, z_\mu, \delta_\mu) \in \mathcal{B}_\mu$, we define the function $x_\mu \mapsto T^{\phi_\mu, x_\mu}$, $M_\mu \to C^\infty(M_1 \times \cdots M_{\mu-1} \times M_{\mu+1} \times \cdots \times M_\nu)'$ by*

$$\langle T(\phi_1 \otimes \cdots \otimes \phi_\nu), \psi_1 \otimes \cdots \otimes \psi_\nu \rangle$$
$$=: \int_{M_\mu} \langle T^{\phi_\mu, x_\mu}(\otimes_{\mu' \neq \mu} \phi_{\mu'}), \otimes_{\mu' \neq \mu} \psi_{\mu'} \rangle \psi_\mu(x_\mu) \, dx_\mu.$$

T^{ϕ_μ, x_μ} *is a priori only defined as a distribution in the* x_μ *variable, but we assume it agrees with a* C^∞ *function in that variable. Furthermore, we assume that for every ordered multi-index* α, *the operator*

$$\mathrm{Vol}\left(B_{\left(X^\mu, \hat{d}^\mu\right)}\left(x_\mu, \delta_\mu\right)\right) \delta_\mu^{t_\mu}\left(\delta_\mu X_{x_\mu}^\mu\right)^\alpha T^{\phi_\mu, x_\mu},$$

which maps $C^\infty\left(\prod_{\mu' \neq \mu} M_{\mu'}\right) \to C^\infty\left(\prod_{\mu' \neq \mu} M_{\mu'}\right)$, *is a product singular integral operator of order* $(t_1, \ldots, t_{\mu-1}, t_{\mu+1}, \ldots, t_\nu)$ *on the* $\nu - 1$ *factor space* $M_1 \times \cdots \times M_{\mu-1} \times M_{\mu+1} \times \cdots \times M_\nu$. *Finally for every continous semi-norm,* $|\cdot|$, *on the space of product kernels of order* $(t_1, \ldots, t_{\mu-1}, t_{\mu+1}, \ldots, t_\nu)$ *on* $M_1 \times \cdots \times M_{\mu-1} \times M_{\mu+1} \times \cdots \times M_\nu$, *every ordered multi-index* α, *and every bounded set of bump functions* $\mathcal{B}_\mu$ *on* M_μ, *we define a semi-norm* $|\cdot|_{\alpha, \mathcal{B}_\mu}$ *on product singular integrals of order* t *by*

$$|T|_{\alpha, \mathcal{B}_\mu} := \sup_{\substack{(\phi_\mu, z_\mu, \delta_\mu) \in \mathcal{B}_\mu \\ x_\mu \in M_\mu}} \left|\mathrm{Vol}\left(B_{\left(X^\mu, \hat{d}^\mu\right)}\left(x_\mu, \delta_\mu\right)\right) \delta_\mu^{t_\mu}\left(\delta_\mu X_{x_\mu}^\mu\right)^\alpha T^{\phi_\mu, x_\mu}\right|,$$

which we assume to be finite. We do the same for the transpose of T *in the* μ *variable, where we define* $z_\mu \mapsto T^{z_\mu, \psi_\mu}$ *reversing the roles of* x_μ, z_μ *and* ϕ_μ, ψ_μ; *thereby obtaining another semi-norm.*

We give the space of product singular integrals of order $(t_1, \ldots, t_\nu)$ *the coarsest topology such that the above semi-norms are continuous.*

The above definitions are justified by the following proposition.

PROPOSITION 4.1.9. *Definitions 4.1.4 and 4.1.8 are equivalent.*

We discuss the proof of Proposition 4.1.9 later. First, we introduce two more ways of viewing product singular integral operators. These are in line with (ii) and (iii) of Theorem 2.0.29. As in Chapter 3, for $j = (j_1, \ldots, j_\nu) \in [0, \infty)^\nu$, we define $2^{-j} = \left(2^{-j_1}, \ldots, 2^{-j_\nu}\right) \in (0, 1]^\nu$. As a result, it makes sense to write $2^{-j} X$ which denotes the list of vector fields $2^{-j \cdot d_1} X_1, \ldots, 2^{-j \cdot d_q} X_q$.

DEFINITION 4.1.10. *We say* $\mathcal{E} \subset C^\infty(M \times M) \times (0, 1]^\nu$ *is a bounded set of pre-elementary operators if:* $\forall \alpha, \beta, \exists C, \forall (E, 2^{-j}) \in \mathcal{E}$,

$$\left|\left(2^{-j} X_x\right)^\alpha \left(2^{-j} X_z\right)^\beta E(x, z)\right| \leq C \frac{\left(1 + 2^j \cdot \rho(x, z)\right)^{-m}}{\mathrm{Vol}\left(B\left(x, 2^{-j} + \rho(x, z)\right)\right)}.$$

DEFINITION 4.1.11. *We define the set of bounded sets of elementary operators,* $\mathcal{G}$, *to be the largest set of subsets of* $C^\infty(M \times M) \times (0, 1]^\nu$ *such that for all* $\mathcal{E} \in \mathcal{G}$,

- $\mathcal{E}$ *is a bounded set of pre-elementary operators.*

- *Let* $e = (1, \ldots, 1) \in \mathbb{N}^\nu$. *We write* $\deg(\alpha) \leq e$ *to denote the inequality holding coordinatewise. We assume* $\forall (E, 2^{-j}) \in \mathcal{E}$,

$$E = \sum_{\deg(\alpha), \deg(\beta) \leq e} 2^{-(2e - \deg(\alpha) - \deg(\beta)) \cdot j}\left(2^{-j} X\right)^\alpha E_{\alpha, \beta}\left(2^{-j} X\right)^\beta,$$

where $\left\{ \left(E_{\alpha,\beta}, 2^{-j} \right) \mid \left(E, 2^{-j} \right) \in \mathcal{E} \right\} \in \mathcal{G}.$

We call elements $\mathcal{E} \in \mathcal{G}$ *bounded sets of elementary operators.*

THEOREM 4.1.12. *Fix* $t \in \left(-Q_1^1, \infty \right) \times \cdots \times \left(-Q_1^\nu, \infty \right)$*, and let* $T : C^\infty \left(M \right) \to C^\infty \left(M \right)$*. The following are equivalent:*

(i) T *is a product singular integral operator of order* t *as in Definition 4.1.4.*

(ii) T *is a product singular integral operator of order* t *as in Definition 4.1.8.*

(iii) *For every bounded set of elementary operators* $\mathcal{E}$,

$$\left\{ \left(2^{-j \cdot t} TE, 2^{-j} \right) \mid \left(E, 2^{-j} \right) \in \mathcal{E} \right\}$$

is a bounded set of elementary operators.

(iv) *There is a bounded set of elementary operators* $\left\{ \left(E_j, 2^{-j} \right) \mid j \in \mathbb{N} \right\}$ *such that* $T = \sum_{j \in \mathbb{N}} 2^{j \cdot t} E_j$*. (Every such sum converges in the topology of bounded convergence as operators* $C^\infty \left(M \right) \to C^\infty \left(M \right)$*; this can be seen just as in Lemma 2.0.28.)*

Furthermore, (iii) and (iv) are equivalent for any $t \in \mathbb{R}^\nu$.

DEFINITION 4.1.13. *We say* $T : C^\infty \left(M \right) \to C^\infty \left(M \right)$ *is a product singular integral operator of order* $t \in \mathbb{R}^\nu$ *if either of the equivalent conditions (iii) or (iv) of Theorem 4.1.12 holds.*

The proof of Theorem 4.1.12 is largely similar to the proof of Theorem 2.0.29. Because of this we only outline the main new ideas. As in Theorem 2.0.29, the main idea is to use the spectral theorem to provide a decomposition of the identity. For this, we may use the decomposition in each factor as guaranteed by Theorem 2.6.6. We use the following lemma.

LEMMA 4.1.14. *For each* μ*, let* $\mathcal{E}_\mu \subset C^\infty \left(M_\mu \times M_\mu \right) \times (0, 1]$ *be a bounded set of elementary operators as in Definition 2.0.21. Then, the set*

$$\left\{ \left(E_1 \otimes \cdots \otimes E_\nu, \left(2^{-j_1}, \ldots, 2^{-j_\nu} \right) \right) \mid \left(E_1, 2^{-j_1} \right) \in \mathcal{E}_1, \ldots, \left(E_\nu, 2^{-j_\nu} \right) \in \mathcal{E}_\nu \right\}$$

is a bounded set of elementary operators as in Definition 4.1.11.

PROOF. This is immediate from the definitions.   □

COROLLARY 4.1.15. *There is a bounded set of elementary operators*

$$\left\{ \left(E_j, 2^{-j} \right) \mid j \in \mathbb{N}^\nu \right\}$$

such that $I = \sum_{j \in \mathbb{N}^\nu} E_j$*, where* $I : C^\infty \left(M \right) \to C^\infty \left(M \right)$ *is the identity operator.*

PROOF. We apply Theorem 2.6.6 to each factor M_μ to decompose the identity operator

$$I : C^\infty (M_\mu) \to C^\infty (M_\mu)$$

as $I = \sum_{j_\mu \in \mathbb{N}} E^\mu_{j_\mu}$, where $\left\{ \left(E^\mu_{j_\mu}, 2^{-j_\mu} \right) \mid j_\mu \in \mathbb{N} \right\}$ is a bounded set of elementary operators (cf. Lemma 2.7.1). Setting $E_{(j_1,\ldots,j_\nu)} := E^1_{j_1} \otimes \cdots \otimes E^\nu_{j_\nu}$, we have $I = \sum_{j \in \mathbb{N}^\nu} E_j$, and the result follows by Lemma 4.1.14. $\qquad\square$

We also need the following proposition. For $j = (j_1,\ldots,j_\nu)$, $k = (k_1,\ldots,k_\nu)$, we write $j \wedge k = (j_1 \wedge k_1,\ldots,j_\nu \wedge k_\nu)$: the coordinatewise minimum of j and k. We have

PROPOSITION 4.1.16. *Let $\mathcal{E}$ be a bounded set of elementary operators. Then, for every N, the set*

$$\left\{ \left(2^{N|j_1-j_2|} E_1 E_2, 2^{-j_1} \right), \left(2^{N|j_1-j_2|} E_1 E_2, 2^{-j_2} \right), \left(2^{N|j_1-j_2|} E_1 E_2, 2^{-j_1 \wedge j_2} \right) \mid \right.$$
$$\left. \left(E_1, 2^{-j_1} \right), \left(E_2, 2^{-j_2} \right) \in \mathcal{E} \right\}$$

is a bounded set of elementary operators.

COMMENTS ON THE PROOF. The proof follows from the same methods as the proof of Proposition 2.7.2. We prove an essentially more general version of this proposition in Section 5.4. Thus, we leave the details to the reader. $\qquad\square$

COMMENTS ON THE PROOF OF THEOREM 4.1.12. (iii)$\Rightarrow$(iv) follows immediately from Corollary 4.1.15, by writing $T = TI = \sum_{j \in \mathbb{N}^\nu} 2^{j \cdot t} \left(2^{-j \cdot t} T E_j \right)$. (iv)$\Rightarrow$(iii) follows easily from Proposition 4.1.16 just as in the proof of Theorem 2.0.29. (iv)$\Rightarrow$(i) and (iv)$\Rightarrow$(ii) both follow from straightforward estimates as in the proof of (iii)$\Rightarrow$(i) of Theorem 2.0.29. (i)$\Rightarrow$(iv) works in the same way as (i)$\Rightarrow$(iii) of Theorem 2.0.29, using Corollary 4.1.15. (ii)$\Rightarrow$(iv) follows in the same way, but here we use the fact that in Corollary 4.1.15, the decomposition $I = \sum_{j \in \mathbb{N}^\nu} E_j$ uses elementary operators of the form $E_j = E^1_{j_1} \otimes \cdots \otimes E^\nu_{j_\nu}$; this is important since the Definition 4.1.8 only gives estimates for T when acting on bump functions which are elementary tensor products. $\qquad\square$

COROLLARY 4.1.17. *Suppose T and S are product singular integral operators of order $t \in \mathbb{R}^\nu$ and $s \in \mathbb{R}^\nu$, respectively. Then TS is a product singular integral operator of order $t + s$.*

PROOF. This follows immediately from (iii) of Theorem 4.1.12. $\qquad\square$

On each factor M_μ of M we have a sub-Laplacian: $\mathcal{L}_\mu := (W^\mu_1)^* W^\mu_1 + \cdots + \left(W^\mu_{r_\mu} \right)^* W^\mu_{r_\mu}$. As shown in Lemma 2.6.3, $\mathcal{L}_\mu$ is an essentially self-adjoint operator, and we identify it with its unique self-adjoint extension, which we think of as an operator on $L^2(M)$. Since $(I + \mathcal{L}_\mu)^{-1}$ are commuting, bounded, self-adjoint operators on $L^2(M)$, we may associate to $\mathcal{L}_1, \ldots, \mathcal{L}_\nu$ a joint spectral decomposition $E(\lambda_1, \ldots, \lambda_\nu)$. For a Borel measurable function $m : [0, \infty)^\nu \to \mathbb{C}$ we define the operator

$$m(\mathcal{L}_1, \ldots, \mathcal{L}_\nu) := \int m(\lambda_1, \ldots, \lambda_\nu) \, dE(\lambda_1, \ldots, \lambda_\nu).$$

THEOREM 4.1.18. *Let $\lambda_0^\mu > 0$ be the least nonzero eigenvalue of $\mathcal{L}_\mu$ (this is possible by Lemma 2.6.4). Fix $t \in \mathbb{R}^\nu$ and let $m : [0, \infty)^\nu \to \mathbb{C}$ be a function with*

$$m\Big|_{(\lambda_0^1/2, \infty) \times \cdots \times (\lambda_0^\nu/2, \infty)} \in C^\infty \text{ and satisfy}$$

$$\sup_{\lambda_\mu \in (\lambda_0^\mu/2, \infty)} \lambda_1^{-t_1} \cdots \lambda_\nu^{-t_\nu} \left| (\lambda_1 \partial_{\lambda_1})^{a_1} \cdots (\lambda_\nu \partial_{\lambda_\nu})^{a_\nu} m(\lambda_1, \ldots, \lambda_\nu) \right| < \infty,$$

$\forall a_1, \ldots, a_\nu \in \mathbb{N}$. Then, $m(\mathcal{L}_1, \ldots, \mathcal{L}_\nu)$ is a product singular integral operator of order $2t$.

It is not hard to prove Theorem 4.1.18 by a reprise of the proof of Theorem 2.6.6 in each factor. This can be seen in a bit more of an abstract way, using tensor products. We include the main part of this, more abstract, approach because it yields an enlightening way to view the decomposition (iv) in Theorem 4.1.12. The key theorem is the product analog of Lemma 2.6.12. Indeed, we have

LEMMA 4.1.19. *Suppose $\mathcal{B} \subset \mathcal{S}(\mathbb{R}^\nu)$ is a bounded set. Then*

$$\left\{ \left(m\left(s_1^2 \mathcal{L}_1, s_2^2 \mathcal{L}_2, \ldots, s_\nu^2 \mathcal{L}_\nu \right), (s_1, \ldots, s_\nu) \right) \mid s_1, \ldots, s_\nu \in (0, 1], m \in \mathcal{B} \right\}$$

is a bounded set of pre-elementary operators.

PROOF. We define a Fréchet space corresponding to pre-elementary operators. This Fréchet space consists of those functions $f : (0, 1]^\nu \to C^\infty(M \times M)$ satisfying, for every α, β, and m,

$$\sup_{j \in [0, \infty)^\nu} \left| \left(2^{-j} X_x \right)^\alpha \left(2^{-j} X_z \right)^\beta f\left(2^{-j} \right) (x, z) \right| \le C_{\alpha, \beta, m} \frac{\left(1 + 2^j \cdot \rho(x, z) \right)^{-m}}{\mathrm{Vol}\left(B\left(x, 2^{-j} + \rho(x, z) \right) \right)}.$$

We give F the coarsest topology such that for each α, β, and m, the least $C_{\alpha, \beta, m}$ defines a continous semi-norm. The proof of Lemma 2.6.12 actually shows that the ν-linear map $\mathcal{S}(\mathbb{R})^\nu \to F$ given by

$$(m_1, \ldots, m_\nu) \mapsto \left[(s_1, \ldots, s_\nu) \mapsto m_1\left(s_1^2 \mathcal{L}_1 \right) \otimes \cdots \otimes m_\nu\left(s_\nu^2 \mathcal{L}_\nu \right) \right]$$

is continuous. The universal property applies to show that this maps extends to a continuous map $\left[\mathcal{S}(\mathbb{R}) \widehat{\otimes} \cdots \widehat{\otimes} \mathcal{S}(\mathbb{R}) \cong \mathcal{S}(\mathbb{R}^\nu) \right] \to F$ (see Appendix A.2 for more details). The result follows. $\qquad \square$

From here the proof of Theorem 4.1.18 follows from a simple modification of the proof of Theorem 2.6.6, which we leave to the reader.

Remark 4.1.20   The proof of Lemma 4.1.19 offers a particularly transparent example of a general idea to which we often turn. Techniques for studying singular integrals in the single-parameter theory when working with something like Definition 2.0.16 often do not translate directly over to the multi-parameter situation when using definitions like Definition 4.1.4; in this particular case this is because the singular integrals in Definition 4.1.4 correspond to the *injective* tensor product of those covered in Definition 2.0.16. However, techniques when dealing with elementary operators often do transfer over. This is because elementary operators in the multi-parameter setting are closely related to nuclear spaces (e.g., $\mathcal{S}(\mathbb{R}^n)$ and $\mathcal{S}_0(\mathbb{R}^n)$) and the tensor products that appear are closely related to the projective tensor product, and are in a setting where the universal property (or something similar) can be exploited.

4.1.1   Non-isotropic Sobolev spaces

In analogy to the theory developed in the single-parameter case in Section 2.10, we develop multi-parameter non-isotropic Sobolev spaces which are adapted to the product singular integral operators. To do this, we use the sub-Laplacians on each factor. Note, for $(s_1, \ldots, s_\nu) \in \mathbb{R}^\nu$, $(I + \mathcal{L}_1)^{s_1/2} \cdots (I + \mathcal{L}_\nu)^{s_\nu/2}$ is a product singular integral operator of order $(s_1, \ldots, s_\nu)$ by Theorem 4.1.18. For $p \in (1, \infty)$, $s = (s_1, \ldots, s_\nu) \in \mathbb{R}^\nu$, let NL_s^p be the completion of $C^\infty(M)$ under the norm

$$\|f\|_{\mathrm{NL}_s^p} := \left\| (I + \mathcal{L}_1)^{s_1/2} \cdots (I + \mathcal{L}_\nu)^{s_\nu/2} f \right\|_{L^p}.$$

In particular $L^p = \mathrm{NL}_0^p$. The main theorem concerning these spaces is

THEOREM 4.1.21.   *Suppose $T : C^\infty(M) \to C^\infty(M)$ is a product singular integral operator of order $t \in \mathbb{R}^\nu$. Then, T extends to a bounded operator $\mathrm{NL}_s^p \to \mathrm{NL}_{s-t}^p$, $1 < p < \infty$, $s \in \mathbb{R}^\nu$.*

To prove Theorem 4.1.21, we begin with the case $s = t = 0$.

PROPOSITION 4.1.22.   *Let $T : C^\infty(M) \to C^\infty(M)$ be a product singular integral operator of order 0. Then T extends to a bounded operator $L^p \to L^p$ $(1 < p < \infty)$.*

We turn to outlining the proof of Proposition 4.1.22. The ideas used here are used more generally in Section 5.8, though the situation here is simpler. Once this situation is well understood, the technical details that appear in later situations become less distracting.

In the proof of Proposition 4.1.22, we see a phenomenon which did not occur in the single-parameter case: product singular integral operators are not, in general, weak-type $(1, 1)$. The Marcinkiewicz interpolation theorem is, therefore, not of direct use to us. However, *indirectly* it is quite useful; see Remark 4.1.24.

The main idea, which is used several times in the sequel, is to use the single-parameter Calderón-Zygmund theory to characterize L^p with a square function. Indeed, on each factor M_μ, we consider the identity operator, $I_\mu : C^\infty(M_\mu) \to C^\infty(M_\mu)$, which is a Calderón-Zygmund singular integral operator of order 0. In light of Theorem 2.0.29, there is a bounded set of elementary operators $\{(D_j^\mu, 2^{-j}) \mid j \in \mathbb{N}\}$ with $I_\mu = \sum_{j \in \mathbb{N}} D_j^\mu$. For $j = (j_1, \ldots, j_\nu) \in \mathbb{N}^\nu$, let

$$D_j := D_{j_1}^1 \otimes \cdots \otimes D_{j_\nu}^\nu : C^\infty(M) \to C^\infty(M).$$

Thus, $I : C^\infty(M) \to C^\infty(M)$ can be decomposed $I = I_1 \otimes \cdots \otimes I_\nu = \sum_{j \in \mathbb{N}^\nu} D_j$. For $j \in \mathbb{Z}^\nu \setminus \mathbb{N}^\nu$, we define $D_j := 0$.

PROPOSITION 4.1.23. *For* $1 < p < \infty$,

$$\|f\|_{L^p(M)} \approx \left\| \left(\sum_{j \in \mathbb{N}^\nu} |D_j f|^2 \right)^{\frac{1}{2}} \right\|_{L^p(M)}, \quad \forall f \in C^\infty(M).$$

PROOF. Let $\{\epsilon_j^\mu\}_{j \in \mathbb{N}, 1 \le \mu \le \nu}$ be a sequence of i.i.d. random variables of mean 0 taking values ± 1. For $j = (j_1, \ldots, j_\nu) \in \mathbb{N}^\nu$ set $\epsilon_j = \epsilon_{j_1}^1 \cdots \epsilon_{j_\nu}^\nu$, so that $\{\epsilon_j\}_{j \in \mathbb{N}^\nu}$ is also a sequence of i.i.d. random variables of mean 0 taking values ± 1. Proposition 2.10.8 and the Khintchine inequality (Theorem 2.10.10) shows, for each μ,

$$\|f\|_{L^p(M_\mu)} \approx \left\| \left(\sum_{j_\mu \in \mathbb{N}} |D_{j_\mu}^\mu f|^2 \right)^{\frac{1}{2}} \right\|_{L^p(M_\mu)} \approx \left(\mathbb{E} \left\| \sum_{j_\mu = 0}^\infty \epsilon_{j_\mu}^\mu D_{j_\mu}^\mu f \right\|_{L^p(M_\mu)}^p \right)^{\frac{1}{p}}.$$

Applying this ν times and using the Khintchine inequality gives

$$\|f\|_{L^p(M)} \approx \left(\mathbb{E} \left\| \sum_{j \in \mathbb{N}^\nu} \epsilon_j D_j f \right\|_{L^p(M)}^p \right)^{\frac{1}{p}} \approx \left\| \left(\sum_{j \in \mathbb{N}^\nu} |D_j f|^2 \right)^{\frac{1}{2}} \right\|_{L^p(M)},$$

completing the proof. □

Remark 4.1.24 Note that the proof of Proposition 4.1.23 uses Proposition 2.10.8, which in turn uses the L^p boundedness of Calderón-Zygmund singular integral operators of order 0. The L^p boundedness of Calderón-Zygmund operators of order 0 follows from the fact that they are weak-type $(1,1)$ and the Marcinkiewicz interpolation theorem. Hence, Proposition 4.1.23 uses the Marcinkiewicz interpolation theorem in an indirect way.

Corresponding to the product singular integrals, there is a product maximal function:

$$\mathcal{M}f(x) = \sup_{\delta \in (0,\infty)^\nu} \frac{1}{\text{Vol}(B(x,\delta))} \int_{B(x,\delta)} |f(y)| \, dy.$$

THEOREM 4.1.25. $\|\mathcal{M}f\|_{L^p} \lesssim \|f\|_{L^p}$, *for* $1 < p \le \infty$.

PROOF. On each factor, M_μ, we have the maximal function

$$M_\mu f(x_\mu) = \sup_{\delta_\mu > 0} \frac{1}{\mathrm{Vol}\left(B_{\left(X^\mu, \hat{d}^\mu\right)}(x_\mu, \delta_\mu)\right)} \int_{B_{\left(X^\mu, \hat{d}^\mu\right)}(x_\mu, \delta_\mu)} |f(y_\mu)| \, dy_\mu.$$

Theorem 2.9.1 shows $\|M_\mu f\|_{L^p(M_\mu)} \lesssim \|f\|_{L^p(M_\mu)}$, $1 < p \le \infty$. We lift M_μ to functions on M by letting M_μ act in only the μth variable. As a consequence, $\|M_\mu f\|_{L^p(M)} \lesssim \|f\|_{L^p(M)}$, $1 < p \le \infty$. It is immediate to verify that we have the pointwise inequality $Mf(x) \le M_1 M_2 \cdots M_\nu f(x)$. The result follows. $\qquad \square$

We introduce some vector valued operators. In analogy to the theory in Section 2.9, we use the spaces $L^p(\ell^q(\mathbb{N}^\nu))$; these are defined in the same way, but with the sequences indexed by $\mathbb{N}^\nu$ instead of $\mathbb{N}$.

LEMMA 4.1.26. *Suppose $\mathcal{E}$ is a bounded set of pre-elementary operators, and let $\{(E_j, 2^{-j}) \mid j \in \mathbb{N}^\nu\} \subset \mathcal{E}$. Define the vector valued operator $\mathcal{T}\{f_j\}_{j \in \mathbb{N}^\nu} = \{E_j f_j\}_{j \in \mathbb{N}^\nu}$. Then, for $1 < p < \infty$, there is a constant $C_{p,\mathcal{E}}$ such that*

$$\|\mathcal{T}\|_{L^p(\ell^2(\mathbb{N}^\nu)) \to L^p(\ell^2(\mathbb{N}^\nu))} \le C_{p,\mathcal{E}}.$$

The proof of Lemma 4.1.26 is nearly identical to the proof of Lemma 2.9.2. We outline the proof as several lemmas, and for each lemma refer the reader to the corresponding version of that lemma in the proof of Lemma 2.9.2. Those proofs can be translated over with only simple modifications to prove the results in this setting.

LEMMA 4.1.27. *Suppose $\mathcal{E}$ is a bounded set of pre-elementary operators. Then, for $1 \le p \le \infty$,*

$$\sup_{(E, 2^{-j}) \in \mathcal{E}} \|E\|_{L^p \to L^p} < \infty.$$

PROOF. This follows just as in Lemma 2.9.4. $\qquad \square$

LEMMA 4.1.28. *Let $\mathcal{E}$ be a bounded set of pre-elementary operators. Then, there is a constant $C_\mathcal{E}$ such that for $(E, 2^{-j}) \in \mathcal{E}$, we have the pointwise estimate $|Ef(x)| \le C_\mathcal{E} Mf(x)$.*

PROOF. This follows just as in the proof of Lemma 2.9.3; here, we decompose the integral by decomposing

$$M = B(x, 2^{-j}) \bigcup \left(\bigcup_{k=1}^{\infty} \left[B\left(x, 2^k 2^{-j}\right) \setminus B\left(x, 2^{k-1} 2^{-j}\right) \right] \right).$$

Indeed, we have

$$
\begin{aligned}
|Ef(x)| &\lesssim \int \frac{\left(1 + 2^j \cdot \rho(x,y)\right)^{-1}}{\mathrm{Vol}\left(B\left(x, 2^{-j} + \rho(x,y)\right)\right)} |f(y)| \, dy \\
&\lesssim \int_{B(x,2^{-j})} \frac{1}{\mathrm{Vol}\left(B\left(x, 2^{-j}\right)\right)} |f(y)| \, dy \\
&\quad + \sum_{k=1}^{\infty} \int_{B(x, 2^k 2^{-j}) \setminus B(x, 2^{k-1} 2^{-j})} \frac{\left(2^j \cdot \rho(x,y)\right)^{-1}}{\mathrm{Vol}\left(B\left(x, \rho(x,y)\right)\right)} |f(y)| \, dy \\
&\lesssim \sum_{k=0}^{\infty} 2^{-k} \mathrm{Vol}\left(B\left(x, 2^k 2^{-j}\right)\right)^{-1} \int_{B(x, 2^k 2^{-j})} |f(y)| \, dy \\
&\lesssim \mathcal{M} f(x),
\end{aligned}
$$

as desired. □

LEMMA 4.1.29. *Let $\mathcal{E}$ be a bounded set of pre-elementary operators, and let*

$$
\left\{ (E_j, 2^{-j}) \mid j \in \mathbb{N}^\nu \right\} \subset \mathcal{E}.
$$

Define the vector valued operator $\mathcal{T}\{f_j\}_{j \in \mathbb{N}^\nu} := \{E_j f_j\}_{j \in \mathbb{N}^\nu}$. Then, there is a constant $C_\mathcal{E}$ such that for $1 \le p < \infty$,

$$
\|\mathcal{T}\|_{L^p(\ell^p(\mathbb{N}^\nu)) \to L^p(\ell^p(\mathbb{N}^\nu))} \le C_\mathcal{E}.
$$

PROOF. This follows just as in Lemma 2.9.5, using Lemma 4.1.27. □

LEMMA 4.1.30. *Let $\mathcal{E}$ be a bounded set of pre-elementary operators, and let*

$$
\left\{ (E_j, 2^{-j}) \mid j \in \mathbb{N}^\nu \right\} \subset \mathcal{E}.
$$

Define the vector valued operator $\mathcal{T}\{f_j\}_{j \in \mathbb{N}^\nu} := \{E_j f_j\}_{j \in \mathbb{N}^\nu}$. Then, for $1 < p < \infty$, there is a constant $C_{p,\mathcal{E}}$ such that

$$
\|\mathcal{T}\|_{L^p(\ell^\infty(\mathbb{N}^\nu)) \to L^p(\ell^\infty(\mathbb{N}^\nu))} \le C_{p,\mathcal{E}}.
$$

PROOF. This follows just as in Lemma 2.9.6, using Theorem 4.1.25. □

PROOF OF LEMMA 4.1.26. Using the above lemmas, this follows just as in the proof of Lemma 2.9.2. □

PROPOSITION 4.1.31. *Suppose $\mathcal{E}$ is a bounded set of elementary operators and let $\left\{ (E_j, 2^{-j}), (F_j, 2^{-j}) \mid j \in \mathbb{N}^\nu \right\} \subset \mathcal{E}$. For $j \in \mathbb{Z}^\nu \setminus \mathbb{N}^\nu$ set $F_j = 0$. For $l \in \mathbb{Z}^\nu$ define the vector valued operator $\mathcal{T}_l\{f_j\}_{j \in \mathbb{N}^\nu} = \{E_j F_{j+l} f_j\}_{j \in \mathbb{N}^\nu}$. Then for $1 < p < \infty$, and $N \in \mathbb{N}$, there is a constant $C_{p,N,\mathcal{E}}$ such that $\|\mathcal{T}_l\|_{L^p(\ell^2(\mathbb{N}^\nu)) \to L^p(\ell^2(\mathbb{N}^\nu))} \le C_{p,N,\mathcal{E}} 2^{-N|l|}$.*

PROOF. This follows just as in Proposition 2.9.7, using Lemma 4.1.26 and Proposition 4.1.16. $\qquad\square$

For each $1 \le \mu \le \nu$ and $L_\mu \in \mathbb{N}$, define $U^\mu_{L_\mu} = \sum_{\substack{j \in \mathbb{N} \\ |l_\mu| \le L_\mu}} D^\mu_{j_\mu} D^\mu_{j_\mu + l_\mu}$. For $L = (L_1, \ldots, L_\nu) \in \mathbb{N}^\nu$ define $U_L = U^1_{L_1} \otimes \cdots \otimes U^\nu_{L_\nu}$.

PROPOSITION 4.1.32. *Fix p, $1 < p < \infty$. There is an $L = L(p)$ such that $U_L : L^p(M) \to L^p(M)$ is an isomorphism.*

PROOF. For each μ, there is $L_\mu = L_\mu(p)$ such that $U^\mu_{L_\mu} : L^p(M_\mu) \to L^p(M_\mu)$ is an isomorphism by Proposition 2.10.11. Setting $L = (L_1, \ldots, L_\nu)$, the result follows. $\qquad\square$

PROOF OF PROPOSITION 4.1.22. Fix p, $1 < p < \infty$ and let T be a product singular integral operator of order 0. Let $L = L(p)$ be as in Proposition 4.1.32 so that $U_L : L^p \to L^p$ is an isomorphism; it suffices to show TU_L is bounded on L^p. For $k \in \mathbb{Z}^\nu$ define the vector valued operator $\mathcal{T}_k \{f_j\}_{j \in \mathbb{N}^\nu} = \{D_j T D_{j+k} f_j\}_{j \in \mathbb{N}^\nu}$. Since $\{(TD_j, 2^{-j}) \mid j \in \mathbb{N}^\nu\}$ is a bounded set of elementary operators, Proposition 4.1.31 shows $\|\mathcal{T}_k\|_{L^p(\ell^2(\mathbb{N}^\nu)) \to L^p(\ell^2(\mathbb{N}^\nu))} \lesssim 2^{-|k|}$. We have, using Proposition 4.1.23 and the triangle inequality,

$$
\begin{aligned}
\|TU_L f\|_{L^p} &\approx \left\| \left(\sum_{j \in \mathbb{N}^\nu} \left| \sum_{\substack{k \in \mathbb{Z}^\nu \\ l_\mu \le L_\mu, \forall \mu}} D_j T D_{j+k} D_{j+k+l} f \right|^2 \right)^{\frac{1}{2}} \right\|_{L^p} \\
&\le \sum_{\substack{k \in \mathbb{Z}^\nu \\ l_\mu \le L_\mu, \forall \mu}} \left\| \left(\sum_{j \in \mathbb{N}^\nu} |D_j T D_{j+k} D_{j+k+l} f|^2 \right)^{\frac{1}{2}} \right\|_{L^p} \\
&= \sum_{\substack{k \in \mathbb{Z}^\nu \\ l_\mu \le L_\mu, \forall \mu}} \left\| \mathcal{T}_k \{D_{j+k+l} f\}_{j \in \mathbb{N}^\nu} \right\|_{L^p(\ell^2(\mathbb{N}^\nu))} \\
&\lesssim \sum_{\substack{k \in \mathbb{Z}^\nu \\ l_\mu \le L_\mu, \forall \mu}} 2^{-|k|} \left\| \{D_j f\}_{j \in \mathbb{N}^\nu} \right\|_{L^p(\ell^2(\mathbb{N}^\nu))} \\
&\approx \left\| \left(\sum_{j \in \mathbb{N}^\nu} |D_j f|^2 \right)^{\frac{1}{2}} \right\|_{L^p} \\
&\approx \|f\|_{L^p},
\end{aligned}
$$

completing the proof. $\qquad\square$

PROOF OF THEOREM 4.1.21. Let T be a product singular integral of order t. We wish to show

$$\left\| (I + \mathcal{L}_1)^{(s_1 - t_1)/2} \cdots (I + \mathcal{L}_\nu)^{(s_\nu - t_\nu)/2} T f \right\|_{L^p}$$
$$\lesssim \left\| (I + \mathcal{L}_1)^{s_1/2} \cdots (I + \mathcal{L}_\nu)^{s_\nu/2} f \right\|_{L^p}.$$

Setting $g = (I + \mathcal{L}_1)^{s_1/2} \cdots (I + \mathcal{L}_\nu)^{s_\nu/2} f$ this is equivalent to showing

$$\left\| (I + \mathcal{L}_1)^{(s_1 - t_1)/2} \cdots (I + \mathcal{L}_\nu)^{(s_\nu - t_\nu)/2} T (I + \mathcal{L}_1)^{-s_1/2} \cdots (I + \mathcal{L}_\nu)^{-s_\nu/2} g \right\|_{L^p}$$
$$\lesssim \| g \|_{L^p}.$$

$$(4.5)$$

Theorem 4.1.18 shows that $(I + \mathcal{L}_1)^{(s_1 - t_1)/2} \cdots (I + \mathcal{L}_\nu)^{(s_\nu - t_\nu)/2}$ is a product singular integral operator of order $s - t$, while $(I + \mathcal{L}_1)^{-s_1/2} \cdots (I + \mathcal{L}_\nu)^{-s_\nu/2}$ is a product singular integral operator of order $-s$. Corollary 4.1.17 then implies

$$(I + \mathcal{L}_1)^{(s_1 - t_1)/2} \cdots (I + \mathcal{L}_\nu)^{(s_\nu - t_\nu)/2} T (I + \mathcal{L}_1)^{-s_1/2} \cdots (I + \mathcal{L}_\nu)^{-s_\nu/2}$$

is a product singular integral operator of order 0, and (4.5) follows from Proposition 4.1.22. $\qquad \square$

We close this section with two more characterizations of the space NL_s^p, which are analogous to Theorems 2.10.6 and 2.10.18.

THEOREM 4.1.33. *Fix $p \in (1, \infty)$ and $s \in \mathbb{R}^\nu$. Then,*

$$\| f \|_{\mathrm{NL}_s^p} \approx \left\| \left(\sum_{j \in \mathbb{N}^\nu} |2^{j \cdot s} D_j f|^2 \right)^{\frac{1}{2}} \right\|_{L^p},$$

where the implicit constants depend on p, s, and the particular choice of decompositions $I_\mu = \sum_{j_\mu \in \mathbb{N}} D_{j_\mu}^\mu$.

PROOF. Let $\{ \epsilon_j^\mu \}_{j \in \mathbb{N}, 1 \le \mu \le \nu}$ be a sequence of i.i.d. random variables of mean 0 taking values ± 1. For $j = (j_1, \dots, j_\nu) \in \mathbb{N}^\nu$ set $\epsilon_j = \epsilon_{j_1}^1 \cdots \epsilon_{j_\nu}^\nu$ so that $\{ \epsilon_j \}_{j \in \mathbb{N}^\nu}$ is also a sequence of i.i.d. random variables of mean 0 taking values ± 1. For each μ, Theorem 2.10.6 and the Khintchine inequality (Theorem 2.10.10) show

$$\left\| (I + \mathcal{L}_\mu)^{s_\mu/2} f \right\|_{L^p(M_\mu)} \approx \left\| \left(\sum_{j_\mu \in \mathbb{N}} \left| 2^{j_\mu s_\mu} D_{j_\mu}^\mu f \right|^2 \right)^{\frac{1}{2}} \right\|_{L^p(M_\mu)}$$

$$\approx \left(\mathbb{E} \left\| \sum_{j_\mu \in \mathbb{N}} \epsilon_{j_\mu}^\mu 2^{j_\mu s_\mu} D_{j_\mu}^\mu f \right\|_{L^p(M_\mu)}^p \right)^{\frac{1}{p}}.$$

Applying the above in each factor, and using the Khintchine inequality again, we have

$$
\left\|(I+\mathcal{L}_1)^{s_1/2}\cdots(I+\mathcal{L}_\nu)^{s_\nu/2} f\right\|_{L^p(M)} \approx \left(\mathbb{E}\left\|\sum_{j\in\mathbb{N}^\nu}\epsilon_j 2^{j\cdot s} D_j f\right\|_{L^p(M)}^p\right)^{\frac{1}{p}}
$$

$$
\approx \left\|\left(\sum_{j\in\mathbb{N}^\nu}\left|2^{j\cdot s} D_j f\right|^2\right)^{\frac{1}{2}}\right\|_{L^p(M)},
$$

completing the proof. $\square$

THEOREM 4.1.34. *Fix $p \in (1, \infty)$ and $N = (N_1, \ldots, N_\nu) \in \mathbb{N}^\nu$. Then,*

$$
\|f\|_{\mathrm{NL}_N^p} \approx \sum_{|\alpha_\mu|\leq N_\mu}\left\|(W^1)^{\alpha_1}\cdots(W^\nu)^{\alpha_\nu} f\right\|_{L^p}.
$$

PROOF. This follows by applying Theorem 2.10.18 in each factor. $\square$

4.1.2 Further reading and references

The product type maximal function considered in Theorem 4.1.25 takes its roots in the maximal function of Jessen, Marcinkiewicz, and Zygmund [JMZ35], and the proof of Theorem 4.1.25, in fact, uses the same methods as that reference.

Singular integrals of product type, like the ones here, were first considered in the work of Fefferman and Stein [FS82]. Another important early work on product type operators is the work of Journé [Jou85]. From here many works followed (see, for instance, [Fef87, CF85, RS92, MRS95]). The particular product type singular integrals we study (at least the operators of order $0 \in \mathbb{R}^\nu$) appeared first in the work of Nagel and Stein [NS04], who proved the L^p boundedness of such operators, but did not address the question of whether or not the operators form an algebra. Parts (iii) and (iv) of Theorem 4.1.12 seem to be new. Another recent related paper is [HLY07].

4.2 FLAG KERNELS ON GRADED GROUPS AND BEYOND

For our next example, we begin by working on a graded, nilpotent Lie group G (see Definition 2.13.3). As discussed in Section 2.13, if $\mathfrak{g}$ is the Lie algebra of G, then $\exp : \mathfrak{g} \to G$ is a diffeomorphism. Let $q_0 = \dim G$. Since $\mathbb{R}^{q_0} \cong \mathfrak{g} \cong G$, we may identify G with $\mathbb{R}^{q_0}$. Example 2.13.6 shows that this graded group has a dilation structure. As in the definition of a graded group (see Definition 2.13.3), we decompose $\mathfrak{g} = \oplus_{\mu=1}^\nu V_\mu$, where $[V_{\mu_1}, V_{\mu_2}] \subseteq V_{\mu_1+\mu_2}$. We identify each V_μ with $\mathbb{R}^{v_\mu}$, where $v_\mu = \dim V_\mu$. This decomposes $\mathbb{R}^{q_0} = \mathbb{R}^{v_1} \oplus \cdots \oplus \mathbb{R}^{v_\nu}$. We write $t = (t_1, \ldots, t_\nu) \in \mathbb{R}^{q_0}$. Notice, our dilations are defined in such a way that $\delta(t_1, t_2, \ldots, t_\nu) = (\delta t_1, \delta^2 t_2, \ldots, \delta^\nu t_\nu)$, and if $Z \in V_\mu$, Z is homogeneous of degree μ under these dilations.

Our first goal is to define a class of ν-parameter, left invariant operators on G. Since these operators are left invariant, they are of the form $f \mapsto f * K$ for some distribution K. We turn to defining these distributions. The class of distributions we study does not depend on the group structure, and so we work more generally on $\mathbb{R}^{q_0}$, where q_0 is arbitrary. More precisely, we assume we are given a decomposition $\mathbb{R}^{q_0} = \mathbb{R}^{v_1} \otimes \cdots \otimes \mathbb{R}^{v_\nu}$, and we are given natural numbers $0 < e_1 < e_2 < \cdots < e_\nu \in \mathbb{N}$. Our dilations are defined to be, for $\delta > 0$, $\delta(t_1, \ldots, t_\nu) = (\delta^{e_1} t_1, \ldots, \delta^{e_\nu} t_\nu)$. On each factor $\mathbb{R}^{v_\mu}$ we define a homogeneous norm $|t_\mu|_\mu = |t_\mu|^{\frac{1}{e_\mu}}$, where the norm on the right is the usual Euclidean norm. Notice, in the case of a graded, nilpotent Lie group, the norm $\sum_{\mu=1}^{\nu} |t_\mu|_\mu$ defines a homogeneous norm as in Definition 2.13.9. For each μ, let $Q_\mu = e_\mu v_\mu$: the homogeneous dimension of $\mathbb{R}^{v_\mu}$ under these dilations. Let $Q = (Q_1, \ldots, Q_\nu)$.

DEFINITION 4.2.1. *Fix $s_1, \ldots, s_\nu \in \mathbb{R}$ with $s_\mu > -Q_\mu$. We define the locally convex topological vector space of flag kernels of order $(s_1, \ldots, s_\nu)$ recursively on ν. The space is a subspace of $C_0^\infty(\mathbb{R}^{q_0})'$. When $\nu = 0$ (and therefore $q_0 = 0$) it is defined to be $\mathbb{C}$ with the usual topology. For $\nu > 0$ it is the space of all distributions $K \in C_0^\infty(\mathbb{R}^{q_0})'$ such that*

- *(Growth Condition) K is C^∞ away from $t_1 = 0$, and $\forall \alpha = (\alpha_1, \ldots, \alpha_\nu) \in \mathbb{N}^{v_1} \times \cdots \times \mathbb{N}^{v_\nu} = \mathbb{N}^{q_0}$, there is a constant C_α such that*

$$|\partial_t^\alpha K(t)| \leq C_\alpha \prod_{\mu=1}^{\nu} \left[|t_1|_1 + \cdots + |t_\mu|_\mu\right]^{-Q_\mu - e_\mu |\alpha_\mu| - s_\mu}.$$

We define the least possible C_α to be a continuous semi-norm.

- *(Cancellation Condition) For any μ, any bounded set $\mathcal{B} \subset C_0^\infty(\mathbb{R}^{v_\mu})$, any $R > 0$, and any $\phi \in \mathcal{B}$, we assume that the distribution*

$$K_{\phi,R} \in C_0^\infty(\mathbb{R}^{v_1} \times \cdots \times \mathbb{R}^{v_{\mu-1}} \times \mathbb{R}^{v_{\mu+1}} \times \cdots \times \mathbb{R}^{v_\nu})'$$

defined by

$$K_{\phi,R}(t_1, \ldots, t_{\mu-1}, t_{\mu+1}, \ldots, t_\nu) = R^{-s_\mu} \int K(t) \, \phi(R^{e_\mu} t_\mu) \, dt_\mu$$

is a flag kernel of order $(s_1, \ldots, s_{\mu-1}, s_{\mu+1}, \ldots, s_\nu)$ on the space $\mathbb{R}^{v_1} \times \cdots \times \mathbb{R}^{v_{\mu-1}} \times \mathbb{R}^{v_{\mu+1}} \times \cdots \times \mathbb{R}^{v_\nu}$ and for any continuous semi-norm $|\cdot|$ on the locally convex space of flag kernels of order $(s_1, \ldots, s_{\mu-1}, s_{\mu+1}, \ldots, s_\nu)$ on the space $\mathbb{R}^{v_1} \times \cdots \times \mathbb{R}^{v_{\mu-1}} \times \mathbb{R}^{v_{\mu+1}} \times \cdots \times \mathbb{R}^{v_\nu}$, we define a continuous semi-norm on the space of flag kernels of order $(s_1, \ldots, s_\nu)$ on $\mathbb{R}^{v_1} \times \cdots \times \mathbb{R}^{v_\nu}$ by $\sup_{\phi \in \mathcal{B}, R > 0} |K_{\phi,R}|$, which we assume to be finite.

We give the space of flag kernels of order $(s_1, \ldots, s_\nu)$ the coarsest topology such that the above semi-norms are continuous.

Given a distribution $K \in C_0^\infty (\mathbb{R}^{q_0})'$, we may think of $K \in C_0^\infty (G)'$ by identifying G with $\mathfrak{g} \cong \mathbb{R}^{q_0}$ via the exponential map. For such a distribution, we define a left-invariant operator $\mathrm{Op}\,(K) : C_0^\infty (G) \to C^\infty (G)$ by $\mathrm{Op}\,(K) f := f * K$, and the convolution is taken in the sense of the group structure on G. The operators $\mathrm{Op}\,(K)$, where K is a flag kernel, have recently received interest; see [NRS01, NRSW12, Głó10, Głó12b, Głó12a]. We state two of the main theorems concerning these operators. The first is due to Nagel, Ricci, Stein, and Wainger [NRSW12], while the second is due to Nagel, Ricci, and Stein [NRS01].

THEOREM 4.2.2. *Suppose K_1 and K_2 are flag kernels of order $t = (t_1, \ldots, t_\nu)$ and $s = (s_1, \ldots, s_\nu)$, respectively, with $t_\mu, s_\mu, t_\mu + s_\mu > -Q_\mu, \forall \mu$. Then there is a flag kernel K_3 of order $t + s$ such that $\mathrm{Op}\,(K_1)\,\mathrm{Op}\,(K_2) = \mathrm{Op}\,(K_3)$.*

THEOREM 4.2.3. *Suppose K is a flag kernel of order 0, then $\mathrm{Op}\,(K)$ extends to a bounded operator $\mathrm{Op}\,(K) : L^p \to L^p \ (1 < p < \infty)$.*

Instead of discussing the proofs of the above theorems directly, we move to a more general situation. Indeed, we leave the setting of translation invariant operators, and even leave the group setting altogether. We again work on $\mathbb{R}^{q_0}$ which we assume to be decomposed $\mathbb{R}^{q_0} = \mathbb{R}^{v_1} \times \cdots \times \mathbb{R}^{v_\nu}$, and we are given natural numbers $0 < e_1 < e_2 < \cdots < e_\nu \in \mathbb{N}$. We define dilations by, for $\delta > 0$,

$$\delta\,(t_1, \ldots, t_\nu) = (\delta^{e_1} t_1, \ldots, \delta^{e_\nu} t_\nu). \tag{4.6}$$

We assume we are given vector fields $Z_1, \ldots, Z_{q_0}$ on $\mathbb{R}^{q_0}$, which span the tangent space to each point of $\mathbb{R}^{q_0}$. Further, we assume each Z_j is homogeneous of degree $0 \ne \hat{d}_j \in \mathbb{N}$ with respect to the above dilations, in the sense that $Z_j f\,(\delta x) = \delta^{\hat{d}_j}\,(Z_j f)\,(\delta x)$. We assign to Z_j the single-parameter formal degree $\hat{d}_j$, and we assume $[Z_j, Z_k] = \sum_{\hat{d}_j + \hat{d}_k = \hat{d}_l} c_{j,k}^l Z_l$ for some *constants* $c_{j,k}^l \in \mathbb{R}$.[2]

Due to the homogeneity and spanning of $Z_1, \ldots, Z_{q_0}$, $\hat{d}_1, \ldots, \hat{d}_{q_0}$ take precisely ν different values: $0 < e_1 < e_2 < \ldots < e_\nu \in \mathbb{N}$. We create ν lists of vector fields with single-parameter formal degrees given by, for each μ, $\left(X^\mu, \hat{d}^\mu\right) = \left(X_1^\mu, \hat{d}_1^\mu\right), \ldots, \left(X_{q_\mu}^\mu, \hat{d}_{q_\mu}^\mu\right)$ is an enumeration of

$$\left\{ \left(Z_l, \hat{d}_l\right) \mid \hat{d}_l \ge e_\mu \right\};$$

so that for $\mu \ge \mu'$ the list $\left(X^{\mu'}, \hat{d}^{\mu'}\right)$ *contains* the list $\left(X^\mu, \hat{d}^\mu\right)$. We let $(X, d) = (X_1, d_1), \ldots, (X_q, d_q)$ be the list of vector fields with ν-parameter formal degrees induced by these ν lists of vector fields with single-parameter formal degrees, as in Chapter 3. With these choices, we have the crucial identity, for $j \in \mathbb{R}^\nu$,

$$\left[2^{-j \cdot d_k} X_k, 2^{-j \cdot d_l} X_l\right] = \sum_{m=1}^q c_{k,l}^{m,j} 2^{-j \cdot d_m} X_m, \tag{4.7}$$

---

[2] In particular, if $\hat{d}_j + \hat{d}_k > e_\nu$, then we assume $[Z_j, Z_k] = 0$.

where $c_{k,l}^{m,j} \in \mathbb{R}$ with $\sup_j \left| c_{k,l}^{m,j} \right| < \infty$. As in Chapter 3, for $j \in \mathbb{R}^\nu$ we denote by $2^{-j} X$ the list of vector fields $2^{-j \cdot d_1} X_1, \ldots, 2^{-j \cdot d_q} X_q$. Corresponding to these vector fields with formal degrees, we also obtain the balls $B_{(X,d)} \left(x, 2^{-j} \right)$, $x \in \mathbb{R}^{q_0}$, $j \in \mathbb{R}^\nu$ as in Chapter 3.

Remark 4.2.4 In light of (4.7), the vector fields (X, d) satisfy the crucial hypothesis needed in Chapter 3. Because of this, all the results from that chapter apply in this special case. In that chapter, though, we restricted attention to balls of the form $B_{(X,d)} \left(x, 2^{-j} \right)$ where x was in a small neighborhood of 0 and 2^{-j} was small. The balls in this chapter are homogeneous, in the sense that for $\delta > 0$ if we use the dilations (4.6) we have $\delta B_{(X,d)} \left(x, 2^{-j} \right) = B_{(X,d)} \left(\delta x, \delta 2^{-j} \right)$. This can be used to show that all of the results from Chapter 3 stated for x and 2^{-j} small actually hold for all x and 2^{-j} because they are homogeneous in the appropriate way.

Remark 4.2.5 Let $j \in \mathbb{R}^\nu$. Define $k \in \mathbb{R}^\nu$ recursively by $k_1 = j_1$ and for $\mu > 1$,

$$k_\mu = \begin{cases} j_\mu & \text{if } j_\mu \leq k_{\mu-1}, \\ k_{\mu-1} & \text{otherwise.} \end{cases}$$

It is easy to see that there is a constant $C > 0$ such that

$$B_{(X,d)} \left(x, 2^{-j} \right) \subseteq B_{(X,d)} \left(x, 2^{-k} \right) \subseteq B_{(X,d)} \left(x, C 2^{-j} \right).$$

Thus, when considering these balls, the main point is to consider balls of radius 2^{-j} where $j_1 \geq j_2 \geq \cdots \geq j_\nu$.

DEFINITION 4.2.6. *In light of Remark 4.2.5, we define* $\mathcal{F} \subset (0, \infty)^\nu$ *to be those* $2^{-j} \in (0, \infty)^\nu$ *with* $j_1 \geq j_2 \geq \cdots \geq j_\nu$. *We define* $\mathcal{F}_\mathbb{Z} = \left\{ 2^{-j} \in \mathcal{F} \mid j \in \mathbb{Z}^\nu \right\}$.

DEFINITION 4.2.7. *For* $j \in \mathbb{R}^\nu$ *we define a metric*

$$\rho_{2^{-j}} (x, y) = \inf \left\{ r > 0 \mid y \in B_{(X,d)} \left(x, r 2^{-j} \right) \right\}.$$

In light of Remark 4.2.5, for any $j \in \mathbb{R}^\nu$, $\rho_{2^{-j}}$ *is equivalent to some* $\rho_{2^{-k}}$, *where* $2^{-k} \in \mathcal{F}$. *We define*

$$B_{2^{-j}} (x, r) = \left\{ y \mid \rho_{2^{-j}} (x, y) < r \right\} = B_{(X,d)} \left(x, r 2^{-j} \right).$$

Remark 4.2.8 Note that the definitions of $\rho_{2^{-j}}$ and $B_{2^{-j}}$ are slightly different from those in Chapter 3. This is due to the fact that in this section, we may work with points which are arbitrarily far apart, in light of Remark 4.2.4. The definitions in Definition 4.2.7 are the appropriate definitions to use to make the results from Chapter 3 carry over to this section, in the sense of Remark 4.2.4.

DEFINITION 4.2.9. *We say* $\mathcal{E} \subset C^\infty \left(\mathbb{R}^{q_0} \times \mathbb{R}^{q_0} \right) \times (0, \infty)^\nu$ *is a bounded set of pre-elementary operators if* $\forall \alpha, \beta, m, \exists C, \forall \left(E, 2^{-j} \right) \in \mathcal{E}$,

$$\left| \left(2^{-j} X_x \right)^\alpha \left(2^{-j} X_z \right)^\beta E (x, z) \right| \leq C \frac{\left(1 + \rho_{2^{-j}} (x, z) \right)^{-m}}{\mathrm{Vol} \left(B_{2^{-j}} \left(x, 1 + \rho_{2^{-j}} (x, z) \right) \right)}.$$

DEFINITION 4.2.10. *We define the set of bounded sets of elementary operators, $\mathcal{G}$, to be the largest set of subsets of $C^\infty\left(\mathbb{R}^{q_0} \times \mathbb{R}^{q_0}\right) \times (0,\infty)^\nu$ such that for all $\mathcal{E} \in \mathcal{G}$,*

- *$\mathcal{E}$ is a bounded set of pre-elementary operators.*

- *For all $\left(E, 2^{-j}\right) \in \mathcal{E}$, and $\forall k \in \mathbb{R}^\nu$ with $k \leq j$ (coordinatewise), we may write*

$$
E = \begin{cases} 2^{-|j-k|_\infty} \sum_{|\alpha|=1} \left(2^{-k}X\right)^\alpha E_{L,\alpha,k} & \text{and} \\ 2^{-|j-k|_\infty} \sum_{|\alpha|=1} E_{R,\alpha,k} \left(2^{-k}X\right)^\alpha, \end{cases} \tag{4.8}
$$

where $\left\{ \left(E_{L,\alpha,k}, 2^{-j}\right), \left(E_{R,\alpha,k}, 2^{-j}\right) \mid \left(E, 2^{-j}\right) \in \mathcal{E}, k \leq j, |\alpha| = 1 \right\} \in \mathcal{G}$.

We call elements $\mathcal{E} \in \mathcal{G}$ bounded sets of elementary operators.

For $S \subseteq \{1,\ldots,\nu\}$, we recall the space $\mathcal{S}_0^S$ from Definition A.2.16. It is the closed subspace of $\mathcal{S}\left(\mathbb{R}^{q_0}\right)$ consisting of those $f \in \mathcal{S}\left(\mathbb{R}^{q_0}\right)$ such that for all $\mu \in S$ and all α, $\int t_\mu^\alpha f(t)\, dt_\mu = 0$.

THEOREM 4.2.11. *Fix $s \in \mathbb{R}^\nu$. For an operator $T : \mathcal{S}_0^{\{\nu\}} \to C^\infty\left(\mathbb{R}^{q_0}\right)'$, the following are equivalent*

(i) *For every bounded set of elementary operators $\mathcal{E} \subset C^\infty\left(G \times G\right) \times \mathcal{F}$, the set $\left\{ \left(2^{-j \cdot s} T E, 2^{-j}\right) \mid \left(E, 2^{-j}\right) \in \mathcal{E} \right\}$ is a bounded set of elementary operators.*

(ii) *For every bounded set of elementary operators $\mathcal{E} \subset C^\infty\left(G \times G\right) \times (0,\infty)^\nu$, the set $\left\{ \left(2^{-j \cdot s} T E, 2^{-j}\right) \mid \left(E, 2^{-j}\right) \in \mathcal{E} \right\}$ is a bounded set of elementary operators.*

(iii) *There is a bounded set of elementary operators $\left\{ \left(E_j, 2^{-j}\right) \mid 2^{-j} \in \mathcal{F}_{\mathbb{Z}} \right\}$ such that $T = \sum_{2^{-j} \in \mathcal{F}_{\mathbb{Z}}} 2^{j \cdot s} E_j$. See Proposition 4.2.17 for a sense in which this sum converges.*

(iv) *There is a bounded set of elementary operators $\left\{ \left(E_j, 2^{-j}\right) \mid j \in \mathbb{Z}^\nu \right\}$ such that $T = \sum_{j \in \mathbb{N}^\nu} 2^{j \cdot s} E_j$. See Proposition 4.2.17 for a sense in which this sum converges.*

DEFINITION 4.2.12. *We say $T : \mathcal{S}_0^{\{\nu\}} \to C^\infty\left(\mathbb{R}^{q_0}\right)'$ is a flag type singular integral operator of order $s \in \mathbb{R}^\nu$ if any of the equivalent conditions from Theorem 4.2.11 hold.*

Remark 4.2.13 Noticeably absent from Theorem 4.2.11 is a definition in terms of "growth conditions" and "cancellation conditions." This omission will be partially rectified when we turn to the more general situation later; see Remark 5.1.18.

Other than Theorem 4.2.11, the two main theorems concerning flag type singular integral operators are.

THEOREM 4.2.14. *Suppose $T, S : \mathcal{S}_0^{\{\nu\}} \to C^\infty\left(\mathbb{R}^{q_0}\right)'$ are flag type singular integral operators of order t and s, respectively. Then, the composition TS can be defined, and is a flag type singular integral operator of order $t + s$.*

PROOF. This follows immediately from (ii) of Theorem 4.2.11.   □

THEOREM 4.2.15. *Suppose* $T : \mathcal{S}_0^{\{\nu\}} \to C^\infty\left(\mathbb{R}^{q_0}\right)'$ *is a flag type singular integral operator of order* 0, *then* $T : L^p \to L^p$ $(1 < p < \infty)$.

Before we discuss the proofs of Theorems 4.2.11 and 4.2.15, we relate them to the operators $\mathrm{Op}\,(K)$ where K is a flag kernel on a graded, nilpotent Lie group as in Definition 4.2.1. Indeed, we consider the case when $\mathbb{R}^{q_0} \cong G$ with the dilations $\delta\,(t_1, t_2, \ldots, t_\nu) = \left(\delta t_1, \delta^2 t_2, \ldots, \delta^\nu t_\nu\right)$ discussed at the beginning of this section. We decompose $\mathfrak{g} = V_1 \oplus \cdots \oplus V_\nu$, and consider the vector fields in $\mathfrak{g}$ as left invariant vector fields on G. The vector fields in V_μ are homogeneous vector fields of degree μ with respect to the above dilations. We pick a basis for $\mathfrak{g}$, $Z_1, \ldots, Z_{q_0}$ where each Z_l is an element of some V_μ and therefore homogeneous of degree μ. From this basis, we proceed as in the beginning of this section to create a list of vector fields with ν parameter formal degrees $(X, d) = (X_1, d_1), \ldots, (X_q, d_q)$, where each d_j is nonzero in only one component, and each X_j is homogeneous of degree $|d_j|$. We obtain an algebra of flag type singular integrals from Definition 4.2.12.

THEOREM 4.2.16. *Let* K *be a flag kernel of order* $(s_1, \ldots, s_\nu)$ *with* $s_\mu > -Q_\mu$. *Then,* $\mathrm{Op}\,(K)$ *is a flag type singular integral operator of order* $(s_1, \ldots, s_\nu)$ *in the sense of (iii) of Theorem 4.2.11.*

Before we prove the above results, we discuss a technical point: the sense in which the sums in Theorem 4.2.11 converge.

PROPOSITION 4.2.17. *Let* $\left\{\left(E_j, 2^{-j}\right) \mid j \in \mathbb{Z}\right\}$ *be a bounded set of elementary operators. Then, for* $s \in \mathbb{R}^\nu$ *and* $f \in \mathcal{S}_0^{\{\nu\}}$, *the sum* $\sum_{j\in\mathbb{Z}} 2^{j\cdot s} E_j f(x)$ *converges uniformly in* x.

To prove Proposition 4.2.17, we need the following lemmas.

LEMMA 4.2.18. *Let* $\mathcal{E}$ *be a bounded set of pre-elementary operators. Then,*

$$\sup_{(E, 2^{-j}) \in \mathcal{E}} \sup_{y \in \mathbb{R}^{q_0}} \int |E\,(x, y)|\ dx < \infty, \qquad \sup_{(E, 2^{-j}) \in \mathcal{E}} \sup_{x \in \mathbb{R}^{q_0}} \int |E\,(x, y)|\ dy < \infty.$$

PROOF. By symmetry (see Proposition 3.2.2), it suffices to verify

$$\sup_{(E, 2^{-j}) \in \mathcal{E}} \sup_{x \in \mathbb{R}^{q_0}} \int |E\,(x, y)|\ dy < \infty.$$

We have

$$\int |E\,(x, y)|\ dy \lesssim \int \frac{\left(1 + \rho_{2^{-j}}\,(x, y)\right)^{-1}}{\mathrm{Vol}\,\left(B_{2^{-j}}\,(x, 1 + \rho_{2^{-j}}\,(x, y))\right)}\ dy$$

$$\lesssim \sum_{k=0}^{\infty} \int_{2^k \le 1 + \rho_{2^{-j}}(x, y) \le 2^{k+1}} 2^{-k}\mathrm{Vol}\,\left(B_{2^{-j}}\,(x, 2^k)\right)^{-1}\ dy$$

$$\lesssim \sum_{k=0}^{\infty} 2^{-k} \int_{B_{2^{-j}}(x, 2^k)} \mathrm{Vol}\,\left(B_{2^{-j}}\,(x, 2^k)\right)^{-1}$$

$$\lesssim 1,$$

completing the proof.   □

Remark 4.2.19   For each μ, recall the vector fields with *single*-parameter degrees $\left(X^\mu, \hat{d}^\mu\right)$. It therefore makes sense to consider the list of vector fields

$$2^{-j_\mu} X^\mu = 2^{-j_\mu \hat{d}_1^\mu} X_1^\mu, \dots, 2^{-j_\mu \hat{d}_{q_\mu}^\mu} X_{q_\mu}^\mu, \text{ for } j_\mu \in \mathbb{R}.$$

In fact, this list (along with several instances of the zero vector field) is the same as the list $2^{-j'} X$ provided $j' = (\infty, \dots, \infty, j_\mu, j_\mu, \dots, j_\mu)$ where the first non-∞ element is in the μth coordinate.

LEMMA 4.2.20. *For $j, k \in \mathbb{R}^\nu$, and for every α,*

$$\left(2^{-j \wedge k} X\right)^\alpha = \sum_{|\beta|+|\gamma| \le |\alpha|} \left(2^{-j} X\right)^\beta c_{\alpha,\beta,\gamma}^{j,k} \left(2^{-k} X\right)^\gamma,$$

where the $c_{\alpha,\beta,\gamma}^{j,k}$ are constants with $\sup_{j,k} \left| c_{\alpha,\beta,\gamma}^{j,k} \right| < \infty$.

PROOF. The proof of this result is a simpler reprise of the induction used to prove Proposition 3.2.22.   □

PROOF OF PROPOSITION 4.2.17. Let $\left(E_j, 2^{-j}\right) \in \mathcal{E}$ and fix $f \in \mathcal{S}_0^{\{\nu\}}$. We show, for every N, $|E_j f(x)| \lesssim 2^{-N|j|_\infty}$ and the result follows. Let $j^0 = j \wedge 0$ (where $0 = (0, \dots, 0) \in \mathbb{R}^\nu$). There are two possibilities: either there is a μ with $-j_\mu^0 = |j|_\infty$, or there is a μ with $j_\mu = |j|_\infty = \left|j - j^0\right|_\infty$.

We begin with the case where there is a μ with $-j_\mu^0 = |j|_\infty$. Consider the vector fields $\left(X^\nu, \hat{d}^\nu\right)$. Each vector field X_j^ν is homogeneous of the highest possible degree. Therefore, each such vector field is of the form $X_j^\nu = \sum_{|\beta|=1} c_\beta \partial_{t_\nu}^\beta$. Moreover the X_j^ν vector fields span all vector fields of the form $\partial_{t_\nu}^\beta$. Using this, and Lemma A.2.18, we see that for every N, we may write

$$f = \sum_{|\beta|=N} \left(X^\nu\right)^\beta f_\beta,$$

where $f_\beta \in \mathcal{S}_0^{\{\nu\}}$. $\left(X^\nu, \hat{d}^\nu\right)$ is a sublist of the list $\left(X^\mu, \hat{d}^\mu\right)$, and so we may write

$$f = \sum_{|\beta|=N} \left(X^\mu\right)^\beta f_\beta.$$

Notice, for $|\beta| = N$, $\left(X^\mu\right)^\beta = 2^{c_\beta N j_0^\mu} \left(2^{-j_\mu} X^\mu\right)^\beta$, where $c_\beta \ge 1$. $\left(2^{-j_\mu} X^\mu\right)^\beta = \left(2^{-j} X\right)^{\beta'}$ for some β' with $|\beta'| = |\beta|$, and we therefore have

$$E_j f = \sum_{|\beta|=N} 2^{-c_\beta N|j|_\infty} E_j \left(2^{-j} X\right)^\beta f_\beta.$$

Since $\left\{ \left(E_j \left(2^{-j} X \right)^\beta, 2^{-j} \right) \mid j \in \mathbb{Z}^\nu \right\}$ is a bounded set of pre-elementary operators, and since $f_\beta \in \mathcal{S} \subset L^\infty$, Lemma 4.2.18 applies to show $|E_j f(x)| \lesssim 2^{-N|j|_\infty}$, completing the proof in this case.

Now suppose there is a μ with $j_\mu = |j|_\infty = |j - j^0|_\infty$. We apply the definition of elementary operators N times, with $k = j^0$, to write

$$E_j = 2^{-N|j|_\infty} \sum_{|\alpha|=N} E_{j,R,\alpha} \left(2^{-j^0} X \right)^\alpha,$$

where $\left\{ \left(E_{j,R,\alpha}, 2^{-j} \right) \mid j \in \mathbb{Z} \right\}$ is a bounded set of elementary operators. Applying Lemma 4.2.20, we may write

$$E_j f = 2^{-N|j|_\infty} \sum_{|\alpha|=N} \sum_{|\beta|+|\gamma|\le N} E_{j,R,\alpha} \left(2^{-j} X \right)^\beta c_{\beta,\gamma} X^\gamma f,$$

where $|c_{\beta,\gamma}| \lesssim 1$. For each γ, $X^\gamma f \in \mathcal{S} \subset L^\infty$, and for each β,

$$\left\{ \left(E_{j,R,\alpha} \left(2^{-j} X \right)^\beta, 2^{-j} \right) \mid j \in \mathbb{Z}^\nu \right\}$$

is a bounded set of elementary operators, and Lemma 4.2.18 applies to show that $\left| E_{j,R,\alpha} \left(2^{-j} X \right)^\beta c_{\beta,\gamma} X^\gamma f(x) \right| \lesssim 1$. The result follows. $\square$

We turn to outlining the proofs of Theorems 4.2.11 and 4.2.16; we outline the proof of Theorem 4.2.15 in Section 4.1.1. We begin with Theorem 4.2.16, and so continue to work in the group setting. Thus, until we complete the proof of Theorem 4.2.16, we are, in particular, restricting attention to the group case where the single-parameter dilations are given by $\delta(t_1, t_2, \ldots, t_\nu) = (\delta t_1, \delta^2 t_2, \ldots, \delta^\nu t_\nu)$.

Since the definition of flag kernels does not use the group structure, it turns out (perhaps surprisingly) that we will be able to use the Euclidean Fourier transform to help prove Theorem 4.2.16. For this, we write ξ_μ to be the dual variable to t_μ. The next result is Theorem 2.3.9 of [NRS01], which we state without proof, though the proof is very similar to the proof of Theorem 5.2.15.

DEFINITION 4.2.21. *For a multi-index* $\alpha = (\alpha_1, \ldots, \alpha_\nu) \in \mathbb{N}^{\upsilon_1} \times \cdots \times \mathbb{N}^{\upsilon_\nu} = \mathbb{N}^{q_0}$ *we define* $\deg(\alpha) = \sum_{\mu=1}^\nu \mu |\alpha_\mu| g_\mu \in \mathbb{N}^\nu$; *here* $g_\mu \in \mathbb{N}^\nu$ *equals 1 in the μth coordinate and 0 in all other coordinates.*

PROPOSITION 4.2.22. *Fix* $s = (s_1, \ldots, s_\nu) \in \mathbb{R}^\nu$, $s_\mu > -Q_\mu$. *Then* $K \in \mathcal{S}(\mathbb{R}^{q_0})'$ *is a flag kernel if and only if the Fourier transform* $\widehat{K}$ *of* K *is given by a function which satisfies*

$$\left| \partial_\xi^\alpha \widehat{K}(\xi) \right| \le C_\alpha \left(|\xi_1|_1 + \cdots + |\xi_\nu|_\nu \right)^{s_1 - \deg(\alpha)_1} \times \cdots$$
$$\times \left(|\xi_{\nu-1}|_{\nu-1} + |\xi_\nu|_\nu \right)^{s_{\nu-1} - \deg(\alpha)_{\nu-1}} |\xi_\nu|_\nu^{s_\nu - \deg(\alpha)_\nu}. \tag{4.9}$$

DEFINITION 4.2.23. *For* $t = (t_1, \ldots, t_\nu) \in \mathbb{R}^{q_0}$ *and* $j \in \mathbb{R}^\nu$, *we define* $2^{-j} t = (2^{-j_1} t_1, 2^{-2j_2} t_2, \ldots, 2^{-\nu j_\nu} t_\nu)$. *For* $x \in G$, *we define* $2^{-j} x$ *by pushing forward these dilations via the exponential map* $\exp : (\mathbb{R}^{q_0} \cong \mathfrak{g}) \xrightarrow{\sim} G$. *For a function* $f(t)$, *we write* $f^{(2^j)}$ *to denote the function* $2^{j \cdot Q} f(2^j t)$. *Note that* $f^{(2^j)}$ *is defined so that* $\int f^{(2^j)}(t) \, dt = \int f(t) \, dt$.

LEMMA 4.2.24. *Suppose* $\widehat{K}(\xi)$ *is a function defined for* $\xi_\nu \neq 0$ *and* $s \in \mathbb{R}^\nu$, *the following are equivalent.*

(a) $\widehat{K}$ *satisfies* (4.9) *(note, we do not restrict to the case* $s_\mu > -Q_\mu$*)*.

(b) *For* $2^{-j} \in \mathcal{F}$, *let* $S(j) = \{\mu \mid \mu = \nu \text{ or } j_\mu > j_{\mu+1}\}$. *There is a bounded set* $\{\varsigma_j \mid 2^{-j} \in \mathcal{F}_{\mathbb{Z}}\} \subset \mathcal{S}(\mathbb{R}^{q_0})$ *with* $\varsigma_j \in \mathcal{S}_0^{S(j)}$, *and* $\widehat{K}(\xi) = \sum_{2^{-j} \in \mathcal{F}_{\mathbb{Z}}} 2^{j \cdot s} \widehat{\varsigma_j}(2^{-j}\xi)$. *Here,* $\widehat{\varsigma_j}$ *is the Fourier transform of* ς_j *and* $2^{-j}\xi$ *is defined in terms of the* ν-*parameter dilations on* $\mathbb{R}^{q_0}$ *from Definition 4.2.23.*

PROOF. (a)$\Rightarrow$(b): For each μ, let $\phi_\mu \in C_0^\infty(\mathbb{R}^{q_\mu})$ be a function which equals 1 on a neighborhood of 0. For $S \subset \{1, \ldots, \nu\}$ define

$$\psi_\mu^S(\xi_\mu) = \begin{cases} \phi_\mu(\xi_\mu) - \phi_\mu(2^\mu \xi_\mu) & \text{if } \mu \in S, \\ \phi_\mu(\xi_\mu) & otherwise. \end{cases}$$

Here, $2^\mu \xi_\mu$ is defined by the usual multiplication, and does not reference any dilations. We define $\psi_S(\xi_1, \ldots, \xi_\nu) = \psi_1^S(\xi_1) \cdots \psi_\nu^S(\xi_\nu)$. Note that the inverse Fourier transform of ψ_S is an element of $\mathcal{S}_0^S$. Also, $\sum_{2^{-j} \in \mathcal{F}_{\mathbb{Z}}} \psi_{S(j)}(2^{-j}\xi) \equiv 1$, where 2^{-j} is defined in terms of the dilations from Definition 4.2.23. Define $\widehat{\varsigma_j}(2^{-j}\xi) := 2^{-j \cdot s} \psi_{S(j)}(2^{-j}\xi) \widehat{K}(\xi)$, so that $\widehat{K}(\xi) = \sum_{2^{-j} \in \mathcal{F}_{\mathbb{Z}}} 2^{j \cdot s} \widehat{\varsigma_j}(2^{-j}\xi)$. Our assumption on $\widehat{K}$ and the support of ψ_S imply that $\{\widehat{\varsigma_j} \mid 2^{-j} \in \mathcal{F}_{\mathbb{Z}}\} \subset \mathcal{S}(\mathbb{R}^{q_0})$ is a bounded set. We let ς_j be the inverse Fourier transform of $\widehat{\varsigma_j}$. Since the inverse Fourier transform of $\psi_{S(j)}$ is an element of $\mathcal{S}_0^{S(j)}$, the same is true of ς_j. This completes the proof of (a)$\Rightarrow$(b).

(b)$\Rightarrow$(a): Since $\partial_{\xi_\mu}^{\alpha_\mu} \widehat{\varsigma_j}(2^{-j}\xi) = 2^{-j_\mu \deg(\alpha)_\mu}(\partial_{\xi_\mu}^{\alpha_\mu} \widehat{\varsigma_j})(2^{-j}\xi)$, (4.9) for arbitrary α follows from (4.9) for $\alpha = 0$, by changing the value of s. Because $\{\varsigma_j \mid 2^{-j} \in \mathcal{F}_{\mathbb{Z}}\} \subset \mathcal{S}(\mathbb{R}^q)$ is a bounded set and because $\varsigma_j \in \mathcal{S}_0^{S(j)}$, $\widehat{\varsigma_j}$ satisfies the estimates, for every m and α,

$$\left| \partial_\xi^\alpha \widehat{\varsigma_j}(\xi) \right| \lesssim \left[\prod_{\mu \in S(j)} |\xi_\mu|_\mu^m \left(1 + |\xi_\mu|_\mu \right)^{-2m} \right] \prod_{\mu \notin S(j)} \left(1 + |\xi_\mu|_\mu \right)^{-m}.$$

See Example A.2.17 for more details on such estimates. Consider, for $\mu < \nu$ and $j_{\mu+1}$

fixed, we have for m sufficiently large by summing in j_μ

$$\sum_{j_\mu > j_{\mu+1}} \left[\prod_{\mu'=1}^{\mu} \left(|\xi_{\mu'}|_{\mu'} + \cdots + |\xi_{\mu-1}|_{\mu-1} + 2^{j_\mu} \right)^{s_{\mu'}} \right]$$
$$\times \left| 2^{-j_\mu \mu} \xi_\mu \right|_\mu^m \left(1 + \left| 2^{-j_\mu \mu} \xi_\mu \right|_\mu \right)^{-2m}$$
$$\lesssim \prod_{\mu'=1}^{\mu} \left(|\xi_{\mu'}|_{\mu'} + \cdots + |\xi_\mu|_\mu + 2^{j_{\mu+1}} \right)^{s_{\mu'}},$$

and we clearly have, for m sufficiently large,

$$\left[\prod_{\mu'=1}^{\mu} \left(|\xi_{\mu'}|_{\mu'} + \cdots + |\xi_{\mu-1}|_{\mu-1} + 2^{j_{\mu+1}} \right)^{s_{\mu'}} \right] \left(1 + \left| 2^{-j_{\mu+1}\mu} \xi_\mu \right|_\mu \right)^{-m}$$
$$\lesssim \prod_{\mu'=1}^{\mu} \left(|\xi_{\mu'}|_{\mu'} + \cdots + |\xi_\mu|_\mu + 2^{j_{\mu+1}} \right)^{s_{\mu'}}.$$

Writing $j = (j_1, \ldots, j_\nu)$, and applying the above inequalities in the sum over j_1, we have

$$\left| \widehat{K}(\xi) \right|$$
$$\lesssim \sum_{j \in \mathcal{F}_{\mathbb{Z}}} 2^{j \cdot s} \left[\prod_{\mu \in S(j)} \left| 2^{-\mu j_\mu} \xi_\mu \right|_\mu^m \left(1 + \left| 2^{-\mu j_\mu} \xi_\mu \right|_\mu \right)^{-2m} \right] \prod_{\mu \notin S(j)} \left(1 + \left| 2^{-\mu j_\mu} \xi_\mu \right|_\mu \right)^{-m}$$
$$\lesssim \sum_{j_2 \geq \cdots \geq j_\nu} 2^{j_2 s_2 + \cdots + j_\nu s_\nu} \left(|\xi_1|_1 + 2^{j_2} \right)^{s_1}$$
$$\times \left(\left[\prod_{\mu \in S(j_2, j_2, j_3, \ldots, j_\nu)} \left| 2^{-\mu j_\mu} \xi_\mu \right|_\mu^m \left(1 + \left| 2^{-\mu j_\mu} \xi_\mu \right|_\mu \right)^{-2m} \right] \right.$$
$$\left. \times \prod_{\substack{\mu \notin S(j_2, j_2, j_3, \ldots, j_\nu) \\ \mu \neq 1}} \left(1 + \left| 2^{-\mu j_\mu} \xi_\mu \right|_\mu \right)^{-m} \right).$$

Repeatedly applying the above in equalities, next for the sum over j_2, then j_3, etc., we obtain

$$\left| \widehat{K}(\xi) \right| \lesssim \prod_{\mu'=1}^{\nu} \left(|\xi_{\mu'}|_{\mu'} + \cdots + |\xi_\nu|_\nu \right)^{s_{\mu'}},$$

as desired. $\qquad \square$

LEMMA 4.2.25. *Let $2^{-j} \in \mathcal{F}$ and let $\Phi_j : G \to G$ be the map given by $\Phi_j(x) = 2^{-j} x$. Let $Y_1, \ldots, Y_q$ be the pullbacks via Φ_j of $2^{-j \cdot d_1} X_1, \ldots, 2^{-j \cdot d_q} X_q$ ($Y_1, \ldots, Y_q$*

depend on j, but we are suppressing the dependance). Then, $Y_1, \ldots, Y_q$ are C^∞ vector fields uniformly in j,[3] and span the tangent space at every point, uniformly in j.[4]

PROOF. We consider the *single*-parameter dilations on G. Recall the basis vector fields $Z_1, \ldots, Z_{q_0}$: each X_l is equal to some Z_k. Because each Z_l is homogeneous of degree $\hat{d}_l$, $Z_1, \ldots, Z_{q_0}$ are linearly independent, and Z_l is nonzero at 0, we may choose coordinates $x_1, \ldots, x_{q_0}$ on G such that, for $\delta > 0$, $\delta(x_1, \ldots, x_{q_0}) = \left(\delta^{\hat{d}_1} x_1, \ldots, \delta^{\hat{d}_{q_0}} x_{q_0} \right)$ and

$$ Z_l = \partial_{x_l} + \sum_{|\deg(\alpha)| + \hat{d}_l = \hat{d}_k} x^\alpha \partial_{x_k}, $$

where $|\cdot|$ denotes the ℓ^1 norm (since we are thinking of $\deg(\alpha)$ as a multi-index). Each X_{l_0} is equal to some Z_l, where $|d_{l_0}| = \hat{d}_l$. Consider, then, the pullback via Φ_j of $2^{-j \cdot d_{l_0}} X_{l_0}$ is equal to

$$ \partial_{x_l} + \sum_{\substack{|\deg(\alpha)| + |d_{l_0}| = |d_{k_0}| \\ \alpha \neq 0}} 2^{-j \cdot d_{l_0} - j \cdot \deg(\alpha) + j \cdot d_{k_0}} x^\alpha \partial_{x_k}, $$

where k_0 is to k as l_0 is to l. Recall that d_{l_0} is nonzero in only one component, and if it is nonzero in the μth component, it is equal to μ–and $\deg(\alpha)$ is a linear combination of elements of the form d_m with coefficients in $\mathbb{N}$. Thus, if $|d_{l_0}| + |\deg(\alpha)| = |d_{k_0}|$, then d_{k_0} must be nonzero in only a coordinate which is strictly greater than the coordinates where d_{l_0} and $\deg(\alpha)$ are nonzero. Using that $2^{-j} \in \mathcal{F}$, we have in this case that $2^{-j \cdot d_l - j \cdot \deg(\alpha) + j \cdot d_k} \leq 1$. The result follows. $\square$

Remark 4.2.26 The map Φ_j used in Lemma 4.2.25 is essentially the map Φ from Theorem 2.2.22 applied to the vector fields $2^{-j \cdot d_1} X_1, \ldots, 2^{-j \cdot d_q} X_q$. Here, though, we were able to consider all $2^{-j} \in \mathcal{F}$, not merely those with 2^{-j} small.

LEMMA 4.2.27. *There is a constant C such that $\forall 2^{-j} \in \mathcal{F}$,*

$$ 2^j B_{(X,d)} \left(0, C^{-1} 2^{-j} \right) \subseteq B^{q_0}(1) \subseteq 2^j B_{(X,d)} \left(0, C 2^{-j} \right). $$

PROOF. Let $Y_1, \ldots, Y_q$ be the pullbacks via Φ_j as given in Lemma 4.2.25. Let $\tilde{d}_l = |d_l|_1$. Note that $2^j B_{(X,d)} \left(0, C^{-1} 2^{-j} \right) = B_{(Y,\tilde{d})} \left(0, C^{-1} \right)$ and $2^j B_{(X,d)} \left(0, C 2^{-j} \right) = B_{(Y,\tilde{d})} \left(0, C \right)$. The result now follows from Lemma 4.2.25. $\square$

DEFINITION 4.2.28. *Let $\|\cdot\|$ be the homogeneous norm on $\mathfrak{g}$ given in Example 2.13.10. We push this forward via the exponential map to define a homogeneous norm on G, which we also denote by $\|\cdot\|$. In particular, if $\delta > 0$ and δx (for $x \in G$) denotes our single-parameter dilations, we have $\|\delta x\| = \delta \|x\|$.*

---

[3] In the sense that, for any compact set K, the norm $\|Y_l\|_{C^m(K)}$ is bounded independent of j.

[4] In the sense that $|\det_{q_0 \times q_0} (Y_1| \cdots |Y_q)|$ is bounded away from 0, uniformly in j.

LEMMA 4.2.29. *For* $2^{-j} \in \mathcal{F}$, $\left\| 2^j \left(y^{-1}x \right) \right\| \approx \rho_{2-j}(y, x)$, *for* $x, y \in G$.

PROOF. ρ_{2-j} is defined in terms of the balls $B_{(X,d)}(x, \delta)$, which are defined in terms of left invariant vector fields. It follows that $\rho_{2-j}(y, x) = \rho_{2-j}\left(0, y^{-1}x \right)$. Thus, it suffices to prove the result for $y = 0$. Both the left-hand side and the right-hand side are homogeneous under the *single*-parameter dilations on G; and so it suffices to prove the result for the case when $\rho_{2-j}(0, y) = 1$. This case follows from Lemma 4.2.27. □

LEMMA 4.2.30. *For* $2^{-j} \in \mathcal{F}$, $\mathrm{Vol}\left(B_{(X,d)}\left(x, 2^{-j} \right) \right) \approx 2^{-j \cdot Q}$.

PROOF. This follows immediately from Lemma 4.2.27. □

LEMMA 4.2.31. *Suppose* $\mathcal{B} \subset \mathcal{S}\left(\mathbb{R}^{q_0} \right)$ *is bounded. The set*

$$\left\{ \left(\mathrm{Op}\left(\varsigma^{\left(2^j \right)} \right), 2^{-j} \right) \,\middle|\, \varsigma \in \mathcal{B}, 2^{-j} \in \mathcal{F} \right\}$$

is a bounded set of pre-elementary operators.

PROOF. Notice, $\mathrm{Op}\left(\varsigma^{\left(2^j \right)} \right)(x, y) = 2^{j \cdot Q} \varsigma\left(2^j \left(y^{-1}x \right) \right)$. Using, by the definition of Schwartz space, that $\left| \varsigma(x) \right| \lesssim (1 + \|x\|)^{-m}$, for every m, the bound

$$\left| \mathrm{Op}\left(\varsigma^{\left(2^j \right)} \right)(x, y) \right| \lesssim \frac{(1 + \rho_{2-j}(x, y))^{-m}}{\mathrm{Vol}\left(B_{2-j}(x, 1 + \rho_{2-j}(x, y)) \right)}$$

follows from Lemmas 4.2.29 and 4.2.30. For derivatives, note that if X_l^R is the right invariant vector field such that $X_l^R(0) = X_l(0)$, then we have

$$2^{-j \cdot d_l} X_{l,x} \mathrm{Op}\left(\varsigma^{\left(2^j \right)} \right)(x, y) = \mathrm{Op}\left((X_l \varsigma)^{\left(2^j \right)} \right)(x, y)$$

and

$$2^{-j \cdot d_l} X_{l,y} \mathrm{Op}\left(\varsigma^{\left(2^j \right)} \right)(x, y) = -\mathrm{Op}\left(\left(X_l^R \varsigma \right)^{\left(2^j \right)} \right)(x, y).$$

Since X_l and X_l^R take bounded subsets of Schwartz space to bounded subsets of Schwartz space, the estimate

$$\left| \left(2^{-j} X_x \right)^\alpha \left(2^{-j} X_z \right)^\beta \mathrm{Op}\left(\varsigma^{\left(2^j \right)} \right)(x, y) \right| \lesssim \frac{(1 + \rho_{2-j}(x, y))^{-m}}{\mathrm{Vol}\left(B_{2-j}(x, 1 + \rho_{2-j}(x, y)) \right)}$$

follows from the case when $|\alpha| = 0 = |\beta|$, and this completes the proof. □

LEMMA 4.2.32. *Let* $S(j) \subseteq \{1, \dots, \nu\}$ *be as in Lemma 4.2.24. Suppose* $\mathcal{B} \subset \mathcal{S}\left(\mathbb{R}^{q_0} \right)$ *is a bounded set. The set*

$$\left\{ \left(\mathrm{Op}\left(\varsigma^{\left(2^j \right)} \right), 2^{-j} \right) \,\middle|\, 2^{-j} \in \mathcal{F}, \varsigma \in \mathcal{B} \bigcap \mathcal{S}_0^{S(j)} \right\}$$

is a bounded set of elementary operators.

PROOF. Lemma 4.2.31 shows that the relevant set is a bounded set of pre-elementary operators. Our goal is to, therefore, show that we may "pull out" derivatives as in Definition 4.2.10. I.e., we need to verify (4.8).

For each μ, recall the vector fields with *single*-parameter dilations $\left(X^\mu, \hat{d}^\mu\right)$. It therefore makes sense to consider the list of vector fields $2^{-j_\mu} X^\mu$ as in Remark 4.2.19.

Fix $2^{-j} \in \mathcal{F}$ and $\varsigma \in \mathcal{B} \cap \mathcal{S}_0^{S(j)}$, and let $j' = (\infty, \ldots, \infty, j_\mu, j_\mu, \ldots, j_\mu)$ as in Remark 4.2.19, so that the first component of j' which is not ∞ is in the μth component. We show, for every $1 \leq \mu \leq \nu$,

$$
\begin{aligned}
\mathrm{Op}\left(\varsigma^{(2^j)}\right) &= \sum_{|\alpha|=1} \mathrm{Op}\left(\tilde{\varsigma}_{\alpha,\mu}^{(2^j)}\right) \left(2^{-j_\mu} X^\mu\right)^\alpha \\
&= \sum_{|\alpha|=1} \mathrm{Op}\left(\tilde{\varsigma}_{\alpha,\mu}^{(2^j)}\right) \left(2^{-j'} X\right)^\alpha,
\end{aligned}
\tag{4.10}
$$

where $\tilde{\varsigma}_{\alpha,\mu} \in \mathcal{S}_0^{S(j)}$ and

$$
\left\{ \tilde{\varsigma}_{\alpha,\mu} \mid 2^{-j} \in \mathcal{F}, \varsigma \in \mathcal{B} \cap \mathcal{S}_0^{S(j)}, |\alpha| = 1, 1 \leq \mu \leq \nu \right\} \subset \mathcal{S}\left(\mathbb{R}^{q_0}\right)
$$

is a bounded set.

Before we prove (4.10), we describe how it completes the proof of the lemma. For $|\alpha| = 1$, $\left(2^{-j_\mu} X^\mu\right)^\alpha = 2^{-|j-k|_\infty} c \left(2^{-k} X\right)^\beta$ for some $|\beta| = 1$, where $0 < c \leq 1$. Thus, we see, by (4.10),

$$
\begin{aligned}
\mathrm{Op}\left(\varsigma^{(2^j)}\right) &= \sum_{|\alpha|=1} \mathrm{Op}\left(\tilde{\varsigma}_{\alpha,\mu}^{(2^j)}\right) \left(2^{-j_\mu} X^\mu\right)^\alpha \\
&= \sum_{|\alpha|=1} 2^{-|j-k|_\infty} \mathrm{Op}\left(\tilde{\varsigma}_{\alpha,\mu,0}^{(2^j)}\right) \left(2^{-k} X\right)^\alpha,
\end{aligned}
$$

where $\tilde{\varsigma}_{\alpha,\mu,0} = c \tilde{\varsigma}_{\alpha,\mu}$ for some $0 < c \leq 1$. This shows that we may "pull out" derivatives on the right as in (4.8). To see that we may "pull out" derivatives on the left, merely take the adjoint on the result on the right.

Thus it suffices to prove (4.10), to which we now turn. Fix $2^{-j} \in \mathcal{F}$ and $\varsigma \in \mathcal{B} \cap \mathcal{S}_0^{S(j)}$. Let $1 \leq \mu \leq \nu$. We claim that it suffices to prove (4.10) for $\mu \in S(j)$. Indeed, if $\mu \notin S(j)$, then replace μ with the smallest $\mu' \geq \mu$ with $\mu' \in S(j)$. We must have $j_{\mu'} = j_\mu$ and every operator of the form $\left(2^{-j_{\mu'}} X^{\mu'}\right)^\alpha$ is of the form $\left(2^{-j_\mu} X^\mu\right)^\beta$ for some β. Thus, (4.10) for μ follows from the same result for μ'. We may, therefore, assume $\mu \in S(j)$.

Out of the list $(X_1, d_1), \ldots, (X_q, d_q)$ we pick a list of basis vector fields,

$$
(X_1, d_1), \ldots, (X_{q_0}, d_{q_0})
$$

which are chosen so that if X_j appears in the list, then d_j is nonzero in the highest possible component; i.e., if X_j is homogeneous of degree μ, then d_j is nonzero in only the μth component. We are considering the operator

$$\mathrm{Op}\left(\varsigma^{(2^j)}\right) f\left(x\right) = \int f\left(e^{t_1 2^{-j \cdot d_1} X_1 + \cdots + t_{q_0} 2^{-j \cdot d_{q_0}} X_{q_0}} x\right) \varsigma\left(At\right)\, dt,$$

where A is some invertible matrix (depending on the exact choice of basis $X_1, \ldots, X_{q_0}$) which fixes the homogeneous subspaces of $\mathfrak{g}$, and where we have used the identification of G with its Lie algebra via the exponential map. Since $\varsigma\left(At\right)$ is of the same form as $\varsigma\left(t\right)$, we prove the result for $A = I$.

Because $\varsigma \in \mathcal{S}_0^{S(j)}$ and $\mu \in S\left(j\right)$, Lemma A.2.18 applies to show

$$\varsigma = \sum_{|\alpha_\mu|=1} \partial_{t_\mu}^{\alpha_\mu} \varsigma_{\alpha_\mu},$$

where $\left\{\varsigma_{\alpha_\mu} \,\middle|\, 2^{-j} \in \mathcal{F}, \varsigma \in \mathcal{B} \cap \mathcal{S}_0^{S(j)}, \mu \in S\left(j\right), |\alpha_\mu| = 1\right\} \subset \mathcal{S}\left(\mathbb{R}^{q_0}\right)$ is a bounded set. Integration by parts shows

$$\mathrm{Op}\left(\varsigma^{(2^j)}\right) f\left(x\right) = - \sum_{|\alpha_\mu|=1} \int \left[\partial_{t_\mu}^{\alpha_\mu} f\left(e^{t_1 2^{-j \cdot d_1} X_1 + \cdots + t_{q_0} 2^{-j \cdot d_{q_0}} X_{q_0}} x\right)\right] \varsigma_{\alpha_\mu}\left(t\right)\, dt.$$

We use the Baker-Campbell-Hausdorff formula to compute

$$\partial_{t_\mu}^{\alpha_\mu} f\left(e^{t_1 2^{-j \cdot d_1} X_1 + \cdots + t_{q_0} 2^{-j \cdot d_{q_0}} X_{q_0}} x\right). \tag{4.11}$$

Recall, since G is nilpotent, the sum in the Baker-Campbell-Hausdorff formula is finite, and yields an exact expression, not merely an asymptotic estimate. Also, we have $\left[2^{-j \cdot d_l} X_l, 2^{-j \cdot d_m} X_m\right] = \sum_{r=1}^{q_0} c_r 2^{-j \cdot d_r} X_r$, where $|c_r| \lesssim 1$–this uses the fact that $2^{-j} \in \mathcal{F}$ and our choice of $(X_1, d_1), \ldots, (X_{q_0}, d_{q_0})$. Since we are taking ∂_{t_μ} the Baker-Campbell-Hausdorff formula will only involve vector fields which are homogeneous of degree $\geq \mu$. We therefore have that (4.11) is a linear combination (with constant coefficients which are in absolute value $\lesssim 1$) of terms of the form

$$t^\gamma \left(\left(2^{-j}X\right)^\beta f\right) \left(e^{t_1 2^{-j \cdot d_1} X_1 + \cdots + t_{q_0} 2^{-j \cdot d_{q_0}} X_{q_0}} x\right),$$

where γ is an explicit multi-index depending on β, and $\left(2^{-j}X\right)^\beta$ only involves those vector fields (X_j, d_j) where d_j is nonzero in a coordinate $\geq \mu$. For such β, $\left(2^{-j}X\right)^\beta = c\left(2^{-j_\mu}X^\mu\right)^{\beta'}$ for some $0 < c \leq 1$ and some β'. Since $t^\gamma \varsigma_\alpha$ is of the same form as ς_α, (4.10) follows with $\tilde{\varsigma}_\alpha$ a sum of terms of the form $t^\gamma \varsigma_\alpha$. This completes the proof. $\square$

PROOF OF THEOREM 4.2.16. Let K be a flag kernel of order $(s_1, \ldots, s_\nu)$ with $s_\mu > -Q_\mu$. Proposition 4.2.22 shows $\widehat{K}$, the Fourier transform of K, satisfies (4.9). Lemma 4.2.24 shows that $\widehat{K}\left(\xi\right) = \sum_{2^{-j} \in \mathcal{F}_{\mathbb{Z}}} 2^{j \cdot s} \widehat{\varsigma_j}\left(2^{-j}\xi\right)$, where $\left\{\varsigma_j \,\middle|\, 2^{-j} \in \mathcal{F}_{\mathbb{Z}}\right\} \subset \mathcal{S}\left(\mathbb{R}^{q_0}\right)$ is a bounded set and $\varsigma_j \in \mathcal{S}_0^{S(j)}$. Taking the inverse Fourier transform, we have $K = \sum_{2^{-j} \in \mathcal{F}_{\mathbb{Z}}} 2^{j \cdot s} \varsigma_j^{(2^j)}$; i.e., $\mathrm{Op}\left(K\right) = \sum_{2^{-j} \in \mathcal{F}_{\mathbb{Z}}} 2^{j \cdot s} \mathrm{Op}\left(\varsigma_j^{(2^j)}\right)$. Lemma

4.2.32 shows that $\left\{ \left(\mathrm{Op}\left(\varsigma_j^{\left(2^j \right)} \right), 2^{-j} \right) \,\middle|\, j \in \mathcal{F}_{\mathbb{Z}} \right\}$ is a bounded set of elementary operators, which completes the proof. □

We now turn to the proof of Theorem 4.2.11. The proof of Theorem 4.2.11 is largely similar to the proof of the more general result in Chapter 5 (Theorem 5.1.12), so we merely outline the proof here. We leave the group situation and work in the more general setting of Theorem 4.2.11. In this more general setting, there is an analog of the Theorem 4.2.16 with nearly the same proof. Recall the basis vector fields $Z_1, \ldots, Z_{q_0}$. Each vector field is homogeneous of degree e_μ for some μ. Let Z^μ denote the list of those vector fields which are homogeneous of degree e_μ.

THEOREM 4.2.33. *Let K be a flag kernel, corresponding to the dilations* (4.6), *of order* $(s_1, \ldots, s_\nu)$ *with* $s_\mu > -Q_\mu$. *Define an operator*

$$Tf(x) = \int f\left(e^{t_1 \cdot Z^1 + \cdots + t_\nu \cdot Z^\nu} x \right) K(t_1, \ldots, t_\nu)\, dt.$$

Then, T is a flag type singular integral of order $(s_1, \ldots, s_\nu)$ in the sense of (iii) of Theorem 4.2.11.

For the proof of Theorem 4.2.33, we need a generalization of Lemma 4.2.31. Indeed, we have the following lemma.

LEMMA 4.2.34. *Suppose $\mathcal{B} \subset \mathcal{S}(\mathbb{R}^{q_0})$ is bounded. For each $\varsigma \in \mathcal{B}$ and $2^{-j} \in \mathcal{F}$ define an operator $E = E\left(\varsigma, 2^{-j} \right)$ by*

$$Ef(x) = \int f\left(e^{t_1 \cdot Z^1 + \cdots + t_\nu \cdot Z^\nu} x \right) \varsigma^{\left(2^j \right)}(t)\, dt.$$

The set $\left\{ \left(E\left(\varsigma, 2^{-j} \right), 2^{-j} \right) \,\middle|\, \varsigma \in \mathcal{B}, 2^{-j} \in \mathcal{F} \right\}$ is a bounded set of pre-elementary operators.

COMMENTS ON THE PROOF. The proof is a simpler version of the proof of the essentially more general result Lemma 5.3.13. □

COMMENTS ON THE PROOF OF THEOREM 4.2.33. With Lemma 4.2.34 in place of Lemma 4.2.31, the proof is largely the same as the proof of Theorem 4.2.16. See also the essentially more general result Theorem 5.1.37. □

It is a corollary of Theorem 4.2.33 which is the key to the proof of Theorem 4.2.11. Namely:

COROLLARY 4.2.35. *The identity operator is a flag type singular integral of order 0, in the sense of (iii) of Theorem 4.2.11.*

PROOF. Let $K = \delta_0$, the Dirac δ function at 0. It is easy to verify that K is a flag kernel of order 0. The result follows from Theorem 4.2.33. □

Another important result we need for the proof of Theorem 4.2.11 is the following proposition.

PROPOSITION 4.2.36. *Let $\mathcal{E}$ be a bounded set of elementary operators. Then, for every N, the set*

$$\left\{ \left(2^{N|j_1-j_2|}E_1E_2, 2^{-j_1}\right), \left(2^{N|j_1-j_2|}E_1E_2, 2^{-j_2}\right), \left(2^{N|j_1-j_2|}E_1E_2, 2^{-j_1 \wedge j_2}\right) \mid \right.$$

$$\left. (E_1, 2^{-j_1}), (E_2, 2^{-j_2}) \in \mathcal{E} \right\}$$

is a bounded set of elementary operators.

COMMENTS ON THE PROOF. The proof is a simpler reprise of the essentially more general result Theorem 5.4.1. □

PROOF OF THEOREM 4.2.11. (ii)⇒(i) and (iii)⇒(iv) are trivial.

Suppose (i) holds. Apply Corollary 4.2.35 to write $I = \sum_{2^{-j} \in \mathcal{F}_{\mathbb{Z}}} E_j$, where $\{(E_j, 2^{-j}) \mid 2^{-j} \in \mathcal{F}_{\mathbb{Z}}\}$ is a bounded set of elementary operators. We have $T = TI = \sum_{2^{-j} \in \mathcal{F}_{\mathbb{Z}}} 2^{j \cdot s}\left(2^{-j \cdot s}TE_j\right)$, showing that (iii) holds.

Suppose (iv) holds; so that $T = \sum_{j \in \mathbb{Z}^\nu} 2^{j \cdot s}E_j$, where $\{(E_j, 2^{-j}) \mid j \in \mathbb{Z}^\nu\}$ is a bounded set of elementary operators. Let $\mathcal{E}$ be a bounded set of elementary operators, and let $(E, 2^{-k}) \in \mathcal{E}$. Define $\widetilde{E}_j = 2^{|j-k|}2^{(j-k)\cdot s}E_j E$. Proposition 4.2.36 shows that $\left\{ \left(\widetilde{E}_j, 2^{-k}\right) \mid 2^{-j} \in \mathbb{Z}^\nu, (E, 2^{-k}) \in \mathcal{E}\right\}$ is a bounded set of elementary operators. Since $2^{-k \cdot s}TE = \sum_{j \in \mathbb{Z}^\nu} 2^{-|j-k|}\widetilde{E}_j$, it follows that $\left\{ \left(2^{-k \cdot s}TE, 2^{-k}\right) \mid (E, 2^{-k}) \in \mathcal{E}\right\}$ is a bounded set of elementary operators, and therefore (ii) holds. □

4.2.1 Non-isotropic Sobolev spaces

To discuss the L^p ($1 < p < \infty$) theory of flag type singular integral operators (e.g., Theorem 4.2.15) we introduce non-isotropic Sobolev spaces. To do so, we introduce a Littlewood-Paley square function. Recall the lists of vector fields with single-parameter formal degrees $\left(X^\mu, \hat{d}^\mu\right) = \left(X_1^\mu, \hat{d}_1^\mu\right), \ldots, \left(X_{q_\mu}^\mu, \hat{d}_{q_\mu}^\mu\right)$; these were defined in such a way that if $\mu \geq \mu'$, the list $\left(X^{\mu'}, \hat{d}^{\mu'}\right)$ contains the list $\left(X^\mu, \hat{d}^\mu\right)$. Define single-parameter dilations on $\mathbb{R}^{q_\mu}$ by, for $\delta_\mu > 0$, $\delta_\mu\left(t_1, \ldots, t_{q_\mu}\right) = \left(\delta_\mu^{\hat{d}_1^\mu}t_1, \ldots, \delta_\mu^{\hat{d}_{q_\mu}^\mu}t_{q_\mu}\right)$.

For $j_\mu \in \mathbb{R}$, $\varsigma : \mathbb{R}^{q_\mu} \to \mathbb{C}$, we use these dilations to define $\varsigma^{(2^{j_\mu})}$ in the usual way: $\varsigma^{(2^{j_\mu})}(t_\mu) = 2^{j_\mu(\sum_l \hat{d}_l^\mu)}\varsigma\left(2^{j_\mu}t_\mu\right)$.

LEMMA 4.2.37. *Let $\delta_0 \in C_0^\infty\left(\mathbb{R}^{q_\mu}\right)'$ be the Dirac δ function at 0. There is a bounded set $\left\{\varsigma_{j_\mu,\mu} \mid j_\mu \in \mathbb{Z}\right\} \subset \mathcal{S}_0\left(\mathbb{R}^{q_\mu}\right)$ such that $\delta_0 = \sum_{j_\mu \in \mathbb{Z}} \varsigma_{j_\mu}^{(2^{j_\mu})}$, with convergence in distribution.*

PROOF. This follows from the $s = 0$ case of (i)$\Rightarrow$(iii) of Theorem 2.13.15 (with the group $G = \mathbb{R}^n$, and with the above nonstandard dilations instead of with the standard Euclidean ones). $\qquad\square$

For each μ we apply Lemma 4.2.37 to decompose $\delta_0(t_\mu) = \sum_{j_\mu \in \mathbb{Z}} \varsigma_{j_\mu,\mu}^{(2^{j_\mu})}(t_\mu)$, where $t_\mu \in \mathbb{R}^{q_\mu}$ and $\{\varsigma_{j_\mu,\mu} \mid j_\mu \in \mathbb{Z}\} \subset \mathcal{S}_0(\mathbb{R}^{q_\mu})$ is a bounded set. For $j_\mu \in \mathbb{Z}$, $1 \le \mu \le \nu$, we define

$$D_{j_\mu}^\mu f(x) = \int_{\mathbb{R}^{q_\mu}} f\left(e^{t \cdot X^\mu} x\right) \varsigma_{j_\mu,\mu}^{(2^{j_\mu})}(t)\, dt.$$

For $j = (j_1, \ldots, j_\nu) \in \mathbb{Z}^\nu$, we define $D_j = D_{j_1}^1 \cdots D_{j_\nu}^\nu$.

This leads us to the following non-isotropic Sobolev spaces. For $1 < p < \infty$, $s \in \mathbb{R}^\nu$, and $f \in \mathcal{S}_0^{\{\nu\}}$, we define

$$\|f\|_{\mathrm{NL}_s^p} := \left\| \left(\sum_{j \in \mathbb{Z}^\nu} \left| 2^{j \cdot s} D_j f \right|^2 \right)^{\frac{1}{2}} \right\|_{L^p}, \tag{4.12}$$

and we define the Banach space NL_s^p to be the completion of $\mathcal{S}_0^{\{\nu\}}$ in the above norm.[5]

The relevant properties of the D_j are in the next two propositions.

PROPOSITION 4.2.38. $\{(D_j, 2^{-j}) \mid j \in \mathbb{Z}^\nu\}$ is a bounded set of elementary operators.

COMMENTS ON THE PROOF. The proof is similar to Lemma 4.2.32. See Proposition 5.3.9 for an essentially more general situation. $\qquad\square$

PROPOSITION 4.2.39. For $1 < p < \infty$,

$$\|f\|_{\mathrm{NL}_0^p} \approx \|f\|_{L^p}.$$

COMMENTS ON THE PROOF. This is a simpler version of Theorem 3.4.1 (where we may take U_0 to be the whole space, in light of the homogeneity of everything involved). $\qquad\square$

We also have the following theorem.

THEOREM 4.2.40. Let $T : \mathcal{S}_0^{\{\nu\}} \to C^\infty(\mathbb{R}^{q_0})'$ be a flag type singular integral of order $s \in \mathbb{R}^\nu$. Then,

$$\|Tf\|_{\mathrm{NL}_{s_0}^p} \lesssim \|f\|_{\mathrm{NL}_{s_0+s}^p},$$

$1 < p < \infty$, $s_0 \in \mathbb{R}^\nu$.

COMMENTS ON THE PROOF. The methods used to prove Theorem 5.1.23, with only obvious modifications, work in this situation as well, yielding the result. $\qquad\square$

PROOF OF THEOREM 4.2.15. This follows immediately by combining Proposition 4.2.39 and Theorem 4.2.40. $\qquad\square$

---

[5] It can be seen, in a manner similar to Proposition 4.2.17, that (4.12) is finite for every $f \in \mathcal{S}_0^{\{\nu\}}$.

4.2.2 Further reading and references

The study of flag type singular integral operators began with the work of Müller, Ricci, and Stein [MRS95, MRS96] on certain spectral multipliers on the Heisenberg group. The study of these flag kernels was later taken up by Nagel, Ricci, and Stein [NRS01], who introduced Definition 4.2.1, and proved Proposition 4.2.22, and an analog of Lemma 4.2.24 (all of this was done in the case $s = 0$). Furthermore, they proved results like Theorem 4.2.2 (when $s = t = 0$), and Theorem 4.2.3, though for both they assumed that the underlying group had a special form. Namely, that it was a product of groups, and the flag structure was adapted to this product structure. They used these results to study the $\Box_b$ complex on certain quadratic CR submanifolds of $\mathbb{C}^n$.

The restriction that Nagel, Ricci, and Stein assumed on the underlying group was removed in a following work of Nagel, Ricci, Stein, and Wainger [NRSW12]. They proved results like Theorem 4.2.2 (when $s = t = 0$), and Theorem 4.2.3, with no additional restrictions on the underlying graded nilpotent Lie group. Our approach follows their ideas. Another approach to the same results was put forth by Głowacki [Gło10, Gło12b, Gło12a]. Two other related papers are the work of Han and Lu [HL08] and the work of Yang [Yan09].

The L^p boundedness of operators given by convolutions with flag kernels of order 0, and more generally the L^p boundedness of the operators covered in Theorem 4.2.33 (when $s = 0$) follows from a more general theory developed by the author and Stein concerning "multi-parameter singular Radon transforms" [SS11, Str12, SS13, SS12].

All of the above papers (except for the work of the author and Stein on multi-parameter singular Radon transforms) are concerned with translation invariant operators on a group. It seems that Definition 4.2.12 is new. This highlights one of the advantages of the theory we develop in Chapter 5: current techniques for multi-parameter singular integrals are often closely tied to an underlying group structure. The theory in Chapter 5 eschews this. Thus, not only do we obtain generalizations like Definition 4.2.12, but many different more complicated ideas immediately fall under the theory in Chapter 5; concepts which are not amenable to the ideas which are tied to a translation invariant setting.

4.3 LEFT AND RIGHT INVARIANT OPERATORS

In the previous examples in this chapter, the algebra of operators being discussed was (in part) defined by estimates on the Schwartz kernel. Such kernel estimates are at the heart of the Calderón-Zygmund paradigm. However, when we move to the general situation in Chapter 5, we do not use estimates on the Schwartz kernel (see, however, Remarks 4.3.13 and 5.1.18). In this example, we discuss an algebra of singular integrals which are not defined in terms of estimates on the Schwartz kernel. The algebra that follows was originally defined in [Str08], and in that paper it is argued that it seems likely that Schwartz kernel estimates are too weak to encapsulate any key part of the algebra.

The setting is a stratified[6] Lie group G. Thus, the Lie algebra of G, $\mathfrak{g}$, can be decomposed $\mathfrak{g} = V_1 \oplus V_2 \oplus \cdots \oplus V_m$, where $[V_1, V_k] = V_{k+1}$ for $1 \leq k < m$ and $[V_1, V_m] = \{0\}$. Since G is stratified, it is graded, and we may apply the theory discussed in Section 2.13. Thus, if K is a Calderón-Zygmund kernel of order $s \in \mathbb{R}$ (in the sense of Definitions 2.13.14 and 2.13.16), we have a theory concerning left invariant operators of the form

$$\mathrm{Op}_L\,(K)\,f = f * K.$$

A completely analogous theory can be developed for right invariant operators of the form

$$\mathrm{Op}_R\,(K)\,f = K * f.$$

The operators of the form $\mathrm{Op}_L\,(K)$ form an algebra: if K_1 is a Calderón-Zygmund kernel of order s_1 and K_2 is a Calderón-Zygmund kernel of order s_2, then

$$\mathrm{Op}_L\,(K_1)\,\mathrm{Op}_L\,(K_2) = \mathrm{Op}_L\,(K_1 * K_2)\,,$$

where $K_1 * K_2$ is a Calderón-Zygmund kernel of order $s_1 + s_2$. Similarly,

$$\mathrm{Op}_R\,(K_1)\,\mathrm{Op}_R\,(K_2) = \mathrm{Op}_R\,(K_2 * K_1)\,,$$

where $K_2 * K_1$ is also a Calderón-Zygmund kernel of order $s_1 + s_2$. Also, if K is a Calderón-Zygmund kernel of order 0, then $\mathrm{Op}_L\,(K), \mathrm{Op}_R\,(K) : L^p \to L^p, 1 < p < \infty$.

Notice, by the associativity of convolution,

$$\mathrm{Op}_L\,(K_1)\,\mathrm{Op}_R\,(K_2)\,f = (K_2 * f) * K_1 = K_2 * (f * K_1) = \mathrm{Op}_R\,(K_2)\,\mathrm{Op}_L\,(K_1)\,f,$$

and it follows that the operators $\mathrm{Op}_L\,(K_1)$ and $\mathrm{Op}_R\,(K_2)$ commute. We have

$$\mathrm{Op}_L\,(K_1)\,\mathrm{Op}_R\,(K_2)\,\mathrm{Op}_L\,(K_3)\,\mathrm{Op}_R\,(K_4) = \mathrm{Op}_L\,(K_1 * K_3)\,\mathrm{Op}_R\,(K_4 * K_1)\,,$$

and so operators of the form $\mathrm{Op}_L\,(K_1)\,\mathrm{Op}_R\,(K_2)$ are closed under composition. Moreover, if K_1 and K_2 are both Calderón-Zygmund kernels of order 0, then

$$\mathrm{Op}_L\,(K_1)\,\mathrm{Op}_R\,(K_2) : L^p \to L^p, \quad 1 < p < \infty.$$

The goal, in this section, is to present a natural, intrinsically defined algebra containing both operators of the form $\mathrm{Op}_L\,(K_1)$ and operators of the form $\mathrm{Op}_R\,(K_2)$ that preserves the above properties.

Since G is a stratified Lie group (and therefore graded), we define dilations on $\mathfrak{g}$ as in Example 2.13.6, giving G the structure of a homogeneous group. For each l, $1 \leq l \leq m$, pick a basis for V_l, $X^{1,l}, \ldots, X^{p_l,l}$, where $p_l = \dim(V_l)$. Notice that under the above dilations, $X^{p_l,l}$ is homogeneous of degree l. We may think of $X^{p_l,l}$

---

[6]The ideas in this section also work in the more general case of a graded Lie group; see the general theory from Chapter 5. The case of a stratified Lie group already contains the main ideas we wish to discuss.

as either a left invariant vector field, or as a right invariant vector field. We denote these two choices by $X_L^{p_l,l}$ and $X_R^{p_l,l}$. We define two sets of vector fields with single-parameter formal degrees:

$$\left\{ \left(X_L^{j,l}, l \right) \mid 1 \le l \le m, 1 \le j \le p_l \right\},$$

$$\left\{ \left(X_R^{j,l}, l \right) \mid 1 \le l \le m, 1 \le j \le p_l \right\}.$$

Using these two lists of vector fields with single-parameter degrees, we create one list of vector fields with 2-parameter degrees as in the introduction to this chapter: $(X, d) = (X_1, d_1), \ldots, (X_q, d_q)$, $0 \ne d_j \in \mathbb{N}^2$, each d_j nonzero in only one component. We have, for $\delta \in [0, \infty)^2$,

$$\left[\delta^{d_j} X_j, \delta^{d_k} X_k \right] = \sum_l c_{j,k}^l \delta^{d_l} X_l,$$

where $c_{j,k}^l$ are *constants*, which are independent of δ. Indeed, there are two possibilities: either X_j and X_k are both left invariant (respectively, right invariant) vector fields, and then the formula follows by the choice of the vector fields and degrees. If one is left invariant and the other right invariant, then $[X_j, X_k] = 0$, and the above formula is trivial.

From (X, d), we obtain the 2-parameter balls $B_{(X,d)} \left(x, 2^{-j} \right)$, where $j \in \mathbb{R}^2$ (as usual, $2^{-(j_1, j_2)} = \left(2^{-j_1}, 2^{-j_2} \right)$).

Remark 4.3.1 Just as in Remark 4.2.4, there is a single-parameter homogeneity in all of our definitions. Namely, all of the definitions that follow behave well under the single-parameter dilations defined in Example 2.13.6. Using the same argument as in Remark 4.2.4, all of the results from Chapter 3 stated for x and 2^{-j} small actually have an appropriate analog in this setting for all x and 2^{-j} because they are homogeneous in the appropriate way.

For $j \in \mathbb{R}^\nu$ and $\delta > 0$, define $B_{2^{-j}} (x, \delta) := B_{(X,d)} \left(x, \delta 2^{-j} \right)$. Define

$$\rho_{2^{-j}} (x, y) = \inf \left\{ \delta > 0 \mid y \in B_{2^{-j}} (x, \delta) \right\}.$$

We now turn to the definition of the algebra of operators. In line with how we proceed in Chapter 5, we present four different definitions, which all turn out to be equivalent. We denote these four a priori different spaces, for $s \in \mathbb{R}^2$, as $\mathcal{A}_1^s$, $\mathcal{A}_2^s$, $\mathcal{A}_3^s$, and $\mathcal{A}_4^s$. The main theorem in this context (Theorem 4.3.18) is that $\mathcal{A}_1^s = \mathcal{A}_2^s = \mathcal{A}_3^s = \mathcal{A}_4^s$, and we denote their common value by $\mathcal{A}^s$.

Remark 4.3.2 The algebra of operators defined in this section is on one extreme of the possible sorts of multi-parameter geometries that we consider. Indeed, in Section 4.1 when we considered the product theory, we had $M = M_1 \times \cdots \times M_\nu$ and on each factor M_μ we were given a Carnot-Carathéodory geometry. Informally, the ν different geometries did not "overlap" at all. In this section, we are given 2 geometries (the left invariant geometry, and the right invariant geometry), that "overlap" completely.

The flag situation discussed in Section 4.2 is between these two extremes. The goal in Chapter 5 is to present one definition that incorporates and generalizes all these possibilities.

DEFINITION 4.3.3. *We say $\mathcal{B} \subset C_0^\infty(G) \times G \times (0, \infty)^2$ is a bounded set of bump functions if $\forall M$, $\exists C_M$, $\forall (\phi, x, 2^{-j}) \in \mathcal{B}$,*

- $\operatorname{supp}(\phi) \subset B_{(X,d)}(x, 2^{-j})$.
- $\sum_{|\alpha| \leq M} \left\| (2^{-j}X)^\alpha \phi \right\|_{C^0(G)} \leq C_M \operatorname{Vol}\left(B_{(X,d)}(x, 2^{-j})\right)^{-1}$.

DEFINITION 4.3.4. *For an ordered multi-index α which is a list of elements of $\{1, \ldots, q\}$ we let $\deg(\alpha) = \sum_{l=1}^q k_l d_l \in \mathbb{N}^2$, where k_l is the number of times l appears in the list α.*

DEFINITION 4.3.5. *For $M \geq 0$, we say $\mathcal{T}_M \subset C^\infty(G) \times G \times (0, \infty)^2$ is a bounded set of test functions of order M, if $\forall (\psi, x, 2^{-j}) \in \mathcal{T}_M$, $\forall \alpha$ with $\deg(\alpha) = (M, M)$, $\exists \phi_\alpha \in C_0^\infty(G)$ with $\psi = \sum_{\deg(\alpha)=(M,M)} (2^{-j}X)^\alpha \phi_\alpha$, and*

$$\left\{ (\phi_\alpha, x, 2^{-j}) \mid (\psi, x, 2^{-j}) \in \mathcal{T}_M, \deg(\alpha) = (M, M) \right\}$$

a bounded set of bump functions.

Remark 4.3.6 Note that test functions are bump functions. Moreover, a test function is a high order derivative (in both left and right invariant vector fields) of a bump function.

DEFINITION 4.3.7. *Let $\mathcal{S}_N(G)$ denote the space of those $f \in \mathcal{S}(G)$ such that*

$$\int t^\alpha f(t)\, dt = 0, \quad \forall |\alpha| \leq N.$$

$\mathcal{S}_N(G)$ *is a closed subspace of $\mathcal{S}(G)$ and we give it the induced Fréchet topology.*

Remark 4.3.8 Note that $\bigcap_N \mathcal{S}_N(G) = \mathcal{S}_0(G)$. Also, if $\mathcal{T}_M$ is a bounded set of test functions of order M, then $\forall (\psi, x, 2^{-j}) \in \mathcal{T}_M$, $\psi \in \mathcal{S}_M(G)$.

DEFINITION 4.3.9. *For $j \in \mathbb{R}^2$ and $r > 0$, we define*

$$B_{2^{-j}}(x, r) := B_{(X,d)}\left(x, r2^{-j}\right)$$

and

$$\rho_{2^{-j}}(x, y) = \inf\left\{ r > 0 \mid y \in B_{2^{-j}}(x, r) \right\}.$$

Remark 4.3.10 Just as in Remark 4.2.8, the definitions of $\rho_{2^{-j}}(x, y)$ and $B_{2^{-j}}(x, r)$ are slightly different from those in Chapter 3. This is for the same reason as discussed in Remarks 4.2.8 and 4.2.4: the definitions here are the appropriate analog to use to make the results from Chapter 3 carry over to this section, as discussed in Remark 4.2.4.

DEFINITION 4.3.11. *Let* $T : \mathcal{S}_N(G) \to \mathcal{S}_N(G)'$ *for some* N. *For* $s \in \mathbb{R}^2$, *we say* $T \in \mathcal{A}_1^s$ *if* $\forall L, m, \exists M \geq N, \forall \mathcal{T}_M$ *bounded sets of test functions of order* M, $\exists C$, $\forall (\psi_1, x, 2^{-j}), (\psi_2, y, 2^{-k}) \in \mathcal{T}_M$,

$$|\langle \psi_1, T, \psi_2 \rangle| \leq C 2^{j \wedge k \cdot s} 2^{-L|j-k|} \frac{(1 + \rho_{2^{-j \wedge k}}(x, y))^{-m}}{\mathrm{Vol}\left(B_{2^{-j \wedge k}}(x, 1 + \rho_{2^{-j \wedge k}}(x, y))\right)}.$$

DEFINITION 4.3.12. *Let* $T : \mathcal{S}_N(G) \to C^\infty(G)$, *for some* N. *For* $s \in \mathbb{R}^2$, *we say* $T \in \mathcal{A}_2^s$ *if* $\forall m, \exists M \geq N, \forall \mathcal{T}_M$ *bounded sets of test functions of order* M, $\forall \alpha, \exists C$, $\forall (\psi, z, 2^{-j}) \in \mathcal{T}_M$,

$$\left|(2^{-j} X)^\alpha T\psi(x)\right| \leq C 2^{s \cdot j} \frac{(1 + \rho_{2^{-j}}(x, z))^{-m}}{\mathrm{Vol}\left(B_{2^{-j}}(x, 1 + \rho_{2^{-j}}(x, z))\right)},$$

with the same estimate for T *replaced with* T^*.

Remark 4.3.13   The definitions of $\mathcal{A}_1$ and $\mathcal{A}_2$ are analogs of the characterizations of singular integrals given in Corollary 2.7.21. Thus, they can be thought of as combining "cancellation conditions" with an analog of "growth conditions," as discussed in Section 2.7.1.

DEFINITION 4.3.14. *We say* $\mathcal{E} \subset C^\infty(G \times G) \times (0, \infty)^\nu$ *is a bounded set of pre-elementary operators if* $\forall \alpha, \beta, \exists C, \forall (E, 2^{-j}) \in \mathcal{E}$,

$$\left|(2^{-j} X_x)^\alpha (2^{-j} X_z)^\beta E(x, z)\right| \leq C \frac{(1 + \rho_{2^{-j}}(x, z))^{-m}}{\mathrm{Vol}\left(B_{2^{-j}}(x, 1 + \rho_{2^{-j}}(x, z))\right)}.$$

DEFINITION 4.3.15. *The set of bounded sets of elementary operators*, $\mathcal{G}$, *is the largest set of subsets of* $C^\infty(G \times G) \times (0, \infty)^2$ *such that* $\forall \mathcal{E} \in \mathcal{G}$,

- $\mathcal{E}$ *is a bounded set of pre-elementary operators.*

- $\forall (E, 2^{-j}) \in \mathcal{E}$,

$$E = \begin{cases} \sum_{\deg(\alpha)=(1,1)} E_{R,\alpha} \left(2^{-j} X\right)^\alpha & and \\ \sum_{\deg(\alpha)=(1,1)} \left(2^{-j} X\right)^\alpha E_{L,\alpha}, \end{cases}$$

where $\left\{(E_{R,\alpha}, 2^{-j}), (E_{L,\alpha}, 2^{-j}) \mid (E, 2^{-j}) \in \mathcal{E}, \deg(\alpha) = (1, 1)\right\} \in \mathcal{G}$.

We call elements $\mathcal{E} \in \mathcal{G}$ *bounded sets of elementary operators.*

DEFINITION 4.3.16. *Let* $T : \mathcal{S}_0(G) \to \mathcal{S}_0(G)$. *For* $s \in \mathbb{R}^2$, *we say* $T \in \mathcal{A}_3^s$ *if for every bounded set of elementary operators* $\mathcal{E}$, *we have*

$$\left\{(2^{-j \cdot s} T E, 2^{-j}) \mid (E, 2^{-j}) \in \mathcal{E}\right\}$$

is a bounded set of elementary operators.

DEFINITION 4.3.17. *Let $T : \mathcal{S}_0(G) \to \mathcal{S}_0(G)$. For $s \in \mathbb{R}^2$, we say $T \in \mathcal{A}_4^s$ if there exists a bounded set of elementary operators $\{(E_j, 2^{-j}) \mid j \in \mathbb{Z}^2\}$ such that $T = \sum_{j \in \mathbb{Z}^2} 2^{j \cdot s} E_j$ (every such sum converges in the topology of bounded convergence as operators $\mathcal{S}_0(G) \to \mathcal{S}_0(G)$).*

We turn to stating the main results about these operators. We do not discuss the proofs of any of the results that follow. They all follow by the same methods as the more general setting in Chapter 5. The reader is also referred to [Str08] for proofs of many of the following results.

THEOREM 4.3.18. *For $s \in \mathbb{R}^2$, $\mathcal{A}_1^s = \mathcal{A}_2^s = \mathcal{A}_3^s = \mathcal{A}_4^s$ in the following sense. $\mathcal{A}_1^s = \mathcal{A}_2^s$ and $\mathcal{A}_3^s = \mathcal{A}_4^s$. Moreover, if $T \in \mathcal{A}_1^s = \mathcal{A}_2^s$, then $T|_{\mathcal{S}_0(G)} \in \mathcal{A}_3^s = \mathcal{A}_4^s$. Finally, if $T \in \mathcal{A}_3^s = \mathcal{A}_4^s$ and if S is a continuous extension of T to $\mathcal{S}_N(G)$ for some N, then $S \in \mathcal{A}_1^s = \mathcal{A}_2^s$. We denote the common value of these spaces as $\mathcal{A}^s$.*

Remark 4.3.19 Let us explain in a little more detail what we mean by extending an operator $T : \mathcal{S}_0(G) \to \mathcal{S}_0(G)$ to $\mathcal{S}_N(G)$ for some N. Since T is continuous $\mathcal{S}_0(G) \to \mathcal{S}_0(G)$, for each semi-norm $|\cdot|_1$ on $\mathcal{S}_0(G)$, there is a continuous semi-norm $|\cdot|_2$ such that

$$|Tf|_1 \lesssim |f|_2.$$

Thus, if $|\cdot|_1$ is a norm on $\mathcal{S}(G)$, it makes sense to define Tf for f in the completion of $\mathcal{S}_0(G)$ with respect to $|\cdot|_2$. This completion necessarily contains $\mathcal{S}_N(G)$ for N sufficiently large.

THEOREM 4.3.20. *Let K be Calderón-Zygmund kernel of order $s \in \mathbb{R}$ (in the sense of Definitions 2.13.14 and 2.13.16). Then, $\mathrm{Op}_L(K) \in \mathcal{A}^{(s,0)}$ and $\mathrm{Op}_R(K) \in \mathcal{A}^{(0,s)}$.*

Let $I : \mathcal{S}_0(G) \to \mathcal{S}_0(G)$ denote the identity operator. Clearly $I \in \mathcal{A}_2^0$, and it follows from Theorem 4.3.18 that $I \in \mathcal{A}_4^0$. Thus, there is a bounded set of elementary operators $\{(D_j, 2^{-j}) \mid j \in \mathbb{Z}^2\}$ such that $I = \sum_{j \in \mathbb{Z}^2} D_j$. For $f \in \mathcal{S}_0(G)$, $s \in \mathbb{R}^2$, and $1 < p < \infty$ we define the norm

$$\|f\|_{\mathrm{NL}_s^p} := \left\| \left(\sum_{j \in \mathbb{Z}^2} \left| 2^{j \cdot s} D_j f \right|^2 \right)^{\frac{1}{2}} \right\|_{L^p},$$

which is finite for $f \in \mathcal{S}_0(G)$. We let NL_s^p be the completion of $\mathcal{S}_0(G)$ in the norm $\|\cdot\|_{\mathrm{NL}_s^p}$.

THEOREM 4.3.21. *(i) NL_s^p does not depend on the choice of the decomposition $I = \sum_{j \in \mathbb{Z}^2} D_j$. Indeed, if $\left\{ \left(\tilde{D}_j, 2^{-j} \right) \mid j \in \mathbb{Z}^2 \right\}$ is another bounded set of elementary operators with $I = \sum_{j \in \mathbb{Z}^2} \tilde{D}_j$, we have for $f \in \mathcal{S}_0(G)$,*

$$\|f\|_{\mathrm{NL}_s^p} \approx \left\| \left(\sum_{j \in \mathbb{Z}^2} \left| 2^{j \cdot s} \tilde{D}_j f \right|^2 \right)^{\frac{1}{2}} \right\|_{L^p}.$$

(ii) $\mathrm{NL}_0^p = L^p$, $1 < p < \infty$. *More precisely, for $f \in \mathcal{S}_0\,(G)$, we have*

$$\|f\|_{\mathrm{NL}_0^p} \approx \|f\|_{L^p}\,.$$

Since $\mathcal{S}_0\,(G)$ is dense in L^p ($1 < p < \infty$; see Remark 1.1.15), it follows that $\mathrm{NL}_0^p = L^p$.

(iii) Let $T \in \mathcal{A}^s$. Then $T : \mathrm{NL}_{s_0+s}^p \to \mathrm{NL}_{s_0}^p$. I.e., for $f \in \mathcal{S}_0\,(G)$,

$$\|Tf\|_{\mathrm{NL}_{s_0}^p} \lesssim \|f\|_{\mathrm{NL}_{s_0+s}^p}\,.$$

In particular, notice that (ii) combined with (iii) shows that if $T \in \mathcal{A}^0$, then T extends to a bounded operator $L^p \to L^p$ ($1 < p < \infty$).

In this setting, we have two sub-Laplacians:

$$\mathcal{L}_L = \left(X_L^{1,1}\right)^* X_L^{1,1} + \cdots + \left(X_L^{1,p_1}\right)^* X_L^{1,p_1},$$

$$\mathcal{L}_R = \left(X_R^{1,1}\right)^* X_R^{1,1} + \cdots + \left(X_R^{1,p_1}\right)^* X_R^{1,p_1}.$$

Since G is stratified, $X_L^{1,1}, \ldots, X_L^{1,p_1}$ satisfies Hörmander's condition, and the same holds for $X_R^{1,1}, \ldots, X_R^{1,p_1}$. $\mathcal{L}_L$ is left invariant, and $\mathcal{L}_R$ is right invariant, and it follows that these two operators commute. They are also essentially self-adjoint and we identify them with their unique self-adjoint extensions. It makes sense to consider multipliers of the form $m\,(\mathcal{L}_L, \mathcal{L}_R)$. We have the following theorem.

THEOREM 4.3.22. *Suppose $m\,(\lambda_1, \lambda_2) : (0, \infty)^2 \to \mathbb{C}$ is a C^∞ function satsifying for some $t_1, t_2 \in \mathbb{R}$,*

$$\sup_{\lambda_1, \lambda_2 \in (0,\infty)} \lambda_1^{-t_1} \lambda_2^{-t_2} \left|(\lambda_1 \partial_{\lambda_1})^a (\lambda_2 \partial_{\lambda_2})^b m\,(\lambda_1, \lambda_2)\right| < \infty, \quad \forall a, b \in \mathbb{N}.$$

Then, $m\,(\mathcal{L}_1, \mathcal{L}_2) \in \mathcal{A}^{(2t_1, 2t_2)}$.

We do not prove Theorem 4.3.22 here, and merely remark that the methods used in Section 2.6 can be easily modified to prove the theorem.

The algebra $\mathcal{A}$ arises when studying an example due to Kohn [Koh05] of a sum of squares of complex vector fields which satisfy the complex version of Hörmander's condition, and which is hypoelliptic but not subelliptic, to which we now turn.

4.3.1   An example of Kohn

The algebra in the previous section arises when studying an example of Kohn's concerning sums of squares of complex vector fields [Koh05]. The setting is $\mathbb{H}^1$, the three dimensional Heisenberg group–which is a stratified Lie group. As a manifold, $\mathbb{H}^1$ is diffeomorphic to $\mathbb{R}^3$, and we assign coordinates $(x, y, t) \in \mathbb{R}^3$. The left invariant vector fields are spanned by

$$X_L = \frac{\partial}{\partial x} + 2y\frac{\partial}{\partial t}, \quad Y_L = \frac{\partial}{\partial y} - 2x\frac{\partial}{\partial t}, \quad [X_L, Y_L] = -4\frac{\partial}{\partial t},$$

while the right invariant vector fields are spanned by

$$X_R = \frac{\partial}{\partial x} - 2y\frac{\partial}{\partial t}, \quad Y_R = \frac{\partial}{\partial y} + 2x\frac{\partial}{\partial t}, \quad [X_R, Y_R] = 4\frac{\partial}{\partial t}.$$

Define $Z_L = \frac{1}{2}(X_L - iY_L)$ and $\overline{Z}_L = \frac{1}{2}(X_L + iY_L)$; also let $z = x + iy$. Fix $k \in \mathbb{N}$, and consider the operator

$$\mathcal{L}_k = \overline{Z}_L^* \overline{Z}_L + \left(\overline{z}^k Z_L\right)^* \overline{z}^k Z_L.$$

Notice that the vector fields $\overline{Z}_L$ and $\overline{z}^k Z_L$ satisfy the complex version of Hörmander's condition: the iterated commutators of $\overline{Z}_L$ and $\overline{z}^k Z_L$ span the complexified tangent space at every point. However, unlike the analogous situation for real vector fields as studied by Hörmander ([Hör67]–see Section 2.6), the operator $\mathcal{L}_k$ is not subelliptic. More precisely, Kohn [Koh05] showed

THEOREM 4.3.23. $\mathcal{L}_k$ *is hypoelliptic on a neighborhood of* $0 \in \mathbb{H}^1$. *Moreover,* $\mathcal{L}_k$ *loses precisely* $k - 1$ *derivatives in* L^2 *Sobolev spaces. This means that if* $\mathcal{L}_k u \in L_s^2$ *on a neighborhood of* 0, *then* $u \in L_{s-k+1}^2$ *on a neighborhood of* 0, *and one can say no better in general. Here,* L_s^2 *denotes the standard* L^2 *Sobolev space of order* $s \in \mathbb{R}$.

Remark 4.3.24   It follows from Theorem 4.3.23 that Hörmander's theorem concerning the sub-Laplacian does not hold with complex vector fields in place of real vector fields, in the sense that $\mathcal{L}_k$ is not subelliptic. However, $\mathcal{L}_k$ is hypoelliptic, and one might wonder if all such sums of squares of complex vector fields satisfying Hörmander's condition must be hypoelliptic. This is not the case, as was shown by Christ [Chr05]. Indeed, [Chr05] shows that if we consider the manifold $\mathbb{H}^1 \times \mathbb{R}$ with new variable s, the operator $\mathcal{L}_k + \partial_s^* \partial_s$ is not hypoelliptic for $k \geq 1$.

Theorem 4.3.23 extends from L^2 Sobolev spaces to L^p Sobolev spaces ($1 < p < \infty$).

THEOREM 4.3.25. $\mathcal{L}_k$ *loses precisely* $k - 1$ *derivatives in* L^p *Sobolev spaces* ($1 < p < \infty$) *in the sense of Theorem 4.3.23.*

Remark 4.3.26   Theorem 4.3.25 extends to a slightly larger class of operators. See [Str10].

The proof of Theorem 4.3.25 involves constructing a parametrix for $\mathcal{L}_k$. This parametrix is an element of the algebra $\mathcal{A}$ developed above. In fact, we are able to say more. We return to this point later; see Section 5.11.4.

4.3.2   Further reading and references

An extrinsic study of an algebra containing both the right and left invariant Calderón-Zygmund operators on a nilpotent Lie group has been taken up before; see, e.g., the work of Kisil [Kis95]. However, it seems that [Str08] was the first paper to develop an intrinsic theory corresponding to such operators.

Kohn was originally motivated to study $\mathcal{L}_k$ [Koh05] by a question of Siu, who wanted to use estimates of such operators to explicitly construct certain varieties useful for studying the $\bar{\partial}$-Neumann problem.

Kohn's example yields an entirely new phenomenon, but there were previous results concerning partial differential operators which are hypoelliptic but not subelliptic. This began with results concerning operators $\mathcal{L}$ where u is always smoother then $\mathcal{L}u$, but not by enough to make $\mathcal{L}$ subelliptic. This includes work of Bell and Mohammed [BM95], Christ [Chr01], Morimoto [Mor78, Mor87], and Kohn [Koh98, Koh02]. The first example of hypoellipticity with no gain seems to be due to Stein [Ste82b], who studied a specific operator on the Heisenberg group. Stein's ideas were extended by Heller in his thesis [Hel86], where he studied certain left invariant partial differential operators on the Heisenberg group which were hypoelliptic but lost derivatives like Kohn's operator (though these operators are not closely related to Kohn's example).

After Kohn's work [Koh05], several authors furthered his results. In an appendix to Kohn's paper, Derridj and Tartakoff prove that Kohn's example is analytic hypoelliptic. Soon after Kohn's paper appeared, Bove, Derridj, Kohn, and Tartakoff [BDKT06] used the same methods to prove hypoellipticity, analytic hypoellipticity, and optimal L^2 Sobolev regularity for a larger class of operators. The first parametrix for a (simplified) operator like Kohn's example was developed by Parenti and Parmeggiani [PP07] who studied the L^2 theory of the parametrix; this was based on their theory concerning more general operators which were hypoelliptic but not subelliptic [PP05]. The L^2 theory of the parameterix developed by Parenti and Parmeggiani was extended to L^p $(1 < p < \infty)$ by the author [Str06]. These methods were extended, also by the author, to Kohn's more complicated example in [Str10]. Recently, Chinni also studied the parametrix for Kohn's operator [Chi11], using the FBI transform.

As discussed in Remark 4.3.24, Christ showed that a sum of squares of complex vector fields whose commutators span the complexified tangent space at every point might not be hypoelliptic at all [Chr05]. Some other papers closely related to Kohn's work are: the work of Khanh, Pinton, and Zampieri [KPZ12], the work of Chinni [Chi12, Chi11], the work of Journé and Trépreau [JT06], and the work of Bove, Mughetti, and Tartakoff [BMT10].

4.4   CARNOT-CARATHÉODORY AND EUCLIDEAN GEOMETRIES

Let M be a compact manifold. Let $W_1, \ldots, W_r$ be vector fields satisfying Hörmander's condition on M. As in Section 2.1, we use this list of vector fields to create a list of vector fields with single-parameter degrees, which we denote by $\left(X^1, \hat{d}^1\right) :=$ $\left(X_1^1, \hat{d}_1^1\right), \ldots, \left(X_{q_1}^1, \hat{d}_{q_1}^1\right)$. Corresponding to this list of vector fields, we obtain the single-parameter balls $B_{\left(X^1, \hat{d}^1\right)}(x, \delta)$. The theory in Chapter 2 applied to give us an algebra of singular integrals associated to the balls $B_{(X,d)}(x, \delta)$. We write the set of Calderón-Zygmund operators of order $s \in \mathbb{R}$, with respect to these vector fields as in Definition 2.0.30, as $\mathcal{B}_1^s$. In particular $\mathcal{L}^{-1} \in \mathcal{B}_1^{-2}$, where $\mathcal{L} = W_1^* W_1 + \cdots + W_r^* W_r$ is the sub-Laplacian (see Corollary 2.6.7).

Let $X_1^2, \ldots, X_{q_2}^2$ be a list of vector fields spanning the tangent space at each point of M. Assign to each of these vector fields the formal degree 1, thereby creating a list of vector fields with formal degrees $(X^2, 1) := (X_1^2, 1), \ldots, (X_{q_2}^2, 1)$. If M is given the structure of a Riemannian manifold, then the balls $B_{(X^2,1)}(x, \delta)$ are comparable to the balls in the Riemannian metric. We let $\mathcal{B}_2^s$ denote the algebra of singular integrals associated to these vector fields. These are the classical Calderón-Zygmund singular integrals corresponding to the usual Euclidean geometry. In particular, if $a(x, D)$ is a standard pseudodifferential operator of order s, then $a(x, D) \in \mathcal{B}_2^s$.

Remark 4.4.1   In the above, we used the notion of a pseudodifferential operator of order s on a compact manifold, whereas in Definition 1.3.3 only pseudodifferential operators on $\mathbb{R}^n$ were defined. For a compact manifold M, an operator $T : C^\infty(M) \to C^\infty(M)'$ is said to be a pseudodifferential operator of order $s \in \mathbb{R}$ if the following holds. For each point $x \in M$, we assume there is a neighborhood U_x of x which is diffeomorphic to an open neighborhood of $\mathbb{R}^n$ via a map $\phi_x : U_x \to \mathbb{R}^n$, and a function $\psi_x \in C_0^\infty(U_x)$ which equals 1 on a neighborhood of x such that $\phi_x^\# \psi_x T \psi_x (\phi_x^{-1})^\#$ is a pseudodifferential operator of order s, where $\phi_x^\#$ denotes the pullback via ϕ_x. We assume, in addition, that T is pseudolocal: i.e., the Schwartz kernel of T is C^∞ off the diagonal.

Putting these two lists of vector fields together, as in the introduction to this chapter, we obtain a list of vector fields with 2-parameter formal degrees:

$$(X, d) = (X_1, d_1), \ldots, (X_q, d_q).$$

These vector fields satisfy (4.1). Indeed, we claim for $\delta \in [0, 1]^2$,

$$\left[\delta^{d_j} X_j, \delta^{d_k} X_k\right] = \sum_{l=1}^q c_{j,k}^{l,\delta} \delta^{d_l} X_l, \tag{4.13}$$

with $\left\{c_{j,k}^{l,\delta} \mid \delta \in [0, 1]^\nu\right\} \subset C^\infty(M)$ bounded. If both X_j and X_k are from the list $(X^1, \hat{d}^1)$, then this follows immediately from Corollary 2.1.4. Otherwise, at least one of the vector fields is from the list $(X^2, \hat{d}^2)$. Write $[X_j, X_k] = \sum_{l=1}^{q_2} c_{j,k}^l X_l^2$, $c_{j,k}^l \in C^\infty$ (this uses that X^2 spans the tangent space). We have

$$\left[\delta^{d_j} X_j, \delta^{d_k} X_k\right] = \sum_{l=1}^{q_2} \left(\delta^{d_j + d_k - (0,1)} c_{j,k}^l\right) \delta_2 X_l^2.$$

Since at least one of X_j and X_k is from the list $(X^2, 1)$, $d_j + d_k - (0, 1)$ is nonnegative in every component. (4.13), and therefore (4.1), follows.

Corresponding to the balls $B_{(X,d)}(x, \delta)$, and as a special case of the operators defined in Chapter 5, we define an algebra of operators $\mathcal{A}^s$, where $s \in \mathbb{R}^2$ (see Section 5.11.3 for details). $\mathcal{B}_1^{s_1}$ is *not* a subset of $\mathcal{A}^{(s_1,0)}$, nor is $\mathcal{B}_2^{s_2}$ a subset of $\mathcal{A}^{(0,s_2)}$. We also do not have either of the reverse containments. However, many important operators from $\mathcal{B}_1$ and $\mathcal{B}_2$ are elements of $\mathcal{A}$. For instance, $\mathcal{L}^{-1} \in \mathcal{A}^{(-2,0)}$. Also, if $a\,(x,D)$ is a pseudodifferential operator of order s_2, then $a\,(x,D) \in \mathcal{A}^{(0,s_2)}$. In particular, by the algebra properties, $a\,(x,D)\,\mathcal{L}^{-1}, \mathcal{L}^{-1}a\,(x,D) \in \mathcal{A}^{(-2,s_2)}$. Thus, the algebra $\mathcal{A}^s$ gives an intrinsic way to understand compositions of pseudodifferential operators and $\mathcal{L}^{-1}$.

We return to a discussion of this setting in Section 5.11.3.

4.4.1   The $\overline{\partial}$-Neumann problem

In [CNS92], Chang, Nagel, and Stein studied the $\overline{\partial}$-Neumann problem on bounded, smooth, weakly pseudoconvex domains of finite type in $\mathbb{C}^2$. Their work involves composing operators from the algebra $\mathcal{B}_1^s$ with standard pseudodifferential operators. As it turns out, the operators they study are elements of the algebra $\mathcal{A}$ described above. We turn to briefly discussing the setup of that paper; we use the same notation so that the reader can easily turn to that paper for further information.

The setting is the same as in Section 2.11.1. We therefore are given $\Omega \subset \mathbb{C}^2$ a smoothly bounded, weakly pseudoconvex domain of finite type. Let U be a neighborhood of $\partial\Omega$ and let $r : U \to \mathbb{R}$ be a defining function so that

- $\Omega \cap U = \{z \in \mathbb{C}^2 \mid r\,(z) > 0\}$,
- $\nabla r\,(z) \neq 0$ for $r\,(z) = 0$.

On $\mathbb{C}^2$ we have $\overline{\partial}$ which takes (p,q) forms to $(p, q+1)$ forms. Let

$$\Box = \overline{\partial}\,\overline{\partial}^* + \overline{\partial}^*\,\overline{\partial}.$$

Corresponding to r define a complex vector field $\mathcal{R} := \sum_{j=1}^{2} \frac{\partial r}{\partial \overline{z}_j} \frac{\partial}{\partial \overline{z}_j}$. The $\overline{\partial}$-Neumann problem is: given a (p,q) form f find a (p,q) form u such that

$$
\begin{aligned}
\Box u &= f \text{ on } \Omega, \\
i_{\mathcal{R}} u &= 0 \text{ on } \partial\Omega, \\
i_{\mathcal{R}} \overline{\partial} u &= 0 \text{ on } \partial\Omega.
\end{aligned}
\tag{4.14}
$$

Here, $i_{\mathcal{R}}$ denotes the interior product with respect to the vector field $\mathcal{R}$; see the proof of Lemma 2.2.30 for more details on this notation. We consider the $\overline{\partial}$-Neumann problem for $(0,1)$ forms. In what follows, we only discuss (4.14) up to C^∞ error terms, which is enough to understand the regularity of the exact solution to (4.14).

Remark 4.4.2   For more on the setting described here, see [CNS92, GS77]. There is one piece of notation which is slightly different in [CNS92]: instead of taking the interior product of vector fields with forms, [CNS92] uses an interior product of u and $\overline{\partial}u$ with the $(0,1)$-form $\overline{\partial}r$. This less standard notation is defined on [GS77, page 72], and is equivalent to the notation we have used above.

Following ideas of Greiner and Stein [GS77], Chang, Nagel, and Stein reduced the $\overline{\partial}$-Neumann problem to inverting an operator on the boundary $\partial\Omega$. To describe this, we introduce two classical operators: a Poisson operator P and a Green's operator G. Let R denote the operator which is restriction to $\partial\Omega$.

P maps $(0,1)$ forms on $\partial\Omega$ to $(0,1)$ forms on Ω and satisfies modulo C^∞ terms

$$\square \circ P = 0 \quad \text{on } \Omega,$$
$$R \circ P = I \quad \text{on } \partial\Omega.$$

G satisfies, modulo C^∞ error terms

$$\square \circ G = I \quad \text{on } \Omega,$$
$$R \circ G = 0 \quad \text{on } \partial\Omega.$$

We look for a solution to (4.14) of the form

$$u = P(u_b) + G(f),$$

where u_b is a $(0,1)$ form on $\partial\Omega$ to be determined.[7] Modulo C^∞ terms, $\square u = f$. The second line of (4.14) is also automatically satisfied as $i_R u_b = 0$ for any $(0,1)$ form on the three dimensional CR-manifold $\partial\Omega$. Thus, the goal is to choose u_b so that the last part of (4.14) is satisfied. Define

$$\square^+ u_b := i_R \bar{\partial} P(u_b) \big|_{\partial\Omega}.$$

Once we solve the equation $\square^+ u_b = -i_R \bar{\partial} G(f) \big|_{\partial\Omega}$, then from the properties of G and P, we have solved (4.14) modulo C^∞ terms. Hence, the goal is to find an inverse of $\square^+$, modulo infinitely smoothing operators.

$\square^+$ maps $(0,1)$ forms on $\partial\Omega$ to $(0,1)$ forms on $\partial\Omega$. Since $\Omega \subset \mathbb{C}^2$, we may identify $(0,1)$ forms on $\partial\Omega$ with scalar valued functions. Once this identification is done, $\square^+$ is a standard pseudodifferential operator of order 1, whose principal symbol can be explicitly computed.

We now turn to discussing the geometry on $\partial\Omega$. Actually, there are two geometries that arise. The first is the usual geometry where $\partial\Omega$ is thought of as a Riemannian submanifold of $\mathbb{C}^2$. Corresponding to this geometry we have the classical Calderón-Zygmund theory of singular integrals, and we refer to this algebra of singular integrals as $\mathcal{B}_2$. In particular, standard pseudodifferential operators are elements of this algebra, and if $a(x, D)$ is a pseudodifferential operator of order s_2, then $a(x, D) \in \mathcal{B}_2^{s_2}$.

The second geometry is the Carnot-Carathéodory geometry from Section 2.11.1. As in Chapter 2, we obtain an algebra of singular integrals corresponding to this geometry, which we refer to as $\mathcal{B}_1$.

We have the Kohn Laplacian $\square_b^1$ acting on $(0,1)$ forms. As in Section 2.11.1, we identify $(0,1)$ forms with functions, and think of $\square_b^1$ as a second order partial differential operator. Also as described in that section, if S is the orthogonal projection onto the L^2 null space of $\square_b^1$, then $S \in \mathcal{B}_1^0$, and there is $K \in \mathcal{B}_1^{-2}$ with $\square_b^1 K = K\square_b^1 = I - S$.

We put these two geometries together to create an algebra of 2-parameter operators $\mathcal{A}^{(s_1, s_2)}$ as described above. We will see $K \in \mathcal{A}^{(-2,0)}$.

In [CNS92], standard pseudodifferential operators $\square^-$, Γ^+, and Q^+ of orders 1, 0, and -1 are defined. Note, $\square^- \in \mathcal{A}^{(0,1)}$, $\Gamma^+ \in \mathcal{A}^{(0,0)}$, and $Q^+ \in \mathcal{A}^{(0,-1)}$. Define $P_0 = \square^- K\Gamma^+ + Q^+ \in \mathcal{A}^{(-2,1)}$. In [CNS92] it is shown that $\square^+ P_0 = I + E$, where E is an error term which "smooths." From this we will show

---

[7] I.e., u_b is a $(0,1)$ form on the three dimensional CR-manifold $\partial\Omega$.

THEOREM 4.4.3. *Let P satisfy $\Box^+ P \equiv I$ mod $C^\infty (\partial\Omega \times \partial\Omega)$, then $P \in \mathcal{A}^{(-2,1)}$.*

In fact, we will see that the above theorem is a way of essentially restating some of the results from [CNS92], and puts the regularity results proved there for the $\bar{\partial}$-Neumann problem into the framework of this monograph. We return to these points in Section 5.11.3.

4.4.2   Further reading and references

One of the original, and main, motivations for studying singular integrals with two underlying geometries, where one is a Carnot-Carathéodory geometry and one is the Euclidean geometry, comes from questions involving the $\bar{\partial}$-Neumann problem as discussed in this section. Such operators first arose in the work of Greiner and Stein [GS77]. It was they who elucidated the usefulness of using both elliptic pseudodifferential operators and operators which are intrinsically associated to the Carnot-Carathéodory geometry on the boundary. Greiner and Stein's work was restricted to the "strictly pseudoconvex" case. Chang, Nagel, and Stein extended Greiner and Stein's results (at least in two complex dimensions) to the "finite type" case [CNS92]. It is the work of Chang, Nagel, and Stein which we follow, here, though the same comments hold in higher dimensions in the strictly pseudoconvex case as studied by Greiner and Stein.

Motivated by these concerns, there have been some recent advances in the theory of singular integrals with two underlying geometries, like the ones discussed here. One is recent work by Nagel, Ricci, Stein, and Wainger who built on their work on flag kernels (see Section 4.2 and [NRSW12]) to develop an algebra of singular integrals containing both the left invariant Calderón-Zygmund operators on a graded group, and the standard psuedodifferential operators. Stein and Yung have recently developed an algebra of generalized pseudodifferential operators which contains the operator P from Theorem 4.4.3 in the special case when the vector fields W_1 and W_2 from Section 2.11.1 satisfy Hörmander's condition of order 2. I.e., when W_1, W_2, and $[W_1, W_2]$ span the tangent space to each point of $\partial\Omega$: the "strictly pseudoconvex" case. To achieve this, they build on work of Nagel and Stein [NS79]. A motivation of these references, and of this section, is to give an intrinsic understanding of the operators which were only studied extrinsically by Greiner and Stein, and by Chang, Nagel, and Stein.

Chapter Five

Multi-parameter Singular Integrals II: General Theory

In Chapter 4, we presented several examples of multi-parameter singular integrals that arise naturally. We now turn to the main purpose of this monograph: to present an intrinsically defined algebra of "multi-parameter singular integral operators" which includes and generalizes all of the ideas from Chapter 4.

The setting is the same as Chapter 3. That is, we are given ν lists of vector fields on an open set $U \subseteq \mathbb{R}^\nu$ with single-parameter formal degrees: for each $1 \le \mu \le \nu$, we have $\left(X^\mu, \hat{d}^\mu \right) := \left(X_1^\mu, \hat{d}_1^\mu \right), \dots, \left(X_{q_\mu}^\mu, \hat{d}_{q_\mu}^\mu \right); 0 \ne \hat{d}_j^\mu \in \mathbb{N}$. We assume that $X_1^1, \dots, X_{q_1}^1, \dots, \dots, X_1^\nu, \dots, X_{q_\nu}^\nu$ span the tangent space at every point of U. From these ν lists of vector fields with formal degrees, we obtain one list of vector fields with ν parameter formal degrees[1] $(X, d) = (X_1, d_1), \dots, (X_q, d_q), 0 \ne d_j \in \mathbb{N}^\nu$, each d_j nonzero in only one component, as in the introduction to Chapter 3. We obtain the balls $B_{(X,d)}(x, \delta)$ for $\delta \in (0, 1]^\nu$, which are open sets.

We fix a distinguished point in U, which we call $0 \in U$. We let $U_1 \Subset U_2 \Subset U$ be relatively compact open subsets of U. **We assume all the assumptions from Section 3.1**, with these choices. In particular, Theorem 3.1.2 applies in this situation. Let $\xi_6 > 0$ be as in Theorem 3.3.1 and let $0 < \xi_0 \le \xi_6/2$ be such that all of the results in Section 3.2 hold. As in Section 3.2, define $\Omega := B_{(X,d)}\left(0, \frac{\xi_0}{2}(1, \dots, 1) \right)$. We will be considering operators $T : C^\infty(\Omega) \to C_0^\infty(\Omega)$. Throughout this chapter, unless otherwise mentioned, we use all these definitions and assumptions.

From (3.4) and (3.5), we obtain (for $j \in [0, \infty)^\nu$) the metric $\rho_{2^{-j}}(x, y)$ and the balls $B_{2^{-j}}(x, r)$ ($r > 0$).

We will define a filtered algebra of operators $\mathcal{A}^s$ ($s \in \mathbb{R}^\nu$). In particular, if $T \in \mathcal{A}^t$ and $S \in \mathcal{A}^s$, we have $TS \in \mathcal{A}^{t+s}$. We also define appropriate non-isotropic Sobolev spaces which interact well with this algebra of operators.

Remark 5.0.4   Since we are considering operators $T : C^\infty(\Omega) \to C_0^\infty(\Omega)$, if we identify T with its Schwartz kernel we have $T \in C^\infty(\Omega \times \Omega)'$. Thus, $\text{supp}(T) \Subset \Omega \times \Omega$. See Appendix A for details on this. Since Ω is a small open set, we are considering only operators with small support. In Chapter 4, we considered many examples where we did not restrict to operators with small support. In each of these examples, there was an appropriate "compactness" (as in Sections 4.1 and 4.4) or homogeneity (as in Sections 4.2 and 4.3) which allowed us to avoid restricting to operators with

---

[1]Recall, we assign to X_j^μ the formal degree $d_j^\mu \in \mathbb{N}^\nu$, where d_j^μ equals $\hat{d}_j^\mu$ in the μth component and 0 in all other components. (X, d) is then an enumeration of all such vector fields with formal degrees, $1 \le \mu \le \nu, 1 \le j \le q_\mu$.

small support. In general, we cannot do this, due to problems like the ones discussed in Remark 2.15.35. In each of the examples from Chapter 4, the methods in this chapter work for operators without small support. Thus, the ideas here generalize all the examples from Chapter 4.

Remark 5.0.5 In this chapter, Vol $(\cdot)$ and dx refer to Lebesgue measure on $\mathbb{R}^n$.

5.1 THE MAIN RESULTS

We start by defining the algebra of operators, $\mathcal{A}^s$. We present four different definitions. The main theorem (Theorem 5.1.12) concerning these definitions is that they are all equivalent. Thus, for each $s \in \mathbb{R}^\nu$, we define four sets of operators $\mathcal{A}_1^s$, $\mathcal{A}_2^s$, $\mathcal{A}_3^s$, and $\mathcal{A}_4^s$. We show $\mathcal{A}_1^s = \mathcal{A}_2^s = \mathcal{A}_3^s = \mathcal{A}_4^s$, and refer to the common value as $\mathcal{A}^s$.

Recall, for $j = (j_1, \ldots, j_\nu), k = (k_1, \ldots, k_\nu) \in [0, \infty)^\nu$ we write $j \leq k$ if $j_\mu \leq k_\mu, \forall \mu$. Also, $j \wedge k = (j_1 \wedge k_1, \ldots, j_\nu \wedge k_\nu)$ and $2^{-j} = \left(2^{-j_1}, \ldots, 2^{-j_\nu}\right) \in (0, 1]^\nu$.

DEFINITION 5.1.1. We say $\mathcal{B} \subset C_0^\infty(\Omega) \times \Omega \times (0, 1]^\nu$ is a bounded set of bump functions if $\forall M, \exists C_M, \forall (\phi, x, 2^{-j}) \in \mathcal{B}$,

- $\operatorname{supp}(\phi) \subset B_{2^{-j}}(x, 1)$,

- $\sum_{|\alpha| \leq M} \left\| (2^{-j}X)^\alpha \phi \right\|_{C^0(\Omega)} \leq C_M \operatorname{Vol}(B_{2^{-j}}(x, 1))^{-1}$.

DEFINITION 5.1.2. For $M \geq 0$, we say $\mathcal{T}_M \subset C_0^\infty(\Omega) \times \Omega \times (0, 1]^\nu$ is a bounded set of test functions of order M if $\forall (\psi, x, 2^{-j}) \in \mathcal{T}_M, \forall l \leq j \ (l \in [0, \infty)^\nu)$,

$$\psi = 2^{-|j-l|_\infty M} \sum_{|\alpha| \leq M} \left(2^{-l}X\right)^\alpha \phi_{l,\alpha},$$

where $\left\{ \left(\phi_{l,\alpha}, x, 2^{-j} \right) \mid (\psi, x, 2^{-j}) \in \mathcal{T}_M, l \leq j, |\alpha| \leq M \right\}$ is a bounded set of bump functions.

Remark 5.1.3 Note that a bounded set of test functions is a bounded set of bump functions (merely take $l = j$ in the definition of test functions). One should think of test functions as a high derivative of bump functions. Indeed, if $\mathcal{B}$ is a bounded set of bump functions and if α is an ordered multi-index so that X^α contains at least M terms from each of the lists $X^1, \ldots, X^\nu$, then $\left\{ \left((2^{-j}X)^\alpha \phi, x, 2^{-j}\right) \mid (\phi, x, 2^{-j}) \in \mathcal{B} \right\}$ is a bounded set of test functions of order M. Test functions are slightly more versatile than this, though. For instance, if $\mathcal{T}_M$ is a bounded set of test functions of order M and if $\eta \in C^\infty(\Omega)$ is a fixed function, then $\left\{ (\eta\psi, x, 2^{-j}) \mid (\psi, x, 2^{-j}) \in \mathcal{T}_M \right\}$ is also a bounded set of test functions of order M.

DEFINITION 5.1.4. For $s \in \mathbb{R}^\nu$ and $T : C^\infty(\Omega) \to C_0^\infty(\Omega)$, we say $T \in \mathcal{A}_1^s$ if $\forall L, m, \exists M, \forall \mathcal{T}_M$ bounded sets of test functions of order M, $\exists C, \forall (\psi_1, x, 2^{-j})$, $(\psi_2, y, 2^{-k}) \in \mathcal{T}_M$,

$$|\langle \psi_1, T\psi_2 \rangle_{L^2}| \leq C 2^{j \wedge k \cdot s} 2^{-L|j-k|} \frac{(1 + \rho_{2^{-j \wedge k}}(x, y))^{-m}}{\operatorname{Vol}(B_{2^{-j \wedge k}}(x, 1 + \rho_{2^{-j \wedge k}}(x, y)))}.$$

DEFINITION 5.1.5. *For $s \in \mathbb{R}^\nu$ and $T : C^\infty (\Omega) \to C_0^\infty (\Omega)$, we say $T \in \mathcal{A}_2^s$ if $\forall m$, $\exists M$, $\forall \mathcal{T}_M$ bounded sets of test functions of order M, $\forall \alpha$, $\exists C$, $\forall (\psi, z, 2^{-j}) \in \mathcal{T}_M$,*

$$\left| \left(2^{-j} X \right)^\alpha T \psi (x) \right| \leq C 2^{j \cdot s} \frac{\left(1 + \rho_{2^{-j}} (x, z) \right)^{-m}}{\mathrm{Vol} \left(B_{2^{-j}} \left(x, 1 + \rho_{2^{-j}} (x, z) \right) \right)},$$

and the same estimate holds for T replaced by T^* (the formal L^2 adjoint of T).

DEFINITION 5.1.6. *We say $\mathcal{E} \subset C_0^\infty (\Omega \times \Omega) \times (0, 1]^\nu$ is a bounded set of pre-elementary operators if:[2]*

(i)
$$\bigcup_{(E, 2^{-j}) \in \mathcal{E}} \mathrm{supp} \, (E) \Subset \Omega \times \Omega.$$

(ii) *$\forall m, M, \exists C, \forall (E, 2^{-j}) \in \mathcal{E}$,*

$$\sum_{|\alpha|, |\beta| \leq M} \left| \left(2^{-j} X_x \right)^\alpha \left(2^{-j} X_z \right)^\beta E (x, z) \right| \leq C \frac{\left(1 + \rho_{2^{-j}} (x, z) \right)^{-m}}{\mathrm{Vol} \left(B_{2^{-j}} \left(x, 1 + \rho_{2^{-j}} (x, z) \right) \right)}.$$

DEFINITION 5.1.7. *We define the **set of bounded sets** of generalized elementary operators, $\mathcal{G}$, to be the largest **set of subsets** of $C_0^\infty (\Omega \times \Omega) \times (0, 1]^\nu \times (0, 1]^\nu$ such that $\forall \mathcal{E} \in \mathcal{G}$:*

(i) *$\forall \left(E, 2^{-j}, 2^{-k} \right) \in \mathcal{E}$, $k \leq j$.*

(ii) *$\left\{ (E, 2^{-j}) \mid (E, 2^{-j}, 2^{-k}) \in \mathcal{E} \right\}$ is a bounded set of pre-elementary operators.*

(iii) *$\forall \left(E, 2^{-j}, 2^{-k} \right) \in \mathcal{E}$, $\forall k \leq l \leq j$, we have*

$$E = \begin{cases} 2^{-|j-l|_\infty} \sum_{|\alpha| \leq 1} E_{\alpha, l}^1 \left(2^{-l} X \right)^\alpha & and \\ 2^{-|j-l|_\infty} \sum_{|\alpha| \leq 1} \left(2^{-l} X \right)^\alpha E_{\alpha, l}^2, \end{cases} \tag{5.1}$$

where

$$\left\{ \left(E_{\alpha, l}^1, 2^{-j}, 2^{-l} \right), \left(E_{\alpha, l}^2, 2^{-j}, 2^{-l} \right) \mid (E, 2^{-j}, 2^{-k}) \in \mathcal{E}, k \leq l \leq j, |\alpha| \leq 1 \right\}$$
$$\in \mathcal{G}.$$

We call elements of $\mathcal{G}$ bounded sets of generalized elementary operators.

DEFINITION 5.1.8. *We say $\mathcal{E} \subset C_0^\infty (\Omega \times \Omega) \times (0, 1]^\nu$ is a bounded set of elementary operators if $\mathcal{E} \times \left\{ 2^{-0} \right\}$ is a bounded set of generalized elementary operators (here $2^{-0} = (1, 1, \ldots, 1) \in (0, 1]^\nu$).*

---

[2] Recall the notation $A \Subset B$ means that A is a relatively compact subset of B.

Remark 5.1.9 To refer to (iii) of Definition 5.1.7, we say that we can "pull out" deriva-tives from generalized elementary operators.

DEFINITION 5.1.10. *Fix $s \in \mathbb{R}^\nu$ and $T : C^\infty(\Omega) \to C_0^\infty(\Omega)$. We say $T \in \mathcal{A}_3^s$ if for every bounded set of elementary operators $\mathcal{E}$,*

$$\left\{ \left(2^{-j \cdot s} T E, 2^{-j} \right) \mid \left(E, 2^{-j} \right) \in \mathcal{E} \right\}$$

is a bounded set of elementary operators.

DEFINITION 5.1.11. *Fix $s \in \mathbb{R}^\nu$ and $T : C^\infty(\Omega) \to C_0^\infty(\Omega)$. We say $T \in \mathcal{A}_4^s$ if for each $j \in \mathbb{N}^\nu$ there is $E_j \in C_0^\infty(\Omega \times \Omega)$ such that $\left\{ \left(E_j, 2^{-j} \right) \mid j \in \mathbb{N}^\nu \right\}$ is a bounded set of elementary operators and*

$$T = \sum_{j \in \mathbb{N}^\nu} 2^{j \cdot s} E_j,$$

with the sum taken in the sense of distribution. We will see (Remark 5.3.3) that every such sum converges in the sense of distribution. In fact, every such sum converges in the topology of bounded convergence as operators $C^\infty(\Omega) \to C_0^\infty(\Omega)$ (Lemma 5.3.2).

The main result concerning these theorems is the following.

THEOREM 5.1.12. *$\mathcal{A}_1^s = \mathcal{A}_2^s = \mathcal{A}_3^s = \mathcal{A}_4^s$, and we denote their common value by* $\mathcal{A}^s$.

COROLLARY 5.1.13. *If $T \in \mathcal{A}^{s_1}$ and $S \in \mathcal{A}^{s_2}$, then $TS \in \mathcal{A}^{s_1 + s_2}$.*

PROOF. This follows immediately by using $\mathcal{A}^s = \mathcal{A}_3^s$. □

COROLLARY 5.1.14. *If $T \in \mathcal{A}^s$, then $T^* \in \mathcal{A}^s$.*

PROOF. This follows immediately by using $\mathcal{A}^s = \mathcal{A}_2^s$. □

Remark 5.1.15 It is easy to see that if $s \le t$, then $\mathcal{A}^s \subseteq \mathcal{A}^t$. This combined with Corollary 5.1.13 can be rephrased as saying $\mathcal{A} := \bigcup_{s \in \mathbb{R}^\nu} \mathcal{A}^s$ forms a *filtered algebra*.

Remark 5.1.16 The definition of $\mathcal{A}$ is invariant under diffeomorphisms. Indeed, if $\Psi : \Omega \to \Omega'$ is a diffeomorphism, and if we define X_j' to be the push forward of X_j via Ψ, then we obtain a new algebra of operators supported on $\Omega' \times \Omega'$. Namely, the class $\mathcal{A}$ associated to the vector fields with formal degrees $(X_1', d_1), \ldots, (X_q', d_q)$; call this class $\mathcal{A}_0$ to avoid confusion–the class associated to $(X_1, d_1), \ldots, (X_q, d_q)$ we continue to refer to as $\mathcal{A}$. Let $\Psi^\#$ denote the pullback via Ψ. Then, $T \in \mathcal{A}_0^s$ if and only if $\left(\Psi^\# \right)^{-1} T \Psi^\# \in \mathcal{A}^s$. This follows immediately from the definitions.

Remark 5.1.17 Many of our definitions involve bounds of the form

$$\frac{\left(1 + \rho_{2^{-j}}(x, z) \right)^{-m}}{\mathrm{Vol}\left(B_{2^{-j}}(x, 1 + \rho_{2^{-j}}(x, z)) \right)},$$

which may not seem symmetric in x and z. However, in light of Proposition 3.2.2, we have

$$\frac{(1 + \rho_{2-j}(x, z))^{-m}}{\mathrm{Vol}\,(B_{2-j}(x, 1 + \rho_{2-j}(x, z)))} \approx \frac{(1 + \rho_{2-j}(z, x))^{-m}}{\mathrm{Vol}\,(B_{2-j}(z, 1 + \rho_{2-j}(z, x)))},$$

and our assumptions are therefore symmetric in x and z

Remark 5.1.18   Unlike the classical definitions in the single-parameter setting (e.g., Definition 2.0.16), none of the definitions of $\mathcal{A}^s$ involve "growth conditions." However, one can think of $\mathcal{A}_1^s$ and $\mathcal{A}_2^s$ as both incorporating a sort of "growth condition," in the sense that they are analogs of the characterizations given in Corollary 2.7.21.

Though we have not made it precise, the definitions of $\mathcal{A}_1^s$, $\mathcal{A}_2^s$, $\mathcal{A}_3^s$, and $\mathcal{A}_4^s$ each give $\mathcal{A}^s$ a locally convex topology in the usual way. It is a consequence of our proofs that all of these topologies are actually the same. This topology is not central to our results, however it does have some uses. In Section 5.7, we define this locally convex topology, and prove that $\mathcal{A}^s$ is complete with respect to this topology. In fact, we will see that $\mathcal{A}^s$ is an LF space (see Definition A.1.24) and is therefore complete (Remark A.2.4).

Remark 5.1.19   As in the rest of this monograph, what is most important to us is not the topology itself, but the bounded subsets of the topology. We remark now on two ways to understand these bounded subsets. Let $\mathcal{B} \subset \mathcal{A}^s$. Then, the following are equivalent:

(i) $\mathcal{B} \subset \mathcal{A}^s$ is a bounded set.

(ii) $\bigcup_{T \in \mathcal{B}} \mathrm{supp}\,(T) \Subset \Omega \times \Omega$ and for every bounded set of elementary operators $\mathcal{E}$, the set
$$\left\{ (2^{-j \cdot s} T E, 2^{-j}) \mid (E, 2^{-j}) \in \mathcal{E}, T \in \mathcal{B} \right\}$$
is a bounded set of elementary operators.

(iii) For each $T \in \mathcal{B}$ and each $j \in \mathbb{N}^\nu$ there is an operator $E_j^T \in C_0^\infty (\Omega \times \Omega)$ such that $\left\{ (E_j^T, 2^{-j}) \mid j \in \mathbb{N}^\nu, T \in \mathcal{B} \right\}$ is a bounded set of elementary operators and $T = \sum_{j \in \mathbb{N}^\nu} 2^{j \cdot s} E_j^T$.

See Section 5.7 for more details.

Remark 5.1.20   The concept of elementary operators in this chapter factors through the more complicated notion of generalized elementary operators. In previous chapters, we did not use these generalized elementary operators. It remains unclear whether or not it is possible to work with a simpler notion of elementary operators than we have introduced here. The main issue is that it is necessary to decompose the identity operator $I = \sum_{j \in \mathbb{N}^\nu} E_j$, where $\left\{ E_j \mid j \in \mathbb{N}^\nu \right\}$ is a bounded set of elementary operators. We do this in Section 5.3, but we only know how to do this (in full generality) with the more complicated notion of elementary operators from this chapter. See Remark 5.3.16 for more details.

5.1.1   Non-isotropic Sobolev spaces

We now define non-isotropic Sobolev spaces which are adapted to the vector fields $(X, d) = (X_1, d_1), \ldots, (X_q, d_q)$. To do this, we look to Theorem 2.10.6 for our motivation.

Fix a neighborhood of $0 \in \mathbb{R}^\nu$, Ω_0, with $\Omega_0 \Subset \Omega$. We will be restricting attention to functions supported in Ω_0. Fix $\psi_0 \in C_0^\infty(\Omega)$ with $\psi_0 \equiv 1$ on a neighborhood of the closure of Ω_0. We identify ψ_0 with the operator given by multiplication with ψ_0. It is easy to see $\psi_0 \in \mathcal{A}_2^0$. Thus, by Theorem 5.1.12, $\psi_0 \in \mathcal{A}_4^0$ and there is a bounded set of elementary operators $\{(D_j, 2^{-j}) \mid j \in \mathbb{N}^\nu\}$ such that $\psi_0 = \sum_{j \in \mathbb{N}^\nu} D_j$.

For $1 < p < \infty$ and $s \in \mathbb{R}^\nu$ define the semi-norm

$$\|f\|_{\mathrm{NL}_s^p} := \left\| \left(\sum_{j \in \mathbb{N}^\nu} |2^{j \cdot s} D_j f|^2 \right)^{\frac{1}{2}} \right\|_{L^p}.$$

Since $\sum_{j \in \mathbb{N}^\nu} D_j f = \psi_0 f$, for $f \in C_0^\infty(\Omega_0)$ this is a genuine norm (as opposed to merely a semi-norm). We will be considering only $f \in C_0^\infty(\Omega_0)$ in what follows; the space NL_s^p should be thought of as the completion of $C_0^\infty(\Omega_0)$ in the above norm.

Remark 5.1.21   The operators in $\mathcal{A}$ are maps $C^\infty(\Omega) \to C_0^\infty(\Omega)$ and therefore if $T \in \mathcal{A}^s$, we have $\mathrm{supp}(T) \Subset \Omega \times \Omega$. Thus, if we are considering a fixed T and take Ω_0 large enough that $\mathrm{supp}(T) \Subset \Omega_0 \times \Omega_0$, then considering only functions supported in Ω_0 is no real restriction.

PROPOSITION 5.1.22.   *The above norm is well-defined. More specifically, let $\widetilde{\psi}_0 \in C_0^\infty(\Omega)$ satisfy $\widetilde{\psi}_0 \equiv 1$ on a neighborhood of the closure of Ω_0 and let*

$$\left\{ \left(\widetilde{D}_j, 2^{-j} \right) \mid j \in \mathbb{N}^\nu \right\}$$

be a bounded set of elementary operators with $\widetilde{\psi}_0 = \sum_{j \in \mathbb{N}^\nu} \widetilde{D}_j$. Then, for $f \in C_0^\infty(\Omega_0)$, we have

$$\left\| \left(\sum_{j \in \mathbb{N}^\nu} |2^{j \cdot s} D_j f|^2 \right)^{\frac{1}{2}} \right\|_{L^p} \approx \left\| \left(\sum_{j \in \mathbb{N}^\nu} |2^{j \cdot s} \widetilde{D}_j f|^2 \right)^{\frac{1}{2}} \right\|_{L^p},$$

where the implicit constants depend on p ($1 < p < \infty$), $s \in \mathbb{R}^\nu$, and the particular choices of ψ_0, $\widetilde{\psi}_0$, D_j, and $\widetilde{D}_j$.

THEOREM 5.1.23.   *Let $T \in \mathcal{A}^s$. Then for $f \in C_0^\infty(\Omega_0)$,*

$$\|Tf\|_{\mathrm{NL}_{s_0}^p} \lesssim \|f\|_{\mathrm{NL}_{s_0+s}^p},$$

for $1 < p < \infty$, $s_0 \in \mathbb{R}^\nu$.

Remark 5.1.24 It is easy to see (using $\mathcal{A}^s = \mathcal{A}_2^s$) that $\psi_0 X_j \in \mathcal{A}^{d_j}$. Thus, for $f \in C_0^\infty(\Omega_0)$, we have

$$\|X_j f\|_{\mathrm{NL}_{s_0}^p} \lesssim \|f\|_{\mathrm{NL}_{s_0+d_j}^p}, \quad 1 < p < \infty, \quad s_0 \in \mathbb{R}^\nu.$$

PROPOSITION 5.1.25. NL_0^p *agrees with* L^p *$(1 < p < \infty)$. More specifically, for* $f \in C_0^\infty(\Omega_0)$,

$$\|f\|_{\mathrm{NL}_0^p} \approx \|f\|_{L^p}.$$

COROLLARY 5.1.26. *Let* $T \in \mathcal{A}^0$. *Then* $T : L^p \to L^p$ *$(1 < p < \infty)$.*

PROOF. Choose Ω_0 so that $\mathrm{supp}(T) \Subset \Omega_0 \times \Omega_0$. Now the result follows by combining Theorem 5.1.23 and Proposition 5.1.25. $\qquad\qquad\square$

We now discuss the relationship between these non-isotropic Sobolev spaces, and the standard isotropic Sobolev spaces. For $r \in \mathbb{R}$ let L_r^p denote the standard L^p Sobolev space of order r, with norm $\|\cdot\|_{L_r^p}$ (see Definition 1.3.10). Recall, we have assumed that $X_1, \ldots, X_q$ span the tangent space. Pick out l of these vector fields which span the tangent space at every point of Ω. Without loss of generality, renumber the vector fields with formal degrees, so that these l are $(X_1, d_1), \ldots, (X_l, d_l)$. Notice $l \geq n$.

PROPOSITION 5.1.27. *Let* $e = (e_1, \ldots, e_\nu) \in \mathbb{N}^\nu$ *be defined by* $e_\mu = \max_{1 \leq j \leq l} d_j^\mu$ *(i.e., the maximum of the μth coordinates of the d_j). Then, for* $r \in \mathbb{N}$, $1 < p < \infty$,

$$\|f\|_{L_r^p} \lesssim \|f\|_{\mathrm{NL}_{re}^p}, \quad f \in C_0^\infty(\Omega_0).$$

Fix ν_0, $1 \leq \nu_0 \leq \nu$. Consider the last ν_0 lists of vector fields

$$\left(X_1^{\nu-\nu_0+1}, \hat{d}_1^{\nu-\nu_0+1}\right), \ldots, \left(X_{q_{\nu-\nu_0+1}}^{\nu-\nu_0+1}, \hat{d}_{q_{\nu-\nu_0+1}}^{\nu-\nu_0+1}\right), \ldots,$$

$$\ldots, \left(X_1^\nu, \hat{d}_1^\nu\right), \ldots, \left(X_{q_\nu}^\nu, \hat{d}_{q_\nu}^\nu\right).$$

These ν_0 lists of vector fields satisfy all the assumptions of this chapter with ν replaced by ν_0, except perhaps that

$$\left\{X_j^\mu \mid \nu_0 + 1 \leq \mu \leq \nu, 1 \leq j \leq q_\mu\right\}$$

might not span the tangent space. Despite the possibility that these vector fields might not span the tangent space, we will see that we may define non-isotropic Sobolev spaces, as above, to create ν_0-parameter non-isotropic Sobolev space in terms of these ν_0 lists of vector fields,[3] whose norm we denote by $\|f\|_{\mathrm{NL}_s^p}$, $1 < p < \infty$, $s \in \mathbb{R}^{\nu_0}$.

There is another ν_0 parameter set of non-isotropic Sobolev norms; namely the ones associated to the norms $\|f\|_{\mathrm{NL}_{(0,s)}^p}$, where $s \in \mathbb{R}^{\nu_0}$ and $0 \in \mathbb{R}^{\nu-\nu_0}$. These are the norms defined in terms of the original ν lists of vector fields. These turn out to be the same spaces:

---

[3] See Remark 5.8.7 for the precise definition of the norm when the vector fields do not span the tangent space.

THEOREM 5.1.28. *For $f \in C_0^\infty(\Omega_0)$,*

$$\|f\|_{\mathrm{NL}_s^p} \approx \|f\|_{\mathrm{NL}_{(0,s)}^p},$$

where the implicit constants depend on $p \in (1, \infty)$ and $s \in \mathbb{R}^{\nu_0}$.

COROLLARY 5.1.29. *If the last[4] list of vector fields $\left(X_1^\nu, \hat{d}_1^\nu\right), \ldots, \left(X_{q_\nu}^\nu, \hat{d}_{q_\nu}^\nu\right)$ is given by the Euclidean vector fields $(\partial_{x_1}, 1), \ldots, (\partial_{x_n}, 1)$, we have for $s \in \mathbb{R}$,*

$$\|f\|_{\mathrm{NL}_{(0,s)}^p} \approx \|f\|_{L_s^p}, \quad f \in C_0^\infty(\Omega_0),$$

where L_s^p denotes the standard isotropic L^p Sobolev space of order s; here $(0, s) \in \mathbb{R}^{\nu-1} \times \mathbb{R} = \mathbb{R}^\nu$.

Remark 5.1.30   In Theorem 5.1.28 (respectively, Corollary 5.1.29), we have taken the first $\nu - \nu_0$ (respectively, $\nu - 1$) coordinates to be 0 in the parameter t of NL_t^p. However, similar results hold with any $\nu - \nu_0$ coordinates chosen to be 0. This is because our assumptions and definitions are invariant under permutations of the parameters.

5.1.2   Multi-parameter pseudodifferential operators

Motivated by the results in Section 2.14, we define a notion of "pseudodifferential operators" adapted to this situation. These pseudodifferential operators are elements of $\mathcal{A}$, and in fact make up many of our main examples of operators in $\mathcal{A}$. Though we refer to these operators as pseudodifferential operators, we do not develop a calculus of these operators akin to the calculus of pseudodifferential operators. That is, we have no analog of Theorem 1.3.6 for these operators.

To define our notion of a pseudodifferential operator, we need to introduce a class of product kernels. We begin with an informal introduction to these ideas, and then turn to making them precise. Recall, $q = q_1 + \cdots + q_\nu$. We write $t = (t_1, \ldots, t_\nu) \in \mathbb{R}^{q_1} \times \cdots \times \mathbb{R}^{q_\nu} = \mathbb{R}^q$. The product kernels we consider are distributions in $C_0^\infty(\Omega \times \mathbb{R}^q)'$, which we write as $K(x, t)$, where $x \in \Omega$ and $t \in \mathbb{R}^q$. These distributions are, moreover, C_0^∞ functions in the x variable.[5] Define $Q_\mu = \sum_{j=1}^{q_\mu} \hat{d}_j^\mu$, and define single-parameter dilations on $\mathbb{R}^{q_\mu}$ by, for $R > 0$,

$$R\left(t^1, \ldots, t^{q_\mu}\right) = \left(R^{\hat{d}_1^\mu} t^1, \ldots, R^{\hat{d}_{q_\mu}^\mu} t^{q_\mu}\right). \tag{5.2}$$

Notice that Q_μ is the homogeneous dimension of $\mathbb{R}^{q_\mu}$ under these dilations.

For $t_\mu = \left(t_\mu^1, \ldots, t_\mu^{q_\mu}\right) \in \mathbb{R}^{q_\mu}$, we write $\|t_\mu\|$ for a choice of a smooth homogeneous norm on $\mathbb{R}^{q_\mu}$. I.e., $\|t_\mu\|$ is smooth away from $t_\mu = 0$, satisfies $\|Rt_\mu\| = R\|t_\mu\|$ for $R > 0$ (where Rt_μ is defined in terms of the dilations on $\mathbb{R}^{q_\mu}$ above), $\|t_\mu\| \geq 0$,

---

[4]It is not essential that we pick out the *last* list of vector fields; see Remark 5.1.30.

[5]More specifically, the product kernels we consider lie in the space $\left[C_0^\infty(\Omega)\right] \hat{\otimes} \left[\mathcal{S}(\mathbb{R}^q)'\right] \cong C_0^\infty(\Omega; \mathcal{S}(\mathbb{R}^q)')$. See Appendix A.2 for more details on this notation.

and $\|t_\mu\| = 0 \Leftrightarrow t_\mu = 0$. Any two choices for the homogeneous norm are equivalent for our purposes. For instance, we can take

$$\|t_\mu\| = \left(\sum_{j=1}^{q_\mu} |t_\mu^j|^{\frac{2(Q_\mu!)}{d_j^\mu}} \right)^{\frac{1}{2(Q_\mu!)}}. \tag{5.3}$$

The product kernels $K(x,t)$ of order $m = (m_1, \ldots, m_\nu) \in (-Q_1, \infty) \times \cdots \times (-Q_\nu, \infty)$ we study are supported in $\Omega \times B^q(a)$ for some small $a > 0$ and satisfy estimates like

$$|K(x, t_1, \ldots, t_\nu)| \lesssim \|t_1\|^{-Q_1 - m_1} \cdots \|t_\nu\|^{-Q_\nu - m_\nu}.$$

K also satisfies appropriate estimates for derivatives and action on certain "bump" functions (see Definition 5.1.32). For arbitrary $m \in \mathbb{R}^\nu$ we use a different type of characterization (see Proposition 5.1.35).

Let $\gamma(x,t)$ be given by either of the following formulas:

$$\gamma(x, t_1, \ldots, t_\nu) = e^{t_1 \cdot X^1} \cdots e^{t_\nu \cdot X^\nu} x, \tag{5.4}$$

$$\gamma(x, t) = e^{t \cdot X} x = e^{t_1 \cdot X^1 + \cdots + t_\nu \cdot X^\nu} x, \tag{5.5}$$

where X^μ denotes the list of vector fields $X_1^\mu, \ldots, X_{q_\mu}^\mu$ and X denotes the list of vector fields $X_1, \ldots, X_q$. We study operators of the form

$$Tf(x) = \int_{\mathbb{R}^q} f(\gamma(x,t)) \, \psi(\gamma(x,t)) \, K(x,t) \, dt, \tag{5.6}$$

where $\psi \in C_0^\infty(\Omega)$ and $K(x,t)$ is a "product kernel of order $m \in (-Q_1, \infty) \times \cdots \times (-Q_\nu, \infty)$" with $\operatorname{supp}(K) \subset \Omega \times B^q(a)$ (and $a > 0$ is some small number to be chosen later). We will show $T \in \mathcal{A}^m$.

Example 5.1.31 *Suppose that* $\left(X_1^1, \hat{d}_1^1 \right), \ldots, \left(X_{q_1}^1, \hat{d}_{q_1}^1 \right)$ *is given by the Euclidean vector fields* $(\partial_{x_1}, 1), \ldots, (\partial_{x_n}, 1)$. *If we then take* $K(x,t) = K_1(x, t_1) \otimes \delta_0(t_2) \otimes \cdots \otimes \delta_0(t_\nu)$, *where* K_1 *is a "Calderón-Zygmund kernel of order* $m \in \mathbb{R}$," *we see that the operator in (5.6) is precisely a standard pseudodifferential operator of order m which is supported in* $\Omega \times \Omega$ *(see Remark 1.3.14, and use the expression* $x - z = e^{-z \cdot \nabla} x$). *It then follows that, in this case, standard pseudodifferential operators of order m are actually elements of* $\mathcal{A}^{\hat{m}}$, *where* $\hat{m} \in \mathbb{R}^\nu$ *is m in the first component, and 0 in all other components.*

We turn to a more formal presentation of these results. We begin by introducing product kernels in the way in which they are usually presented. Once that is done, we introduce a more general definition (Definition 5.1.34).

DEFINITION 5.1.32. *The space of product kernels of order* $m = (m_1, \ldots, m_\nu) \in (-Q_1, \infty) \times \cdots \times (-Q_\nu, \infty)$ *is a locally convex topological vector space consisting* of certain distributions $K(x, t_1, \ldots, t_\nu) \in C^\infty(\Omega \times \mathbb{R}^q)'$. The space is defined recursively. For $\nu = 0$, it is defined to be $C_0^\infty(\Omega)$, with the usual locally convex topology. We assume that we have defined locally convex topological vector spaces of product kernels of any order up to $\nu - 1$ factors, and we define it for ν factors. The space of product kernels is the space of distributions $K \in C^\infty(\Omega \times \mathbb{R}^q)'$ such that the following two types of semi-norms are finite:

(i) (Growth Condition) For each continuous semi-norm $|\cdot|$ on $C_0^\infty(\Omega)$ and each $\alpha = (\alpha_1, \ldots, \alpha_\nu) \in \mathbb{N}^{q_1} \times \cdots \times \mathbb{N}^{q_\nu} = \mathbb{N}^q$ we assume that there is a constant $C = C(\alpha, |\cdot|)$ such that for $t_1 \neq 0, \ldots, t_\nu \neq 0$,

$$\left| \partial_{t_1}^{\alpha_1} \cdots \partial_{t_\nu}^{\alpha_\nu} K(\cdot, t) \right| \leq C \, \|t_1\|^{-Q_1 - m_1 - \alpha_1 \cdot \hat{d}^1} \cdots \|t_\nu\|^{-Q_\nu - m_\nu - \alpha_\nu \cdot \hat{d}^\nu}.$$

We define a semi-norm to be the least possible C. In particular, we assume $K(x, t)$ is a C^∞ function of (x, t) when $t_1 \neq 0, \ldots, t_\nu \neq 0$.

(ii) (Cancellation Condition) Given $1 \leq \mu \leq \nu$, $R > 0$, and a bounded set $\mathcal{B} \subset C_0^\infty(\mathbb{R}^{q_\mu})$ for $\phi \in \mathcal{B}$ we define

$$K_{\phi, R}(x, t_1, \ldots, t_{\mu-1}, t_{\mu+1}, \ldots, t_\nu) = R^{-m_\mu} \int K(x, t) \, \phi(Rt_\mu) \, dt_\mu,$$

which defines a distribution[6]

$$K_{\phi, R} \in C_0^\infty(\Omega \times \mathbb{R}^{q_1} \times \cdots \times \mathbb{R}^{q_{\mu-1}} \times \mathbb{R}^{q_{\mu+1}} \times \cdots \times \mathbb{R}^{q_\nu})'.$$

We assume that this distribution is a product kernel of order

$$(m_1, \ldots, m_{\mu-1}, m_{\mu+1}, \ldots, m_\nu).$$

Let $\|\cdot\|$ be a continuous semi-norm on the space of $\nu - 1$ factor product kernels of order $(m_1, \ldots, m_{\mu-1}, m_{\mu+1}, \ldots, m_\nu)$. We define a semi-norm on ν factor product kernels of order m by $\|K\| := \sup_{\phi \in \mathcal{B}, R > 0} \|K_{\phi, R}\|$, which we assume to be finite.

We give the space of product kernels of order m the coarsest topology such that all of the above semi-norms are continuous.

Remark 5.1.33 Notice that if K is a product kernel, then $\{x \mid \exists t, (x, t) \in \operatorname{supp}(K)\} \Subset \Omega$.

While Definition 5.1.32 is the most commonly used definition for product kernels, it will be of use to us to introduce a more general way of viewing these kernels. Given our single-parameter dilations (5.2) on each $\mathbb{R}^{q_\mu}$, we may define ν-parameter dilations on $\mathbb{R}^q = \mathbb{R}^{q_1} \times \cdots \times \mathbb{R}^{q_\nu}$ by, for $j = (j_1, \ldots, j_\nu) \in \mathbb{R}^\nu$, $2^j(t_1, \ldots, t_\nu) =$

---

[6]This defines a distribution in the standard way: for $\psi \in C_0^\infty(\Omega \times \mathbb{R}^{q_1} \times \cdots \times \mathbb{R}^{q_{\mu-1}} \times \mathbb{R}^{q_{\mu+1}} \times \cdots \times \mathbb{R}^{q_\nu})$, we define $\int K_{\phi, R} \psi :=$ $\int K_{\phi, R} \psi(x, t_1, \ldots, t_{\mu-1}, t_{\mu+1}, \ldots, t_\nu) \, \phi(Rt_\mu) \, dx \, dt$.

$\left(2^{j_1}t_1,\ldots,2^{j_\nu}t_\nu\right)$, and $2^{j_\mu}t_\mu$ is defined by the single-parameter dilations on $\mathbb{R}^{q_\mu}$ (i.e., is defined by (5.2)). For $j = (j_1,\ldots,j_\nu) \in \mathbb{R}^\nu$ and a function $f(t)$, we define $f^{(2^j)}(t) := 2^{j_1 Q_1 + \cdots + j_\nu Q_\nu} f(2^j t)$, so that $\int f^{(2^j)} = \int f$. Also, for $E \subset \{1,\ldots,\nu\}$, we use the space $\mathcal{S}_0^E$ from Definition A.2.16: $\mathcal{S}_0^E$ consists of those $f \in \mathcal{S}(\mathbb{R}^q)$ such that $\int t_\mu^\alpha f(t)\, dt_\mu = 0$, for all $\mu \in E$, $\alpha \in \mathbb{N}^{q_\mu}$.

We have the following analog of Definition 2.14.2.

DEFINITION 5.1.34. *For a distribution* $K \in C_0^\infty(\mathbb{R}^q)'$, $s \in \mathbb{R}^\nu$, *and* $a > 0$, *we say* $K \in \mathrm{PK}(s,a)$ *if there is* $\eta \in C_0^\infty(B^q(a))$ *and a bounded set* $\{\varsigma_j \mid j \in \mathbb{N}^\nu\} \subset \mathcal{S}(\mathbb{R}^q)$ *with* $\varsigma_j \in \mathcal{S}_0^{\{\mu \mid j_\mu \neq 0\}}$ *such that*

$$K = \eta \sum_{j \in \mathbb{N}^\nu} \varsigma_j^{(2^j)}. \tag{5.7}$$

We will see (Lemma 5.2.1) that every sum of the form (5.7) converges in distribution. Also, we show that $\mathrm{PK}(s,a)$ is a vector space (Remark 5.2.7) and give it a complete locally convex topology (in fact, we give it an LF topology–see Definition 5.2.11). With this locally convex topology, it makes sense to consider the tensor product $C_0^\infty(\Omega) \widehat{\otimes} \mathrm{PK}(s,a) \cong C_0^\infty(\Omega; \mathrm{PK}(s,a))$ (see Appendix A.2). We have the following result.

PROPOSITION 5.1.35. *Let* $K \in C_0^\infty(\Omega \times \mathbb{R}^q)'$ *be supported in* $\Omega \times B^q(a)$ *and let* $m \in (-Q_1, \infty) \times \cdots \times (-Q_\nu, \infty)$. *If* K *is a product kernel of order* m, *then* $K \in C_0^\infty(\Omega) \widehat{\otimes} \mathrm{PK}(m,a)$.

DEFINITION 5.1.36. *Let* $a > 0$ *be a small number to be chosen later. We say* $T : C^\infty(\Omega) \to C_0^\infty(\Omega)$ *is a pseudodifferential operator of order* $m \in \mathbb{R}^\nu$ *if there is* $\psi \in C_0^\infty(\Omega)$ *and* $K \in C_0^\infty(\Omega) \widehat{\otimes} \mathrm{PK}(m,a)$ *such that*

$$Tf(x) = \int f(\gamma(x,t))\, \psi(\gamma(x,t))\, K(x,t)\, dt, \tag{5.8}$$

where γ *is given by either (5.4) or (5.5).*

THEOREM 5.1.37. *If* $a > 0$ *is sufficiently small, and if* T *is a pseudodifferential operator of order* $m \in \mathbb{R}^\nu$, *then* $T \in \mathcal{A}^m$. *More precisely, we prove* $T \in \mathcal{A}_4^m$.

Let $K(x,t_1,\ldots,t_\nu) \in C_0^\infty(\Omega) \widehat{\otimes} \mathrm{PK}(m,a)$ as above, and let T be given by (5.8) so that $T \in \mathcal{A}^m$. Suppose we add a $\nu + 1$st list of vector fields to our ν lists of vector fields $\left(X_1^{\nu+1}, \hat{d}_1^{\nu+1}\right), \ldots, \left(X_{q_{\nu+1}}^{\nu+1}, \hat{d}_{q_{\nu+1}}^{\nu+1}\right)$; and suppose that these $\nu + 1$ lists of vector fields still satisfy the assumptions of this chapter (with ν replaced by $\nu + 1$). Notice if we replace K by $\widetilde{K} := K \otimes \delta_0(t_{\nu+1})$, then $\widetilde{K} \in C_0^\infty(\Omega) \widehat{\otimes} \mathrm{PK}((m,0),a) \subset C_0^\infty(\Omega \times \mathbb{R}^{q+q_{\nu+1}})'$, and T is *also* given by (5.8) with K replaced by $\widetilde{K}$. Applying Theorem 5.1.37 we have $T \in \mathcal{A}^{(m,0)}$, where $\mathcal{A}^{(m,0)}$ is the algebra of $\nu + 1$ parameter operators defined in terms of the $\nu + 1$ lists of vector fields. Theorem 5.1.23 then shows that T has good mapping properties on the $\nu + 1$ parameter non-isotropic Sobolev spaces as defined in Section 5.1.1. Namely, for $s \in \mathbb{R}^\nu$, $t \in \mathbb{R}$, $T : \mathrm{NL}_{(s,t)}^p \to \mathrm{NL}_{(s-m,t)}^p$.

In particular, we see in Section 5.11.1 that one may always take this $\nu + 1$st list to be the Euclidean vector fields $(\partial_{x_1}, 1), \ldots, (\partial_{x_n}, 1)$. In that case, if we take $K \in C_0^\infty(\Omega) \widehat{\otimes} \mathrm{PK}(0, a)$, we see by the above discussion and Corollary 5.1.29 that $T : L_t^p \to L_t^p$ $(1 < p < \infty, t \in \mathbb{R})$, where L_t^p denotes the standard isotropic Sobolev space of order t.

5.1.3 Adding parameters

The idea of the algebra $\mathcal{A}$ is that it involves singular integrals which have ν different "underlying geometries." This raises the following question. Suppose we take a sublist of our vector fields (X, d) in the following way. Fix μ_0, $1 \leq \mu_0 < \nu$, and let (X', d') denote the list of vector fields with ν parameter formal degrees

$$(X', d') := \left(X_1^1, d_1^1 \right), \ldots, \left(X_{q_1}^1, d_{q_1}^1 \right), \ldots, \ldots, (X_1^{\mu_0}, d_1^{\mu_0}), \ldots, \left(X_{q_{\mu_0}}^{\mu_0}, d_{q_{\mu_0}}^{\mu_0} \right).$$

Each of the above formal degrees is zero in the last $\nu - \mu_0$ components, so we may think of these vector fields in (X', d') as having μ_0 parameter formal degrees. We assume that the vector fields in X' span the tangent space at each point of Ω. I.e., we assume that $X_1^1, \ldots, X_{q_1}^1, \ldots, \ldots, X_1^{\mu_0}, \ldots, X_{q_{\mu_0}}^{\mu_0}$ span the tangent space at each point of Ω. Using that (X, d) satisfies the assumptions of this chapter, it is easy to verify that (X', d') satisfies the assumptions of this chapter, and we therefore obtain a corresponding μ_0 parameter algebra of singular integrals; call it $\mathcal{A}_0^s$, $s \in \mathbb{R}^{\mu_0}$. Let $\mathcal{A}^t$ $(t \in \mathbb{R}^\nu)$ denote the algebra corresponding to (X, d).

In general, one has neither $\mathcal{A}_0^s \subseteq \mathcal{A}^{(s,0)}$ nor $\mathcal{A}^{(s,0)} \subseteq \mathcal{A}_0^s$. For instance, $\mathcal{A}^{(s,0)}$ contains some operators corresponding to the geometries $\left(X^{\mu_0+1}, \hat{d}^{\mu_1+1} \right), \ldots, \left(X^\nu, \hat{d}^\nu \right)$ (see Section 4.4 for a particularly transparent example). We already have some examples of operators in $\mathcal{A}_0^s \cap \mathcal{A}^{(s,0)}$, namely, the pseudodifferential operators of order s in $\mathcal{A}_0^s$ are also elements of $\mathcal{A}^{(s,0)}$ as discussed at the end of Section 5.1.2.

However, if we are given an operator in $\mathcal{A}_0^s$ which is not a pseudodifferential operator, how might we tell if this operator is an element of $\mathcal{A}^{(s,0)}$? We now present a sufficient condition for this to be the case. For this, we need some new notation. Let (X'', d'') denote the list of vector fields with formal degrees

$$\left(X_1^{\mu_0+1}, d_1^{\mu_0+1} \right), \ldots, \left(X_{q_{\mu_0+1}}^{\mu_0+1}, d_{q_{\mu_0+1}}^{\mu_0+1} \right), \ldots, \ldots, (X_1^\nu, d_1^\nu), \ldots, \left(X_{q_\nu}^\nu, d_{q_\nu}^\nu \right).$$

For an ordered multi-index, we write

$$\deg(\alpha) = \sum_{l=1}^q k_l d_l \in \mathbb{N}^\nu, \tag{5.9}$$

where $k_l \in \mathbb{N}$ denotes the number of times l appears in the list α. Notice $\left(2^{-j} X \right)^\alpha = 2^{-j \cdot \deg(\alpha)} X^\alpha$. When we raise X'' to an ordered multi-index $(X'')^\alpha$ we can consider this as an ordered multi-index involving only the indices appearing in X'', so that the following equation holds $(X'')^\alpha = X^\alpha$ and it makes sense to write $\deg(\alpha)$ for such an ordered multi-index.

We begin by stating a corollary of the theorem we prove.

COROLLARY 5.1.38. *Let $T \in \mathcal{A}_0^s$ and suppose $\forall \alpha$,*

$$(X'')^\alpha T = \sum_{\deg(\beta) \leq \deg(\alpha)} S_{\alpha,\beta} (X'')^\beta, \tag{5.10}$$

with $S_{\alpha,\beta} \in \mathcal{A}_0^s$. Assume that the same holds for T replaced by T^*. Then, $T \in \mathcal{A}^{(s,0)}$.

We now state the more general theorem. For $j \in [0,\infty)^\nu$, we separate j into two variables $j = (j_1, j_2) \in [0,\infty)^{\mu_0} \times [0,\infty)^{\nu-\mu_0}$; so that $\left(2^{-j_2} X''\right)^\alpha = \left(2^{-j} X\right)^\alpha$ for appropriate multi-indices as discussed above.

THEOREM 5.1.39. *Let $T \in \mathcal{A}_0^s$. Suppose $\forall \alpha$, $\exists M$, $\forall j \in [0,\infty)^\nu$,*

$$\left(2^{-j_2} X''\right)^\alpha T = \sum_{|\beta| \leq M} S_{j,\alpha,\beta} \left(2^{-j} X\right)^\beta, \tag{5.11}$$

where $\left\{ S_{j,\alpha,\beta} \mid |\beta| \leq M, j \in [0,\infty)^\nu \right\} \subset \mathcal{A}_0^s$ is a bounded set. Also assume the same condition holds with T replaced by T^*. Then $T \in \mathcal{A}^{(s,0)}$.

PROOF OF COROLLARY 5.1.38 GIVEN THEOREM 5.1.39. Multiply both sides of (5.10) by $2^{-j \cdot \deg(\alpha)}$ to obtain (5.11). The result follows from Theorem 5.1.39. □

5.1.4 Pseudolocality

In general, the operators in $\mathcal{A}^s$ might not be pseudolocal; i.e., for $T \in \mathcal{A}^s$, $T(x,y)$ is not necessarily C^∞ for $x \neq y$. When $\nu = 1$, these operators are pseudolocal; in fact, the algebra $\mathcal{A}$ consists exactly of the operators from Definitions 2.15.15 and 2.15.21, which are clearly pseudolocal. When $\nu > 1$ we have seen several examples of operators which are not pseudolocal: for instance, the product theory from Section 4.1 and the flag theory from Section 4.2. However, we have the following theorem which gives a condition implying pseudolocality.

THEOREM 5.1.40. *Suppose for each μ, $1 \leq \mu \leq \nu$, the vector fields $X_1^\mu, \ldots, X_{q_\mu}^\mu$ span the tangent space at every point of Ω. Then if $T \in \mathcal{A}^s$, $T(x,y)$ is C^∞ for $x \neq y$.*

5.2 SCHWARTZ SPACE AND PRODUCT KERNELS

We turn to discussing the product kernels and the space PK (s,a) defined in Section 5.1.2. The key to fully understanding these kernels lies in an understanding of the space $\mathcal{S}_0^E$ ($E \subseteq \{1, \ldots, \nu\}$) from Definition A.2.16.

We recall the dilations from Section 5.1.2. There we decomposed $\mathbb{R}^q = \mathbb{R}^{q_1} \times \cdots \times \mathbb{R}^{q_\nu}$. On each $\mathbb{R}^{q_\mu}$ we defined dilations by, for $\delta_\mu > 0$, $\delta_\mu (t_1, \ldots, t_q) = \left(\delta_\mu^{\hat{d}_1^\mu} t_1, \ldots, \delta_\mu^{\hat{d}_{q_\mu}^\mu} t_{q_\mu} \right)$. This defines ν-parameter dilations on $\mathbb{R}^q = \mathbb{R}^{q_1} \times \cdots \times \mathbb{R}^{q_\nu}$ by, for $\delta = (\delta_1, \ldots, \delta_\nu) \in [0,\infty)^\nu$, $\delta (t_1, \ldots, t_\nu) = (\delta_1 t_1, \ldots, \delta_\nu t_\nu)$, where $t_\mu \in \mathbb{R}^{q_\mu}$ and $\delta_\mu t_\mu$ is defined by the above single-parameter dilations.

Letting $Q_\mu = \sum_{j=1}^{q_\mu} \hat{d}_j^\mu$, for $\varsigma : \mathbb{R}^q \to \mathbb{C}$ and $j = (j_1, \ldots, j_\nu) \in \mathbb{R}^\nu$, we define $\varsigma^{(2^j)}(t) = 2^{j_1 Q_1 + \cdots + j_\nu Q_\nu} \varsigma(2^j t)$, so that $\int \varsigma^{(2^j)} = \int \varsigma$.

We extend these dilations to the gradient ∂_t. Indeed, for each μ and $\delta_\mu > 0$, we write $\delta_\mu \left(\partial_{t_1}, \ldots, \partial_{t_{q_\mu}} \right) = \left(\delta_\mu^{\hat{d}_1^\mu} \partial_{t_1}, \ldots, \delta_\mu^{\hat{d}_{q_\mu}^\mu} \partial_{t_{q_\mu}} \right)$. For each μ we also have a Laplacian in the t_μ variable given by $\triangle_\mu = -\partial_{t_\mu} \cdot \partial_{t_\mu}$. For $\delta_\mu > 0$, we write $\delta_\mu \triangle_\mu := -\left(\delta_\mu \partial_{t_\mu} \right) \cdot \left(\delta_\mu \partial_{t_\mu} \right)$, where $\delta_\mu \partial_{t_\mu}$ is defined by the above single-parameter dilations. Notice, this is defined in such a way that, for $j = (j_1, \ldots, j_\nu) \in \mathbb{R}^\nu$, $(\triangle_\mu \varsigma)^{(2^j)} = (2^{-j_\mu} \triangle_\mu) \varsigma^{(2^j)}$.

LEMMA 5.2.1. *Let $\left\{ \varsigma_j \mid j \in \mathbb{N}^\nu \right\} \subset \mathcal{S}(\mathbb{R}^q)$ be a bounded set, with $\varsigma_j \in \mathcal{S}_0^{\{\mu \mid j_\mu \neq 0\}}$. Then for any $s \in \mathbb{R}^\nu$, the sum*

$$\sum_{j \in \mathbb{N}^\nu} 2^{j \cdot s} \varsigma_j^{(2^j)}$$

converges in the sense of tempered distributions.

PROOF. Let $\varsigma \in \mathcal{S}(\mathbb{R}^q)$. We show, for every N,

$$\left| \int \varsigma_j^{(2^j)}(t) \varsigma(t) \, dt \right| \lesssim 2^{-2N|j|_\infty}, \tag{5.12}$$

and the result will follow.

If $j = 0$, (5.12) is trivial. We therefore assume $j > 0$. Take μ so that $j_\mu = |j|_\infty > 0$. Note that $\varsigma_j \in \mathcal{S}_0^{\{\mu\}}$. We use Example A.2.17 to see that $\varsigma_j = \triangle_\mu^N \tilde{\varsigma}_j$, where $\left\{ \tilde{\varsigma}_j \mid j \in \mathbb{N}^\nu, j_\mu \neq 0 \right\} \subset \mathcal{S}(\mathbb{R}^q)$ is a bounded set. We have

$$\int \varsigma_j^{(2^j)}(t) \varsigma(t) \, dt = \int \tilde{\varsigma}_j^{(2^j)}(t) \left(2^{-j_\mu} \triangle_\mu \right)^N \varsigma(t) \, dt.$$

But $\left\{ 2^{2N|j|_\infty} \left(2^{-j_\mu} \triangle_\mu \right)^N \varsigma(t) \mid j \in \mathbb{N}^\nu, j \neq 0, j_\mu = |j|_\infty \right\} \subset \mathcal{S}(\mathbb{R}^q)$ is a bounded set. (5.12) follows, completing the proof. $\square$

PROPOSITION 5.2.2. *Let $s \in \mathbb{R}^\nu$ and $a > 0$. For a distribution $K \in \mathcal{S}(\mathbb{R}^q)'$, the following are equivalent:*

(i) $K \in \mathrm{PK}(s, a)$.

(ii) $\mathrm{supp}(K) \subset B^q(a)$ *and there is a bounded set $\left\{ \tilde{\varsigma}_j \mid j \in \mathbb{N}^\nu \right\} \subset \mathcal{S}(\mathbb{R}^q)$ with $\tilde{\varsigma}_j \in \mathcal{S}_0^{\{\mu \mid j_\mu \neq 0\}}$ and $K = \sum_{j \in \mathbb{N}^\nu} 2^{j \cdot s} \tilde{\varsigma}_j^{(2^j)}$.*

To prove Proposition 5.2.2, we need several lemmas.

LEMMA 5.2.3. *Let $\mathcal{B} \subset \mathcal{S}(\mathbb{R}^q)$ be a bounded set. For $j_1, j_2 \in \mathbb{R}^\nu$ and $\varsigma_1, \varsigma_2 \in \mathcal{B}$ define $\varsigma_3^{(2^{j_1 \wedge j_2})} := \varsigma_1^{(2^{j_1})} * \varsigma_2^{(2^{j_2})}$, where $*$ denotes the usual Euclidean convolution on $\mathbb{R}^q$. Then,*

$$\left\{ \varsigma_3 \mid \varsigma_1, \varsigma_2 \in \mathcal{B}, j_1, j_2 \in \mathbb{R}^\nu \right\} \subset \mathcal{S}(\mathbb{R}^q)$$

is a bounded set.

PROOF. Let $\hat{\varsigma}$ denote the Fourier transform of ς. By definition, $\hat{\varsigma}_3 \left(2^{-j_1 \wedge j_2}\xi\right) = \hat{\varsigma}_1 \left(2^{-j_1}\xi\right)\hat{\varsigma}_2 \left(2^{-j_2}\xi\right)$. The result now follows easily. $\qquad\square$

LEMMA 5.2.4. *Fix $a > 0$. Let $\mathcal{B}_1 \subset \mathcal{S}\left(\mathbb{R}^q\right)$ and $\mathcal{B}_2 \subset C_0^\infty\left(B^q\left(a\right)\right)$ be bounded sets. For $j_1 = \left(j_1^1,\ldots,j_1^\nu\right), j_2 = \left(j_2^1,\ldots,j_2^\nu\right) \in [0,\infty)^\nu$, $\eta \in \mathcal{B}_2$, and $\varsigma_1,\varsigma_2 \in \mathcal{B}_1$ with*

$$\varsigma_1 \in \mathcal{S}_0^{\left\{\mu | j_1^\mu \neq 0\right\}}, \quad \varsigma_2 \in \mathcal{S}_0^{\left\{\mu | j_2^\mu \neq 0\right\}},$$

define $\varsigma_3^{\left(2^{j_1 \wedge j_2}\right)} := \varsigma_1^{\left(2^{j_1}\right)} * \left(\eta\varsigma_2^{\left(2^j\right)}\right)$. Then, for every N, the set

$$\left\{ 2^{N|j_1-j_2|}\varsigma_3 \;\middle|\; \eta \in \mathcal{B}_2, j_1, j_2 \in [0,\infty)^\nu, \varsigma_1, \varsigma_2 \in \mathcal{B}_1, \right. \tag{5.13}$$
$$\left. \varsigma_1 \in \mathcal{S}_0^{\left\{\mu | j_1^\mu \neq 0\right\}}, \varsigma_2 \in \mathcal{S}_0^{\left\{\mu | j_2^\mu \neq 0\right\}} \right\} \subset \mathcal{S}\left(\mathbb{R}^q\right)$$

is a bounded set and $\varsigma_3 \in \mathcal{S}_0^{\left\{\mu | j_1^\mu \wedge j_2^\mu \neq 0\right\}}$.

PROOF. Since $\varsigma_1 \in \mathcal{S}_0^{\left\{\mu | j_1^\mu \neq 0\right\}}$, it follows $\varsigma_3 \in \mathcal{S}_0^{\left\{\mu | j_1^\mu \neq 0\right\}} \subseteq \mathcal{S}_0^{\left\{\mu | j_1^\mu \wedge j_2^\mu \neq 0\right\}}$. Thus, we need only prove the set in (5.13) is bounded. If $j_1 = j_2$, the result follows from Lemma 5.2.3, and so we assume $|j_1 - j_2| > 0$. Pick μ so that $|j_1^\mu - j_2^\mu| = |j_1 - j_2|_\infty$. There are two cases: either $j_1^\mu > j_2^\mu$ or $j_2^\mu > j_1^\mu$. We prove the result for the case $j_2^\mu > j_1^\mu$, the reverse case is similar. Since $j_2^\mu > j_1^\mu \geq 0$, we have $\varsigma_2 \in \mathcal{S}_0^{\{\mu\}}$. Fix $M = M\left(N\right)$ large. Applying Example A.2.17, we see that we may write $\varsigma_2 = \triangle_\mu^M \tilde{\varsigma}_2$, where $\left\{\tilde{\varsigma}_2 \;\middle|\; \varsigma_2 \in \mathcal{B}_1 \cap \mathcal{S}_0^{\{\mu\}}\right\}$ is a bounded set. Thus, we have

$$\varsigma_1^{\left(2^{j_1}\right)} * \left(\eta\varsigma_2^{\left(2^{j_2}\right)}\right) = \varsigma_1^{\left(2^{j_1}\right)} * \left(\eta\left(2^{-j_2^\mu}\triangle_\mu\right)^M \tilde{\varsigma}_2^{\left(2^{j_2}\right)}\right).$$

We now integrate by parts so that $\left(2^{-j_2^\mu}\triangle_\mu\right)^M$ acts on η and $\varsigma_1^{\left(2^{j_1}\right)}$. If more than half of the derivatives land on η, we get a large power of $2^{-j_2^\mu} \leq 2^{-|j_1-j_2|_\infty}$. If more than half land on $\varsigma_1^{\left(2^{j_1}\right)}$ we get a large power of $2^{-|j_1-j_2|_\infty}$. Either way, we have a large power of $2^{-|j_1-j_2|_\infty}$ times a sum of terms of the form covered in Lemma 5.2.3. The result follows. $\qquad\square$

LEMMA 5.2.5. *There is a bounded set $\left\{\psi_j \;\middle|\; j \in \mathbb{N}^\nu\right\} \subset \mathcal{S}\left(\mathbb{R}^q\right)$ with $\psi_j \in \mathcal{S}_0^{\left\{\mu | j_\mu \neq 0\right\}}$ and such that $\delta_0 = \sum_{j \in \mathbb{N}^\nu} \psi_j^{\left(2^j\right)}$, where δ_0 denotes the Dirac δ function at $0 \in \mathbb{R}^q$ and the convergence is taken in the sense of tempered distributions.*

PROOF. We decompose δ_0 on the Fourier transform side. Indeed, let $\hat{\phi} \in C_0^\infty\left(\mathbb{R}^q\right)$ equal 1 on a neighborhood of 0. Define

$$\hat{\psi}_j\left(\xi\right) = \sum_{\substack{p=(p_1,\ldots,p_\nu)\in\{0,1\}^\nu \\ j_\mu \geq p_\mu}} (-1)^{p_1+\cdots+p_\nu}\,\hat{\phi}\left(2^p\xi\right).$$

Notice, $\sum_{j\in\mathbb{N}^\nu} \widehat{\psi}_j \left(2^{-j}\xi\right) \equiv 1$. Also notice $\left\{\widehat{\psi}_j \mid j \in \mathbb{N}^\nu\right\} \subset \mathcal{S}\left(\mathbb{R}^q\right)$ is a finite set, and therefore bounded. Letting ψ_j denote the inverse Fourier transform of $\widehat{\psi}_j$, we have $\sum_{j\in\mathbb{N}^\nu} \psi_j^{(2^j)} = \delta_0$ and $\left\{\psi_j \mid j \in \mathbb{N}^\nu\right\} \subset \mathcal{S}\left(\mathbb{R}^q\right)$ is bounded. Finally, if $j_\mu > 0$, $\widehat{\psi}_j(\xi_1,\ldots,\xi_\nu)$ vanishes on a neighborhood of 0 in the ξ_μ variable. It follows that $\varsigma_j \in \mathcal{S}_0^{\{\mu\mid j_\mu\neq0\}}$, which completes the proof. $\qquad\square$

LEMMA 5.2.6. *Fix $a > 0$. Suppose $\mathcal{B}_1 \subset \mathcal{S}\left(\mathbb{R}^q\right)$ and $\mathcal{B}_2 \subset C_0^\infty\left(B^q\left(a\right)\right)$ are bounded sets. For $j \in [0,\infty)^\nu$, $\varsigma \in \mathcal{B}_1 \cap \mathcal{S}_0^{\{\mu\mid j_\mu\neq0\}}$, $\eta \in \mathcal{B}_2$, we may write*

$$\eta\varsigma^{(2^j)} = \sum_{\substack{k\leq j \\ k\in\mathbb{N}^\nu}} \varsigma_{j,k}^{(2^k)},$$

where, for every N,

$$\mathcal{B}_N := \left\{2^{N|j-k|}\varsigma_{j,k} \mid j \in [0,\infty)^\nu, k \leq j, k \in \mathbb{N}^\nu, \varsigma \in \mathcal{B}_1 \cap \mathcal{S}_0^{\{\mu\mid j_\mu\neq0\}}, \eta \in \mathcal{B}_2\right\}$$
$$\subset \mathcal{S}\left(\mathbb{R}^q\right)$$

is a bounded set, and $\varsigma_{j,k} \in \mathcal{S}_0^{\{\mu\mid k_\mu\neq0\}}$.

PROOF. We apply Lemma 5.2.5 to write $\delta_0 = \sum_{k\in\mathbb{N}^\nu} \psi_k^{(2^k)}$, where ψ_k is as in that lemma. Define $\widetilde{\varsigma}_{j,k}^{(2^{j\wedge k})} := \psi_k^{(2^k)} * \left(\eta\varsigma^{(2^j)}\right)$, so that $\sum_{k\in\mathbb{N}^\nu} \widetilde{\varsigma}_{j,k}^{(2^{j\wedge k})} = \eta\varsigma^{(2^j)}$, and by Lemma 5.2.4 for every N,

$$\left\{2^{N|j-k|}\widetilde{\varsigma}_{j,k} \mid j \in [0,\infty)^\nu, k \in \mathbb{N}^\nu, \varsigma \in \mathcal{B}_1 \cap \mathcal{S}_0^{\{\mu\mid j_\mu\neq0\}}, \eta \in \mathcal{B}_2\right\} \subset \mathcal{S}\left(\mathbb{R}^q\right)$$

is a bounded set with $\widetilde{\varsigma}_{j,k} \in \mathcal{S}_0^{\{\mu\mid k_\mu\neq0\}}$. For $k \in \mathbb{N}^\nu$ with $k \leq j$, define

$$\varsigma_{j,k}^{(2^k)} := \sum_{\substack{l\in\mathbb{N}^\nu \\ j\wedge l=k}} \widetilde{\varsigma}_{j,l}^{(2^k)}.$$

But with the above properties of $\widetilde{\varsigma}_{j,k}$, it follows that $\varsigma_{j,k}$ satisfy the conclusions of the lemma. $\qquad\square$

PROOF OF PROPOSITION 5.2.2. (ii)$\Rightarrow$(i): Let $K \in \mathcal{S}\left(\mathbb{R}^q\right)'$ be as in (ii) so that $\operatorname{supp}\left(K\right) \subset B^q\left(a\right)$ and there is a bounded set $\left\{\widetilde{\varsigma}_j \mid j \in \mathbb{N}^\nu\right\} \subset \mathcal{S}\left(\mathbb{R}^q\right)$ with $\widetilde{\varsigma}_j \in \mathcal{S}_0^{\{\mu\mid j_\mu\neq0\}}$ and $K = \sum_{j\in\mathbb{N}^\nu} \widetilde{\varsigma}_j^{(2^j)}$. Since $\operatorname{supp}\left(K\right)$ is closed, it is a compact subset of $B^q\left(a\right)$. Let $\eta \in C_0^\infty\left(B^q\left(a\right)\right)$ equal 1 on a neighborhood of $\operatorname{supp}\left(K\right)$. We have $K = \eta K = \eta \sum_{j\in\mathbb{N}^\nu} \widetilde{\varsigma}_j^{(2^j)}$, which shows $K \in \operatorname{PK}\left(s,a\right)$.

(i)$\Rightarrow$(ii): Let $K \in \mathrm{PK}\,(s,a)$ so that there is $\eta \in C_0^\infty\,(B^q\,(a))$ and $\{\varsigma_j \mid j \in \mathbb{N}^\nu\} \subset S\,(\mathbb{R}^q)$ a bounded set with $\varsigma_j \in S_0^{\{\mu \mid j_\mu \neq 0\}}$ and $K = \eta \sum_{j \in \mathbb{N}^\nu} \varsigma_j^{(2^j)}$. For each j apply Lemma 5.2.6 to decompose $\eta \varsigma_j^{(2^j)} = \sum_{k \le j} \varsigma_{j,k}^{(2^k)}$, where for every N,

$$\left\{ 2^{N|j-k|} \varsigma_{j,k} \mid j \in \mathbb{N}^\nu, k \le j \right\} \subset S\,(\mathbb{R}^q)$$

is bounded, and $\varsigma_{j,k} \in S_0^{\{\mu \mid k_\mu \neq 0\}}$. For $k \in \mathbb{N}^\nu$ define $\widetilde{\varsigma}_k := \sum_{j \ge k} 2^{j \cdot s - k \cdot s} \varsigma_{j,k}$, so that $\{\widetilde{\varsigma}_k \mid k \in \mathbb{N}^\nu\} \subset S\,(\mathbb{R}^q)$ is a bounded set, $\widetilde{\varsigma}_k \in S_0^{\{\mu \mid k_\mu \neq 0\}}$, and $K = \sum_{k \in \mathbb{N}^\nu} 2^{k \cdot s} \widetilde{\varsigma}_k^{(2^k)}$. This is the desired decomposition to show that (ii) holds. $\square$

Remark 5.2.7   It follows immediately from Proposition 5.2.2 that PK (s,a) is a vector space.

DEFINITION 5.2.8.   *For $0 < b < a$ and $s \in \mathbb{R}^\nu$, define the vector space $\mathrm{PK}_0\,(s,b)$ to be those $K \in \mathrm{PK}\,(s,a)$ such that $\mathrm{supp}\,(K) \subseteq \overline{B^q}\,(b)$. In light of Proposition 5.2.2, we can write $K = \sum_{j \in \mathbb{N}^\nu} 2^{j \cdot s} \widetilde{\varsigma}_j^{(2^j)}$, where $\{\widetilde{\varsigma}_j \mid j \in \mathbb{N}^\nu\} \subset S\,(\mathbb{R}^q)$ is a bounded set and $\widetilde{\varsigma}_j \in S_0^{\{\mu \mid j_\mu \neq 0\}}$. For each continuous semi-norm $|\cdot|$ on $S\,(\mathbb{R}^q)$ we define a semi-norm on $\mathrm{PK}_0\,(s,b)$ by $|K| := \inf \sup_{j \in \mathbb{N}^\nu} |\widetilde{\varsigma}_j|$, where the infimum is taken over all such representations of K. We give $\mathrm{PK}_0\,(s,b)$ the coarsest topology in which all such semi-norms are continuous.*

LEMMA 5.2.9.   *For $0 < b < a$ and $s \in \mathbb{R}^q$, $\mathrm{PK}_0\,(s,b)$ is a Fréchet space.*

To prove Lemma 5.2.9, we need a slightly different decomposition of elements of $\mathrm{PK}_0\,(s,b)$. For this, we use the following lemma, which is a slight modification of Lemma 5.2.4.

LEMMA 5.2.10.   *Fix $a > 0$. Let $\mathcal{B} \subset S\,(\mathbb{R}^q)$ be a bounded set. For $j_1 = (j_1^1, \ldots, j_1^\nu)$, $j_2 = (j_2^1, \ldots, j_2^\nu) \in [0,\infty)^\nu$ and $\varsigma_1, \varsigma_2 \in \mathcal{B}$ with*

$$\varsigma_1 \in S_0^{\{\mu \mid j_1^\mu \neq 0\}}, \quad \varsigma_2 \in S_0^{\{\mu \mid j_2^\mu \neq 0\}},$$

define $\varsigma_3^{(2^{j_1})} := \varsigma_1^{(2^{j_1})} * \varsigma_2^{(2^{j_2})}$. Then, for every N, the set

$$\left\{ 2^{N|j_1 - j_2|} \varsigma_3 \mid j_1, j_2 \in [0,\infty)^\nu, \varsigma_1, \varsigma_2 \in \mathcal{B}, \right.$$

$$\left. \varsigma_1 \in S_0^{\{\mu \mid j_1^\mu \neq 0\}}, \varsigma_2 \in S_0^{\{\mu \mid j_2^\mu \neq 0\}} \right\} \subset S\,(\mathbb{R}^q) \tag{5.14}$$

is a bounded set and $\varsigma_3 \in S_0^{\{\mu \mid j_1^\mu \neq 0\}}$.

PROOF. That $\varsigma_3 \in \mathcal{S}_0^{\{\mu | j_1^\mu \neq 0\}}$ is clear from the definitions. A simple reprise of the proof of Lemma 5.2.4 shows that for every M,

$$\left\{ 2^{M|j_1-j_2|} \varsigma_3^{\left(2^{j_1-j_1 \wedge j_2}\right)} \,\middle|\, j_1, j_2 \in [0,\infty)^\nu, \varsigma_1, \varsigma_2 \in \mathcal{B}, \right.$$
$$\left. \varsigma_1 \in \mathcal{S}_0^{\{\mu | j_1^\mu \neq 0\}}, \varsigma_2 \in \mathcal{S}_0^{\{\mu | j_2^\mu \neq 0\}} \right\} \subset \mathcal{S}(\mathbb{R}^q)$$

is a bounded set. Let $\|\cdot\|$ be a continuous semi-norm on $\mathcal{S}(\mathbb{R}^q)$. By taking M large depending on N and $\|\cdot\|$ and using that the above set is bounded, it is easy to see that $\sup \left\| 2^{N|j_1-j_2|} \varsigma_3 \right\| < \infty$, where the supremum is taken over the relevant parameters which vary. This is equivalent to showing that for every N, (5.14) is a bounded set. $\square$

PROOF SKETCH OF LEMMA 5.2.9. Definition 5.2.8 gives a countable family of semi-norms defining the topology of $\mathrm{PK}_0(s,b)$. Thus, the content of the lemma is that $\mathrm{PK}_0(s,b)$ is complete. As in Sections 2.8 and 5.7, the main difficulty is the infimum defining the semi-norms in Definition 5.2.8. To avoid this, we pick a particular way of decomposing $K \in \mathrm{PK}_0(s,b)$, and we do this in such a way that the topology can be equivalently defined by this particular decomposition.

Fix $K \in \mathrm{PK}_0(s,b)$. Write $K = \sum_{j\in\mathbb{N}^\nu} \varsigma_j^{(2^j)}$, with $\{\varsigma_j \mid j \in \mathbb{N}^\nu\} \subset \mathcal{S}(\mathbb{R}^q)$ a bounded set with $\varsigma_j \in \mathcal{S}_0^{\{\mu | j_\mu \neq 0\}}$. Let ψ_j be as in Lemma 5.2.5, so that $\delta_0 = \sum_{j\in\mathbb{N}^\nu} \psi_j^{(2^j)}$. Set

$$\tilde{\varsigma}_j^{(2^j)} := 2^{-j\cdot s} K * \psi_j^{(2^j)} = \sum_{k\in\mathbb{N}^\nu} 2^{-(j-k)\cdot s} \varsigma_k^{(2^k)} * \psi_j^{(2^j)},$$

so that $K = \sum_{j\in\mathbb{N}^\nu} 2^{j\cdot s} \tilde{\varsigma}_j^{(2^j)}$. Lemma 5.2.10 shows that $\{\tilde{\varsigma}_j \mid j \in \mathbb{N}^\nu\} \subset \mathcal{S}(\mathbb{R}^q)$ is a bounded set and $\tilde{\varsigma}_j \in \mathcal{S}_0^{\{\mu | j_\mu \neq 0\}}$. Furthermore, it follows by the same proof that for every continuous semi-norm $|\cdot|_1$ on $\mathcal{S}(\mathbb{R}^q)$ there is a continuous semi-norm $|\cdot|_2$ on $\mathcal{S}(\mathbb{R}^q)$ such that if $K = \sum_{j\in\mathbb{N}^\nu} 2^{j\cdot s} \varsigma_j^{(2^j)}$ with $\{\varsigma_j \mid j \in \mathbb{N}^\nu\} \subset \mathcal{S}(\mathbb{R}^q)$ a bounded set and $\varsigma_j \in \mathcal{S}_0^{\{\mu | j_\mu \neq 0\}}$, then

$$\sup_{j\in\mathbb{N}^\nu} |\tilde{\varsigma}_j|_1 \lesssim \sup_{j\in\mathbb{N}^\nu} |\varsigma_j|_2.$$

Taking the infimum over all such representations of K, we have that for every continuous semi-norm $|\cdot|_1$ on $\mathcal{S}(\mathbb{R}^q)$, there is a continuous semi-norm $|\cdot|_2$ on $\mathrm{PK}_0(s,b)$ such that

$$\sup_{j\in\mathbb{N}^\nu} |\tilde{\varsigma}_j|_1 \lesssim |K|_2.$$

Thus, for each continuous semi-norm $|\cdot|$ on $\mathcal{S}(\mathbb{R}^q)$ we obtain a continuous semi-norm on $\mathrm{PK}_0(s,b)$, given by $|K| := \sup_j |\tilde{\varsigma}_j|_j$, where $\tilde{\varsigma}_j$ is as chosen above. It follows that this family of semi-norms generates the topology of $\mathrm{PK}_0(s,b)$.

Now suppose $\{K_n\}_{n\in\mathbb{N}} \subset PK_0(s,b)$ is a Cauchy sequence. Decompose each K_n as $K_n = \sum_{j\in\mathbb{N}^\nu} 2^{j\cdot s} \widetilde{\varsigma}_{n,j}^{(2^j)}$ as above. Because $\{K_n\}_{n\in\mathbb{N}}$ is Cauchy, it follows that for each j, $\{\widetilde{\varsigma}_{n,j}\}_{n\in\mathbb{N}} \subset \mathcal{S}(\mathbb{R}^q)$ is Cauchy uniformly in j. Thus, $\widetilde{\varsigma}_{n,j} \to \widetilde{\varsigma}_{\infty,j}$ in $\mathcal{S}(\mathbb{R}^q)$ uniformly for $j \in \mathbb{N}^\nu$, where $\{\widetilde{\varsigma}_{\infty,j} \mid j \in \mathbb{N}^\nu\} \subset \mathcal{S}(\mathbb{R}^q)$ is a bounded set and $\widetilde{\varsigma}_{\infty,j} \in \mathcal{S}_0^{\{\mu|j_\mu\neq 0\}}$. Set $K_\infty = \sum_{j\in\mathbb{N}^\nu} 2^{j\cdot s} \widetilde{\varsigma}_{\infty,j}^{(2^j)} \in PK_0(s,b)$. We have $K_n \to K_\infty$ in $PK_0(s,b)$ proving that $PK_0(s,b)$ is complete, and therefore a Fréchet space. $\qquad\square$

Note that if $0 < b_1 < b_2 < a$, Proposition 5.2.2 shows that $PK_0(s,b_1) \hookrightarrow PK_0(s,b_2)$. Furthermore, it follows directly from the definitions that the topology on $PK_0(s,b_1)$ is exactly the topology $PK_0(s,b_1)$ inherits as a subspace of $PK_0(s,b_2)$.

DEFINITION 5.2.11. *We define a locally convex topology on* $PK(s,a)$ *by seeing it as* $PK(s,a) = \varinjlim PK_0(s,b)$ *where the inductive limit is taken over* $0 < b < a$. *See Appendix A.1 for the definition of an inductive limit.*

Remark 5.2.12 Pick a countable sequence $b_1 < b_2 < \cdots < a$ with $\lim_{j\to\infty} b_j = a$. We have $PK(s,a) = \varinjlim PK_0(s,b_j)$. This sees $PK(s,a)$ as an LF space (Definition A.1.24) and we therefore have $PK(s,a)$ is complete (Remark A.2.4).

We now turn to Proposition 5.1.35. To study this, we introduce a class of product kernels which do not depend on the variable $x \in \Omega$.

DEFINITION 5.2.13. *The space of product kernels of order* $m = (m_1,\ldots,m_\nu) \in (-Q_1,\infty) \times \cdots \times (-Q_\nu,\infty)$ *is a locally convex topological vector space made of distributions* $K(t_1,\ldots,t_\nu) \in C^\infty(\mathbb{R}^q)'$. *The space is defined recursively. For* $\nu = 0$, *it is defined to be* $\mathbb{C}$, *with the usual locally convex topology. We assume that we have defined locally convex topological vector spaces of product kernels of any order up to* $\nu - 1$ *factors, and we define it for* ν *factors. The space of product kernels is the space of distributions* $K \in C^\infty(\mathbb{R}^q)'$ *such that the following two types of semi-norms are finite:*

(i) *(Growth Condition) For each multi-index* $\alpha = (\alpha_1,\ldots,\alpha_\nu) \in \mathbb{N}^{q_1} \times \cdots \times \mathbb{N}^{q_\nu} = \mathbb{N}^q$ *we assume that there is a constant* $C = C(\alpha)$ *such that*

$$\left| \partial_{t_1}^{\alpha_1} \cdots \partial_{t_\nu}^{\alpha_\nu} K(t) \right| \leq C \|t_1\|^{-Q_1-m_1-\alpha_1\cdot\hat{d}^1} \cdots \|t_\nu\|^{-Q_\nu-m_\nu-\alpha_\nu\cdot\hat{d}^\nu}.$$

We define a semi-norm to be the least possible C. *Here* $\|t_\mu\|$ *is the homogeneous norm on* $\mathbb{R}^{q_\mu}$ *(see (5.3)).*

(ii) *(Cancellation Condition) Given* $1 \leq \mu \leq \nu$, $R > 0$, *and a bounded set* $\mathcal{B} \subset C_0^\infty(\mathbb{R}^{q_\mu})$ *for* $\phi \in \mathcal{B}$ *we define*

$$K_{\phi,R}(t_1,\ldots,t_{\mu-1},t_{\mu+1},\ldots,t_\nu) = R^{-m_\mu} \int K(t)\,\phi(Rt_\mu)\,dt_\mu,$$

which defines a distribution

$$K_{\phi,R} \in C_0^\infty\left(\mathbb{R}^{q_1} \times \cdots \times \mathbb{R}^{q_{\mu-1}} \times \mathbb{R}^{q_{\mu+1}} \times \cdots \times \mathbb{R}^{q_\nu}\right)'.$$

We assume that this distribution is a product kernel of order

$$(m_1, \ldots, m_{\mu-1}, m_{\mu+1}, \ldots, m_\nu).$$

Let $\|\cdot\|$ be a continuous semi-norm on the space of $\nu - 1$ factor product kernels of order $(m_1, \ldots, m_{\mu-1}, m_{\mu+1}, \ldots, m_\nu)$. We define a semi-norm on ν factor product kernels of order m by $\|K\| := \sup_{\phi \in \mathcal{B}, R>0} \|K_{\phi, R}\|$, which we assume to be finite.

We give the space of product kernels of order m the coarsest topology such that all of the above semi-norms are continuous.

We prove the following result.

PROPOSITION 5.2.14. *Fix $a > 0$ and $m = (m_1, \ldots, m_\nu) \in (-Q_1, \infty) \times \cdots \times (-Q_\nu, \infty)$. Every product kernel of order m (in the sense of Definition 5.2.13) supported in $B^q(a)$ is an element of $\mathrm{PK}(m, a)$.*

PROOF OF PROPOSITION 5.1.35 GIVEN PROPOSITION 5.2.14. It is easy to see that the locally convex topological vector space of product kernels of order m from Definition 5.1.32 is equal to the completed tensor product of the nuclear space $C_0^\infty(\Omega)$ with the space of product kernels of order m from Definition 5.2.13. From here, Proposition 5.1.35 follows from Proposition 5.2.14. See Appendix A.2 for more details on these tensor products. □

The key to the proof of Proposition 5.2.14 is the following theorem.

THEOREM 5.2.15. *For a tempered distribution $K \in \mathcal{S}(\mathbb{R}^q)'$ and*

$$m = (m_1, \ldots, m_\nu) \in (-Q_1, \infty) \times \cdots \times (-Q_\nu, \infty)$$

the following are equivalent:

(i) *K is a product kernel of order m in the sense of Definition 5.2.13.*

(ii) *The Fourier transform, $\widehat{K}$, of K is given by a function which satisfies*

$$\left| \partial_{\xi_1}^{\alpha_1} \cdots \partial_{\xi_\nu}^{\alpha_\nu} \widehat{K}(\xi_1, \ldots, \xi_\nu) \right| \lesssim \|\xi_1\|^{m_1 - \alpha_1 \cdot \hat{d}^1} \cdots \|\xi_\nu\|^{m_\nu - \alpha_\nu \cdot \hat{d}^\nu}, \qquad (5.15)$$

where $\|\xi_\mu\|$ is the homogeneous norm on $\mathbb{R}^{q_\mu}$ (see (5.3)).

PROOF. We begin with (i)⇒(ii). The proof proceeds by induction on ν. The base case is $\nu = 0$, in which there is nothing to prove. Now assume $\nu \geq 1$, and we have the result for all smaller ν.

Let $K_{\xi_1}(x_2, \ldots, x_\nu) = \int K(x) e^{-2\pi i x_1 \cdot \xi_1} \, dx_1$ denote the partial Fourier transform in the x_1 variable. A priori, K is just a tempered distribution in the ξ_1 variable (in the sense that $K_{\xi_1}(x_2, \ldots, x_\nu) \in \mathcal{S}(\mathbb{R}^{q_1})' \widehat{\otimes} \mathcal{S}(\mathbb{R}^{q-q_1})'$). We will show that K_{ξ_1} is in fact a C^∞ function in the ξ_1-variable away from $\xi_1 = 0$ (in the sense that

$K_{\xi_1}(x_2,\ldots,x_\nu) \in C^\infty(\xi_1 \neq 0)\,\widehat{\otimes}\,\mathcal{S}(\mathbb{R}^{q-q_1})')$, and for every continuous semi-norm $|\cdot|$ on product kernels of order $(m_2,\ldots,m_\nu)$ on $\mathbb{R}^{q_2}\times\cdots\times\mathbb{R}^{q_\nu}$, we have

$$\sup_{1\le|\xi_1|\le 2}\left|\partial_{\xi_1}^\alpha K_{\xi_1}(x_2,\ldots,x_\nu)\right| \le C_\alpha, \tag{5.16}$$

where C_α can be chosen to depend only on α and $\|K\|$, where $\|\cdot\|$ is a finite sum of the semi-norms on product kernels of order $m = (m_1,\ldots,m_\nu)$ from Definition 5.2.13, and the choice of $\|\cdot\|$ depends on α.

First, let us see why this yields (ii). Indeed, for $R > 0$, if we replace K by the distribution $K^R(x_1,\ldots,x_\nu) = R^{m_1+Q_1}K(Rx_1,x_2,\ldots,x_\nu)$, then $\{K^R \mid R > 0\}$ is a bounded set of product kernels of order m. If we take the partial Fourier transform in the x_1 variable of K^R, we obtain $R^{m_1}K_{R^{-1}\xi_1}(x_2,\ldots,x_\nu)$. From this scaling, (5.16), and the inductive hypothesis, (5.15) follows. The only remaining issue is whether or not K_{ξ_1} has a part which is supported on $\xi_1 = 0$. That is, we now know that K_{ξ_1} equals a function which satisfies (5.15), plus a term which is supported at 0 in the ξ_1 variable. The only distributions supported at 0 in the ξ_1 variable are finite linear combinations of derivatives of $\delta_0(\xi_1)$. I.e., such distributions are finite sums of the form $\partial_{\xi_1}^{\alpha_1}\delta_0(\xi_1)\otimes L(x_2,\ldots,x_\nu)$, where L is a tempered distribution–each term in this finite sum is homogeneous of degree $\ge Q_1$ in the ξ_1 variable. Because $\{K^R \mid R > 0\}$ is a bounded set of tempered distributions, the same is true for $R^{m_1}K_{R^{-1}\xi_1}(x_2,\ldots,x_\nu)$. Since $m_1 > -Q_1$, by taking $R \to \infty$, we see that if there were any terms supported on $\xi_1 = 0$, then $\{R^{m_1}K_{R^{-1}\xi_1}(x_2,\ldots,x_\nu) \mid R > 0\}$ would not be a bounded set of tempered distributions, which achieves a contradiction.

Thus, to prove (i)$\Rightarrow$(ii) it suffices to prove (5.16). Now consider ξ_1 with $1 \le |\xi_1| \le 2$. We are considering

$$\partial_{\xi_1}^{\alpha_1}K_{\xi_1}(x_2,\ldots,x_\nu) = \partial_{\xi_1}^{\alpha_1}\int K(x)\,e^{-2\pi i x_1\cdot\xi_1}\,dx_1$$
$$= \int(-2\pi i x_1)^{\alpha_1}K(x)\,e^{-2\pi i x_1\cdot\xi_1}\,dx_1.$$

Define $K_{\xi_1,\alpha_1} := \partial_{\xi_1}^{\alpha_1}K_{\xi_1}(x_2,\ldots,x_\nu)$. Let $\phi \in C_0^\infty(B^{q_1}(2))$ equal 1 on a neighborhood of the closure of $B^{q_1}(1)$.

We decompose

$$K_{\xi_1,\alpha_1}(x_2,\ldots,x_\nu) := \int(-2\pi i x_1)^{\alpha_1}\,\phi(x_1)\,K(x)\,e^{-2\pi i x_1\cdot\xi_1}\,dx_1$$
$$+ \int(-2\pi i x_1)^{\alpha_1}\,(1-\phi(x_1))\,K(x)\,e^{-2\pi i x_1\cdot\xi_1}\,dx_1,$$

and we estimate these two terms separately. The desired estimates for

$$\int(-2\pi i x_1)^{\alpha_1}\,\phi(x_1)\,K(x)\,e^{-2\pi i x_1\cdot\xi_1}\,dx_1$$

follow immediately from the cancellation condition (recall, $1 \leq |\xi_1| \leq 2$). For the second term, we integrate by parts to see

$$\int (-2\pi i x_1)^{\alpha_1} (1 - \phi(x_1)) K(x) e^{-2\pi i x_1 \cdot \xi_1} \, dx_1$$
$$= (-2\pi i |\xi_1|)^{-2L} \int \triangle_{x_1}^L [(-2\pi i x_1)^\alpha (1 - \phi(x_1)) K(x)] e^{-2\pi i x_1 \cdot \xi_1} \, dx_1.$$

We take $L = L(\alpha, m)$ large. If any of the derivatives land on $1 - \phi(x_1)$, then the resulting function is supported on $B^{q_1}(1)$ and we may again use the cancellation condition as before. Otherwise, all but at most $|\alpha|$ derivatives land on K. The growth condition shows that the resulting distribution falls off quickly (by taking L large) as $x_1 \to \infty$, and therefore the integral converges absolutely and yields (5.16), completing the proof.

We turn to (ii)$\Rightarrow$(i). Again, we proceed by induction on ν, the base case $\nu = 0$ being trivial. The growth condition follows just as in the above argument. Indeed, notice that the "cancellation condition" holds for $\widehat{K}$ trivially, and so the above proof goes through unchanged to prove the growth condition for K. We turn to the cancellation condition. By symmetry of our assumptions, we prove only the cancellation condition for $\mu = 1$. Let $\mathcal{B} \subset C_0^\infty(\mathbb{R}^{q_1})$ be a bounded set. Let $R > 0$ and $\phi \in \mathcal{B}$. We have

$$R^{-m_1} \int K(x) \phi(Rx_1) \, dx_1$$
$$= \int \widehat{K}(\xi) R^{-m_1 - Q_1} \widehat{\phi}(R^{-1}\xi_1) e^{-2\pi i(\xi_2 \cdot x_2 + \cdots + \xi_\nu \cdot x_n u)} \, d\xi.$$

The right-hand side is the inverse Fourier transform in the variables $\xi_2, \ldots, \xi_\nu$ of the tempered distribution

$$\int \widehat{K}(\xi) R^{-m_1 - Q_1} \widehat{\phi}(R^{-1}\xi_1) \, d\xi_1 = \int R^{-m_1} \widehat{K}(R\xi_1, \xi_2, \ldots, \xi_\nu) \widehat{\phi}(\xi_1) \, d\xi_1.$$

Since $\left\{ \widehat{\phi} \mid \phi \in \mathcal{B} \right\} \subset \mathcal{S}(\mathbb{R}^{q_1})$ is a bounded set, it follows that the right-hand side of the above equation satisfies the $\nu - 1$ factor version of (5.15), uniformly in ϕ. The result follows by our inductive hypothesis. $\qquad \square$

COROLLARY 5.2.16. *Let $K \in \mathcal{S}(\mathbb{R}^q)'$ be a tempered distribution and let $m = (m_1, \ldots, m_\nu) \in (-Q_1, \infty) \times \cdots \times (-Q_\nu, \infty)$. The following are equivalent:*

(i) *K is a product kernel of order m in the sense of Definition 5.2.13.*

(ii) *There is a bounded set $\{\varsigma_j \mid j \in \mathbb{Z}^\nu\} \subset \mathcal{S}_0^{\{1, \ldots, \nu\}}$ such that $K = \sum_{j \in \mathbb{Z}^\nu} 2^{j \cdot m} \varsigma_j^{(2^j)}$. (We will see in the proof that every such sum converges in the sense of tempered distributions.)*

PROOF. We begin with (i)$\Rightarrow$(ii). For each μ, let $\phi_\mu \in C_0^\infty(B^{q_\mu}(2))$ be equal to 1 on a neighborhood of the closure of $B^{q_\mu}(1)$. Define $\psi_\mu(\xi_\mu) = \phi_\mu(\xi_\mu) - \phi_\mu(2\xi_\mu)$, where $2\xi_\mu$ is defined in terms of the single-parameter dilations on $\mathbb{R}^{q_\mu}$. For $\xi_\mu \neq 0$, we have $\sum_{j_\mu \in \mathbb{Z}} \psi_\mu \left(2^{-j_\mu} \xi_\mu\right) = 1$. Notice that $\psi_\mu \left(2^{-j_\mu} \xi_\mu\right)$ is supported on $\|\xi_\mu\| \approx 2^{j_\mu}$. Define

$$\widehat{\varsigma_j}\left(2^{-j}\xi\right) = 2^{-j \cdot m} \psi_1 \left(2^{-j_1} \xi_1\right) \cdots \psi_\nu \left(2^{-j_\nu} \xi_\nu\right) \widehat{K}\left(\xi\right).$$

Clearly we have $\sum_{j \in \mathbb{Z}^\nu} 2^{j \cdot m} \widehat{\varsigma_j}\left(2^{-j}\xi\right) = \widehat{K}$, in the sense of tempered distributions, and taking the inverse Fourier transform shows $\sum_{j \in \mathbb{Z}^\nu} 2^{j \cdot m} \varsigma_j^{\left(2^j\right)} = K$. By the bounds in Theorem 5.2.15, $\left\{\widehat{\varsigma_j} \mid j \in \mathbb{Z}^\nu\right\} \subset \mathcal{S}\left(\mathbb{R}^q\right)$ is bounded, and the same is therefore true for $\left\{\varsigma_j \mid j \in \mathbb{Z}^\nu\right\} \subset \mathcal{S}\left(\mathbb{R}^q\right)$. Since each $\psi_\mu \left(\xi_\mu\right)$ vanishes on a neighborhood of 0, we have $\varsigma_j \in \mathcal{S}_0^{\{1,\ldots,\nu\}}$, completing the proof.

For the converse, suppose $\left\{\varsigma_j \mid j \in \mathbb{Z}^\nu\right\} \subset \mathcal{S}_0^{\{1,\ldots,\nu\}}$ is a bounded set. We then have, for every α and L,

$$\left|\partial_\xi^\alpha \widehat{\varsigma_j}\left(\xi\right)\right| \lesssim \|\xi_1\|^L \cdots \|\xi_\nu\|^L \left(1 + \|\xi_1\|\right)^{-2L} \cdots \left(1 + \|\xi_\nu\|\right)^{-2L}.$$

See Example A.2.17 for more details on estimates like this. From this it is immediate to see that $\sum_{j \in \mathbb{Z}^\nu} 2^{j \cdot m} \widehat{\varsigma_j}\left(2^{-j}\xi\right)$ converges in the sense of tempered distributions to a function satisfying (5.15). In light of Theorem 5.2.15, this completes the proof. $\square$

PROOF OF PROPOSITION 5.2.14. Fix $m \in (-Q_1, \infty) \times \cdots \times (-Q_\nu, \infty)$ and let K be a product kernel of order m (in the sense of Definition 5.2.13) supported in $B^q\left(a\right)$. Let $\eta \in C_0^\infty\left(B^q\left(a\right)\right)$ be equal to 1 on a neighborhood of supp (K). Applying Corollary 5.2.16, we may write $K = \sum_{j \in \mathbb{Z}^\nu} 2^{j \cdot m} \widetilde{\varsigma_j}^{\left(2^j\right)}$, where $\left\{\widetilde{\varsigma_j} \mid j \in \mathbb{Z}^\nu\right\} \subset \mathcal{S}_0^{\{1,\ldots,\nu\}}$ is a bounded set. We then have $K = \eta K = \sum_{j \in \mathbb{Z}^\nu} \eta \widetilde{\varsigma_j}^{\left(2^j\right)}$. For $j \in (\mathbb{N} \setminus \{0\})^\nu$ define $\varsigma_j := \widetilde{\varsigma_j}$. For each μ, let $\eta_\mu \in C_0^\infty\left(\mathbb{R}^{q_\mu}\right)$ be such that $\eta_\mu\left(t_\mu\right)\eta\left(t\right) = \eta\left(t\right)$. Now suppose $j \in \mathbb{N}^\nu$, with at least one coordinate equal to 0. Set $E = \left\{\mu \mid j_\mu \neq 0\right\} \neq \{1,\ldots,\nu\}$. Let

$$2^{j \cdot m} \varsigma_j^{\left(2^j\right)}\left(t\right) := \sum_{\substack{k \in \mathbb{Z}^\nu \\ k_\mu \leq 0 \text{ for } \mu \in E^c \\ k_\mu = j_\mu \text{ for } \mu \in E}} \left[\prod_{\mu \in E^c} \eta_\mu\left(t_\mu\right)\right] 2^{k \cdot m} \widetilde{\varsigma_k}^{\left(2^k\right)}\left(t\right).$$

I.e.,

$$\varsigma_j\left(t\right) = \sum_{\substack{k \in \mathbb{Z}^\nu \\ k_\mu \leq 0 \text{ for } \mu \in E^c \\ k_\mu = j_\mu \text{ for } \mu \in E}} \left[\prod_{\mu \in E^c} \eta_\mu\left(t_\mu\right)\right] 2^{(k-j) \cdot m} \widetilde{\varsigma_k}^{\left(2^{k-j}\right)}\left(t\right).$$

Notice,

$$2^{(k-j) \cdot m} \widetilde{\varsigma_k}^{\left(2^{k-j}\right)}\left(t\right) = 2^{(k_1-j_1)(m_1+Q_1)+\cdots+(k_\nu-j_\nu)(m_\nu+Q_\nu)} \widetilde{\varsigma_k}\left(2^{k-j}t\right).$$

Since $k_\mu = j_\mu$ for $\mu \in E$, the scaling of t on the right-hand side only affects those t_μ for $\mu \in E^c$. Also notice, for $\mu \in E^c$, $j_\mu = 0$.

We claim $\{\varsigma_j \mid j \in \mathbb{N}^\nu\} \subset \mathcal{S}(\mathbb{R}^q)$ is a bounded set. This is clear for $j \in (\mathbb{N} \setminus \{0\})^\nu$, and so we consider only j where some coordinates are 0. Let E be as above and notice that $\varsigma_j(t) \neq 0$ only when $t_\mu \in \mathrm{supp}\,(\eta_\mu)$ for $\mu \in E^c$, and therefore has compact support in the t_μ variable (for $\mu \in E^c$). We have, for every L,

$$|\partial_t^\alpha \varsigma_j(t)| \lesssim \sum_{k_\mu \leq 0 \text{ for } \mu \in E^c} 2^{k_1(m_1+Q_1)+\cdots+k_\nu(m_\nu+Q_\nu)} \prod_{\mu \in E} (1+|t_\mu|)^{-L}$$

$$\lesssim \prod_{\mu \in E} (1+|t_\mu|)^{-L}.$$

This combined with the compact support of ς_j in the t_μ variable for $\mu \in E^c$ shows $\{\varsigma_j \mid j \in \mathbb{N}^\nu\} \subset \mathcal{S}(\mathbb{R}^q)$ is a bounded set. That $\varsigma_j \in \mathcal{S}_0^E$ follows from the fact that all $\widetilde{\varsigma}_k \in \mathcal{S}_0^{\{1,\ldots,\nu\}}$ and the formula for ς_j.

We have

$$K = \eta K = \sum_{j \in \mathbb{Z}^\nu} \eta \widetilde{\varsigma}_j^{(2^j)} = \sum_{j \in \mathbb{N}^\nu} \eta \varsigma_j^{(2^j)},$$

completing the proof. $\qquad\square$

One technical result we will need in studying pseudodifferential operators is the following decomposition.

PROPOSITION 5.2.17. *Fix* $a > 0$, $M \in \mathbb{N}$ *and let* $\mathcal{B}_1 \subset \mathcal{S}(\mathbb{R}^q)$ *and* $\mathcal{B}_2 \subset C_0^\infty(B^q(a))$ *be bounded sets. Let* $j \in [0,\infty)^\nu$, $\varsigma \in \mathcal{B}_1 \cap \mathcal{S}_0^{\{\mu|j_\mu \neq 0\}}$, *and* $\eta \in \mathcal{B}_2$. *In what follows* $\alpha_\mu \in \mathbb{N}^{q_\mu}$ *is a multi-index in the* t_μ *variable, and* $\alpha = (\alpha_1, \ldots, \alpha_\nu) \in \mathbb{N}^q$ *is a multi-index in the* t *variable. There exists*

$$\{\gamma_{k,\alpha} \mid k \leq j, k \in \mathbb{N}^\nu, |\alpha_\mu| = M \text{ when } k_\mu \neq 0, |\alpha_\mu| = 0 \text{ when } k_\mu = 0\}$$
$$\subset C_0^\infty(B^q(a))$$

such that if we set

$$\varsigma_k := \sum_{\substack{\alpha \in \mathbb{N}^q \\ |\alpha_\mu| = M \text{ when } k_\mu \neq 0 \\ |\alpha_\mu| = 0 \text{ when } k_\mu = 0}} \partial_t^\alpha \gamma_{k,\alpha},$$

we have

$$\eta(t)\varsigma^{(2^j)}(t) = \sum_{\substack{k \leq j \\ k \in \mathbb{N}^\nu}} \eta(t)\varsigma_k^{(2^k)}(t).$$

Furthermore, for every $N \in \mathbb{N}$, *the following set is bounded:*

$$\Big\{2^{N|j-k|}\gamma_{k,\alpha} \mid j \in [0,\infty)^\nu, k \leq j, k \in \mathbb{N}^\nu, \varsigma \in \mathcal{B}_1 \cap \mathcal{S}_0^{\{\mu|j_\mu \neq 0\}},$$

$$|\alpha_\mu| = M \text{ when } k_\mu \neq 0, |\alpha_\mu| = 0 \text{ when } k_\mu = 0, \eta \in \mathcal{B}_2\Big\} \subset C_0^\infty(B^q(a)).$$

To prove Proposition 5.2.17, we need a lemma.

LEMMA 5.2.18. *Fix $a > 0$, and let $\mathcal{B}_1 \subset \mathcal{S}(\mathbb{R}^q)$ and $\mathcal{B}_2 \subset C_0^\infty(B^q(a))$ be bounded sets. Let $j \in [0, \infty)^\nu$, $\varsigma \in \mathcal{B}_1$, and $\eta \in \mathcal{B}_2$. Then, there exists $\{\gamma_k\}_{\substack{k \in \mathbb{N}^\nu \\ k \leq j}} \subset C_0^\infty(B^q(a))$ such that*

$$
\eta(t)\, \varsigma^{(2^j)}(t) = \sum_{\substack{k \leq j \\ k \in \mathbb{N}^\nu}} \eta(t)\, \gamma_k^{(2^k)}(t).
$$

Furthermore, for every $N \in \mathbb{N}$, the set

$$
\left\{ 2^{N|j-k|}\gamma_k \mid j \in [0,\infty)^\nu, k \leq j, k \in \mathbb{N}^\nu, \varsigma \in \mathcal{B}_1, \eta \in \mathcal{B}_2 \right\} \tag{5.17}
$$
$$
\subset C_0^\infty(B^q(a))
$$

is bounded.

PROOF. Let $\eta' \in C_0^\infty(B^q(a))$ equal 1 on a neighborhood of $\cup_{\eta \in \mathcal{B}_2} \operatorname{supp}(\eta)$. For $k \in \mathbb{N}^\nu$ let

$$
\delta_k(t) := \sum_{\substack{p \in \{0,1\}^\nu \\ k+p \leq j}} (-1)^{\sum_\mu p_\mu}\, \eta'\left(2^{k+p}t\right),
$$

so that $\eta'(t) = \sum_{0 \leq k \leq j} \delta_k(t)$, and $\delta_k(t) = 0$ if $|2^{k_\mu}t_\mu|$ is small enough (independent of k, j) for any μ such that $k_\mu \neq 0$. Define

$$
\gamma_k^{(2^k)}(t) := \delta_k(t)\, \varsigma^{(2^j)}.
$$

Notice, for any N, the set in (5.17) is bounded in light of the support of δ_k and the rapid decrease of ς. Since $\eta\varsigma^{(2^j)} = \eta\eta'\varsigma^{(2^j)} = \sum_{k \leq j} \eta\gamma_k^{(2^k)}$, this completes the proof. $\square$

PROOF OF PROPOSITION 5.2.17. Let $j \in [0, \infty)^\nu$, $\varsigma \in \mathcal{B}_1 \cap \mathcal{S}_0^{\{\mu | j_\mu \neq 0\}}$, and $\eta \in \mathcal{B}_2$. We prove the result for M replaced by $2M$ (of course the result for M follows from the result for $2M$, so this is sufficient). Define $E_0 := \{\mu | j_\mu \neq 0\}$. Because $\varsigma \in \mathcal{S}_0^{E_0}$, we may use the arguments from Example A.2.17 to see

$$
\varsigma = \left[\prod_{\mu \in E_0} \Delta_\mu^M \right] \tilde{\varsigma},
$$

where $\left\{ \tilde{\varsigma} \mid \varsigma \in \mathcal{B}_1 \cap \mathcal{S}_0^{E_0} \right\} \subset \mathcal{S}(\mathbb{R}^q)$ is a bounded set. Let $\eta' \in C_0^\infty(B^q(a))$ be such that η' equals 1 on a neighborhood of the closure of $\cup_{\eta \in \mathcal{B}_2} \operatorname{supp}(\eta)$. We apply Lemma 5.2.18 to $\eta'\tilde{\varsigma}^{(2^j)}$ to write

$$
\eta'(t)\, \tilde{\varsigma}^{(2^j)}(t) = \sum_{\substack{k \leq j \\ k \in \mathbb{N}^\nu}} \eta'(t)\, \tilde{\gamma}_k^{(2^k)}(t),
$$

where for every N,

$$\left\{ 2^{N|j-k|}\widetilde{\gamma}_k \mid j \in [0,\infty)^\nu , k \leq j, k \in \mathbb{N}^\nu , \varsigma \in \mathcal{B}_1 \cap \mathcal{S}_0^{E_0} \right\} \subset C_0^\infty (B^q(a))$$

is a bounded set. Consider,

$$\eta_\varsigma^{(2^j)} = \eta \left[\prod_{\mu \in E_0} \left(2^{-j_\mu} \triangle_\mu \right)^M \right] \varsigma^{(2^j)}$$

$$= \sum_{\substack{k \leq j \\ k \in \mathbb{N}^\nu}} \eta \left[\prod_{\mu \in E_0} \left(2^{-j_\mu} \triangle_\mu \right)^M \right] \eta' \widetilde{\gamma}_k^{(2^k)}$$

$$= \sum_{\substack{k \leq j \\ k \in \mathbb{N}^\nu}} \eta \left[\prod_{\mu \in E_0} \left(2^{-j_\mu} \triangle_\mu \right)^M \right] \widetilde{\gamma}_k^{(2^k)}.$$

We expand $\prod_{\mu \in E_0} \triangle_\mu^M$ as

$$\prod_{\mu \in E_0} \triangle_\mu^M = \sum_{\substack{\alpha \in \mathbb{N}^q \\ |\alpha_\mu|=2M \text{ when } \mu \in E_0 \\ |\alpha_\mu|=0 \text{ when } \mu \notin E_0}} c_\alpha \partial_t^\alpha,$$

where c_α is a constant depending on α. Consider,

$$\left(2^{-j} \partial_t \right)^\alpha \widetilde{\gamma}_k^{(2^k)} = \left(2^{(k-j)\cdot e_\alpha} \partial_t^\alpha \widetilde{\gamma}_k \right)^{(2^k)},$$

where e_α is a vector depending on α. Setting $\gamma_{k,\alpha} = 2^{(k-j)\cdot e_\alpha} c_\alpha \widetilde{\gamma}_k$, the result follows.
$\square$

5.3 PSEUDODIFFERENTIAL OPERATORS AND $\mathcal{A}_3 \subseteq \mathcal{A}_4$

In this section, we prove Theorem 5.1.37: pseudodifferential operators are elements of $\mathcal{A}_4$. As a simple consequence, we obtain

THEOREM 5.3.1. *For $s \in \mathbb{R}^\nu$, $\mathcal{A}_3^s \subseteq \mathcal{A}_4^s$.*

First we show how Theorem 5.3.1 follows from Theorem 5.1.37. For this, we need to understand in what sense sums of elementary operators converge. This is contained in the next lemma.

LEMMA 5.3.2. *Let $\left\{ (E_j, 2^{-j}) \mid j \in \mathbb{N}^\nu \right\}$ be a bounded set of elementary operators. Fix $s \in \mathbb{R}^\nu$. The sum*

$$\sum_{j \in \mathbb{N}^\nu} 2^{j \cdot s} E_j \tag{5.18}$$

converges in the topology of bounded convergence as operators $C^\infty(\Omega) \to C_0^\infty(\Omega)$. See Definition A.1.28 for the definition of this topology.

Remark 5.3.3   As an immediate consequence of Lemma 5.3.2, (5.18) also converges in distribution.

To prove Lemma 5.3.2, we begin with a simpler lemma.

LEMMA 5.3.4.   *Let $\mathcal{E}$ be a bounded set of pre-elementary operators, and let $\mathcal{B} \subset C^\infty(\Omega)$ be a bounded set. Then, for every ordered multi-index α,*

$$\sup_{f\in\mathcal{B}} \sup_{(E,2^{-j})\in\mathcal{E}} \sup_{x\in\Omega} \left|\left(2^{-j}X\right)^\alpha Ef(x)\right| < \infty.$$

PROOF.   Since $\mathcal{E}$ is a bounded set of pre-elementary operators, we have

$$\bigcup_{(E,2^{-j})\in\mathcal{E}} \operatorname{supp}(E) \Subset \Omega \times \Omega.$$

Let $\psi \in C_0^\infty(\Omega)$ be such that $\psi \otimes \psi \equiv 1$ on $\operatorname{supp}(E)$, $\forall (E, 2^{-j}) \in \mathcal{E}$. For $(E, 2^{-j}) \in \mathcal{E}$, $f \in \mathcal{B}$, we have $Ef = E\psi f$. By the definition of pre-elementary operators, $\forall (E, 2^{-j}) \in \mathcal{E}$,

$$\left|\left(2^{-j}X_x\right)^\alpha E(x,z)\right| \lesssim \frac{(1+\rho_{2^{-j}}(x,z))^{-1}}{\operatorname{Vol}(B_{2^{-j}}(x, 1+\rho_{2^{-j}}(x,z)))}.$$

Applying Lemma 3.2.14, we have

$$\sup_{x\in\Omega} \sup_{(E,2^{-j})\in\mathcal{E}} \int \left|\left(2^{-j}X_x\right)^\alpha E(x,z)\right| dz < \infty.$$

Thus, we have

$$\sup_{\substack{f\in\mathcal{B}\\(E,2^{-j})\in\mathcal{E}\\x\in\Omega}} \left|\left(2^{-j}X\right)^\alpha Ef(x)\right| = \sup_{\substack{f\in\mathcal{B}\\(E,2^{-j})\in\mathcal{E}\\x\in\Omega}} \left|\left(2^{-j}X\right)^\alpha E\psi f(x)\right|$$

$$\leq \left(\sup_{\substack{x\in\Omega\\(E,2^{-j})\in\mathcal{E}}} \int \left|\left(2^{-j}X_x\right)^\alpha E(x,z)\right| dz\right) \left(\sup_{\substack{f\in\mathcal{B}\\z\in\Omega}} |\psi(z)f(z)|\right)$$

$$< \infty,$$

completing the proof.   □

PROOF OF LEMMA 5.3.2.   Fix a bounded set $\mathcal{B} \subset C^\infty(\Omega)$. The result follows once we show for every ordered multi-index α,

$$\sup_{j\in\mathbb{N}} \sup_{f\in\mathcal{B}} \sup_{x\in\Omega} 2^{|j|_\infty} \left|X^\alpha 2^{j\cdot s} E_j f(x)\right| < \infty.$$

That this is sufficient uses the fact that the vector fields $X_1, \ldots, X_q$ span the tangent space at every point. By repeated applications of the definition of elementary operators (using $l = 0$ in the definition), we have for any N,

$$E_j = 2^{-N|j|_\infty} \sum_{|\beta| \leq N} E_{j,\beta} X^\beta,$$

where $\left\{ \left(E_{j,\beta}, 2^{-j} \right) \mid j \in \mathbb{N}^\nu, |\beta| \leq N \right\}$ is a bounded set of elementary operators (and therefore a bounded set of pre-elementary operators). Let $\deg(\alpha)$ be as in (5.9). We have

$$\sup_{\substack{j \in \mathbb{N}, f \in \mathcal{B} \\ x \in \Omega}} 2^{|j|_\infty} \left| X^\alpha 2^{j \cdot s} E_j f(x) \right|$$

$$\leq \sum_{|\beta| \leq N} \sup_{\substack{j \in \mathbb{N}, f \in \mathcal{B} \\ x \in \Omega}} 2^{(-N+1)|j|_\infty} 2^{|j|_\infty (|s + \deg(\alpha)|_1)} \left| \left(2^{-j} X \right)^\alpha E_{j,\beta} X^\beta f \right|.$$

Taking $N = 1 + |s + \deg(\alpha)|_1$, it follows from Lemma 5.3.4 (applied with f replaced by $X^\beta f$) that the right-hand side of the above equation is finite, which completes the proof. $\qquad\square$

PROOF OF THEOREM 5.3.1 GIVEN THEOREM 5.1.37. Suppose $s \in \mathbb{R}^\nu$ and $T \in \mathcal{A}_3^s$. Since $T : C^\infty(\Omega) \to C_0^\infty(\Omega)$, we have $\operatorname{supp}(T) \Subset \Omega \times \Omega$. Let $\psi \in C_0^\infty(\Omega)$ be such that $\psi \otimes \psi \equiv 1$ on a neighborhood of $\operatorname{supp}(T)$; and therefore (if we identify ψ with the operator given by multiplication by ψ) $T\psi = T$. Let $\psi_1 \in C_0^\infty(\Omega)$ equal 1 on a neighborhood of $\operatorname{supp}(\psi)$. We may write ψf as

$$\psi(x) f(x) = \int f\left(e^{t \cdot X} x \right) \psi\left(e^{t \cdot X} x \right) \psi_1(x) \otimes \delta_0(t) \ dt.$$

It is evident that $\psi_1(x) \otimes \delta_0(t)$ is a product kernel of order 0 in the sense of Definition 5.1.32, and by Proposition 5.1.35 we have $\psi_1(x) \otimes \delta_0(t) \in C_0^\infty(\Omega) \widehat{\otimes} \mathrm{PK}(0, a)$ for any $a > 0$. Theorem 5.1.37 applies to show that $\psi \in \mathcal{A}_4^0$; i.e., there is a bounded set of elementary operators $\left\{ \left(E_j, 2^{-j} \right) \mid j \in \mathbb{N}^\nu \right\}$ such that $\psi = \sum_{j \in \mathbb{N}^\nu} E_j$. We have

$$T = T\psi$$

$$= T \sum_{j \in \mathbb{N}^\nu} E_j$$

$$= \sum_{j \in \mathbb{N}^\nu} T E_j$$

$$= \sum_{j \in \mathbb{N}^\nu} 2^{j \cdot s} \left(2^{-j \cdot s} T E_j \right),$$

where to move T past the sum, we have used that $T : C^\infty(\Omega) \to C_0^\infty(\Omega)$ is continuous and applied Lemma 5.3.2. By the assumption $T \in \mathcal{A}_3^s$, we have

$$\left\{ \left(2^{-j \cdot s} T E_j, 2^{-j} \right) \mid j \in \mathbb{N}^\nu \right\}$$

is a bounded set of elementary operators, which completes the proof that $T \in \mathcal{A}_4^s$. $\qquad\square$

We devote the rest of this section to the proof of Theorem 5.1.37. In all of the results that follow, $a > 0$ is some small number to be chosen in the proofs. In Theorem 5.1.37, we are considering operators of the form

$$Tf(x) = \int_{\mathbb{R}^q} f(\gamma(x,t)) \, \psi(\gamma(x,t)) \, K(x,t) \, dt,$$

where $\psi \in C_0^\infty(\Omega)$, and $K \in C_0^\infty(\Omega) \widehat{\otimes} \mathrm{PK}(s,a)$ for some $s \in \mathbb{R}^\nu$. Here, γ can be given by either (5.4) or (5.5). The proof of Theorem 5.1.37 for the two choices of γ are nearly identical. We present the proof when γ is given by (5.5); i.e., $\gamma(x,t) = e^{t \cdot X} x$. The other case follows by the same arguments.

By the universal property, we need only consider the case when $K(x,t)$ is an elementary tensor product–i.e., when $K(x,t)$ is given by $\kappa(x) \otimes K(t)$, where $\kappa \in C_0^\infty(\Omega)$ and $K(t) \in \mathrm{PK}(s,a)$, and we henceforth assume this form for $K(x,t)$. By the definition of $\mathrm{PK}(s,a)$, there is $\eta \in C_0^\infty(B^q(a))$ and $\{\varsigma_j \mid j \in \mathbb{N}^\nu\} \subset \mathcal{S}(\mathbb{R}^q)$ a bounded set with $\varsigma_j \in \mathcal{S}_0^{\{\mu | j_\mu \neq 0\}}$ such that $K(t) = \sum_{j \in \mathbb{N}^\nu} \eta(t) 2^{j \cdot s} \varsigma_j^{(2^j)}$. Define

$$E_j f(x) = \int f(e^{t \cdot X} x) \, \psi(e^{t \cdot X} x) \, \kappa(x) \, \varsigma_j^{(2^j)}(t) \, dt,$$

so that $T = \sum_{j \in \mathbb{N}^\nu} 2^{j \cdot s} E_j$. We show

PROPOSITION 5.3.5. $\{(E_j, 2^{-j}) \mid j \in \mathbb{N}^\nu\}$ *is a bounded set of elementary operators.*

Note that Proposition 5.3.5 completes the proof that $T \in \mathcal{A}_4^s$.

Remark 5.3.6 We have ignored one technical point in the above. In order to use the universal property to reduce to the case when K is given by an elementary tensor product, we need that the bilinear map $(\kappa, K) \mapsto T$ is a continuous map $C_0^\infty(\Omega) \times \mathrm{PK}(s,a) \to \mathcal{A}_4^s$ (this also requires that $\mathcal{A}_4^s$ is complete; see Section 5.7). Our proof demonstrates the desired continuity. We do not specifically say in each result that follows that the corresponding maps are continuous in the proper sense, as that would needlessly complicate the statements of each result. We therefore only remark here that all of the results that follow are continuous in the appropriate sense, and that the proofs of those results demonstrate that fact.

Remark 5.3.7 In the proof of Theorem 5.3.1, we used Theorem 5.1.37 to write (for $\psi \in C_0^\infty(\Omega)$) $\psi = \sum_{j \in \mathbb{N}^\nu} E_j$, where $\{(E_j, 2^{-j}) \mid j \in \mathbb{N}^\nu\}$ is a bounded set of elementary operators. In this case, we used $K = \psi_1(x) \otimes \delta_0$, for some $\psi_1 \in C_0^\infty(\Omega)$ which is an elementary tensor product. Thus, we do not need to use the topology of $\mathcal{A}_4^s$ as in Remark 5.3.6 to decompose the operator given by multiplication by ψ into elementary operators. This becomes an important point in Section 5.7, where we use such a decomposition to prove that $\mathcal{A}_4^s$ is complete.

Our goal is to prove Proposition 5.3.5, from which Theorem 5.1.37 follows. We do this by proving a slightly more general result. For this, we introduce a new definition.

DEFINITION 5.3.8. *We say*

$$\mathcal{I} \subset \mathcal{S}(\mathbb{R}^q) \times (0,1]^\nu \times C^\infty(\Omega \times B^q(a)) \times C_0^\infty(B^q(a)) \times C_0^\infty(\Omega) \times C_0^\infty(\Omega)$$

is a bounded set of pseudodifferential operator data if:

- $\forall (\varsigma, 2^{-j}, \kappa, \eta, \psi_1, \psi_2) \in \mathcal{I}, \varsigma \in \mathcal{S}_0^{\{\mu | j_\mu \neq 0\}}$.

- $\{\varsigma \mid (\varsigma, 2^{-j}, \kappa, \eta, \psi_1, \psi_2) \in \mathcal{I}\} \subset \mathcal{S}(\mathbb{R}^q)$ *is a bounded set.*

- $\{\kappa \mid (\varsigma, 2^{-j}, \kappa, \eta, \psi_1, \psi_2) \in \mathcal{I}\} \subset C^\infty(\Omega \times B^q(a))$ *is a bounded set.*

- $\{\eta \mid (\varsigma, 2^{-j}, \kappa, \eta, \psi_1, \psi_2) \in \mathcal{I}\} \subset C_0^\infty(B^q(a))$ *is a bounded set.*

- $\{\psi_1, \psi_2 \mid (\varsigma, 2^{-j}, \kappa, \eta, \psi_1, \psi_2) \in \mathcal{I}\} \subset C_0^\infty(\Omega)$ *is a bounded set.*

PROPOSITION 5.3.9. *Let $\mathcal{I}$ be a bounded set of pseudodifferential operator data. For each $\mathcal{D} = (\varsigma, 2^{-j}, \kappa, \eta, \psi_1, \psi_2) \in \mathcal{I}$ we define the operator $E(\mathcal{D})$ by*

$$E(\mathcal{D}) f(x) = \psi_1(x) \int f(e^{t \cdot X} x) \psi_2(e^{t \cdot X} x) \kappa(x,t) \eta(t) \varsigma^{(2^j)}(t) \, dt. \qquad (5.19)$$

Then $\{(E(\mathcal{D}), 2^{-j}) \mid \mathcal{D} \in \mathcal{I}\}$ is a bounded set of elementary operators.

Remark 5.3.10 Proposition 5.3.9 also holds with γ given by (5.4) instead of (5.5). The proof is nearly identical.

PROOF OF PROPOSITION 5.3.5 GIVEN PROPOSITION 5.3.9. Let κ, η, ς_j, and ψ be as in Proposition 5.3.5. Let $\psi_1(x) \in C_0^\infty(\Omega)$ be such that $\psi_1 \equiv 1$ on a neighborhood of supp(κ). Applying Proposition 5.3.9 with ψ_2 replaced by ψ, and $\kappa(x)$ independent of t, Proposition 5.3.5 follows. $\square$

Proposition 5.3.9 follows immediately from a more general proposition concerning generalized elementary operators. To make sense of this, we must weaken the restrictions on κ in the definition of pseudodifferential operator data.

DEFINITION 5.3.11. *We say*

$$\mathcal{I} \subset \mathcal{S}(\mathbb{R}^q) \times (0,1]^\nu \times (0,1]^\nu \times C^\infty(\Omega \times B^q(a)) \times C_0^\infty(B^q(a)) \times C_0^\infty(\Omega) \times C_0^\infty(\Omega)$$

is a bounded set of generalized pseudodifferential operator data if

- $\forall (\varsigma, 2^{-j}, 2^{-k}, \kappa, \eta, \psi_1, \psi_2) \in \mathcal{I}, k \le j$.

- $\forall (\varsigma, 2^{-j}, 2^{-k}, \kappa, \eta, \psi_1, \psi_2) \in \mathcal{I}, \varsigma \in \mathcal{S}_0^{\{\mu | j_\mu \neq 0\}}$.

- $\{\varsigma \mid (\varsigma, 2^{-j}, 2^{-k}, \kappa, \eta, \psi_1, \psi_2) \in \mathcal{I}\} \subset \mathcal{S}(\mathbb{R}^q)$ *is a bounded set.*

- $\forall M, \exists C, N, \forall \left(\varsigma, 2^{-j}, 2^{-k}, \kappa, \eta, \psi_1, \psi_2 \right) \in \mathcal{I}$,

$$
\sup_{\substack{x \in \Omega \\ t \in B^q(a)}} \sum_{|\alpha|, |\beta| \le M} \left| \left(2^{-k} \partial_t \right)^{\beta} \left(2^{-k} X \right)^{\alpha} \left(1 + \left| 2^k t \right|^2 \right)^{-N} \kappa \left(x, t \right) \right| \le C,
$$

where $2^k t$ and $2^{-k} \partial_t$ are defined as in the beginning of Section 5.2.

- $\left\{ \eta \mid \left(\varsigma, 2^{-j}, 2^{-k}, \kappa, \eta, \psi_1, \psi_2 \right) \in \mathcal{I} \right\} \subset C_0^\infty \left(B^q \left(a \right) \right)$ is a bounded set.

- $\left\{ \psi_1, \psi_2 \mid \left(\varsigma, 2^{-j}, 2^{-k}, \kappa, \eta, \psi_1, \psi_2 \right) \in \mathcal{I} \right\} \subset C_0^\infty \left(\Omega \right)$ is a bounded set.

PROPOSITION 5.3.12. *Let $\mathcal{I}$ be a bounded set of generalized pseudodifferential operator data. For $\mathcal{D} = \left(\varsigma, 2^{-j}, 2^{-k}, \kappa, \eta, \psi_1, \psi_2 \right) \in \mathcal{I}$ define the operator $E \left(\mathcal{D} \right)$ by*

$$
E \left(\mathcal{D} \right) f \left(x \right) = \psi_1 \left(x \right) \int f \left(e^{t \cdot X} x \right) \psi_2 \left(e^{t \cdot X} x \right) \kappa \left(x, t \right) \eta \left(t \right) \varsigma^{\left(2^j \right)} \left(t \right) \, dt. \qquad (5.20)
$$

Then $\left\{ \left(E \left(\mathcal{D} \right), 2^{-j}, 2^{-k} \right) \mid \mathcal{D} \in \mathcal{I} \right\}$ is a bounded set of generalized elementary operators.

PROOF OF PROPOSITION 5.3.9 GIVEN PROPOSITION 5.3.12. Let $\mathcal{I}$ be a bounded set of pseudodifferential operator data and let E be the operator in Proposition 5.3.9. Take $\psi_3 \in C_0^\infty \left(\Omega \right)$ such that $\psi_3 \equiv 1$ on a neighborhood of the closure of

$$
\bigcup_{\left(\varsigma, 2^{-j}, \kappa, \eta, \psi_1, \psi_2 \right) \in \mathcal{I}} \operatorname{supp} \left(\psi_1 \right) \Subset \Omega
$$

and $\eta' \in C_0^\infty \left(B^q \left(a \right) \right)$ so that $\eta' \equiv 1$ on a neighborhood of the closure of

$$
\bigcup_{\left(\varsigma, 2^{-j}, \kappa, \eta, \psi_1, \psi_2 \right) \in \mathcal{I}} \operatorname{supp} \left(\eta \right) \Subset B^q \left(a \right).
$$

Notice in the definition of E we may replace κ with $\tilde{\kappa} \left(x, t \right) := \psi_3 \left(x \right) \kappa \left(x, t \right) \eta' \left(t \right)$ and obtain the same operator. From here, the result follows by applying Proposition 5.3.12 in the case $k = 0$, with κ replaced by $\tilde{\kappa}$. $\qquad \square$

To prove Proposition 5.3.12, we begin with a few technical lemmas.

LEMMA 5.3.13. *Let $\mathcal{I}$ be a bounded set of generalized pseudodifferential operator data. For $\mathcal{D} = \left(\varsigma, 2^{-j}, 2^{-k}, \kappa, \eta, \psi_1, \psi_2 \right) \in \mathcal{I}$ define $E \left(\mathcal{D} \right)$ by (5.20). Then $\left\{ \left(E \left(\mathcal{D} \right), 2^{-j} \right) \mid \mathcal{D} \in \mathcal{I} \right\}$ is a bounded set of pre-elementary operators.*

PROOF. Because of the support of ψ_1 and ψ_2 in the definition of $E \left(\mathcal{D} \right)$, we have

$$
\bigcup_{\mathcal{D} \in \mathcal{I}} \operatorname{supp} \left(E \left(\mathcal{D} \right) \right) \subseteq \bigcup_{\left(\varsigma, 2^{-j}, 2^{-k}, \kappa, \eta, \psi_1, \psi_2 \right) \in \mathcal{I}} \operatorname{supp} \left(\psi_1 \right) \times \operatorname{supp} \left(\psi_2 \right) \Subset \Omega \times \Omega.
$$

Thus, the result will follow once we show $\forall m$, M, $\exists C$, $\forall \mathcal{D} \in \mathcal{I}$,

$$\sum_{|\alpha|,|\beta| \leq M} \left| \left(2^{-j} X_x \right)^\alpha \left(2^{-j} X_z \right)^\beta E \left(\mathcal{D} \right) (x, z) \right| \leq C \frac{\left(1 + \rho_{2-j} (x, z) \right)^{-m}}{\mathrm{Vol} \left(B_{2-j} \left(x, 1 + \rho_{2-j} (x, z) \right) \right)}.$$

Fix $\mathcal{D} = \left(\varsigma, 2^{-j}, 2^{-k}, \kappa, \eta, \psi_1, \psi_2 \right) \in \mathcal{I}$ and define $E = E \left(\mathcal{D} \right)$ as in (5.20). Fix M and take $N = N \left(M \right)$ so large if we define $\tilde{\kappa} \left(x, t \right) := \left(1 + \left| 2^k t \right|^2 \right)^{-N} \kappa \left(x, t \right)$, we have

$$\sum_{|\alpha|,|\beta| \leq M} \left| \left(2^{-k} \partial_t \right)^\beta \left(2^{-k} X \right)^\alpha \tilde{\kappa} \left(x, t \right) \right| \lesssim 1.$$

Define $\tilde{\varsigma} := \left(1 + \left| 2^{k-j} t \right|^2 \right)^N \varsigma \left(t \right)$ (notice that $\tilde{\varsigma}$ is of the same form as ς, since $k \geq j$). We have

$$E f \left(x \right) = \psi_1 \left(x \right) \int f \left(e^{t \cdot X} x \right) \psi_2 \left(e^{t \cdot X} x \right) \tilde{\kappa} \left(x, t \right) \eta \left(t \right) \tilde{\varsigma}^{\left(2^j \right)} \left(t \right) \, dt.$$

We apply Lemma 5.2.18 to $\eta \tilde{\varsigma}^{\left(2^j \right)}$ to obtain $\eta \tilde{\varsigma}^{\left(2^j \right)} = \sum_{\substack{l \leq j \\ l \in \mathbb{N}^\nu}} \eta \varsigma_l^{\left(2^l \right)}$, where for every M_0, $\left\{ 2^{M_0 |j-l|} \varsigma_l \mid \mathcal{D} \in \mathcal{I}, l \leq j \right\} \subset C_0^\infty \left(B^q \left(a \right) \right)$ is a bounded set. Defining

$$E_l f \left(x \right) = \psi_1 \left(x \right) \int f \left(e^{t \cdot X} x \right) \psi_2 \left(e^{t \cdot X} x \right) \tilde{\kappa} \left(x, t \right) \eta \left(t \right) \varsigma_l^{\left(2^l \right)} \left(t \right) \, dt,$$

we have $E = \sum_{l \leq j} E_l$. We apply Proposition 2.2.28 to E_l (taking $\left(Z, d \right)$ in that theorem to be $\left(2^{-l} X, \hat{d} \right)$ where $\hat{d}_r = |d_r|_1$), to obtain for $a > 0$ sufficiently small and every M_0,

$$\sum_{|\alpha|,|\beta| \leq M} \left| \left(2^{-l} X_x \right)^\alpha \left(2^{-l} X_z \right)^\beta E_l \left(x, z \right) \right| \lesssim 2^{-M_0 |j-l|} \frac{\chi_{\left\{ \rho_{2-l} (x,z) < \xi_2 \right\}}}{\mathrm{Vol} \left(B_{2-l} \left(x, z \right) \right)}.$$

Using that M is fixed, we see (by changing M_0)

$$\sum_{|\alpha|,|\beta| \leq M} \left| \left(2^{-j} X_x \right)^\alpha \left(2^{-j} X_z \right)^\beta E_l \left(x, z \right) \right| \lesssim 2^{-M_0 |j-l|} \frac{\chi_{\left\{ \rho_{2-l} (x,z) < \xi_2 \right\}}}{\mathrm{Vol} \left(B_{2-l} \left(x, z \right) \right)},$$

for every M_0. Thus, we have

$$\sum_{|\alpha|,|\beta| \leq M} \left| \left(2^{-j} X_x \right)^\alpha \left(2^{-j} X_z \right)^\beta E \left(x, z \right) \right| \lesssim \sum_{l \leq j} 2^{-M_0 |j-l|} \frac{\chi_{\left\{ \rho_{2-l} (x,z) < \xi_2 \right\}}}{\mathrm{Vol} \left(B_{2-l} \left(x, z \right) \right)},$$

for every M_0. The result now follows from Lemma 3.2.16, by taking $M_0 = M_0 \left(m \right)$ sufficiently large. $\qquad \square$

LEMMA 5.3.14. *Let $\kappa \in C^\infty \left(\Omega \times B^q\left(a\right)\right)$ be a function such that $\forall M, \exists C_M, N_M,$ such that*

$$\sup_{\substack{x \in \Omega \\ t \in B^q(a)}} \sum_{|\alpha|,|\beta| \leq M} \left| \left(2^{-k}\partial_t\right)^\beta \left(2^{-k}X\right)^\alpha \left(1 + |2^k t|^2\right)^{-N_M} \kappa\left(x,t\right) \right| \leq C_M.$$

Then, $\forall M, \exists C'_M = C'_M\left(C_M, N_M\right), N'_M = N'_M\left(N_M\right)$ such that

$$\sup_{\substack{x \in \Omega \\ t \in B^q(a)}} \sum_{|\alpha|,|\beta| \leq M} \left| \left(2^{-k}\partial_t\right)^\beta \left(2^{-k}X\right)^\alpha \left(1 + |2^k t|^2\right)^{-N'_M} \kappa\left(e^{t \cdot X}x,t\right) \right| \leq C'_M.$$

PROOF. Notice, for $|\alpha| = 1$,

$$\left(2^{-k}X\right)^\alpha \kappa\left(e^{t \cdot X}x,t\right) = \left(2^{-k}\partial_s\right)^\alpha \Big|_{s=0} \kappa\left(e^{t \cdot X}e^{s \cdot X}x,t\right).$$

Thus to prove the lemma, it suffices to show that, for any p, $1 \leq p \leq q$,

$$2^{-k \cdot d_p}\partial_{t_p}\kappa\left(e^{t \cdot X}x,t\right) =: \kappa_1\left(e^{t \cdot X}x,t\right),$$

$$2^{-k \cdot d_p}\partial_{s_p}\Big|_{s=0}\kappa\left(e^{t \cdot X}e^{s \cdot X}x,t\right) =: \kappa_2\left(e^{t \cdot X}x,t\right),$$

where κ_1 and κ_2 both satisfy the hypotheses of the lemma. This proves the result for $|\alpha| = 1$ and the result for all α follows by a simple induction. By taking $a > 0$ sufficiently small, that both κ_1 and κ_2 are of the right form follows from Proposition 3.2.10.   □

LEMMA 5.3.15. *The class of operators in Proposition 5.3.12 is closed under adjoints. More precisely, let $\mathcal{I}$ be a bounded set of generalized pseudodifferential operator data. For each $\mathcal{D} = \left(\varsigma, 2^{-j}, 2^{-k}, \kappa, \eta, \psi_1, \psi_2\right) \in \mathcal{I}$ let $E = E\left(\mathcal{D}\right)$ be given by (5.20). Then E^* is of the same form as E. That is $E^* = E\left(\mathcal{D}^*\right)$ (as in (5.20)) where $\mathcal{D}^* = \left(\varsigma^*, 2^{-j}, 2^{-k}, \kappa^*, \eta^*, \psi_1^*, \psi_2^*\right)$ for some $\varsigma^*, \kappa^*, \eta^*, \psi_1^*,$ and ψ_2^* with $\left\{\mathcal{D}^* \,\middle|\, \mathcal{D} \in \mathcal{I}\right\}$ a bounded set of generalized pseudodifferential operator data.*

PROOF. E is given by

$$Ef\left(x\right) = \psi_1\left(x\right) \int f\left(e^{t \cdot X}x\right) \psi_2\left(e^{t \cdot X}x\right) \kappa\left(x,t\right) \eta\left(t\right) \varsigma^{\left(2^j\right)}\left(t\right) \, dt.$$

Using that $\det d_x e^{t \cdot X}x\big|_{t=0} = 1$, if we take $a > 0$ sufficiently small and apply a change of variables as in Appendix B.3, we have

$$E^*f\left(x\right) = \overline{\psi_2\left(x\right)} \int f\left(e^{-t \cdot X}x\right) \overline{\psi_1\left(e^{-t \cdot X}x\right)} \kappa\left(e^{-t \cdot X}x,t\right) J\left(x,t\right) \eta\left(t\right) \varsigma^{\left(2^j\right)}\left(t\right) \, dt,$$

where $J\left(x,t\right)$ is the Jacobian in the x variable of the mapping $x \mapsto e^{t \cdot X}x$, and therefore $J\left(x,t\right)$ is C^∞. Changing variables $u = -t$, we have

$$E^*f\left(x\right) = \overline{\psi_2\left(x\right)}\left(-1\right)^q \int f\left(e^{u \cdot X}x\right) \overline{\psi_1\left(e^{u \cdot X}x\right)}$$

$$\overline{\kappa\left(e^{u \cdot X}x,-u\right)} J\left(-u,x\right) \eta\left(-u\right) \varsigma^{\left(2^j\right)}\left(-u\right) \, du.$$

Setting $\kappa^* = \overline{\kappa\left(e^{t\cdot X}x, -t\right)}J\left(-t, x\right)$, $\psi_1^* = (-1)^q\,\overline{\psi_2}$, $\psi_2^* = \overline{\psi_1}$, $\eta^*\left(t\right) = \overline{\eta\left(-t\right)}$, $\varsigma^*\left(t\right) = \overline{\varsigma\left(-t\right)}$, and $\mathcal{D}^* = \left(\varsigma^*, 2^{-j}, 2^{-k}, \kappa^*, \eta^*, \psi_1^*, \psi_2^*\right)$, we have $E^* = E\left(\mathcal{D}^*\right)$. That κ^* satisfies the right estimates follows from Lemma 5.3.14. It is clear that ψ_1^*, ψ_2^*, η^*, and ς^* satisfy all the right properties to show $\left\{\mathcal{D}^* \mid \mathcal{D} \in \mathcal{I}\right\}$ is a bounded set of generalized pseudodifferential operator data, completing the proof. $\qquad\square$

PROOF OF PROPOSITION 5.3.12. Lemma 5.3.13 shows that

$$\left\{\left(E\left(\mathcal{D}\right), 2^{-j}\right) \mid \mathcal{D} = \left(\varsigma, 2^{-j}, 2^{-k}, \kappa, \eta, \psi_1, \psi_2\right) \in \mathcal{I}\right\}$$

is a bounded set of pre-elementary operators. Thus we need to show that we may "pull out" derivatives as in the definition of generalized elementary operators.

Let $E = E\left(\mathcal{D}\right)$, where $\mathcal{D} = \left(\varsigma, 2^{-j}, 2^{-k}, \kappa, \eta, \psi_1, \psi_2\right) \in \mathcal{I}$ and let $k \le l \le j$. We wish to show that there is a finite set $\mathcal{F}$ (whose size depends only on q) such that

$$E = \begin{cases} 2^{-|j-l|_\infty} \sum_{r\in\mathcal{F}} \sum_{|\alpha|\le 1} E_{\alpha,l,r}^1 \left(2^{-l}X\right)^\alpha & \text{and} \\ 2^{-|j-l|_\infty} \sum_{r\in\mathcal{F}} \sum_{|\alpha|\le 1} \left(2^{-l}X\right)^\alpha E_{\alpha,l,r}^2, \end{cases} \tag{5.21}$$

where $E_{\alpha,l,r}^1$ and $E_{\alpha,l,r}^2$ are "of the same form" as E with k replaced by l. That is, $E_{\alpha,l,r}^1 = E\left(\mathcal{D}_{\alpha,l,r}^1\right)$ and $E_{\alpha,l,r}^2 = E\left(\mathcal{D}_{\alpha,l,r}^2\right)$ as in (5.20), where

$$\left\{\mathcal{D}_{\alpha,l,r}^1, \mathcal{D}_{\alpha,l,r}^2 \mid \mathcal{D} \in \mathcal{I}, k \le l \le j, |\alpha| \le 1, r \in \mathcal{F}\right\}$$

is a bounded set of generalized pseudodifferential operator data, and in $\mathcal{D}_{\alpha,l,r}^1$ and $\mathcal{D}_{\alpha,l,r}^2$, the third coordinate is 2^{-l} instead of 2^{-k} and the second coordinate remains 2^{-j}. The result will then follow.

Note that it suffices to prove only the first line of (5.21). Indeed, if we have the first line, we may apply it to E^* (using Lemma 5.3.15) to obtain

$$E^* = 2^{-|j-l|_\infty} \sum_{r\in\mathcal{F}} \sum_{|\alpha|\le 1} E_{\alpha,l,r}^1 \left(2^{-l}X\right)^\alpha,$$

where $E_{\alpha,l,r}^1$ is of the proper form. Taking adjoints we obtain

$$E = 2^{-|j-l|_\infty} \sum_{r\in\mathcal{F}} \sum_{|\alpha|\le 1} \left[\left(2^{-l}X\right)^\alpha\right]^* \left[E_{\alpha,l,r}^1\right]^*.$$

Noting that $\left[\left(2^{-l}X\right)^\alpha\right]^* = f_{l,\alpha}\left(x\right) + \left(2^{-l}X\right)^\alpha$ (for $|\alpha| = 1$), where for every m,

$$\sup_{l\in[0,\infty)^\nu} \|f_{l,\alpha}\|_{C^m(\Omega)} < \infty,$$

the second line of (5.21) follows by another application of Lemma 5.3.15.

We turn to proving the first line of (5.21). Take μ_0 such that $|j - l|_\infty = j_{\mu_0} - l_{\mu_0}$. Notice, if $j_{\mu_0} = l_{\mu_0}$, then $|j - l|_\infty = 0$ and there is nothing to prove. We therefore assume $j_{\mu_0} > l_{\mu_0}$. Since $j_{\mu_0} > l_{\mu_0} \ge 0$, we have $\mu_0 \in \left\{\mu \mid j_\mu \ne 0\right\}$. By assumption

$\varsigma \in S_0^{\{\mu|j_\mu \neq 0\}}$ and Lemma A.2.18 shows that we may write $\varsigma = \sum_{v=1}^{q_{\mu_0}} \partial_{t_{\mu_0}^v} \varsigma_v$, where $\varsigma_v \in S_0^{\{\mu|j_\mu \neq 0\}}$ and $\{\varsigma_v \mid \mathcal{D} \in \mathcal{I}\} \subset S(\mathbb{R}^q)$ is a bounded set. Integrating by parts, we obtain

$$Ef(x) = \sum_{v=1}^{q_\mu} -2^{(l-j)\cdot d_v^{\mu_0}} \psi_1(x) \int 2^{-l\cdot d_v^{\mu_0}} \partial_{t_{\mu_0}^v} \left[f\left(e^{t\cdot X}x\right) \psi_2\left(e^{t\cdot X}x\right) \kappa(x,t) \eta(t) \right]$$

$$\varsigma_v^{(2^j)}(t)\, dt.$$

We expand the derivative by the product rule. When $2^{-l\cdot d_v^{\mu_0}} \partial_{t_{\mu_0}^v}$ lands on $\psi_2\left(e^{t\cdot X}x\right)$, $\kappa(x,t)$, or $\eta(t)$ we obtain an expression of the same form. Using that $(j-l)\cdot d_v^{\mu_0} \geq |j-l|_\infty$, it follows that these terms are of the proper form.

All that is left to consider is the case when $2^{-l\cdot d_v^{\mu_0}} \partial_{t_{\mu_0}^v}$ lands on $f\left(e^{t\cdot X}x\right)$. For these terms, we apply Proposition 3.2.10 to see that this yields

$$\sum_{p=1}^q -2^{(l-j)\cdot d_v^{\mu_0}} \psi_1(x) \int \left(2^{-l\cdot d_p} X_p f\right)\left(e^{t\cdot X}x\right) \psi_2\left(e^{t\cdot X}x\right) c_{p,v}^l(t,x)$$

$$\kappa(x,t)\eta(t)\varsigma_v^{(2^j)}(t)\, dt,$$

where for every M, there is a constant C independent of l such that

$$\sum_{|\alpha|,|\beta| \leq M} \left| \left(2^{-l}\partial_t\right)^\beta \left(2^{-l}X\right)^\alpha c_{p,v}^l(t,x) \left(1+|2^l t|^2\right)^{-\frac{1}{2}} \right| \leq C.$$

Thus, $c_{p,v}^l(t,x)\kappa(x,t)$ satisfies all the assumptions that κ did, but with k replaced by l. Using that $(j-l)\cdot d_v^{\mu_0} \geq |j-l|_\infty$, we see that the above equation is a sum of terms of the form $2^{-|j-l|_\infty} E_\alpha \left(2^{-l}X\right)^\alpha$, where $|\alpha| = 1$ and E_α is of the same form as E but with k replaced by l. This completes the proof. $\qquad \square$

Remark 5.3.16 The reader might note that we defined elementary operators in terms of "generalized elementary operators" in this chapter, but in the examples we covered in earlier chapters, this was not necessary. The reason is the lack of regularity guaranteed on $c_{p,v}^l(t,x)$ by Proposition 3.2.10. If we could assume that $c_{p,v}^l(t,x)$ were, for instance, C^∞ uniformly in any relevant parameters, then we would not need to use these more complicated generalized elementary operators. In the examples covered in earlier chapters, the relevant functions are, in fact, C^∞ uniformly in the relevant parameters.

5.4 ELEMENTARY OPERATORS AND $\mathcal{A}_4 \subseteq \mathcal{A}_3$

This section is devoted to studying the composition of elementary operators. Informally, if we have two elementary operators $\left(E_1, 2^{-j_1}\right), \left(E_2, 2^{-j_2}\right)$ then if $E_3 = E_1 E_2$ we have for any fixed N,

$$\left(2^{N|j_1-j_2|}E_3, 2^{-j_1}\right), \quad \left(2^{N|j_1-j_2|}E_3, 2^{-j_2}\right), \quad \text{and} \quad \left(2^{N|j_1-j_2|}E_3, 2^{-j_1 \wedge j_2}\right)$$

are all elementary operators, uniformly in the relevant parameters. More precisely, we have

THEOREM 5.4.1. *Suppose $\mathcal{E}$ is a bounded set of elementary operators. Then, for every N, the set*

$$\mathcal{E}_N := \left\{ \left(2^{N|j_1-j_2|} E_1 E_2, 2^{-j_1} \right), \left(2^{N|j_1-j_2|} E_1 E_2, 2^{-j_2} \right), \right.$$
$$\left. \left(2^{N|j_1-j_2|} E_1 E_2, 2^{-j_1 \wedge j_2} \right) \, \middle| \, (E_1, 2^{-j_1}), (E_2, 2^{-j_2}) \in \mathcal{E} \right\}$$

is a bounded set of elementary operators.

Before we prove Theorem 5.4.1 we show how it implies $\mathcal{A}_4 \subseteq \mathcal{A}_3$.

THEOREM 5.4.2. *For $s \in \mathbb{R}^\nu$, $\mathcal{A}_4^s \subseteq \mathcal{A}_3^s$.*

PROOF. Let $T \in \mathcal{A}_4^s$, so that there is a bounded set of elementary operators

$$\left\{ (E_j, 2^{-j}) \mid j \in \mathbb{N}^\nu \right\}$$

such that $T = \sum_{j \in \mathbb{N}^\nu} 2^{j \cdot s} E_j$. Fix a bounded set of elementary operators $\mathcal{E}$. For $(F, 2^{-k}) \in \mathcal{E}$ we have

$$TF = \sum_{j \in \mathbb{N}^\nu} 2^{j \cdot s} E_j F = \sum_{j \in \mathbb{N}^\nu} 2^{k \cdot s} 2^{-|k-j|} \widetilde{E}_{j,k},$$

where $\widetilde{E}_{j,k} = 2^{(j-k) \cdot s + |k-j|} E_j F$ and therefore $\left\{ \left(\widetilde{E}_{j,k}, 2^{-k} \right) \mid (F, 2^{-k}) \in \mathcal{E} \right\}$ is a bounded set of elementary operators by taking $N = N(s)$ large in Theorem 5.4.1. Thus,

$$\left\{ \left(2^{-k \cdot s} TF, 2^{-k} \right) \mid (F, 2^{-k}) \in \mathcal{E} \right\}$$

is a bounded set of elementary operators, verifying $T \in \mathcal{A}_3^s$ and completing the proof.
□

We devote the rest of the section to the proof of Theorem 5.4.1. Theorem 5.4.1 follows by taking $l_1 = l_2 = 0$ in the next theorem.

THEOREM 5.4.3. *Suppose $\mathcal{E}$ is a bounded set of generalized elementary operators. Then, for every N, the set*

$$\mathcal{E}_N := \left\{ \left(2^{N|j_1-j_2|} E_1 E_2, 2^{-j_1}, 2^{-l_1 \vee l_2} \right), \left(2^{N|j_1-j_2|} E_1 E_2, 2^{-j_2}, 2^{-l_1 \vee l_2} \right), \right.$$
$$\left(2^{N|j_1-j_2|} E_1 E_2, 2^{-j_1 \wedge j_2}, 2^{-l_1 \vee l_2} \right) \, \middle| \, (E_1, 2^{-j_1}, 2^{-l_1}), (E_2, 2^{-j_2}, 2^{-l_2}) \in \mathcal{E},$$
$$\left. l_1 \vee l_2 \le j_1 \wedge j_2 \right\}$$

is a bounded set of generalized elementary operators.

To prove Theorem 5.4.3, we need a number of technical lemmas.

LEMMA 5.4.4. *Let $\mathcal{E}$ be a bounded set of pre-elementary operators and let c be a bounded measurable function on Ω. Then, for every m, M, there exists $C = C\left(\mathcal{E}, m, M, \|c\|_{L^\infty}\right)$ such that for every $\left(E_1, 2^{-j_1}\right), \left(E_2, 2^{-j_2}\right) \in \mathcal{E}$, we have*

$$\sum_{\substack{|\alpha_1|,|\alpha_2|\leq M \\ |\beta_1|,|\beta_2|\leq M}} \left| \left(2^{-j_1} X_x\right)^{\alpha_1} \left(2^{-j_2} X_z\right)^{\beta_1} \left[E_1 \left(2^{-j_1} X\right)^{\alpha_2} c \left(2^{-j_2} X\right)^{\beta_2} E_2\right](x,z)\right|$$

$$\leq C \frac{\left(1 + \rho_{2^{-j_1 \wedge j_2}}(x,z)\right)^{-m}}{\mathrm{Vol}\left(B_{2^{-j_1 \wedge j_2}}\left(x, 1 + \rho_{2^{-j_1 \wedge j_2}}(x,z)\right)\right)},$$

(5.22)

where we have identified c with the operator given by multiplication by c.

PROOF. The left-hand side of (5.22) can be re-written as

$$\sum_{\substack{|\alpha_1|,|\alpha_2|\leq M \\ |\beta_1|,|\beta_2|\leq M}} \left| \left[\left(2^{-j_1} X\right)^{\alpha_1} E_1 \left(2^{-j_1} X\right)^{\alpha_2} c \left(2^{-j_2} X\right)^{\beta_2} E_2 \left(\left(2^{-j_2} X\right)^{\beta_1}\right)^*\right](x,z)\right|.$$

The definition of pre-elementary operators shows, for every m,

$$\left| \left[\left(2^{-j_1} X\right)^{\alpha_1} E_1 \left(2^{-j_1} X\right)^{\alpha_2}\right](x,z)\right| \lesssim \frac{\left(1 + \rho_{2^{-j_1}}(x,z)\right)^{-m}}{\mathrm{Vol}\left(B_{2^{-j_1}}\left(x, 1 + \rho_{2^{-j_1}}(x,z)\right)\right)},$$

$$\left| \left[\left(2^{-j_2} X\right)^{\beta_2} E_2 \left[\left(2^{-j_2} X\right)^{\beta_1}\right]^*\right](x,z)\right| \lesssim \frac{\left(1 + \rho_{2^{-j_2}}(x,z)\right)^{-m}}{\mathrm{Vol}\left(B_{2^{-j_2}}\left(x, 1 + \rho_{2^{-j_2}}(x,z)\right)\right)}.$$

The result now follows by Proposition 3.2.19. □

LEMMA 5.4.5. *Let $\mathcal{E}$ be a bounded set of generalized elementary operators. Then for every m, M, L, there exists $C = C\left(\mathcal{E}, m, M, L\right)$ such that for every*

$$\left(E_1, 2^{-j_1}, 2^{-l_1}\right), \left(E_2, 2^{-j_2}, 2^{-l_2}\right) \in \mathcal{E}$$

with $l_1 \vee l_2 \leq j_1 \wedge j_2$, we have

$$\sum_{|\alpha|,|\beta|\leq M} \left| \left(2^{-j_1 \wedge j_2} X_x\right)^{\alpha} \left(2^{-j_1 \wedge j_2} X_z\right)^{\beta} \left[E_1 E_2\right](x,z)\right|$$

$$\leq C 2^{-L|j_1 - j_2|} \frac{\left(1 + \rho_{2^{-j_1 \wedge j_2}}(x,z)\right)^{-m}}{\mathrm{Vol}\left(B_{2^{-j_1 \wedge j_2}}\left(x, 1 + \rho_{2^{-j_1 \wedge j_2}}(x,z)\right)\right)}.$$

PROOF. We will show, for every N,

$$\sum_{|\alpha|,|\beta|\leq M} \left| \left(2^{-j_1} X_x\right)^{\alpha} \left(2^{-j_2} X_z\right)^{\beta} \left[E_1 E_2\right](x,z)\right|$$

$$\leq C 2^{-N|j_1 - j_2|_\infty} \frac{\left(1 + \rho_{2^{-j_1 \wedge j_2}}(x,z)\right)^{-m}}{\mathrm{Vol}\left(B_{2^{-j_1 \wedge j_2}}\left(x, 1 + \rho_{2^{-j_1 \wedge j_2}}(x,z)\right)\right)}.$$

(5.23)

The result will then follow by taking $N = N(M, L)$ sufficiently large. There are two possibilities: either $|j_1 - j_2|_\infty = |j_1 - j_1 \wedge j_2|_\infty$ or $|j_1 - j_2|_\infty = |j_2 - j_1 \wedge j_2|_\infty$. We prove (5.23) for the first case, the second case is similar and we leave it to the reader. Hence, we are in the case when $|j_1 - j_2|_\infty = |j_1 - j_1 \wedge j_2|_\infty$. Applying the definition of generalized elementary operators N times (taking $l = j_1 \wedge j_2$ in Definition 5.1.7) we have

$$E_1 = \sum_{|\alpha| \leq N} 2^{-N|j_1 - j_1 \wedge j_2|_\infty} E_\alpha \left(2^{-j_1 \wedge j_2} X\right)^\alpha,$$

where $\left\{\left(E_\alpha, 2^{-j}, 2^{-j_1 \wedge j_2}\right) \mid \left(E_1, 2^{-j_1}, 2^{-l_1}\right) \in \mathcal{E}, l_1 \leq j_2\right\}$ is a bounded set of generalized elementary operators. Applying Proposition 3.2.22, we have

$$E_1 = \sum_{|\alpha| \leq N} 2^{-N|j_1 - j_2|_\infty} \sum_{|\beta| + |\gamma| \leq |\alpha|} E_\alpha \left(2^{-j_1} X\right)^\beta c_{\beta, \gamma} \left(2^{-j_2} X\right)^\gamma, \tag{5.24}$$

where $c_{\beta, \gamma}$ is as in Proposition 3.2.22–though the important property is that $\|c_{\beta, \gamma}\|_{L^\infty}$ is bounded uniformly in j_1, j_2. Plugging (5.24) into the left-hand side of (5.23), (5.23) follows from Lemma 5.4.4, which completes the proof. □

LEMMA 5.4.6. *Let $\mathcal{E}$ be a bounded set of generalized elementary operators. Then, for every m, M, L there exists $C = C(\mathcal{E}, m, M, L)$ such that for every*

$$\left(E_1, 2^{-j_1}, 2^{-l_1}\right), \left(E_2, 2^{-j_2}, 2^{-l_2}\right) \in \mathcal{E}$$

with $l_1 \vee l_2 \leq j_1 \wedge j_2$, we have

$$\sum_{|\alpha|, |\beta| \leq M} \left| \left(2^{-j_1 \wedge j_2} X_x\right)^\alpha \left(2^{-j_1 \wedge j_2} X_z\right)^\beta [E_1 E_2](x, z)\right|$$
$$\leq C 2^{-L|j_1 - j_2|} \frac{\left(1 + \rho_{2^{-j_1 \wedge j_2}}(x, z)\right)^{-m}}{\mathrm{Vol}\left(B_{2^{-j_1 \wedge j_2}}(x, 1 + \rho_{2^{-j_1 \wedge j_2}}(x, z))\right)}, \tag{5.25}$$

$$\sum_{|\alpha|, |\beta| \leq M} \left| \left(2^{-j_1} X_x\right)^\alpha \left(2^{-j_1} X_z\right)^\beta [E_1 E_2](x, z)\right|$$
$$\leq C 2^{-L|j_1 - j_2|} \frac{\left(1 + \rho_{2^{-j_1}}(x, z)\right)^{-m}}{\mathrm{Vol}\left(B_{2^{-j_1}}(x, 1 + \rho_{2^{-j_1}}(x, z))\right)}, \tag{5.26}$$

$$\sum_{|\alpha|, |\beta| \leq M} \left| \left(2^{-j_2} X_x\right)^\alpha \left(2^{-j_2} X_z\right)^\beta [E_1 E_2](x, z)\right|$$
$$\leq C 2^{-L|j_1 - j_2|} \frac{\left(1 + \rho_{2^{-j_2}}(x, z)\right)^{-m}}{\mathrm{Vol}\left(B_{2^{-j_2}}(x, 1 + \rho_{2^{-j_2}}(x, z))\right)}. \tag{5.27}$$

PROOF. (5.25) is just a restatement of Lemma 5.4.5. The proofs of (5.26) and (5.27) are similar, and so we focus only on (5.26). Fix m. Lemma 3.2.5 shows we may take $L_0 = L_0\left(m\right)$ so large

$$2^{-L_0|j_1-j_2|}\frac{\left(1+\rho_{2^{-j_1\wedge j_2}}\left(x,z\right)\right)^{-m}}{\mathrm{Vol}\left(B_{2^{-j_1\wedge j_2}}\left(x,1+\rho_{2^{-j_1\wedge j_2}}\left(x,z\right)\right)\right)}$$

$$\lesssim \frac{\left(1+\rho_{2^{-j_1}}\left(x,z\right)\right)^{-m}}{\mathrm{Vol}\left(B_{2^{-j_1}}\left(x,1+\rho_{2^{-j_1}}\left(x,z\right)\right)\right)}.$$

Applying (5.25) with L replaced by $L + L_0$, we have

$$\sum_{|\alpha|+|\beta|\leq M}\left|\left(2^{-j_1}X_x\right)^\alpha\left(2^{-j_1}X_z\right)^\beta\left[E_1E_2\right]\left(x,z\right)\right|$$

$$\leq \sum_{|\alpha|+|\beta|\leq M}\left|\left(2^{-j_1\wedge j_2}X_x\right)^\alpha\left(2^{-j_1\wedge j_2}X_z\right)^\beta\left[E_1E_2\right]\left(x,z\right)\right|$$

$$\lesssim 2^{-(L+L_0)|j_1-j_2|}\frac{\left(1+\rho_{2^{-j_1\wedge j_2}}\left(x,z\right)\right)^{-m}}{\mathrm{Vol}\left(B_{2^{-j_1\wedge j_2}}\left(x,1+\rho_{2^{-j_1\wedge j_2}}\left(x,z\right)\right)\right)}$$

$$\lesssim 2^{-L|j_1-j_2|}\frac{\left(1+\rho_{2^{-j_1}}\left(x,z\right)\right)^{-m}}{\mathrm{Vol}\left(B_{2^{-j_1}}\left(x,1+\rho_{2^{-j_1}}\left(x,z\right)\right)\right)},$$

completing the proof. $\square$

PROOF OF THEOREM 5.4.3. Let $\mathcal{E}$ be a bounded set of generalized elementary operators, and define $\mathcal{E}_N$ as in the statement of the theorem. Notice that

$$\bigcup_{\left(E,2^{-j},2^{-l}\right)\in\mathcal{E}}\mathrm{supp}\left(E\right)\Subset\Omega\times\Omega$$

and therefore

$$\bigcup_{\left(E_1,2^{-j_1},2^{-l_1}\right),\left(E_2,2^{-j_2},2^{-l_2}\right)\in\mathcal{E}}\mathrm{supp}\left(E_1E_2\right)\Subset\Omega\times\Omega.$$

Combining this with Lemma 5.4.6 shows that $\mathcal{E}_N$ is a bounded set of pre-elementary operators, for every N. Our goal is to show that we may "pull out" derivatives as in the definition of generalized elementary operators.

We begin by showing

$$\left\{\left(2^{N|j_1-j_2|}E_1E_2,2^{-j_1\wedge j_2},2^{-l_1\vee l_2}\right)\,\middle|\,\left(E_1,2^{-j_1},2^{-l_1}\right)\in\mathcal{E},\right.$$

$$\left.\left(E_2,2^{-j_2},2^{-l_2}\right)\in\mathcal{E},l_1\vee l_2\leq j_1\wedge j_2\right\} \tag{5.28}$$

is a bounded set of generalized elementary operators. Let

$$\left(E_1,2^{-j_1},2^{-l_1}\right),\left(E_2,2^{-j_2},2^{-l_2}\right)\in\mathcal{E}$$

with $l_1 \vee l_2 \leq j_1 \wedge j_2$. Fix k with $l_1 \vee l_2 \leq k \leq j_1 \wedge j_2$. Applying the definition of generalized elementary operators, we have

$$2^{N|j_1-j_2|} E_1 E_2 = \begin{cases} 2^{-|j_1-k|_\infty} \sum_{|\alpha|\leq 1} \left(2^{-k}X\right)^\alpha 2^{N|j_1-j_2|} E_{1,\alpha,k} E_2 & \text{and} \\ 2^{-|j_2-k|_\infty} \sum_{|\alpha|\leq 1} 2^{N|j_1-j_2|} E_1 E_{2,\alpha,k} \left(2^{-k}X\right)^\alpha, \end{cases}$$

where

$$\left\{ \left(E_{1,\alpha,k}, 2^{-j_1}, 2^{-k}\right), \left(E_{2,\alpha,k}, 2^{-j_2}, 2^{-k}\right) \; \middle| \right.$$
$$\left. \left(E_1, 2^{-j_1}, 2^{-l_1}\right), \left(E_2, 2^{-j_2}, 2^{-l_2}\right) \in \mathcal{E}, l_1 \vee l_2 \leq k \leq j_1 \wedge j_2, |\alpha| \leq 1 \right\}$$

is a bounded set of generalized elementary operators. Since $|j_1 - k|_\infty, |j_2 - k|_\infty \geq |j_1 \wedge j_2 - k|_\infty$ this shows that $E_1 E_2$ can be written as $2^{-|j_1 \wedge j_2 - k|_\infty}$ times an appropriate sum of derivatives of terms of the same form with $l_1 \vee l_2$ replaced by k. This completes the proof that (5.28) is a bounded set of generalized elementary operators.

We now wish to show

$$\left\{ \left(2^{N|j_1-j_2|} E_1 E_2, 2^{-j_1}, 2^{-l_1 \vee l_2}\right) \; \middle| \; \left(E_1, 2^{-j_1}, 2^{-l_1}\right) \in \mathcal{E}, \right.$$
$$\left. \left(E_2, 2^{-j_2}, 2^{-l_2}\right) \in \mathcal{E}, l_1 \vee l_2 \leq j_1 \wedge j_2 \right\} \tag{5.29}$$

$$\left\{ \left(2^{N|j_1-j_2|} E_1 E_2, 2^{-j_2}, 2^{-l_1 \vee l_2}\right) \; \middle| \; \left(E_1, 2^{-j_1}, 2^{-l_1}\right) \in \mathcal{E}, \right.$$
$$\left. \left(E_2, 2^{-j_2}, 2^{-l_2}\right) \in \mathcal{E}, l_1 \vee l_2 \leq j_1 \wedge j_2 \right\} \tag{5.30}$$

are bounded sets of generalized elementary operators. The proofs for (5.29) and (5.30) are similar, and we prove only the result for (5.29).

Fix k with $l_1 \vee l_2 \leq k \leq j_1$. We apply the definition of generalized elementary operators at the scale $k \wedge j_2$ to see

$$2^{N|j_1-j_2|} E_1 E_2$$
$$= \begin{cases} 2^{-|j_1-k\wedge j_2|_\infty} \sum_{|\alpha|\leq 1} \left(2^{-k\wedge j_2}X\right)^\alpha 2^{N|j_1-j_2|} E_{1,\alpha,k} E_2 & \text{and} \\ 2^{-|j_2-k\wedge j_2|_\infty} \sum_{|\alpha|\leq 1} 2^{N|j_1-j_2|} E_1 E_{2,\alpha,k} \left(2^{-k\wedge j_2}X\right)^\alpha, \end{cases} \tag{5.31}$$

where

$$\left\{ \left(E_{1,\alpha,k}, 2^{-j_1}, 2^{-k\wedge j_2}\right), \left(E_{2,\alpha,k}, 2^{-j_2}, 2^{-k\wedge j_2}\right) \; \middle| \right.$$
$$\left. \left(E_1, 2^{-j_1}, 2^{-l_1}\right), \left(E_2, 2^{-j_2}, 2^{-l_2}\right) \in \mathcal{E}, l_1 \vee l_2 \leq k \leq j_1, |\alpha| \leq 1 \right\}$$

is a bounded set of generalized elementary operators.

Both the top and bottom terms of (5.31) work in the same way, and so we focus only on the bottom. Because $k \leq j_1$, we have

$$2^{-|j_2-k\wedge j_2|_\infty} 2^{N|j_1-j_2|} \left(2^{-k\wedge j_2}X\right)^\alpha = c\left(\alpha\right) 2^{-|j_1-k|_\infty} 2^{(N+e(\alpha))|j_1-j_2|} \left(2^{-k}X\right)^\alpha,$$

where $c(\alpha)$ is some constant with $0 < c(\alpha) \leq 1$ and $e(\alpha)$ is some constant. Thus, we have

$$2^{N|j_1-j_2|} E_1 E_2 = 2^{-|j_1-k|_\infty} \sum_{|\alpha|\leq 1} 2^{(N+e(\alpha))|j_1-j_2|} c(\alpha) E_1 E_{2,\alpha,k} \left(2^{-k}X\right)^\alpha.$$

This shows that $2^{N|j_1-j_2|} E_1 E_2$ is $2^{-|j_1-k|_\infty}$ times an appropriate sum of derivatives of operators of the same form, with $l_1 \vee l_2$ replaced by $j_2 \wedge k$, and with a different choice of N. Since $k \leq j_1$, we have $j_2 \wedge k \leq j_1 \wedge j_2$. This completes the proof that (5.29) is a bounded set of generalized elementary operators.   □

5.5 $\mathcal{A}_4 \subseteq \mathcal{A}_2 \subseteq \mathcal{A}_1$

In this section, we prove the containments $\mathcal{A}_4^s \subseteq \mathcal{A}_2^s \subseteq \mathcal{A}_1^s$. These are the easiest containments to prove. We begin with $\mathcal{A}_4^s \subseteq \mathcal{A}_2^s$.

LEMMA 5.5.1.   *Let $\mathcal{E}$ be a bounded set of elementary operators. Then $\forall L$, $\exists M$, $\forall \alpha$, $\forall T_M$ a bounded set of test functions of order M, $\forall m$, $\exists C$, $\forall (E, 2^{-k}) \in \mathcal{E}$, $\forall (\psi, z, 2^{-j}) \in T_M$,*

$$\left|\left(2^{-j}X\right)^\alpha E\psi(x)\right| \leq C 2^{-L|j-k|} \frac{\left(1 + \rho_{2^{-j\wedge k}}(x,z)\right)^{-m}}{\mathrm{Vol}\left(B_{2^{-j\wedge k}}(x, 1 + \rho_{2^{-j\wedge k}}(x,z))\right)}.$$

PROOF.   Take $M = M(L)$ and $\tilde{N} = \tilde{N}(L, |\alpha|)$ large to be chosen later, and let $(E, 2^{-k}) \in \mathcal{E}$ and $(\psi, z, 2^{-j}) \in T_M$. Applying the definition of elementary operators, we may write

$$E = 2^{-\tilde{N}|k-j\wedge k|_\infty} \sum_{|\beta|\leq \tilde{N}} F_\beta \left(2^{-j\wedge k}X\right)^\beta,$$

where $\left\{(F_\beta, 2^{-k}, 2^{-j\wedge k}) \mid (E, 2^{-k}) \in \mathcal{E}, j \in [0, \infty)^\nu\right\}$ is a bounded set of generalized elementary operators. $\left\{(F_\beta, 2^{-k}) \mid (E, 2^{-k}) \in \mathcal{E}, j \in [0, \infty)^\nu\right\}$ is therefore a bounded set of pre-elementary operators, and we have for every m and R,

$$\sum_{|\alpha|,|\gamma|\leq R} \left|\left(2^{-k}X_x\right)^\alpha \left(2^{-k}X_y\right)^\gamma F_\beta(x,y)\right|$$

$$\lesssim \frac{\left(1 + \rho_{2^{-k}}(x,y)\right)^{-m}}{\mathrm{Vol}\left(B_{2^{-k}}(x, 1 + \rho_{2^{-k}}(x,y))\right)}. \tag{5.32}$$

Similarly, using the definition of test functions, we may write

$$\psi = 2^{-M|j-j\wedge k|_\infty} \sum_{|\beta|\leq M} \left(2^{-j\wedge k}X\right)^\beta \phi_{\beta,j\wedge k},$$

where $\left\{ \left(\phi_{\beta,z,j\wedge k}, 2^{-j} \right) \mid \left(\psi, z, 2^{-j} \right) \in \mathcal{T}_M, k \in [0, \infty)^\nu \right\}$ is a bounded set of bump functions, and therefore, for every m and R,

$$
\sum_{|\alpha| \leq R} \left| \left(2^{-j} X \right)^\alpha \phi_{\beta,j\wedge k} (y) \right| \lesssim \frac{\chi_{\{y \in B_{2^{-j}}(z,1)\}}}{\text{Vol}\left(B_{2^{-j}} (z, 1) \right)}
$$

$$
\lesssim \frac{\left(1 + \rho_{2^{-j}} (y, z) \right)^{-m}}{\text{Vol}\left(B_{2^{-j}} (y, 1 + \rho_{2^{-j}} (y, z)) \right)}.
$$

(5.33)

For $|\beta_1| \leq \widetilde{N}$ and $|\beta_2| \leq M$, we apply Proposition 3.2.22 to write

$$
\left(2^{-j\wedge k} X \right)^{\beta_1} \left(2^{-j\wedge k} X \right)^{\beta_2} = \sum_{|\gamma_1| + |\gamma_2| \leq |\beta_1| + |\beta_2|} \left(2^{-k} X \right)^{\gamma_1} c_{\gamma_1,\gamma_2}^{\beta_1,\beta_2,j,k} \left(2^{-j} X \right)^{\gamma_2},
$$

where $\left\| c_{\gamma_1,\gamma_2}^{\beta_1,\beta_2,j,k} \right\|_{L^\infty} \lesssim 1$ and the implicit constant depends on M and $\widetilde{N}$ but not on j or k.

Consider,

$$
\left| \left(2^{-j} X \right)^\alpha E\psi (x) \right|
$$

$$
= \sum_{|\beta_1| \leq \widetilde{N}, |\beta_2| \leq M} 2^{-\widetilde{N}|k-j\wedge k|_\infty} 2^{-M|j-j\wedge k|_\infty}
$$

$$
\times \left| \left(2^{-j} X \right)^\alpha F_{\beta_1} \left(2^{-j\wedge k} X \right)^{\beta_1} \left(2^{-j\wedge k} X \right)^{\beta_2} \phi_{\beta_2,j\wedge k} (x) \right|
$$

$$
= \sum_{\substack{|\beta_1| \leq \widetilde{N}, |\beta_2| \leq M \\ |\gamma_1| + |\gamma_2| \leq |\beta_1| + |\beta_2|}} 2^{-\widetilde{N}|k-j\wedge k|_\infty} 2^{-M|j-j\wedge k|_\infty}
$$

$$
\times \left| \left(2^{-j} X \right)^\alpha F_{\beta_1} \left(2^{-k} X \right)^{\gamma_1} c_{\gamma_1,\gamma_2}^{\beta_1,\beta_2,j,k} \left(2^{-j} X \right)^{\gamma_2} \phi_{\beta_2,j\wedge k} (x) \right|.
$$

By taking $M = M(L)$ and $\widetilde{N} = \widetilde{N}(L, |\alpha|)$ large, we have

$$
2^{-\widetilde{N}|k-j\wedge k|_\infty} 2^{-M|j-j\wedge k|_\infty} \left(2^{-j} X \right)^\alpha = c 2^{-L|j-k|} \left(2^{-k} X \right)^\alpha,
$$

where $0 < c \leq 1$ is some constant. We therefore have

$$
\left| \left(2^{-j} X \right)^\alpha E\psi (x) \right| \leq \sum_{\substack{|\beta_1| \leq \widetilde{N}, |\beta_2| \leq M \\ |\gamma_1| + |\gamma_2| \leq |\beta_1| + |\beta_2|}} 2^{-L|j-k|}
$$

$$
\times \left| \left(2^{-k} X \right)^\alpha F_{\beta_1} \left(2^{-k} X \right)^{\gamma_1} c_{\gamma_1,\gamma_2}^{\beta_1,\beta_2,j,k} \left(2^{-j} X \right)^{\gamma_2} \phi_{\beta_2,j\wedge k} (x) \right|.
$$

(5.34)

In light of (5.32), we have for every m

$$
\left| \left[\left(2^{-k} X \right)^\alpha F_{\beta_1} \left(2^{-k} X \right)^{\gamma_1} \right] (x, y) \right| \lesssim \frac{\left(1 + \rho_{2^{-k}} (x, y) \right)^{-m}}{\text{Vol}\left(B_{2^{-k}} (x, 1 + \rho_{2^{-k}} (x, y)) \right)},
$$

and in light of (5.33), we have for every m

$$\left| c_{\gamma_1,\gamma_2}^{\beta_1,\beta_2,j,k}(y)\left(2^{-j}X\right)^{\gamma_2}\phi_{\beta_2,j\wedge k}(x)\right| \lesssim \frac{(1+\rho_{2^{-j}}(y,z))^{-m}}{\mathrm{Vol}\left(B_{2^{-j}}(y,1+\rho_{2^{-j}}(y,z))\right)}.$$

Plugging the above two inequalities into (5.34) and applying Proposition 3.2.19, we have

$$\left|\left(2^{-j}X\right)^{\alpha}E\psi(x)\right| \lesssim 2^{-L|j-k|}\frac{(1+\rho_{2^{-j\wedge k}}(x,z))^{-m}}{\mathrm{Vol}\left(B_{2^{-j\wedge k}}(x,1+\rho_{2^{-j\wedge k}}(x,z))\right)},$$

completing the proof. $\square$

LEMMA 5.5.2. *Let $\mathcal{E}$ be a bounded set of elementary operators. Then,*

$$\left\{\left(E^*,2^{-j}\right)\mid \left(E,2^{-j}\right)\in\mathcal{E}\right\}$$

is a bounded set of elementary operators.

PROOF. Using Proposition 3.2.2, this follows immediately from the definitions. $\square$

THEOREM 5.5.3. *For $s\in\mathbb{R}^{\nu}$, $\mathcal{A}_4^s\subseteq\mathcal{A}_2^s$.*

PROOF. Let $T\in\mathcal{A}_4^s$, so that $T=\sum_{k\in\mathbb{N}^{\nu}}2^{k\cdot s}E_k$ and $\left\{\left(E_k,2^{-k}\right)\mid k\in\mathbb{N}^{\nu}\right\}$ is a bounded set of elementary operators.. Fix m and take $M=M(m,s)$ and $L=L(m,s)$ large. Let $\mathcal{T}_M$ be a bounded set of test functions of order M. Consider, for $\left(\psi,z,2^{-j}\right)\in\mathcal{T}_M$ we have by Lemma 5.5.1,

$$\left|\left(2^{-j}X\right)^{\alpha}T\psi(x)\right| \le \sum_{k\in\mathbb{N}^{\nu}}2^{k\cdot s}\left|\left(2^{-j}X\right)^{\alpha}E_k\psi(x)\right|$$

$$\lesssim \sum_{k\in\mathbb{N}^{\nu}}2^{k\cdot s}2^{-L|k-j|}\frac{(1+\rho_{2^{-j\wedge k}}(x,z))^{-m}}{\mathrm{Vol}\left(B_{2^{-j\wedge k}}(x,1+\rho_{2^{-j\wedge k}}(x,z))\right)}. \qquad (5.35)$$

By taking $L=L(m,s)$ sufficiently large, we have by Lemma 3.2.5

$$2^{k\cdot s}2^{-L|k-j|}\frac{(1+\rho_{2^{-j\wedge k}}(x,z))^{-m}}{\mathrm{Vol}\left(B_{2^{-j\wedge k}}(x,1+\rho_{2^{-j\wedge k}}(x,z))\right)}$$

$$\lesssim 2^{j\cdot s}2^{-|j-k|}\frac{(1+\rho_{2^{-j}}(x,z))^{-m}}{\mathrm{Vol}\left(B_{2^{-j}}(x,1+\rho_{2^{-j}}(x,z))\right)}.$$

Plugging this into (5.35), we have

$$\left|\left(2^{-j}X\right)^{\alpha}T\psi(x)\right| \lesssim \sum_{k\in\mathbb{N}^{\nu}}2^{j\cdot s}2^{-|j-k|}\frac{(1+\rho_{2^{-j}}(x,z))^{-m}}{\mathrm{Vol}\left(B_{2^{-j}}(x,1+\rho_{2^{-j}}(x,z))\right)}$$

$$\lesssim 2^{j\cdot s}\frac{(1+\rho_{2^{-j}}(x,z))^{-m}}{\mathrm{Vol}\left(B_{2^{-j}}(x,1+\rho_{2^{-j}}(x,z))\right)},$$

which is the desired bound.

The same proof works with T replaced by T^*, since (in light of Lemma 5.5.2) $\left\{ \left(E_k^*, 2^{-k} \right) \mid k \in \mathbb{N}^\nu \right\}$ is a bounded set of elementary operators. This completes the proof. $\qquad\square$

THEOREM 5.5.4. *For $s \in \mathbb{R}^\nu$, $\mathcal{A}_2^s \subseteq \mathcal{A}_1^s$.*

PROOF. Let $T \in \mathcal{A}_2^s$. Fix L and m and take $M = M(L, m, s)$ large. Let $\mathcal{T}_M$ be a bounded set of test functions of order M, and let $\left(\psi_1, x, 2^{-j} \right), \left(\psi_2, z, 2^{-k} \right) \in \mathcal{T}_M$. We wish to show

$$|\langle \psi_1, T\psi_2 \rangle| \lesssim 2^{-L|j-k|} 2^{s \cdot j \wedge k} \frac{\left(1 + \rho_{2^{-j \wedge k}} (x, z) \right)^{-m}}{\mathrm{Vol}\left(B_{2^{-j \wedge k}} \left(x, 1 + \rho_{2^{-j \wedge k}} (x, z) \right) \right)}.$$

There are two cases: either $|j - k|_\infty = |j - j \wedge k|_\infty$ or $|j - k|_\infty = |k - j \wedge k|_\infty$. We deal with the former case first, so that $|j - k|_\infty = |j - j \wedge k|_\infty$.

Apply Proposition 3.2.19 to obtain $\widetilde{m} = \widetilde{m}(m)$ so that

$$\int_\Omega \frac{\left(1 + \rho_{2^{-j}} (x, y) \right)^{-\widetilde{m}}}{\mathrm{Vol}\left(B_{2^{-j}} \left(x, 1 + \rho_{2^{-j}} (x, y) \right) \right)} \frac{\left(1 + \rho_{2^{-k}} (y, z) \right)^{-\widetilde{m}}}{\mathrm{Vol}\left(B_{2^{-k}} \left(y, 1 + \rho_{2^{-k}} (y, z) \right) \right)} \, dy$$

$$\leq C \frac{\left(1 + \rho_{2^{-j \wedge k}} (x, z) \right)^{-m}}{\mathrm{Vol}\left(B_{2^{-j \wedge k}} \left(x, 1 + \rho_{2^{-j \wedge k}} (x, z) \right) \right)},$$

$\forall x, z \in \Omega, j, k \in [0, \infty)^\nu$.

We may write

$$\psi_1 = 2^{-M|j-j \wedge k|_\infty} \sum_{|\beta| \leq M} \left(2^{-j \wedge k} X \right)^\beta \phi_{\beta, j \wedge k},$$

where $\left\{ \left(\phi_{\beta, j \wedge k}, x, 2^{-j} \right) \mid \left(\psi_1, x, 2^{-j} \right) \in \mathcal{T}_M, k \in [0, \infty)^\nu, |\beta| \leq M \right\}$ is a bounded set of bump functions. Thus, for every R and m'

$$\sum_{|\alpha| \leq R} \left| \left(2^{-j} X \right)^\alpha \phi_{\beta, j \wedge k} (y) \right| \lesssim \frac{\chi_{\left\{ y \in B_{2^{-j}} (x, 1) \right\}}}{\mathrm{Vol}\left(B_{2^{-j}} (x, 1) \right)}$$

$$\lesssim \frac{\left(1 + \rho_{2^{-j}} (x, y) \right)^{-m'}}{\mathrm{Vol}\left(B_{2^{-j}} \left(x, 1 + \rho_{2^{-j}} (x, y) \right) \right)}. \tag{5.36}$$

We apply Proposition 3.2.22 to write, for $|\beta| \leq M$,

$$\left(2^{-j \wedge k} X \right)^\beta = \sum_{|\beta|_1 + |\beta_2| \leq M} \left(2^{-k} X \right)^{\beta_2} c_{\beta_1, \beta_2}^{\beta, j, k} \left(2^{-j} X \right)^{\beta_1}.$$

For $|\beta_2| \leq M$, let

$$F_{\beta_2, \psi_2, z, 2^{-k}} (y) := 2^{-k \cdot s} \left[\left(2^{-k} X \right)^{\beta_2} \right]^* T\psi_2 (y).$$

By the assumption that $T \in \mathcal{A}_2^s$, we have (if $M = M\,(m)$ is sufficiently large)

$$\left| F_{\beta_2, \psi_2, z, 2^{-k}}\,(y) \right| \lesssim \frac{(1 + \rho_{2^{-k}}\,(y, z))^{-\widetilde{m}}}{\mathrm{Vol}\,(B_{2^{-k}}\,(y, 1 + \rho_{2^{-k}}\,(y, z)))}. \tag{5.37}$$

Combining the above, we have

$$|\langle \psi_1, T\psi_2 \rangle| \leq 2^{-M|j - j \wedge k|_\infty + k \cdot s} \sum_{\substack{|\beta| \leq M \\ |\beta_1| + |\beta_2| \leq |\beta|}} \left| \left\langle c_{\beta_1, \beta_2}^{\beta, j, k} \left(2^{-j} X \right)^{\beta_1} \phi_{\beta, j \wedge k}, F_{\beta_2, \psi_2, z, 2^{-k}} \right\rangle \right|.$$

Plugging (5.36) and (5.37) into the above equation we see

$$|\langle \psi_1, T\psi_2 \rangle|$$
$$\lesssim 2^{-M|j - j \wedge k|_\infty + k \cdot s}$$
$$\times \int_\Omega \frac{(1 + \rho_{2^{-j}}\,(x, y))^{-\widetilde{m}}}{\mathrm{Vol}\,(B_{2^{-j}}\,(x, 1 + \rho_{2^{-j}}\,(x, y)))} \frac{(1 + \rho_{2^{-k}}\,(y, z))^{-\widetilde{m}}}{\mathrm{Vol}\,(B_{2^{-k}}\,(y, 1 + \rho_{2^{-k}}\,(y, z)))}\, dy$$
$$\lesssim 2^{-M|j - j \wedge k|_\infty + k \cdot s} \frac{(1 + \rho_{2^{-j \wedge k}}\,(x, z))^{-m}}{\mathrm{Vol}\,(B_{2^{-j \wedge k}}\,(x, 1 + \rho_{2^{-j \wedge k}}\,(x, z)))},$$

where the last line follows from our choice of $\widetilde{m}$. We are considering the case when $|j - j \wedge k|_\infty = |j - k|_\infty$ and therefore $|j - j \wedge k|_\infty \approx |j - k|$. Taking $M = M\,(L, m, s)$ sufficiently large, we have

$$|\langle \psi_1, T\psi_2 \rangle| \lesssim 2^{-L|j - k| + s \cdot j \wedge k} \frac{(1 + \rho_{2^{-j \wedge k}}\,(x, z))^{-m}}{\mathrm{Vol}\,(B_{2^{-j \wedge k}}\,(x, 1 + \rho_{2^{-j \wedge k}}\,(x, z)))},$$

which is the desired bound.

If $|j - k|_\infty = |k - j \wedge k|_\infty$ then we may reverse the roles of j and k in the above proof by replacing T with T^*. Since the assumptions of $\mathcal{A}_2^s$ are symmetric in T and T^*, this completes the proof. $\qquad\square$

Once we have completed the proof of $\mathcal{A}_1 = \mathcal{A}_2 = \mathcal{A}_3 = \mathcal{A}_4$ in Section 5.6, we will have shown that $\mathcal{A}_1$ and $\mathcal{A}_2$ can be re-written in a way which seems slightly stronger, but is actually equivalent. We present this for $\mathcal{A}_2$, and similar comments hold for $\mathcal{A}_1$.

THEOREM 5.5.5. *Fix* $s \in \mathbb{R}^\nu$. $\forall m$, $\exists M$, $\forall \mathcal{T}_M$ *bounded sets of test functions of order* M, $\forall \alpha$, $\forall T \in \mathcal{A}_2^s$, $\exists C$, $\forall\,(\psi, z, 2^{-j}) \in \mathcal{T}_M$,

$$\left| \left(2^{-j} X \right)^\alpha T\psi\,(x) \right| \leq C 2^{s \cdot j} \frac{(1 + \rho_{2^{-j}}\,(x, z))^{-m}}{\mathrm{Vol}\,(B_{2^{-j}}\,(x, 1 + \rho_{2^{-j}}\,(x, z)))}.$$

Remark 5.5.6 The idea in Theorem 5.5.5 is that M depends on s and m but *not* on the particular $T \in \mathcal{A}_2^s$.

PROOF OF THEOREM 5.5.5. Let $T \in \mathcal{A}_2^s$. Once we have completed the proof of $\mathcal{A}_1^s = \mathcal{A}_2^s = \mathcal{A}_3^s = \mathcal{A}_4^s$, we will have (in particular) $\mathcal{A}_2^s \subseteq \mathcal{A}_4^s$, and so $T \in \mathcal{A}_4^s$ and there is a bounded set of elementary operators $\left\{ \left(E_j, 2^{-j} \right) \mid j \in \mathbb{N}^\nu \right\}$ such that

$$T = \sum_{j \in \mathbb{N}^\nu} 2^{j \cdot s} E_j.$$

The proof of $\mathcal{A}_4^s \subseteq \mathcal{A}_2^s$ picks M depending on s and m but not on the particular bounded set of elementary operators $\left\{ \left(E_j, 2^{-j} \right) \mid j \in \mathbb{N}^\nu \right\}$ which completes the proof. $\qquad \square$

5.6 $\mathcal{A}_1 \subseteq \mathcal{A}_4$

In this section, we prove the final containment: $\mathcal{A}_1^s \subseteq \mathcal{A}_4^s$. To do this, we need some additional information about the elementary operators associated to the pseudodifferential operators studied in Section 5.3. Let $\mathcal{I}$ be a bounded set of pseudodifferential operator data as in Definition 5.3.8. For $\mathcal{D} := \left(\varsigma, 2^{-j}, \kappa, \eta, \psi_1, \psi_2 \right) \in \mathcal{I}$, we define $E \left(\mathcal{D} \right)$ as in (5.19). I.e.,

$$E \left(\mathcal{D} \right) f \left(x \right) = \psi_1 \left(x \right) \int f \left(e^{t \cdot X} x \right) \psi_2 \left(e^{t \cdot X} x \right) \kappa \left(x, t \right) \eta \left(t \right) \varsigma^{\left(2^j \right)} \left(t \right) \, dt. \qquad (5.38)$$

In light of Proposition 5.3.9, $\left\{ \left(E \left(\mathcal{D} \right), 2^{-j} \right) \mid \mathcal{D} = \left(\varsigma, 2^{-j}, \kappa, \eta, \psi_1, \psi_2 \right) \in \mathcal{I} \right\}$ is a bounded set of elementary operators. For a finite set $\mathcal{F} \subset [0, \infty)^\nu$, we write

$$\operatorname{diam} \left\{ \mathcal{F} \right\} := \max_{j, k \in \mathcal{F}} |j - k|.$$

PROPOSITION 5.6.1. *Fix $s \in \mathbb{R}^\nu$ and $T \in \mathcal{A}_1^s$. Let $\mathcal{I}$ be a bounded set of pseudodifferential operator data. For $l = 1, 2, 3, 4$, let $\mathcal{D}_l = \left(\varsigma_l, 2^{-j_l}, \kappa_l, \eta_l, \psi_{l,1}, \psi_{l,1} \right) \in \mathcal{I}$. We define*

$$E \left(\mathcal{D}_1, \mathcal{D}_2, \mathcal{D}_3, \mathcal{D}_4 \right) := 2^{-s \cdot j_1 \wedge j_2 \wedge j_3 \wedge j_4} E \left(\mathcal{D}_1 \right) E \left(\mathcal{D}_2 \right) T E \left(\mathcal{D}_3 \right) E \left(\mathcal{D}_4 \right).$$

We have, for every N,

$$\left\{ \left(2^{N \operatorname{diam} \{ j_1, j_2, j_3, j_4 \}} E \left(\mathcal{D}_1, \mathcal{D}_2, \mathcal{D}_3, \mathcal{D}_4 \right), 2^{-j_1 \wedge j_2 \wedge j_3 \wedge j_4} \right) \mid \mathcal{D}_1, \mathcal{D}_2, \mathcal{D}_3, \mathcal{D}_4 \in \mathcal{I} \right\}$$

is a bounded set of elementary operators.

THEOREM 5.6.2. *For $s \in \mathbb{R}^\nu$, $\mathcal{A}_1^s \subseteq \mathcal{A}_4^s$.*

PROOF OF THEOREM 5.6.2 GIVEN PROPOSITION 5.6.1. Fix $T \in \mathcal{A}_1^s$. Using that $\operatorname{supp} \left(T \right) \Subset \Omega \times \Omega$, we may take $\psi \in C_0^\infty \left(\Omega \right)$ such that $\psi \otimes \psi \equiv 1$ on a neighborhood of $\operatorname{supp} \left(T \right)$. We identify ψ with the operator given by multiplication by ψ, and we have $\psi^2 T \psi^2 = T$. As we saw in the proof of Theorem 5.3.1, there exists a bounded set of pseudodifferential operator data $\{\mathcal{D}_j \mid j \in \mathbb{N}^\nu\}$ such that $\psi = \sum_{j \in \mathbb{N}^\nu} E(\mathcal{D}_j)$. We therefore have

$$T = \psi^2 T \psi^2$$

$$= \sum_{j_1, j_2, j_3, j_4 \in \mathbb{N}^\nu} E(\mathcal{D}_{j_1}) E(\mathcal{D}_{j_2}) T E(\mathcal{D}_{j_3}) E(\mathcal{D}_{j_4})$$

$$= \sum_{j_1, j_2, j_3, j_4 \in \mathbb{N}^\nu} 2^{-\operatorname{diam}\{j_1, j_2, j_3, j_4\}} 2^{s \cdot j_1 \wedge j_2 \wedge j_3 \wedge j_4}$$

$$\times \left(2^{\operatorname{diam}\{j_1, j_2, j_3, j_4\}} 2^{-s \cdot j_1 \wedge j_2 \wedge j_3 \wedge j_4} E(\mathcal{D}_1) E(\mathcal{D}_2) T E(\mathcal{D}_3) E(\mathcal{D}_4) \right)$$

$$=: \sum_{j_1, j_2, j_3, j_4 \in \mathbb{N}^\nu} 2^{-\operatorname{diam}\{j_1, j_2, j_3, j_4\}} 2^{s \cdot j_1 \wedge j_2 \wedge j_3 \wedge j_4} \widetilde{E}_{j_1, j_2, j_3, j_4},$$

where, by Proposition 5.6.1, $\left\{ \left(\widetilde{E}_{j_1, j_2, j_3, j_4}, 2^{-j_1 \wedge j_2 \wedge j_3 \wedge j_4} \right) \mid j_1, j_2, j_3, j_4 \in \mathbb{N}^\nu \right\}$ is a bounded set of elementary operators (here we have taken $N = 1$ in that proposition). For $k \in \mathbb{N}^\nu$ define

$$\widetilde{E}_k := \sum_{\substack{j_1 \wedge j_2 \wedge j_3 \wedge j_4 = k \\ j_1, j_2, j_3, j_4 \in \mathbb{N}^\nu}} 2^{-\operatorname{diam}\{j_1, j_2, j_3, j_4\}} \widetilde{E}_{j_1, j_2, j_3, j_4},$$

so that $\left\{ \left(\widetilde{E}_k, 2^{-k} \right) \mid k \in \mathbb{N}^\nu \right\}$ is a bounded set of elementary operators. We have $T = \sum_{k \in \mathbb{N}^\nu} 2^{s \cdot k} \widetilde{E}_k$, completing the proof that $T \in \mathcal{A}_4^s$. $\qquad\square$

The rest of this section is devoted to the proof of Proposition 5.6.1. In fact, we prove the following more general version.

PROPOSITION 5.6.3. *Fix $s \in \mathbb{R}^\nu$ and $T \in \mathcal{A}_1^s$. Let $\mathcal{I}_1$ be a bounded set of generalized pseudodifferential operator data and let $\mathcal{I}_2$ be a bounded set of pseudodifferential operator data. For each*

$$\mathcal{D}_1 = \left(\varsigma_1, 2^{-j_1}, 2^{-k_1}, \kappa_1, \eta_1, \psi_{1,1}, \psi_{1,2} \right) \in \mathcal{I}_1,$$

$$\mathcal{D}_4 = \left(\varsigma_4, 2^{-j_4}, 2^{-k_4}, \kappa_4, \eta_4, \psi_{4,1}, \psi_{4,2} \right) \in \mathcal{I}_1,$$

and each

$$\mathcal{D}_2 = \left(\varsigma_2, 2^{-j_2}, \kappa_2, \eta_2, \psi_{2,1}, \psi_{2,2} \right), \mathcal{D}_3 = \left(\varsigma_3, 2^{-j_3}, \kappa_3, \eta_3, \psi_{3,1}, \psi_{3,2} \right) \in \mathcal{I}_2,$$

define $E(\mathcal{D}_l)$, $l = 1, 2, 3, 4$, as in (5.38). We restrict attention to the case when $k_1 \vee k_4 \le j_1 \wedge j_2 \wedge j_3 \wedge j_4$. Let

$$E_{j_1, j_2, j_3, j_4} := 2^{-s \cdot j_1 \wedge j_2 \wedge j_3 \wedge j_4} E(\mathcal{D}_1) E(\mathcal{D}_2) T E(\mathcal{D}_3) E(\mathcal{D}_4).$$

Then, for every N,

$$\left\{ \left(2^{N \operatorname{diam}\{j_1, j_2, j_3, j_4\}} E_{j_1, j_2, j_3, j_4}, 2^{-j_1 \wedge j_2 \wedge j_3 \wedge j_4}, 2^{-k_1 \vee k_4} \right) \mid \right.$$

$$\left. \mathcal{D}_1, \mathcal{D}_4 \in \mathcal{I}_1, \mathcal{D}_2, \mathcal{D}_3 \in \mathcal{I}_2, k_1 \vee k_4 \le j_1 \wedge j_2 \wedge j_3 \wedge j_4 \right\} \tag{5.39}$$

is a bounded set of generalized elementary operators.

Remark 5.6.4 Recall, the definition of bounded sets of generalized pseudodifferential operator data used a small number $a > 0$. As before, we are allowed to shrink this number, so long as how large it is depends only on (X, d) and Ω, not on any particular operator we are considering.

Notice that Proposition 5.6.1 follows directly from Proposition 5.6.3 by taking $k_1 = k_4 = 0$. Thus, we focus on proving Proposition 5.6.3. To do this, we introduce a new kind of elementary operator which will only be used in this section.

DEFINITION 5.6.5. *For $M \geq 0$ we say $\mathcal{E} \subset C_0^\infty (\Omega \times \Omega) \times (0, 1]^\nu \times (0, 1]^\nu$ is a bounded set of generalized left elementary operators of type 2 of depth M if*

- $$\bigcup_{(E, 2^{-j}, 2^{-k}) \in \mathcal{E}} \operatorname{supp}(E) \Subset \Omega \times \Omega.$$

- $\forall (E, 2^{-j}, 2^{-k}) \in \mathcal{E}$, $\operatorname{supp}(E) \subset \{(x, z) \mid z \in B_{2^{-j}}(x, 1)\}$.

- $\forall L, \exists C_L, \forall (E, 2^{-j}, 2^{-k}) \in \mathcal{E}$,
$$\sum_{|\alpha|, |\beta| \leq L} \left| (2^{-j} X_x)^\alpha (2^{-j} X_z)^\beta E(x, z) \right| \leq C_L \operatorname{Vol}(B_{2^{-j}}(x, 1))^{-1}.$$

- *If $M \geq 1$, then $\forall (E, 2^{-j}, 2^{-k}) \in \mathcal{E}$, $\forall k \leq l \leq j$,*
$$E = 2^{-|j-l|_\infty} \sum_{|\alpha| \leq 1} (2^{-l} X)^\alpha E_{\alpha, l}, \tag{5.40}$$

where $\{(E_{\alpha,l}, 2^{-j}, 2^{-l}) \mid (E, 2^{-j}, 2^{-l}) \in \mathcal{E}, k \leq l \leq j, |\alpha| \leq 1\}$ is a bounded set of generalized left elementary operators of type 2 of depth $M - 1$.

If we instead say $\mathcal{E}$ is a bounded set of generalized right elementary operators of type 2 of depth M, we replace (5.40) with
$$E = 2^{-|j-l|_\infty} \sum_{|\alpha| \leq 1} E_{\alpha, l} (2^{-l} X)^\alpha.$$

DEFINITION 5.6.6. *We say $\mathcal{E} \subset C_0^\infty (\Omega \times \Omega) \times (0, 1]^\nu$ is a bounded set of left (respectively, right) elementary operators of type 2 of depth M if $\mathcal{E} \times \{2^{-0}\}$ is a bounded set of generalized left (respectively, right) elementary operators of type 2 of depth M.*

Remark 5.6.7 If $\mathcal{E}$ is a bounded set of generalized left elementary operators of type 2 of depth M, then so is
$$\left\{ \left(E (2^{-j} X)^\alpha, 2^{-j}, 2^{-k} \right) \mid (E, 2^{-j}, 2^{-k}) \in \mathcal{E} \right\},$$

for every ordered multi-index α; a similar result holds for bounded sets of generalized right elementary operators of type 2 of depth M if we replace $E\left(2^{-j}X\right)^{\alpha}$ with $\left(2^{-j}X\right)^{\alpha}E$.

LEMMA 5.6.8. *If $\mathcal{E}$ is a bounded set of right elementary operators of type 2 of depth M, then $\{\left(E\left(x,\cdot\right),x,2^{-j}\right) \mid x \in \Omega, \left(E,2^{-j}\right) \in \mathcal{E}\}$ is a bounded set of test functions of order M. If, instead, $\mathcal{E}$ is a bounded set of left elementary operators of type 2 of depth M, then $\{\left(E\left(\cdot,x\right),x,2^{-j}\right) \mid x \in \Omega, \left(E,2^{-j}\right) \in \mathcal{E}\}$ is a bounded set of test functions of order M.*

PROOF. This is immediate from the definitions.   □

PROPOSITION 5.6.9. *Suppose $T \in \mathcal{A}_1^s$. Fix α,β ordered multi-indices and fix $N,m \in \mathbb{N}$. There exists M such that if $\mathcal{E}_L$ and $\mathcal{E}_R$ are bounded sets of left and right (respectively) elementary operators of type 2 of depth M, we have $\exists C, \forall\left(E_1,2^{-j_1}\right) \in \mathcal{E}_R, \left(E_2,2^{-j_2}\right) \in \mathcal{E}_L,$*

$$\left|\left(2^{-j_1\wedge j_2}X_x\right)^{\alpha}\left(2^{-j_1\wedge j_2}X_z\right)^{\beta}\left[E_1TE_2\right](x,z)\right|$$
$$\leq C2^{-N|j_1-j_2|}2^{s\cdot j_1\wedge j_2}\frac{\left(1+\rho_{2^{-j_1\wedge j_2}}(x,z)\right)^{-m}}{\mathrm{Vol}\left(B_{2^{-j_1\wedge j_2}}\left(x,1+\rho_{2^{-j_1\wedge j_2}}(x,z)\right)\right)}.$$

PROOF. Notice

$$\left(2^{-j_1\wedge j_2}X\right)^{\alpha}=2^{c_1(\alpha)|j_1-j_2|}\gamma_1\left(\alpha\right)\left(2^{-j_1}X\right)^{\alpha},$$

where $\gamma_1\left(\alpha\right)$ is a constant with $0<\gamma_1\left(\alpha\right)\leq 1$; and similarly,

$$\left(2^{-j_1\wedge j_2}X\right)^{\beta}=2^{c_2(\beta)|j_1-j_2|}\gamma_2\left(\beta\right)\left(2^{-j_2}X\right)^{\beta},$$

with $0<\gamma_2\left(\beta\right)\leq 1$. Thus, by changing N, it suffices to instead show

$$\left|\left(2^{-j_1}X_x\right)^{\alpha}\left(2^{-j_2}X_z\right)^{\beta}\left[E_1TE_2\right](x,z)\right|$$
$$\leq C2^{-N|j_1-j_2|}2^{s\cdot j_1\wedge j_2}\frac{\left(1+\rho_{2^{-j_1\wedge j_2}}(x,z)\right)^{-m}}{\mathrm{Vol}\left(B_{2^{-j_1\wedge j_2}}\left(x,1+\rho_{2^{-j_1\wedge j_2}}(x,z)\right)\right)}. \tag{5.41}$$

We have

$$\left(2^{-j_1}X_x\right)^{\alpha}\left(2^{-j_2}X_z\right)^{\beta}\left[E_1TE_2\right](x,z)=\left[\left(2^{-j_1}X\right)^{\alpha}E_1TE_2\left(\left(2^{-j_2}X\right)^{\beta}\right)^*\right].$$

By Remark 5.6.7, we have

$$\mathcal{E}_R':=\left\{\left(\left(2^{-j_x}X\right)^{\alpha}E_1,2^{-j_1}\right) \mid \left(E_2,2^{-j_1}\right) \in \mathcal{E}_R\right\}$$

is a bounded set of right elementary operators of type 2 of depth M, and

$$\mathcal{E}_L':=\left\{\left(E_2\left(\left(2^{-j_2}X\right)^{\beta}\right)^*,2^{-j_2}\right) \mid \left(E_2,2^{-j_2}\right) \in \mathcal{E}_L\right\}$$

is a bounded set of left elementary operators of type 2 of depth M. Thus, by replacing $\mathcal{E}_L$ and $\mathcal{E}_R$ with $\mathcal{E}'_L$ and $\mathcal{E}'_R$ it suffices to consider only the case $|\alpha| = |\beta| = 0$ in (5.41). I.e., we wish to show if M is sufficiently large (depending on N and m), for $\left(E_1, 2^{-j_1}\right) \in \mathcal{E}'_R$ and $\left(E_2, 2^{-j_2}\right) \in \mathcal{E}'_L$,

$$\left|[E_1 T E_2](x, z)\right|$$

$$\lesssim 2^{-N|j_1 - j_2|} 2^{s \cdot j_1 \wedge j_2} \frac{\left(1 + \rho_{2^{-j_1 \wedge j_2}}(x, z)\right)^{-m}}{\mathrm{Vol}\left(B_{2^{-j_1 \wedge j_2}}\left(x, 1 + \rho_{2^{-j_1 \wedge j_2}}(x, z)\right)\right)}. \tag{5.42}$$

But,

$$[E_1 T E_2](x, z) = \left\langle \overline{E_1(x, \cdot)}, T E_2(\cdot, z) \right\rangle. \tag{5.43}$$

Lemma 5.6.8 shows that

$$\left\{ \left(E_1(x, \cdot), x, 2^{-j_1}\right) \mid \left(E_1, 2^{-j_1}\right) \in \mathcal{E}'_R, x \in \Omega \right\}$$

and

$$\left\{ \left(E_2(\cdot, z), z, 2^{-j_2}\right) \mid \left(E_2, 2^{-j_2}\right) \in \mathcal{E}'_R, z \in \Omega \right\}$$

are bounded sets of test functions of order M. Combining this with (5.43), (5.42) follows from the definition of $\mathcal{A}^s_1$. □

Proposition 5.6.3 involves the action of T on generalized elementary operators, not generalized elementary operators of type 2 which are the sort of operators covered in Proposition 5.6.9. To be able to apply Proposition 5.6.9, we decompose the operators appearing in Proposition 5.6.3 by using Proposition 5.2.17.

PROPOSITION 5.6.10. *Let $\mathcal{I}_1$ be a bounded set of generalized pseudodifferential operator data and let $\mathcal{I}_2$ be a bounded set of pseudodifferential operator data. For each*

$$\mathcal{D}_1 = \left(\varsigma_1, 2^{-j_1}, 2^{-k_0}, \kappa_1, \eta_1, \psi_{1,1}, \psi_{1,2}\right) \in \mathcal{I}_1,$$

$$\mathcal{D}_2 = \left(\varsigma_2, 2^{-j_2}, \kappa_2, \eta_2, \psi_{2,1}, \psi_{2,2}\right) \in \mathcal{I}_2,$$

with $k_0 \leq j_1 \wedge j_2$, define $E_1 = E_1(\mathcal{D}_1)$ and $E_2 = E_2(\mathcal{D}_2)$ as in (5.38). Then, for every M, we may decompose

$$E_1 E_2 = \sum_{\substack{j \leq j_1 \wedge j_2 \\ j \in \mathbb{N}^\nu}} \widetilde{E}_j,$$

where

$$\left\{ \left(2^{M(|j_1 - j| + |j_2 - j|)} \widetilde{E}_j, 2^j\right) \mid \mathcal{D}_1 \in \mathcal{I}_1, \mathcal{D}_2 \in \mathcal{I}_2, k \leq j_1 \wedge j_2, k \in \mathbb{N}^\nu \right\}$$

is a bounded set of right elementary operators of type 2 of depth M (here, $\widetilde{E}_j$ depends on M). A similar decomposition holds for $E_2 E_1$, where one obtains left elementary operators of type 2 of depth M, instead of right elementary operators of type 2 of depth M.

We turn to the proof of Proposition 5.6.10, which requires a preliminary definition and lemma. We prove only the result for $E_1 E_2$–the result for $E_2 E_1$ follows by a simple reprise, or by taking adjoints of the results for $E_1 E_2$. Consider the operators E_1 and E_2 in Proposition 5.6.10. Notice we may, without loss of generality, take $k_0 = j_1 \wedge j_2$. Fix M as in Proposition 5.6.10, and take $M' = M'(M)$ large to be chosen later. We apply Proposition 5.2.17 with M replaced by M' to $\eta_1 \varsigma_1^{(2^{j_1})}$ and $\eta_2 \varsigma_2^{(2^{j_2})}$ to obtain

$$\{\gamma_{1,k,\alpha} \mid k \le j_1, k \in \mathbb{N}^\nu, |\alpha_\mu| = M' \text{ when } k_\mu \ne 0, |\alpha_\mu| = 0 \text{ when } k_\mu = 0\}$$
$$\subset C_0^\infty (B^q(a))$$

and

$$\{\gamma_{2,k,\alpha} \mid k \le j_2, k \in \mathbb{N}^\nu, |\alpha_\mu| = M' \text{ when } k_\mu \ne 0, |\alpha_\mu| = 0 \text{ when } k_\mu = 0\}$$
$$\subset C_0^\infty (B^q(a))$$

such that for every N, the following sets are bounded

$$\left\{ 2^{N|j_1 - k|} \gamma_{1,k,\alpha} \,\middle|\, \mathcal{D}_1 \in \mathcal{I}_1, k \le j_1, k \in \mathbb{N}^\nu, |\alpha_\mu| = M' \text{ when } k_\mu \ne 0, \right.$$
$$\left. |\alpha_\mu| = 0 \text{ when } k_\mu = 0 \right\} \subset C_0^\infty (B^q(a)),$$

$$\left\{ 2^{N|j_2 - k|} \gamma_{2,k,\alpha} \,\middle|\, \mathcal{D}_2 \in \mathcal{I}_2, k \le j_2, k \in \mathbb{N}^\nu, |\alpha_\mu| = M' \text{ when } k_\mu \ne 0, \right.$$
$$\left. |\alpha_\mu| = 0 \text{ when } k_\mu = 0 \right\} \subset C_0^\infty (B^q(a)),$$

and if we set (for $l = 1, 2$)

$$\varsigma_{l,k} := \sum_{\substack{\alpha \in \mathbb{N}^q \\ |\alpha_\mu| = M \text{ when } k_\mu \ne 0 \\ |\alpha_\mu| = 0 \text{ when } k_\mu = 0}} \partial_t^\alpha \gamma_{l,k,\alpha},$$

then we have

$$\eta_l(t) \varsigma_l^{(2^{j_l})}(t) = \sum_{\substack{k \le j_l \\ k \in \mathbb{N}^\nu}} \eta_l(t) \varsigma_{l,k}^{(2^k)}(t).$$

This decomposition of $\eta_1 \varsigma_1^{(2^{j_1})}$ and $\eta_2 \varsigma_2^{(2^{j_2})}$ gives rise to a decomposition of $E_1 E_2$, namely,

$$E_1 E_2 f(x) = \sum_{\substack{k_1 \le j_1 \\ k_2 \le j_2}} \psi_{1,1}(x) \int f\left(e^{t_2 \cdot X} e^{t_1 \cdot X} x\right) \psi_{2,2}\left(e^{t_2 \cdot X} e^{t_1 \cdot X} x\right) \kappa(x, t_1, t_2)$$

$$\times \kappa_1(x, t_1) \varsigma_{1,k_1}^{(2^{k_1})}(t_1) \varsigma_{2,k_2}^{(2^{k_2})}(t_2) \, dt_1 dt_2,$$

$$(5.44)$$

where

$$\kappa\left(x, t_{1}, t_{2}\right)=\kappa_{2}\left(e^{t_{1}} x, t_{2}\right) \eta_{1}\left(t_{1}\right) \eta_{2}\left(t_{2}\right) \psi_{1,2}\left(e^{t_{1} \cdot X} x\right) \psi_{2,1}\left(e^{t_{1} \cdot X} x\right)$$
$$\in C^{\infty}\left(\Omega \times B^{q}(a) \times B^{q}(a)\right).$$

Here, we are careful to not include κ_{1} in κ, as it satisfies less regularity. (5.44) motivates the following, ad hoc definition.

DEFINITION 5.6.11. *We say*

$$\mathcal{I} \subset C_{0}^{\infty}\left(B^{q}(a)\right) \times C_{0}^{\infty}\left(B^{q}(a)\right) \times C^{\infty}\left(\Omega \times B^{q}(a)\right) \times C^{\infty}\left(\Omega \times B^{q}(a) \times B^{q}(a)\right)$$
$$\times C_{0}^{\infty}(\Omega) \times C_{0}^{\infty}(\Omega) \times(0,1]^{\nu} \times(0,1]^{\nu} \times(0,1]^{\nu} \times(0,1]^{\nu} \times(0,1]^{\nu}$$

is a bounded set of generalized right pseudodifferential operator data of type 2 of depth M' if the following holds.

(i) $\forall\left(\varsigma_{1}, \varsigma_{2}, \kappa_{1}, \kappa, \psi_{1}, \psi_{2}, 2^{-l}, 2^{-k_{1}}, 2^{-k_{2}}, 2^{-j_{1}}, 2^{-j_{2}}\right) \in \mathcal{I}, l \leq k_{1} \wedge k_{1}, k_{1} \leq j_{1}, k_{2} \leq j_{2}.$

(ii) *For $\alpha \in \mathbb{N}^{q}$, we write $\alpha=\left(\alpha_{1}, \ldots, \alpha_{\nu}\right) \in \mathbb{N}^{q_{1}} \times \cdots \times \mathbb{N}^{q_{\nu}}$. We assume, for $m=1,2$, and $\forall\left(\varsigma_{1}, \varsigma_{2}, \kappa_{1}, \kappa, \psi_{1}, \psi_{2}, 2^{-l}, 2^{-k_{1}}, 2^{-k_{2}}, 2^{-j_{1}}, 2^{-j_{2}}\right) \in \mathcal{I},$*

$$\varsigma_{m}=\sum_{\substack{\alpha \in \mathbb{N}^{q} \\ |\alpha_{\mu}|=M' \text{ if } k_{\mu} \neq 0 \\ \alpha_{\mu}=0 \text{ otherwise}}} \partial_{t}^{\alpha} \gamma_{m, \alpha},$$

where, for every N,

$$\left\{2^{N\left|j_{m}-k_{m}\right|} \gamma_{m, \alpha} \mid\left(\varsigma_{1}, \varsigma_{2}, \kappa_{1}, \kappa, \psi_{1}, \psi_{2}, 2^{-l}, 2^{-k_{1}}, 2^{-k_{2}}, 2^{-j_{1}}, 2^{-j_{2}}\right) \in \mathcal{I}\right\}$$
$$\subset C_{0}^{\infty}\left(B^{q}(a)\right)$$

is a bounded set, for $m=1,2$.

(iii) $\left\{\psi_{1}, \psi_{2} \mid\left(\varsigma_{1}, \varsigma_{2}, \kappa_{1}, \kappa, \psi_{1}, \psi_{2}, 2^{-l}, 2^{-k_{1}}, 2^{-k_{2}}, 2^{-j_{1}}, 2^{-j_{2}}\right) \in \mathcal{I}\right\} \subset C_{0}^{\infty}(\Omega)$ is *a bounded set.*

(iv) $\forall L, \exists C, N, \forall\left(\varsigma_{1}, \varsigma_{2}, \kappa_{1}, \kappa, \psi_{1}, \psi_{2}, 2^{-l}, 2^{-k_{1}}, 2^{-k_{2}}, 2^{-j_{1}}, 2^{-j_{2}}\right) \in \mathcal{I}, \forall x \in \Omega,$
$\forall t_{1}, t_{2} \in B^{q}(a),$

$$\sum_{|\alpha|,\left|\beta_{1}\right|,\left|\beta_{2}\right| \leq L}\left|\left(2^{-l} X_{x}\right)^{\alpha}\left(2^{-l} \partial_{t_{1}}\right)^{\beta_{1}}\left(2^{-l} \partial_{t_{2}}\right)^{\beta_{2}}\right.$$
$$\left.\left[\left(1+\left|2^{l} t_{1}\right|^{2}+\left|2^{l} t_{2}\right|^{2}\right)^{-N} \kappa\left(x, t_{1}, t_{2}\right)\right]\right| \leq C,$$

where $2^{-l} \partial_{t}$ and $2^{l} t$ are defined as in the beginning of Section 5.2.

(v) $\forall L, \exists C, N, \forall \left(\varsigma_1, \varsigma_2, \kappa_1, \kappa, \psi_1, \psi_2, 2^{-l}, 2^{-k_1}, 2^{-k_2}, 2^{-j_1}, 2^{-j_2}\right) \in \mathcal{I}, \forall x \in \Omega,$
$\forall t \in B^q(a),$

$$\sum_{|\alpha|, |\beta| \leq L} \left| \left(2^{-j_1 \wedge j_2} X_x\right)^\alpha \left(2^{-j_1 \wedge j_2} \partial_t\right)^\beta \left(1 + |2^{j_1 \wedge j_2} t|^2\right)^{-N} \kappa_1(x,t) \right| \leq C.$$

LEMMA 5.6.12. *Fix M. There exists $M' = M'(M)$ such that the following holds. Let $\mathcal{I}$ be a bounded set of generalized right pseudodifferential operator data of type 2 of order M'. For each*

$$\mathcal{D} := \left(\varsigma_1, \varsigma_2, \kappa_1, \kappa, \psi_1, \psi_2, 2^{-l}, 2^{-k_1}, 2^{-k_2}, 2^{-j_1}, 2^{-j_2}\right) \in \mathcal{I},$$

define the operator $E(\mathcal{D})$ by

$$E(\mathcal{D}) = \psi_1(x) \int f\left(e^{t_2 \cdot X} e^{t_1 \cdot X} x\right) \psi_2\left(e^{t_2 \cdot X} e^{t_1 \cdot X} x\right) \kappa(x, t_1, t_2) \kappa_1(x, t_1)$$
$$\times \varsigma_1^{(2^{k_1})}(t_1) \varsigma_2^{(2^{k_2})}(t_2) \, dt_1 dt_2.$$

Then,

$$\left\{ \left(2^{M(|j_1 - k_1| + |j_2 - k_2| + |k_1 - k_2|)} E(\mathcal{D}), 2^{-k_1 \wedge k_2}, 2^{-l}\right) \mid \mathcal{D} \in \mathcal{I} \right\}$$

is a bounded set of generalized right elementary operators of type 2 of depth M.

PROOF. In light of (ii) of Definition 5.6.11, we have

$$\{\varsigma_1, \varsigma_2 \mid \mathcal{D} \in \mathcal{I}\} \subset C_0^\infty(B^q(a))$$

is a bounded set. It follows (by taking $a > 0$ sufficiently small), that

$$\mathrm{supp}\,(E(\mathcal{D})) \subseteq \left\{(x,z) \mid z \in B_{2^{-k_1 \wedge k_2}}(x,1)\right\}.$$

By the support of ψ_1 and ψ_2 it is clear that

$$\bigcup_{\mathcal{D} \in \mathcal{I}} \mathrm{supp}\,(E(\mathcal{D})) \Subset \Omega \times \Omega.$$

For notational simplicity, we consider only the special case of (ii) when $\varsigma_1 = \partial_t^{\alpha_1} \gamma_1$ and $\varsigma_2 = \partial_t^{\alpha_2} \gamma_2$, where $|\alpha_{1,\mu}| = M'$ if $k_{1,\mu} \neq 0$ and $|\alpha_{2,\mu}| = M'$ if $k_{2,\mu} \neq 0$. In general, ς_1 and ς_2 are actually sums of such terms, so it suffices to consider only this case.

Fix $l \leq l' \leq k_1 \wedge k_2$. Our first goal is to verify

$$E(\mathcal{D}) = 2^{-|k_1 \wedge k_2 - l'|_\infty} \sum_{|\beta| \leq 1} E_{\beta, l'} \left(2^{-l'} X\right)^\beta, \tag{5.45}$$

where $E_{\alpha, l'}$ is a sum of terms of the same form as $E(\mathcal{D})$ but with l replaced by l' and M' replaced by $M' - 1$ (in this, we are considering only the case $M' \geq 1$).

(5.45) is trivial if $|k_1 \wedge k_2 - l'|_\infty = 0$, and so we assume $|k_1 \wedge k_2 - l'|_\infty > 0$. Let $k = k_1 \wedge k_2$, and pick μ so that $k_\mu - l'_\mu = |k - l'|_\infty$. Notice that $k_{2,\mu} - l'_\mu \geq |k - l'|_\infty > 0$ and it follows that $k_{2,\mu} > 0$. Since $|a_{2,\mu}| = M' \geq 1$, there exists a multi-index $\delta \in \mathbb{N}^{q_\mu}$ with $|\delta| = 1$ so that $\partial_{t_2}^{\alpha_2} = \partial_{t_2}^{\delta} \partial_{t_2}^{\alpha'_2}$. Integrating by parts, we see

$$E(\mathcal{D}) f(x) = -2^{(l'_\mu - k_{2,\mu})\delta \cdot \hat{d}^\mu} \psi_1(x) \int \left[\left(2^{-l'_\mu} \partial_{t_2^\mu}\right)^\delta F(t_1, t_2, x) \right] \kappa_1(t_1, x)$$
$$\varsigma_1^{(2^{k_1})}(t_1) \left(\partial_{t_2}^{\alpha'_2} \gamma_2\right)^{(2^{k_2})}(t_2) \, dt_1 dt_2,$$

where

$$F(t_1, t_2, x) = f\left(e^{t_2 \cdot X} e^{t_1 \cdot X} x\right) \psi_2\left(e^{t_2 \cdot X} e^{t_1 \cdot X} x\right) \kappa(x, t_1, t_2).$$

We compute $\left(2^{-l'_\mu} \partial_{t_2^\mu}\right)^\delta F(t_1, t_2, x)$ by the product rule. When $\left(2^{-l'_\mu} \partial_{t_2^\mu}\right)^\delta$ lands on $\psi_2\left(e^{t_2 \cdot X} e^{t_1 \cdot X} x\right) \kappa(x, t_1, t_2)$, we obtain a sum of terms of the same form, therefore obtain operators of the form (5.45) with, in fact, $|\beta| = 0$–this uses the fact that $\left(k_{2,\mu} - l'_\mu\right) \left(\delta \cdot \hat{d}^\mu\right) \geq |k - l'|_\infty$. All that remains to complete the proof of (5.45) is to consider the case when $\left(2^{-l'} \partial_{t_2^\mu}\right)^\delta$ lands on $f\left(e^{t_2 \cdot X} e^{t_1 \cdot X} x\right)$.

By Proposition 3.2.10, we have

$$\left(2^{-l'} \partial_{t_2^\mu}\right)^\delta f\left(e^{t_2 \cdot X} e^{t_1 \cdot X} x\right) = \sum_{r=1}^q c_r^{l'}(x, t_1, t_2) \left(2^{-l' \cdot d_r} X_r f\right) \left(e^{t_2 \cdot X} e^{t_1 \cdot X} x\right),$$

where, for every L,

$$\sum_{|\alpha|, |\beta_1|, |\beta_2| \leq L} \left| \left(2^{-l'} \partial_{t_1}\right)^{\beta_1} \left(2^{-l'} \partial_{t_2}\right)^{\beta_2} \left(2^{-l'} X\right)^\alpha \right.$$
$$\left. \left[\left(1 + \left|2^{l'} t_1\right|^2 + \left|2^{l'} t_2\right|^2\right)^{-1} c_r^{l'}(x, t_1, t_2) \right] \right| \lesssim 1.$$

From here, it follows immediately that $c_r^{l'} \kappa$ satisfies all the hypotheses of κ with l replaced by l' and N replaced by $N + 1$. (5.45) follows, again using the fact that $\left(k_{2,\mu} - l'_\mu\right) \left(\delta \cdot \hat{d}^\mu\right) \geq |k - l'|_\infty$.

Thus, to complete the proof of the lemma, we need to show, for every L,

$$\sum_{|\beta_1|, |\beta_2| \leq L} \left| \left(2^{-k} X_x\right)^{\beta_1} \left(2^{-k} X_z\right)^{\beta_2} E(\mathcal{D})(x, z) \right| \lesssim \frac{2^{-M(|j_1 - k_1| + |j_2 - k_2| + |k_1 - k_2|)}}{\mathrm{Vol}(B_{2^{-k}}(x, 1))}.$$

By our assumptions, we may replace ς_1 and ς_2 with $2^{M|j_1 - k_1|} \varsigma_1$ and $2^{M|j_2 - k_2|} \varsigma_2$, respectively, and obtain functions of the same form. Thus, it suffices to show

$$\sum_{|\beta_1|, |\beta_2| \leq L} \left| \left(2^{-k} X_x\right)^{\beta_1} \left(2^{-k} X_z\right)^{\beta_2} E(\mathcal{D})(x, z) \right| \lesssim \frac{2^{-M|k_1 - k_2|}}{\mathrm{Vol}(B_{2^{-k}}(x, 1))}. \qquad (5.46)$$

Fix L for which we wish to show (5.46). Take $M'' = M''(M', L)$ large to be chosen later. Set

$$\widetilde{\kappa}(x, t_1, t_2) := \psi_2\left(e^{t_2 \cdot X} e^{t_1 \cdot X} x\right) \kappa_1(x, t_1) \kappa(x, t_1, t_2).$$

On the support of $\varsigma_1^{(2^{k_1})}(t_1)\, \varsigma_2^{(2^{k_2})}(t_2)$, $\left|2^{j_1 \wedge j_2} t_1\right| \lesssim 2^{c_0 |j_1 - k_1|}$ and $\left|2^{j_1 \wedge j_2} t_2\right| \lesssim 2^{c_0 |j_2 - k_2|}$, where $c_0 = \max_{1 \le j \le q} |d_q|$. Therefore, using the assumptions on κ and κ_1, we have for such t_1, t_2,

$$\sum_{|\alpha|, |\beta_1|, |\beta_2| \le M''} \left| \left(2^{-j_1 \wedge j_2} \partial_{t_1}\right)^{\beta_1} \left(2^{-j_1 \wedge j_2} \partial_{t_2}\right)^{\beta_2} \left(2^{-j_1 \wedge j_2} X\right)^{\alpha} \widetilde{\kappa}(x, t_1, t_2)\right|$$
$$\lesssim 2^{c|j_1 - k_1| + c|j_2 - k_2|},$$

where $c = c(M'')$ is independent of the element of $\mathcal{I}$ we are considering.

If we replace ς_1 and ς_2 with $2^{c|j_1 - k_1|} \varsigma_1$ and $2^{c|j_2 - k_2|} \varsigma_2$ (which we may do, without disrupting our hypotheses), we may replace $\widetilde{\kappa}$ with $2^{-c|j_1 - k_1| - c|j_2 - k_2|} \widetilde{\kappa}$ and obtain the same operator. This new choice of $\widetilde{\kappa}$ then satisfies

$$\sum_{|\alpha|, |\beta_1|, |\beta_2| \le M''} \left| \left(2^{-j_1 \wedge j_2} \partial_{t_1}\right)^{\beta_1} \left(2^{-j_1 \wedge j_2} \partial_{t_2}\right)^{\beta_2} \left(2^{-j_1 \wedge j_2} X\right)^{\alpha} \widetilde{\kappa}(x, t_1, t_2)\right| \tag{5.47}$$
$$\lesssim 1.$$

Fix a point $x_0 \in \Omega$, we will prove (5.46) for $x = x_0$ (uniformly in x_0). Let $\Phi = \Phi_{x_0, 2^{-k}} : B^n(\eta) \to B_{(X,d)}(x_0, 2^{-k})$ be the map given in Theorem 3.1.2. Let Y_r be the pullback of $2^{-k \cdot d_r} X_r$ via Φ, as in that theorem.

Let $\Phi^\# g = g \circ \Phi$. By the definition of $E(\mathcal{D})$, $E(\mathcal{D})(x, z)$ is supported on those (x, z) with $z \in B_{(X,d)}(x, \xi 2^{-k})$, where ξ can be as small as we like by taking $a > 0$ small. Thus, bounds for the Schwartz kernel of E for $x = x_0$ (like (5.46)) are equivalent to bounds for the Schwartz kernel of $F := \Phi^\# E \left(\Phi^\#\right)^{-1}$. Since $|\det d\Phi| \approx \mathrm{Vol}\left(B_{(X,d)}(x_0, 2^{-k})\right) \approx \mathrm{Vol}\left(B_{2^{-k}}(x_0, 1)\right)$, it is easy to see that (5.46) (with $x = x_0$) follows from

$$\sum_{|\alpha|, |\beta| \le L} \left| Y_u^\alpha Y_v^\beta F(u, v)\right| \lesssim 2^{-M|k_1 - k_2|}, \tag{5.48}$$

where we are actually only interested in $u = 0$ (since $\Phi(0) = x_0$), and therefore may restrict attention to u small. Using that $\|Y_r\|_{C^m} \lesssim 1$, for every m, (5.48) follows from

$$\|F(u, v)\|_{C^L} \lesssim 2^{-M|k_1 - k_2|}, \tag{5.49}$$

where, again, we may restrict attention to u small. We conclude the proof of the lemma by proving (5.49).

To compute F, notice by a change of variables

$$E(\mathcal{D}) f(x) = \psi_1(x) \int f\left(e^{t_2 \cdot (2^{-k} X)} e^{t_1 \cdot (2^{-k} X)} x\right) \widetilde{\kappa}(x, 2^{-k} t_1, 2^{-k} t_2)$$
$$\times \varsigma_1^{(2^{k_1 - k})}(t_1)\, \varsigma_2^{(2^{k_2 - k})}(t_2)\, dt_1 dt_2.$$

We therefore have

$$Fg\left(u\right) = \psi_1\left(\Phi\left(u\right)\right)\int g\left(e^{t_2 \cdot Y}e^{t_1 \cdot Y}u\right)\widetilde{\kappa}\left(\Phi\left(u\right), 2^{-k}t_1, 2^{-k}t_2\right)$$
$$\times \varsigma_1^{\left(2^{k_1 - k}\right)}\left(t_1\right)\varsigma_2^{\left(2^{k_2 - k}\right)}\left(t_2\right)\, dt_1 dt_2.$$

Let $\widehat{\kappa}\left(u, t_1, t_2\right) := \widetilde{\kappa}\left(\Phi\left(u\right), 2^{-k}t_1, 2^{-k}t_2\right)$. (5.47) shows

$$\sum_{|\alpha|, |\beta_1|, |\beta_2| \leq M''}\left|\partial_{t_1}^{\beta_1}\partial_{t_2}^{\beta_2}Y^{\alpha}\widehat{\kappa}\left(u, t_1, t_2\right)\right| \lesssim 1.$$

From (3.2) and (3.3), this implies

$$\|\widehat{\kappa}\|_{C^{M''}\left(B^n\left(\eta\right)\times B^n\left(\eta\right)\right)} \lesssim 1.$$

We separate the proof of (5.49) into two cases. Either $k_1 = k_2$ or $|k_1 - k_2| > 0$. We deal with the more difficult case $|k_1 - k_2| > 0$ first. Recall, $k = k_1 \wedge k_2$. Either $|k_1 - k|_\infty = |k_1 - k_2|_\infty$ or $|k_2 - k|_\infty = |k_1 - k_2|_\infty$. We prove (5.49) when $|k_1 - k|_\infty = |k_1 - k_2|_\infty$; the other case is similar and we leave the details to the reader. Take μ such that $k_{1,\mu} - k_\mu = |k_1 - k|_\infty$.

Recall, we have separated $t_1 \in \mathbb{R}^q$ into ν variables $t_1 = \left(t_1^1, \ldots, t_1^\nu\right) \in \mathbb{R}^{q_1} \times \cdots \times \mathbb{R}^{q_\nu}$. We use the above choice of μ to separate t_1 into two variables: $t_1 = \left(t_1', t_1^\mu\right)$, where $t_1^\mu \in \mathbb{R}^{q_\mu}$ is as above, and $t_1' \in \mathbb{R}^{q - q_\mu}$ is the rest of the variables. In the t_1^μ variable we have single-parameter dilations as before. In the t_1' variable we have $\nu - 1$ parameter dilations, induced by the dilations on each of the variables $t_1^1, \ldots, t_1^{\mu-1}, t_1^{\mu+1}, \ldots, t_1^\nu$. Let $k', k_1' \in [0, \infty)^{q - q_\mu}$ denote k and k_1 (respectively) with out the μth coordinate. Using the above dilations, it makes sense to write $2^{k'}t_1'$ and $2^{k_{1,\mu}}t_1^\mu$, and similarly with k in place of k_1. For a function $\eta_1\left(t_1'\right)$ we define

$$\eta_1^{\left(2^{k'}\right)}\left(t_1'\right) = 2^{k_1 Q_1 + \cdots + k_{\mu-1}Q_{\mu-1} + k_{\mu+1}Q_{\mu+1} + \cdots + k_\nu Q_\nu}\eta_1\left(2^{k'}t_1'\right)$$

so that $\int \eta_1^{\left(2^{k'}\right)} = \int \eta_1$ (see Section 5.2 for more details). Also, for $\eta_2\left(t_1^\mu\right)$ we define $\eta_2^{\left(2^{k_\mu}\right)}\left(t_1^\mu\right) = 2^{Q_\mu k_\mu}\eta_2\left(2^{k_\mu}t_1^\mu\right)$ so that $\int \eta_2^{\left(2^{k_\mu}\right)} = \int \eta_2$. Combining these definitions, we have $\left(\eta_1 \otimes \eta_2\right)^{\left(2^k\right)} = \eta_1^{\left(2^{k'}\right)} \otimes \eta_2^{\left(2^{k_\mu}\right)}$.

Because $C_0^\infty\left(B^{q-q_\mu}\left(a\right) \times B^{q_\mu}\left(a\right)\right) \cong C_0^\infty\left(B^{q-q_\mu}\left(a\right)\right)\widehat{\otimes}C_0^\infty\left(B^{q_\mu}\left(a\right)\right)$, it suffices to prove the result when $\gamma_1\left(t_1\right)$ is of the form $\gamma_1\left(t_1\right) = \gamma_{1,1}\left(t_1'\right)\gamma_{1,2}\left(t_1^\mu\right)$, where $\gamma_{1,1} \in C_0^\infty\left(B^{q-q_\mu}\left(a\right)\right)$ and $\gamma_{1,2} \in C_0^\infty\left(B^{q_\mu}\left(a\right)\right)$ (see Appendix A.2 for more details). Notice, then,

$$\varsigma_1\left(t_1\right) = \partial_{t_1'}^{\alpha_1'}\gamma_{1,1}\left(t_1'\right)\partial_{t_1^\mu}^{\alpha_1^\mu}\gamma_{1,2}\left(t_1^\mu\right),$$

where $|\alpha_1^\mu| = M'$. Let $\varsigma_{1,1}\left(t_1'\right) = \partial_{t_1'}^{\alpha_1'}\gamma_{1,1}\left(t_1'\right)$. We have

$$Fg\left(u\right) = \psi_1\left(\Phi\left(u\right)\right)\int g\left(e^{t_2 \cdot Y}e^{t_1 \cdot Y}u\right)\widehat{\kappa}\left(u, t_1, t_2\right)\varsigma_{1,1}^{\left(2^{k_1' - k'}\right)}\left(t_1'\right)\varsigma_2^{\left(2^{k_2 - k}\right)}\left(t_2\right)$$
$$\times \left[\partial_{t_1^\mu}^{\alpha_1^\mu}\gamma_{1,2}\right]^{\left(2^{k_{1,\mu} - k_\mu}\right)}\left(t_1^\mu\right)\, dt_1 dt_2.$$

(3.3) shows $|\det_{n \times n} Y| \approx 1$, and so we may pick n of the vector fields (call them $Y_{r_1}, \ldots, Y_{r_n}$) for which $|\det (Y_{r_1}(0)| \cdots |Y_{r_n}(0))| \approx 1$. Each of these n vector fields is multiplied by a coordinate of t_1 and by a coordinate of t_2 in the exponentials $e^{t_2 \cdot Y} e^{t_1 \cdot Y} u$. Since $k_{2,\mu} - k_\mu = 0$, and for each $\mu' \neq \mu$ either $k'_{1,\mu'} - k'_{\mu'} = 0$ or $k_{2,\mu'} - k_{\mu'} = 0$, each of $Y_{r_1}, \ldots, Y_{r_n}$ is multiplied by a coordinate of either t'_1 or t_2 (in $e^{t_2 \cdot Y} e^{t_1 \cdot Y} u$) which is *not scaled* in either the expression $2^{k'_1 - k'} t'_1$ or the expression $2^{k_2 - k} t_2$. Let $s \in \mathbb{R}^n$ denote these unscaled n coordinates, taken from either t'_1 or t_2, which are multiplied by $Y_{r_1}, \ldots, Y_{r_n}$. Notice,

$$\left| \det \frac{\partial}{\partial s} e^{t_1 \cdot Y_1} e^{t_2 \cdot Y_2} u \big|_{t_1 = t_2 = u = 0} \right| \approx 1.$$

By a change of variables in the s variable as in Appendix B.3, we see by taking $a > 0$ sufficiently small, we have

$$Fg(u) = \psi_1 (\Phi(u)) \int f(v) K(u, t_1^\mu, v) \, dv \, \left[\partial_{t_1^\mu}^{\alpha_\mu^1} \gamma_{1,2} \right]^{(2^{k_{1,\mu} - k_\mu})} (t_1^\mu) \, dt_1^\mu,$$

where $\| K(u, t_1^\mu, v) \|_{C^{M''}} \lesssim 1$.

Integrating by parts in the t_1^μ variable, and using that $|\alpha_\mu^1| = M'$ and $|k_{1,\mu} - k_\mu| = |k_1 - k_2|_\infty \approx |k_1 - k_2|$, we see by taking $M' = M'(M)$ sufficiently large

$$Fg(u) = 2^{-M|k_1 - k_2|} \int f(v) \widetilde{K}(u, v) \, dv,$$

where $\left\| \widetilde{K} \right\|_{C^{M'' - M'}} \lesssim 1$. Taking M'' so that $M'' - M' \geq L$, (5.49) follows and completes the proof in this case.

All that remains is to consider the case when $k_1 = k_2$, and again it suffices to prove (5.49). Since $k_1 = k_2$ we may assume $M = 0$. The difference is that, in this case, we are not guaranteed that $k_{1,\mu} > 0$ or $k_{2,\mu} > 0$ and so there may be no derivatives on ς_1 or ς_2 with which to integrate by parts. However, since $M = 0$, we do not need the gain $2^{-M|k_1 - k_2|}$, and a simpler reprise of the above proof goes through, without the integration by parts, to prove (5.49) and complete the proof. $\qquad \square$

COMPLETION OF THE PROOF OF PROPOSITION 5.6.10. (5.44) shows $E_1 E_2$ can be decomposed into a sum of terms satisfying the hypotheses of Lemma 5.6.12 with $l = 0$: $E_1 E_2 = \sum_{k_1 \leq j_1, k_2 \leq j_2} E_{k_1, k_2}$. The conclusion of Lemma 5.6.12, in this case, says that for any fixed M, we may do this decomposition in such a way that

$$\left\{ \left(2^{M|j_1 - k_2| + M|j_2 - k_2| + M|k_1 - k_2|} E_{k_1, k_2}, 2^{-k_1 \wedge k_2} \right) \; \Big| \right.$$

$$\left. \mathcal{D}_1 \in \mathcal{I}_1, \mathcal{D}_2 \in \mathcal{I}_2, k_1 \leq j_1, k_2 \leq j_2, k_1, k_2 \in \mathbb{N}^\nu \right\}$$

is a bounded set of right elementary operators of type 2 depth M. Setting $\widetilde{E}_j = \sum_{k_1 \wedge k_2 = j} E_{k_1, k_2}$, we have $E_1 E_2 = \sum_{j \leq j_1 \wedge j_2} \widetilde{E}_j$ and the $\widetilde{E}_j$ satisfy the conclusions of the proposition. The corresponding result for $E_2 E_1$ can be achieved by taking adjoints of the result for $E_1 E_2$; cf. Lemma 5.3.15. $\qquad \square$

PROOF OF PROPOSITION 5.6.3. Let $k_1 \vee k_4 \leq l \leq j_1 \wedge j_2 \wedge j_3 \wedge j_4$. We saw in the proof of Proposition 5.3.12 that

$$E\left(\mathcal{D}_1\right) = \sum_{|\alpha|\leq 1} 2^{-|j_1-l|_\infty} \left(2^{-l}X\right)^\alpha E_{1,l,\alpha},$$

and

$$E\left(\mathcal{D}_4\right) = \sum_{|\alpha|\leq 1} 2^{-|j_4-l|_\infty} E_{4,l,\alpha} \left(2^{-l}X\right)^\alpha,$$

where $E_{1,l,\alpha}$ is a sum of terms of the same form as $E\left(\mathcal{D}_1\right)$ with k_1 replaced by l and $E_{4,l,\alpha}$ is a sum of terms of the same form as $E\left(\mathcal{D}_4\right)$ with k_4 replaced by l. Combining these, we have

$$E\left(\mathcal{D}_1\right) E\left(\mathcal{D}_2\right) TE\left(\mathcal{D}_3\right) E\left(\mathcal{D}_4\right) =$$
$$\begin{cases} \sum_{|\alpha|\leq 1} 2^{-|j_1-l|_\infty} \left(2^{-l}X\right)^\alpha E_{1,l,\alpha} E\left(\mathcal{D}_2\right) TE\left(\mathcal{D}_3\right) E\left(\mathcal{D}_4\right) & \text{and} \\ \sum_{|\alpha|\leq 1} 2^{-|j_4-l|_\infty} E\left(\mathcal{D}_1\right) E\left(\mathcal{D}_2\right) TE\left(\mathcal{D}_3\right) E_{4,l,\alpha} \left(2^{-l}X\right)^\alpha. \end{cases}$$

Since $|j_1 - l|_\infty, |j_4 - l|_\infty \geq |j_1 \wedge j_2 \wedge j_3 \wedge j_4 - l|_\infty$, this shows

$$E\left(\mathcal{D}_1\right) E\left(\mathcal{D}_2\right) TE\left(\mathcal{D}_3\right) E\left(\mathcal{D}_4\right)$$

is an appropriate sum of derivatives of terms of the same form but with $k_1 \vee k_2$ replaced by l. I.e., we may "pull out" derivatives as in the definition of elementary operators.

Thus, to complete the proof of the proposition, it suffices to show that for every N, (5.39) is a bounded set of pre-elementary operators. By the support of $\psi_{1,1}$ and $\psi_{4,2}$ it is clear that

$$\bigcup_{\mathcal{D}_1,\mathcal{D}_4\in\mathcal{I}_1,\mathcal{D}_2,\mathcal{D}_3\in\mathcal{I}_2} \text{supp}\left(E_{j_1,j_2,j_3,j_4}\right) \Subset \Omega \times \Omega.$$

The proof will then be complete once we prove the appropriate estimates on E_{j_1,j_2,j_3,j_4}. More specifically, let $j_0 = j_1 \wedge j_2 \wedge j_3 \wedge j_4$. We wish to show $\forall N, L, m, \exists C,$

$$\sum_{|\alpha|,|\beta|\leq L} \left|\left(2^{-j_0}X_x\right)^\alpha \left(2^{-j_0}X_z\right)^\beta E_{j_1,j_2,j_3,j_4}(x,z)\right|$$

$$\leq C2^{-N\text{diam}\{j_1,j_2,j_3,j_4\}} 2^{s\cdot j_0} \frac{\left(1+\rho_{2^{-j_0}}(x,z)\right)^{-m}}{\text{Vol}\left(B_{2^{-j_0}}\left(x,1+\rho_{2^{-j_0}}(x,z)\right)\right)}.$$
(5.50)

To prove (5.50) take $M' = M'(N,L,m)$ large. Let $M = M(M',m)$ be so large that Proposition 5.6.9 applies to show that if $\mathcal{E}_L$ and $\mathcal{E}_R$ are bounded sets of left and right (respectively) elementary operators of type 2 of depth M and if $|\alpha|, |\beta| \leq L$, we have $\forall \left(E_1, 2^{-l_1}\right) \in \mathcal{E}_R, \left(E_2, 2^{-l_2}\right) \in \mathcal{E}_L,$

$$\left|\left(2^{-l_1\wedge l_2}X_x\right)^\alpha \left(2^{-l_1\wedge l_2}X_z\right)^\beta \left[E_1 TE_2\right](x,z)\right|$$

$$\lesssim 2^{-M'|l_1-l_2|} 2^{s\cdot l_1\wedge l_2} \frac{\left(1+\rho_{2^{-l_1\wedge l_2}}(x,z)\right)^{-m}}{\text{Vol}\left(B_{2^{-l_1\wedge l_2}}\left(x,1+\rho_{2^{-l_1\wedge l_2}}(x,z)\right)\right)}.$$
(5.51)

We apply Proposition 5.6.10 with this choice of M to decompose

$$E\left(\mathcal{D}_1\right) E\left(\mathcal{D}_2\right) = \sum_{k \leq j_1 \wedge j_2} E_k^R, \quad E\left(\mathcal{D}_3\right) E\left(\mathcal{D}_4\right) = \sum_{k \leq j_3 \wedge j_4} E_k^L,$$

where

$$\left\{ \left(2^{M\left(|j_1 - k| + |j_2 - k|\right)} E_k^R, 2^{-k}\right) \mid \mathcal{D}_1 \in \mathcal{I}_1, \mathcal{D}_2 \in \mathcal{I}_2, k \leq j_1 \wedge j_2, k \in \mathbb{N}^\nu \right\}$$

is a bounded set of right elementary operators of type 2 of depth M and

$$\left\{ \left(2^{M\left(|j_3 - k| + |j_4 - k|\right)} E_k^L, 2^{-k}\right) \mid \mathcal{D}_3 \in \mathcal{I}_2, \mathcal{D}_4 \in \mathcal{I}_1, k \leq j_3 \wedge j_4, k \in \mathbb{N}^\nu \right\}$$

is a bounded set of left elementary operators of type 2 of depth M.

Notice, if we take $c\left(L\right)$ large enough, we have for $|\alpha|, |\beta| \leq L$,

$$2^{-c(L)|j_0 - l_1 \wedge l_2|} \left(2^{-j_0} X_x\right)^\alpha \left(2^{-j_0} X_z\right)^\beta \tag{5.52}$$
$$= \gamma\left(\alpha, \beta, j_0, l_1, l_2\right) \left(2^{-l_1 \wedge l_2} X_x\right)^\alpha \left(2^{-l_1 \wedge l_2} X_z\right)^\beta,$$

where $\gamma\left(\alpha, \beta, j_0, l_1, l_2\right)$ is a constant with $0 < \gamma\left(\alpha, \beta, j_0, l_1, l_2\right) \leq 1$.

We use (5.51) and (5.52) to see

$$\sum_{|\alpha|, |\beta| \leq L} \left| \left(2^{-j_0} X_x\right)^\alpha \left(2^{-j_0} X_z\right)^\beta \left[E\left(\mathcal{D}_1\right) E\left(\mathcal{D}_2\right) T E\left(\mathcal{D}_3\right) E\left(\mathcal{D}_4\right)\right](x, z) \right|$$

$$\leq \sum_{\substack{l_1 \leq j_1 \wedge j_2 \\ l_2 \leq j_3 \wedge j_4}} \sum_{|\alpha|, |\beta| \leq L} \left| \left(2^{-j_0} X_x\right)^\alpha \left(2^{-j_0} X_z\right)^\beta \left[E_{l_1}^R T E_{l_2}^L\right](x, z) \right|$$

$$\leq \sum_{\substack{l_1 \leq j_1 \wedge j_2 \\ l_2 \leq j_3 \wedge j_4}} \sum_{|\alpha|, |\beta| \leq L} 2^{c(L)|j_0 - l_1 \wedge l_2|} \left| \left(2^{-l_1 \wedge l_2} X_x\right)^\alpha \left(2^{-l_1 \wedge l_2} X_z\right)^\beta \left[E_{l_1}^R T E_{l_2}^L\right](x, z) \right|$$

$$\lesssim \sum_{\substack{l_1 \leq j_1 \wedge j_2 \\ l_2 \leq j_3 \wedge j_4}} \sum_{|\alpha|, |\beta| \leq L} 2^{-M'|l_1 - l_2| + c(L)|j_0 - l_1 \wedge l_2| + s \cdot l_1 \wedge l_2 - M|j_1 - l_1|}$$

$$\times 2^{-M|j_2 - l_1| - M|j_3 - l_2| - M|j_4 - l_2|} \frac{\left(1 + \rho_{2^{-l_1 \wedge l_2}}(x, z)\right)^{-m}}{\mathrm{Vol}\left(B_{2^{-l_1 \wedge l_2}}\left(x, 1 + \rho_{2^{-l_1 \wedge l_2}}(x, z)\right)\right)}$$

$$\lesssim 2^{-N \mathrm{diam}\{j_1, j_2, j_3, j_4\}} 2^{s \cdot j_1 \wedge j_2 \wedge j_3 \wedge j_4}$$

$$\times \sum_{\substack{l_1 \leq j_1 \wedge j_2 \\ l_2 \leq j_3 \wedge j_4}} 2^{-M_0 |j_0 - l_1| - M_0 |j_0 - l_2|} \frac{\left(1 + \rho_{2^{-l_1 \wedge l_2}}(x, z)\right)^{-m}}{\mathrm{Vol}\left(B_{2^{-l_1 \wedge l_2}}\left(x, 1 + \rho_{2^{-l_1 \wedge l_2}}(x, z)\right)\right)},$$

where M_0 can be as large as we like by taking M' and M large. Setting $k = l_1 \wedge l_2$ and summing, this shows

$$\sum_{|\alpha|, |\beta| \leq L} \left| \left(2^{-j_0} X_x\right)^\alpha \left(2^{-j_0} X_z\right)^\beta \left[E\left(\mathcal{D}_1\right) E\left(\mathcal{D}_2\right) T E\left(\mathcal{D}_3\right) E\left(\mathcal{D}_4\right)\right](x, z) \right|$$

$$\lesssim 2^{-N \mathrm{diam}\{j_1, j_2, j_3, j_4\}} 2^{s \cdot j_1 \wedge j_2 \wedge j_3 \wedge j_4} \sum_{k \leq j_0} 2^{-M_0 |j_0 - k|} \frac{\left(1 + \rho_{2^{-k}}(x, z)\right)^{-m}}{\mathrm{Vol}\left(B_{2^{-k}}\left(x, 1 + \rho_{2^{-k}}(x, z)\right)\right)}.$$

By taking $M_0 = M_0(m)$ sufficiently large, (5.50) follows from Proposition 3.2.6, completing the proof. □

5.7 THE TOPOLOGY

In this section we turn to defining a topology on $\mathcal{A}_4^s$. This topology gives $\mathcal{A}_4^s$ the structure of an LF space (see Definition A.1.24), and therefore $\mathcal{A}_4^s$ is complete (see Remark A.2.4). We used this completeness in Section 5.3 (see Remark 5.3.6).

DEFINITION 5.7.1. For $j, k \in [0, \infty)^\nu$ with $k \le j$, we say $E \in C_0^\infty(\Omega \times \Omega)$ is a $2^{-j}, 2^{-k}$ generalized elementary operator if $\{(E, 2^{-j}, 2^{-k})\}$ is a bounded set of generalized elementary operators. We say E is a 2^{-j} elementary operator if E is a $2^{-j}, 2^{-0}$ generalized elementary operator.

DEFINITION 5.7.2. For each $j, k \in [0, \infty)^\nu$ with $k \le j$ and $M, N \in \mathbb{N}$ we define a norm, $|\cdot|_{2^{-j}, 2^{-k}, N, M}$ on $2^{-j}, 2^{-k}$ generalized elementary operators recursively on M as follows.

- $M = 0$: $|E|_{2^{-j}, 2^{-k}, N, 0}$ is defined to be the least C such that

$$\sum_{|\alpha|, |\beta| \le N} \left| (2^{-j} X_x)^\alpha (2^{-j} X_z)^\beta E(x, z) \right| \le C \frac{(1 + \rho_{2^{-j}}(x, z))^{-N}}{\mathrm{Vol}\,(B_{2^{-j}}(x, 1 + \rho_{2^{-j}}(x, z)))}.$$

- $M \ge 1$: For $k \le l \le j$, $l \in [0, \infty)^\nu$, we write

$$E = \begin{cases} 2^{-|j-l|_\infty} \sum_{|\alpha| \le 1} E_{\alpha,l}^1 (2^{-l} X)^\alpha & and \\ 2^{-|j-l|_\infty} \sum_{|\alpha| \le 1} (2^{-l} X)^\alpha E_{\alpha,l}^2, \end{cases} \tag{5.53}$$

where

$$\left\{ (E_{\alpha,l}^1, 2^{-j}, 2^{-l}), (E_{\alpha,l}^2, 2^{-j}, 2^{-l}) \mid k \le l \le j, |\alpha| \le 1 \right\}$$

is a bounded set of generalized elementary operators. We define

$$|E|_{2^{-j}, 2^{-k}, N, M} := |E|_{2^{-j}, 2^{-k}, N, M-1}$$
$$+ \sup_{k \le l \le j} \inf \sum_{|\alpha| \le 1} \left[|E_{\alpha,l}^1|_{2^{-j}, 2^{-l}, N, M-1} + |E_{\alpha,l}^2|_{2^{-j}, 2^{-l}, N, M-1} \right], \tag{5.54}$$

where the infimum is taken over all representations of the form (5.53).

DEFINITION 5.7.3. Fix a compact set $K \Subset \Omega$. For $s \in \mathbb{R}^\nu$, we define the vector space $\mathcal{A}_4^s(K)$ to consist of those $T \in \mathcal{A}_4^s = \mathcal{A}^s$ with $\mathrm{supp}\,(T) \subseteq K \times K$. For each $N \in \mathbb{N}$, we define a semi-norm, $|\cdot|_{s,N}$, on $\mathcal{A}_4^s(K)$ by writing

$$T = \sum_{j \in \mathbb{N}^\nu} 2^{j \cdot s} E_j, \tag{5.55}$$

where $\left\{ \left(E_j, 2^{-j} \right) \mid j \in \mathbb{N}^\nu \right\}$ *is a bounded set of elementary operators. We define*

$$|T|_{s,N} := \inf \sup_{j \in \mathbb{N}^\nu} |E_j|_{2^{-j}, 2^{-0}, N, N}, \tag{5.56}$$

where the infimum is taken over all representations of T of the form (5.55). We give $\mathcal{A}_4^s$ the coarsest topology such that all of the above semi-norms are continuous.

PROPOSITION 5.7.4. $\mathcal{A}_4^s(K)$ *is a Fréchet space.*

Proposition 5.7.4 would be straightforward were it not for the infima in (5.54) and (5.56). For instance, if we are able to pick a particular way to decompose T as in (5.55) and then a particular way to decompose each E_j as in (5.54) and so forth, so that the infima could be avoided, then it would easily follow that $\mathcal{A}_4^s(K)$ is a Fréchet space. In fact, we will be able to choose such a particular decomposition. The key is the next lemma.

LEMMA 5.7.5. *Let E_0 be a 2^{-j} elementary operator. Let $\mathcal{E}_1$ be a bounded set of elementary operators and $\mathcal{E}_2$ be a bounded set of generalized elementary operators. Then $\forall \left(E_1, 2^{-j_1}, 2^{-k_1} \right), \left(E_4, 2^{-j_4}, 2^{-k_4} \right) \in \mathcal{E}_2, \left(E_2, 2^{-j_2} \right), \left(E_3, 2^{-j_3} \right) \in \mathcal{E}_1$ if we define $j = j_1 \wedge j_2 \wedge j_3 \wedge j_4$ and if we have $k_1 \vee k_2 \le j$ and set*

$$E = E_1 E_2 E_0 E_3 E_4$$

then $\forall N, \exists M$ with

$$|E|_{2^{-j}, 2^{-k_1 \vee k_2}, N, 0} \le C_{\mathcal{E}} |E_0|_{2^{-j_0}, 2^{-0}, M, M} \, 2^{-N \mathrm{diam}\{j_0, j_1, j_2, j_3, j_4\}},$$

where $\mathrm{diam}\{j_0, j_1, j_2, j_3, j_4\} = \max_{0 \le l_1, l_2 \le 5} |j_{l_1} - j_{l_2}|$, and $C_{\mathcal{E}}$ depends on $\mathcal{E}$ but not on E_0.

COMMENTS ON THE PROOF. To obtain

$$|E|_{2^{-j}, 2^{-k_1 \vee k_2}, N, 0} \le C 2^{-N \mathrm{diam}\{j_0, j_1, j_2, j_3, j_4\}}$$

for just some constant C, follows by several applications of Theorem 5.4.3. That C is of the form $C_{\mathcal{E}} |E_0|_{2^{-j_0}, 2^{-0}, M, M}$ follows by keeping track of constants in the proof of Theorem 5.4.3, which we leave to the reader. $\qquad\square$

Let $T \in \mathcal{A}_4^s(K)$, and fix $\psi \in C_0^\infty(\Omega)$ with $\psi \equiv 1$ on a neighborhood of K; we identify ψ with the operator given by multiplication by ψ. Decompose $\psi = \sum_{j \in \mathbb{N}^\nu} D_j$, where $\left\{ \left(D_j, 2^{-j} \right) \mid j \in \mathbb{N}^\nu \right\}$ is a bounded set of elementary operators (see Remark 5.3.7). Note that

$$T = \psi^2 T \psi^2 = \sum_{j_1, j_2, j_3, j_4 \in \mathbb{N}^\nu} D_{j_1} D_{j_2} T D_{j_3} D_{j_4}.$$

For $j \in \mathbb{N}^\nu$ define

$$\widetilde{E}_j := 2^{-j \cdot s} \sum_{j_1 \wedge j_2 \wedge j_3 \wedge j_4 = j} D_{j_1} D_{j_2} T D_{j_3} D_{j_4}.$$

Theorem 5.4.1 can be used to easily show that $\left\{ \left(\tilde{E}_j, 2^{-j} \right) \mid j \in \mathbb{N}^\nu \right\}$ is a bounded set of elementary operators, and we have

$$T = \sum_{j \in \mathbb{N}^\nu} 2^{j \cdot s} \tilde{E}_j. \tag{5.57}$$

This is our chosen decomposition in (5.55).

Suppose we take any decomposition as in (5.55): $T = \sum_{j \in \mathbb{N}^\nu} 2^{j \cdot s} E_j$. We have

$$\tilde{E}_j := 2^{(j_0 - j) \cdot s} \sum_{j_1 \wedge j_2 \wedge j_3 \wedge j_4 = j} \sum_{j_0 \in \mathbb{N}^\nu} D_{j_1} D_{j_2} E_{j_0} D_{j_3} D_{j_4}.$$

Lemma 5.7.5 shows that $\forall N, \exists M,$

$$\left| \tilde{E}_j \right|_{2^{-j}, 2^{-0}, N, 0} \lesssim \sup_{j_0 \in \mathbb{N}^\nu} |E_{j_0}|_{2^{-j_0}, 2^{-0}, M, M}.$$

Taking the infimum over all such representations of T and the supremum over $j \in \mathbb{N}^\nu$, we get

$$\sup_{j \in \mathbb{N}^\nu} \left| \tilde{E}_j \right|_{2^{-j}, 2^{-0}, N, 0} \lesssim |T|_{s, M}.$$

Next, we discuss "pulling out" derivatives. We write, for $0 \le l \le j,$

$$D_j = \begin{cases} 2^{-|j-l|_\infty} \sum_{|\alpha| \le 1} D^1_{j, \alpha, l} \left(2^{-l} X \right)^\alpha & \text{and} \\ 2^{-|j-l|_\infty} \sum_{|\alpha| \le 1} \left(2^{-l} X \right)^\alpha D^2_{j, \alpha, l}, \end{cases}$$

where

$$\left\{ \left(D^1_{j, \alpha, l}, 2^{-j}, 2^{-l} \right), \left(D^2_{j, \alpha, l}, 2^{-j}, 2^{-l} \right) \mid 0 \le l \le j, |\alpha| \le 1 \right\}$$

is a bounded set of generalized elementary operators.

We have, for $l \le j,$ and $T = \sum_{j \in \mathbb{N}^\nu} 2^{j \cdot s} E_j,$

$$\tilde{E}_j = \sum_{|\alpha| \le 1} 2^{(j_0 - j) \cdot s} \sum_{j_1 \wedge j_2 \wedge j_3 \wedge j_4 = j} \sum_{j_0 \in \mathbb{N}^\nu} 2^{-|j_4 - l|_\infty} D_{j_1} D_{j_2} E_{j_0} D_{j_3} D^1_{j_4, \alpha, l} \left(2^{-j} X \right)^\alpha$$

$$=: 2^{-|j-l|_\infty} \sum_{|\alpha| \le 1} \tilde{E}^1_{j, \alpha, l} \left(2^{-l} X \right)^\alpha$$

and

$$\tilde{E}_j = \sum_{|\alpha| \le 1} 2^{(j_0 - j) \cdot s} \sum_{j_1 \wedge j_2 \wedge j_3 \wedge j_4 = j} \sum_{j_0 \in \mathbb{N}^\nu} 2^{-|j_1 - l|_\infty} \left(2^{-j} X \right)^\alpha D^2_{j_1, \alpha, l} D_{j_2} E_{j_0} D_{j_3} D_{j_4}$$

$$=: 2^{-|j-l|_\infty} \sum_{|\alpha| \le 1} \left(2^{-l} X \right)^\alpha \tilde{E}^2_{j, \alpha, l}.$$

Lemma 5.7.5 shows for $\forall N, \exists M,$

$$\left| \tilde{E}_j \right|_{2^{-j}, 2^{-l}, N, 0} + \sup_{l \le j} \sum_{|\alpha| \le 1} \left[\left| \tilde{E}^1_{j, \alpha, l} \right|_{2^{-j}, 2^{-l}, N, 0} + \left| \tilde{E}^2_{j, \alpha, l} \right|_{2^{-j}, 2^{-l}, N, 0} \right]$$

$$\lesssim \sup_{j \in \mathbb{N}^\nu} |E_j|_{2^{-j}, 2^{-0}, M, M}.$$

Taking the infimum over all such representations of T, we get

$$\left|\widetilde{E}_j\right|_{2^{-j},2^{-l},N,0} + \sup_{l\leq j}\sum_{|\alpha|\leq 1}\left[\left|\widetilde{E}_{j,\alpha,l}^1\right|_{2^{-j},2^{-l},N,0} + \left|\widetilde{E}_{j,\alpha,l}^2\right|_{2^{-j},2^{-l},N,0}\right] \lesssim |T|_{s,M}.$$

Inductively continuing this process, "pulling out" more and more derivatives from the D_j, we see that we can define the topology on $\mathcal{A}_4^s(K)$ equivalently in terms of the $\widetilde{E}_j$, and thereby removing the infima from (5.54) and (5.56).

PROOF OF PROPOSITION 5.7.4. Clearly $\mathcal{A}_4^s(K)$ is a locally convex topological vector space whose topology is given by a countable family of semi-norms. The only thing to verify is that this space is complete. Let $\{T_n\} \subset \mathcal{A}_4^s(K)$ be a Cauchy sequence. For each n we decompose

$$T_n = \sum_{j\in\mathbb{N}^\nu} 2^{j\cdot s}\widetilde{E}_j^n,$$

where $\widetilde{E}_j^n$ are the elementary operators from the above chosen decomposition. By the above remarks, we have $\forall N$, $\exists M$,

$$\left|\widetilde{E}_j^n - \widetilde{E}_j^m\right|_{2^{-j},2^{-0},N,0} \lesssim |T_n - T_m|_{s,M}.$$

Thus $\left\{\widetilde{E}_j^n\right\}$ is Cauchy in the $|\cdot|_{2^{-j},2^{-0},N,0}$ norm (uniformly in j) and it follows that $\widetilde{E}_j^n$ converges (uniformly in j) to some $\widetilde{E}_j^\infty$ in the $|\cdot|_{2^{-j},2^{-0},N,0}$ norm for every N. Furthermore we showed that we can canonically pull out derivatives so that for instance, $\forall l \leq j$,

$$\widetilde{E}_j^n = \begin{cases} 2^{-|j-l|_\infty}\sum_{|\alpha|\leq 1}\widetilde{E}_{j,\alpha,l}^{n,1}\left(2^{-l}X\right)^\alpha & \text{and} \\ 2^{-|j-l|_\infty}\sum_{|\alpha|\leq 1}\left(2^{-l}X\right)^\alpha\widetilde{E}_{j,\alpha,l}^{n,2}, \end{cases}$$

and $\forall N$, $\exists M$ with

$$\sup_{l\leq j}\sum_{|\alpha|\leq 1}\left[\left|\widetilde{E}_{j,\alpha,l}^{n,1} - \widetilde{E}_{j,\alpha,l}^{m,1}\right|_{2^{-j},2^{-l},N,0} + \left|\widetilde{E}_{j,\alpha,l}^{n,2} - \widetilde{E}_{j,\alpha,l}^{m,2}\right|_{2^{-j},2^{-l},N,0}\right]$$

$$\lesssim |T_n - T_m|_{s,M}.$$

Thus, the $\widetilde{E}_{j,\alpha,l}^{n,1}$ and $\widetilde{E}_{j,\alpha,l}^{n,2}$ converge (uniformly in j) to some $\widetilde{E}_{j,\alpha,l}^{\infty,1}$ and $\widetilde{E}_{j,\alpha,l}^{\infty,2}$ in the $|\cdot|_{2^{-j},2^{-0},N,0}$ norm for every N. We have

$$\widetilde{E}_j^\infty = \begin{cases} 2^{-|j-l|_\infty}\sum_{|\alpha|\leq 1}\widetilde{E}_{j,\alpha,l}^{\infty,1}\left(2^{-l}X\right)^\alpha & \text{and} \\ 2^{-|j-l|_\infty}\sum_{|\alpha|\leq 1}\left(2^{-l}X\right)^\alpha\widetilde{E}_{j,\alpha,l}^{\infty,2}. \end{cases}$$

Inductively continuing this process, we have that $\left\{\left(\widetilde{E}_j^\infty, 2^{-j}\right) \mid j \in \mathbb{N}^\nu\right\}$ is a bounded set of elementary operators and $T_\infty := \sum_{j\in\mathbb{N}^\nu} 2^{j\cdot s}\widetilde{E}_j^\infty \in \mathcal{A}_4^s(K)$. Finally, we have $T_n \to T_\infty$ in the topology of $\mathcal{A}_4^s(K)$. This proves that $\mathcal{A}_4^s(K)$ is complete. $\qquad\square$

It is immediate from the definitions that (as sets) $\mathcal{A}_4^s = \bigcup_{K \Subset \Omega} \mathcal{A}_4^s(K)$, where the union is taken over compact subsets of Ω. It is also clear that if $K_1 \subset K_2$ are two compact subsets of Ω, then the inclusion $\mathcal{A}_4^s(K_1) \hookrightarrow \mathcal{A}_4^s(K_2)$ is continuous; and moreover the topology on $\mathcal{A}_4^s(K_1)$ is the same as the subspace topology where $\mathcal{A}_4^s(K_1)$ is thought of as a subset of $\mathcal{A}_4^s(K_2)$. This defines an inductive system (indexed by compact subsets of Ω), and we give $\mathcal{A}_4^s$ a topology by setting $\mathcal{A}_4^s = \varinjlim \mathcal{A}_4^s(K)$. See Definition A.1.21 and Example A.1.23 for more details.

With this topology, $\mathcal{A}_4^s$ is an LF space (see Definition A.1.24). Indeed, let $K_1 \subsetneq K_2 \subsetneq \cdots \subsetneq \Omega$ be a countable increasing sequence of compact subsets of Ω with $\bigcup_n K_n = \Omega$. Then, it is easy to see that we have $\mathcal{A}_4^s = \varinjlim \mathcal{A}_4^s(K_n)$, which proves $\mathcal{A}_4^s$ is an LF space.

We now turn to discussing Remark 5.1.19 which says that for $\mathcal{B} \subset \mathcal{A}^s$, the following are equivalent:

(i) $\mathcal{B} \subset \mathcal{A}^s$ is a bounded set.

(ii) $\bigcup_{T \in \mathcal{B}} \mathrm{supp}\,(T) \Subset \Omega \times \Omega$ and for every bounded set of elementary operators $\mathcal{E}$, the set
$$\left\{ \left(2^{-j \cdot s} T E, 2^{-j}\right) \mid (E, 2^{-j}) \in \mathcal{E}, T \in \mathcal{B} \right\}$$
is a bounded set of elementary operators.

(iii) For each $T \in \mathcal{B}$ and each $j \in \mathbb{N}^\nu$ there is an operator $E_j^T \in C_0^\infty(\Omega \times \Omega)$ such that $\left\{ \left(E_j^T, 2^{-j}\right) \mid j \in \mathbb{N}^\nu, T \in \mathcal{B} \right\}$ is a bounded set of elementary operators and $T = \sum_{j \in \mathbb{N}^\nu} 2^{j \cdot s} E_j^T$.

Due to Proposition A.1.25, a subset $\mathcal{B} \subset \mathcal{A}_4^s$ is bounded if and only if $\mathcal{B} \subset \mathcal{A}_4^s(K)$ for some K and is a bounded subset of $\mathcal{A}_4^s(K)$. Suppose $\mathcal{B} \subset \mathcal{A}_4^s(K)$ is bounded. By decomposing each $T \in \mathcal{B}$ in the way set out in (5.57), $T = \sum_{j \in \mathbb{N}^\nu} 2^{j \cdot s} \widetilde{E}_j^T$, we see that $\left\{ \left(\widetilde{E}_j^T, 2^{-j}\right) \mid T \in \mathcal{B}, j \in \mathbb{N}^\nu \right\}$ is a bounded set of elementary operators. Conversely, if we have a subset $\mathcal{B} \subset \mathcal{A}_4^s(K)$ such that for each $T \in \mathcal{B}$ we have a decomposition $T = \sum_{j \in \mathbb{N}^\nu} 2^{j \cdot s} E_j^T$ with $\left\{ \left(E_j^T, 2^{-j}\right) \mid T \in \mathcal{B}, j \in \mathbb{N}^\nu \right\}$ a bounded set of elementary operators then it is easy to see that $\mathcal{B}$ is a bounded subset of $\mathcal{A}_4^s(K)$. These comments prove (i)$\Leftrightarrow$(iii).

For (ii)$\Rightarrow$(iii) let $\mathcal{B}$ be as in (ii). Take $\psi \in C_0^\infty(\Omega)$ with $\psi \otimes \psi \equiv 1$ on a neighborhood of the closure of $\bigcup_{T \in \mathcal{B}} \mathrm{supp}\,(T) \Subset \Omega \times \Omega$. Let
$$\left\{ \left(D_j, 2^{-j}\right) \mid j \in \mathbb{N}^\nu \right\}$$
be a bounded set of elementary operators with $\psi = \sum_{j \in \mathbb{N}^\nu} D_j$. Writing
$$T = \sum_{j \in \mathbb{N}^\nu} 2^{j \cdot s} E_j^T,$$
where $E_j^T = 2^{-j \cdot s} T D_j$, (iii) follows.

Finally, we turn to (iii)$\Rightarrow$(ii). Let $\mathcal{B}$ be as in (iii) so that for each $T \in \mathcal{B}$ we can write $T = \sum_{j \in \mathbb{N}^\nu} 2^{j \cdot s} E_j^T$ where $\left\{ \left(E_j^T, 2^{-j}\right) \mid j \in \mathbb{N}^\nu, T \in \mathcal{B} \right\}$ is a bounded set of elementary operators. Part of the definition of a bounded set of elementary operators shows

$$\bigcup_{T \in \mathcal{B}, j \in \mathbb{N}^\nu} \operatorname{supp}\left(E_j^T\right) \Subset \Omega,$$

and it follows that $\bigcup_{T \in \mathcal{B}} \operatorname{supp}(T) \Subset \Omega$. Let $\mathcal{E}$ be a bounded set of elementary operators. For $\left(E, 2^{-j}\right) \in \mathcal{B}$ define

$$\widetilde{E} := 2^{-j \cdot s} T E = \sum_{k \in \mathbb{N}^\nu} 2^{(k-j) \cdot s} E_k^T E.$$

It follows from Theorem 5.4.1 that $\left\{\left(\widetilde{E}, 2^{-j}\right) \mid (E, 2^{-j}) \in \mathcal{E}, T \in \mathcal{B}\right\}$ is a bounded set of elementary operators, which completes the proof of (ii).

Remark 5.7.6   Each of the definitions $\mathcal{A}_1^s$, $\mathcal{A}_2^s$, $\mathcal{A}_3^s$, and $\mathcal{A}_4^s$ give rise to a locally convex topology on $\mathcal{A}^s$ in the usual way; we have only outlined this for $\mathcal{A}_4^s$. Our proof methods show that all of the topologies on $\mathcal{A}^s$ are the same. This gives rise to a surprising fact: even though it is not at all clear that the topologies induced by $\mathcal{A}_1^s$, $\mathcal{A}_2^s$, and $\mathcal{A}_3^s$ give rise to LF spaces (a priori there seem to be too many semi-norms), our methods do in fact show that these actually are LF spaces. We have seen simpler versions of this before; see Remarks 1.1.6 and 2.8.2.

5.8   NON-ISOTROPIC SOBOLEV SPACES

We now prove the results from Section 5.1.1 concerning the non-isotropic Sobolev spaces NL_s^p, $1 < p < \infty$, $s \in \mathbb{R}^\nu$. As in that section, we fix a neighborhood, Ω_0, of 0 with $\Omega_0 \Subset \Omega$. We are concerned with the norm $\|f\|_{\mathrm{NL}_s^p}$ for f with $\operatorname{supp}(f) \subset \Omega_0$. We make a number of choices in defining this norm, but all of these choices give rise to equivalent norms (Proposition 5.1.22).

We recall the definition of $\|f\|_{\mathrm{NL}_s^p}$. Let $\psi_0 \in C_0^\infty(\Omega)$ with $\psi_0 \equiv 1$ on a neighborhood of the closure of Ω_0. We identify ψ_0 with the operator given by multiplication by ψ_0. It is easy to check that $\psi_0 \in \mathcal{A}_2^0 = \mathcal{A}^0$ and therefore $\psi_0 \in \mathcal{A}_4^0 = \mathcal{A}^0$. Thus, there is a bounded set of elementary operators $\left\{\left(D_j, 2^{-j}\right) \mid j \in \mathbb{N}^\nu\right\}$ with $\psi_0 = \sum_{j \in \mathbb{N}^\nu} D_j$. For $1 < p < \infty$ and $s \in \mathbb{R}^\nu$, we define

$$\|f\|_{\mathrm{NL}_s^p} := \left\|\left(\sum_{j \in \mathbb{N}^\nu} \left|2^{j \cdot s} D_j f\right|^2\right)^{\frac{1}{2}}\right\|_{L^p}.$$

Throughout this section, if we have a sequence of operators, E_j, indexed by $j \in \mathbb{N}^\nu$ (e.g., D_j, $j \in \mathbb{N}^\nu$), we extend this sequence to $\mathbb{Z}^\nu$ by setting $E_j = 0$ for $j \in \mathbb{Z}^\nu \setminus \mathbb{N}^\nu$.

We begin by describing a technical result which is at the heart of many of the proofs in this section. For this we introduce some new notation. For $p \in (1, \infty)$, $s \in \mathbb{R}^\nu$, $\mathcal{E}$ a bounded set of elementary operators, and $f \in C_0^\infty(\Omega)$, we define the semi-norm

$$\|f\|_{p,s,\mathcal{E}} := \sup_{\{(E_j, 2^{-j}) \mid j \in \mathbb{N}^\nu\} \subseteq \mathcal{E}} \left\| \left(\sum_{j \in \mathbb{N}^\nu} |2^{j \cdot s} E_j f|^2 \right)^{\frac{1}{2}} \right\|_{L^p}.$$

If we take $\mathcal{E} = \{(D_j, 2^{-j}) \mid j \in \mathbb{N}^\nu\}$, then we have $\|\cdot\|_{p,s,\mathcal{E}} = \|\cdot\|_{\mathrm{NL}_s^p}$. The next proposition shows that $\|\cdot\|_{\mathrm{NL}_s^p}$ is the largest such norm.

PROPOSITION 5.8.1. *Let $p \in (1, \infty)$, $s \in \mathbb{R}^\nu$, and let $\mathcal{E}$ be a bounded set of elementary operators. Then there is $C = C(p, s, \mathcal{E})$ such that for all $f \in C_0^\infty(\Omega_0)$,*

$$\|f\|_{p,s,\mathcal{E}} \leq C \|f\|_{\mathrm{NL}_s^p}.$$

Remark 5.8.2 Notice that in Proposition 5.8.1, we are only considering $f \in C_0^\infty(\Omega_0)$. This is important, as $\sum_{j \in \mathbb{N}^\nu} D_j = \psi_0$, and so the norm $\|f\|_{\mathrm{NL}_s^p}$ may not "see" any part of f which is supported outside $\mathrm{supp}(\psi_0)$.

To prove Proposition 5.8.1, we need several preliminary results.

LEMMA 5.8.3. *Let $\mathcal{E}$ be a bounded set of pre-elementary operators. There exists C such that for every $(E, 2^{-j}) \in \mathcal{E}$, $\|E\|_{L^p \to L^p} \leq C$, $1 \leq p \leq \infty$.*

PROOF. For $p = 1, \infty$ this follows immediately from Lemma 3.2.14. Interpolation yields the full result. $\square$

LEMMA 5.8.4. *Suppose $\mathcal{E}$ is a bounded set of pre-elementary operators, and let $\{(E_j, 2^{-j}) \mid j \in \mathbb{N}^\nu\} \subseteq \mathcal{E}$. Define the vector valued operator*

$$\mathcal{T}\{f_j\}_{j \in \mathbb{N}^\nu} = \{E_j f_j\}_{j \in \mathbb{N}^\nu}.$$

Then, $\|\mathcal{T}\|_{L^p(\ell^2(\mathbb{N}^\nu)) \to L^p(\ell^2(\mathbb{N}^\nu))} \leq C_{p,\mathcal{E}}$, $1 < p < \infty$.

PROOF. We need only prove the result for $1 < p \leq 2$, and the result for $2 < p < \infty$ then follows by duality. Lemma 5.8.3 shows that $\sup_{(E, 2^{-j}) \in \mathcal{E}} \|E\|_{L^p \to L^p} < \infty$. Notice,

$$\left\| \mathcal{T}\{f_j\}_{j \in \mathbb{N}^\nu} \right\|_{L^p(\ell^p(\mathbb{N}^\nu))} = \left\| \left(\sum_{j \in \mathbb{N}^\nu} |E_j f_j|^p \right)^{\frac{1}{p}} \right\|_{L^p}$$

$$= \left(\sum_{j \in \mathbb{N}^\nu} \|E_j f_j\|_{L^p}^p \right)^{\frac{1}{p}}$$

$$\lesssim \left(\sum_{j \in \mathbb{N}^\nu} \|f_j\|_{L^p}^p \right)^{\frac{1}{p}}$$

$$= \left\| \{f_j\}_{j \in \mathbb{N}^\nu} \right\|_{L^p(\ell^p(\mathbb{N}^\nu))},$$

and so $\|\mathcal{T}\|_{L^p(\ell^p(\mathbb{N}^\nu)) \to L^p(\ell^p(\mathbb{N}^\nu))} \lesssim 1$. Thus, by interpolation, to prove the result it suffices to show for $1 < p \le 2$,

$$\|\mathcal{T}\|_{L^p(\ell^\infty(\mathbb{N}^\nu)) \to L^p(\ell^\infty(\mathbb{N}^\nu))} \le C'_{p,\varepsilon}, \tag{5.58}$$

for some $C'_{p,\varepsilon}$. Notice, we are using that $1 < p \le 2$, as for $p > 2$ interpolating $L^p(\ell^p(\mathbb{N}^\nu))$ with $L^p(\ell^\infty(\mathbb{N}^\nu))$ cannot yield $L^p(\ell^2(\mathbb{N}^\nu))$.

We recall the maximal operator from Section 3.3:

$$\mathcal{M}f(x) = \psi(x) \sup_{\substack{\delta \in (0,1]^\nu \\ |\delta| < \xi_6}} \frac{1}{\text{Vol}\left(B_{(X,d)}(x,\delta)\right)} \int_{B_{(X,d)}(x,\delta)} |f(y)| \, \psi(y) \, dy;$$

see Section 3.3 for more details. Theorem 3.3.1 gives, for $1 < p < \infty$, $\|\mathcal{M}f\|_{L^p} \le C_p \|f\|_{L^p}$. Proposition 3.3.3 shows that we have the pointwise inequality, $\forall (E, 2^{-j}) \in \mathcal{E}$, $|Ef(x)| \lesssim \mathcal{M}f(x)$, where the implicit constant is independent of $(E, 2^{-j}) \in \mathcal{E}$.

Consider, for $1 < p < \infty$,

$$\begin{aligned}
\left\| \mathcal{T} \{f_j\}_{j \in \mathbb{N}^\nu} \right\|_{L^p(\ell^\infty(\mathbb{N}^\nu))} &= \left\| \sup_{j \in \mathbb{N}^\nu} |E_j f_j(x)| \right\|_{L^p} \\
&\lesssim \left\| \mathcal{M} \sup_{j \in \mathbb{N}^\nu} |f_j| \right\|_{L^p} \\
&\lesssim \left\| \sup_{j \in \mathbb{N}^\nu} |f_j| \right\|_{L^p} \\
&= \left\| \{f_j\}_{j \in \mathbb{N}^\nu} \right\|_{L^p(\ell^\infty(\mathbb{N}^\nu))}.
\end{aligned}$$

This establishes (5.58) and completes the proof. $\qquad\square$

PROPOSITION 5.8.5. *Suppose $\mathcal{E}$ is a bounded set of elementary operators, and let $\{(E_j, 2^{-j}) \mid j \in \mathbb{N}^\nu\}, \{(F_j, 2^{-j}) \mid j \in \mathbb{N}^\nu\} \subseteq \mathcal{E}$. For $k \in \mathbb{Z}^\nu$ define the vector valued operator*

$$\mathcal{T}_k \{f_j\}_{j \in \mathbb{N}^\nu} := \{E_j F_{j+k} f_j\}_{j \in \mathbb{N}^\nu}.$$

Then, for every N, and $1 < p < \infty$,

$$\|\mathcal{T}_k\|_{L^p(\ell^2(\mathbb{N}^\nu)) \to L^p(\ell^2(\mathbb{N}^\nu))} \le C_{p,\varepsilon,N} 2^{-N|k|}.$$

PROOF. Fix N, and define $G_{j,k} := 2^{N|k|} E_j F_{j+k}$. Theorem 5.4.1 shows

$$\left\{ (G_{j,k}, 2^{-j}) \mid j \in \mathbb{N}^\nu, k \in \mathbb{Z}^\nu, (E_j, 2^{-j}), (F_{j+k}, 2^{-j-k}) \in \mathcal{E} \right\}$$

is a bounded set of elementary operators.

For $k \in \mathbb{Z}^\nu$, define a vector valued operator

$$\mathcal{S}_k \{f_j\}_{j \in \mathbb{N}^\nu} := \{G_{j,k} f_j\}_{j \in \mathbb{N}^\nu}.$$

Lemma 5.8.4 shows $\|\mathcal{S}_k\|_{L^p(\ell^2(\mathbb{N}^\nu)) \to L^p(\ell^2(\mathbb{N}^\nu))} \lesssim 1$. Since $\mathcal{S}_k = 2^{N|k|} \mathcal{T}_k$, this completes the proof. $\qquad\square$

PROOF OF PROPOSITION 5.8.1. Fix $s \in \mathbb{R}^\nu$. For $j, k, l \in \mathbb{Z}^\nu$ with $|l| \geq |k|$, define a new operator by

$$F_{j,k,l} := 2^{-k \cdot s + |k| + |l|} D_{j+k} D_{j+k+l}.$$

Define

$$\mathcal{E}' := \mathcal{E} \bigcup \left\{ \left(F_{j,k,l}, 2^{-j-k} \right) \mid j, k, l \in \mathbb{Z}^\nu, j + k \in \mathbb{N}^\nu, |l| \geq |k| \right\}.$$

Theorem 5.4.1 shows $\mathcal{E}'$ is a bounded set of elementary operators (this uses $|l| \geq |k|$). Since $\|f\|_{p,s,\mathcal{E}} \leq \|f\|_{p,s,\mathcal{E}'}$, it suffices to prove the result with $\mathcal{E}$ replaced by $\mathcal{E}'$.

Let $\left\{ \left(E_j, 2^{-j} \right) \mid j \in \mathbb{N}^\nu \right\} \subseteq \mathcal{E}'$. For $f \in C_0^\infty (\Omega_0)$, we have $f = \psi_0^2 f$. Thus, we have

$$
\left\| \left(\sum_{j \in \mathbb{N}^\nu} |2^{j \cdot s} E_j f|^2 \right)^{\frac{1}{2}} \right\|_{L^p} = \left\| \left(\sum_{j \in \mathbb{N}^\nu} |2^{j \cdot s} E_j \psi_0^2 f|^2 \right)^{\frac{1}{2}} \right\|_{L^p}
$$

$$
= \left\| \left(\sum_{j \in \mathbb{N}^\nu} \left| \sum_{k,l \in \mathbb{Z}^\nu} 2^{j \cdot s} E_j D_{j+k} D_{j+k+l} f \right|^2 \right)^{\frac{1}{2}} \right\|_{L^p} \tag{5.59}
$$

$$
\leq \sum_{k,l \in \mathbb{Z}^\nu} \left\| \left(\sum_{j \in \mathbb{N}^\nu} |2^{j \cdot s} E_j D_{j+k} D_{j+k+l} f|^2 \right)^{\frac{1}{2}} \right\|_{L^p},
$$

where the last line follows by the triangle inequality.

Fix M large to be chosen later. We separate the right-hand side of (5.59) into three terms,

$$
(I) := \sum_{\substack{k,l \in \mathbb{Z}^\nu \\ |l| \leq M}} \left\| \left(\sum_{j \in \mathbb{N}^\nu} |2^{j \cdot s} E_j D_{j+k} D_{j+k+l} f|^2 \right)^{\frac{1}{2}} \right\|_{L^p},
$$

$$
(II) := \sum_{\substack{k,l \in \mathbb{Z}^\nu \\ |l| > M \\ |k| \geq |l|}} \left\| \left(\sum_{j \in \mathbb{N}^\nu} |2^{j \cdot s} E_j D_{j+k} D_{j+k+l} f|^2 \right)^{\frac{1}{2}} \right\|_{L^p},
$$

$$
(III) := \sum_{\substack{k,l \in \mathbb{Z}^\nu \\ |l| > M \\ |l| > |k|}} \left\| \left(\sum_{j \in \mathbb{N}^\nu} |2^{j \cdot s} E_j D_{j+k} D_{j+k+l} f|^2 \right)^{\frac{1}{2}} \right\|_{L^p}.
$$

Notice that the right-hand side of (5.59) is exactly $(I) + (II) + (III)$. We bound each of these terms separately. In what follows, implicit constants may depend on

$p \in (1, \infty)$, $s \in \mathbb{R}^\nu$, and $\mathcal{E}'$, but not on k, l, M, f, or the particular choice of $\left\{ (E_j, 2^{-j}) \mid j \in \mathbb{N}^\nu \right\} \subseteq \mathcal{E}'$.

We begin with (I). Define a vector valued operator, for $k \in \mathbb{Z}^\nu$,

$$\mathcal{T}_k^1 \{f_j\}_{j \in \mathbb{N}^\nu} := \left\{ 2^{|k| - k \cdot s} E_j D_{j+k} f_j \right\}_{j \in \mathbb{N}^\nu}.$$

Proposition 5.8.5 shows $\left\| \mathcal{T}_k^1 \right\|_{L^p(\ell^2(\mathbb{N}^\nu)) \to L^p(\ell^2(\mathbb{N}^\nu))} \lesssim 1$. We have

$$(I) = \sum_{\substack{k,l \in \mathbb{Z}^\nu \\ |l| \le M}} 2^{-l \cdot s - |k|} \left\| \left(\sum_{j \in \mathbb{N}^\nu} \left| \left(2^{|k| - k \cdot s} E_j D_{j+k} \right) \left(2^{(j+k+l) \cdot s} D_{j+k+l} \right) f \right|^2 \right)^{\frac{1}{2}} \right\|_{L^p}$$

$$= \sum_{\substack{k,l \in \mathbb{Z}^\nu \\ |l| \le M}} 2^{-l \cdot s - |k|} \left\| \mathcal{T}_k^1 \left\{ 2^{(j+k+l) \cdot s} D_{j+k+l} f \right\}_{j \in \mathbb{N}^\nu} \right\|_{L^p(\ell^2(\mathbb{N}^\nu))}$$

$$\lesssim \sum_{\substack{k,l \in \mathbb{Z}^\nu \\ |l| \le M}} 2^{-l \cdot s - |k|} \|f\|_{\mathrm{NL}_s^p}$$

$$\lesssim 2^{M|s|} \|f\|_{\mathrm{NL}_s^p}.$$

We now bound (II). For $k, l \in \mathbb{Z}^\nu$ with $|k| \ge |l|$ define the vector valued operator

$$\mathcal{T}_{k,l}^2 \{f_j\}_{j \in \mathbb{N}^\nu} := \left\{ 2^{|k| + |l| - (k+l) \cdot s} E_j D_{j+k} f_j \right\}_{j \in \mathbb{N}^\nu}.$$

Proposition 5.8.5 shows $\left\| \mathcal{T}_{k,l}^2 \right\|_{L^p(\ell^2(\mathbb{N}^\nu)) \to L^p(\ell^2(\mathbb{N}^\nu))} \lesssim 1$ (this uses $|k| \ge |l|$). We have

(II)

$$= \sum_{\substack{k,l \in \mathbb{Z}^\nu \\ |l| > M \\ |k| \ge |l|}} 2^{-|k| - |l|} \left\| \left(\sum_{j \in \mathbb{N}^\nu} \left| \left(2^{|k| + |l| - (k+l) \cdot s} E_j D_{j+k} \right) \left(2^{(j+k+l) \cdot s} D_{j+k+l} \right) f \right|^2 \right)^{\frac{1}{2}} \right\|_{L^p}$$

$$= \sum_{\substack{k,l \in \mathbb{Z}^\nu \\ |l| > M \\ |k| \ge |l|}} 2^{-|k| - |l|} \left\| \mathcal{T}_{k,l}^2 \left\{ 2^{(j+k+l) \cdot s} D_{j+k+l} f \right\}_{j \in \mathbb{N}^\nu} \right\|_{L^p(\ell^2(\mathbb{N}^\nu))}$$

$$\lesssim \sum_{\substack{k,l \in \mathbb{Z}^\nu \\ |l| > M \\ |k| \ge |l|}} 2^{-|k| - |l|} \|f\|_{\mathrm{NL}_s^p}$$

$$\lesssim 2^{-M} \|f\|_{\mathrm{NL}_s^p}.$$

We now bound (III). Define a vector valued operator

$$\mathcal{T}^3 \{f_j\}_{j \in \mathbb{N}^\nu} := \{E_j f_j\}_{j \in \mathbb{N}^\nu}.$$

By Lemma 5.8.4, $\left\|\mathcal{T}^3\right\|_{L^p(\ell^2(\mathbb{N}^\nu))\to L^p(\ell^2(\mathbb{N}^\nu))} \lesssim 1$. Consider,

(III)

$$= \sum_{\substack{k,l\in\mathbb{Z}^\nu \\ |l|>M \\ |l|>|k|}} 2^{-|k|-|l|} \left\|\left(\sum_{j\in\mathbb{N}^\nu} \left|E_j\left(2^{(j+k)\cdot s}2^{-k\cdot s+|k|+|l|}D_{j+k}D_{j+k+l}\right)f\right|^2\right)^{\frac{1}{2}}\right\|_{L^p}$$

$$= \sum_{\substack{k,l\in\mathbb{Z}^\nu \\ |l|>M \\ |l|>|k|}} 2^{-|k|-|l|} \left\|\mathcal{T}^3\left\{2^{(j+k)\cdot s}F_{j,k,l}f\right\}_{j\in\mathbb{N}^\nu}\right\|_{L^p(\ell^2(\mathbb{N}^\nu))}$$

$$\lesssim \sum_{\substack{k,l\in\mathbb{Z}^\nu \\ |l|>M \\ |l|>|k|}} 2^{-|k|-|l|} \left\|\left\{2^{(j+k)\cdot s}F_{j,k,l}f\right\}_{j\in\mathbb{N}^\nu}\right\|_{L^p(\ell^2(\mathbb{N}^\nu))}$$

$$\leq \sum_{\substack{k,l\in\mathbb{Z}^\nu \\ |l|>M \\ |l|>|k|}} 2^{-|k|-|l|} \|f\|_{p,s,\mathcal{E}'}$$

$$\lesssim 2^{-M} \|f\|_{p,s,\mathcal{E}'}.$$

Plugging these estimates into (5.59), we have

$$\left\|\left(\sum_{j\in\mathbb{N}^\nu}|2^{j\cdot s}E_jf|^2\right)^{\frac{1}{2}}\right\|_{L^p} \leq (I)+(II)+(III) \lesssim 2^{M|s|}\|f\|_{\mathrm{NL}^p_s} + 2^{-M}\|f\|_{p,s,\mathcal{E}'}.$$

Taking the supremum over all $\{(E_j, 2^{-j}) \mid j\in\mathbb{N}^\nu\} \subseteq \mathcal{E}'$ we have that there exists $C = C(p, s, \mathcal{E}')$ (but independent of M and f) such that

$$\|f\|_{p,s,\mathcal{E}'} \leq C2^{M|s|}\|f\|_{\mathrm{NL}^p_s} + C2^{-M}\|f\|_{p,s,\mathcal{E}'}.$$

Taking M so large $C2^{-M} \leq \frac{1}{2}$, we have

$$\|f\|_{p,s,\mathcal{E}'} \leq 2C2^{M|s|}\|f\|_{\mathrm{NL}^p_s},$$

completing the proof. $\qquad\square$

PROOF OF PROPOSITION 5.1.22. The assumptions of Proposition 5.1.22 are symmetric in D_j and $\widetilde{D}_j$, and it therefore suffices to prove, for $f\in C_0^\infty(\Omega_0)$,

$$\left\|\left(\sum_{j\in\mathbb{N}^\nu}|2^{j\cdot s}D_jf|^2\right)^{\frac{1}{2}}\right\|_{L^p} \gtrsim \left\|\left(\sum_{j\in\mathbb{N}^\nu}|2^{j\cdot s}\widetilde{D}_jf|^2\right)^{\frac{1}{2}}\right\|_{L^p}, \tag{5.60}$$

as the reverse inequality then follows by symmetry. (5.60) follows immediately from Proposition 5.8.1. $\qquad\square$

LEMMA 5.8.6. *Let $T \in \mathcal{A}^s$ and let $\mathcal{E}$ be a bounded set of elementary operators. Then,*

$$\left\{ \left(2^{-j \cdot s} ET, 2^{-j} \right) \mid \left(E, 2^{-j} \right) \in \mathcal{E} \right\}$$

is a bounded set of elementary operators.

PROOF. In light of Lemma 5.5.2, it suffices to show

$$\left\{ \left(2^{-j \cdot s} T^* E^*, 2^{-j} \right) \mid \left(E, 2^{-j} \right) \in \mathcal{E} \right\}$$

is a bounded set of elementary operators.

Lemma 5.5.2 shows $\left\{ \left(E^*, 2^{-j} \right) \mid \left(E, 2^{-j} \right) \in \mathcal{E} \right\}$ is a bounded set of elementary operators, while Corollary 5.1.14 shows $T^* \in \mathcal{A}^s$. The result now follows from $T^* \in \mathcal{A}^s = \mathcal{A}_3^s$. $\qquad\square$

PROOF OF THEOREM 5.1.23. Let $T \in \mathcal{A}^s$, $p \in (1, \infty)$ and $s_0 \in \mathbb{R}^\nu$. Define $E_j := 2^{-j \cdot s} D_j T$, so that $\left\{ \left(E_j, 2^{-j} \right) \mid j \in \mathbb{N}^\nu \right\}$ is a bounded set of elementary operators by Lemma 5.8.6. Then, for $f \in C_0^\infty (\Omega_0)$, we have

$$
\|Tf\|_{\mathrm{NL}_{s_0}^p} = \left\| \left(\sum_{j \in \mathbb{N}^\nu} \left| 2^{j \cdot s_0} D_j T f \right|^2 \right)^{\frac{1}{2}} \right\|_{L^p}
$$

$$
= \left\| \left(\sum_{j \in \mathbb{N}^\nu} \left| 2^{j \cdot (s_0 + s)} E_j f \right|^2 \right)^{\frac{1}{2}} \right\|_{L^p}
$$

$$
\lesssim \|f\|_{\mathrm{NL}_{s_0 + s}^p},
$$

where the last line follows by Proposition 5.8.1. $\qquad\square$

In light of Proposition 5.1.22, we may choose any $\psi_0 \in C_0^\infty (\Omega)$ with $\psi_0 \equiv 1$ on Ω_0 and any bounded set of elementary operators $\left\{ \left(D_j, 2^{-j} \right) \mid j \in \mathbb{N}^\nu \right\}$ with $\sum_{j \in \mathbb{N}^\nu} D_j = \psi_0$, to define the space NL_s^p. For a convenient choice of the D_j, we turn to the theory developed in Section 3.4. Take U_1 in that section to be Ω and U_0 to be Ω_0. For each $1 \leq \mu \leq \nu$ and $j_\mu \in \mathbb{N}$ let $D_{j_\mu}^\mu$ be the operator of the same name defined in that section. For $j = (j_1, \ldots, j_\nu) \in \mathbb{N}^\nu$, we take $D_j = D_{j_1}^1 \cdots D_{j_\nu}^\nu$. It follows from the definitions that $\sum_{j \in \mathbb{N}^\nu} D_j = \psi_\nu^2$, where $\psi_\nu \in C_0^\infty (\Omega)$ equals 1 on a neighborhood of the closure of Ω_0. $\left\{ \left(D_j, 2^{-j} \right) \mid j \in \mathbb{N}^\nu \right\}$ is of the form covered in Proposition 5.3.9 (with γ given by (5.4); see Remark 5.3.10) and is therefore a bounded set of elementary operators. In light of Proposition 5.1.22, we have for $f \in C_0^\infty (\Omega_0)$,

$$
\|f\|_{\mathrm{NL}_s^p} \approx \left\| \left(\sum_{j \in \mathbb{N}^\nu} \left| 2^{j \cdot s} D_j f \right|^2 \right)^{\frac{1}{2}} \right\|_{L^p}, \tag{5.61}
$$

with this choice of D_j, for $1 < p < \infty$ and $s \in \mathbb{R}^\nu$.

PROOF OF PROPOSITION 5.1.25. In light of (5.61), Theorem 3.4.1 shows for $1 < p < \infty$, and $f \in C_0^\infty(\Omega_0)$,

$$\|f\|_{\mathrm{NL}_0^p} \approx \|f\|_{L^p},$$

completing the proof of Proposition 5.1.25. □

We now turn to Theorem 5.1.28. For this, we continue to use the particular choice of the D_js from the preceding paragraphs, so that $D_j = D_{j_1}^1 \cdots D_{j_\nu}^\nu$, and

$$\|f\|_{\mathrm{NL}_s^p} \approx \left\| \left(\sum_{j \in \mathbb{N}^\nu} \left| 2^{j \cdot s} D_j f \right|^2 \right)^{\frac{1}{2}} \right\|_{L^p}. \tag{5.62}$$

Remark 5.8.7 Fix ν_0. Consider the ν_0 lists of vector fields

$$\left(X^{\nu - \nu_0 + 1}, \hat{d}^{\nu - \nu_0 + 1} \right), \dots, \left(X^\nu, \hat{d}^\nu \right).$$

Due to (5.62), for $s = (s_{\nu - \nu_0 + 1}, \dots, s_\nu) \in \mathbb{R}^{\nu_0}$ it makes sense to define ν_0 parameter non-isotropic Sobolev spaces in terms of these ν_0 lists of vector fields as

$$\left\| \left(\sum_{j_1, \dots, j_\nu \in \mathbb{N}} \left| 2^{j_\nu - \nu_0 + 1 s_{\nu - \nu_0 + 1} + \cdots + j_\nu s_\nu} D_{j_\nu - \nu_0 + 1}^{\nu - \nu_0 + 1} \cdots D_{j_\nu}^\nu f \right|^2 \right)^{\frac{1}{2}} \right\|_{L^p}.$$

In light of Remark 5.8.7, we may restate Theorem 5.1.28 as the following proposition.

PROPOSITION 5.8.8. *Fix $1 \le \nu_0 \le \nu$ and let $s = (s_{\nu - \nu_0 + 1}, \dots, s_\nu) \in \mathbb{R}^{\nu_0}$. Then, for $1 < p < \infty$,*

$$\left\| \left(\sum_{j_1, \dots, j_\nu \in \mathbb{N}} \left| 2^{j_\nu - \nu_0 + 1 s_{\nu - \nu_0 + 1} + \cdots + j_\nu s_\nu} D_{j_1}^1 \cdots D_{j_\nu}^\nu f \right|^2 \right)^{\frac{1}{2}} \right\|_{L^p}$$

$$\approx \left\| \left(\sum_{j_\nu - \nu_0 + 1, \dots, j_\nu \in \mathbb{N}} \left| 2^{j_\nu - \nu_0 + 1 s_{\nu - \nu_0 + 1} + \cdots + j_\nu s_\nu} D_{j_\nu - \nu_0 + 1}^{\nu - \nu_0 + 1} \cdots D_{j_\nu}^\nu f \right|^2 \right)^{\frac{1}{2}} \right\|_{L^p}.$$

PROOF. Let $\left\{ \epsilon_j^\mu \right\}_{j \in \mathbb{N}, 1 \le \mu \le \nu}$ be i.i.d. random variables of mean 0 taking values ± 1. For $j = (j_1, \dots, j_\nu) \in \mathbb{N}^\nu$ define $\epsilon_j = \epsilon_{j_1}^1 \epsilon_{j_2}^2 \cdots \epsilon_{j_\nu}^\nu$. Notice $\{\epsilon_j\}_{j \in \mathbb{N}^\nu}$ are i.i.d. random variables of mean 0 taking values ± 1. The Khintchine inequality (Theorem

2.10.10) applies to show

$$
\left\| \left(\sum_{j_1,\ldots,j_\nu \in \mathbb{N}} \left| 2^{j_\nu - \nu_1 + 1 s_\nu - \nu_0 + 1 + \cdots + j_\nu s_\nu} D_{j_1}^1 \cdots D_{j_\nu}^\nu f \right|^2 \right)^{\frac{1}{2}} \right\|_{L^p}
$$

$$
\approx \left(\mathbb{E} \left\| \sum_{j_1,\ldots,j_\nu \in \mathbb{N}} \epsilon_{j_1}^1 \cdots \epsilon_{j_\nu}^\nu 2^{j_\nu - \nu_1 + 1 s_\nu - \nu_0 + 1 + \cdots + j_\nu s_\nu} D_{j_1}^1 \cdots D_{j_\nu}^\nu f \right\|_{L^p}^p \right)^{\frac{1}{p}}
$$

$$
= \left(\mathbb{E} \left\| \sum_{j_1 \in \mathbb{N}} \epsilon_{j_1}^1 D_{j_1}^1 \sum_{j_2,\ldots,j_\nu \in \mathbb{N}} \epsilon_{j_2}^2 \cdots \epsilon_{j_\nu}^\nu 2^{j_\nu - \nu_1 + 1 s_\nu - \nu_0 + 1 + \cdots + j_\nu s_\nu} D_{j_2}^2 \cdots D_{j_\nu}^\nu f \right\|_{L^p}^p \right)^{\frac{1}{p}}
$$

$$
\approx \left(\mathbb{E} \left\| \sum_{j_2,\ldots,j_\nu \in \mathbb{N}} \epsilon_{j_2}^2 \cdots \epsilon_{j_\nu}^\nu 2^{j_\nu - \nu_1 + 1 s_\nu - \nu_0 + 1 + \cdots + j_\nu s_\nu} D_{j_2}^2 \cdots D_{j_\nu}^\nu f \right\|_{L^p}^p \right)^{\frac{1}{p}}
$$

$$
\approx \left\| \left(\sum_{j_2,\ldots,j_\nu \in \mathbb{N}} \left| 2^{j_\nu - \nu_1 + 1 s_\nu - \nu_0 + 1 + \cdots + j_\nu s_\nu} D_{j_2}^2 \cdots D_{j_\nu}^\nu f \right|^2 \right)^{\frac{1}{2}} \right\|_{L^p},
$$

where in the second-to-last line we applied (3.23) just as in the proof of Theorem 3.4.1, and in the last line we have used the Khintchine inequality (Theorem 2.10.10). Repeating this argument ν_0 times, we obtain

$$
\left\| \left(\sum_{j_1,\ldots,j_\nu \in \mathbb{N}} \left| 2^{j_\nu - \nu_1 + 1 s_\nu - \nu_0 + 1 + \cdots + j_\nu s_\nu} D_{j_1}^1 \cdots D_{j_\nu}^\nu f \right|^2 \right)^{\frac{1}{2}} \right\|_{L^p}
$$

$$
\approx \left(\mathbb{E} \left\| \sum_{j_\nu - \nu_0 + 1,\ldots,j_\nu \in \mathbb{N}} \epsilon_{j_\nu - \nu_0 + 1}^{\nu - \nu_0 + 1} \cdots \epsilon_{j_\nu}^\nu 2^{j_\nu - \nu_1 + 1 s_\nu - \nu_0 + 1 + \cdots + j_\nu s_\nu} D_{j_\nu - \nu_0 + 1}^{\nu - \nu_0 + 1} \cdots D_{j_\nu}^\nu f \right\|_{L^p}^p \right)^{\frac{1}{p}}
$$

$$
\approx \left\| \left(\sum_{j_\nu - \nu_0 + 1,\ldots,j_\nu \in \mathbb{N}} \left| 2^{j_\nu - \nu_0 + 1 s_\nu - \nu_0 + 1 + \cdots + j_\nu s_\nu} D_{j_\nu - \nu_0 + 1}^{\nu - \nu_0 + 1} \cdots D_{j_\nu}^\nu f \right|^2 \right)^{\frac{1}{2}} \right\|_{L^p},
$$

where the last line follows from the Khintchine inequality (Theorem 2.10.10). This completes the proof. □

PROOF OF PROPOSITION 5.1.27. Take $X_1,\ldots,X_l$ as in the statement of Proposition 5.1.27, and let $\widetilde{X}$ denote the list of vector fields $(X_1,\ldots,X_l)$. Notice, for each $k = 1,\ldots,n$, $\partial_{x_k} = \sum_{j=1}^l c_k^j X_j$, where $c_j^k \in C^\infty$. We let $\psi_1 \in C_0^\infty(\Omega)$ equal 1 on a neighborhood of the closure of Ω_0. Consider, for $f \in C_0^\infty(\Omega_0)$, and $1 < p < \infty$,

$$\|f\|_{L_r^p} \approx \sum_{|\alpha| \le r} \|\psi_1 \partial_x^\alpha f\|_{L^p}$$

$$\approx \sum_{|\alpha| \le r} \left\|\psi_1 \tilde{X}^\alpha f\right\|_{\mathrm{NL}_0^p}$$

$$\lesssim \|f\|_{\mathrm{NL}_{re}^p},$$

where in the last line we have used $\psi_1 X^\alpha \in \mathcal{A}^{\deg(\alpha)}$ and $\deg(\alpha)$ is as in (5.9). □

LEMMA 5.8.9. *Suppose we consider the single-parameter case $\nu = 1$, and consider the single-parameter vector fields $(\partial_{x_1}, 1), \ldots, (\partial_{x_n}, 1)$. Corresponding to this list of vector fields with single-parameter formal degrees, we obtain a single-parameter family of non-isotropic Sobolev spaces NL_s^p, $s \in \mathbb{R}$ as in Section 5.1.1. Then, for $f \in C_0^\infty(\Omega_0)$, and $s \in \mathbb{R}$, $1 < p < \infty$,*

$$\|f\|_{L_s^p} \approx \|f\|_{\mathrm{NL}_s^p}.$$

PROOF. Let $a(x, D)$ be a standard pseudodifferential operator of order s, with $\mathrm{supp}(a(x, D)) \subset \Omega \times \Omega$ and such that for $f \in C_0^\infty(\Omega_0)$,

$$\|f\|_{L_s^p} \approx \|a(x, D) f\|_{L^p}.$$

In light of Example 5.1.31, $a(x, D) \in \mathcal{A}^s$, and it follows that

$$\|f\|_{L_s^p} \approx \|a(x, D) f\|_{L^p} \approx \|a(x, D) f\|_{\mathrm{NL}_0^p} \lesssim \|f\|_{\mathrm{NL}_s^p}.$$

Conversely, let $b(x, D)$ be a standard pseudodifferential operator of order $-s$ (with $\mathrm{supp}(b(x, D)) \subset \Omega \times \Omega$) such that there is a standard pseudodifferential operator $a(x, D)$ (also supported in $\Omega \times \Omega$) such that $b(x, D) a(x, D) f = f$ for $\mathrm{supp}(f) \subset \Omega_0$, we have

$$\|f\|_{\mathrm{NL}_s^p} = \|b(x, D) a(x, D) f\|_{\mathrm{NL}_s^p} \lesssim \|a(x, D) f\|_{\mathrm{NL}_0^p} \approx \|a(x, D) f\|_{L^p} \lesssim \|f\|_{L_s^p},$$

completing the proof. □

PROOF OF COROLLARY 5.1.29. In light of Lemma 5.8.9, Corollary 5.1.29 is an immediate consequence of Theorem 5.1.28. □

5.9 ADDING PARAMETERS

In this section, we prove Theorem 5.1.39. As in the setup of Theorem 5.1.39, we have the original ν parameter list of vector fields (X, d) and the μ_0 parameter sublist (X', d')–where we assume the vector fields in X' span the tangent space at every point. We therefore obtain two algebras: $\mathcal{A}_0^s$ ($s \in \mathbb{R}^{\mu_0}$) corresponding to (X', d') and $\mathcal{A}^s$ ($s \in \mathbb{R}^\nu$) corresponding to (X, d). Theorem 5.1.39 gives a sufficient condition for an operator in $\mathcal{A}_0^s$ to be an element of $\mathcal{A}^{(s,0)}$. For clarity, in this section, when we refer to bounded sets of test functions and elementary operators, we will specify whether we are using the list (X', d') or the list (X, d) by referring to bounded sets of (X', d')-test functions or bounded sets of (X, d)-test functions, and similarly for bounded sets of elementary operators. The metrics ρ_{2-j} and the balls B_{2-j} are defined in terms of (X', d') if $j \in [0, \infty)^{\mu_0}$ and are defined in terms of (X, d) if $j \in [0, \infty)^{\nu}$.

Remark 5.9.1   The balls $B_{(X,d)}(x, (\delta_1, \ldots, \delta_\nu))$ are closely related to the balls

$$B_{(X',d')}(x, (\delta_1, \ldots, \delta_{\mu_0})).$$

Indeed we have the identity

$$B_{(X',d')}(x, (\delta_1, \ldots, \delta_{\mu_0})) = B_{(X,d)}(x, (\delta_1, \ldots, \delta_{\mu_0}, 0, \ldots, 0)),$$

which can be immediately seen from the definitions.

LEMMA 5.9.2. *$\forall m$, there exist 2-admissible constants $C = C(m)$, $m' = m'(m)$ such that for all $j \in [0, \infty)^{\mu_0}$ and $k = (k_1, k_2) \in [0, \infty)^{\mu_0} \times [0, \infty)^{\nu-\mu_0}$, and all $x, z \in \Omega$,*

$$\int_\Omega \frac{(1 + \rho_{2-j}(x, y))^{-m'}}{\mathrm{Vol}\,(B_{2-j}(x, 1 + \rho_{2-j}(x, y)))} \frac{(1 + \rho_{2-k}(y, z))^{-m'}}{\mathrm{Vol}\,(B_{2-k}(y, 1 + \rho_{2-k}(y, z)))}\, dy$$

$$\leq C \frac{(1 + \rho_{2-(j \wedge k_1, k_2)}(x, z))^{-m}}{\mathrm{Vol}\,(B_{2-(j \wedge k_1, k_2)}(x, 1 + \rho_{2-(j \wedge k_1, k_2)}(x, z)))}.$$

COMMENTS ON THE PROOF. Using Remark 5.9.1, this is a reprise of the proof of Proposition 3.2.19–no new ideas are required.   □

As in Section 5.1.3, let (X'', d'') be the list of those vector fields with formal degrees which are in the list (X, d) but not in the list (X', d'). We may think of the formal degrees d'' as $\nu - \mu_0$ parameter formal degrees, since they are zero in the first μ_0 components.

LEMMA 5.9.3. *Let $j \in [0, \infty)^{\mu_0}$ and $k = (k_1, k_2) \in [0, \infty)^{\mu_0} \times [0, \infty)^{\nu-\mu_0}$. Then,*

$$\left(2^{-(j \wedge k_1, k_2)} X\right)^\alpha = \sum_{|\beta| + |\gamma| \leq |\alpha|} \left(2^{-j} X'\right)^\beta c_{\beta,\gamma}^{\alpha,j,k} \left(2^{-k} X\right)^\gamma,$$

where $\left\| c_{\beta,\gamma}^{\alpha,j,k} \right\|_{L^\infty} \lesssim 1$, and the implicit constant may depend on α, but not on j or k. We also have

$$\left(2^{-(j,k_2)} X\right)^\alpha = \sum_{|\beta| + |\gamma| \leq |\alpha|} c_{\beta,\gamma}^{\alpha,j,k_2} \left(2^{-j} X'\right)^\beta \left(2^{-k_2} X''\right)^\gamma,$$

with $\left\| c_{\beta,\gamma}^{\alpha,j,k_2} \right\|_{L^\infty} \lesssim 1$, and again the implicit constant is independent of j, k_2.

COMMENTS ON THE PROOF. Noting that $(j \wedge k_1, k_2) = (j, \infty, \dots, \infty) \wedge k$, the first part of the lemma is a reprise of Proposition 3.2.22, where we allow some of the coordinates of j to be ∞, and the proof is otherwise unchanged.

A similar proof works for the second part, where we consider

$$(j, k_2) = (j, \infty, \dots, \infty) \wedge (\infty, \dots, \infty, k_2),$$

and we must also use the regularity of the functions $c_{\beta,\gamma}^{\alpha,j,k_2}$ guaranteed in Proposition 3.2.22 to commute them to the front of the expression. We leave the remaining details to the reader. $\qquad\square$

LEMMA 5.9.4. *Let $\mathcal{E}$ be a bounded set of (X', d')-elementary operators. Then, $\forall L_1, m, \exists M, \forall \mathcal{T}_M$ bounded sets of (X, d)-test functions of order M, $\forall \alpha, \forall L_2, \exists C,$ $\forall (E, 2^{-j}) \in \mathcal{E}$ $(j \in [0, \infty)^{\mu_0})$, $\forall (\psi, z, 2^{-(k_1,k_2)}) \in \mathcal{T}_M$ $(k_1 \in [0, \infty)^{\mu_0}, k_2 \in [0, \infty)^{\nu-\mu_0})$, we have*

$$\left| (2^{-j} X')^\alpha E\psi(x) \right|$$

$$\leq C 2^{-L_1|k_1 - j\wedge k_1| - L_2|j - j\wedge k_1|} \frac{(1 + \rho_{2^{-(j\wedge k_1, k_2)}}(x, z))^{-m}}{\mathrm{Vol}\left(B_{2^{-(j\wedge k_1, k_2)}}(x, 1 + \rho_{2^{-(j\wedge k_1, k_2)}}(x, z))\right)}.$$

PROOF. Applying the definition of bounded sets of (X', d')-elementary operators, for $L_2' = L_2'(L_2)$ large to be chosen later and for $(E, 2^{-j}) \in \mathcal{E}$, we have

$$E = \sum_{|\beta| \leq L_2'} 2^{-L_2'|j - j\wedge k_1|_\infty} E_{k_1, \beta} \left(2^{-j\wedge k_1} X'\right)^\beta,$$

where $\left\{ (E_{k_1,\beta}, 2^{-j}) \mid (E, 2^{-j}) \in \mathcal{E}, |\beta| \leq L_2', k_1 \in [0, \infty)^{\mu_0} \right\}$ is a bounded set of (X', d')-pre-elementary operators.

Applying the definition of bounded sets of (X, d)-test functions of order M, we have for $(\psi, z, 2^{-(k_1, k_2)}) \in \mathcal{T}_M$,

$$\psi = \sum_{|\beta| \leq M} 2^{-M|k_1 - j\wedge k_1|_\infty} \left(2^{-(j\wedge k_1, k_2)} X\right)^\beta \phi_{j,\beta},$$

where $\left\{ (\phi_{j,\beta}, z, 2^{-k}) \mid (\psi, z, 2^{-k}) \in \mathcal{T}_M, |\beta| \leq M, j \in [0, \infty)^{\mu_0} \right\}$ is a bounded set of (X, d)-bump functions.

Combining the above two equations, we have

$$(2^{-j} X')^\alpha E\psi(x)$$

$$= 2^{-M|k_1 - j\wedge k_1|_\infty - L_2'|j - j\wedge k_1|_\infty}$$

$$\times \sum_{|\beta_1| \leq L_2', |\beta_2| \leq M} (2^{-j} X')^\alpha E_{k_1, \beta_1} \left(2^{-j\wedge k_1} X'\right)^{\beta_1} \left(2^{-(j\wedge k_1, k_2)} X\right)^{\beta_2} \phi_{j, \beta_2}$$

$$=: 2^{-M|k_1 - j\wedge k_1|_\infty - L_2'|j - j\wedge k_1|_\infty} \sum_{|\beta| \leq L_2' + M} (2^{-j} X')^\alpha E_\beta \left(2^{-(j\wedge k_1, k_2)} X\right)^\beta \phi_\beta.$$

In the above, E_β and ϕ_β depend on $\left(E, 2^{-j}\right) \in \mathcal{E}$ and $\left(\psi, z, 2^{-k}\right) \in \mathcal{T}_M$, but we suppress this dependance. The important part is that we have the following (uniform) estimates. For every m, L',

$$
\sum_{|\gamma_1|+|\gamma_2| \leq L'} \left| \left(2^{-j} X_x'\right)^{\gamma_1} \left(2^{-j} X_y'\right)^{\gamma_2} E_\beta (x, y) \right|
$$
$$
\lesssim \frac{\left(1 + \rho_{2^{-j}} (x, y)\right)^{-m}}{\operatorname{Vol}\left(B_{2^{-j}} \left(x, 1 + \rho_{2^{-j}} (x, y)\right)\right)}, \tag{5.63}
$$

$$
\sum_{|\gamma| \leq L'} \left(2^{-k} X\right)^\gamma \phi_\beta (y) \lesssim \frac{\chi_{\{y \in B_{2^{-k}} (z,1)\}}}{\operatorname{Vol}\left(B_{2^{-k}} (z, 1)\right)}
$$
$$
\lesssim \frac{\left(1 + \rho_{2^{-k}} (y, z)\right)^{-m}}{\operatorname{Vol}\left(B_{2^{-k}} \left(y, 1 + \rho_{2^{-k}} (y, z)\right)\right)}. \tag{5.64}
$$

Applying Lemma 5.9.3, we see

$$
\left(2^{-j} X'\right)^\alpha E_\beta \left(2^{-(j \wedge k_1, k_2)} X\right)^\beta \phi_\beta
$$
$$
= \sum_{|\gamma_1|+|\gamma_2| \leq |\beta|} \left(2^{-j} X'\right)^\alpha E_\beta \left(2^{-j} X'\right)^{\gamma_1} c_{\gamma_1,\gamma_2}^\beta \left(2^{-k} X\right)^{\gamma_2} \phi_\beta, \tag{5.65}
$$

where $\left\| c_{\gamma_1,\gamma_2}^\beta \right\|_{L^\infty} \lesssim 1$. Using (5.63) and (5.64), we have for every m,

$$
\left| \left[\left(2^{-j} X'\right)^\alpha E_\beta \left(2^{-j} X'\right)^{\gamma_1} \right] (x, y) \right| \lesssim \frac{\left(1 + \rho_{2^{-j}} (x, y)\right)^{-m}}{\operatorname{Vol}\left(B_{2^{-j}} \left(x, 1 + \rho_{2^{-j}} (x, y)\right)\right)},
$$

$$
\left| c_{\gamma_1,\gamma_2}^\beta (y) \left(2^{-k} X\right)^{\gamma_2} \phi_\beta (y) \right| \lesssim \frac{\left(1 + \rho_{2^{-k}} (y, z)\right)^{-m}}{\operatorname{Vol}\left(B_{2^{-k}} \left(y, 1 + \rho_{2^{-k}} (y, z)\right)\right)}.
$$

Plugging these two estimates into (5.65) and applying Lemma 5.9.2 shows

$$
\left| \left(2^{-j} X'\right)^\alpha E\psi (x) \right|
$$
$$
\leq C 2^{-M|k_1 - j \wedge k_1|_\infty - L_2'|j - j \wedge k_1|_\infty} \frac{\left(1 + \rho_{2^{-(j \wedge k_1, k_2)}} (x, z)\right)^{-m}}{\operatorname{Vol}\left(B_{2^{-(j \wedge k_1, k_2)}} \left(x, 1 + \rho_{2^{-(j \wedge k_1, k_2)}} (x, z)\right)\right)}.
$$

Taking $L_2' = L_2' (L_2)$ and $M = M (L_1)$ large yields the result. $\qquad\square$

COROLLARY 5.9.5. *Let $\mathcal{E}$ be a bounded set of (X', d')-elementary operators. $\forall L$, $\forall m$, $\exists M$, $\forall \mathcal{T}_M$ bounded sets of (X, d)-test functions of order M, $\forall \alpha$, $\exists C$, $\forall \left(E, 2^{-j}\right) \in \mathcal{E}$ ($j \in [0, \infty)^{\mu_0}$), $\forall \left(\psi, z, 2^{-(k_1, k_2)}\right) \in \mathcal{T}_M$ ($k_1 \in [0, \infty)^{\mu_0}$, $k_2 \in [0, \infty)^{\nu - \mu_0}$), we have*

$$
\left| \left(2^{-k_1} X'\right)^\alpha E\psi (x) \right| \leq C 2^{-L|j - k_1|} \frac{\left(1 + \rho_{2^{-(j \wedge k_1, k_2)}} (x, z)\right)^{-m}}{\operatorname{Vol}\left(B_{2^{-(j \wedge k_1, k_2)}} \left(x, 1 + \rho_{2^{-(j \wedge k_1, k_2)}} (x, z)\right)\right)}.
$$

PROOF. This follows from Lemma 5.9.4 by taking $L_2 = L_2(\alpha, L)$ large and taking $L_1 = L$. $\qquad\square$

PROPOSITION 5.9.6. *Fix* $s \in \mathbb{R}^{\mu_0}$ *and let* $\mathcal{B} \subset \mathcal{A}_0^s$ *be a bounded set (see Remark 5.1.19). Then,* $\forall m,$ $\exists M,$ $\forall \mathcal{T}_M$ *bounded sets of* (X, d)*-test functions of order* $M,$ $\forall \alpha,$ $\exists C,$ $\forall T \in \mathcal{B},$ $\forall (\psi, z, 2^{-k}) \in \mathcal{T}_M,$

$$\left| \left(2^{-k_1} X' \right)^\alpha T\psi(x) \right| \le C 2^{k \cdot (s,0)} \frac{(1 + \rho_{2^{-k}}(x, z))^{-m}}{\mathrm{Vol}\left(B_{2^{-k}}(x, 1 + \rho_{2^{-k}}(x, z)) \right)}.$$

PROOF. Using $\mathcal{A} = \mathcal{A}_4$, we may decompose $T = \sum_{j \in \mathbb{N}^{\mu_0}} 2^{j \cdot s} E_j^T$, where

$$\left\{ \left(E_j^T, 2^{-j} \right) \mid j \in \mathbb{N}^{\mu_0}, T \in \mathcal{B} \right\}$$

is a bounded set of (X', d')-elementary operators (this is equivalent to $\mathcal{B} \subset \mathcal{A}_4^s$ being bounded–see Remark 5.1.19). Fix m and take $L' = L'(m)$ large, and let $L = L(L', m, s)$ be large. We see, applying Corollary 5.9.5 and taking M sufficiently large, for $(\psi, z, 2^{-k}) \in \mathcal{T}_M,$

$$\left| \left(2^{-k_1} X' \right)^\alpha T\psi(x) \right| \le \sum_{j \in \mathbb{N}^{\mu_0}} \left| \left(2^{-k_1} X' \right)^\alpha 2^{j \cdot s} E_j \psi(x) \right|$$

$$\lesssim \sum_{j \in \mathbb{N}^{\mu_0}} 2^{-L|j - k_1|} 2^{j \cdot s} \frac{(1 + \rho_{2^{-(j \wedge k_1, k_2)}}(x, z))^{-m}}{\mathrm{Vol}\left(B_{2^{-(j \wedge k_1, k_2)}}(x, 1 + \rho_{2^{-(j \wedge k_1, k_2)}}(x, z)) \right)}$$

$$\lesssim \sum_{j' \in \mathbb{N}^\nu} 2^{-L'|j' - k|} 2^{k \cdot (s,0)} \frac{(1 + \rho_{2^{-j' \wedge k}}(x, z))^{-m}}{\mathrm{Vol}\left(B_{2^{-j' \wedge k}}(x, 1 + \rho_{2^{-j' \wedge k}}(x, z)) \right)}$$

$$\lesssim 2^{k \cdot (s,0)} \frac{(1 + \rho_{2^{-k}}(x, z))^{-m}}{\mathrm{Vol}\left(B_{2^{-k}}(x, 1 + \rho_{2^{-k}}(x, z)) \right)},$$

where in the last line we applied Proposition 3.2.6. $\qquad\square$

LEMMA 5.9.7. *Suppose* $\mathcal{T}_M$ *is a bounded set of* (X, d)*-functions of order* M*. Then, for every* $\beta,$

$$\left\{ \left(\left(2^{-k} X \right)^\beta \psi, x, 2^{-k} \right) \mid (\psi, x, 2^{-k}) \in \mathcal{T}_M \right\}$$

is a bounded set of (X, d)*-test functions of order* M*.*

PROOF. We proceed by induction on $|\beta|$. The base case $|\beta| = 0$ is trivial. By our inductive hypothesis, it suffices to prove the result for $|\beta| = 1$. We, henceforth, assume $\left(2^{-k} X \right)^\beta = 2^{-k \cdot d_r} X_r$, for some r.

Let $(\psi, x, 2^{-k}) \in \mathcal{T}_M$. Let $l \in [0, \infty)^\nu$ with $l \le k$. By the definition of $\mathcal{T}_M$, we may write

$$\psi = 2^{-|k-l|_\infty M} \sum_{|\alpha| \le M} \left(2^{-l} X \right)^\alpha \phi_{l,\alpha},$$

where $\left\{ \left(\phi_{l,\alpha}, x, 2^{-j} \right) \mid (\psi, x, 2^{-k}) \in \mathcal{T}_M, l \leq k, |\alpha| \leq M \right\}$ is a bounded set of (X, d)-bump functions.

Notice, $2^{-k \cdot d_r} X_r \left(2^{-l} X \right)^\alpha = \left(2^{-l} X \right)^\alpha 2^{-k \cdot d_r} X_r + \left[2^{-k \cdot d_r} X_r, \left(2^{-l} X \right)^\alpha \right]$. Furthermore, since $l \leq k$, $2^{-k \cdot d_r} = b 2^{-l \cdot d_r}$ for some $0 < b \leq 1$. Finally, by repeated applications of Lemma 3.2.8, we have

$$\left[2^{-l \cdot d_r} X_r, \left(2^{-l} X \right)^\alpha \right] = \sum_{|\gamma| \leq |\alpha| - 1} \left(2^{-l} X \right)^\gamma c_{l,\alpha,\gamma}$$

where for every γ', $\left| \left(2^{-l} X \right)^{\gamma'} c_{l,\alpha,\gamma} \right| \lesssim 1$. We therefore have

$$2^{-k \cdot d_r} X_r \psi = 2^{-|k-l|_\infty M} \sum_{|\alpha| \leq M} \left(2^{-l} X \right)^\alpha 2^{-k \cdot d_r} X_r \phi_{l,\alpha}$$

$$+ 2^{-|k-l|_\infty M} \sum_{|\alpha| \leq M} b \sum_{|\gamma| \leq |\alpha| - 1} \left(2^{-l} X \right)^\gamma c_{l,\alpha,\gamma} \phi_{l,\alpha}.$$

Because $\left\{ \left(\phi_{l,\alpha}, x, 2^{-k} \right) \mid (\psi, x, 2^{-k}) \in \mathcal{T}_M, l \leq k, |\alpha| \leq M \right\}$ is a bounded set of (X, d)-bump functions, the same is true of

$$\left\{ \left(2^{-k \cdot d_r} X_r \phi_{l,\alpha}, x, 2^{-k} \right) \mid (\psi, x, 2^{-k}) \in \mathcal{T}_M, l \leq k, |\alpha| \leq M \right\}$$

and

$$\left\{ \left(b c_{l,\alpha,\gamma} \phi_{l,\alpha}, x, 2^{-k} \right) \mid (\psi, x, 2^{-k}) \in \mathcal{T}_M, l \leq k, |\alpha| \leq M, |\gamma| \leq |\alpha| \right\}.$$

This completes the proof. □

PROOF OF THEOREM 5.1.39. Suppose $T \in \mathcal{A}_0^s$ satisfies the hypotheses of Theorem 5.1.39. We wish to show $T \in \mathcal{A}^{(s,0)}$. To do so, we show $T \in \mathcal{A}_2^{(s,0)}$. Fix m and take $M = M(m)$ large, and let $\mathcal{T}_M$ be a bounded set of (X, d) test functions of order M. Fix α and consider, for $k = (k_1, k_2) \in [0, \infty)^\nu$, if we apply Lemma 5.9.3,

$$\left(2^{-k} X \right)^\alpha = \sum_{|\beta| + |\gamma| \leq |\alpha|} c_{\beta,\gamma}^{\alpha,k} \left(2^{-k_1} X' \right)^\beta \left(2^{-k_2} X'' \right)^\gamma,$$

where $\left\| c_{\beta,\gamma}^{\alpha,k} \right\|_{L^\infty} \lesssim 1$. Consider, then, by our hypotheses, for $(\psi, z, 2^{-k}) \in \mathcal{T}_M$,

$$\left(2^{-k} X \right)^\alpha T\psi = \sum_{|\beta| + |\gamma| \leq |\alpha|} c_{\beta,\gamma}^{\alpha,k} \left(2^{-k_1} X' \right)^\beta \left(2^{-k_2} X'' \right)^\gamma T$$

$$= \sum_{|\beta| + |\gamma| \leq |\alpha|} \sum_{|\delta| \leq M_0(|\gamma|)} c_{\beta,\gamma}^{\alpha,k} \left(2^{-k_1} X' \right)^\beta S_{k,\gamma,\delta} \left(2^{-k} X \right)^\delta \psi(x),$$

where $M_0(|\gamma|)$ is the M from the statement of the theorem, and

$$\left\{ S_{k,\gamma,\delta} \mid k \in [0, \infty)^\nu, |\delta| \leq M_0 \right\} \subset \mathcal{A}_0^s$$

is a bounded set. It then follows by combining Lemma 5.9.7 and Proposition 5.9.6 that

$$\left|\left(2^{-k}X\right)^{\alpha} T\psi\left(x\right)\right| \lesssim 2^{k\cdot(s,0)} \frac{\left(1+\rho_{2^{-k}}\left(x,z\right)\right)^{-m}}{\mathrm{Vol}\left(B_{2^{-k}}\left(x,1+\rho_{2^{-k}}\left(x,z\right)\right)\right)},$$

for every m, provided $M = M(m)$ is sufficiently large. The same bound with T^* in place of T holds, since our assumptions are symmetric in T and T^*. This proves $T \in \mathcal{A}^{(s,0)} = \mathcal{A}_2^{(s,0)}$. $\qquad\square$

5.10   PSEUDOLOCALITY

In this section, we prove Theorem 5.1.40. I.e., we assume that each list $X_1^{\mu}, \ldots, X_{q_{\mu}}^{\mu}$ spans the tangent space at every point–for the rest of the section we assume this. We then prove that all operators in $\mathcal{A}^s$ are pseudolocal: if $T \in \mathcal{A}^s$, then $T(x,y)$ is C^{∞} for $x \neq y$.

To show $T(x,y)$ is C^{∞} for $x \neq y$ it suffices to show for all ordered multi-indices α_1, α_2 we have

$$X_x^{\alpha_1} X_y^{\alpha_2} T(x,y)$$

is bounded on a neighborhood of each fixed $x \neq y$. Let $\psi \in C_0^{\infty}(\Omega)$ be such that $\psi \otimes \psi \equiv 1$ on a neighborhood of supp (T). Because $\psi X^{\alpha_1} \in \mathcal{A}^{\deg(\alpha_1)}$ and $\psi X^{\alpha_2} \in \mathcal{A}^{\deg(\alpha_2)}$ (where deg is as in (5.9)) we have $X^{\alpha_1} T (X^{\alpha_2})^* \in \mathcal{A}^{s+\deg(\alpha_1)+\deg(\alpha_2)}$. Thus, to prove $X_x^{\alpha_1} X_y^{\alpha_2} T(x,y)$ is bounded on a neighborhood of a fixed $x \neq y$, it suffices to consider the case when $|\alpha_1| = |\alpha_2| = 0$, by changing the choice of s.

Fix $x \neq y$. Using $T \in \mathcal{A}^s = \mathcal{A}_4^s$, we may write $T = \sum_{j \in \mathbb{N}^{\nu}} 2^{j \cdot s} E_j$, where $\left\{\left(E_j, 2^{-j}\right) \mid j \in \mathbb{N}^{\nu}\right\}$ is a bounded set of elementary operators. We wish to bound

$$|T(x,y)| \leq \sum_{j \in \mathbb{N}^{\nu}} 2^{j \cdot s} |E_j(x,y)|.$$

Our assumptions are symmetric in the parameters $1, \ldots, \nu$, and so it suffices to bound

$$\sum_{\substack{j \in \mathbb{N}^{\nu} \\ j_1 \geq j_2 \geq \cdots \geq j_{\nu}}} 2^{j \cdot s} |E_j(x,y)|.$$

For $j \in \mathbb{N}^{\nu}$ with $j_1 \geq j_2 \geq \cdots \geq j_{\nu}$, define $l = l(j) \in \mathbb{N}^{\nu}$ by $l(j) = (j_{\nu}, j_{\nu}, \ldots, j_{\nu})$.

LEMMA 5.10.1. *Let* $j_1 \geq j_2 \geq \cdots \geq j_{\nu}$. $\forall N$, $\exists M = M(s,N)$, $\forall m$,

$$2^{j \cdot s} |E_j(x,y)| \lesssim 2^{Mj_{\nu}} 2^{-N|j-l|} \frac{\left(1+\rho_{2^{-l}}\left(x,z\right)\right)^{-m}}{\mathrm{Vol}\left(B_{2^{-l}}\left(x,1+\rho_{2^{-l}}\left(x,z\right)\right)\right)},$$

where $l = l(j)$ *as above and the implicit constants are independent of* j.

PROOF. We use the fact that $X_1^\nu, \ldots, X_{q_\nu}^\nu$ span the tangent space to each point of Ω. Because of this, for each α, we may write

$$\left(2^{-l}X\right)^\alpha = \sum_{|\beta| \leq |\alpha|} 2^{C(\alpha)j_\nu} f_{\alpha,\beta,j}(x) \left(2^{-j_\nu}X^\nu\right)^\beta,$$

where $C(\alpha)$ is some constant depending on α and $\{f_{\alpha,\beta,j} \mid j \in \mathbb{N}^\nu, j_1 \geq j_2 \geq \cdots \geq j_\nu, |\beta| \leq |\alpha|\} \subset C^\infty(\Omega)$ is a bounded set. This follows from the fact that each X_j^μ can be written as a C^∞ linear combination of the X^ν.

Take $\widetilde{N} = \widetilde{N}(N,s)$ large. Applying the definition of elementary operators, we have

$$E_j = 2^{-\widetilde{N}|j-l|_\infty} \sum_{|\alpha| \leq \widetilde{N}} \left(2^{-l}X\right)^\alpha E_{j,\alpha},$$

where $\left\{\left(E_{j,\alpha}, 2^{-j}\right) \mid j \in \mathbb{N}^\nu, j_1 \geq j_2 \geq \cdots \geq j_\nu, |\alpha| \leq \widetilde{N}\right\}$ is a bounded set of pre-elementary operators. We therefore have

$$2^{j \cdot s} |E_j(x,z)| \leq 2^{-\widetilde{N}|j-l|_\infty} 2^{j \cdot s} \sum_{|\alpha| \leq \widetilde{N}} \left|\left(2^{-l}X\right)^\alpha E_{j,\alpha}(x,z)\right|$$

$$\lesssim 2^{-\widetilde{N}|j-l|_\infty} 2^{j \cdot s} \sum_{|\alpha| \leq \widetilde{N}} \sum_{|\beta| \leq |\alpha|} 2^{C(\alpha)j_\nu} \left|\left(2^{-j_\nu}X^\nu\right)^\beta E_{j,\alpha}(x,z)\right|$$

$$\lesssim 2^{-(\widetilde{N}-|s|_1)|j-l|_\infty} 2^{Mj_\nu} \frac{(1+\rho_{2^{-j}}(x,z))^{-m}}{\mathrm{Vol}\left(B_{2^{-j}}(x, 1+\rho_{2^{-j}}(x,z))\right)},$$

where $M = \sup_{|\alpha| \leq \widetilde{N}} C(\alpha) + |s|_1$. Using that $\rho_{2^{-l}}(x,z) \leq \rho_{2^{-j}}(x,z)$ we have $(1+\rho_{2^{-j}}(x,z))^{-m} \leq (1+\rho_{2^{-l}}(x,z))^{-m}$. Using Lemma 3.2.4, for some fixed D,

$$2^{-D|j-l|}\mathrm{Vol}\left(B_{2^{-j}}(x, 1+\rho_{2^{-j}}(x,z))\right)^{-1} \lesssim \mathrm{Vol}\left(B_{2^{-l}}(x, 1+\rho_{2^{-l}}(x,z))\right)^{-1}.$$

Putting these estimates together, we have

$$2^{j \cdot s} |E_j(x,z)| \lesssim 2^{-(\widetilde{N}-|s|)|j-l|_\infty + D|j-l|} 2^{Mj_\nu} \frac{(1+\rho_{2^{-l}}(x,z))^{-m}}{\mathrm{Vol}\left(B_{2^{-l}}(x, 1+\rho_{2^{-l}}(x,z))\right)},$$

which completes the proof, by taking $\widetilde{N}$ sufficiently large. $\qquad\square$

Theorem 5.1.40 follows from the discussion at the start of this section and the following proposition.

PROPOSITION 5.10.2. *Let* $s \in \mathbb{R}^\nu$. *There exists* $M = M(s)$ *such that*

$$\sum_{\substack{j \in \mathbb{N}^\nu \\ j_1 \geq j_2 \geq \cdots \geq j_\nu}} 2^{j \cdot s} |E_j(x,z)| \lesssim \rho_{2^{-0}}(x,z)^{-M} \mathrm{Vol}\left(B_{2^{-0}}(x, \rho_{2^{-0}}(x,z))\right)^{-1},$$

where $2^{-0} = \left(2^{-0}, \ldots, 2^{-0}\right) = (1, \ldots, 1) \in (0,1]^\nu$.

PROOF. Take $N = 1$ in Lemma 5.10.1 and let M be the corresponding $M = M(s, N)$. Let $m = m(M)$ be a large number, to be chosen later. In what follows, $l = (j_\nu, \ldots, j_\nu)$. We use the easily seen fact that $\rho_{2-l}(x, z) = 2^{j_\nu} \rho_{2-0}(x, z)$.

$$\sum_{\substack{j \in \mathbb{N}^\nu \\ j_1 \geq j_2 \geq \cdots \geq j_\nu}} 2^{j \cdot s} |E_j(x, z)|$$

$$\lesssim \sum_{\substack{j \in \mathbb{N}^\nu \\ j_1 \geq j_2 \geq \cdots \geq j_\nu}} 2^{-|j-l|} 2^{M j_\nu} \frac{(1 + \rho_{2-l}(x, z))^{-m}}{\mathrm{Vol}(B_{2-l}(x, 1 + \rho_{2-l}(x, z)))}$$

$$\lesssim \sum_{j_\nu \in \mathbb{N}} 2^{M j_\nu} \frac{(1 + 2^{j_\nu} \rho_{2-0}(x, z))^{-m}}{\mathrm{Vol}(B_{2-0}(x, 2^{-j_\nu}(1 + 2^{j_\nu} \rho_{2-0}(x, z))))}.$$

For $x, z \in \Omega$ fixed with $x \neq z$, we separate the above sum into two sums. For the first, we sum over those j_ν such that $2^{j_\nu} \rho_{2-0}(x, z) \leq 1$. This sum is comparable to

$$\sum_{2^{j_\nu} \rho_{2-0}(x,z) \leq 1} 2^{M j_\nu} \mathrm{Vol}(B_{2-0}(x, 2^{-j}))^{-1},$$

which is bounded termwise by a geometric series, and is therefore bounded by a constant times its largest term:

$$\rho_{2-0}(x, z)^{-M} \mathrm{Vol}(B_{2-0}(x, \rho_{2-0}(x, z)))^{-1},$$

which is the desired estimate.

For the second sum, we sum of those j_ν with $2^{j_\nu} \rho_{2-0}(x, z) \geq 1$. This sum is comparable to:

$$\sum_{2^{j_\nu} \rho_{2-0}(x,z) \geq 1} 2^{M j_\nu} \frac{2^{-m j_\nu} \rho_{2-0}(x, z)^{-m}}{\mathrm{Vol}(B_{2-0}(x, \rho_{2-0}(x, z)))},$$

which, if $m > M$, is bounded termwise by a geometric series, and is therefore bounded by a constant times its largest term:

$$\rho_{2-0}(x, z)^{-M} \mathrm{Vol}(B_{2-0}(x, \rho_{2-0}(x, z)))^{-1},$$

which is the desired estimate. This completes the proof. □

Remark 5.10.3 The above estimates on the Schwartz kernel of T are certainly not optimal. In particular, we did not specify M in the estimate. More seriously, even if the optimal M were found, this does not yield the optimal estimate. Even in simple special cases, it seems to be somewhat involved to deduce the optimal estimates (see, e.g., [Str08, Section 9.1]). Moreover, even if the optimal estimates were obtained, we do not know of any use for these estimates: in particular, we know of no way to simplify any of the definitions $\mathcal{A}_1$, $\mathcal{A}_2$, $\mathcal{A}_3$, or $\mathcal{A}_4$ using Schwartz kernel estimates.

5.10.1   Operators on a compact manifold

The definition of $\mathcal{A}$ behaves well under diffeomorphisms (Remark 5.1.16). It is therefore desirable to define these operators on a manifold. Unfortunately, the theory described in this chapter only deals with the composition of operators which are both defined on a particular small open set, and a natural way to extend these definitions (in general) to operators which are not defined on a small set is not clear. One situation where it is clear is when the operators are pseudolocal and we are working on a compact manifold.

Let M be a compact manifold. Suppose we are given ν lists of vector fields $\left(X^1, \hat{d}^1\right), \ldots, \left(X^\nu, \hat{d}^\nu\right)$ as in this chapter, and from them we obtain the list of vector fields with ν parameter degrees $(X, d) = (X_1, d_1), \ldots, (X_q, d_q)$. We assume all the assumptions from Section 3.1, where we take $U_1 = U_2 = U = M$. We assume, in addition, that $each\ list\ X^\mu$ spans the tangent space at each point of M. As we saw above, in this case, the operators in our algebra $\mathcal{A}^s$ are pseudolocal.

For each x in M there is a neighborhood $\Omega_x = B_{(X,d)}\left(x, \frac{\xi_0}{2}(1, \ldots, 1)\right)$ of x_0 so that we may define an algebra of operators, $\mathcal{A}_x$, whose Schwartz kernels are supported in $\Omega_x \times \Omega_x$. Here our assumptions imply that ξ_0 may be chosen independent of $x \in M$.

Pick a finite cover of M consisting of balls of the form $B_{(X,d)}\left(x, \frac{\xi_0}{6}(1, \ldots, 1)\right)$, call these balls $\Omega_{x_1,0}, \ldots, \Omega_{x_L,0}$. Notice, if $\Omega_{x_j,0} \cap \Omega_{x_k,0} \neq \emptyset$ then $\Omega_{x_j,0} \cup \Omega_{x_k,0} \subset \Omega_{x_j}$, where $\Omega_{x_j} = B_{(X,d)}\left(x, \frac{\xi_0}{2}(1, \ldots, 1)\right)$. Take a partition of unity, $\psi_1, \ldots, \psi_L$, with $\psi_j \in C_0^\infty\left(\Omega_{x_j,0}\right)$. In what follows, we identify ψ_j with the operator given by multiplication by ψ_j.

DEFINITION 5.10.4.   *Let $s \in \mathbb{R}^\nu$ and $T : C^\infty(M) \to C^\infty(M)$. We say $T \in \mathcal{A}^s$ under the following conditions:*

- *If $\Omega_{x_j,0} \cap \Omega_{x_k,0} \neq \emptyset$, we assume $\psi_j T \psi_k \in \mathcal{A}^s_{x_j}$. It is easy to see that this is equivalent to $\psi_j T \psi_k \in \mathcal{A}^s_{x_k}$.*

- *If $\Omega_{x_j,0} \cap \Omega_{x_k,0} = \emptyset$, we assume $\psi_j T \psi_k \in C^\infty(M \times M)$.*

The following results follow immediately from the corresponding results in Section 5.1.

THEOREM 5.10.5.   • *If $T \in \mathcal{A}^t$ and $S \in \mathcal{A}^s$, then $TS \in \mathcal{A}^{s+t}$.*

- *If $T \in \mathcal{A}^0$, then $T : L^p(M) \to L^p(M)$.*

DEFINITION 5.10.6.   *For $1 < p < \infty$, we say $f \in \mathrm{NL}^p_s$ if $\psi_j f \in \mathrm{NL}^p_s$ for every j, where the later space is defined for functions supported in $\Omega_{x_j,0}$ as in Section 5.1.1.*

THEOREM 5.10.7.   *If $T \in \mathcal{A}^s$, then $T : \mathrm{NL}^p_{s_0} \to \mathrm{NL}^p_{s_0-s}$.*

Remark 5.10.8   It is easy to see that $\mathcal{A}^s$, as defined in Definition 5.10.4, depends only on the manifold M and the list of vector field (X, d). It does not depend on, for instance, the chosen partition of unity, or the chosen finite cover $\Omega_{x_j,0}$, etc.

5.11 EXAMPLES

In this section, we present several special cases of the theory developed in this chapter. For some of these examples, we further discuss some of the algebras brought up in Chapter 4. Other examples are new.

5.11.1 Euclidean vector fields

The Euclidean geometry induced by $(\partial_{x_1}, 1), \ldots, (\partial_{x_n}, 1)$ on $\mathbb{R}^n$ has a very special property with respect to our assumptions. Indeed, suppose we have a list of vector fields with ν parameter formal degrees $(X, d) = (X_1, d_1), \ldots, (X_q, d_q)$ induced by ν lists of vector fields with single-parameter formal degrees $\left(X^1, \hat{d}^1\right), \ldots, \left(X^\nu, \hat{d}^\nu\right)$ satisfying all the assumptions of this chapter. We, therefore, obtain a ν parameter algebra of operators $\mathcal{A}^s$ ($s \in \mathbb{R}^\nu$) as in this chapter.

Suppose we wish to add another list of vector fields with single-parameter formal degrees $\left(X^{\nu+1}, \hat{d}^{\nu+1}\right)$ to our original ν lists of vector fields to obtain a $\nu+1$ parameter algebra of operators. Of course, we need to verify that our assumptions still hold with this new list added. However, if we take this list to be the Euclidean vector fields $(\partial_{x_1}, 1), \ldots, (\partial_{x_n}, 1)$ then our assumptions automatically hold.

Indeed, to see this, note that we have

$$\left[X_j^\mu, \partial_{x_k}\right] = \sum_{l=1}^{n} c_{j,k}^l \partial_{x_l},$$

where $c_{j,k}^l \in C^\infty$. It follows that

$$\left[\delta_\mu^{\hat{d}_j^\mu} X_j^\mu, \delta_{\nu+1}\partial_{x_k}\right] = \sum_{l=1}^{n} \left(\delta_\mu^{\hat{d}_j^\mu} c_{j,k}^l\right) \left(\delta_{\nu+1}\partial_{x_l}\right),$$

where, of course, $\delta_\mu^{\hat{d}_j^\mu} c_{j,k}^l \in C^\infty$ uniformly for $\delta_\mu \in (0, 1]$. Thus, if we take the $\nu + 1$st list to be the Euclidean vector fields, then all of the assumptions of this chapter hold, and we obtain a $\nu + 1$ parameter algebra of singular integrals, call it $\mathcal{B}^s$, $s \in \mathbb{R}^{\nu+1}$.

As commented in Example 5.1.31, standard pseudodifferential operators of order m which are supported in $\Omega \times \Omega$ are elements of $\mathcal{B}^{(0,\ldots,0,m)}$. Furthermore, as seen in Corollary 5.1.29, elements of $\mathcal{B}^{(0,\ldots,0,m)}$ extend to bounded operators $L_r^p \to L_{r-m}^p$ ($1 < p < \infty$), where L_r^p denotes the standard L^p Sobolev space of order $r \in \mathbb{R}$.

Not every element of $\mathcal{A}^s$ is an element of $\mathcal{B}^{(s,0)}$. However, Corollary 5.1.38 gives a convenient sufficient condition for an element of $\mathcal{A}^s$ to be an element of $\mathcal{B}^{(s,0)}$. In particular, if $T \in \mathcal{A}^0$, and the assumptions of that corollary hold, then $T \in \mathcal{B}^0$ and $T : L_r^p \to L_r^p$ ($1 < p < \infty, r \in \mathbb{R}$).

Furthermore, if $T \in \mathcal{A}^s$ and the assumptions of Corollary 5.1.38 hold, then we have an intrinsic understanding of the composition of T with a standard pseudodifferential operator of order m (supported in $\Omega \times \Omega$): it is an element of $\mathcal{B}^{(s,m)}$.

5.11.2 Hörmander vector fields and other geometries

We suppose we are given ν lists of vector fields $\left(X^1, \hat{d}^1\right), \dots, \left(X^\nu, \hat{d}^\nu\right)$ satisfying the assumptions of this chapter. We suppose, in addition, that the first[7] list of vector fields is "generated" by vector fields satisfying Hörmander's condition.

That is, suppose we are given C^∞ vector fields $W_1, \dots, W_r$ (defined on the open set U from Section 3.1). We assume that $W_1, \dots, W_r$ satisfy Hörmander's condition of order m on U (see Definition 2.0.1). We assign to $W_1, \dots, W_r$ the formal degree 1, and recursively if Y_1 has formal degree e and Y_2 has formal degree f, we assign to $[Y_1, Y_2]$ the formal degree $e + f$. We assume $\left(X_1^1, \hat{d}_1^1\right), \dots, \left(X_{q_1}^1, \hat{d}_{q_1}^1\right)$ is an enumeration of the above generated vector fields with formal degrees $\le m$. By our hypothesis, $X_1^1, \dots, X_{q_1}^1$ span the tangent space at every point.

Because we are assuming the list of vector fields with ν parameter formal degrees, (X, d), satisfies the hypotheses of this chapter, we obtain an algebra of ν parameter singular integrals $\mathcal{A}^s$, $s \in \mathbb{R}^\nu$.

Let $\mathcal{L} = W_1^* W_1 + \dots + W_r^* W_r$, the sub-Laplacian. Let $\psi_1, \psi_2 \in C_0^\infty(\Omega)$. From the theory in Chapter 2, we know that if $\mathcal{L}^{-1}$ is a fundamental solution for $\mathcal{L}$, then $\psi_1 \mathcal{L}^{-1} \psi_2$ is a Calderón-Zygmund singular integral operator of order -2, associated to the vector fields with formal degrees $\left(X^1, \hat{d}^1\right)$. The main theorem of this section is the following.

THEOREM 5.11.1. *Let $\psi_1, \psi_2 \in C_0^\infty(\Omega)$ and let $\mathcal{L}^{-1}$ be a fundamental solution for $\mathcal{L}$. Then, $\psi_1 \mathcal{L}^{-1} \psi_2 \in \mathcal{A}^{(-2,0,\dots,0)}$.*

To prove Theorem 5.11.1, we look back to the pseudodifferential operators of Section 2.14. Let $\psi_1, \psi_2 \in C_0^\infty(\Omega)$ be as in Theorem 5.11.1 and let $\psi \in C_0^\infty(\Omega)$ equal 1 on a neighborhood of $\mathrm{supp}(\psi_1) \cup \mathrm{supp}(\psi_2)$. Looking back at the proof of Theorem 2.14.28, we see that there is a Calderón-Zygmund kernel $K(t)$ of order -2, as in Definition 2.13.14, and $\eta \in C_0^\infty(B^q(a))$ where $a > 0$ is small, such that if we define

$$T_0 f(x) := \psi(x) \int f\left(e^{t_1 \cdot X^1} x\right) \psi(x) \eta(t_1) K(t_1) \, dt,$$

then $T_0 \mathcal{L} \equiv \psi^2$, modulo a Calderón-Zygmund operator of order -1, defined with respect to the list $\left(X^1, \hat{d}^1\right)$.[8]

If we set $\widetilde{K}(t_1, \dots, t_\nu) = K(t_1) \otimes \delta_0(t_2) \otimes \dots \otimes \delta_0(t_\nu)$, we have that $\widetilde{K}$ is a product kernel of order $(-2, 0, \dots, 0)$ as in Definition 5.2.13. Theorem 5.1.37 shows $T_0 \in \mathcal{A}^{(-2,0,\dots,0)}$.

Define R by $T_0 \mathcal{L} = \psi^2 + R$. We know that R is a Calderón-Zygmund operator of order -1. In addition, since $T_0 \in \mathcal{A}^{(-2,\dots,0)}$, $T_0 \mathcal{L} \in \mathcal{A}^{(0,0,\dots,0)}$ and it follows that

---

[7]It is not important that we single out the *first* list of vector fields.

[8]Here we are working with Calderón-Zygmund operators, as in Definition 2.0.16, supported on $\Omega \times \Omega$, instead of supported on a compact manifold, but this is not an essential difference. Indeed, this is merely the theory developed in this chapter in the case $\nu = 1$.

$R \in \mathcal{A}^0$. Fix $N \in \mathbb{N}$. Define

$$T_N = \sum_{j=0}^{N} T_0 \left(-R\right)^j \in \mathcal{A}^{(-2,0,\dots,0)}.$$

Let $\psi_0 \in C_0^\infty(\Omega)$ equal 1 on a neighborhood of $\mathrm{supp}(\psi_1) \cup \mathrm{supp}(\psi_2)$ but with $\psi \equiv 1$ on a neighborhood of $\mathrm{supp}(\psi_0)$. Notice, $\psi_0 T_0 \mathcal{L} \psi_0 = \psi_0^2 + R_N$, where R_N is a Calderón-Zygmund operator of order $-N$. We have

PROPOSITION 5.11.2. *Let $\mathcal{L}^{-1}$ be a fundamental solution for $\mathcal{L}$. Then, for every L there is an N such that $\psi_1 T_N \psi_2 = \psi_1 \mathcal{L}^{-1} \psi_2 + E_N$, where $E_N \in C_0^L(\Omega \times \Omega)$.*

PROOF. Fix L' large. If $N = N(L')$ is sufficiently large, $R_N \in C_0^{L'}(\Omega \times \Omega)$ by the definition of Calderón-Zygmund operators. Consider, then $\psi_0 \left(T_N - \mathcal{L}^{-1}\right) \mathcal{L}\psi_0 \equiv 0 \mod C_0^{L'}(\Omega \times \Omega)$. By taking $L' = L'(L)$ sufficiently large, the subellipticity of $\mathcal{L}$ completes the proof. $\square$

The next theorem does not use the fact that the first list of vector fields was generated by Hörmander vector fields: it holds in the general setting of this chapter.

THEOREM 5.11.3. *Fix $s \in \mathbb{R}^\nu$. Suppose $T : C^\infty(\Omega) \to C_0^\infty(\Omega)$ is such that for every L, there exists $S_L \in C_0^L(\Omega \times \Omega)$ with $T \equiv S_L \mod \mathcal{A}^s$. Then, $T \in \mathcal{A}^s$.*

PROOF. We show $T \in \mathcal{A}_2^s$, and to do this we use Theorem 5.5.5. Fix m and take $M_0 = M_0(m, s)$ to be the $M(m, s)$ from Theorem 5.5.5. Notice, for $x, z \in \Omega$,
$$\frac{\left(1 + \rho_{2^{-0}}(x, z)\right)^{-m}}{\mathrm{Vol}\left(B_{2^{-0}}\left(x, 1 + \rho_{2^{-0}}(x, z)\right)\right)} \approx 1.$$ Take N so large for $j \in [0, \infty)^\nu$,

$$2^{-N|j|_\infty} \approx 2^{-N|j|_\infty} \frac{\left(1 + \rho_{2^{-0}}(x, z)\right)^{-m}}{\mathrm{Vol}\left(B_{2^{-0}}(x, 1 + \rho_{2^{-0}}(x, z))\right)}$$
$$\lesssim 2^{j \cdot s} \frac{\left(1 + \rho_{2^{-j}}(x, z)\right)^{-m}}{\mathrm{Vol}\left(B_{2^{-j}}(x, 1 + \rho_{2^{-j}}(x, z))\right)},$$

where we have used Lemma 3.2.5. Let $\mathcal{T}_M$ be a bounded set of test functions of order $M := M_0 \vee N$. We wish to estimate $\left(2^{-j}X\right)^\alpha T\psi(x)$ for $\left(\psi, z, 2^{-j}\right) \in \mathcal{T}_M$. Take $L = M + |\alpha|$ and write $T = S_L + A_L$ where $A_L \in \mathcal{A}^s$ and $S_L \in C_0^L(\Omega \times \Omega)$. By Theorem 5.5.5, we have

$$\left|\left(2^{-j}X\right)^\alpha A_L \psi(x)\right| \lesssim 2^{s \cdot j} \frac{\left(1 + \rho_{2^{-j}}(x, z)\right)^{-m}}{\mathrm{Vol}\left(B_{2^{-j}}(x, 1 + \rho_{2^{-j}}(x, z))\right)},$$

and so it suffices to prove the corresponding estimate for S_L. Using the definition of test functions, we have
$$\psi = 2^{-M|j|_\infty} \sum_{|\beta| \le M} X^\beta \phi_\beta,$$

where $\left\{ \left(\phi_\beta, z, 2^{-j} \right) \mid \left(\psi, z, 2^{-j} \right) \in \mathcal{T}_M, |\beta| \le M \right\}$ is a bounded set of bump functions. Thus,

$$\left(2^{-j} X \right)^\alpha S_L \psi \left(x \right) \le 2^{-N|j|_\infty} \sum_{|\beta| \le M} \left| X^\alpha S_L X^\beta \phi_\beta \right|.$$

$X^\alpha S_L X^\beta \in C_0^0 \left(\Omega \times \Omega \right)$ and it follows that $\left| X^\alpha S_L X^\beta \phi_\beta \right| \lesssim 1$. Thus,

$$\left| \left(2^{-j} X \right)^\alpha S_L \psi \left(x \right) \right| \lesssim 2^{-N|j|_\infty} \lesssim 2^{j \cdot s} \frac{\left(1 + \rho_{2^{-j}} \left(x, z \right) \right)^{-m}}{\mathrm{Vol} \left(B_{2^{-j}} \left(x, 1 + \rho_{2^{-j}} \left(x, z \right) \right) \right)},$$

completing the proof that $T \in \mathcal{A}_2^s$. $\square$

PROOF OF THEOREM 5.11.1.   Theorem 5.11.1 follows by combining Proposition 5.11.2 and Theorem 5.11.3 and using $T_N \in \mathcal{A}^{(-2,0,\ldots,0)}$. $\square$

Remark 5.11.4   Combining the results of this section with the discussion from Section 5.11.1, we see that if $\nu = 2$ and we take the first list of vector fields to be generated by Hörmander vector fields, and the second list of vector fields to be the Euclidean vector fields, then we obtain a two parameter algebra of singular integrals containing both the standard pseudodifferential operators and the fundamental solutions to $\mathcal{L}$. We discuss this more in the next section.

5.11.3   Carnot-Carathéodory and Euclidean geometries

In this section, we return to the example discussed in Section 4.4 (and also the setting discussed in Remark 5.11.4). The setting is a **compact manifold** M. We are given two lists of vector fields with single-parameter formal degrees. The first $\left(X^1, \hat{d}^1 \right)$ is generated by vector fields satisfying Hörmander's condition, $W_1, \ldots, W_r$, as in Section 5.11.2. The second list of vector fields induces the usual Euclidean geometry on M.[9] To obtain this, we take $X_1^2, \ldots, X_{q_2}^2$ to be the list of vector fields spanning the tangent space at each point of M. We define $\left(X^2, \hat{d}^2 \right) := \left(X_1^2, 1 \right), \ldots, \left(X_{q_2}^2, 1 \right)$. Just as in Section 5.11.1, we have that the two lists of vector fields satisfy the hypotheses of this chapter.

Since *each* of the lists of vector fields (X^1 and X^2) span the tangent space to each point of M, the operators we consider are pseudolocal (see Section 5.1.4). Using the theory in Section 5.10.1, we obtain an algebra $\mathcal{A}^s$ ($s \in \mathbb{R}^2$) of operators supported on $M \times M$; i.e., we do not need to restrict attention to operators with small support.

Just as in Section 5.11.2, if $\mathcal{L} = W_1^* W_1 + \cdots + W_r^* W_r$ is the sub-Laplacian, and if we define $\mathcal{L}^{-1}$ as in Section 2.6, then $\mathcal{L}^{-1} \in \mathcal{A}^{(-2,0)}$. Also, using the discussion from Section 5.11.1, we see that standard pseudodifferential operators of order s are elements of $\mathcal{A}^{(0,s)}$. Combining these two together, we have an intrinsic understanding

---

[9] By this we mean that the balls are comparable to the balls associated to any Riemannian metric.

of the composition $a(x, D)\mathcal{L}^{-1}$ where $a(x, D)$ is a pseudodifferential operator of order s: the composition is an element of $\mathcal{A}^{(-2,s)}$.

Among other things, combining the above results with Corollary 5.1.29, we see that $W_iW_j\mathcal{L}^{-1} : L^p_s \to L^p_s$ ($1 < p < \infty$, $s \in \mathbb{R}$), where L^p_s denotes the standard isotropic L^p Sobolev space of order s–this boundedness was originally proved by Rothschild and Stein [RS76] by different methods.[10]

We now turn to the special case covered in Section 4.4.1, and we discuss the proof of Theorem 4.4.3. We take all the same notation as Section 4.4.1, and so we are in a special case of the above setting, and we have an algebra of singular integrals $\mathcal{A}^s$, $s \in \mathbb{R}^2$.

Recall the algebra of operators $\mathcal{B}^s_1$, $s \in \mathbb{R}$ defined in terms of the Hörmander vector fields on M; as in Section 4.4.1 there are operators $K \in \mathcal{B}^{-2}_1$ and $S \in \mathcal{B}^0_1$ such that $\Box^1_b K = K\Box^1_b = I - S$.

LEMMA 5.11.5. *Let T be a differential operator of order m. Then, TK can be written as a finite sum of terms K_jT_j, where $K_j \in \mathcal{B}^{-2}_1$ and T_j is a differential operator of order $\leq m$.*

COMMENTS ON THE PROOF. This is an immediate corollary of Theorem 6.7 of [CNS92], and we refer the reader there for the proof. □

PROPOSITION 5.11.6. $K \in \mathcal{A}^{(-2,0)}$.

PROOF. Lemma 5.11.5 gives precisely the hypotheses of Corollary 5.1.38 in this situation. The conclusion of Corollary 5.1.38 is the statement of the proposition. □

Define $P_0 = \Box^- K\Gamma^+ + Q^+$ as in Section 4.4.1. Since $K \in \mathcal{A}^{(-2,0)}$ and pseudodifferential operators of order m are in $\mathcal{A}^{(0,m)}$, we see $P_0 \in \mathcal{A}^{(-2,1)}$. We state the next result of Chang, Nagel, and Stein without proof.

THEOREM 5.11.7 (Theorem 4.7 of [CNS92]). $\Box^+ P_0 = I + E$, *where there exists $\epsilon > 0$ such that $E : L^2_s \to L^2_{s+\epsilon}$ for every $s \in \mathbb{R}$ and L^2_s denotes the standard L^2 Sobolev space of order s.*

LEMMA 5.11.8. *Let E be as in Theorem 5.11.7. Then $E \in \mathcal{A}^0$.*

PROOF. It suffices to show $\Box^+ P_0 \in \mathcal{A}^0$, since $E = \Box^+ P_0 - I$. $\Box^+ Q^+$ is a standard pseudodifferential operator of order 0, and so $\Box^+ Q^+ \in \mathcal{A}^0$. We have $K\Gamma^+ \in \mathcal{A}^{(-2,0)}$ and so to complete the proof it suffices to show $\Box^+\Box^- \in \mathcal{A}^{(2,0)}$. This follows immediately from the formula in Lemma 4.4 of [CNS92], completing the proof. □

For each N define $P_N = \sum_{j=0}^{N-1} P_0(-E)^N \in \mathcal{A}^{(-2,1)}$. Notice, $\Box^+ P_N = I + E_N$, where $E_N : L^2_s \to L^2_{s+N\epsilon}$ for every $s \in \mathbb{R}$.

THEOREM 5.11.9. *Let P satisfy $\Box^+ P \equiv I \mod C^\infty(M \times M)$. Then, $P \in \mathcal{A}^{(-2,1)}$.*

---

[10]It also follows from this argument that standard pseudodifferential operators are bounded on the non-isotropic Sobolev spaces associated to $W_1, \ldots, W_r$ (see Section 2.10). This can also be proved directly using the calculus of pseudodifferential operators from Section 1.3.

PROOF. Because $E_N : L^2_s \to L^2_{s+N\epsilon}$ for every $s \in \mathbb{R}$, we have $\forall L, \exists N$, with $E_N \in C^L (M \times M)$. From here the proof proceeds just as in Theorem 5.11.1. $\square$

5.11.4 An Example of Kohn

In this section, we return to the setting of Section 4.3.1, and we discuss a parametrix for Kohn's hypoelliptic operator $\mathcal{L}_k$. Away from $z = 0$, $\mathcal{L}_k$ is known to be maximally hypoellitpic, and so we are only concerned with $\mathcal{L}_k$ for z near 0. $\mathcal{L}_k$ is translation invariant in t, and it therefore suffices to understand the operator for t near 0. Thus, we fix a small open neighborhood, Ω, of $0 \in \mathbb{H}^1$, and we wish to study $\mathcal{L}_k$ on Ω.

There are three geometries that come into play. The left invariant geometry associated to the vector fields

$$\left(X^1, \hat{d}^1\right) := (X_L, 1), (Y_L, 1), (T, 2),$$

where $T = \frac{\partial}{\partial t}$ and X_L, Y_L are as in Section 4.3.1. The right invariant geometry associated to the vector fields

$$\left(X^2, \hat{d}^2\right) := (X_R, 1), (Y_R, 1), (T, 2).$$

Finally, the Euclidean geometry associated to the vector fields

$$\left(X^3, \hat{d}^3\right) := \left(\frac{\partial}{\partial x}, 1\right), \left(\frac{\partial}{\partial y}, 1\right), \left(\frac{\partial}{\partial t}, 1\right).$$

Because left invariant vector fields commute with right invariant vector fields, and because of the special nature of the Euclidean vector fields outlined in Section 5.11.1, it is immediate to verify that these three lists of vector fields satisfy the assumptions of this chapter. We, therefore, obtain an algebra of operators on Ω, denoted by $\mathcal{A}^s$, where $s \in \mathbb{R}^3$.

We now outline the construction of the parametrix for $\mathcal{L}_k$ developed in [Str10]. For further details and more general operators, we refer the reader to that reference.

There are three types of operators which arise in the constriction of the parametrix for $\mathcal{L}_k$. If K is a Calderón-Zygmund kernel of order $s > -4$ as in Definition 2.13.14, then we obtain two operators given by convolution with K. A left invariant operator $\text{Op}_L (K) f = f * K$ and a right invariant operator $\text{Op}_R (K) f = K * f$. If $\psi_1, \psi_2 \in C^\infty_0 (\Omega)$, it is immediate to verify that $\psi_1 \text{Op}_L (K) \psi_2$ is a pseudodifferential operator in the sense of Section 5.1.2, and therefore $\psi_1 \text{Op}_L (K) \psi_2 \in \mathcal{A}^{(s,0,0)}$. Similarly, $\psi_1 \text{Op}_R (K) \psi_2 \in \mathcal{A}^{(0,s,0)}$. We write L^s for the space of operators of the form $\psi_1 \text{Op}_L (K) \psi_2$ (for any choice of K, ψ_1, and ψ_2) and R^s for the space of operators of the form $\psi_1 \text{Op}_R (K) \psi_2$. The third type of operators which arises comes directly as pseudodifferential operators with respect to the left invariant vector fields $\left(X^1, \hat{d}^1\right)$. Indeed, if $K (x, y)$ is a C^∞ function in the x variable taking values in Calderón-Zygmund kernels of order $s > -4$ (again, as in Definition 2.13.14), then for $\psi_1, \psi_2 \in C^\infty_0 (\Omega)$, we may define an operator

$$Tf (\xi_1) = \psi_1 (\xi_1) \int f (\xi_2) \psi_2 (\xi_2) K \left(\xi_1, \xi_2^{-1} \xi_1\right) d\xi_2.$$

It is easy to see that T is a pseudodifferential operator of order s with respect to the list $\left(X^1, \hat{d}^1\right)$ (in the sense of Section 5.1.2) and we therefore have $T \in \mathcal{A}^{(s,0,0)}$. We denote the space of such operators T as P_L^s.

Fix $\psi_0 \in C_0^\infty(\Omega)$ with $\psi_0 \equiv 1$ on a neighborhood of 0. In [Str10] it is shown that there are operators $U_{-2} \in P_L^{-2}$, $R_0, R_0' \in R^0$, $L_0, L_0' \in L^0$ such that if we define

$$S = U_{-2} + \frac{\partial^{k-1}}{\partial t^{k-1}} R_0 L_0 + R_0' L_0' U_{-2} \in \mathcal{A}^{(0,0,k-1)},$$

we have $\psi_0 S \mathcal{L}_k \psi_0 = \psi_0^2 + E$, where $E \in \mathcal{A}^{(-1,0,0)}$.

For each $N \in \mathbb{N}$ define

$$S_N = \sum_{j=0}^{N} S(-E)^j \in \mathcal{A}^{(0,0,k-1)}.$$

We have $\psi_0 S_N \mathcal{L}_k \psi_0 = \psi_0^2 + E_N$, where $E_N \in \mathcal{A}^{(-N-1,0,0)}$.

Because X_L, Y_L, T span the tangent space, for every s there is N such that $E_N : L_{-s}^2 \to L_s^2$. Thus, for every R there is N such that $E_N \in C_0^R(\Omega \times \Omega)$. From here, the next theorem follows just as in Theorem 5.11.1.

THEOREM 5.11.10. *Let $\mathcal{L}_k^{-1}$ be a fundamental solution for $\mathcal{L}_k$. Then, for $\psi_1, \psi_2 \in C_0^\infty(\Omega)$, $\psi_1 \mathcal{L}_k^{-1} \psi_2 \in \mathcal{A}^{(0,0,k-1)}$.*

COROLLARY 5.11.11. *Let $1 < p < \infty$ and suppose $\mathcal{L}_k u \in L_s^p$ on a neighborhood of $0 \in \mathbb{H}^1$. Then, $u \in L_{s-k+1}^p$ on a neighborhood of 0.*

PROOF. Using Theorem 5.11.10 this follows immediately from Corollary 5.1.29. $\qquad\square$

5.11.5   The product theory of singular integrals

In Section 4.1, we outlined the theory of "product singular integrals." It is evident that the definitions there are closely related to the definitions in this chapter, but it is perhaps not obvious that the theory there is essentially a special case of the theory in this chapter. In this section, we discuss this fact.

We take the same setting as Section 4.1. There we are given ν compact manifolds $M_1, \ldots, M_\nu$ and we consider operators on the compact manifold $M = M_1 \times \cdots \times M_\nu$. On each factor M_μ, we are given vector fields $W_1^\mu, \ldots, W_{r_\mu}^\mu$ satisfying Hörmander's condition, and we use these to generate a list of vector fields with formal degrees $\left(X^\mu, \hat{d}^\mu\right) = \left(X_1^\mu, \hat{d}_1^\mu\right), \ldots, \left(X_{q_\mu}^\mu, \hat{d}_{q_\mu}^\mu\right)$ as in Section 2.1. We think of each X_j^μ as a vector field on M. Since $\left[X_j^\mu, X_k^{\mu'}\right] = 0$ for $\mu \neq \mu'$ it is evident that these ν lists of vector fields satisfy the hypotheses of this chapter–see (4.2).

Fix $x_0 \in M$. Applying the theory of this chapter, there is a small open set Ω, containing x_0, and an algebra of operators $\mathcal{A}^s$ ($s \in \mathbb{R}^\nu$): for $T \in \mathcal{A}^s$, supp $(T) \subset \Omega \times \Omega$.

THEOREM 5.11.12. *If $T \in \mathcal{A}^s$, then T is a product singular integral operator of order s in the sense of Definition 4.1.13. Conversely, if T is a product singular integral operator of order s and if $\psi_1, \psi_2 \in C_0^\infty(\Omega)$, then $\psi_1 T \psi_2 \in \mathcal{A}^s$.*

The difficulty in Theorem 5.11.12 is that the definition of bounded sets of elementary operators in this chapter (Definition 5.1.8) is, a priori,[11] less restrictive than the corresponding definition for product singular integral operators (Definition 4.1.11). Indeed, one part of Theorem 5.11.12 is easy:

PROPOSITION 5.11.13. *If T is a product singular integral operator of order s and if $\psi_1, \psi_2 \in C_0^\infty(\Omega)$, then $\psi_1 T \psi_2 \in \mathcal{A}^s$.*

To prove Proposition 5.11.13 we use the next lemma.

LEMMA 5.11.14. *Suppose $\mathcal{E}$ is a bounded set of elementary operators in the sense of Definition 4.1.11. Let $\psi_1, \psi_2 \in C_0^\infty(\Omega)$. Then*

$$\left\{ \left(\psi_1 E \psi_2, 2^{-j} \right) \mid \left(E, 2^{-j} \right) \in \mathcal{E} \right\}$$

is a bounded set of elementary operators in the sense of Definition 5.1.8.

PROOF. This follows easily from the definitions. ☐

PROOF OF PROPOSITION 5.11.13. If T is a product singular integral operator of order s, then in light of (iv) of Theorem 4.1.12, we may write $T = \sum_{j \in \mathbb{N}^\nu} 2^{j \cdot s} E_j$, where $\left\{ \left(E_j, 2^{-j} \right) \mid j \in \mathbb{N}^\nu \right\}$ is a bounded set of elementary operators in the sense of Definition 4.1.11. Thus, $\psi_1 T \psi_2 = \sum_{j \in \mathbb{N}^\nu} 2^{j \cdot s} \psi_1 E_j \psi_2$, where $\left\{ \left(\psi_1 E_j \psi_2, 2^{-j} \right) \mid j \in \mathbb{N}^\nu \right\}$ is a bounded set of elementary operators in the sense of Definition 5.1.8 (by Lemma 5.11.14). This shows $T \in \mathcal{A}_4^s$ and completes the proof. ☐

We now turn to the other part of Theorem 5.11.12, which we again state as a proposition.

PROPOSITION 5.11.15. *If $T \in \mathcal{A}^s$, then T is a product singular integral operator of order s in the sense of Definition 4.1.13.*

We separate the proof of Proposition 5.11.15 into a few lemmas.

LEMMA 5.11.16. *Let $\mathcal{E}_1$ be a bounded set of elementary operators in the sense of Definition 4.1.11 and let $\mathcal{E}_2$ be a bounded set of elementary operators in the sense of Definition 5.1.8. Then, for every N, the set*

$$\left\{ \left(2^{N \operatorname{diam}\{j_1, j_2, j_3\}} E_1 E_2 E_3, 2^{-j_2} \right) \mid \left(E_1, 2^{-j_1} \right), \left(E_3, 2^{-j_3} \right) \in \mathcal{E}_1, \left(E_2, 2^{-j_2} \right) \in \mathcal{E}_2 \right\}$$

is a bounded set of elementary operators in the sense of Definition 4.1.11. Here,

$$\operatorname{diam}\{j_1, j_2, j_3\} = \max_{1 \le k, l \le 3} |j_k - j_l|.$$

---

[11] We will see that Definition 5.1.8 is, in fact, not less restrictive than Definition 4.1.11, but this is not obvious and requires proof.

COMMENTS ON THE PROOF. This can be shown using all the same ideas as Theorem 5.4.1. $\qquad\square$

LEMMA 5.11.17. *Suppose $\mathcal{E}$ is a bounded set of elementary operators in the sense of Definition 5.1.8. Then $\mathcal{E}$ is a bounded set of elementary operators in the sense of Definition 4.1.11.*

PROOF. We apply Corollary 4.1.15 to write $I = \sum_{j \in \mathbb{N}^\nu} E_j$, where

$$\left\{ \left(E_j, 2^{-j} \right) \mid j \in \mathbb{N}^\nu \right\}$$

is a bounded set of elementary operators in the sense of Definition 4.1.11. For

$$\left(E, 2^{-j_2} \right) \in \mathcal{E}$$

we have

$$E = IEI = \sum_{j_1, j_3 \in \mathbb{N}^\nu} E_{j_1} E E_{j_3}$$

$$= \sum_{j_1, j_3 \in \mathbb{N}^\nu} 2^{-\operatorname{diam}\{j_1, j_2, j_3\}} \left(2^{\operatorname{diam}\{j_1, j_2, j_3\}} E_{j_1} E E_{j_3} \right).$$

Lemma 5.11.16 shows

$$\left\{ \left(\left(2^{\operatorname{diam}\{j_1, j_2, j_3\}} E_{j_1} E E_{j_3} \right), 2^{-j_2} \right) \mid j_1, j_3 \in \mathbb{N}^\nu, \left(E, 2^{-j_2} \right) \in \mathcal{E} \right\}$$

is a bounded set of elementary operators in the sense of Definition 4.1.11, and it follows that $\mathcal{E}$ is a bounded set of elementary operators in the sense of Definition 4.1.11. $\qquad\square$

PROOF OF PROPOSITION 5.11.15. Since $T \in \mathcal{A}^s = \mathcal{A}_4^s$, we have

$$T = \sum_{j \in \mathbb{N}^\nu} 2^{j \cdot s} E_j,$$

where $\left\{ \left(E_j, 2^{-j} \right) \mid j \in \mathbb{N}^\nu \right\}$ is a bounded set of elementary operators in the sense of Definition 5.1.8. Lemma 5.11.17 shows $\left\{ \left(E_j, 2^{-j} \right) \mid j \in \mathbb{N}^\nu \right\}$ is a bounded set of elementary operators in the sense of Definition 4.1.11. Theorem 4.1.12 then shows that T is a product singular integral operator of order s. $\qquad\square$

5.12 SOME GENERALIZATIONS

In the setup to this chapter, we were given vector fields with single-parameter formal degrees $\left(X_1^\mu, \hat{d}_1^\mu \right), \dots, \left(X_{q_\mu}^\mu, \hat{d}_{q_\mu}^\mu \right) 1 \le \mu \le \nu$ and $0 \ne \hat{d}_j^\mu \in \mathbb{N}$. Sometimes, applications arise when $\hat{d}_j^\mu \in (0, \infty)$. Our theory can be easily adapted to this case. Indeed, set

$$\epsilon_0 = \min_{\substack{1 \le \mu \le \nu \\ 1 \le j \le q_\mu}} \hat{d}_j^\mu.$$

If, in Definitions 5.1.2 and 5.1.7 we replace $2^{-|j-l|_\infty}$ with $2^{-\epsilon_0|j-l|_\infty}$, then all of our results generalize to this case with the same proofs.

In fact, one could replace ϵ_0 with any ϵ, $0 < \epsilon \le \epsilon_0$ and obtain the same algebra of operators. We can take this one step further and use the following, seemingly weaker, definitions.

DEFINITION 5.12.1. *For $M, N \ge 0$, we say $\mathcal{T}_M^N \subset C_0^\infty(\Omega) \times \Omega \times (0,1]^\nu$ is a bounded set of test functions of order M, N if $\forall\, (\psi, x, 2^{-j}) \in \mathcal{T}_M^N$, $\forall l \le j$ (here, $l \in [0, \infty)^\nu$)*

$$\psi = 2^{-|j-l|N} \sum_{|\alpha| \le M} \left(2^{-l} X\right)^\alpha \phi_{l,\alpha}$$

where $\left\{\left(\phi_{l,\alpha}, x, 2^{-j}\right) \mid (\psi, x, 2^{-j}) \in \mathcal{T}_M^N, l \le j, |\alpha| \le M\right\}$ is a bounded set of bump functions. Here, bounded sets of bump functions are defined as in Definition 5.1.1.

DEFINITION 5.12.2. *For $s \in \mathbb{R}^\nu$ and $T : C^\infty(\Omega) \to C_0^\infty(\Omega)$, we say $T \in \mathcal{B}_1^s$ if $\forall L, m, \exists N, \forall M, \forall \mathcal{T}_M^N$ bounded sets of test functions of order M, N, $\exists C, \forall\, (\psi_1, x, 2^{-j})$, $(\psi_2, y, 2^{-k}) \in \mathcal{T}_M^N$,*

$$\left|\langle \psi_1, T\psi_2\rangle_{L^2}\right| \le C 2^{j \wedge k \cdot s} 2^{-L|j-k|} \frac{(1 + \rho_{2^{-j \wedge k}}(x, y))^{-m}}{\mathrm{Vol}\left(B_{2^{-j \wedge k}}\left(x, 1 + \rho_{2^{-j \wedge k}}(x, y)\right)\right)}.$$

DEFINITION 5.12.3. *For $s \in \mathbb{R}^\nu$ and $T : C^\infty(\Omega) \to C_0^\infty(\Omega)$, we say $T \in \mathcal{B}_2^s$ if $\forall m, \exists N, \forall M, \forall \mathcal{T}_M^N$ bounded sets of test functions of order M, N, $\forall \alpha, \exists C$, $\forall\, (\psi, z, 2^{-j}) \in \mathcal{T}_M^N$,*

$$\left|\left(2^{-j} X\right)^\alpha T\psi(x)\right| \le 2^{s \cdot j} \frac{(1 + \rho_{2^{-j}}(x, z))^{-m}}{\mathrm{Vol}\left(B_{2^{-j}}\left(x, 1 + \rho_{2^{-j}}(x, z)\right)\right)},$$

and the same result holds for T replaced by T^* (the formal L^2 adjoint of T).

DEFINITION 5.12.4. *We define the set of bounded sets of generalized elementary operators of order $\epsilon > 0$, $\mathcal{G}_\epsilon$, to be the largest set of subsets of $C_0^\infty(\Omega \times \Omega) \times (0,1]^\nu \times (0,1]^\nu$ such that $\forall \mathcal{E} \in \mathcal{G}_\epsilon$:*

(i) $\forall\, (E, 2^{-j}, 2^{-k}) \in \mathcal{E}$, $k \le j$.

(ii) $\left\{\left(E, 2^{-j}\right) \mid (E, 2^{-j}, 2^{-k}) \in \mathcal{E}\right\}$ *is a bounded set of pre-elementary operators. Here, bounded sets of pre-elementary operators are defined as in Definition 5.1.6.*

(iii) $\forall\, (E, 2^{-j}, 2^{-k}) \in \mathcal{E}$, $\forall k \le l \le j$, *we have*

$$E = \begin{cases} 2^{-\epsilon|j-l|_\infty} \sum_{|\alpha| \le 1} E_{\alpha,l}^1 \left(2^{-l} X\right)^\alpha & and \\ 2^{-\epsilon|j-l|_\infty} \sum_{|\alpha| \le 1} \left(2^{-l} X\right)^\alpha E_{\alpha,l}^2, \end{cases} \tag{5.66}$$

where

$$\left\{\left(E_{\alpha,l}^1, 2^{-j}, 2^{-l}\right), \left(E_{\alpha,l}^2, 2^{-j}, 2^{-l}\right) \mid (E, 2^{-j}, 2^{-k}) \in \mathcal{E}, k \le l \le j, |\alpha| \le 1\right\}$$
$$\in \mathcal{G}_\epsilon.$$

We call elements of $\mathcal{G}_\epsilon$ bounded sets of generalized elementary operators of order ϵ.

DEFINITION 5.12.5. *We say $\mathcal{E} \subset C_0^\infty (\Omega \times \Omega) \times (0,1]^\nu$ is a bounded set of elementary operators of order $\epsilon > 0$ if $E \times \{2^{-0}\}$ is a bounded set of generalized elementary operators of order ϵ. We say $\mathcal{E}$ is a bounded set of elementary operators, if there is an $\epsilon > 0$ such that $\mathcal{E}$ is a bounded set of elementary operators of order ϵ.*

DEFINITION 5.12.6. *Fix $s \in \mathbb{R}^\nu$ and $T : C^\infty (\Omega) \to C_0^\infty (\Omega)$. We say $T \in \mathcal{B}_3^s$ if for every bounded set of elementary operators $\mathcal{E}$,*

$$\left\{ \left(2^{-j \cdot s} T E, 2^{-j}\right) \mid (E, 2^{-j}) \in \mathcal{E} \right\}$$

is a bounded set of elementary operators.

DEFINITION 5.12.7. *Fix $s \in \mathbb{R}^\nu$ and $T : C^\infty (\Omega) \to C_0^\infty (\Omega)$. We say $T \in \mathcal{B}_4^s$ if for each $j \in \mathbb{N}^\nu$ there is $E_j \in C_0^\infty (\Omega \times \Omega)$ such that $\left\{ (E_j, 2^{-j}) \mid j \in \mathbb{N}^\nu \right\}$ is a bounded set of elementary operators and*

$$T = \sum_{j \in \mathbb{N}^\nu} 2^{j \cdot s} E_j,$$

with the sum taken in the sense of distribution.

THEOREM 5.12.8. *$\mathcal{B}_1^s = \mathcal{B}_2^s = \mathcal{B}_3^s = \mathcal{B}_4^s$, and if we define $\mathcal{A}^s$ as in this chapter (with the use of ϵ_0 as remarked at the start of this section), then we have $\mathcal{B}_1^s = \mathcal{B}_2^s = \mathcal{B}_3^s = \mathcal{B}_4^s = \mathcal{A}^s$.*

COMMENTS ON THE PROOF. The proof that $\mathcal{B}_1^s = \mathcal{B}_2^s = \mathcal{B}_3^s = \mathcal{B}_4^s$ follows just as the corresponding proof for $\mathcal{A}_1^s = \mathcal{A}_2^s = \mathcal{A}_3^s = \mathcal{A}_4^s$. It is evident that $\mathcal{B}_2^s \subseteq \mathcal{A}_2^s$ and $\mathcal{A}_4^s \subseteq \mathcal{B}_4^s$ showing equality with $\mathcal{A}^s$. $\qquad \square$

5.13 CLOSING REMARKS

The main setting of this chapter involves vector fields paired with formal degrees $(X_1, d_1), \ldots, (X_q, d_q)$ satisfying certain hypotheses. A key assumption was that each d_j was nonzero in only one component. This played a fundamental role in many of our estimates, as can be seen explicitly in Section 3.2. Nevertheless, we did not use this assumption in proving the L^p boundedness of the maximal function in Section 3.3; nor did we need this assumption to make use of the quantitative Frobenius theorem from Section 2.2.

It is not hard to see that the assumption that each d_j is nonzero in only one component is essentially necessary for operators like the ones we have defined to form an algebra–for instance, it was necessary to prove Lemma 3.2.18. However, the question of L^p $(1 < p < \infty)$ boundedness is more subtle.

One issue that arises is that if some of the d_j are nonzero in more than one component, then the four definitions $\mathcal{A}_1, \mathcal{A}_2, \mathcal{A}_3, \mathcal{A}_4$ are, in general, not equivalent. It is therefore not obvious what a "singular integral operator" should be in this context. Nevertheless, there is still some hope for an L^p boundedness theorem, as can be seen by the following special case. We work on the three-dimensional Heisenberg group $\mathbb{H}^1$, which (as a manifold) is diffeomorphic to $\mathbb{R}^3$, and we use coordinates (x, y, t) on $\mathbb{H}^1$. Let $\{\varsigma_j \mid j \in \mathbb{Z}^2\} \subset \mathcal{S}(\mathbb{R}^3)$ be a bounded set such that

$$\int x^\alpha t^\beta \varsigma_j (x, y, t) \ dx \ dt = 0, \quad \int y^\alpha t^\beta \varsigma_j (x, y, t) \ dy \ dt = 0,$$

$\forall \alpha, \beta, j$. Define a distribution

$$K(x, y, t) = \sum_{(j_1, j_2) \in \mathbb{Z}^2} 2^{2j_1 + 2j_2} \varsigma_j \left(2^{j_1} x, 2^{j_2} y, 2^{j_1 + j_2} t \right).$$

Then, the operator $f \mapsto f * K$ is bounded on L^p $(1 < p < \infty)$, where the convolution is taken in the sense of $\mathbb{H}^1$. See [SS11, Str12, SS13] for more details on this. Let X, Y be the left invariant vector fields on $\mathbb{H}^1$ from Section 4.3.1, and define $T = [X, Y]$. The operator $f \mapsto f * K$ is morally a singular integral operator with respect to this 2-parameter Carnot-Carathéodory geometry given by the vector fields with formal degrees $(X, (1, 0)), (Y, (0, 1)), (T, (1, 1))$–which satisfy the main assumptions of this chapter (except, of course, not all of the formal degrees are nonzero in only one component). In this case, the operator is most closely related to the definition $\mathcal{A}_4^0$. It would be interesting to generalize this sort of result in a manner analogous to the more general definitions in this monograph.

In another direction, one might hope to generalize the singular integrals discussed in this paper to a setting where the balls are more general than Carnot-Carathéodory balls. For instance, in the single-parameter setting, there is the theory of singular integrals on spaces of homogeneous type. In that context, one is given a Borel measure and a family of single-parameter balls $B(x, \delta)$ satisfying certain axioms. In that setting, the Calderón-Zygmund theory of singular integrals is well known (see, e.g., [CW71, Ste93]). If one is instead given *multi*-parameter balls $B(x, \delta_1, \ldots, \delta_\nu)$ then it makes sense to ask the following questions:

- What should the definition of a singular integral corresponding to the multi-parameter balls be?

- What axioms (akin to the axioms of a space of homogeneous type) should one assume on the balls $B(x, \delta_1, \ldots, \delta_\nu)$ to be able to develop a theory of singular integrals?

We do not know of a way to generalize the definitions from this chapter to give any sort of reasonable answer to the first question. If possible, such an answer could be quite interesting. The second question is also open, but it seems likely that one would need to assume something which implies a result like Lemma 3.2.18 to carry out any theory like the one developed here.

Appendix A

Functional Analysis

Throughout this monograph, we use terminology and theorems from functional analysis. This appendix is designed to be a brief introduction to the relevant theory. The presentation here relies heavily on the books by Trèves [Trè67] and Rudin [Rud91]. We separate the discussion into two parts. In the first part, we discuss the basic definitions involved in the category of locally convex topological vector spaces; including the notions of categorical limits. In particular, we discuss the important space $C_0^\infty(\Omega)$, where $\Omega \subseteq \mathbb{R}^n$ is open. In the second part, we introduce the injective and projective tensor products of locally convex topological vector spaces.

A.1 LOCALLY CONVEX TOPOLOGICAL VECTOR SPACES

DEFINITION A.1.1. *A topological vector space is a vector space V with a Hausdorff topology such that the maps*

$$(v, w) \mapsto v + w, \quad V \times V \to V$$

$$(c, v) \mapsto cv, \quad \mathbb{C} \times V \to V$$

are continuous.

Remark A.1.2 Note that we have assumed that topological vector spaces are *Hausdorff*. This is not always assumed in the literature; however, every example we are concerned with is Hausdorff.

DEFINITION A.1.3. *A topological vector space V is said to be locally convex if there is a basis for the topology of V consisting of convex sets. In such a situation we say V is a "locally convex topological vector space" or LCTVS.*

Remark A.1.4 In what follows, it is best to think of LCTVS as a category whose object are locally convex topological vector spaces and whose morphisms are continuous linear maps.

DEFINITION A.1.5. *Let V be a vector space. A function $|\cdot| : V \to [0, \infty)$ is said to be a semi-norm on V if*

$$|v + w| \leq |v| + |w|, \quad \forall v, w \in V,$$

$$|cv| = |c| \, |v|, \quad \forall c \in \mathbb{C}, v \in V.$$

If V is a topological vector space, a semi-norm $|\cdot|$ is said to be a continuous semi-norm if $|\cdot| : V \to [0, \infty)$ is continuous.

DEFINITION A.1.6. *A family, $\mathcal{P}$, of semi-norms on a vector space V is said to be separating if for every $0 \neq v \in V$, there is $|\cdot| \in \mathcal{P}$ with $|v| \neq 0$.*

Given a separating family of semi-norms on a vector space V, there is a unique coarsest topology on V with respect to which all of these semi-norms are continuous. It is straightforward to verify that this topology makes V an LCTVS. Thus, when defining a topology on a vector space, we will often do so by defining a family of semi-norms which we take to be continuous. See [Rud91, Chapter 1] for more details.

Conversely, if V is an LCTVS, and we take $\mathcal{P}$ to be the collection of all continuous semi-norms on V, then the topology on V is the coarsest topology with respect to which all these semi-norms are continuous. Thus, every LCTVS can be defined by giving a vector space and a separating family of semi-norms on the space (see [Trè67, Proposition 7.5]).

DEFINITION A.1.7. *A metric $d : V \times V \to [0, \infty)$ is said to be invariant if $d(v + y, w + y) = d(v, w)$, $\forall v, w, y \in V$.*

Remark A.1.8 Suppose we have defined a locally convex topology on V by taking the weakest topology corresponding to a *countable* collection of semi-norms $\mathcal{P}$. This topology is induced by an invariant metric. Namely, if we take $|\cdot|_j$, $j \in \mathbb{N}$ to be an enumeration of $\mathcal{P}$, then

$$d(v, w) = \sum_{j \in \mathbb{N}} 2^{-j} \left(|v - w|_j \wedge 1 \right)$$

is an invariant metric inducing the same topology. In fact, an LCTVS having a topology induced by an invariant metric is equivalent to having the topology induced by a countable family of semi-norms.

DEFINITION A.1.9. *A topological vector space is called an F-space if its topology is induced by a complete, invariant metric. An LCTVS is called a Fréchet space if it is an F-space.*

We now turn to three simple examples of Fréchet spaces which we use.

Example A.1.10 *Let $\Omega \subseteq \mathbb{R}^n$ be open. We let $C^\infty(\Omega)$ be the vector space of those functions $f : \Omega \to \mathbb{C}$ which are C^∞. We let $K_1 \subset K_2 \subset K_3 \subset \cdots \subset \Omega$ be a sequence of compact sets with $\cup_j K_j = \Omega$. For each j and each multi-index α, we define a semi-norm*

$$|f|_{j,\alpha} = \sup_{x \in K_j} |\partial_x^\alpha f(x)|.$$

It is easy to verify that the topology so defined is independent of the choice of the sequence of compact sets, $\{K_j\}$, and turns $C^\infty(\Omega)$ into a Fréchet space.

Example A.1.11 *We define Schwartz space, $S(\mathbb{R}^n)$, by*

$$S(\mathbb{R}^n) = \left\{ f \in C^\infty(\mathbb{R}^n) \;\middle|\; \forall \alpha, \beta, \; \sup_{x \in \mathbb{R}^n} |\partial_x^\alpha x^\beta f(x)| < \infty \right\},$$

and we give $S(\mathbb{R}^n)$ the Fréchet topology given by the countable family of semi-norms

$$\|f\|_{\alpha,\beta} := \sup_{x \in \mathbb{R}^n} |\partial_x^\alpha x^\beta f(x)|.$$

Example A.1.12 *Let $K \subset \mathbb{R}^n$ be a compact set. We define $C^\infty(K)$ to be the vector space of those $f \in C^\infty(\mathbb{R}^n)$ with $\text{supp}(f) \subset K$. We define a Fréchet topology on K by giving the countable family of semi-norms: for each multi-index α*

$$|f|_\alpha = \sup_{x \in K} |\partial_x^\alpha f(x)|.$$

Two important theorems we often use for Fréchet spaces are the open mapping theorem and the closed graph theorem. In fact, these hold in the more general context of F-spaces.

THEOREM A.1.13 (The open mapping theorem (see page 47 of [Rud91])). *Suppose V is an F-space, W is a topological vector space, $T : V \to W$ is a continuous linear map, and $T(V) \subseteq W$ is of second category in W. Then, $T(V) = W$, T is an open mapping, and W is an F-space.*

THEOREM A.1.14 (The closed graph theorem (see page 50 of [Rud91])). *Suppose V and W are F-spaces, $T : V \to W$ is linear, and the graph $\{(v, Tv) : v \in V\} \subseteq V \times W$ is closed. Then T is continuous.*

A concept which is of repeated use to us is the notion of a bounded set.

DEFINITION A.1.15. *Let V be a topological vector space, and let $B \subseteq V$. We say B is bounded if for every neighborhood U of 0 in V, there is a $\lambda > 0$ with $B \subseteq \lambda U$.*

Since we usually define the topologies on LCTVS through semi-norms, it is convenient to have a characterization of bounded sets in terms of the semi-norms which generate the topology. This is given in the next proposition.

PROPOSITION A.1.16 (Proposition 14.5 of [Trè67]). *Let $\mathcal{P}$ be a family of continuous semi-norms on the LCTVS V which generate the topology. Then $B \subseteq V$ is bounded if and only if for all $|\cdot| \in \mathcal{P}$,*

$$\sup_{v \in B} |v| < \infty.$$

Another way we define an LCTVS is through categorical limits. We remind the reader of the definitions of projective and inductive limits in an arbitrary category. We only tangentially use these ideas in the text, and the uninterested reader can avoid these topics while still being able to understand our main results and proofs. Recall, the category we have in mind is the category LCTVS, whose objects are locally convex topological spaces, and whose morphisms are continuous linear maps.

DEFINITION A.1.17. *A directed set is a nonempty set I together with a reflexive and transitive binary relation $\leq$, with the additional property that for every $a, b \in I$, there exists $c \in I$ with $a \leq c$ and $b \leq c$.*

DEFINITION A.1.18. *Let $(I, \leq)$ be a directed set. A projective system of objects and morphisms is a family $(V_i)_{i \in I}$ of objects and a family of morphisms $T_{i,j} : V_j \to V_i$ for all $i \leq j$, $i, j \in I$ such that $T_{i,i}$ is the identity and $T_{i,k} = T_{i,j} \circ T_{j,k}$ for all $i \leq j \leq k$.*

DEFINITION A.1.19. *Let V_i, $T_{i,j}$ be a projective system as above. The projective limit of this system (if it exists) is an object V along with morphisms $\pi_i : V \to V_i$ satisfying $\pi_i = T_{i,j} \circ \pi_j$ for $i \leq j$, and such that if (W, ψ_i) is an other such pair there exists a unique morphism $S : W \to V$ making the following diagram commute*

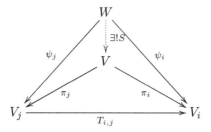

When V exists, it is unique up to isomorphism, and we denote it by $\varprojlim V_i$.

DEFINITION A.1.20. *Let $(I, \leq)$ be a directed set. An inductive system of objects and morphisms is a family $(V_i)_{i \in I}$ of objects and a family of morphisms $T_{i,j} : V_i \to V_j$ for all $i \leq j$, $i, j \in I$ such that $T_{i,i}$ is the identity and $T_{i,k} = T_{j,k} \circ T_{i,j}$ for all $i \leq j \leq k$.*

DEFINITION A.1.21. *Let V_i, $T_{i,j}$ be an inductive system as above. The inductive limit of this system (if it exists) is an object V along with morphisms $\phi_i : V_i \to V$ satisfying $\phi_i = \phi_j \circ T_{i,j}$ for $i \leq j$, and such that if (W, ψ_i) is an other such pair there exists a unique morphism $S : V \to W$ making the following diagram commute*

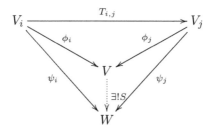

When V exists, it is unique up to isomorphism, and we denote it by $\varinjlim V_i$.

THEOREM A.1.22 (See pages 511–512 of [Trè67]). *In the category LCTVS, projective and inductive limits always exist.*

Example A.1.23 *Let $\Omega \subseteq \mathbb{R}^n$ be open. We let $C_0^\infty(\Omega)$ denote those smooth functions $f : \Omega \to \mathbb{C}$ such that $\mathrm{supp}(f)$ is a compact subset of Ω. For $K \subset \Omega$ compact, recall the space $C^\infty(K)$ from Example A.1.12. If $K_1 \subseteq K_2 \subset \Omega$ with K_1, K_2 compact, we have a natural inclusion $C^\infty(K_1) \hookrightarrow C^\infty(K_2)$. This defines an inductive system (indexed by compact subsets of Ω, ordered by inclusion), and we define the topology on $C_0^\infty(\Omega)$ by $C_0^\infty(\Omega) = \varinjlim C^\infty(K)$.*

Example A.1.23 defined the topology on $C_0^\infty(\Omega)$, and for many purposes this characterization of the topology is quite useful. However, it is a somewhat indirect definition, and understanding certain aspects of the topology from there can be challenging. One useful thing is to notice that if we take $K_1 \subsetneq K_2 \subsetneq K_3 \subsetneq \cdots \subset \Omega$ to be an increasing sequence of compact subsets of Ω with $\cup K_j = \Omega$, then we have an inductive subsystem of the one in Example A.1.23 and it is easy to see $C_0^\infty(\Omega) = \varinjlim C^\infty(K_j)$. This leads us to the following definition.

DEFINITION A.1.24. *Suppose a vector space V is equal to a countable increasing union of subspaces V_j, each V_j is a Fréchet space, and the natural inclusion $V_j \subseteq V_l$ ($l \geq j$) is an isomorphism (i.e., the topology on V_j is the same as the subspace topology on V_j as a subspace of V_l). Then we say V with the inductive limit topology given by identifying V with $\varinjlim V_j$ is an "LF space."*

LF spaces have many convenient properties (see [Trè67] for details). One such property is the following.

PROPOSITION A.1.25 (See Proposition 14.6 of [Trè67]). *Let V be an LF space with a sequence V_j of subspaces as above. Then $B \subseteq V$ is a bounded set if and only if there is a j such that $B \subseteq V_j$ and B is a bounded subset of V_j.*

COROLLARY A.1.26. *Let $\Omega \subseteq \mathbb{R}^n$ be open and let $B \subseteq C_0^\infty(\Omega)$. Then B is a bounded set if and only if the following two properties hold.*

- *There is a compact set $K \Subset \Omega$ such that $\forall f \in B$, $\mathrm{supp}(f) \subseteq K$.*

- *For every multi-index α, $\sup_{x \in \Omega, f \in B} |\partial_x^\alpha f(x)| < \infty$.*

A.1.1 Duals and distributions

For an LCTVS V, we denote by V' the dual of V with the weak dual topology. That is, V' is the vector space of continuous linear maps $\lambda : V \to \mathbb{C}$. The topology on V' is the one induced by the semi-norms given by, for each $v \in V$, $|\lambda|_v := |\lambda(v)|$. There are three spaces whose duals we often use: $C_0^\infty(\Omega)'$, $C^\infty(\Omega)'$, and $\mathcal{S}(\mathbb{R}^n)'$; the elements of these spaces we refer to as distributions, distributions with compact support, and tempered distributions. Here $\Omega \subseteq \mathbb{R}^n$ is open. If we have a squence $\lambda_n \in C_0^\infty(\Omega)'$ and we say $\lambda_n \to \lambda$ "in distribution," we mean $\lambda_n \to \lambda$ in the weak dual topology.

Remark A.1.27 Let us explain the terminology "distribution with compact support" as it relates to $C^\infty(\Omega)'$. Given an element $\lambda \in C^\infty(\Omega)'$, we may define $\mathrm{supp}(\lambda) \subseteq \Omega$ as follows. A point $x \in \Omega$ is *not* in $\mathrm{supp}(\lambda)$ if there is a neighborhood U of x such that for any $f \in C^\infty(\Omega)$ with $\mathrm{supp}(f) \subset U$, we have $\lambda(f) = 0$. It is immediate to see that supp (λ) is a closed subset of Ω. Suppose $\lambda \in C_0^\infty (\Omega)'$ is such that supp (λ) is a compact subset of Ω. We wish to identify λ with an element of $C^\infty (\Omega)'$. Indeed, take $\phi \in C_0^\infty (\Omega)$ such that $\phi \equiv 1$ on a neighborhood of supp (λ). It is easy to see $\forall f \in C_0^\infty (\Omega), \lambda (f) = \lambda (\phi f)$. Thus, for $f \in C^\infty (\Omega)$, we may define $\lambda (f) = \lambda (\phi f)$. This sees λ as an element of $C^\infty (\Omega)'$. In fact *every* element of $C^\infty (\Omega)'$ is of this form: every element of $C^\infty (\Omega)'$ is a "distribution with compact support."

Given a function $\lambda \in L_{\mathrm{loc}}^1 (\Omega)$, we obtain an element of $C_0^\infty (\Omega)$ given by

$$f \mapsto \int \lambda (x) f (x) \ dx.$$

In light of this, we abuse notation, and for any $\lambda \in C_0^\infty (\Omega)'$ we write

$$\lambda (f) = \int \lambda (x) f (x) \ dx.$$

I.e., we use notation as if distributions are functions even when they are not actually given by integration against an L_{loc}^1 function. Moreover, suppose $\lambda \in C_0^\infty (\Omega)'$ and $U \subseteq \Omega$ is an open set such that there is a $\lambda_0 \in L_{\mathrm{loc}}^1 (U)$ and such that $\forall f \in C_0^\infty (U)$, $\int \lambda (x) f (x) \ dx = \int \lambda_0 (x) f (x) \ dx$. Then, for $x \in U$, we identify $\lambda (x)$ with $\lambda_0 (x)$. I.e., when λ is actually given by integration against an L_{loc}^1 function, then we may treat λ as a function. Conversely, if we are given an L_{loc}^1 function, then we often identify it with a distribution.

If V, W are two LCTVS, then the vector space $\mathcal{L} (V, W)$ consisting of continuous linear maps from V to W has several different topologies which can be of use. One we often use is the following.

DEFINITION A.1.28. *For each continuous semi-norm $|\cdot|$ on W and each bounded set $\mathcal{B} \subset V$, we define a semi-norm $|\cdot|'$ on $\mathcal{L} (V, W)$ by*

$$|T|' = \sup_{v \in \mathcal{B}} |Tv| .$$

The coarsest topology on $\mathcal{L} (V, W)$ with respect to which all the above semi-norms are continuous is called the "topology of bounded convergence."

DEFINITION A.1.29. *As a vector space, we may identify V' with $\mathcal{L} (V, \mathbb{C})$. When we give V' the topology of bounded convergence, we also say we have given V' the "strong dual topology."*

Whenever we define singular integrals they are, a priori, operators $C_0^\infty (\Omega) \to C_0^\infty (\Omega)'$ or $C^\infty (\Omega) \to C^\infty (\Omega)'$ or $\mathcal{S} (\mathbb{R}^n) \to \mathcal{S} (\mathbb{R}^n)'$. The Schwartz kernel theorem gives a nice way to deal with these sorts of operators. Notice, given a distribution $\lambda \in C_0^\infty (\Omega \times \Omega)'$, we obtain an operator $T_\lambda : C_0^\infty (\Omega) \to C_0^\infty (\Omega)'$ by

$$\int \phi (x) (T_\lambda \psi) (x) \ dx := \int \phi (x) \psi (y) \lambda (x, y) \ dx \ dy.$$

Similarly, given an element of $C^\infty (\Omega \times \Omega)'$ we obtain a map $C^\infty (\Omega) \to C^\infty (\Omega)'$, and given an element of $\mathcal{S} (\mathbb{R}^n \times \mathbb{R}^n)'$ we obtain a map $\mathcal{S} (\mathbb{R}^n) \to \mathcal{S} (\mathbb{R}^n)'$. The Schwartz kernel theorem states that these are all bijections. In fact, we have the following theorem.

THEOREM A.1.30 (The Schwartz kernel theorem–see Chapter 51 of [Trè67]). *We have the natural bijections*

$$\mathcal{L} \left(C_0^\infty (\Omega), C_0^\infty (\Omega)' \right) \cong C_0^\infty (\Omega \times \Omega)',$$

$$\mathcal{L} \left(C^\infty (\Omega), C^\infty (\Omega)' \right) \cong C^\infty (\Omega \times \Omega)',$$

$$\mathcal{L} \left(\mathcal{S} (\mathbb{R}^n), \mathcal{S} (\mathbb{R}^n)' \right) \cong \mathcal{S} (\mathbb{R}^n \times \mathbb{R}^n)',$$

$$\mathcal{L} \left(C_0^\infty (\Omega)', C_0^\infty (\Omega) \right) \cong C_0^\infty (\Omega \times \Omega). \tag{A.1}$$

Remark A.1.31 The bijections in Theorem A.1.30 are just meant in the algebraic sense. They can be interpreted as topological ismorphisms, provided one uses the right topologies; see [Trè67, Chapter 51] for details on this.

Remark A.1.32 The above version of the Schwartz kernel theorem is actually just four instances of the same general principle concerning tensor products. This perspective is due to Grothendieck. We discuss tensor products briefly in the next section, but refer the reader to [Trè67] for the connection between tensor products and the Schwartz kernel theorem.

Remark A.1.33 In light of Theorem A.1.30, we may think of $\mathcal{L} \left(C_0^\infty (\Omega), C_0^\infty (\Omega)' \right)$ as a space of distributions. Thus if we have a sequence $T_n : C_0^\infty (\Omega) \to C_0^\infty (\Omega)'$, it makes sense to say $T_n \to T$ "converges in distribution," where this means that $T_n \to T$ in the weak dual topology on $C_0^\infty (\Omega \times \Omega)'$. Similar remarks hold for $\mathcal{L} \left(C^\infty (\Omega), C^\infty (\Omega)' \right)$ and $\mathcal{L} \left(\mathcal{S} (\mathbb{R}^n), \mathcal{S} (\mathbb{R}^n)' \right)$.

Remark A.1.34 (A.1) can be informally restated as saying that operators which are "infinitely smoothing" are precisely those whose Schwartz kernels are infinitely smooth.

Remark A.1.35 Up until this point in this appendix, we have been explicit about whether or not a linear map T between two topological vector spaces V and W is assumed to be continuous. For the rest of the monograph, we assume that all linear maps between topological vector spaces are continuous, and do not explicitly state this each time.

In light of Theorem A.1.30, we may identify operators $T : C_0^\infty (\Omega) \to C_0^\infty (\Omega)'$ (and other similar situations) with their Schwartz kernels $T (x, y) \in C_0^\infty (\Omega \times \Omega)'$. We use this identification throughout the monograph. Since we consider T as a distribution we abuse notation, as above, and treat it as a function. Furthermore, given a function $T (x, y) \in L_{\text{loc}}^1 (\Omega \times \Omega)$, we associate to it a distribution $T \in C_0^\infty (\Omega \times \Omega)'$ and therefore we associate to it an operator $T : C_0^\infty (\Omega) \to C_0^\infty (\Omega)'$. Hence, given a function in $L_{\text{loc}}^1 (\Omega \times \Omega)$, we obtain an operator $C_0^\infty (\Omega) \to C_0^\infty (\Omega)'$, and conversely given an operator $C_0^\infty (\Omega) \to C_0^\infty (\Omega)'$ we obtain a distribution which we may often be able to identity with a function.

A.2   TENSOR PRODUCTS

It is often useful, when studying multi-parameter operators, to use the tensor product of two locally convex topological vector spaces. In this section, we discuss various issues which arise when putting a topology on this tensor product. We assume the reader is familiar with the algebraic notion of a tensor product. The exposition here is very brief, and we refer the reader to [Trè67] for more details, proofs, and further reading.

Let V and W be two vector spaces and $V \otimes W$ the tensor product. There is the bilinear inclusion $V \times W \hookrightarrow V \otimes W$ given by $(v, w) \mapsto v \otimes w$. Recall the following universal property which characterizes the tensor product up to isomorphism. Let Z be a vector space and $B : V \times W \to Z$ a bilinear map. Then there is a unique linear map $V \otimes W \to Z$ making the following diagram commute

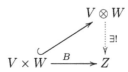

We now move to the category of LCTVS, where the morphisms are given by continuous linear maps. We will mostly be interested in *complete* LCTVS. To introduce the notion of completeness, we need a few definitions.

DEFINITION A.2.1. *Let X be a topological space and $(I, \leq)$ be a directed set (see Definition A.1.17). A set of elements of X indexed by I, $\{x_\alpha\}_{\alpha \in I}$, is called a net. We say $\lim x_\alpha = x$ (where $x \in X$) if for every neighborhood U of x, there is $\beta \in I$ such that $\beta \leq \alpha \Rightarrow x_\alpha \in U$. In this case, we say x_α converges to x, and also write $x_\alpha \to x$.*

DEFINITION A.2.2. *Let V be a topological vector space. We say a net $\{x_\alpha\}_{\alpha \in I} \subset V$ is Cauchy if for every neighborhood U of 0, there is $\beta \in I$ such that $\beta \leq \alpha_1, \alpha_2 \Rightarrow x_{\alpha_1} - x_{\alpha_2} \in U$.*

DEFINITION A.2.3. *We say a topological vector space V is complete if every Cauchy net in V converges to a point in V.*

Remark A.2.4   LF spaces are always complete. See [Trè67, Theorem 13.1].

THEOREM A.2.5 (Theorem 5.2 of [Trè67]). *Let V be a topological vector space. There exits a complete topological vector space $\widehat{V}$ and an inclusion $i : V \hookrightarrow \widehat{V}$ such that:*

(i) *The mapping i is an isomorphism, in the sense that the topology on V is the same as subspace topology V inherits from $\widehat{V}$.*

(ii) *The image of V under i is dense in $\widehat{V}$.*

(iii) *Let W be a complete LCTVS and consider a continuous linear map $V \to W$. Then, there is a unique continuous linear map $\widehat{V} \to W$ making the following*

diagram commute

Furthermore, $\widehat{V}$ is unique up to isomorphism. We call $\widehat{V}$ the completion of V.

From a category theoretic perspective, there is a natural topology to put on $V \otimes W$, namely the following.

DEFINITION A.2.6. *Let V and W be LCTVS. The projective topology on $V \otimes W$ is the finest topology such that the natural bilinear mapping $V \times W \hookrightarrow V \otimes W$ given by $(v, w) \mapsto v \otimes w$ is continuous. We denote by $V \otimes_\pi W$, $V \otimes W$ with this topology, and we denote by $V \widehat{\otimes}_\pi W$ the completion of $V \otimes_\pi W$.*

$V \otimes_\pi W$ and $V \widehat{\otimes}_\pi W$ enjoy the following universal properties.

PROPOSITION A.2.7. *Let V, W, and Z be LCTVS and $B : V \times W \to Z$ a continuous bilinear map. There is a unique continuous linear map $V \otimes_\pi W \to Z$ making the following diagram commute*

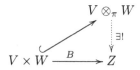

If, in addition, Z is a complete LCTVS then there is a unique continuous map $V \widehat{\otimes}_\pi W \to Z$ making the following diagram commute

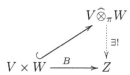

Putting the two universal properties above together we have the following restatement. If V, W, and Z are LCTVS and $B : V \times W \to Z$ is a continuous bilinear map, and if we denote by $\widehat{Z}$ the completion of Z with the natural inclusion $Z \hookrightarrow \widehat{Z}$, then there are unique continuous linear maps $V \otimes_\pi W \to Z$ and $V \widehat{\otimes}_\pi W \to \widehat{Z}$ making the following diagram commute

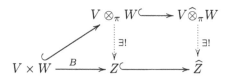

Proposition A.2.7 says that the projective tensor product (respectively, the completion of the projective tensor product) is *the* tensor product in the category of LCTVS (respectively, the category of complete LCTVS).

Unfortunately, in many examples that arise, the topology that $V \otimes W$ comes with is *not* the projective topology. For example, for $1 \leq p \leq \infty$, $L^p([0,1]) \otimes L^p([0,1])$ is a dense subspace of $L^p([0,1] \times [0,1])$, but the induced subspace topology agrees with $L^p([0,1]) \otimes_\pi L^p([0,1])$ if and only if $p = 1$. When $p = 1$, the bilinear map $(f(x), g(y)) \mapsto f(x)g(y)$ induces a canonical isomorphism

$$L^1([0,1]) \widehat{\otimes}_\pi L^1([0,1]) \cong L^1([0,1] \times [0,1]).$$

There is another important topology on $V \otimes W$ which arises: the *injective* topology. We define this topology by giving a family of semi-norms of $V \otimes W$. For continuous linear functionals $\lambda_1 \in V'$, $\lambda_2 \in W'$, there is a natural linear functional $\lambda_1 \otimes \lambda_2 : V \otimes W \to \mathbb{C}$. Let $|\cdot|_1$ be a continuous semi-norm on V and $|\cdot|_2$ be a continuous semi-norm on W. Let $U_1 = \left\{ \lambda \in V' \mid \sup_{|v|_1 = 1} |\lambda(v)| < 1 \right\} \subset V'$, and similarly define $U_2 \subset W'$ where we replace V with W and $|\cdot|_1$ with $|\cdot|_2$. Define a semi-norm on $V \otimes W$ by $|x| = \sup_{\lambda_1 \in U_1, \lambda_2 \in U_2} |\lambda_1 \otimes \lambda_2(x)|$.

DEFINITION A.2.8. *The injective topology on $V \otimes W$ is the coarsest topology in which every semi-norm defined in the above way is continuous. We denote by $V \otimes_\epsilon W$, $V \otimes W$ with this topology. We denote by $V \widehat{\otimes}_\epsilon W$ the completion of $V \otimes_\epsilon W$.*

It is easy to see that the bilinear map $(v, w) \mapsto v \otimes w$ is continuous $V \times W \to V \otimes_\epsilon W$. Thus, this map extends uniquely to inclusions $V \otimes_\pi W \hookrightarrow V \otimes_\epsilon W$ and $V \widehat{\otimes}_\pi W \hookrightarrow V \widehat{\otimes}_\epsilon W$. In other words, we may think of $V \widehat{\otimes}_\pi W$ as a dense subspace of $V \widehat{\otimes}_\epsilon W$ with a finer (but maybe not strictly finer) topology than the usual subspace topology.

In practice, we are often given a topology on $V \otimes W$ as a dense subspace of some space which arises in our application, call this topology $V \otimes_S W$ (where we have used S to denote subspace); we denote the ambient space by $V \widehat{\otimes}_S W$. It is almost always the case that we have the following continuous maps $V \otimes_\pi W \hookrightarrow V \otimes_S W \hookrightarrow V \otimes_\epsilon W$, where $\hookrightarrow$ denotes the identity map $V \otimes W \to V \otimes W$. These maps extend to the completions $V \widehat{\otimes}_\pi W \hookrightarrow V \widehat{\otimes}_S W \hookrightarrow V \widehat{\otimes}_\epsilon W$. For example, if we return to the case of $L^p([0,1])$, we have $L^p([0,1]) \widehat{\otimes}_\pi L^p([0,1]) \hookrightarrow L^p([0,1] \times [0,1]) \hookrightarrow L^p([0,1]) \widehat{\otimes}_\epsilon L^p([0,1])$, where both inclusions are strict if $1 < p < \infty$. This leads us to the following useful notion.

DEFINITION A.2.9. *We say an LCTVS V is nuclear if for every LCTVS W the canonical inclusion $V \widehat{\otimes}_\pi W \hookrightarrow V \widehat{\otimes}_\epsilon W$ is an isomorphism.*

Roughly speaking if V is nuclear, there is only one natural topology to put on $V \otimes W$. In this case, we drop the subscripts on $\otimes$ and $\widehat{\otimes}$, since there is no need to indicate the topology, and we write merely $V \widehat{\otimes} W$ to denote $V \widehat{\otimes}_\pi W \cong V \widehat{\otimes}_\epsilon W$.

Now if we return to the case above of some tensor product given by our application $V \widehat{\otimes}_S W$, if either V or W is nuclear, then we must have that the inclusions $V \widehat{\otimes}_\pi W \hookrightarrow V \widehat{\otimes}_S W \hookrightarrow V \widehat{\otimes}_\epsilon W$ are actually isomorphisms. Thus, $V \widehat{\otimes}_S W$ must agree with the projective tensor product, and therefore enjoys the universal property in Proposition A.2.7. It is therefore useful to know how to recognize nuclear spaces, so that we may easily recognize situations where we may apply the universal property.

PROPOSITION A.2.10. • An LCTVS is nuclear if and only if its completion is nuclear.

• A subspace of a nuclear space is nuclear.

• A product of nuclear spaces is nuclear.

• A (Hausdorff) projective limit of nuclear spaces is nuclear.

• A countable inductive limit of nuclear spaces is nuclear.

• If V and W are nuclear, then so is $V \widehat{\otimes} W$.

• A Fréchet space is nuclear if and only if its strong dual is nuclear.

Proposition A.2.10 shows how to build nuclear spaces out of other nuclear spaces. To use it, we need some useful nuclear spaces as building blocks.

THEOREM A.2.11. The following spaces are nuclear:

• $S(\mathbb{R}^n)$ (Schwartz space–see Definition 1.1.3),

• $S(\mathbb{R}^n)'$ (the space of tempered distributions), given the strong dual topology,

• $C^\infty(K)$ where M is a manifold and $K \Subset M$ is compact,

• $C_0^\infty(M)$, $C^\infty(M)$, $C_0^\infty(M)'$, and $C^\infty(M)'$, where M is a manifold and all duals are given the strong dual topology.

Remark A.2.12 Some of the examples in Theorem A.2.11 can be built out of the other examples using Proposition A.2.10. For instance, $C_0^\infty(M)$ is a countable inductive limit of $C^\infty(K)$ over various choices of $K \Subset M$.

We now discuss a few uses we have for nuclear spaces. Let M be a smooth manifold, and V a *complete* LCTVS. We wish to understand the space $C^\infty(M; V)$–C^∞ functions on M taking values in V.

DEFINITION A.2.13. Let $\Omega \subseteq \mathbb{R}^n$ be an open set, and $f : \Omega \to V$ a function. We say f is differentiable at $x \in \Omega$ if there are vectors $e_1, \ldots, e_n \in V$ so that

$$\lim_{y \to x} |y - x|^{-1} \left(f(y) - f(x) - \sum_{j=1}^n (y_j - x_j) e_j \right) = 0,$$

where the limit is taken in V. We define $\frac{\partial f}{\partial x_j}(x) = e_j$.

DEFINITION A.2.14. *For an open set $\Omega \subseteq \mathbb{R}^n$, and $0 \leq k$, we define $C^k(\Omega; V)$ to be those $f : \Omega \to V$ such that f and all its derivatives to order k are continuous. We give $C^k(\Omega; V)$ the topology of uniform convergence of f and all its derivatives to order up to k on compact subsets of Ω. We define $C^\infty(\Omega; V)$ to be $\bigcap_{k \geq 0} C^k(\Omega; V)$, with the usual (projective limit) topology.*

We may, in the usual way, extend Definition A.2.13 to the manifold M to obtain the space $C^\infty(M; V)$. In a similar manner we may also define the space $C_0^\infty(M; V)$, the details of which we leave to the reader.

Example A.2.15 *Consider the nuclear space $C^\infty(M)$, where M is a manifold. It is not hard to see that $C^\infty(M) \otimes V$ is dense in $C^\infty(M; V)$ and that we have the following inclusions $C^\infty(M) \widehat{\otimes}_\pi V \hookrightarrow C^\infty(M; V) \hookrightarrow C^\infty(M) \widehat{\otimes}_\epsilon V$ (this requires V be complete). However, since $C^\infty(M)$ is nuclear, these must be isomorphisms and we must have, canonically, $C^\infty(M; V) \cong C^\infty(M) \widehat{\otimes}_\pi V$. Similarly, we have $C_0^\infty(M; V) \cong C_0^\infty(M) \widehat{\otimes}_\pi V$, via the canonical map. See [Trè67] for further details.*

We now turn to our next example. Suppose we have decomposed $\mathbb{R}^q$ into factors: $\mathbb{R}^q = \mathbb{R}^{q_1} \times \mathbb{R}^{q_2} \times \cdots \times \mathbb{R}^{q_\nu}$. We write an element $t \in \mathbb{R}^q$ as $t = (t_1, \ldots, t_\nu) \in \mathbb{R}^{q_1} \times \cdots \times \mathbb{R}^{q_\nu}$.

DEFINITION A.2.16. *For $E \subseteq \{1, \ldots, \nu\}$ we define $\mathcal{S}_0^E$ to be the space of those $f \in \mathcal{S}(\mathbb{R}^q)$ such that for every $\mu \in E$ and every $\alpha \in \mathbb{N}^{q_\mu}$,*

$$\int t_\mu^\alpha f(t) \; dt_\mu = 0.$$

It is easy to see that $\mathcal{S}_0^E$ is a closed subspace of $\mathcal{S}(\mathbb{R}^q)$ and we give it the subspace topology, making it a Fréchet space.

Example A.2.17 *Consider the subspace $\mathcal{S}_0(\mathbb{R}^{q_\mu}) \subset \mathcal{S}(\mathbb{R}^{q_\mu})$—the space of Schwartz functions all of whose moments vanish; see Definition 1.1.10. $\mathcal{S}_0(\mathbb{R}^{q_\mu})$ is a subspace of the nuclear space $\mathcal{S}(\mathbb{R}^{q_\mu})$ and is therefore nuclear. Fix $E \subseteq \{1, \ldots, \nu\}$. There is an obvious ν-linear inclusion $\left[\prod_{\mu \in E} \mathcal{S}_0(\mathbb{R}^{q_\mu})\right] \times \left[\prod_{\mu \in E^c} \mathcal{S}(\mathbb{R}^{q_\mu})\right] \to \mathcal{S}_0^E$. From the nuclearity of $\mathcal{S}_0$ and $\mathcal{S}$ it is easy to see that this yields an isomorphism*

$$\left[\widehat{\otimes}_{\mu \in E} \mathcal{S}_0(\mathbb{R}^{q_\mu})\right] \widehat{\otimes} \left[\widehat{\otimes}_{\mu \in E^c} \mathcal{S}(\mathbb{R}^{q_\mu})\right] \cong \mathcal{S}_0^E.$$

For each $\mu \in \{1, \ldots, \nu\}$ and $s_\mu \in \mathbb{R}$, Corollary 1.1.13 gives the isomorphism $\triangle_\mu^{s_\mu} : \mathcal{S}_0(\mathbb{R}^{q_\mu}) \to \mathcal{S}_0(\mathbb{R}^{q_\mu})$. If we let $I_\mu : \mathcal{S}(\mathbb{R}^{q_\mu}) \to \mathcal{S}(\mathbb{R}^{q_\mu})$ denote the identity, then we have the following ν-linear map

$$\left[\otimes_{\mu \in E} \triangle_\mu^{s_\mu}\right] \otimes \left[\otimes_{\mu \in E^c} I_\mu\right] : \left[\prod_{\mu \in E} \mathcal{S}_0(\mathbb{R}^{q_\mu})\right] \times \left[\prod_{\mu \in E^c} \mathcal{S}(\mathbb{R}^{q_\mu})\right] \to \mathcal{S}_0^E.$$

The universal property shows that this extends to a continuous map

$$\left[\otimes_{\mu \in E} \triangle_\mu^{s_\mu}\right] \otimes \left[\otimes_{\mu \in E^c} I_\mu\right] : \mathcal{S}_0^E \to \mathcal{S}_0^E,$$

which is an isomorphism in light of its continuous inverse

$$\left[\otimes_{\mu \in E} \triangle_\mu^{-s_\mu}\right] \otimes \left[\otimes_{\mu \in E^c} I_\mu\right] : \mathcal{S}_0^E \to \mathcal{S}_0^E.$$

We have the next lemma.

LEMMA A.2.18. *Fix $E \subseteq \{1, \ldots, \nu\}$ and $\mu \in E$. Let $\mathcal{B} \subset \mathcal{S}_0^E$ be a bounded set. Then for each $f \in \mathcal{B}$ we may write*

$$f(t_1, \ldots, t_\nu) = \sum_{|\alpha_\mu|=1} \partial_{t_\mu}^{\alpha_\mu} f_\alpha(t_1, \ldots, t_\nu),$$

where $\{f_\alpha | f \in \mathcal{B}, |\alpha| = 1\} \subset \mathcal{S}_0^E$ is a bounded set.

PROOF. For $f \in \mathcal{S}_0^E$ we may write $f = \sum_{|\alpha_\mu|=1} - \left(\partial_{t_\mu}^{\alpha_\mu}\right)^2 \triangle_\mu^{-1} f$. The result immediately follows from the continuity of $\triangle_\mu^{-1} : \mathcal{S}_0^E \to \mathcal{S}_0^E$. $\square$

Appendix B

Three Results from Calculus

In this appendix, we discuss three connected results from calculus, which we use several times:

- Given vector fields $Y_1, \ldots, Y_\nu$, we discuss the function $\Psi(t_1, \ldots, t_\nu, x) := e^{t_1 Y_1 + \cdots + t_\nu Y_\nu} x$. In particular, we discuss how the smoothness of Ψ depends on the smoothness of $Y_1, \ldots, Y_\nu$.

- The inverse function theorem. In particular, we discuss the fact that the inverse function theorem holds "uniformly on compact sets in C^1."

- A certain sort of change of variables which involves the above two points.

B.1 EXPONENTIAL OF VECTOR FIELDS

We begin by recalling the definition of the exponential map from Section 2.1. For a C^1 vector field Y on an open set $U \subseteq \mathbb{R}^n$ and for $x_0 \in U$ we define $E_Y(t) = e^{tY} x_0$ to be the unique solution to the ODE

$$\frac{d}{dt} E_Y(t) = Y(E_Y(t)), \quad E(0) = x_0.$$

By the Picard-Lindelöf theorem, this unique solution always exists for $|t|$ sufficiently small (how small depends on U, the distance from x_0 to ∂U, and the C^1 norm of Y).[1] We then define $e^Y x_0 := E_Y(1)$. It is easy to see that $e^{tY} x_0 = E_Y(t) = E_{tY}(1)$, justifying the notation. This allows us to define

$$e^{t_1 Y_1 + \cdots + t_\nu Y_\nu} x_0$$

for $|t|$ sufficiently small, where $t = (t_1, \ldots, t_\nu)$.

THEOREM B.1.1. *Suppose $Y_1, \ldots, Y_\nu$ are C^m vector fields ($m \geq 1$) defined on an open set $\Omega \subseteq \mathbb{R}^n$. Then the function*

$$u(t, x) = e^{t_1 Y_1 + \cdots + t_\nu Y_\nu} x - x$$

is C^m. Moreover, the C^m norm of u can be bounded by a constant which depends only on upper bounds for n, ν, and the C^m norms of $Y_1, \ldots, Y_\nu$.

---

[1] Actually, the interval of existence can be shown to only depend on the C^0 norm of Y (instead of the C^1 norm), so long as Y is Lipschitz, but we will not need this.

COMMENTS ON THE PROOF. It is perhaps easiest to consider the function

$$v\left(\epsilon, t, x\right) = e^{\epsilon(t_1 Y_1 + \cdots + t_\nu Y_\nu)} x.$$

Then $u\left(t, x\right) = v\left(1, t, x\right) - x$, and v is defined by an ODE in the ϵ variable. That $v - x$ is C^m in all three variables (with C^m norm bounded as in the statement of the theorem) is classical. See [Die60, Chapter X]. □

Remark B.1.2 For a more modern approach to Theorem B.1.1, we refer the reader to [Izz99], the methods of which can be modified to prove Theorem B.1.1.

B.2 THE INVERSE FUNCTION THEOREM

We now turn to the inverse function theorem. The moral of the version we state here is that the inverse function holds "uniformly on compact sets."

THEOREM B.2.1. *Fix an open set $U \subseteq \mathbb{R}^n$ and $x_0 \in U$. Suppose $K \subset C^1\left(U; \mathbb{R}^n\right)$ is a compact set such that $\forall f \in K$, $\det df\left(x_0\right) \neq 0$ (and thus, by compactness, $\left|\det df\left(x_0\right)\right|$ is bounded away from 0, uniformly for $f \in K$). Then, there exist constants $\delta_1, \delta_2 > 0$ such that $\forall f \in K$*

- $f\big|_{B(x_0,\delta_1)}$ *is a C^1 diffeomorphism onto its image.*

- $B\left(f\left(x_0\right), \delta_2\right) \subseteq f\left(B\left(x_0, \delta_1\right)\right).$

Here, $B\left(y, \delta\right)$ is the usual Euclidean ball of radius δ, centered at y.

COMMENTS ON THE PROOF. This follows from a straightforward modification of the proof of the inverse function theorem in [Spi65], by using the Arzelà-Ascoli theorem. □

The usual way that Theorem B.2.1 arises in this monograph is the following. We will have a *bounded set* $\mathcal{B} \subset C^2\left(U; \mathbb{R}^n\right)$, where $U \subseteq \mathbb{R}^n$ is some open set. We will also have some fixed $x_0 \in U$ such that $\inf_{f \in \mathcal{B}} \left|\det f\left(x_0\right)\right| > 0$. Because $\mathcal{B}$ is bounded in C^2, the Arzelà-Ascoli theorem shows that it is pre-compact in C^1. Theorem B.2.1 then applies to the closure, in C^1, of $\mathcal{B}$.

Remark B.2.2 Because, as mentioned above, we will always be applying Theorem B.2.1 to bounded subsets of C^2 (instead of arbitrary pre-compact subsets of C^1), the derivatives of the functions in the set will be, in particular, uniformly Lipschitz. In this case, one can use the version of the inverse function theorem in [HH99] instead of the more general Theorem B.2.1.

B.3   A CHANGE OF VARIABLES

In this section, we discuss a type of change of variables we use several times throughout the monograph. Fix $n, m \in \mathbb{N}$, and suppose we are given a C^∞ function of two variables $\Psi(t_1, t_2) : B^n(\eta) \times B^m(\eta) \to \mathbb{R}^n$, where $\eta > 0$ is some fixed number. Here, $B^d(\eta) \subset \mathbb{R}^d$ denotes the usual Euclidean ball of radius η centered at 0. Suppose $\Psi(0,0) = 0$ and $|\det d_{t_1} \Psi(0,0)| =: \epsilon_0 > 0$.

THEOREM B.3.1. *There is a constant $a > 0$ depending only on upper bounds for $\|\Psi\|_{C^2(B^n(\eta) \times B^m(\eta))}$, n, and m, and lower bounds for ϵ_0 and η such that $\forall \phi \in C_0^\infty(B^{n+m}(a))$,*

$$\int f(\Psi(t_1, t_2)) \phi(t_1, t_2) \, dt_1 \, dt_2 = \int f(u) K(u) \, du$$

where $\|K\|_{C^m}$ can be bounded in terms of upper bounds for $\|\Psi\|_{C^{m+1}}$, $\|\phi\|_{C^m}$, n, and m, and a lower bound for ϵ_0. Furthermore, suppose there is $\eta_2 > 0$ such that $\phi \geq 1$ on $B^{n+m}(\eta_2)$ then there is $a_2 > 0$, depending only on upper bounds for $\|\Psi\|_{C^2}$, n, and m, and lower bounds for ϵ_0 and η_2 such that

$$K(u) \geq c_0, \quad \forall u \in B^n(a_2),$$

where $c_0 > 0$ can be chosen to depend only on an upper bound for $\|\Psi\|_{C^2}$.

PROOF. We may choose $0 < \eta_1 \leq \eta$, depending only on a lower bound for η and ϵ_0 and an upper bound for $\|\Psi\|_{C^2}$ such that for all $(t_1, t_2) \in B^n(\eta_1) \times B^m(\eta_2)$,

$$|\det d_{t_1} \Psi(t_1, t_2)| \geq \epsilon_0/2.$$

The inverse function theorem (Theorem B.2.1) can be used to show that there exist $\delta_1, \delta_2 > 0$, depending only on lower bounds for η_1, and ϵ_0, and an upper bound for $\|\Psi\|_{C^2}$ such that $\forall t_2 \in B^m(\delta_2)$, $\Psi(\cdot, t_2)|_{B^n(\delta_1)}$ is a diffeomorphism onto its image and $B^n(\delta_2) \subseteq \Psi(B^n(\delta_1), t_2)$. (These are not the same δ_1, δ_2 as in the statement of Theorem B.2.1, but this follows easily from that theorem.) Setting $a = \delta_2$ we separate the integral

$$\int f(\Psi(t_1, t_2)) \phi(t_1, t_2) \, dt_1 \, dt_2 = \int_{B^m(\delta_2)} \int_{B^n(\delta_1)} f(\Psi(t_1, t_2)) \phi(t_1, t_2) \, dt_1 \, dt_2.$$

Applying a change of variables in the t_1 variable ($u = \Psi(t_1, t_2)$), the result follows immediately. □

The usual situation in which we apply Theorem B.3.1 is as follows. We are given C^∞ vector fields $Y_1, \ldots, Y_q$ on $B^n(\eta)$, where $\eta > 0$ is some fixed number and we have

$$\sup_{1 \leq l_1 < l_2 < \cdots < l_n \leq q} |\det(Y_{l_1}(0)| \cdots |Y_{l_n}(0))| =: \epsilon_0 > 0, \tag{B.1}$$

where $(Y_{l_1}(0)|\cdots|Y_{l_n}(0))$ denotes the matrix whose columns are given by the vectors $Y_{l_1}(0),\ldots,Y_{l_n}(0)$. (B.1) can be rephrased as saying that $Y_1(0),\ldots,Y_q(0)$ span the tangent space at 0. We will be concerned with integrals like

$$\int f\left(e^{t_1 Y_1+\cdots+t_\nu Y_\nu}0\right)\varsigma(t)\ dt,$$

where $\varsigma\in C_0^\infty(B^n(a))$, and $a>0$ is some number which we may choose, but we wish to only depend on upper bounds for n, q, $\|Y_1\|_{C^2},\ldots,\|Y_q\|_{C^2}$, and lower bounds for ϵ_0 and η. Noting that $\partial_{t_j}\big|_{t=0}e^{t_1 Y_1+\cdots+t_\nu Y_\nu}0 = Y_j(0)$, Theorem B.3.1 (when combined with Theorem B.1.1 and (B.1)) shows that we may choose such an $a>0$ so that we have

$$\int f\left(e^{t_1 Y_1+\cdots+t_\nu Y_\nu}0\right)\varsigma(t)\ dt = \int f(u)K(u)\ du,$$

where K is C^∞ and $\|K\|_{C^m}$ (for $m\geq 2$) can be bounded by a constant which depends only on upper bounds for q, n, and $\|Y_1\|_{C^m},\ldots,\|Y_q\|_{C^m}$, and lower bounds for ϵ_0 and η. Furthermore, if $\varsigma\geq 1$ on $B^n(\eta_2)$, then there exists $a_2>0$ (depending on the same constants as a and on a lower bound for $\eta_2>0$) such that $K(u)\gtrsim 1$ on $B^n(a_2)$.

Appendix C

Notation

In this appendix, we record notation which is used throughout the monograph.

- If we write $T : V \to W$, where V and W are topological vector spaces, we mean that T is a continuous linear map between these topological vector spaces.

- For $j, k \in \mathbb{R}$, we write $j \wedge k$ for the minimum of j and k and $j \vee k$ for the maximum. If, instead, $j = (j_1, \ldots, j_\nu), k = (k_1, \ldots, k_\nu) \in \mathbb{R}^\nu$, then $j \wedge k = (j_1 \wedge k_1, \ldots, j_\nu \wedge k_\nu)$ and $j \vee k = (j_1 \vee k_1, \ldots, j_\nu \vee k_\nu)$.

- If $j = (j_1, \ldots, j_\nu), k = (k_1, \ldots, k_\nu) \in \mathbb{R}^\nu$, we write $j \leq k$ to mean $j_\mu \leq k_\mu$, $\forall \mu$.

- If $j = (j_1, \ldots, j_\nu) \in \mathbb{R}^\nu$, then $2^j := (2^{j_1}, \ldots, 2^{j_\nu}) \in (0, \infty)^\nu$.

- If $\mathcal{F} \subset \mathbb{R}^\nu$ is a finite set, we write $\operatorname{diam}\{\mathcal{F}\} := \max_{j,k \in \mathcal{F}} |j - k|$.

- $A \lesssim B$ means $A \leq CB$, where C is a constant which does not depend on any relevant parameters. $A \approx B$ means $A \lesssim B$ and $B \lesssim A$.

- If U and V are subsets of a topological space, we write $U \Subset V$ if the closure of U is a compact subset of V.

- If $\lambda \in C_0^\infty(\mathbb{R}^n)'$ is a distribution, then we abuse notation, and for $f \in C_0^\infty(\mathbb{R}^n)$ we write $\langle \lambda, f \rangle = \int \lambda(x) f(x) \, dx$. If there is an open set $U \subseteq \mathbb{R}^n$ for which $\langle \lambda, f \rangle$ is given by integration against an L^1_{loc} function, then we identify $\lambda|_U$ with this function.

- We identify operators $T : C_0^\infty(\mathbb{R}^n) \to C_0^\infty(\mathbb{R}^n)'$ with their Schwartz kernels $T(x, y) \in C_0^\infty(\mathbb{R}^n \times \mathbb{R}^n)'$. Conversely, given a distribution $T(x, y) \in C_0^\infty(\mathbb{R}^n \times \mathbb{R}^n)'$, we identify it with an operator $T : C_0^\infty(\mathbb{R}^n) \to C_0^\infty(\mathbb{R}^n)'$. Combining this with the previous point, if we are given a function
$$E \in L^1_{loc}(\mathbb{R}^n \times \mathbb{R}^n)$$
we identify it with a distribution $E \in C_0^\infty(\mathbb{R}^n \times \mathbb{R}^n)'$ and therefore with an operator $E : C_0^\infty(\mathbb{R}^n) \to C_0^\infty(\mathbb{R}^n)'$.

- If $A_1, \ldots, A_k$ are possibly noncommuting operators, and if we denote by $A = (A_1, \ldots, A_k)$, then we use "ordered multi-index notation." Indeed, if α is a list of elements of $\{1, \ldots, k\}$ then it makes sense to write A^α. For instance, if $\alpha = (1, 2, 2, 4, 1)$, then $A^\alpha = A_1 A_2 A_2 A_4 A_1$, and $|\alpha|$ denotes the length of the list (in this case, 5).

- If A is an $n \times q$ matrix, and $n_0 \leq n \wedge q$, we let $\det_{n_0 \times n_0} A$ denote the *vector* whose coordinates are given by the determinants of $n_0 \times n_0$ submatrices of A–the order of the coordinates does not matter.

Bibliography

[Ale94]   G. Alexopoulos, *Spectral multipliers on Lie groups of polynomial growth*, Proc. Amer. Math. Soc. **120** (1994), no. 3, 973–979. MR 1172944 (95j:22016)

[BDKT06]   Antonio Bove, Makhlouf Derridj, Joseph J. Kohn, and David S. Tartakoff, *Sums of squares of complex vector fields and (analytic-) hypoellipticity*, Math. Res. Lett. **13** (2006), no. 5-6, 683–701. MR 2280767 (2007k:35051)

[Bes26]   A. Besicovitch, *Sur la nature des fonctions à carré sommable mésurables*, Fund. Math. **4** (1926), 172–196.

[BM95]   Denis R. Bell and Salah Eldin A. Mohammed, *An extension of Hörmander's theorem for infinitely degenerate second-order operators*, Duke Math. J. **78** (1995), no. 3, 453–475. MR 1334203 (96g:35034)

[BMT10]   Antonio Bove, Marco Mughetti, and David S. Tartakoff, *Gevrey hypoellipticity for an interesting variant of Kohn's operator*, Complex analysis, Trends Math., Birkhäuser/Springer Basel AG, Basel, 2010, pp. 51–73. MR 2885108 (2012k:35067)

[Car09]   C. Carathéodory, *Untersuchungen über die Grundlagen der Thermodynamik*, Math. Ann. **67** (1909), no. 3, 355–386. MR 1511534

[CD99]   Thierry Coulhon and Xuan Thinh Duong, *Riesz transforms for $1 \leq p \leq 2$*, Trans. Amer. Math. Soc. **351** (1999), no. 3, 1151–1169. MR 1458299 (99e:58174)

[CF85]   Sun-Yung A. Chang and Robert Fefferman, *Some recent developments in Fourier analysis and H^p-theory on product domains*, Bull. Amer. Math. Soc. (N.S.) **12** (1985), no. 1, 1–43. MR 766959 (86g:42038)

[CG90]   Lawrence J. Corwin and Frederick P. Greenleaf, *Representations of nilpotent Lie groups and their applications. Part I*, Cambridge Studies in Advanced Mathematics, vol. 18, Cambridge University Press, Cambridge, 1990, Basic theory and examples. MR 1070979 (92b:22007)

[CGGP92]   Michael Christ, Daryl Geller, Paweł Głowacki, and Larry Polin, *Pseudodifferential operators on groups with dilations*, Duke Math. J. **68** (1992), no. 1, 31–65. MR 1185817 (94b:35316)

[Che46]   Claude Chevalley, *Theory of Lie Groups. I*, Princeton Mathematical Se-
            ries, vol. 8, Princeton University Press, Princeton, NJ, 1946. MR 0015396
            (7,412c)

[Chi11]   Gregorio Chinni, *A proof of hypoellipticity for Kohn's operator via
            FBI*, Rev. Mat. Iberoam. **27** (2011), no. 2, 585–604. MR 2848531
            (2012h:35030)

[Chi12]   _____, *Germ hypoellipticity and loss of derivatives*, Proc. Amer. Math.
            Soc. **140** (2012), no. 7, 2417–2427. MR 2898704

[Cho39]   Wei-Liang Chow, *Über Systeme von linearen partiellen Differentialgle-
            ichungen erster Ordnung*, Math. Ann. **117** (1939), 98–105. MR 0001880
            (1,313d)

[Chr91]   Michael Christ, L^p *bounds for spectral multipliers on nilpotent groups*,
            Trans. Amer. Math. Soc. **328** (1991), no. 1, 73–81. MR 1104196
            (92k:42017)

[Chr01]   _____, *Hypoellipticity in the infinitely degenerate regime*, Complex anal-
            ysis and geometry (Columbus, OH, 1999), Ohio State Univ. Math. Res.
            Inst. Publ., vol. 9, de Gruyter, Berlin, 2001, pp. 59–84. MR 1912731
            (2003e:35044)

[Chr05]   _____, *A remark on sums of squares of complex vector fields*, 2005,
            preprint.

[CNS92]   D.-C. Chang, A. Nagel, and E. M. Stein, *Estimates for the $\bar{\partial}$-Neumann
            problem in pseudoconvex domains of finite type in $\mathbb{C}^2$*, Acta Math. **169**
            (1992), no. 3-4, 153–228. MR MR1194003 (93k:32025)

[CW71]   Ronald R. Coifman and Guido Weiss, *Analyse harmonique non-
            commutative sur certains espaces homogènes*, Lecture Notes in Math-
            ematics, Vol. 242, Springer-Verlag, Berlin, 1971, Étude de certaines
            intégrales singulières. MR 0499948 (58 #17690)

[CZ52]   A. P. Calderón and A. Zygmund, *On the existence of certain singular in-
            tegrals*, Acta Math. **88** (1952), 85–139. MR 0052553 (14,637f)

[Die60]   J. Dieudonné, *Foundations of modern analysis*, Pure and Applied Math-
            ematics, Vol. X, Academic Press, New York, 1960. MR 0120319 (22
            #11074)

[DJ84]   Guy David and Jean-Lin Journé, *A boundedness criterion for generalized
            Calderón-Zygmund operators*, Ann. of Math. (2) **120** (1984), no. 2, 371–
            397. MR 763911 (85k:42041)

[Dye70]   Joan L. Dyer, *A nilpotent Lie algebra with nilpotent automorphism group*,
            Bull. Amer. Math. Soc. **76** (1970), 52–56. MR 0249544 (40 #2789)

[Fef87]   Robert Fefferman, *Harmonic analysis on product spaces*, Ann. of Math. (2) **126** (1987), no. 1, 109–130. MR 898053 (90e:42030)

[Fol75]   G. B. Folland, *Subelliptic estimates and function spaces on nilpotent Lie groups*, Ark. Mat. **13** (1975), no. 2, 161–207. MR 0494315 (58 #13215)

[Fol95]   Gerald B. Folland, *Introduction to partial differential equations*, second ed., Princeton University Press, Princeton, NJ, 1995. MR 1357411 (96h:35001)

[Fol99]   ———, *Real analysis*, second ed., Pure and Applied Mathematics (New York), John Wiley & Sons Inc., New York, 1999, Modern techniques and their applications, A Wiley-Interscience Publication. MR 1681462 (2000c:00001)

[FP83]   C. Fefferman and D. H. Phong, *Subelliptic eigenvalue problems*, Conference on harmonic analysis in honor of Antoni Zygmund, Vol. I, II (Chicago, Ill., 1981), Wadsworth Math. Ser., Wadsworth, Belmont, CA, 1983, pp. 590–606. MR 730094 (86c:35112)

[FS74]   G. B. Folland and E. M. Stein, *Estimates for the $\bar{\partial}_b$ complex and analysis on the Heisenberg group*, Comm. Pure Appl. Math. **27** (1974), 429–522. MR 0367477 (51 #3719)

[FS82]   Robert Fefferman and Elias M. Stein, *Singular integrals on product spaces*, Adv. in Math. **45** (1982), no. 2, 117–143. MR 664621 (84d:42023)

[FSC86]   Charles L. Fefferman and Antonio Sánchez-Calle, *Fundamental solutions for second order subelliptic operators*, Ann. of Math. (2) **124** (1986), no. 2, 247–272. MR 855295 (87k:35047)

[Gło10]   Paweł Głowacki, *Composition and L^2-boundedness of flag kernels*, Colloq. Math. **118** (2010), no. 2, 581–585. MR 2602167 (2011h:42015)

[Gło12a]   ———, *Flag kernels of arbitrary order*, 2012, preprint.

[Gło12b]   ———, *L^p-boundedness of flag kernels on homogenous groups*, 2012, preprint.

[Goo76]   Roe W. Goodman, *Nilpotent Lie groups: structure and applications to analysis*, Lecture Notes in Mathematics, Vol. 562, Springer-Verlag, Berlin, 1976. MR 0442149 (56 #537)

[GS77]   P. C. Greiner and E. M. Stein, *Estimates for the $\overline{\partial}$-Neumann problem*, Princeton University Press, Princeton, NJ, 1977, Mathematical Notes, No. 19. MR 0499319 (58 #17218)

[Haa81]   Uffe Haagerup, *The best constants in the Khintchine inequality*, Studia Math. **70** (1981), no. 3, 231–283 (1982). MR 654838 (83m:60031)

[Hel86]   Peter Niels Heller, *Analyticity and regularity for nonhomogeneous opera-tors on the Heisenberg group*, ProQuest LLC, Ann Arbor, MI, 1986, The-sis (Ph.D.)–Princeton University. MR 2634700

[HH99]   John Hamal Hubbard and Barbara Burke Hubbard, *Vector calculus, linear algebra, and differential forms*, Prentice Hall Inc., Upper Saddle River, NJ, 1999, A unified approach. MR 1657732 (99k:00002)

[HJ83]   Andrzej Hulanicki and Joe W. Jenkins, *Almost everywhere summability on nilmanifolds*, Trans. Amer. Math. Soc. **278** (1983), no. 2, 703–715. MR 701519 (85f:22011)

[HL08]   Yongsheng Han and Guozhen Lu, *Discrete Littlewood-Paley-Stein theory and multi-parameter hardy spaces associated with flag singular integrals*, 2008, arXiv preprint arXiv:0801.1701.

[HLL13]   Yongsheng Han, Ji Li, and Guozhen Lu, *Multiparameter Hardy space the-ory on Carnot-Carathéodory spaces and product spaces of homogeneous type*, Trans. Amer. Math. Soc. **365** (2013), no. 1, 319–360. MR 2984061

[HLY07]   Yongsheng Han, Guozhen Lu, and Dachun Yang, *Product theory on spaces of homogeneous type*, 2007, unpublished manuscript.

[HN79]   B. Helffer and J. Nourrigat, *Caracterisation des opérateurs hypoellip-tiques homogènes invariants à gauche sur un groupe de Lie nilpotent gradué*, Comm. Partial Differential Equations **4** (1979), no. 8, 899–958. MR 537467 (81i:35034)

[HN85]   Bernard Helffer and Jean Nourrigat, *Hypoellipticité maximale pour des opérateurs polynômes de champs de vecteurs*, Progress in Mathemat-ics, vol. 58, Birkhäuser Boston Inc., Boston, MA, 1985. MR 897103 (88i:35029)

[Hör65]   Lars Hörmander, *Pseudo-differential operators*, Comm. Pure Appl. Math. **18** (1965), 501–517. MR 0180740 (31 #4970)

[Hör67]   _____, *Hypoelliptic second order differential equations*, Acta Math. **119** (1967), 147–171. MR 0222474 (36 #5526)

[Izz99]   Alexander J. Izzo, *C^r convergence of Picard's successive approxima-tions*, Proc. Amer. Math. Soc. **127** (1999), no. 7, 2059–2063. MR 1486736 (99j:34003)

[JMZ35]   B. Jessen, J. Marcinkiewicz, and A. Zygmund, *Note on the differentiability of multiple integrals*, Funda. Math. **25** (1935), 217–234.

[Jou85]   Jean-Lin Journé, *Calderón-Zygmund operators on product spaces*, Rev. Mat. Iberoamericana **1** (1985), no. 3, 55–91. MR 836284 (88d:42028)

[JT06]   Jean-Lin Journé and Jean-Marie Trépreau, *Hypoellipticité sans sous-elliptcité: le cas des systèmes de n champs de vecteurs complexes en $(n + 1)$ variables*, Seminaire: Equations aux Dérivées Partielles. 2005–2006, Sémin. Équ. Dériv. Partielles, École Polytech., Palaiseau, 2006, pp. Exp. No. XIV, 19. MR 2276079 (2007k:35055)

[Kis95]   Vladimir V. Kisil, *Connection between two-sided and one-sided convolution type operators on non-commutative groups*, Integral Equations Operator Theory **22** (1995), no. 3, 317–332. MR 1337379 (96d:44004)

[KN65]   J. J. Kohn and L. Nirenberg, *An algebra of pseudo-differential operators*, Comm. Pure Appl. Math. **18** (1965), 269–305. MR 0176362 (31 #636)

[Koe02]   Kenneth D. Koenig, *On maximal Sobolev and Hölder estimates for the tangential Cauchy-Riemann operator and boundary Laplacian*, Amer. J. Math. **124** (2002), no. 1, 129–197. MR MR1879002 (2002m:32061)

[Koh78]   J. J. Kohn, *Lectures on degenerate elliptic problems*, Pseudodifferential operator with applications (Bressanone, 1977), Liguori, Naples, 1978, pp. 89–151. MR 660652 (84a:35059)

[Koh98]   _____, *Hypoellipticity of some degenerate subelliptic operators*, J. Funct. Anal. **159** (1998), no. 1, 203–216. MR 1654190 (99m:35035)

[Koh02]   _____, *Superlogarithmic estimates on pseudoconvex domains and CR manifolds*, Ann. of Math. (2) **156** (2002), no. 1, 213–248. MR 1935846 (2003i:32059)

[Koh05]   _____, *Hypoellipticity and loss of derivatives*, Ann. of Math. (2) **162** (2005), no. 2, 943–986, With an appendix by Makhlouf Derridj and David S. Tartakoff. MR 2183286 (2006k:35036)

[KPZ12]   Tran Vu Khanh, Stefano Pinton, and Giuseppe Zampieri, *Loss of derivatives for systems of complex vector fields and sums of squares*, Proc. Amer. Math. Soc. **140** (2012), no. 2, 519–530. MR 2846320

[Lob70]   Claude Lobry, *Contrôlabilité des systèmes non linéaires*, SIAM J. Control **8** (1970), 573–605. MR 0271979 (42 #6860)

[LP31]   J. E. Littlewood and R. E. A. C. Paley, *Theorems on Fourier Series and Power Series*, J. London Math. Soc. **S1-6** (1931), no. 3, 230. MR 1574750

[LP37]   _____, *Theorems on Fourier Series and Power Series (II)*, Proc. London Math. Soc. **S2-42** (1937), no. 1, 52. MR 1577045

[LP38]   _____, *Theorems on Fourier Series and Power Series(III)*, Proc. London Math. Soc. **S2-43** (1938), no. 2, 105. MR 1575588

[Lun92]   Albert T. Lundell, *A short proof of the Frobenius theorem*, Proc. Amer. Math. Soc. **116** (1992), no. 4, 1131–1133. MR 1145422 (93c:58005)

[Mar36]   József Marcinkiewicz, *Sur les series de fourier*, Fund. Math **27** (1936), 38–69.

[Mar39]   ———, *Sur linterpolation dopérations*, CR Acad. Sci. Paris **208** (1939), no. 193, 1272–1273.

[Mel86]   Richard Melrose, *Propagation for the wave group of a positive subelliptic second-order differential operator*, Hyperbolic equations and related topics (Katata/Kyoto, 1984), Academic Press, Boston, MA, 1986, pp. 181–192. MR 925249 (89h:35177)

[Mor78]   Yoshinori Morimoto, *On the hypoellipticity for infinitely degenerate semielliptic operators*, J. Math. Soc. Japan **30** (1978), no. 2, 327–358. MR 494715 (81d:35017)

[Mor87]   ———, *Hypoellipticity for infinitely degenerate elliptic operators*, Osaka J. Math. **24** (1987), no. 1, 13–35. MR 881744 (88m:35030)

[MRS95]   Detlef Müller, Fulvio Ricci, and Elias M. Stein, *Marcinkiewicz multipliers and multi-parameter structure on Heisenberg (-type) groups. I*, Invent. Math. **119** (1995), no. 2, 199–233. MR 1312498 (96b:43005)

[MRS96]   ———, *Marcinkiewicz multipliers and multi-parameter structure on Heisenberg (-type) groups. II*, Math. Z. **221** (1996), no. 2, 267–291. MR 1376298 (97c:43007)

[Mül89]   Detlef Müller, *On Riesz means of eigenfunction expansions for the Kohn-Laplacian*, J. Reine Angew. Math. **401** (1989), 113–121. MR 1018056 (90i:22017)

[Mül04]   ———, *Marcinkiewicz multipliers and multi-parameter structure on heisenberg groups*, 2004, Lecture Notes. Padova.

[NRS01]   Alexander Nagel, Fulvio Ricci, and Elias M. Stein, *Singular integrals with flag kernels and analysis on quadratic CR manifolds*, J. Funct. Anal. **181** (2001), no. 1, 29–118. MR MR1818111 (2001m:22018)

[NRSW89]   A. Nagel, J.-P. Rosay, E. M. Stein, and S. Wainger, *Estimates for the Bergman and Szegő kernels in* $\mathbb{C}^2$, Ann. of Math. (2) **129** (1989), no. 1, 113–149. MR MR979602 (90g:32028)

[NRSW12]   Alexander Nagel, Fulvio Ricci, Elias Stein, and Stephen Wainger, *Singular integrals with flag kernels on homogeneous groups, I*, Rev. Mat. Iberoam. **28** (2012), no. 3, 631–722. MR 2949616

[NS79]   Alexander Nagel and E. M. Stein, *Lectures on pseudodifferential operators: regularity theorems and applications to nonelliptic problems*, Mathematical Notes, vol. 24, Princeton University Press, Princeton, NJ, 1979. MR 549321 (82f:47059)

[NS01] Alexander Nagel and Elias M. Stein, *Differentiable control metrics and scaled bump functions*, J. Differential Geom. **57** (2001), no. 3, 465–492. MR 1882665 (2003i:58003)

[NS04] _____, *On the product theory of singular integrals*, Rev. Mat. Iberoamericana **20** (2004), no. 2, 531–561. MR MR2073131 (2006i:42023)

[NSW85] Alexander Nagel, Elias M. Stein, and Stephen Wainger, *Balls and metrics defined by vector fields. I. Basic properties*, Acta Math. **155** (1985), no. 1-2, 103–147. MR MR793239 (86k:46049)

[PP05] Cesare Parenti and Alberto Parmeggiani, *On the hypoellipticity with a big loss of derivatives*, Kyushu J. Math. **59** (2005), no. 1, 155–230. MR 2134059 (2005m:35049)

[PP07] _____, *A note on Kohn's and Christ's examples*, Hyperbolic problems and regularity questions, Trends Math., Birkhäuser, Basel, 2007, pp. 151–158. MR 2298790 (2007m:35019)

[Puk67] L. Pukánszky, *Leçons sur les représentations des groupes*, Monographies de la Société Mathématique de France, No. 2, Dunod, Paris, 1967. MR 0217220 (36 #311)

[Rie28] Marcel Riesz, *Sur les fonctions conjuguées*, Math. Z. **27** (1928), no. 1, 218–244. MR 1544909

[Roc78] Charles Rockland, *Hypoellipticity on the Heisenberg group-representation-theoretic criteria*, Trans. Amer. Math. Soc. **240** (1978), 1–52. MR 0486314 (58 #6071)

[Rot79] Linda Preiss Rothschild, *A criterion for hypoellipticity of operators constructed from vector fields*, Comm. Partial Differential Equations **4** (1979), no. 6, 645–699. MR 532580 (80e:58040)

[RS76] Linda Preiss Rothschild and E. M. Stein, *Hypoelliptic differential operators and nilpotent groups*, Acta Math. **137** (1976), no. 3-4, 247–320. MR 0436223 (55 #9171)

[RS80] Michael Reed and Barry Simon, *Methods of modern mathematical physics. I*, second ed., Academic Press Inc. [Harcourt Brace Jovanovich Publishers], New York, 1980, Functional analysis. MR 751959 (85e:46002)

[RS92] F. Ricci and E. M. Stein, *Multiparameter singular integrals and maximal functions*, Ann. Inst. Fourier (Grenoble) **42** (1992), no. 3, 637–670. MR 1182643 (94d:42020)

[Rud91] Walter Rudin, *Functional analysis*, second ed., International Series in Pure and Applied Mathematics, McGraw-Hill Inc., New York, 1991. MR 1157815 (92k:46001)

[Sch11] J. Schur, *Bemerkungen zur theorie der beschränkten bilinearformen mit unendlich vielen veränderlichen.*, Journal für die reine und Angewandte Mathematik **140** (1911), 1–28.

[See59] R. T. Seeley, *Singular integrals on compact manifolds*, Amer. J. Math. **81** (1959), 658–690. MR 0110022 (22 #905)

[See65] ———, *Refinement of the functional calculus of Calderón and Zygmund*, Nederl. Akad. Wetensch. Proc. Ser. A 68=Indag. Math. **27** (1965), 521–531. MR 0226450 (37 #2040)

[Sik04] Adam Sikora, *Riesz transform, Gaussian bounds and the method of wave equation*, Math. Z. **247** (2004), no. 3, 643–662. MR 2114433 (2005j:58034)

[Spi65] Michael Spivak, *Calculus on manifolds. A modern approach to classical theorems of advanced calculus*, W. A. Benjamin, Inc., New York-Amsterdam, 1965. MR 0209411 (35 #309)

[SS11] Elias M. Stein and Brian Street, *Multi-parameter singular radon transforms*, Math. Res. Lett. **18** (2011), no. 2, 257–277. MR 2784671

[SS12] ———, *Multi-parameter singular Radon transforms III: Real analytic surfaces*, Adv. Math. **229** (2012), no. 4, 2210–2238. MR 2880220

[SS13] ———, *Multi-parameter singular Radon transforms II: the L^p theory*, 2013, to appear in Adv. Math.

[Ste70a] Elias M. Stein, *Singular integrals and differentiability properties of functions*, Princeton Mathematical Series, No. 30, Princeton University Press, Princeton, NJ, 1970. MR 0290095 (44 #7280)

[Ste70b] ———, *Topics in harmonic analysis related to the Littlewood-Paley theory.*, Annals of Mathematics Studies, No. 63, Princeton University Press, Princeton, NJ, 1970. MR 0252961 (40 #6176)

[Ste82a] ———, *The development of square functions in the work of A. Zygmund*, Bull. Amer. Math. Soc. (N.S.) **7** (1982), no. 2, 359–376. MR 663787 (83i:42001)

[Ste82b] ———, *An example on the Heisenberg group related to the Lewy operator*, Invent. Math. **69** (1982), no. 2, 209–216. MR 674401 (84c:35031)

[Ste93] ———, *Harmonic analysis: real-variable methods, orthogonality, and oscillatory integrals*, Princeton Mathematical Series, vol. 43, Princeton University Press, Princeton, NJ, 1993, with the assistance of Timothy S. Murphy, Monographs in Harmonic Analysis, III. MR 1232192 (95c:42002)

[Ste99] _____, *Calderón and Zygmund's theory of singular integrals*, Harmonic analysis and partial differential equations (Chicago, IL, 1996), Chicago Lectures in Math., Univ. Chicago Press, Chicago, IL, 1999, pp. 1–26. MR 1731194 (2001e:42021)

[Str06] Brian Street, L^p *regularity for Kohn's operator*, Math. Res. Lett. **13** (2006), no. 5-6, 703–711. MR 2280768 (2007i:35026)

[Str08] _____, *An algebra containing the two-sided convolution operators*, Adv. Math. **219** (2008), no. 1, 251–315. MR 2435424 (2009h:43005)

[Str09] _____, *The* $\Box_b$ *heat equation and multipliers via the wave equation*, Math. Z. **263** (2009), no. 4, 861–886. MR 2551602 (2011b:32063)

[Str10] _____, *A parametrix for Kohn's operator*, Forum Math. **22** (2010), no. 4, 767–810. MR 2661448 (2011e:35071)

[Str11] _____, *Multi-parameter Carnot-Carathéodory balls and the theorem of Frobenius*, Rev. Mat. Iberoam. **27** (2011), no. 2, 645–732. MR 2848534

[Str12] _____, *Multi-parameter singular radon transforms I: The* L^2 *theory*, J. Anal. Math. **116** (2012), 83–162. MR 2892618

[Sus73] Héctor J. Sussmann, *Orbits of families of vector fields and integrability of distributions*, Trans. Amer. Math. Soc. **180** (1973), 171–188. MR 0321133 (47 #9666)

[Tit29] E. C. Titchmarsh, *On Conjugate Functions*, Proc. London Math. Soc. **S2-29** (1929), no. 1, 49. MR 1575323

[Trè67] François Trèves, *Topological vector spaces, distributions and kernels*, Academic Press, New York, 1967. MR 0225131 (37 #726)

[TW03] Terence Tao and James Wright, L^p *improving bounds for averages along curves*, J. Amer. Math. Soc. **16** (2003), no. 3, 605–638. MR 1969206 (2004j:42005)

[UB64] André Unterberger and Juliane Bokobza, *Les opérateurs de Calderón-Zygmund précisés*, C. R. Acad. Sci. Paris **259** (1964), 1612–1614. MR 0176360 (31 #635a)

[Wey08] Hermann Weyl, *Singuläre integralgleichungen mit besonderer berücksichtigung des fourierschen integraltheorems: vorgelegt von h. weyl*, W. Fr. Kaestner, 1908.

[Yan09] Dachun Yang, *Besov and Triebel-Lizorkin spaces related to singular integrals with flag kernels*, Rev. Mat. Complut. **22** (2009), no. 1, 253–302. MR 2499336 (2010c:42026)

[ZS60]   Oscar Zariski and Pierre Samuel, *Commutative algebra. Vol. II*, The University Series in Higher Mathematics, D. Van Nostrand Co., Inc., Princeton, NJ-Toronto-London-New York, 1960. MR 0120249 (22 #11006)

Index

Ingram Content Group UK Ltd.
Milton Keynes UK
UKHW020839290523
422449UK00011B/534